Bioethics and Biosafety

Bioethics and Biosafety

M.K. Sateesh
Faculty
Dept. of Microbiology and Biotechnology
Bangalore University
Bangalore

I.K. International Publishing House Pvt. Ltd.

NEW DELHI   •   BANGALORE

Published by

I.K. International Publishing House Pvt. Ltd.
S-25, Green Park Extension
Uphaar Cinema Market
New Delhi–110 016 (India)
E-mail:info@ikinternational.com
Website: www.ikbooks.com

ISBN 978-81-906757-0-3

© 2010 I.K. International Publishing House Pvt. Ltd.

Reprint 2014

Published by Krishan Makhijani for I.K. International Publishing House Pvt. Ltd., S-25, Green Park Extension, Uphaar Cinema Market, New Delhi–110 016 and Printed by Rekha Printers Pvt. Ltd., Okhla Industrial Area, Phase II, New Delhi–110 020.

Dedicated to my beloved and benevolent parents

B. Krishnaiah and Susheela Krishnaiah

To my wife

Sowmya Sateesh

To my son

Yajath S. Gowda

Foreword

The most striking feature of the Earth is the existence of life and the most striking feature of life is its diversity on the Earth. Today's biodiversity is the result of 3.5 billion years of evolution. In the present fast moving biotechnological world, biodiversity and natural ecosystem play a major role in the enhancement of human needs. Human domination of the Earth's ecosystem is reducing diversity of species within many habitats throughout the world. Biotechnology is also contributing for such instances from its varied type of activities. Hence the present need is developing wiser ethical and safety aspects of our ecosystem without compromising the development. Thus, education on bioethics and biosafety is essential at present.

In 1971, the biologist Van Rensselaer Potter first coined the term "bioethics". Potter used the term to refer to a new field devoted to human survival and an improved quality of life. Gradually, the term "bioethics" came to refer to "the broad terrain of the moral problems of the life sciences, ordinarily taken to encompass medicine, biology, and some important aspects of the environmental, population and social sciences. A special field of bioethics is needed because, for better or for worse, the present century will probably be best described as the biological century or the biological age. Not only advances in medical science, but advances in many other fields will likely have biological components as well. We cannot hide our heads in the sand and pretend that difficult questions are not on their way, indeed, we already know about many of them, or at least their broad outlines.

Many people have voiced concern about biotechnology and genetic engineering. Scientists have considered the issue of safety over recent years. Prevention of large-scale loss of biological integrity referring to human health as well as ecology is defined as "biosafety". Biosafety issues, including human safety, environmental safety, food safety, socioeconomic impacts, ethical and moral issues to be clearly followed for any product of biotechnology, either living or non-living products, before their widespread use or application. Thus, biosafety aspects are considered in large number of biotechnology syllabi throughout the world.

This book is an attempt to give information in a single volume much needed by the students, scholars and scientists of different disciplines in biotechnology. I congratulate Dr. M.K. Sateesh for his efforts in bringing out this book. I am confident that the students, teachers and research workers will receive this book.

Dr. T. Prabhu
B.Sc., M.B.B.S., M.D., FICP.,
Chairman, Medical Advisory Board,
LN Lifecare Pvt. Ltd., Bangalore

Preface

Biotechnology is a much sought after subject these days. The reasons behind this scenario are many. Some being, tremendous growth of industries, high pay scales, simulating stories in the media about large-scale genomic sequencing such as human genome sequencing and new studies on genetic diseases. This makes it very challenging for a teacher to sustain the students' enthusiasm while at the same time not to forget the basic moral principles of life and environment. Biotechnology has to be adapted to the wider arena of social and ethical contexts. The focus of the present-day biotechnologists is on goal-oriented approaches, without understanding the sustainable developments that form the environment.

Biosafety and Bioethics introduces the readers to the basic concepts of ethics and safety that are essential for various branches of science, which use different components of the environment for technical procedures. The study of biosafety and bioethics has had a major impact on all the biological sciences and the quality of human life. Significant achievements have been made using many prokaryotic and eukaryotic organisms to understand the principles that govern life. Many such achievements made by biotechnologists have helped in developing new branches of science such as molecular biology, genetic engineering, proteomics and genomics. Thus, the field of biotechnology is a major branch of science encompassing several smaller branches of it. Bioethics is a very complex and difficult subject as it concerns troubling questions like the nature of life and death, the kind of life to be considered worth living, what constitutes murder, how people in very painful circumstances should be treated, what are the responsibilities of one human being to others, and other such. Bioethics is not a wholly independent field. It has to draw a great deal from other ethical discussions.

Today, the medical technology is pushing the limits of what humans can do as also our understanding of such discussions. Medicine, today, forces us to confront the nature of life and death in a manner that we are not accustomed to. This has necessitated the application of ethical principles and arguments to medical and biological issues. Although medical professionals are educated to handle triage that involves technical questions like who has the best chances of survival, neither they nor other members of the society are trained for situations involving ethical triage. In our pluralistic society, we are faced with a diversity of values and ethical ideals, but it is difficult to tell which values need to be employed when and which ethical dilemmas deserve our immediate attention.

One of the results seems to be that we are forced to deal with one crisis after another. Our lack of sound reasoning and coherent values prevents us from engaging in careful and advance deliberations that would make it easier for us to handle new situations. Today, bioethics has become more a matter of crisis management and less of reasoned discourse. A special field of bioethics is needed because, for better or for worse, the coming century will probably be best described as the Biological Century or the Biological Age (in contrast to the Industrial Age that the West so recently experienced). Not only advances in medical science, but advancements in many other fields too will likely involve biological components.

Throughout the world, many people are voicing increasing concern about biotechnology and genetic engineering. Scientists have also considered the issue of safety over the recent years. A special committee of the National Academy of Sciences specifically reviewed the issues on the introduction of genetically engineered organisms using recombinant DNA (rDNA) technology into the environment. It concluded that there is no evidence of any specific hazards existing either in the use of rDNA technique or in the transfer of genes between unrelated organisms, and that the risks associated with the introduction of rDNA engineered organisms are the same kind as those associated with the introduction of unmodified organisms. Thus, at present more emphasis has to be given on using releasing, transporting and handling of potentially dangerous or new organisms in science. While taking any decision, a balanced integration of scientific and social knowledge should be considered.

Biotechnology is one of the most exciting areas of natural sciences and its applications have to reach the common man. Biotechnology has attracted worldwide attention after the successful mapping of the entire human genome in February 2001 by the National Institute of Health and Celera Genomics. In December 2002, the birth of Baby Eve, claimed to be the first cloned human, has only reinforced the global attention on biotechnology. This branch of science is considered a field that can end the world's hunger in a lesser duration. It is no longer a specialized segment of science confined to high-tech laboratories. The subject of biotechnology is now entering into the degree and even pre-university classes. The common man today talks of biotechnology in the same breath as he talks about information technology. Newspapers too are given information on all fields of biotechnology to educate the masses. The Government of India framed a draft of the biotechnology policy in 2005. This further attracted more and more categories of people from all walks of life to know about the role of biotechnology in society.

Biotechnology is an interdisciplinary field encompassing not only biological sciences but also material sciences. Owing to this interdisciplinary nature, biotechnology finds appeal in both science as well as engineering students. The present book is intended to be an introduction to the subject for the students of both streams. An integrated approach to biosafety and bioethics has been adopted in the book. The content has been systematically presented, built up with simple and clear illustrations, and contains both a historical account and the latest information on the subject.

Biosafety and Bioethics has been written to meet the requirements of the students of biotechnology of various universities and engineering institutes. It provides an insight into the basics of bioethics and biosafety in the field of biotechnology.

Under 'biosafety', topics such as socio-economic impact of biotechnology, biosafety regulations, biosafety protocol and its implementation, guidelines for research and release of transgenic organisms, public education of transgenic organisms production, handling and disposal of hazardous material, good manufacturing practices and good laboratory practices are dealt with. It also includes topics such a IPR and their impact on biodiversity, Indian Biodiversity Act, international organizations dealing with IPR and their related issues, traditional knowledge and their exploitation, ethical and safety issues in consumption of GM foods.

Under 'bioethics', topics such as the study of Human Genome Project (HGP), along with the ethical, legal and social issues involved in it, molecular detection of presymptomatic genetic diseases, prenatal diagnosis, fetal sex determination, genetic diversity research and related ethical issues, stem cell research, use of animals and humans in research and testing, organ transplantations, xenotransplantation, human and animal cloning along with related ethical and social issues, have been dealt with. It also covers other topics like the intellectual property rights (IPR), patents, biodiversity, genetically modified foods and crops and labeling of GM products with related ethical issues. This part of the book explains both basic and applied aspect of bioethics.

Though I have tried to make the book comprehensive and correct in all respects, it is possible that some errors might still have been left due to oversight. Any suggestions or constructive criticism from the readers will be deeply appreciated and carefully considered in the preparation of the next edition.

M.K. Sateesh
mk.sateesh@gmail.com

Acknowledgements

I express my heartfelt gratitude to Dr. T. Prabhu, Former Principal, Bangalore Medical College and Former Chief of Laboratories, Medinova Diagnostic Center and Sagar Apollo Hospital, Bangalore for being a true source of inspiration to me and extending his constant encouragement and support. I am also grateful to Prof. H.A. Ranganath, Vice-Chancellor, Bangalore University, Prof. V.S. Ranganathan, Prof. S. Shankara Bhat, Prof. K.A. Raveesha, Dr. R. Ramachandra, Prof. Rangappa, Mysore University, Prof. Geetha Bali, Bangalore University, Dr. S. Shishupala, Kuvempu University. Dr. A.B. Vedamuthy, HOD, Oxford College, Dr. D.V.S.S.R. Prakash, AMC College, Mr. Bhaskar Reddy, Director, CMR Institutions, Mrs. Rashmi Bhaskar Reddy, Dr. Krishne Gowda, Principal, Government Science College, Bangalore, Dr. Ramakrishnaiah, Government Science College, Bangalore, Dr. H. Nagabhushana, Dr. H.N. Ramesh Babu, Dr. Rajeshwari Ramesh Babu, Dr. Y.L. Krishnamurthy, Kuvempu University, Dr. Sumithra K. Urs. U.S.A., Dr. K.P. Guruprasad, Manipal Institute of Biotechnology, Dr. Gangadhar, V.V. Puram College, Mr. Chandrashekara Joshi, Dr. G.R. Jhanardhana, Dr. S. Umesh, Dr. S. Satish, Mysore University, Dr. S. Puttaraju, Karnataka Pollution Control Board, Dr. S. Manjunath, Bangalore University, Dr. L. Rajanna, Bangalore University, Mr. M.K. Girish, President, Rotaract International, Mysore, Mr. S. Satish Kumar, Ku. Samrudhi, Ku. Samskruthi, Mrs. Gayathri Girish, Mr. Sangameshar, Mr. B. Umamahesh, Sigma-Aldrich, Mr. C.R. Naveen Kumar, Biocon India, Late Mr. Krishne Urs, Dr. S.T. Girish, Mrs. Indira Marigowda, Mrs. Shylaja Suresh, Mrs. Pankaja Ramanna, Ms. K.S. Sneha, Mr. K.S. Sandeep, Mr. Sripadhraj, Mr. Prashanth, GCI Solutions, Mr. Santhosh, Oxford College, Mr. Raghunath, Research Scholar, NIMHANS, Mr. Nagendra Prasad, SJCE, Mysore, Late Mr. B. Basavaiah and Late Mr. Chickathayamma, my teachers, my colleagues, my students, friends and all my well wishers for their direct and indirect support. I also thank the Forensic Science Laboratories for providing an environment that made this book possible.

Thanks are also due to Mr. V.N. Narasimha Murthy and my wife, Mrs. Sowmya Sateesh, a software engineer with GCI Solutions Pvt. Ltd., for their help in the form of suggestions after a thorough and critical reading of the manuscript.

I pay may humble regards to my parents, Mr. B. Krishnaiah and Mrs. Susheela Krishnaiah. I thank my wife and my son Yajath S. Gowda for all their patience and co-operation while preparing this book. I also express my thanks to my brother Mr. M.K. Girish, my sister Mrs. M.K. Sandhya Sateesh Kumar, my in-laws and other family members for showing great forbearance and encouraging me during the writing of this book. Without the help of all these people, timely completion of this book would not have been possible.

M.K. Sateesh

Contents

<table><tr><td># 1</td><td></td></tr><tr><td>**CHAPTER**</td><td># Introduction to Bioethics and Biosafety</td></tr></table>

1.1. INTRODUCTION TO BIOETHICS

At present biotechnology discoveries, in particular, and science discoveries, at a large extent, are occurring with unprecedented speed. In biology, there is an information explosion that is almost incomprehensible. Mager states that the amount of information available in biology doubles every three and half years. The species presently inhabiting Earth are the result of over three billion (3×10^9) years of natural selection that likely favoured efficiency, productivity and specialization. These living or biological organisms are the catalysts that capture and transform energy and materials, producing various essentials including food, fuel, fiber and medicines. These species recycle wastes, create pure potable water, and drive global biogeochemical cycles. These cylces, in turn, create and maintain aerobic atmosphere, regulate entire earth's climate through effects on greenhouse gases, and local climate through effects on evapotranspiration, and make the soil fertile. The living organisms also provide other ecosystem goods and services. Along with these major activities, the living organisms present in earth's biodiversity are the source of all crops, livestock, pollinators of crops and of many pharmaceuticals and pesticides. Out of such diversity, only the three crops such as corn, rice, and wheat provide about 60 percent of the human food supply.

The viability of these crops and other food sources depends on the maintenance of high genetic diversity, which can allow, among other things, development of strains that are resistant to emerging and evolving diseases and pests. This also has a greater effect on the entire food chain and food web at the biosphere scale. In the long term, food stability will require development of new crops and animals from what are now wild plants and animals, because disease or pesticide-resistant pests and pathogens will cause the loss of current crops and animals, just as disease caused the loss of chestnut, elm and other tree species from North American forests and deer populations in the Indian forests.

Indiscriminate use of various biological entities has greater impact on the entire processes of the biosphere. Humans, like all other organisms, experience trade-offs. The loss of biodiversity will diminish the capacity of ecosystems to provide society with a stable and sustainable supply of essential goods and services, but many of the very actions that harm biodiversity simultaneously provide valuable societal benefits. There exists a trade-off defining the net benefits that society receives from various ways that human could use and impact nature, but as yet this is poorly defined. This trade-off itself is likely to shift through time in response to the remaining amounts and states of various resources, including biodiversity. The amounts and states of biotic resources have changed rapidly during the past century, as global population increased 3.7-fold and per capita gross domestic product, a reasonable proxy for consumption, increased 4.6-fold. It seems likely that environmental policy that is optimal from a societal perspective would be markedly different now from that of 250 years ago. However, we still use environmental and land-use ethics, codified in law, which were articulated during the era when the human population, at one-tenth its present size, tamed wilderness with axe and ox.

In the present condition of biotechnological applications and implementation, science has much to contribute to dialogues on policy and ethics. Although academic institutions seem to value such contributions less than contributions to peer-reviewed journals, this is shortsighted. Ultimately, society invests in science because advances in scientific knowledge benefit society. The ethics of science cannot eschew involvement in public discourse. Science must contribute in an open, unbiased manner to relevant issues.

The introduction of transgenic organisms, and the transboundary movement of these **genetically modified organisms (GMOs)** or **living modified organisms (LMOs)** caused greater concern worldwide. Because of the emergence of human domination of global ecosystems, society faces new and tough trade-offs. These include trade-offs between the current benefits and the future costs of environmental damage, and between benefits to a few and costs to many. Research is needed to quantify these trade-offs, and the work done so far on biodiversity provides a good start. Additional

work at the interface between ecology and economics is needed to quantify the immediate and long-term costs and benefits of alternative actions.

Humans have achieved greater success in the recent past. The world that will exist in 100 and 1,000 years will, unavoidably, be of human design, whether deliberate or haphazard. The principles that should guide this design must be based on science, much of it done only sketchily to date and on ethics. Ethics should, among other things, apportion costs and benefits between individuals and society as a whole, and between current generations and all future generations. A sustainable world will require an ethic that is ultimately incorporated into culture, and as long lasting as a constitutional bill of rights or as religious commandments. The earth will retain its most striking feature, biodiversity, only if humans have the prescience to do so. This will occur, it seems, only if we realize the extent to which we use biodiversity.

Over the past twenty years, scientists and educators have discussed teaching bioethics or values in conjunction with the regular biology **(Fig. 1.1)**. Large numbers of world's major scientific and science educational organizations

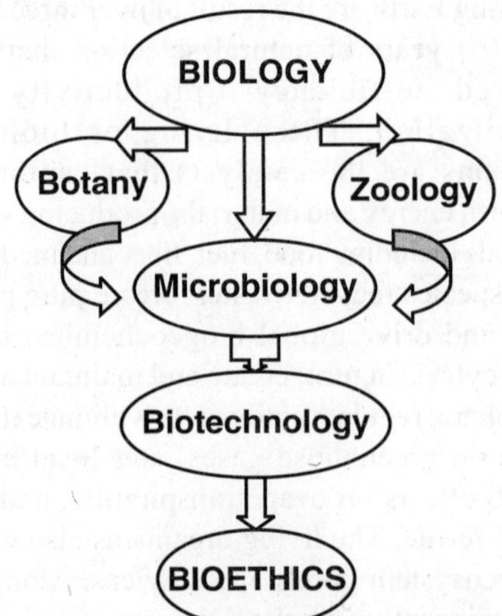

Fig. 1.1. Evolution of different branches of science and their relationship.

have clearly recognized the importance of ethical aspects in the biology. They are in the direction of promoting a better understanding of the integration of these two branches. Bioethics is not only an essential part of the teaching of biology and biotechnology in college, but it is also a unique opportunity to relate the subject content more closely to students' lives, while examining the priorities, which affect the long term survival and sustainability of our planet Earth. This presents the possibility to have a deeper look at what we teach, and perhaps make a shift in emphasis, which will bring about the amalgamation of **biology** and **ethics**.

1.2. NEED OF BIOETHICS

Nowadays, a large number of biotechnological curriculum include bioethics as one of the papers in the study along with other reforms including process and the strategies for collection, compilation and interpretation of data. Today, the realization is that critical thinking skills of ethical analysis are the same as those emphasized in this process-oriented method of teaching science. In fact, the integration of ethical decision-making into biology curricula only reinforces the critical thinking procedures, which are a foundation of the process-oriented teaching reforms. These values or moral activities essential for scientific skills or research at this level are very essential, though the basic values of nonmalfeasance, benevolence, fairness and truthfulness are taught in the lower classes.

Thus, at the present time, the investigation of biology, scientific technology, and ethical issues combine to give out a new branch of science known as **bioethics**. The term "bioethics" was coined by Van Potter in 1971 for this multidisciplinary science.

1.3. DEFINITION OF BIOETHICS

Van Potter also defines bioethics as "biology combined with diverse humanistic knowledge, forging a science that sets a system of medical and environmental priorities for acceptable survival". Later, in 1988, he published a detailed account on bioethics in his book titled **Global Bioethics**.

In a simple term, it can be defined as the humanistic knowledge of scientific methodology to handle living organisms in an acceptable survival or a sustainable society within a healthy ecosystem.

Encyclopedia of Bioethics describes the term "bioethics" as "the broad terrain of the moral problems of the life sciences, ordinarily taken to encompass medicine, biology and some important aspects of the environmental, population and social sciences. The traditional domain of medical ethics would be included within this array, accompanied now by many other topics and problems".

1.4. OTHER DEFINITIONS OF BIOETHICS

- A discipline dealing with the ethical implications of biological research and applications.
- The study of ethical issues raised by the developments in the life science technologies.
- The branch of ethics, philosophy and social commentary that discusses the life sciences and their potential impact on our society.
- The branch of ethics that studies moral values in the biomedical and biological sciences.
- Bioethics is the ethics of biological science and medicine.

1.5. BIOETHICS AND ITS RELATIONS WITH OTHER BRANCHES

This view of bioethics is eminently useful in promoting critical thinking at many levels. It allows for greater accessibility to the content through connectivity rather than stand-alone units. It engages the content and process of real situations, where decisions have real consequences, seldom with risk-free results. Bioethics, generally, obtains its requirements from different branches of sciences, such as biology, genetics, biostatistics philosophy, theology, social sciences, medicine, moral sciences and law **(Fig. 1.2)**. It also promotes a focusing framework that places the biology in a more fully integrated form with the issues, which the student, quickly and precisely, relates to the situation. And now, a word about the new role played by a new professional group known as the bioethicists. The approach that is usually taken is based on biomedical ethics and physician-patient relationships, autonomy of the patient, and informed consent. The question is, is the approach adequate for the kinds of issues that we are facing today? Bioethicists are involved at every level. They are in local committees at universities and national boards, and industry is setting up their ethics boards on these issues. So, they are prominent, they are everywhere, and they are quoted on every major issue.

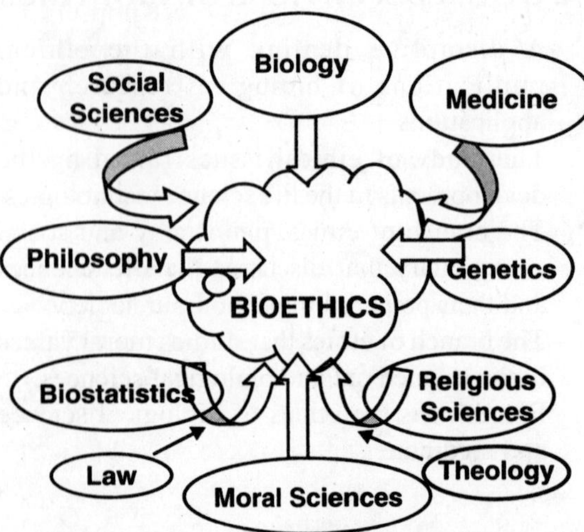

Fig. 1.2. Bioethics draws its requirements from various branches of science and philosophy

1.6. APPLICATIONS OF BIOETHICS

The study of bioethics is thus useful in large number of situations that need proper decision-making based on presuppositions, personal or societal values, moral values and subjective interpretations. They find its applications in dealing with value-rich issues as follows:

* Research using various organisms
* Environmental degradation
* Environmental pollution
* Species diversity
* Fetal tissue transfer
* Organ transplantation
* Xenotransplantation
* Stem cells study and storage
* Cloning
* Genomic studies
* Release of transgenic organisms
* The new reproductive strategies

These are only a few of the many, which could be mentioned at this context. Humankind has always tried to tame the nature in many unacceptable, unethical, and unpredictable ways (or many ways), establishing his supremacy on this mother planet, Earth. During the process of handling the abiotic and biotic environment, human beings have created unimaginative chaos mainly because of total ignorance of ethics or bioethics. Bioethics also gives opportunities in the following ways:

* Creates awareness in students and teachers of the role of science in personal, social, economic, and political realms.
* Problem-solving abilities are in response to human needs.
* Students learn how to use scientific knowledge and skill in their personal, professional, and social lives.
* It is viewed through the impact of technological changes.
* Education becomes a life long process.
* Ambiguity is looked at as an opportunity to evaluate and select the present best solution, and not as a barrier to learning.
* It provides greater opportunity to do meaningful cooperative work.
* To perform and evaluate more than one scientific principles.

Bioethics creates bridges between science, policy and public discussion. It promotes a common understanding of terms like "clone", "embryo", and "stem cell" to help disparate groups discuss the implications of bench science and agree on policy and research conduct.

Stanislaw Lem, a very well-known Polish science-fiction novelist and journalist, at the end of the last century, wrote in one of his assays that twentieth century was the period of totalitarisms, and the twenty-first century will be the time of terrorisms. Biochemistry, in the twentieth century, became a dominant discipline among natural sciences. The vivid development of the methods and ideas permitting to describe the phenomena of life, health and diseases in terms of molecular events caused several traditional disciplines like botany, zoology, evolutionism, even anatomy to adapt the biochemical way of investigating their problems. One may say that the second half of twentieth century has been the time of "molecularisation" of natural sciences. The evidence of this process is the emergence of several new disciplines, which developed as a result of the application of biochemical research to solve the problems, which were investigated previously by morphological methods. In this way it raised disciplines, now known as molecular biology, molecular genetics, immunochemistry, comparative biochemistry, biochemical evolutionism, biochemical pathology and others.

Another story has been the development of technologies applying the achievements of contemporary biochemistry (biotechnology, genetical engineering, biomolecular engineering, biomedicine). This led to the possibility of manipulating and modifying living organisms, by now not only microorganisms, plants, and animals, but also humans. These manipulations and modifications have been used to produce better and more abundant food and pharmaceuticals. It has also helped in the medical treatment of human subjects. However, these possibilities may also be easily used for evil purposes. The awareness of it

caused about thirty years ago the formation of a new discipline, which was the subject of interest for both philosophers and biologists—bioethics. The term "bioterrorism" means not only the use of modified bacteria or isolated toxins to kill people, but also manipulations done on human without the the consent of the subject. If bioethics would be neglected, the biochemical achievements could not be used to treat diseases, prevent sufferings of living beings, and to preserve natural environment. The achievements may serve also the advocates of bioterrorism. This is the reason why the organizers of this Annual Meeting of the Polish Biochemical Society decided to include in the agenda a session entitled "ethical aspects of research in natural sciences—the challenge of 21st century".

The biochemical research is becoming now-a-days more and more expensive. Thus, the subject of research performed by an individual investigator depends not only on the researcher himself, but also on the sponsor who is paying for it. The direction of the development of biochemistry in the twenty-first century will be, therefore, dependent not only on the new scientific ideas, but also on the appreciation by researchers and by the sponsors (governments, military authorities, business people) of the bioethical indications.

The Center for Biomedical Ethics provides education and consultation, and performs research on ethical problems in our health-care system. Community clinicians and researchers can request a panel to explore issues in depth, or call the center with shorter-term problems, such as responding to grant agency concerns.

Magnus is the co-chair of the ethics committee for the Stanford Health Center, where he runs a bedside consultation program after which the bench-side program is modeled. "We're trying to give researchers the same kind of consultation service that we've been giving on the clinical side for a long time," he said.

By the close of the twentieth century, humankind had acquired the power to transform the processes of all living species, including its own. The potential benefits of these powers can be exciting, and the perils ominous. Consider, for example, the recent announcement that scientists have successfully produced cultures of embryonic stem cells. This breakthrough offers the potential to grow any type of human tissue, and may eventually be used to repair damaged hearts, blood vessels, or brains. The very same week, the UK's Sunday Times reported that scientists could theoretically engineer deadly biological organisms to produce "ethno-bombs", which are capable of targeting human victims by ethnic origin.

George Annas, noted bioethicist and lawyer, observed in 1989 that "ethics is generally taken seriously by physicians and scientists only when it either fosters their agenda or does not interfere with it. If it cautions a slower pace or a more deliberate consideration of science's darker side, it is dismissed as fearful of the future, anti-intellectual, or simply uninformed." Given the dizzying pace of technological advancements in genetics and biology, it is not surprising that society is grappling ever more urgently with the social, ethical and legal implications of humankind's ability to decipher and control the genetic blueprint of life. Opinions differ sharply on the implications of new biotechnologies, but nearly everyone agrees that advances in technology are taking place at a rate far faster than social policies can be devised to guide them, or legal systems can evolve to address them. There is growing recognition worldwide that the development of scientific knowledge must be accompanied by public debate on societal choices and the informed participation of citizens. Bioethics attempts to identify the social and cultural implications of breakthroughs in life sciences, to anticipate its applications, and to ensure that progress in the life sciences benefits humanity as a whole. Bioethics acknowledges that there is a distinction between what is scientifically possible and ethically acceptable. What is good for society, what is equitable, and what is safe? Who will decide? These are among the questions that are fueling debate on the applications of biotechnology in health, agriculture and human development.

At the intergovernmental level, the United Nations Educational, Scientific and Cultural Organization (UNESCO) created the International Bioethics Committee in 1993 as the world's only international body to study the implications of human genome research and genetic engineering. In November, 1997, UNESCO adopted a non-binding Universal Declaration on the Human Genome and Human Rights—the first international text on the ethics of genetics research. Since 1993, a growing number of countries have established national bioethics advisory committees to examine bioethics and to provide guidance to national governments. In 1998, the CGIAR adopted a set of ethical principles to guide its work on genetic resources. The Crucible Group recognizes and appreciates the vital contribution of ethical debates. There is concern, however, that the appointment of expert panels and commissions devoted to bioethics should not become a substitute for broad public debate and participation in the review and assessment of new technologies.

1.7. INTRODUCTION TO BIOSAFETY

In the present scenario in biotechnology, genetically modified organisms (GMOs) are a fact of modern agriculture and are here to stay. Genetically modified organisms are also a fact of public preoccupation and opinion, which politicians must take into account. Food and Agriculture Organization (FAO) recognizes the great potential and complications of these new technologies. We need to move carefully, with a full understanding of all the factors involved. Genetically modified organisms are produced in almost all major categories of living organisms that are identified. In particular, we need to assess these organisms in terms of their impact on food security, poverty, safety of living entities on the earth or biosphere, and the sustainability of agriculture.

Will genetically modified organisms increase the amount of food in the world, and make more food accessible to the hungry? Clearly, genetically modified organisms should be seen not in isolation as technical achievements. The public in many countries distrusts genetically modified organisms. They are often seen in the context of globalization and of privatization and even as "antidemocratic" or "meddling with evolution". There are as yet few perceived advantages for the public, because applications of these to organisms have concentrated on reducing costs for producers without direct consumer benefits. In particular, it has been a tactical error of the industry to concentrate on pesticide-resistance as one of the earliest applications, as this has stimulated environmental concerns. The public often confuses the industry with the science. And consumers worry about risk, not about scientific freedom.

A large number of researchers and scientists in both the private and public sectors clearly see genetic modification as a major new set of tools. They are also participants and spectators in a major shift of research from the public to the private sector, which will undoubtedly influence the future direction of research and research investment. As shareholders in the genetically modified organisms debate, scientists must recognize that there is also a substantial public distrust of science. Obviously, the industry looks at genetically modified organisms as opportunities for corporate profit, yet, at the same time, recognizes that public acceptance may be a stumbling block. In turn, governments often lack coherent policies in relation to genetically modified organisms and have not yet developed and implemented adequate regulatory instruments and infrastructures. As a result, in most countries, there is no consensus on how biotechnology and genetically modified organisms, in particular, can focus on the key challenges of the food and agricultural sector. Governments need to be more proactive in addressing these questions, and in this, the role of scientists in public service will be crucial.

It is not possible to talk intelligently about genetically modified organisms if we remain at the level of generalities. Any statement must refer explicitly to specific crops, genes, ecologies and production systems. For this reason, FAO has been conducting a worldwide inventory of agricultural biotechnology applications and products, with special reference to developing countries. We wished to test the hypothesis that biotechnology has been uneven, and that most current biotechnology applications in crop science—particularly GM crops—have not addressed the special needs of developing countries. We have looked specifically at current as well as pipeline research on GM crops. Our sources of information have been varied—informal discussions with scientists as well as formal reports. Our inventory will be available electronically and will be updated regularly. Some of the preliminary results are as follows.

The total area cultivated with genetically modified organism crops in the world currently stands at about 44.2 million hectares. This compares with 11 million hectares only three years ago. About 75 percent of the area planted with GM crops is in industrialized countries. Substantial plantings, so far, largely concern only four crops—soybean, maize, cotton, and canola, and about 16 percent of the total area planted with these major crops is now under GM varieties. Two traits—insect resistance (mainly based on Bt) and herbicide tolerance—dominate.

There are also small areas of potato and papaya, with genes for delayed ripening and virus-resistance inserted. Only seven developing countries commercially cultivate GMO crops, and with the exception of Argentina and China, the areas involved are less than 100,000 hectares. Here again, the dominant crops are soybean and cotton, and the traits are herbicide tolerance and insect resistance. Only in China, one of GM cotton was locally developed and commercialized. Other countries obtained genetic constructs or varieties from industrialized countries. Although several forest tree species, such as conifers, poplar, sweet gum and eucalypts have been transformed using recombinant DNA technology, they have not so far been released for commercial purposes. Tropical fruit tree species seem to have been largely neglected. Current releases are still very narrow in terms of crops, traits, and countries involved. So what is in the pipeline? Throughout the world, several thousand GMO field tests have been conducted or are underway, again mostly in industrialized countries.

Some 200 crops are under field testing in developing countries; the vast majority of these are in Latin America (152), followed by Africa (33), and Asia (19). Many more countries that have already released GMOs are involved in this and many more crop-trait combinations are being investigated, with greater focus on virus resistance, quality, and in some cases, tolerance to abiotic stresses. It can, therefore, be expected that the number of GMOs ready for commercial release in these countries will expand considerably in the next few years, though many important crops, such as pulses, vegetables, and fodder, and industrial crops and certain traits, such as drought and aluminum tolerance, are still almost entirely neglected.

Thus, the present portfolio of GM applications is increasing in the near future. We need to deal rather rapidly with some overall questions to ensure that GM crops make an optimal contribution to world food security, food safety, quality, and sustainability, and remain to the public at large. There are three big problems here. Firstly, directing genomics and related research for meeting these key challenges. Secondly, implementing international instruments and developing international standards of governance, which ensures health and environmental safety, and so command public support. Thirdly, facilitating the access to research and new technologies for developing countries, poor producers, and consumers. A related issue is that if developing countries are to be full partners in the international economy, they need assistance to build up their national legislation and regulations, as well as the relevant skills and infrastructures.

In the transgenics field, GMOs being directed to the right research priorities is a major question raised at present. Despite some hopeful signs, the inventory suggests that this is not yet the case. There are specific technical challenges that are difficult to handle through traditional crop-breeding, where transgenics may offer great possibilities for crops and forestry. These include drought and heat-tolerance, improved nutrient uptake and rooting, biological nitrogen fixation, responses to carbon dioxide, and tolerance to key abiotic stresses, such as salinity and drought. As part of a strategy of diet diversification, genetic modification may help us address protein, vitamin, and iron deficiencies—see the famous example of the golden rice. Unfortunately, most of the traits needed are governed by multiple genes, and it will probably be sometime before farmers and consumers benefit from such research.

The perceived profit potential of GMOs has already changed the direction of investment in research and development in both the public and private sectors. The possible long-term environmental costs of such strategies may be overlooked. If research is to address the challenges of agriculture in the future, we need to put genetic modification in context and realize that it is but one of the many elements of agricultural change. Scientists must not be blinded by the glamour of cutting-edge molecular science for its own sake. It is also worried about the disconnection between the laboratory and the field and the decline of agronomy as an integrative science. Governments need to be vigilant and not let this glamour, or the perception by private industry of major profit opportunities, draw investment away from research in other more traditional fields such as water and soil management or ecology and from public sector research. At the same time, the best science is developed in a climate of intellectual freedom, without much direct government interference. It is a difficult balance to strike. The cost of molecular technologies in plant breeding and the size of markets that are required to recoup the profits lead to the growing use of "hard" intellectual property rights over seeds and planting material and the tools of genetic engineering. This changes the relationship between the public and private sectors, to the detriment of the public sector.

A policy question that governments must take up, in both the national and international contexts, is how to ensure that public research is not the "poor relation". In developing countries, in particular, it is important for the public sector to retain enough capacity, resources and freedom of action to provide the services on which their national private sectors can build. They will also need to build their policy and regulatory capacities with regard to transgenic crops that originate elsewhere, and in this FAO has an "honest broker" role to play. Thus, we need a science that, in a globalized world, addresses local agriculture and local food security and helps to keep the rural areas liveable and attractive to young people in search of employment. The constraints to doing so are more economic than technical. Globalization is not only a question of size, but also of kind. It is inextricably linked to privatization, and major economic restructuring in both developed and developing countries has changed the balance of public and private sectors, and the privatization of knowledge through intellectual property rights. This results in concentration, and there is often little economic incentive for global companies to address local needs. It has been estimated that the world's top ten companies account for about 85 percent of the global market for the seed and agro-chemical industries. In 1998, just four companies controlled 69 percent of the seed market in the USA. Of course, globalization also promises to create opportunities for millions of farmers throughout the world. It can offer opportunities to the poorer countries, and can drive rapid development and local capital accumulation. But it may also exacerbate existing differences among

countries. Developing transgenic crops implies massive investments, and the need for massive returns. The small number of GM technologies currently in use suggests that there is a real danger that the scale of the investment may lead to selective concentration on species, problems of global importance, and concomitant capital inertia. In this context, a sound note of warning about the rising costs of regulation is also essential. In the health industry, where regulation is heaviest, bringing a pharmaceutical to market now costs about \$500 million; bringing a pesticide to market can cost about \$200 million; and it has recently been reported that a GM crop costs about \$30 million to produce, to which regulatory costs can add up to a further \$5–6 million. Which crops can bear such costs in the commercial market? How are we to serve local needs, small farmers and poor consumers, and promote genetic and dietary diversity?

1.8. NEED FOR BIOSAFETY

Most policy and regulatory activities regarding GMOs are national. International agreements have a role to play in harmonizing national regulatory systems, for example, on transboundary questions, and when trade standards are involved. International agreements can also help governments, particularly the smaller governments, to rationalize their national capacities and investments through common action. In relation to GM crops, the major instruments concern risk-assessment and management in three areas:

- Food
- Health
- Environment

GMOs in food and agriculture have been around long enough for us to have some background against which to assess possible risks. There has been as yet no proven effect on human life. Moreover, a recent study of GMO releases into the environment in Britain showed that they had not survived in nature, as, in fact, they only would if they had an evolutionary advantage, which remains an important possibility. But our present knowledge, which is state-of-the-art knowledge, may not always be adequate to assess the risk of GMOs to health or the environment. Of course, lack of evidence of adverse effects is not the same as the knowledge that genetic modification is safe.

FAO deals with environmental and food risk-assessment questions under a more general concept of "biosecurity", namely the application of sanitary and phytosanitary measures (SPS), and a methodology of risk analysis for food and agriculture, including fisheries and forestry.

1.9. ISSUES IN USE OF GENETICALLY MODIFIED ORGANISMS

Genetically modified organisms (GMOs) are created by recombinant DNA technology. These are, at present, used in agriculture, forestry, environmental clean-up programmes, biomedical fields, research and development, and other in-house experiments **(Fig. 1.3)**. Especially, genetically modified microorganisms or genetically engineered microorganisms (GEMs) have better advantage in the introduced ecosystem, and thus they can be considered as a greater risk in the following ways:

1. They can adopt and multiply rapidly in the ecosystem compared to native flora.
2. They utilize the nutrients at a faster rate.
3. They can transfer the genes related to virulence or pathogenesis into native microbial flora.
4. They can produce newer toxins and allergens.
5. They can transfer the new traits to related microbes.

Hence, these organisms create situations, which are unpredictable, unexplained, uncontrolled and unmanageable. But, this is not always true, and can occur with the introduction of unmodified organisms. Thus, a committee of the National Academy of Science, USA, in 1987 has considered that assessment of danger should be made based on the type of the organism that is under study and not the method of engineering. For these reasons, a biosafety protocol was established on January 2000 under the supervision of Convention on Biological Diversity (or Biodiversity Convention), which began in 1992.

1.10. DEFINITION OF BIOSAFETY

Biosafety refers to "efforts to ensure safety in using, transporting, transferring, handling, releasing, and disposing of biological organisms, including genetically modified organisms (GMOs) when they are considered potentially capable of harming human, animal, or plant health or the environment".

In the present biotechnological research and trade, most of the developed countries have stringent laws and regulations that ensure the biosafety of potentially hazardous organisms, both in domestic use and in trade. But large number of developing countries do not have such biosafety regulatory guidelines for safeguarding their life and the environments.

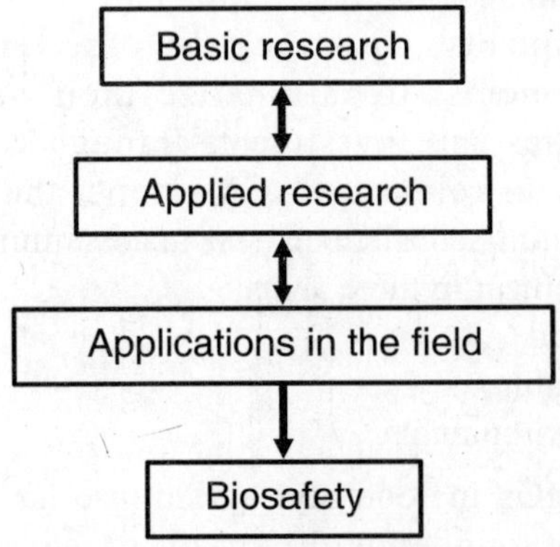

Fig. 1.3. Biosafety is necessary in handling and release of the biohazard or genetically modified organisms.

1.11. APPLICATIONS OF BIOSAFETY

The implementation of biosafety sets forth a number of procedures and rules that will be helpful in the following conditions:

- The protection of animals or plants from pests, diseases, or disease-causing organisms in different situations.
- The protection of human or animal life from risks arising due to additives, contaminants, residual toxins or disease-causing organisms in foods, beverages, or foodstuff.
- Avoiding risk organisms in transboundary movements.
- Helpful for standard-setting in relation to the handling, transport, packaging and identification of GMOs or LMOs.
- Making standards for risk assessment, that is, need for labelling and for several other food safety aspects of GMOs.
- To design an appropriate control measure for the hazardous organism.
- To take a decision for accepting or rejecting the commodities.
- Helpful for identification of the source of contamination or risk.
- To plan the required emergency action plan for treating the danger.
- Segregation and identity preservation.
- The protection of human life or health from risks arising from diseases carried by animals, plants, or plant products, or from pests.
- The prevention or limitation of other damage from pests.

Thus, in a nutshell, it covers the transboundary movement, transit, handling and use of all LMOs (except pharmaceuticals) that may have adverse effects on the conservation, and sustainable use of biological diversity, taking also into account risks to human health. Firstly, it allows for standard setting in relation to the handling, transport, packaging, and identification of LMOs. Secondly, food safety aspects of Genetically Modified Organisms are, at international level, dealt with by the FAO/WHO Codex Alimentarius, which covers all aspects of food safety. Thirdly, objective of preventing the spread and introduction of microbes, plant and plant product pests, including weeds and other species that have indirect effects on both wild and cultivated plants, and to promote appropriate control measures. This also applies to risks associated with LMOs. It is also in the process of establishing practical cooperation with the CBD and its biosafety protocol.

1.12. LEVELS OF BIOSAFETY

Based on the specific risk associated with the organisms, generally four categories of biosafety are recognized. They are named as:

- Biosafety level 1
- Biosafety level 2
- Biosafety level 3
- Biosafety level 4

1.13. CRITERIA USED FOR BIOSAFETY LEVEL

During the decision of the biosafety levels, the following criteria is considered for placing the respective levels from one (Biosafety level 1) to four (Biosafety level 4):

- Source and nature of the introduced DNA
- Recipient organism
- Nature of expressed proteins
- Local environment
- Experimental procedures

The biosafety condition ensures the maximum protection while dealing with any potential hazardous organisms used in different aspects of biotechnology. Hence, greater protection can be planned for the expected hazard.

SUMMARY

In the present information, explosion in the field

of biology and the amount of information available in biology doubles every three and half years. With this increase in information of biological organisms, more and more potential organisms and their multifarious usefulness is identified. These living or biological organisms are the catalysts that capture and transform energy and materials producing various essentials including food, fuel, fiber and medicines. These species recycle wastes, create pure potable water, drive global biogeochemical cycles that created and maintain an aerobic atmosphere, regulate entire earths' climate through effects on greenhouse gases and local climate through effects on evapotranspiration, generate soil fertility. They also provide other ecosystem goods and services. Along with these major activities, the living organisms present in Earth's biodiversity are the source of all crops, livestock, pollinators of crops and of many pharmaceuticals and pesticides. Only the three crops such as corn, rice and wheat provide about 60% of the human food supply. These entire living organisms should be handled, used and exploited in a sustainable manner. They should not be allowed to be endangered and extinct. Hence bioethics is essential for all these activities to be performed in proper way.

In present situation, the realization is that critical thinking skills of ethical analysis are the same as those emphasized in this process-oriented method of teaching science. In fact, the integration of ethical decision making into biology curricula only reinforces the critical thinking procedures that are a foundation of the process-oriented teaching reforms. These values or moral activities essential for scientific skills or research at this level are very essential, even though basic values of nonmalfeasance, benevolence, fairness and truthfulness are taught in the lower classes. The investigation of biology, scientific technology, and ethical issues combine to give out a new branch of science known as bioethics and is coined by Van Potter. The bioethics is defined as "biology combined with diverse humanistic knowledge forging a science that sets a system of medical and environmental priorities for acceptable survival". The study of bioethics is thus useful in large number of situations that needs proper decision-making based on presuppositions, personal or societal values, moral values and subjective interpretations. They find its applications in dealing with value-rich issues such as environmental degradation, environmental pollutions, species diversity, fetal tissue transfer, organ transplantation, xenotransplantation, stem cells study and storage, cloning, genomic studies, release of transgenic organisms or the new reproductive strategies.

The genetically modified organisms (GMOs), created by using recombinant DNA technology, are at present used in agriculture, forestry, environmental clean-up programmes, biomedical fields, research and development, and other in-house experiments. Especially the genetically modified microorganisms or a genetically engineered microorganism (GEMs) has better advantage in the introduced ecosystem and thus they can be considered as greater risk-posing organisms in many ways, because they adopt and multiply rapidly in the ecosystem when compared to native flora, transfer the genes related to virulence or pathogenesis into native microbial flora, and can produced newer toxins, allergens.

Hence, these organisms that are created by the biotechnological processes create situations that are unpredictable, unexplained, uncontrolled and unmanageable conditions and risk. This is not true always and can occur with the introduction of unmodified organisms. A separate committee of the National Academy of Science, U.S.A. has considered that assessment of danger should be made based on the type of the organism that is under study and not the method of engineering. For this reasons, in biosafety protocol is established under the supervision of convention on biological diversity (or biodiversity convention).

The biosafety refers to "efforts to ensure safety in using, transporting, transferring, handling, releasing and disposing of biological organisms including genetically modified organisms (GMOs) when they are considered potentially capable of harming human, animal or plant health or the environment". In the present biotechnological research and trade, most of the developed countries have stringent laws and regulations that ensure the biosafety of potentially hazardous organisms, both in domestic use and in trade. But large number of developing countries do not have such biosafety regulatory guidelines for safeguarding their life and the environment. The implementation of biosafety sets forth a number of procedures and rules that will be helpful in a number of conditions such as protection of animals or plants from pests, diseases, or disease-causing organisms, protection of human or animal life from risks arising from additives, contaminants, residues, toxins or disease-causing organisms in foods, beverages or feedstuffs, in biomedical uses and in environments.

EXERCISE

1. Why is bioethics essential in biological field?
2. Define bioethics.
3. Who coined the word "bioethics" and when?
4. Mention benefits of the bioethics study in the biology.
5. Explain applications of the bioethics.
6. How development in biology is making difficult in considering the ethical values?
7. Discuss the opportunities of the bioethics to students.
8. How does bioethics help the biotechnology in future development?
9. What do you mean by biosafety?
10. Name important points that made the origin of biosafety.
11. What do you mean by biosafety?
12. Why biosafety is essential in the biotechnological development?
13. Name different levels of biosafety that are categorized in the biotechnology.
14. Give an account on different criteria that are used for assigning the levels of biosafety.
15. List out applications of the biosafety.

2 CHAPTER

Human Genome Project and its Ethical Issues

2.1. INTRODUCTION

The Human Genome Project (HGP) is an 13-year effort, which formally initiated in October 1990. The first idea of Human Genome Project came from the discussions held during scientific meetings, which were organized by the US department of energy and other scientific organizations between 1984 and 1986. The project was planned, spanning a period of 15 years, but rapid technological advances accelerated its completion within 13 years, i.e., in 2003. The three billion US dollars funds were earmarked for the sequencing of more than two meters length of human DNA. The goal of the project was to determine the complete sequence of the three billion (3×10^9) DNA subunits (bases), identify all human genes, and make them accessible for further biological study. As a part of the HGP, parallel sequencing was done for selected model organisms, such as the bacterium *E. coli* to help develop the technology and interpret human gene function. The Department of Energy's 'Human Genome Program (HGP)' and the National Institutes of Health's 'National Human Genome Research Institute (NHGRI)' together sponsored the US Human Genome Project. Ari Patrinos, head of the Office of Biological and Environmental Research, directed the Department of Energy's 'Human Genome Program' research. Francis Collins directed the National Institutes of Health's, National Human Genome Research Institute efforts.

The Corporate Genome Project was initiated rather late by Celera Genomics, a company founded by a former NIH scientist, Craig Venter, and funded by Perkin-Elmer, a large instrumentation manufacturer that makes and sells instruments to the government and to the private sector as well.

The Human Genome Project (HGP) and the Corporate Genome Project (CGP) are two very distinctly different entities having different cultures and attitudes. The focus has been around the privacy issue. Celera wanted to retain some of the information or control over the information that it was going to publish about the human genome, because, in fact, its business model was dependent on that fact.

From the inception of this project, due to the huge budget in sequencing of human DNA, many laboratories around the United States received funding from either the Department of Energy (DOE) or the National Institutes of Health (NIH), or both, for Human Genome Project research **(Table 2.1)**. Other researchers at numerous colleges, universities, and laboratories throughout the United States also received DOE and NIH funding for human genome research. At any given time, the DOE Human Genome Program funded about 200 separate principal investigators. Many private companies have also been conducting genome research.

At least 18 countries have established Human Genome Research Programs. Some of the larger programs are in Australia, Brazil, Canada, China, Denmark, European Union, France, Germany, Israel, Italy, Japan, Korea, Mexico, Netherlands, Russia, Sweden, United Kingdom, and the United States. Some developing countries have participated through studies of molecular biology techniques for genome research and studies of organisms that are particularly interesting to their geographical regions. The Human Genome Organization (HUGO) helped to coordinate international collaboration in the genome project.

The Human Genome Project will reap fantastic benefits for humankind. Generations of biologists and researchers have been provided with detailed DNA information, which will be a key leading to the understanding of the structure, organization, and function of DNA in chromosomes. Genome maps of other organisms will provide the basis for comparative studies that are often critical for the understanding of more complex biological systems. Information generated and technologies developed are revolutionizing future biological explorations.

2.2. FACTS OF GENOME

Importance of Genome Sequencing

A genome is the entire DNA in an organism, including its genes. Genes carry all the information, which is required for the synthesis of proteins in the organisms. These proteins

Table 2.1. US human genome project funding ($ Millions).

Financial year	DOE	NIH	US Total
1988	10.7	17.2	27.9
1989	18.5	28.2	46.7
1990	27.2	59.5	86.7
1991	47.4	87.4	134.8
1992	59.4	104.8	164.2
1993	63.0	106.1	169.1
1994	63.3	127.0	190.3
1995	68.7	153.8	222.5
1996	73.9	169.3	243.2
1997	77.9	188.9	266.8
1998	85.5	218.3	303.8
1999	89.9	225.7	315.6
2000	88.9	271.7	360.6
2001	86.4	308.4	394.8
2002	90.1	346.7	434.3
2003	64.2	372.8	437

determine, among other things, how the organism looks, how well its body metabolizes food or fights infection, and sometimes even how it behaves. The human genome is made up of DNA (which has four different chemical building blocks).

DNA is made up of four similar chemicals (called bases and abbreviated A, T, C and G) that are repeated millions or billions of times throughout a genome. The human genome, for example, has 3 billion pairs of bases.

In DNA, the particular order of As, Ts, Cs and Gs is extremely important. The order underlies the life's diversity, even dictating whether an organism is human or another species, such as yeast, rice, or fruit fly, all of which have their own genomes and are themselves the focus of genome projects. Since all organisms are related through similarities in DNA sequences, insights gained from nonhuman genomes often lead to new knowledge about human biology. To get an idea of the size of the human genome present in each of our cells, consider the following analogy: If the DNA sequence of the human genome were compiled in books, the equivalent of 200 volumes the size of a telephone book (at 1000 pages each) would be needed to hold it all **(Fig. 2.1)**.

Storing all this information is a great challenge for computer experts known as bioinformatics specialists. One million bases (called a mega base and abbreviated Mb) of DNA sequence data is roughly equivalent to 1 megabyte of computer data storage space. Since the human genome is 3 billion base pairs long, 3 gigabytes of computer data storage space is needed to store the entire genome. This includes only nucleotide sequence data, and does not include data annotations and other information that can be associated with sequence data.

As time goes on, more annotations will be entered as a result of laboratory findings, literature searches, data analyses, personal communications, automated data-analysis programs, and auto annotators. These annotations associated with the sequence data are likely to dwarf the amount of

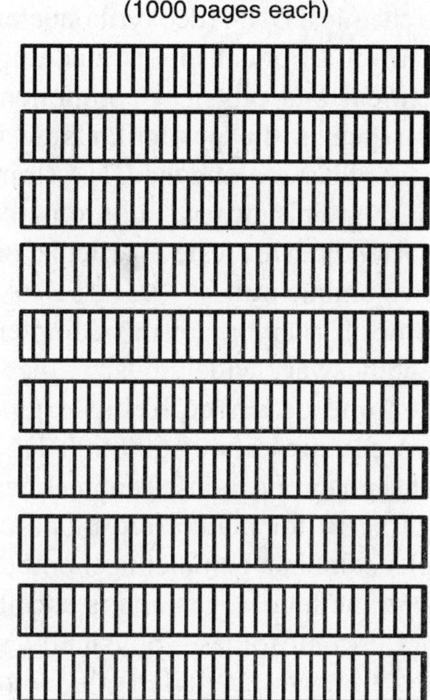

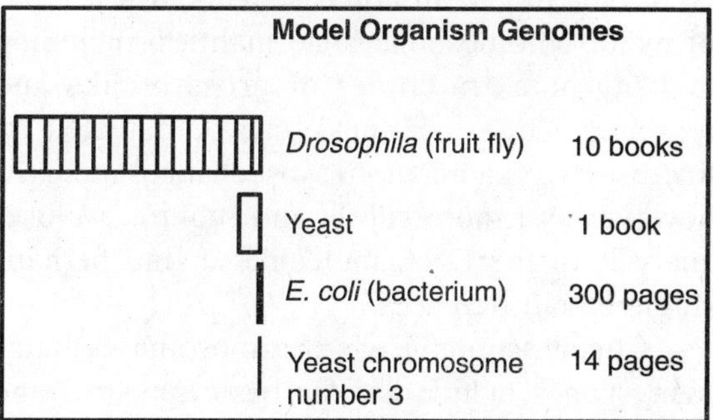

Fig. 2.1. Compiling the DNA sequence from the human genome into books would require 200 volumes, each the size of the 1,000 page Bangalore telephone book.

storage space actually taken up by the initial 3 billion nucleotide sequence. Of course, that's not much of a surprise because the sequence is merely a starting point for a much deeper biological understanding.

Human beings are also similar to other living cells in their basic cell characteristics.

Cells: These are the fundamental working units of every living system. All the instructions needed to direct their activities are contained within the chemical DNA (deoxyribonucleic acid).

DNA from all organisms is made up of the same chemical and physical components. The DNA sequence is the particular side-by-side arrangement of bases along the DNA strand (e.g., ATTCCGGA). This order spells out the exact instructions required to create a particular organism with its own unique traits.

Genome: It is an organism's complete set of DNA. Genomes vary widely in size: the smallest known genome of a free-living organism (a bacterium) contains about 600,000 DNA base pairs, while human and mouse genomes have some 3 billion. Except for mature red blood cells, all human cells contain a complete genome.

DNA in the human genome is arranged into 24 distinct chromosomes—physically separate molecules that range in length from about 50 million to 250 million base pairs. A few types of major chromosomal abnormalities, including missing or extra copies or gross breaks and rejoinings (translocations), can be detected by microscopic examination. Most changes in DNA, however, are more subtle and require a closer analysis of the DNA molecule to find perhaps single-base differences.

Chromosome: Each chromosome contains many genes, which are the basic physical and functional units of heredity. Genes are specific sequences of bases that encode instructions on how to make proteins. Genes comprise only about 2% of the human genome; the remainder consists of noncoding regions, whose functions may include providing chromosomal structural integrity and regulating the protein synthesis. The human genome is estimated to contain 20,000–25,000 genes.

Although genes get a lot of attention, it's the proteins that perform most life functions and even make-up the majority of cellular structures. Proteins are large, complex molecules made up of smaller subunits called amino acids. Chemical properties that distinguish the 20 different amino acids cause the protein chains to fold up into specific three-dimensional structures, which define their particular functions in the cell.

The constellation of all proteins in a cell is called its **proteome**. Unlike the relatively unchanging genome, the dynamic proteome changes from minute to minute in response to tens of thousands of intra- and extracellular environmental signals. The number and identities of other proteins made in the same cell at the same time and with which it associates and reacts specify by the gene sequence and a protein's chemistry and behavior. Studies to explore protein structure and activities, known as **proteomics**, will be the focus of much research for decades to come, and will help elucidate the molecular basis of health and disease.

An average gene consists of 3000 bases, but sizes vary greatly. The largest known human gene 'dystrophin' consists of 2.4 million bases.

The total number of genes is estimated to be 30,000, which is much lower than previous estimates of 80,000 to 140,000 that had been based on extrapolations from gene-rich areas as opposed to a composite of gene-rich and gene-poor areas.

The functions are unknown for over 50% of the discovered genes. Less than 2% of the genome codes for proteins.

Repeated sequences that do not code for proteins ("junk DNA") make-up at least 50% of the human genome. Repetitive sequences **(Fig. 2.2)** are thought to have no direct functions, but they shed light on chromosome structure and dynamics. Over time, these repeats reshape the genome by rearranging it, creating entirely new genes, and modifying and reshuffling existing genes.

During the past 50 million years, a dramatic decrease seems to have occurred in the rate of accumulation of repeats in the human genome.

2.3. GENOME ARRANGEMENT

The human genome's gene-dense "urban centers" are predominantly composed of the DNA building

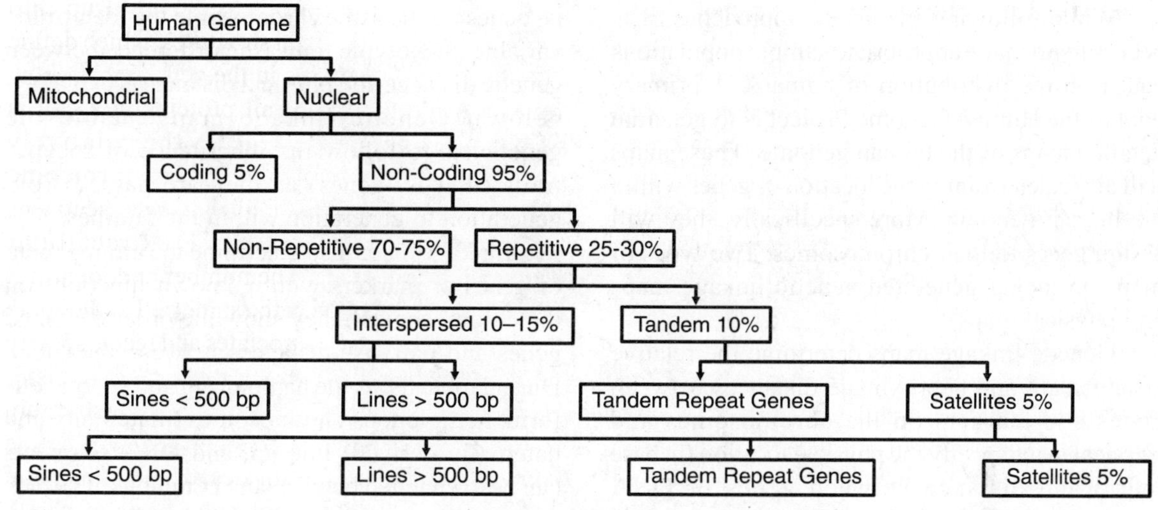

Fig. 2.2. Human genome and nature of DNA.

blocks, G and C. In contrast, the gene-poor "deserts" are rich in the DNA building blocks, A and T. GC- and AT-rich regions can usually be seen through a microscope as light and dark bands on chromosomes.

Genes appear to be concentrated in random areas along the genome, with vast expanses of non-coding DNA in between.

Stretches of upto 30,000 C and G bases repeating over and over often occur adjacent to gene-rich areas, forming a barrier between the genes and the "junk DNA". These CpG islands are believed to help regulate gene activity. Chromosome 1 has most of the genes (2968), and the Y chromosome has the fewest (231).

Scientists have identified about 1.4 million locations where single-base DNA differences i.e., single nucleoride polymorphisms (SNPs) occur in humans. This information promises to revolutionize the processes of finding chromosomal locations for disease-associated sequences and tracing human history.

2.4. GENOME SEQUENCING

Genome sequencing is the term used to describe the laboratory process of reading the order of the four letters of the genetic alphabet (A, C, G, T) along a strand of DNA. The steps involved in such efforts are as follows:

1. Selection of suitable sample materials.
2. Isolation of DNA from the cells, and preparation of large samples of high quality DNA from these cells.
3. Cutting the purified DNA at random sites into a manageable size, overlapping pieces of the DNA sample.
4. Insertion of these DNA pieces into packages for the production of unlimited copies of such selected DNA.
5. Recording the order of bases for each DNA sample piece by using DNA sequencing techniques.
6. Determination of the overlap of each piece, and assembling the sequences to give the final genome of the human.

While following the above approaches, it is necessary to make appropriate sample populations based on the distribution of humans. A primary goal of the Human Genome Project is to generate detailed maps of the human genome. These maps will aid in determining the location of genes within the human genome. More specifically, they will assign genes to their chromosomes. Two types of maps are being generated genetic linkage maps and physical maps.

Genetic linkage maps determine the relative arrangement and approximate distances between genes and markers on the chromosomes and physical maps specify the physical location (in base pairs) and distance between genes or DNA fragments with unknown functions that are mapped to specific regions of the chromosomes.

Maps have different levels of resolution, ranging from low to high. The degree of resolution that is appropriate depends on whether, for example, a large fragment of DNA is to be studied or a more detailed picture of a small DNA region is needed. A human genomic library consists of random DNA fragments, and is used to establish sets of ordered, overlapping cloned DNA fragments or contigs for each chromosome of the genome, In other words, these are high-resolution maps.

After mapping is complete, the DNA must be sequenced to determine the order of all the nucleotide bases of the chromosomes, and the genes in the DNA sequence must be identified. In all phases of the project, a major focus has been on developing instrumentation to increase the speed of data collection and analysis. New, automated technologies are significantly increasing the speed and accuracy of DNA sequencing, while decreasing the cost. Software and database systems manage the data generated from mapping and sequencing projects. Database management systems store and aid in distributing genomic information **(Fig. 2.3)**.

Genetic linkage maps show the order and genetic distance between pairs of linked genes, that is, genes on the same chromosome that determine variable phenotypic traits (the difference between genetic distance and physical distance is explained below). Genetic linkage maps enable the geneticists to follow the inheritance of specific traits (that is, genes) as they are passed from generation to generation within the families.

Linkage maps also determine the arrangement of genes or markers with unknown functions on the chromosomes. They show the order of linked genes and pairwise distances between their loci. During meiosis, as the haploid egg and sperm cells form, homologous chromosomes (maternally and paternally derived) line up, and DNA segments can be exchanged between the homologs. The new combinations of alleles result from this process of homologous recombination. During meiosis, each human chromosome pair is involved, on an average, in 1.5 crossover events. The likelihood of crossing over increases as the distance between the two loci increases. Crossing over between two genes or markers on the same chromosome can sometimes occur if there is enough distance between them. If two genes are very close, they are "linked" and recombination is unlikely to occur between them. Thus, the frequency of recombination is a quantitative index of the linear distance between two genes on a genetic linkage map. Distances are measured in centimorgans (cM), named after the famous geneticist, Thomas Hunt Morgan. If genes (for example, A and B) are separated by recombination 1% of the time, that is, if one out of 100 products of meiosis is recombinant, they are 1 cM apart. A genetic distance of 1 cM represents a physical distance of approximately one million base pairs (1 Mb). Genetic maps are very powerful. An inherited disease gene can be located on the map if a second gene or DNA reference marker is also inherited in individuals with the disease, but is not found in individuals who do not have that disease. Exact chromosomal locations have already been found for many disease genes, including fragile X syndrome, cystic fibrosis and Buntington's disease.

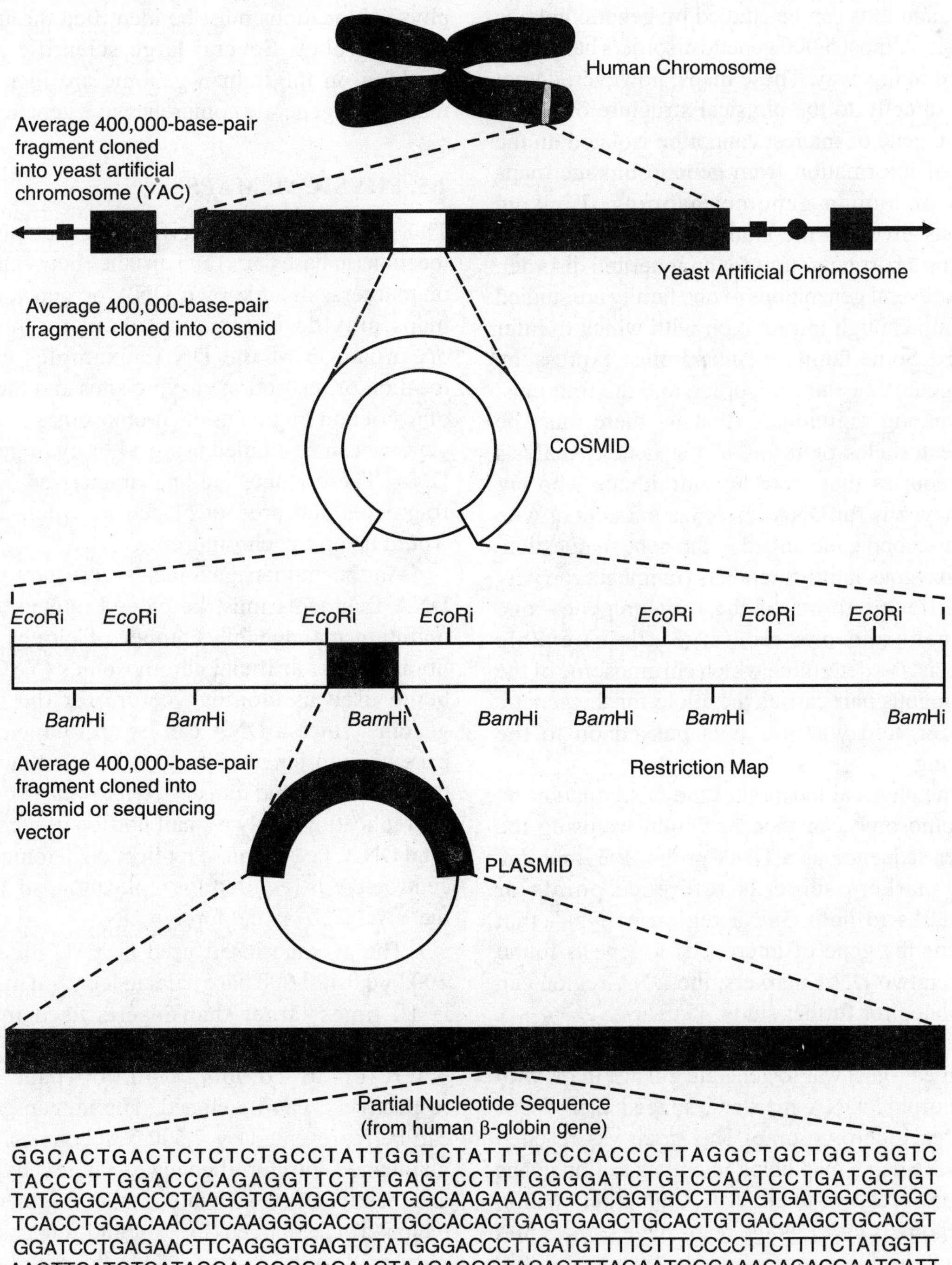

Fig. 2.3. Process of determination of DNA sequence from human chromosome.

Many inherited diseases are caused by single genes, and thus can be studied by genetic linkage analysis. Almost 5,000 genetic disorders have been studied in this way. These maps, however, do not relate directly to the physical structure of DNA, and the gene of interest cannot be isolated on the basis of information from genetic linkage maps alone or human genome mapping. Linkage analysis involves the study of family members carrying a particular trait for an inherited disorder. Often, several generations of one family are studied to obtain enough information with which to infer linkage. Some family members must express the trait (gene) or genetic disorder, and the trait must vary among individuals (that is, there must be different alleles or forms of the gene). Analysis also requires that there be individuals who are heterozygous for DNA reference markers or who have a second gene linked to the gene in question. Heterozygous family members (members carrying two different forms of the trait or gene—one dominant and one recessive allele) enable geneticists to determine which chromosome of the homologous pair carries the allele for the genetic disorder, and whether it is passed on to the offspring.

The physical location of the DNA marker on a chromosome can then be found by using the marker sequence as a DNA probe. Polymorphic DNA markers serve as reference points or landmarks to help find a region of DNA that contains the gene of interest. If a gene is found between two DNA markers, the DNA region can be isolated for further study.

An early goal of the investigators of Human Genome Project was to generate linkage maps with polymorphic DNA markers, spaced 2 to 5 cM along each chromosome. This goal was reached in 1995. Such a map helps scientists to find genes of interest relative to about 1,500 markers within the genome, Once linkage maps have some 3,300 polymorphic DNA markers, each separated by only 1 cM, gene hunting will be much easier. Thus, for polymorphic DNA markers to be valuable, their linkage with a gene must be established, and their physical locations must be identified through the use of probes. Several large scientific groups working on the human genome are identifying markers to generate comprehensive genetic maps.

2.5. PHYSICAL MAPS

The physical maps specify the exact physical location (in base pairs) and distance between genes or markers, or unknown DNA or genes. These maps provide information about the physical organization of the DNA; examples are the location of restriction enzyme sites and the order of restriction fragments of chromosomes. An entire genome can be studied using a library of genomic DNA. These clones are uncharacterized, random fragments and are not placed in order, as they would be on the chromosome.

As the human genome is very large, large DNA fragments must be cloned into vectors to maintain manageable number of clones in the library. Yeast artificial chromosomes (YACs) are being used as cloning vectors for the human genome, since a DNA can be up to one million base pairs in length. Human DNA is attached to the yeast DNA and transferred into yeast host cells for replication. Only a small portion of the yeast's total DNA, i.e., origin of replication, telomere, and centromere is required for replication, so most of the YAC DNA is the foreign DNA.

The average insert used in YAC libraries is 200,000 0 400,000 base pairs in length. This range is 10 times larger than inserts used in other libraries, such as for bacteriophage and cosmids, where up to 20,000 to 40,000 base pairs, respectively, can be cloned. The human genome can be represented by 7,500 YAC clones, and is maintained and amplified in yeast host cells. YACs and their inserts are cut into smaller fragments and recloned or subcloned (for example, into cosmids), so that a detailed map of a YAC clone is obtained. YAC clones are screened by PCR to isolate specific genes of interest. DNA inserts are also analyzed by obtaining restriction maps, identifying polymorphic markers, and/or DNA sequencing. However, without an ordered physical map, i.e., one that refers to actual physical distances in base pairs between landmarks, the location of particular clones cannot be identified.

Another method called fluorescence *in situ* hybridization (FISH) of probes to metaphase chromosomes provides information for constructing low-resolution chromosomal maps. Chromosomal maps are actual physical maps because distances are measured in base pairs. Metaphase chromosomes are spread out on a microscope slide, and a solution containing a fluorescent-tagged DNA probe is added. Under the appropriate conditions, the probe hybridizes to its DNA complement on the chromosome and is detected with a fluorescent microscope **(Figs. 2.4 & 2.5)**. The relative orientation of genes and DNA fragments can be assigned to specific chromosomes, and the gaps between mapped

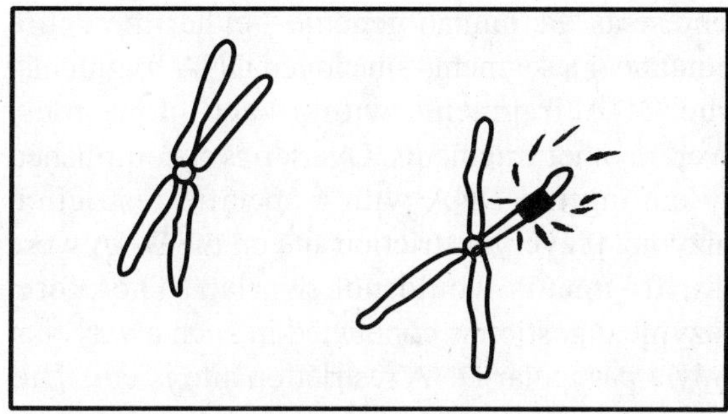

Fig. 2.5. Fluorescence of chromosome position by probes in fluorescence *in situ* hybridization.

cosmids can be bridged. Chromosomal mapping is used to locate genetic markers that are associated with observable traits.

Another type of physical map is the cDNA map, which localizes coding regions (for example exons) to specific chromosome regions or bands. The cDNA molecules are synthesized from an mRNA template. The DNA map is probably one of the most important types of map, since it can identify the chromosomal location of specific genes, whether their functions are known or not. Researchers searching for a specific disease causing gene can use cDNA maps to help locate it after having established a general location by genetic linkage methods.

High-resolution physical maps can be generated by a method that is sometimes called bottom-up mapping. The chromosome is cut into small overlapping fragments, each of which is cloned and the order determined. These fragments form continuous DNA blocks called contigs. The bottom-up method generates a detailed map called a 'contig' map. A library of clones ranging from 10,000 base pairs to 1 Mb is used for mapping. Each clone can be localized to specific regions within chromosomal bands. This "linked" library of overlapping clones comprises a chromosomal segment.

The production of human contig maps requires several steps. First, a library must be made that

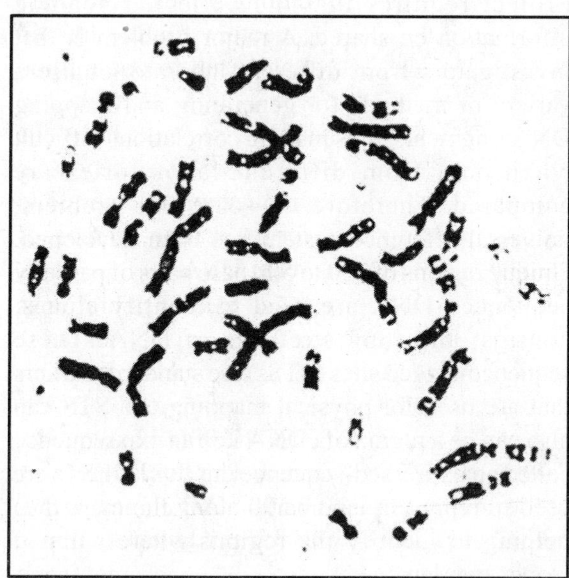

Fig. 2.4. A microscopic preparation (metaphase squash before a karyotype is made) of human chromosomes showing the differences in size and banding patterns of the chromosomes.

represents the human genome—either the entire genome or a segment—in cloned DNA fragments. The DNA fragments within each clone must overlap other fragments. Overlap is accomplished by cutting the DNA with a specific restriction enzyme. If every restriction site on the DNA were cut, fragments would not overlap. Therefore, enzyme digestion is conducted in such a way that only a particular DNA restriction site is cut. This partial digestion randomly leaves many sites uncut, so that overlapping DNA fragments are produced and the order along the chromosomes can be determined.

The order of the clones or contigs can be determined by identifying the overlaps in the DNA fragments. Overlap can be detected when some of the DNA bands are the same i.e., two clones have bands in common. This method of assembling pairs of clones into contigs is difficult and time-consuming.

Automation and sophisticated computer algorithms may increase the efficiency. Different approaches may be used to fill in the gaps that are likely to be present even after researchers generate detailed physical maps. For example, micro-dissection, which is used to physically cut a piece of DNA from a specific region of a chromosome. This chromosomal piece can be cut into smaller fragments by restriction enzymes, cloned, mapped, and sequenced by standard methods.

An alternate method is "chromosome walking", in which a small region at the end of the DNA fragment is used as a probe to screen the library for the adjacent clone. A DNA piece at the end of this second cloned fragment is used as a next probe. This process continues until a complete physical map has been obtained. Since the human genome is divided into chromosomes, chromosome specific libraries can be constructed so that each chromosome has a contig map.

Mapping is simplified if each chromosome is separated from the others before being cut by restriction enzymes and cloned to make libraries. Twenty-four libraries are required: 22 autosomal libraries, and one each for the X and Y chromosomes. The several types of maps range from coarse to fine resolution.

The map with the lowest resolution is the genetic map, which measures the frequency of recombination between linked markers (which can be genes or noncoding DNA). The next level of resolution is the restriction map, on which DNA restriction fragments ranging from 1 to 2 Mb are separated and mapped. The next higher level of resolution is achieved by placing in order 400,000 to 1,000,000 base pair fragments of overlapping clones from libraries of YAC clones. These clones are then further subcloned (with insert sizes of 20,000 to 40,000 base pairs) into other vectors to produce contig maps. Finally, the DNA base sequence map having the finest resolution is determined.

Sequence-Tagged Sites

In the sequencing approaches, Human Genome Project requires that the collected genome information be shared. A major problem is that investigators from different laboratories use a variety of methods for generating and mapping DNA fragments, thus making correlations difficult when data from different laboratories are compared. Therefore, to solve this problem, universal reference system has been developed. Unique regions of 200 to 500 base pairs of partially sequenced DNA are used to identify clones, contigs, and long stretches of DNA. These sequence-tagged sites (STSs) are standard markers that are used for physical mapping. An STS can also can be a region of cDNA i.e., an exp sequence called an expressed-sequence tag (EST). ESTs are used to represent landmarks along the map, thus helping to identify the regions where pairs of clones overlap.

These special sequences constitute a "universal mapping language", enabling everyone to refer to a specific region of the genome by the same name, and enabling investigators to share information and construct compatible maps. A goal is to generate STSs every 100,000 base pairs for each human chromosome. Each chromosome would comprise contigs that are one to two million base pairs in length, and cover over 95% of the chromosome.

Thus, approximately, 30,000 STSs must be identified. To identify an STS, a short DNA fragment is isolated from a chromosome- specific library of clones or from one small fragment within a clone from a contig. A small region, 200 to 400 base pairs in length, is sequenced and compared to all known repeated sequences. Sophisticated computer analyses programs help in identifying unique sequences. PCR can be used to screen a library for clones containing STSs. The PCR amplifies the unique DNA region from the total human genome.

Gel electrophoresis will yield one DNA band if the STS is unique, and more bands on the gel if additional regions are amplified, indicating that the STS is not unique. An STS can also include a repeated sequence (examples: unique sequence GAGAGAGA. . . . . . . .GA unique sequence or unique sequence (GAn) unique sequence) if unique sequences flank the repeat on either side. STSs would then be polymorphic in the region of the repeat, (that is, repeated *n* times) and would have many alleles in a population.

Each individual carries two copies of the STS marker—one on each homologous chromosome. Thus, the inheritance of the STS can be determined. Once located on the physical map, polymorphic STSs aid in aligning the genetic linkage map with the physical map. The use of STSs as probes for contig mapping allows clone overlap to be established. This method is called STS-content mapping. Clones that share an STS overlap belong to the same contig. To be most useful, STSs should be spaced every few thousand base pairs. STS information is stored in computer databases analogous to, for example, the large DNA sequence database called Genbank. As progress is made, information will include PCR primer sequences, PCR reaction conditions, and the size of the

fragment amplified. The amplified PCR fragment can then be used as a probe in hybridizations to isolate clones containing the STS marker.

2.6. FUSION OF GENETIC LINKAGE MAPS AND PHYSICAL MAPS

The use of polymorphic DNA markers for genetic linkage analysis is a powerful method for finding the approximate location of genes causing inherited diseases. If the polymorphic DNA markers are co-inherited or linked closely to the gene of interest, the markers arid gene must be physically close to one another. To be amenable for further analysis, genetic linkage markers must be within two million base pairs (preferably no more than 1 cM) from the disease causing gene. To isolate the disease causing gene, a method called chromosome walking is used. DNA markers are used to link the genetic linkage map to the physical map. Flanking DNA markers are used as a starting point in hybridizations to clones comprising contigs in the disease gene region. Once the two types of maps are integrated, genes for inherited disorders can be identified, isolated and sequenced.

2.7. COMPARISON OF HUMAN WITH OTHER ORGANISMS

Unlike the human's seemingly random distribution of gene-rich areas, many other organisms' genomes are more uniform, with genes spaced evenly throughout.

Humans have on an average three times as many kinds of proteins as the fly or worm, because of mRNA transcript "alternative splicing" and chemical modifications of the proteins. This process can yield different protein products from the same gene.

Humans share most of the same protein families with worms, flies and plants, but the number of gene family members has expanded in humans, especially in proteins involved in development and immunity.

The human genome has a much greater portion (50%) of repeat sequences than the mustard weed (11%), the worm (7%) and the fly (3%).

Although humans appear to have stopped accumulating repeated DNA over 50 million years ago, there seems to be no such decline in rodents. This may account for some of the fundamental differences between hominids and rodents, though gene estimates are similar in these species. Scientists have proposed many theories to explain evolutionary contrasts between humans and other organisms, including those of life span, litter sizes, inbreeding, and genetic drift.

Variations and Mutations

US Human Genome Project Research Goals

The completion of the human DNA sequence in the spring of 2003 coincided with the 50th anniversary of Watson and Crick's description of the fundamental structure of DNA. The analytical power arising from the reference DNA sequences of entire genomes and other genomics resources has jump-started, what some call the "biology century".

The Human Genome Project was marked by accelerated progress. In June, 2000, the rough draft of the human genome was completed a year before its schedule time. In February 2001, special issues of Science and Nature contained the working draft sequence and analyses were published.

The project's first 5-year plan, intended to guide research in financial years 1990-1995, was revised in 1993 due to unexpected progress, and the second plan outlined goals through the FY, 1998. The third and final plan (Science, 23 October 1998) was developed during a series of DOE and NIH workshops. Some 18 countries have participated in the worldwide effort, with significant contributions from the Sanger Center in the United Kingdom, and research centers in Germany, France and Japan.

Difference Between Draft Sequence and Finished Sequence

To generate the high-quality reference sequence, completed in April 2003, an additional sequencing was done to close the gaps and reduce the ambiguities. Further, only a single error was allowed for every 10,000 bases, the agreed-upon standard for the HGP. Investigators believe that a high-quality sequence is critical for recognizing regulatory components of genes that are very important in understanding human biology and disorders such as heart disease, cancer, and diabetes. The genomes have been sequenced completely as shown in the **Table 2.2**.

The small genomes of several viruses and bacteria, and the much larger genomes of three higher organisms have been completely sequenced. They are bakers' or brewers' yeast (*Saccharomyces cerevisiae*), the roundworm (*Caenorhabditis elegans*) and the fruit fly (*Drosophila melanogaster*). In October, 2001, the draft sequence of the pufferfish *Fugu rubripes*, the first vertebrate after the human, was completed, and scientists finished the first genetic sequence of a plant weed *Arabidopsis thaliana*, in December 2000. Many more genomes have been sequenced since then.

Human Genome project is also called **Human Genome Initiative Scientific Research Effort** to analyze the DNA of humans and of several lower organisms. The project began in the United States in 1990 under the sponsorship of the US Department of Energy and the National Institutes of Health. Projects undertaken concurrently in Japan, the United Kingdom, Italy, France, and Russia are coordinated with the American effort through the Human Genome Organization.

The ultimate goal of the project is to identify the chromosomal location of every human gene, and to determine the precise chemical structure of each gene in order to elucidate its function in health and disease. The information gathered is expected to serve as the basic reference for research in human biology and medicine in the 21st century,

Table 2.2. The list of organisms whose genome sequence is completed.

Group	Organism	Genome size	Haploid number
Virus	MS2	4 kb	1
	SV40	5 kb	1
	ϕX174	5 kb	1
	M13	6 kb	1
	λ	50 kb	1
	Herpes simplex	152 kb	1
	T_2, T_4, T_6	165 kb	1
	Smallpox	267 kb	1
Prokaryotes	*Methanococcus jannaschii*	1600 kb	1
	E. coli	4600 kb	1
Eukaryotes	*Borrelia burgdorferi*	910 kb	1
	Saccharomyces cerevisiae	13 Mb	16
	Caenorhabditis elegans	97 Mb	06
	Arabidopsis thaliana	100 Mb	05
	Drosophila melanogaster	180 Mb	04
	Homo sapiens	3000 Mb	23
	Zea mays	4500 Mb	10
	Fugu rubripes	400 Mb	22
	Amphiuma means	90,000 Mb	14

and to provide fundamental insights into the genetic basis of human diseases. The new technologies developed in the course of the project will be applicable in numerous other fields of biomedical endeavour.

Each cell of an organism has a set of chromosomes containing the heritable genetic material that directs its development, i.e., its genome. The genetic material of chromosomes is DNA. Each of the paired strands of the DNA molecule is a linear array of subunits called nucleotides, or bases, of which there are four types—adenine, cytosine, thymine, and guanine. Genes are discrete stretches of nucleotides that carry the information, which is used by the cell to synthesize proteins.

Human genes take up only about 5 to 10% of the DNA. Some of the remaining DNA, which does not code for proteins, may regulate whether or not proteins are made, but the function of most of it is unknown.

This landmark of scientific achievement represented the completion of the first stage of the project. Initial results published by both groups in February 2001 declared that the human genome actually contains only about 30,000 to 40,000 genes, much fewer than originally thought. Two types of maps were constructed—genetic linkage maps and physical maps. Genetic linkage map provides the relative location of genes and other markers on the basis of how frequently genes are inherited together; the closer genes are to each other on a chromosome, the more likely they are to be inherited together. Physical maps locate genes in relation to the presence of known nucleotide sequences that act as landmarks along the length of a chromosome.

One such "marker" used to map the human genome is a sequence-tagged site (STS)—a short sequence of nucleotides that occurs only once throughout the genome. A relatively detailed physical map was needed before sequencing could begin. Sequencing, in which the precise order of the nucleotide sequence is determined, was the most technically challenging part of the project.

DNA sequencing of the nematode worm *Caenorhabditis elegans* and the yeast *Saccharomyces cerevisiae* was completed in 1996. The DNA sequencing of the other organisms was completed in the following order:

(1) *E. coli*-1997.

(2) Fruit fly *(Drosophila melanogaster)* and plant *Arabidopsis thaliana*—2000.

(3) The laboratory mouse (*Mus musculus*) and bacterium *Staphylococcus aureus*—2001.

The rationale for these efforts is that many genes with similar functions in disparate organisms have been conserved in evolution and show surprising similarities. Genes from simpler organisms can thus be used to study human beings.

Another objective of the Human Genome Project is to address the ethical, legal, and social implications of the information obtained. Society will derive the greatest benefit from this knowledge only if it takes measures to prevent abuses, such as invasion of the privacy of an individual's genetic background by employers, insurers, or government agencies, or discrimination based on genetic grounds.

The HGP was the natural culmination of the history of genetics research. In 1911, Alfred Sturtevant, then an undergraduate researcher in the laboratory of Thomas Hunt Morgan, realized that he could—and had to, in order to manage his data—map the location of the fruit fly (*Drosophila melanogaster*) genes, whose mutations Morgan laboratory was tracking over generations. Sturtevant's very first gene map can be likened to the Wright brothers' first flight at Kitty Hawk. In turn, the Human Genome Project can be compared to the Apollo program bringing humanity to the moon.

The hereditary material of all multicellular organisms is the famous double helix of deoxyribonucleic acid (DNA), which contains all of our genes. DNA, in turn, is made up of four chemical bases, pairs of which form the "rungs" of the twisted, ladder-shaped DNA molecules. All genes are made up of stretches of these four bases, which are arranged in different ways and in different lengths. HGP researchers have deciphered the human genome in three major ways—determining the order or "sequence", of all the bases in our genome's DNA, making maps that show the locations of genes in major sections of all our chromosomes, and producing what are called linkage maps, complex versions of the type originated in early Drosophila research, through which inherited traits (such as those for genetic disease) can be tracked over generations.

The HGP has revealed that there are probably somewhere between 30,000 and 40,000 human genes, and their location can be identified now. This ultimate product of the HGP has given the world a resource of detailed information about the structure, organization and function of the complete set of human genes. This information can be thought as the basic set of inheritable "instructions" for the development and functioning of a human being.

The International Human Genome Sequencing Consortium published the first draft of the human genome in the journal 'Nature' in February 2001, with the sequence of the entire genome's three billion base pairs some 90 percent complete. A startling finding of this first draft was that the number of human genes appeared to be significantly fewer than previous estimates, which ranged from 50,000 genes to as many as 140,000. The full sequence was completed and published in April 2003.

How to Sequence

The task of determining the complete sequence of the 3,200,000,000 bases of the human genome (30× the size of the nematode genome) was extremely daunting at the time when the project was formally launched. Several lines of investigation focused on an alternative approach to characterize the human genome. For example, complete genome sequencing may be bypassed by selectively sequencing just expressed sequences, obtained by extracting mRNA from a wide range of human tissues. Large scale expressed sequence tags (EST) projects in both the public and private domain resulted in the collection of huge amount of sequence information on human genes. An international consortium to map the ESTs in the genome, using the genetic map as a framework, resulted in the human gene map of 35,000 genes. This was an important and valuable milestone in HGP. However, the sequence of most of the mRNAs was incomplete, unknown number of genes were missing from the collection, and there was no information available on gene structures.

In contrast, the experience gained from the study of smaller genomes, especially those of the nematode and yeast illustrated the enormous potential to obtain a complete set of genes, gene structures and all other genetic information by determining the complete sequence of the genome. Furthermore, by breaking the task into manageable segments, and using a physical map to co-ordinate the work, it was possible to undertake projects to sequence genomes that were far beyond the capabilities of a simple shotgun approach. For the human genome, therefore, the strategy adopted was to use the landmarks provided by the genetic map, and later the gene map, as a framework to anchor a physical map of overlapping clones which represented all human chromosomes. initially much of the work was done using yeast artificial chromosomes (YACs), a yeast cloning system, which accepts vary large fragments and thus allows a physical map to be built quickly over very large distances. However, the development of new bacterial cloning systems called BACs or PACs (bacteria or P1 derived artificial chromosomes), which were capable of taking large inserts (up to 250 kb) made it possible to make long range maps directly in bacterial clones. This coupled with the greater convenience and stability of bacterial clones compared to YACs, resulted in the choice

of this system for construction of the physical map to provide the tile path of clones for sequencing.

Each BACs or PACs has been sequenced using a random shotgun approach. This approach is essentially the same as was developed for the early whole genome sequencing projects. DNA from the BAC or PAC is broken up randomly into short fragments (typically 1-2kb long), which are sub cloned into plasmids or bacteriophage M13 cloning vector. The resulting sub clones (transformed bacterial colonies) are picked at random, cultured and the sub clone DNA is extracted for use as a sequencing template. A primer (short DNA strand) is hybridized to the template within the vector sequence (which is common to all clones). This provides a starting point for DNA polymerase to synthesize new strands of DNA by incorporating the deoxynucleotide triphosphate (dNTPs), which are the single base precursors of DNA.

Fluorescently labeled analogues for each base are included in the same reaction (dideoxy NTPs or ddNTPs); a different fluorescent tag is used for each of the four bases. These analogues extend the chain in a base-specific manner when they are incorporated (and these are called "chain terminators"). The product of the reaction is a ladder of newly synthesized DNA fragments of increasing size in single base increments. Each fragment in the mixture is terminated at a specific place, which can be identified according to its specific fluorescent label that can be separated on the basis of size by electrophoresis, either through polyacrylamide "slab" gels, or more recently through a viscous liquid matrix held in individual capillaries (capillary gel electrophoresis). The ladder of colored bands thus represents the sequence of the bases in the DNA, and can be read automatically by an automatic fluorescent detector.

The sequence of all the sub clones of a single BAC or PAC is analyzed together. All overlap[s] between sequences are identified, and the individual reads are assembled onto contigs. A consensus sequence is obtained at this stage, and all the assembled sequence data is released on the internet. Much of the human genome sequence is currently in this form as the "working draft". in the second phase, the assembled sequence for each BAC or PAC is re-examined manually—all gaps are closed, all ambiguities are resolved, and every base is checked. This process of finishing results in a handcrafted product of high accuracy (> 99.90%). The finished sequence of each BAC or PAC is submitted to the public databases.

At the time of release of the sequence of each BAC or PAC, either as working draft or in finished form, further analysis is carried out to identify all possible genes and other important biological features. These features are marked up or annotated on the sequence. For the working draft, a fast process has been developed at multiple sites, so that all the genomic sequence annotation is kept up to date and is carried out to the same standard across the entire genome.

These ambitious goals required and will continue to demand a variety of new technologies that have made it possible to rapidly construct a first draft of the human genome and to continue to refine that draft. These techniques include the following:
- DNA Sequencing
- The Employment of Restriction Fragment-Length Polymorphisms (RFLP)
- Yeast Artificial Chromosomes (YAC)
- Bacterial Artificial Chromosomes (BAC)
- The Polymerase Chain Reaction (PCR).
- Electrophoresis

Of course, information is only as good as the ability to use it. Therefore, advanced methods for widely disseminating the information generated by the HGP to scientists, physicians and others, is necessary in order to ensure the most rapid application of research results for the benefit of humanity. Biomedical technology and research are particular beneficiaries of the HGP.

However, the momentous implications for individuals and society for possessing the detailed genetic information made possible by the HGP were recognized from the outset. Another major component of the HGP and an ongoing component of NHGRI is therefore devoted to the analysis of the ethical, legal and social implications (ELSI) of our newfound genetic knowledge, and the subsequent development of policy options for public consideration.

Human Genome Project and its Accomplishment

In addition to the human genome, the Human Genome Project sequenced the genomes of several other organisms, including brewers' yeast, the roundworm, and the fruit fly. In 2002, researchers announced that they had also completed a working draft of the mouse genome. By studying the similarities and differences between human genes and those of other organisms, researchers can discover the functions of particular genes, and identify which genes are critical for life.

The Project's Ethical, Legal, and Social Implications (ELSI) program became the world's largest bioethics program and a model for other ELSI programs worldwide.

Completed in 2003, the Human Genome Project (HGP) was a 13-year project, which was coordinated by the US Department of Energy and the National Institutes of Health. During the early years of the HGP, the Wellcome Trust (UK) became a major partner; additional contributions came from Japan, France, Germany, China, and others. The goals of project were to:
- identify all the, approximately 20,000–25,000 genes in human DNA;
- determine the sequences of the 3 billion chemical base pairs that make-up human DNA;
- store this information in databases;
- improve tools for data analysis;
- transfer related technologies to the private sector; and
- address the ethical, legal, and social issues (ELSI) that may arise from the project.

Although the HGP is finished, analyses of the data will continue for many years. An important feature of the HGP project was the federal government's long-standing dedication to the transfer of technology to the private sector. By licensing technologies to private companies and awarding grants for innovative research, the project catalyzed the multibillion-dollar US biotechnology industry, and fostered the development of new medical applications.

Rapid progress in genome science and a glimpse into its potential applications have spurred observers to predict that biology will be the foremost science of the 21st century. Technology and resources generated by the Human Genome Project and other genomics research are already having a major impact on research across the life sciences. The potential for commercial development of genomics research presents US industry with a wealth of opportunities, and sales of DNA-based products and technologies in the biotechnology industry are projected to exceed $45 billion by 2009 (Consulting Resources Corporation Newsletter, Spring 1999).

Current and Potential Applications of Genome Research Include the Following:

- Molecular medicine
- Energy sources and environmental applications
- Risk assessment
- Bioarchaeology, anthropology, evolution, and human migration
- DNA forensics (identification)
- Agriculture, livestock breeding and bioprocessing
- Molecular medicine
- Improved diagnosis of disease
- Earlier detection of genetic predispositions to disease
- Rational drug design
- Gene therapy and control systems for drugs
- Pharmacogenomics "custom drugs"

Broader applications reaching into many areas of the economy include the following:

- **Clinical medicine:** Many more individualized diagnostics and prognostics, drugs, and other therapies.
- **Agriculture and livestock:** More nutritious and healthier crops and animals.
- **Industrial processes:** Cleaner and more efficient manufacturing in sectors such as chemicals, pulp and paper, textiles, food, fuels, metals, and minerals.
- **Environmental biotechnology:** Biodegradable products, new energy resources, environmental diagnostics and less hazardous cleanup of mixed toxic-waste sites.
- **DNA fingerprinting:** Identification of humans and other animals, plants and microbes; evolutionary and human anthropological studies; and detection of and resistance to harmful agents that might be used in biological warfare.

Technology and resources promoted by the Human Genome Project are beginning to have profound impacts on biomedical research, and promise to revolutionize the wider spectrum of biological research and clinical medicine. Increasingly detailed genome maps have aided researchers seeking genes associated with dozens of genetic conditions, including myotonic dystrophy, fragile X syndrome, neurofibromatosis, diabetes types 1 and 2, inherited colon cancer, Alzheimer's disease and familial breast cancer.

On the horizon is a new era of molecular medicine, characterized less by treating symptoms and more by looking to the most fundamental causes of disease. Rapid and more specific diagnostic tests will make earlier treatment of countless maladies possible. Medical researchers will also be able to devise novel therapeutic regimens based on new classes of drugs, immunotherapy techniques, avoidance of environmental conditions that may trigger disease, and possible augmentation or even replacement of defective genes through gene therapy.

Research Milestones of Human Genome Project

- Mouse genome sequencing consortium publishes draft sequence of mouse genome, December 2002.
- International consortium led by the DOE joint genome institute publishes draft sequence of *Fugu rubripes*, July 2002.
- HGP leaders publish the initial analysis of the working draft of the human genome sequence, February 2001.
- HGP leaders and President Clinton announce the completion of a "working draft" DNA sequence of the human genome, June 26, 2000.
- Press briefing and remarks.
- Human genome news article.
- Human chromosome 22 completed: First human chromosome to be sequenced, December 1999.
- HGP leaders confirm accelerated timetable for draft sequence, October 1999.
- Completion date for "Human genome working draft" accelerated to spring of 2000, March 1999 news release.
- Human genome project 5-year research goals 1998-2003, October 1998; also contains links to previous goals since the project's beginnings.
- Drosophila genome completed, March 2000.
- Draft sequences of human chromosomes 5, 16, 19 completed, April 2000.
- Human chromosome 2 completed, April 2005.
- Human chromosome 4 completed, April 2005.
- Human chromosome X completed, March 2005.
- Human chromosome 16 completed, December 2004.
- Human chromosome 5 completed, September 2004.
- Human chromosome 9 completed, May 2004.
- Human chromosome 10 completed, May 2004.
- Human chromosome 19 completed, March 2004.
- Human chromosome 13 completed, March 2004.
- Human chromosome 6 completed, October 2003.
- Human chromosome 7 completed, July 2003.
- Human chromosome Y completed, June 2003.
- Human chromosome 21 completed, May 2000.
- Human chromosome 14 completed, January 2003.
- Human chromosome 20 completed, December 2001.
- Human genome project completion:1990-2003 (April 2003).
- Human gene count estimates changed to 20,000 to 25,000, October 2004.

In April 2003, researchers announced that the Human Genome Project had completed a high-quality sequence of essentially the entire human genome. This sequence closed the gaps from a working draft of the genome, which was published in 2001. It also identified the locations of many human genes and provided information about their structure and organization. The project made the sequence of the human genome and tools to analyze the data freely available via the Internet.

Upon publication of the majority of the genome in February 2001, Francis Collins, the director of NHGRI, noted that the genome could be thought of in terms of a book with multiple uses: "It's a history book—a narrative of the journey of our species through time. It's a shop manual, with an incredibly detailed blueprint for building every human cell. And it's a transformative textbook of medicine, with insights that will give health care providers immense new powers to treat, prevent and cure disease."

In generating the draft sequence (released in June 2000), scientists determined the order of base pairs in each chromosomal area at least 4 to 5 times ($4\times$ to $5\times$) to ensure data accuracy and to help with reassembling DNA fragments in their original order. This repeated sequencing is known as genome "depth of coverage". Draft sequence data is mostly in the form of 10,000 basepair-sized fragments whose approximate chromosomal locations are known.

In June 2000, the Human Genome Project and Celera Genomics, a privately owned firm founded in 1998, jointly announced the completion of the initial sequencing of the human genome, which is composed of about three billion nucleotide base pairs.

Developing the Tools and Technologies for the Success of HGP

The DOE investments described below helped to make the Human Genome Project a success. Substantial investments by the NIH and the Wellcome Trust in the UK were equally important, however, and should not be overlooked. In most cases, the DOE achievements outlined below were the result of basic research programs. Research is an incremental process that learns from both the success and failures of other research investments, including other agencies and organizations. Furthermore, no single instrument, technology, reagent, or protocol made high-throughput DNA sequencing possible, many contributors were responsible.

DNA Sequencers

Research on capillary-based DNA sequencing contributed to the development of two major DNA sequencing machines—the Perkin-Elmer 3700 and the MegaBace DNA sequencers. The MegaBace DNA sequencer was developed initially with DOE funds by Dr. Richard Mathies at UC, Berkeley. The Perkin-Elmer 3700 was based, in part, on DOE-funded research by Dr. Norman Dovichi at the University of Alberta. These high-throughput instruments are one of the keys to the success of the genome project.

Fluorescent Dyes

DNA sequencing originally used radiolabeled DNA subunits. DOE-funded research contributed to the development of fluorescent dyes, which increased the accuracy and safety of DNA sequencing as well as the ability to automate the procedures.

DNA Cloning Vectors

Before the sequencing of large DNA molecules, they are cut into small pieces and multiplied, or cloned into numerous copies using microbial-based "cloning" vectors. Today, the bacterial artificial chromosome (BAC) is the most commonly used vector for initial DNA amplification before sequencing. These cloning vectors were developed with DOE funds.

BAC-End Sequencing

The widely agreed-upon strategy for sequencing the human genome is based on the use of BACs, which carry fragments of human DNA from known locations in the genome. DOE-funded research at the Institute for Genomic Research in Rockville, Maryland, and at the University of Washington provided the sequencing community with a complete set of over 450,000 BAC-based genetic "markers" corresponding to a sequence tag every 3 to 4 kilobases across the entire human genome. These markers were needed to assemble both the draft and the final human DNA sequence.

Gene Recognition and Assembly Internet Link (GRAIL)

Gene Recognition and Assembly Internet Link (GRAIL) is one of the most widely used computer programs for identifying the potential genes in DNA sequence and for general DNA sequence analysis. This powerful analytical tool was developed with DOE funds by Dr. Ed Uberbacher at Oak Ridge National Laboratory. Although a number of gene-locating tools are available now, GRAIL led the way.

Reducing Costs and Speeding Up Sequencing

The above-mentioned technological developments dramatically decreased DNA sequencing cost, while increasing its speed and efficiency. For example, it took 4 years for the international Human Genome Project to produce the first billion base pairs of sequence, and less than 4 months to produce the second billion base pairs. In the month of January, 2003, the DOE team sequenced 1.5 billion bases. The cost of sequencing has dropped dramatically since the project began, and is still dropping rapidly. Other major factors involved in cost and time reduction were greatly improved sequencing instruments, and efficient biological resources such as the following:

- DOE-funded research on capillary-based DNA sequencing contributed to the development of the two major sequencing machines. The core optical system concept of the Perkin-Elmer 3700 sequencing machine (used by Celera and others) was pioneered with DOE support. The instrumentation concepts that matured as the MegaBACE sequencer were pioneered by Richard Mathies (University of California, Berkeley). The DOE JGI chose this sequencing hardware platform after competitive trials.
- DNA sequencing was originally done with radiolabeled DNA fragments. Nowadays, DOE improvements in fluorescent dyes have decreased the amount of DNA needed and increased the accuracy of sequencing data.
- Bacterial artificial chromosome (BAC) clones, developed in the DOE program, became the preferred starting resource in sequencing procedures because of their superior stability and large size. A critical component of public- and private-sector sequencing, BACs were used to assemble both the draft and final human DNA reference sequences.

- Further extending the usefulness of BACs, the DOE HGP funded the production of sequence tag connectors (STCs) from BAC ends. This early information enabled the selection of optimal BACs for complete sequencing, thus saving time and money. STC use for the HGP was advocated by Craig Venter and Nobelist Hamilton Smith (Celera) and Leroy Hood (now at the Institute for Systems Biology).

A Successful Transformation

These successes transferred much of the repetitive labor from humans to automated machines. In addition, new software for data processing alleviated and sped human decision-making. Over the last decade, advances in instrumentation, automation, and computation have transformed the entire process. Further innovations, however, are still needed for completing many large sequences and increasing the effectiveness of sequencing.

2.8. FUNCTIONAL GENOMICS IS THE FUTURE OF HGP

The genome data grows daily. The new challenge will be to use this vast reservoir of data to explore how DNA and proteins work with each other and the environment to create complex, dynamic living systems. Systematic studies of function on a grand scale-functional genomics-will be the focus of biological explorations in this century and beyond. These explorations will encompass studies in transcriptomics, proteomics, structural genomics, new experimental methodologies, and comparative genomics.

- **Transcriptomics** involves large-scale analysis of messenger RNAs transcribed from active genes to follow when, where and under what conditions genes are expressed.
- Studying protein expression and function or **proteomics** can bring researchers closer to what is actually happening in the cell than gene-expression studies. This capability has applications in drug design.

- **Structural genomics:** Initiatives are being launched worldwide to generate the 3-D structures of one or more proteins from each protein family, thus offering clues to function and biological targets for drug design.
- Experimental methods for understanding the function of DNA sequences and the proteins they encode include **knockout studies** to inactivate genes in living organisms, and monitor any changes that could reveal their functions.
- **Comparative genomics:** Analyzing DNA sequence patterns of humans and well-studied model organisms side-by-side has become one of the most powerful strategies for identifying human genes and interpreting their function.

2.9. FUTURE OF HGP IN THE MEDICINE AND GENETICS

The medical industry is building upon the knowledge, resources, and technologies emanating from the HGP to further understanding of genetic contributions to human health. As a result of this expansion of genomics into human health applications, the field of genomic medicine was born. Genetics is playing an increasingly important role in the diagnosis, monitoring and treatment of diseases.

Diagnosing and Predicting Disease and Disease Susceptibility

All diseases have a genetic component **(Fig. 2.6)**, whether inherited or resulting from the body's response to environmental stresses like viruses or toxins. The success of the HGP has even enabled researchers to pinpoint errors in genes—the smallest units of heredity—that cause or contribute to disease.

The ultimate goal is to use this information to develop new ways to treat, cure, or even prevent the thousands of diseases that afflict humankind.

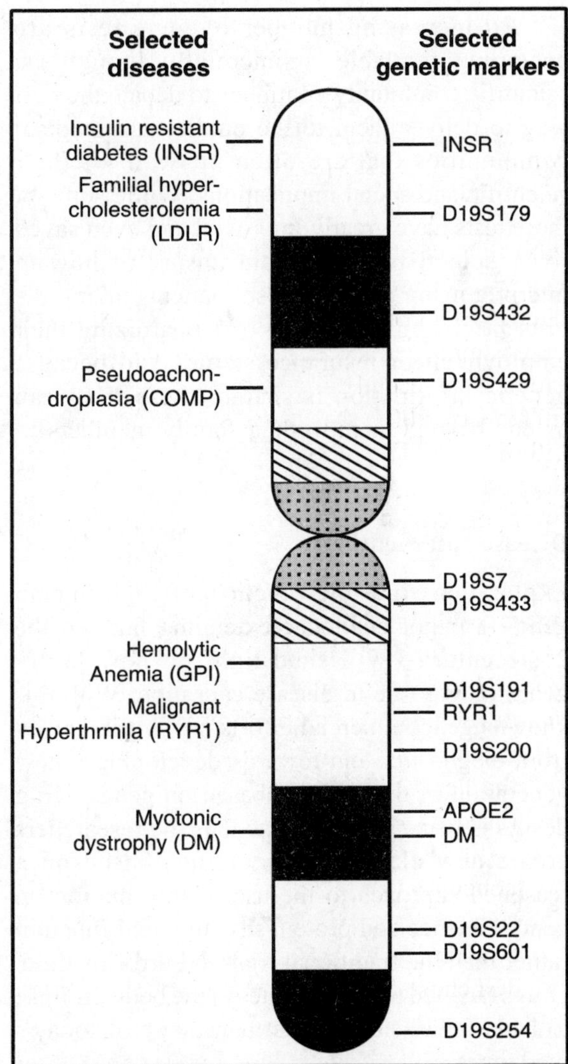

Fig. 2.6. Diagram of human chromosome 19 showing the locations of selected defective genes and genetic markers.

But the road from gene identification to effective treatments is long and fraught with challenges. In the meantime, biotechnology companies are racing ahead with commercialization by designing diagnostic tests to detect errant genes in people suspected of having particular diseases or of being at risk for developing them.

An increasing number of gene tests are becoming available commercially, though the scientific community continues to debate the best way to deliver them to the public, and medical communities that are often unaware of their scientific and social implications. While some of these tests have greatly improved and even saved lives, scientists still remain unsure of how to interpret many of them. Also, patients taking the tests face significant risks of jeopardizing their employment or insurance status. And because genetic information is shared, these risks can extend beyond them to their family members as well.

Disease Intervention

Explorations into the function of each human gene—a major challenge extending far into the 21st century—will shed light on how faulty genes play a role in disease causation. With this knowledge, commercial efforts are shifting away from diagnostics and towards developing a new generation of therapeutics based on genes. Drug design is being revolutionized, as the researchers create new classes of medicines based on a reasoned approach to the use of information on gene sequence and protein structure and function rather than the traditional trial-and-error method. Drugs targeted to specific sites in the body promise to have fewer side effects than many of today's medicines.

Using the genes to treat disease—gene therapy—is the most exciting application of biotechnology. It has captured the imaginations of the public and the biomedical community for good reason. This rapidly developing field holds great potential for treating or even curing genetic and acquired diseases. This technique uses normal genes to replace or supplement a defective gene, or to bolster immunity to disease (e.g., by adding a gene that suppresses the growth of tumor.

Scientific Versus Commercial Goals

The HGP's commitment from the outset was to create a scientific standard (an entire reference genome). Most private-sector human genome sequencing projects, however, focused on gathering just enough DNA to meet their customers' needs—probably in the 95% to 99% range for gene-rich, potentially lucrative regions. Such private data continue being enriched greatly by accurate free public mapping (location) and sequence information. Celera's shotgun sequencing strategy, for example, created millions of tiny fragments that had to be ordered and oriented computationally using HGP research results. Most data at Celera, Incyte, and other genomics information-based companies are proprietary, or available only for a fee. In addition, companies are filing numerous patent applications to stake claims to genes and other potentially important DNA fragments.

2.10. ETHICAL, LEGAL AND SOCIAL IMPLICATIONS (ELSI) OF HGP

The planners of the US Human Genome Project (HGP) recognized that the information gained from mapping and sequencing the human genome would have profound implications for individuals, families, and society. While this information would have the potential to dramatically improve human health, they realized that it would also raise a number of complex ethical, legal and social issues. How should this new genetic information be interpreted and used? Who should have access to it? How can people be protected from the harm that might result from its improper disclosure or use?

To address these issues, the Ethical, Legal and Social Implications (ELSI) Program was established as an integral part of the HGP. The ELSI Program was designed to provide a new approach to scientific research by identifying, analyzing and addressing the ethical, legal and social implications of human genetics research at the same time that the basic scientific issues are being studied. In this way, problem areas can be identified, and solutions developed before the scientific information is integrated into health care practice.

The ELSI Program is viewed as essential to the success of the genome project in the United States, and is supported by federal HGP funds. The National Institutes of Health's 'National Human Genome Research Institute (NHGRI)' has committed 5% of its annual research budget to study ELSI issues. The US Department of Energy Office of Energy Research, NHGRI's partner in the US Human Genome Project, also reserves a portion of its funding for ELSI research and education.

ELSI and its establishments anticipates and addresses the implications of mapping and sequencing of the human genomes for the individuals and society. It also examines the ethical, legal and social consequences of mapping and sequencing the human genome; stimulates public discussion of the issues; and develops policy options, which would assure that the information is used for the benefit of individuals and society.

The Working Group envisioned a program that would anticipate potential problems before they actually occur, and identify possible solutions for the problems. It suggested a number of means for accomplishing these goals. Specifically, it encouraged the research community to explore and gather data on a wide range of issues pertinent to the human genome program that could be used to develop educational programs, policy recommendations or possible legislative solutions. A number of areas for focus were identified, including fairness in the use of genetic information, the impact of knowledge of genetic variation on individuals, and the privacy and confidentiality of genetic information, to name a few.

In 1990, in response to the Working Group's report, the NHGRI established the ELSI Branch (later renamed the ELSI Research Program) in its Division of Extramural Research, and the DOE established an ELSI program in their Office of Energy Research. Since the beginning, these two programs have collaborated closely, including the joint support of the ELSI Working Group, the development of complementary research priority areas, and the co-funding of ELSI activities of mutual interest.

SUMMARY

Humans are higher in the hierarchy of living organisms because of their independent thinking and fantasizing capacity. Thus, understanding the human genome and its contents gives an idea about how simple, single-celled zygote/organisms developed into a complex individual. Human Genome project is also called Human Genome Initiative scientific research effort to analyze the DNA of humans and of several lower organisms. The project began in the United States in 1990 under the sponsorship of the US Department of Energy and the National Institutes of Health. Projects undertaken concurrently in Japan, the United Kingdom, Italy, France, and Russia are coordinated with the American effort through the Human Genome Organization.

The project's ultimate goal is to identify the chromosomal location of every human gene, and to determine each gene's precise chemical structure in order to elucidate its function in health and disease. The information gathered is expected to serve as the basic reference for research in human biology and medicine in the 21st century, and to provide fundamental insights into the genetic basis of human disease. The new technologies developed in the course of the project will be applicable in numerous other fields of biomedical endeavour.

The total number of genes is estimated to be 30,000, which is much lower than previous estimates of 80,000 to 140,000 that had been based on extrapolations from gene-rich areas as opposed to a composite of gene-rich and gene-poor areas.

The functions are unknown for over 50% of the discovered genes. Less than 2% of the human genome codes for functional proteins of the total three billion base pairs in all cells of the body. Repeated sequences that do not code for proteins ("junk DNA") make-up at least 50% of the human genome. Repetitive sequences are thought to have no direct functions, but they shed light on chromosome structure and dynamics. Over time, these repeats reshape the genome by rearranging it, creating entirely new genes, and modifying and reshuffling existing genes. Chromosome 1 has most of the genes (2968) and the Y chromosome has the fewest (231).

Genome sequencing is the term used to describe the laboratory process of reading the order of the four letter of the genetic alphabets (A,C,G,T) along a strand of DNA. The various steps involved in such efforts are as follows: Selection of suitable sample materials for the DNA; isolation of DNA from cells and preparation of large samples of high quality DNA from these cells; cutting the purified DNA at random sites into manageably sized, overlapping pieces of the DNA sample; insertion of these DNA pieces into packages for the production of limitless copies of such selected DNA; recording the order of bases for each DNA sample piece by using DNA sequencing techniques; determination of the overlap of each piece, and assembling the sequences to give the final genome of the human.

While following the above approaches, it is necessary to make appropriate sample populations based on the humans distribution. A primary goal of the Human Genome project is to generate detailed maps of the human genome. These maps will aid in determining the location of genes within the human genome. More specifically, they will assign genes to their chromosomes. Two types of maps are being generated. Genetic linkage maps determine the relative arrangement and approximate distances between genes and markers on the chromosomes; physical maps specify the physical location (in base pairs) and distance between genes or DNA fragments with unknown functions that are mapped to specific regions of the chromosomes.

In the Human Genome Project, importance is also given to the sequencing of other model organisms. DNA sequencing of the nematode worm *Caenorhabditis elegans* and the yeast *Saccharomyces cerevisiae* was completed in 1996, the bacterium *Escherichia coli* in 1997, the fruit fly (*Drosophila melanogaster*) and the plant *Arabidopsis thaliana* in 2000, and the laboratory mouse (*Mus musculus*) and the bacterium *Staphylococcus aureus* in 2001. The rationale behind these findings is that many genes with similar functions in disparate organisms have been conserved in evolution and show surprising similarities. Genes from simpler organisms can thus be used to study their counterparts found in human beings.

Another objective of the Human Genome Project is to address the ethical, legal, and social implications of the information obtained. Society will derive benefit from this knowledge only if it takes measures to prevent abuses, such as invasion of the privacy of an individual's genetic background by employers, insurers, or government agencies, or discrimination based on genetic grounds.

Large number of advancement made in the diverse fields including molecular biology, genetic engineering and sequencing provided a great impetus to the Human Genome Project. These technological developments dramatically decreased the cost of DNA sequencing, while increasing its speed and efficiency. For example, it took four years for the international Human Genome Project to produce the first billion base pairs of sequence, and less than four months to produce the second billion base pairs. In the month of January, 2003, the DOE team sequenced 1.5 billion bases. The cost of sequencing has dropped dramatically since the project began and is still dropping rapidly. Other major factors involved in cost and time reduction were greatly improved sequencing instruments and efficient biological resources.

To generate the high-quality reference sequence completed in April 2003, additional sequencing was done for closing the gaps and reducing the ambiguities. Further only a single error was allowed for every 10,000 bases, the agreed-upon standard for the HGP. Investigators believe that a high-quality sequence is critical for recognizing regulatory components of genes that are very important in understanding human biology and such disorders as heart disease, cancer, and diabetes. In addition to the human genome, the Human Genome Project sequenced the genomes of several other organisms, including brewers' yeast, the roundworm, and the fruit fly. In 2002, researchers announced that they had also completed a working draft of the mouse genome. By studying the similarities and differences between human genes and those of other organisms, researchers can discover the functions of particular genes, and identify the genes critical for life.

The avalanche of genome data is growing day by day. The new challenge will be to use this vast reservoir of data to explore how DNA and proteins work with each other, and the environment to create complex, dynamic living systems. Systematic studies of function on a grand scale— functional genomics—will be the focus of biological explorations in this century and beyond. These explorations will encompass studies in transcriptomics, proteomics, structural genomics, new experimental methodologies and comparative genomics.

The HGP's commitment from the outset was to create a scientific standard (an entire reference genome). Most private-sector human genome sequencing projects, however, focused on gathering just enough DNA to meet their customers' needs—probably in the 95 percent to 99 percent range for gene-rich, potentially lucrative regions. Such private data continues to be enriched greatly by accurate free public mapping (location)

and sequence information. Celera's shotgun sequencing strategy, for example, created millions of tiny fragments that had to be ordered and oriented computationally using HGP research results. Most data at Celera, Incyte and other genomics information-based companies are proprietary or available only for a fee. In addition, companies are filing numerous patent applications to stake claims to genes and other potentially important DNA fragments. This practice has to be reduced and public domain should get more and more patents.

EXERCISE

1. Define human genome project, and describe its impact on the ethical, legal and social issues.
2. Describe in detail the initiation of human genone project, and mention its goals and expectations.
3. What do you mean by ethical, legal and social issues, and how are they associated with the human genome project studies?
4. List out how religious groups are saying against the human genome project in the present circumstances of conserved society.
5. Write a note on the procedures adopted for human genome project and its importance in the field of biology and biotechnology, in particular.
6. Distinguish between the use of the human genome project and model organisms genome projects that are completed.
7. Explain the different types of model organisms used in genome studies. Give an account of their applications in human genome project.
8. Give a brief account of various organisations involved in human genome project, around the world.
9. Critically comment on the ethical, social and legal issues raised by the human genome project.
10. Write about the involvement and methodologies followed by the national institutes of health and department of energy for human genome project.
11. Discuss how human genome project has proved helpful for understanding the genetic diseases and explain its applications and risk involved in the field of medicine.

12. Enumerate the importance of human genome project and its adaptations in understanding the different biological processes of human beings.

13. Give an account of the role played by the Celera genomics, and its approaches to the completion of human genome project.

14. Give a brief account of the advantages of human genome project, and its applications in the field of genomic and proteomics.

15. Explain different types of limitations associated with the draft human genome sequences and gene estimations.

16. What is meant by human cloning? Explain its importance in the present day patent rich gene technology world.

17. Give an account on the arguments for and against that associate with human genome project and their applications.

18. Elucidate the future plans of human genome project and their usage by private companies patents. Also describe its impact on the human research and development.

19. What are the ethical issues associated with describing sequencing of different organisms. Add a note on its impact on human genome project completion?

20. Briefly explain how scientists and researchers are supporting the continued studies on human genome project and technological need of the biology.

21. What is meant by sequence-tagged sites? Write a note mentioning its uses in the human genome project and identification of genes?

22. Describe how different countries are benefiting from the human genome project informations. Given an account of its impact on their society.

3

Molecular Detection of Pre-Symptomatic Genetic Diseases and its Importance in Healthcare

3.1. INTRODUCTION

The initiation of the human genome project in 1990 paved the way for the medical field to utilise some of the benefits provided by it. Later, rapid growth of bioinformatics, genetics, medicine, proteomics and genomics coupled with human genome project was also found to be beneficial for wide varieties of fields. The human genome project finds applications in wide variety of fields ranging from biochemistry to behavioural psychology, paleontology to parasitology, conservation biology to cancer biology, genetics to gene cloning and pharmacology to forensic medicine. Since the discovery of sequencing methods, a large number of gene-based diseases have been known. So, it is always better to get the idea of the genetic status of fetus well in advance, or before the onset of the symptoms in the individuals who are having the defective gene or genes. Human development depends on genetic and environmental factors. A person's genetic composition (genome) is established during conception. The genetic information is carried in the DNA of the chromosomes and mitochondria. Most diseases probably have some genetic component, the extent of which varies. Environmental factors may alter the genetic information through mutation or other structural alteration. Presymptomatic genetic screening may be appropriate for persons with a family history of dominantly inherited disorder (e.g., Huntington's disease, breast cancer). Identifying a definite carrier of the genetic disorder may allow the patient to make decisions (e.g., monitoring in the case of breast cancer, reproductive choices in the case of Huntington's disease or adult polycystic kidney disease). Presymptomatic genetic screening is most appropriate for women or couples whose medical histories or family backgrounds indicate an elevated risk of fetal genetic disorders. Physicians should inform women or couples, who are not having an elevated risk of disorders, why prenatal diagnosis may not be desirable in their case. Then they can legitimately request prenatal diagnosis, provided they understand and accept the risks involved. Now our society has reached a critical juncture in the evolution of genetic technologies. The media has given us a glimpse of what is currently taking place in the laboratories and what may take place in mainstream society in the very near future.

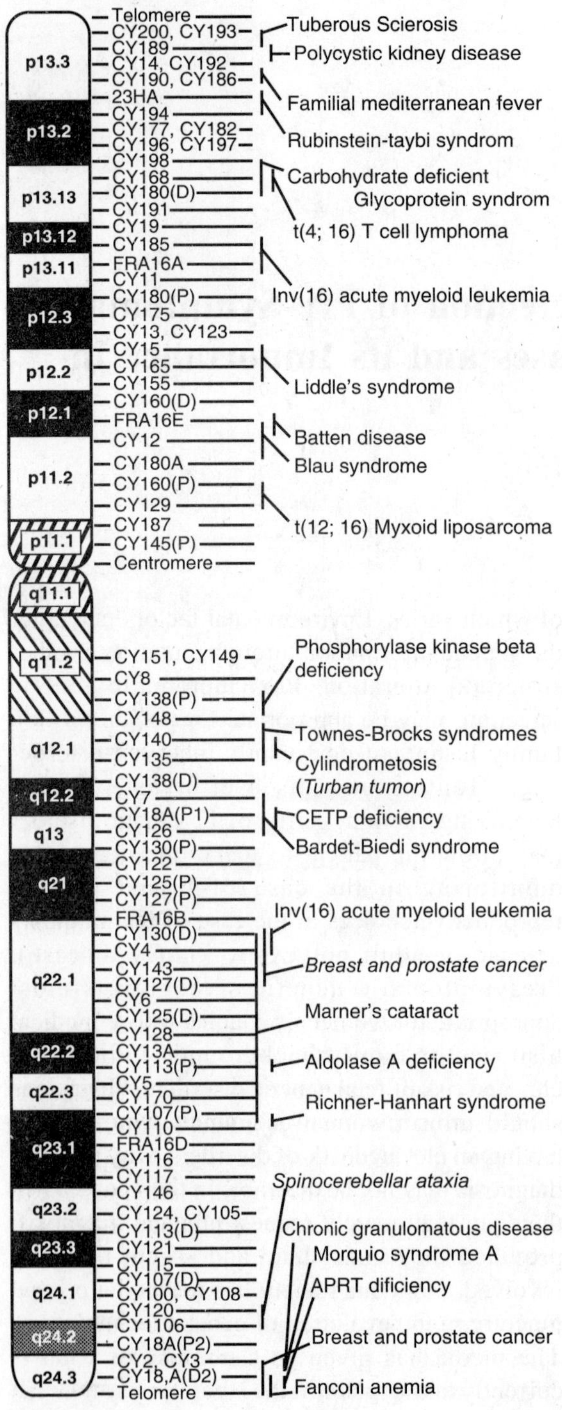

Fig. 3.1. Position of disease genes on chromosome 16 of human. Italicized genes have not yet been cloned.

Currently, prenatal genetic screening is available for serious disorders like Tay-Sachs disease, Down's syndrome and cystic fibrosis **(Fig. 3.1)**. In future, genetic screening may also become available for milder genetic disorders. As the ability to screen for genetic disorders increases, the possibility of genetic selection will also increase. It is not possible to give precise guidelines on when prenatal genetic selection would be ethically acceptable. Many situations in which selection would be possible are only hypothetical at this time, and other factors would affect the acceptability of selection.

The medical industry is building upon the knowledge, resources and technologies emanating from the HGP for further understanding of genetic contributions to human health. As a result of this expansion of genomics into human health applications, the field of genomic medicine was born. Genetics is playing an increasingly important role in the diagnosis, monitoring, and treatment of diseases.

Diagnosing and Predicting Disease with Disease Susceptibility

All diseases have a genetic component, whether inherited or resulting from the body's response to the environmental stresses like viruses or toxins. The success of the HGP has enabled researchers to pinpoint errors in genes—the smallest units of heredity—which cause or contribute to the disease.

The ultimate goal is to use this information to develop new ways to treat, cure, or even prevent the thousands of diseases that afflict humankind. But the road from gene identification to effective treatments is long and fraught with challenges. In the meantime, biotechnology companies are racing ahead with commercialization by designing diagnostic tests to detect errant genes in people suspected of having particular disease or of being at risk for developing them.

An increasing number of gene tests are becoming available commercially, although the scientific community continues to debate the best way to deliver them to the public and medical communities that are often unaware of their scientific and social implications. While some of these tests have greatly improved and even saved lives, scientists remain unsure of how to interpret many of them. Also, patients taking the tests face significant risks of jeopardizing their employment or insurance status. And because genetic information is shared, these risks can extend beyond them to their family members as well.

Disease Intervention

Exploration into the function of each human gene—a major challenge extending far into the 21st century—will shed light on how faulty genes play a role in disease causation. With this knowledge, commercial efforts are shifting away from diagnostics and toward developing a new generation of therapeutics based on genes. Drug design is being revolutionized, as researchers create new classes of medicines based on a reasoned approach to the use of information on gene sequence and protein structure function rather than the traditional trial-and-error method. Drugs targeted to specific sites in the body promise to have fewer side effects than many of today's medicines.

The potential of using genes to treat disease—gene therapy—is the most exciting application of DNA science. It has captured the imaginations of the public and the biomedical community for good reason. This rapidly developing field holds great potential for treating or even curing genetic and acquired diseases using normal genes to replace or supplement a defective gene, or to bolster immunity to the disease (e.g., by adding a gene that suppresses tumor growth).

Both environmental and genetic factors play an important role in the development of disease. Generally, prenatal tests have focused on identifying a limited number of genetic disorders.

However, with the completion of the Human Genome Project, the types of genetic characteristics that can be detected through prenatal technologies are sure to accelerate. Some scientists even go as far as to predict with enthusiasm that with the help of genetic manipulation, parents will be able to assemble their children from the genes listed in a catalog. It is clear that the clock cannot be turned back, and as genetic science develops, it will be possible to detect more and more genes prior to birth. As a result, it is very likely that prenatal technologies will become more and more of a predominant force in the life of a pregnant woman.

At this point, it is important to distinguish between the terms "predisposed genetic condition" and "presymptomatic genetic condition". People who are predisposed to a genetic disease do not have the disease. Rather, they have an increased likelihood that the disease will develop. The Equal Employment Opportunity Commission ("EEOC") provides an excellent explanation of a predisposed genetic condition:

"If your grandmother had diabetes or heart disease, you may have a predisposition to diabetes or heart disease. And if your father has cancer, you may be predisposed to developing cancer. However, the presence of these genetic characteristics do not indicate that an individual has an impairment or a record of an impairment, or necessarily that the individual may develop an impairment in the future."

Breast cancer, heart disease, lung cancer, alcoholism, and obesity are all examples of predisposed genetic conditions. There is a possibility that someone with a predisposition to one of the above conditions may never develop that disease. For example, the carrier of the BRCA1 gene, which predisposes an individual to breast cancer, has around an 85 percent chance of developing breast cancer. The carrier may never develop breast cancer, however, the presence of the mutation indicates an increased risk for developing the disease.

On the other hand, people with presymptomatic genetic conditions will develop the disease if they live long enough. An example of such a condition is Huntington's disease. Although the carrier of the Huntington's mutation may not develop symptoms of the disease for the first forty years of his life, symptoms will eventually develop, regardless of the outside factors. Once the symptoms develop, the carrier can be classified as symptomatic for the disease.

DNA-based tests are among the first commercial applications of the new genetic discoveries. These tests are employed to diagnose a condition or estimate the likelihood for developing it. Another aspect of genetic diseases is the gene's penetrance and expressivity. Some diseases express themselves more consistently than others. For example, the BRCA1 gene causes breast cancer in 85 percent of the people who have it (predisposing them to breast cancer), while the Huntington's gene manifests itself in nearly 100 percent of the people who have it (causing them to be presymptomatic for Huntington's). The ability of a gene to express itself in a person is known as a gene's penetrance. In the earlier example, the BRCA1 gene has 85 percent penetrant, while the Huntington's gene has nearly 100 percent penetrant. The Huntington's gene is so penetrant that people are considered presymptomatic for the disease if they have the gene. However, there are some cases where the Huntington's gene does not express itself until very late in life. Expressivities deal with the degree and manner in which the gene manifests itself once it has penetrated. A gene may express itself differently in one individual relative to another, possibly leading to a more severe genetic condition in that individual.

3.2. DNA AND DISEASE

For all our apparent outward diversity, humans are surprisingly alike at the DNA level **(Fig. 3.2 (a)** **& (b)**. We differ by only one or two tenths of one percent of our DNA—some three to six million bases—yet these tiny DNA variations are responsible for all our physical differences and may influence many of our behaviors as well.

Most DNA variations among individuals are normal, but harmful variations called mutations can cause or contribute to many different diseases and conditions. Depending on their size and location in the DNA, mutations can have devastating effects or none at all. If they occur within genes, the result can be the creation of faulty proteins that function at less-than-normal levels or are completely nonfunctional and result in disease.

All diseases have a genetic basis. We may inherit a particular condition, such as the lung disease cystic fibrosis, or an increased likelihood for developing disorders such as heart disease or colon cancer. We also inherit the particular ability to respond to the environmental stresses. such as viruses, bacteria, and toxins. Understanding how DNA influences every aspect of health eventually will lead to far more effective ways to treat, cure, or even prevent the thousands of diseases that afflict humankind.

3.3. DISEASES AND THEIR GENETIC BASIS

Genetic disorder is a disease caused by the abnormalities in an individual's genetic material (genome). Genetic disorders may occur when there is a mutated gene, which causes the cell to produce aberrant proteins or other gene products that may obstruct efficient bodily performance. Sometimes full segments of DNA may be missing, multiplied, or transposed (found on a different segment of the chromosome), resulting in a missing gene, an extra gene, or a misplaced gene that alters the regular production of gene products, which the cell needs. These mutations may be either inherited or acquired.

Inherited mutations arise when one parent (or both) donates an aberrant gene in the set of

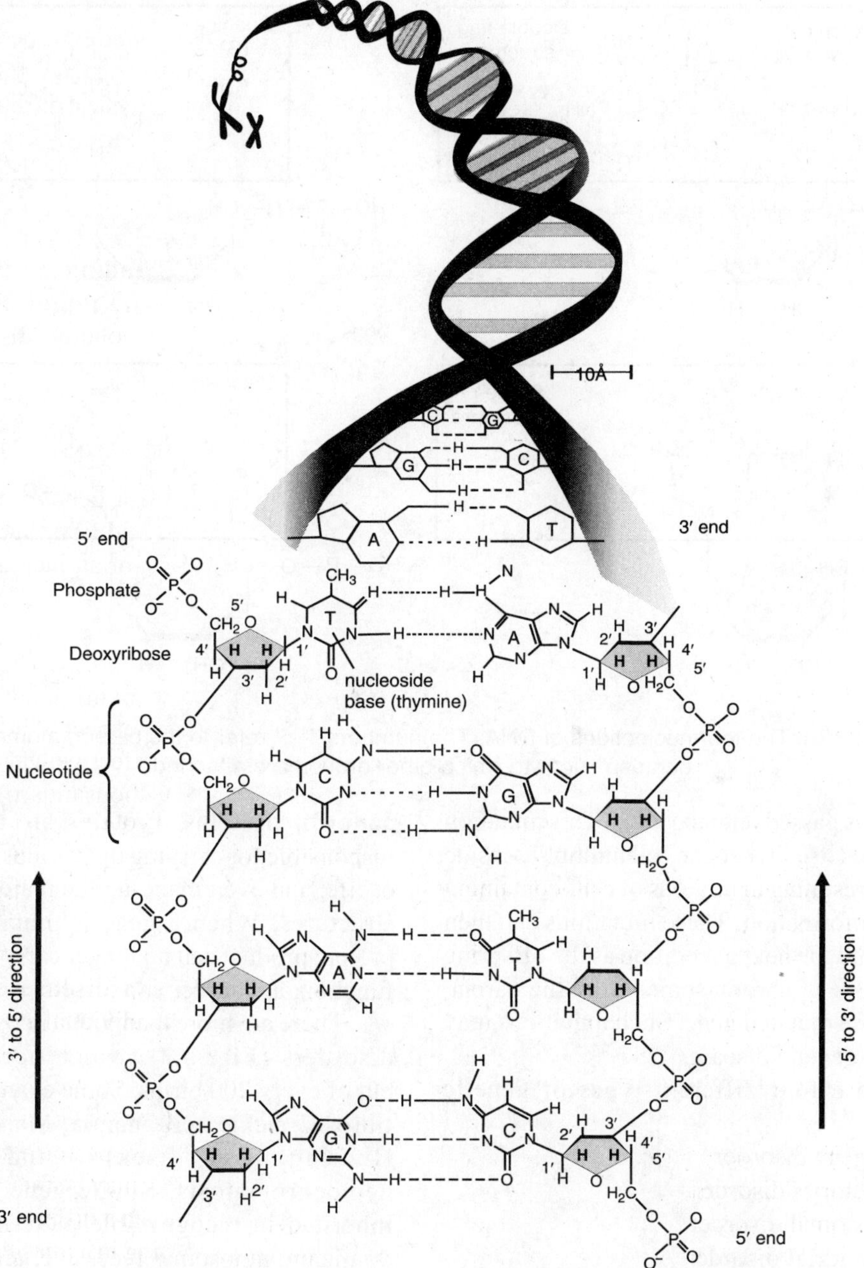

Fig. 3.2(a). The double-stranded DNA showing the nucleotides, each composed of a deoxyribose sugar, a phosphate, and a nitrogen-containing base. The two strands are held together by hydrogen bonds (dashed lines) between pairs of bases. Guanine (G) and cytosine (C) are held together by three hydrogen bonds, and adenine (A) and thymine (T) by two. The two strands are antiparallel, since one strand is 5' to 3' in one direction and the complementary strand is 3' to 5'.

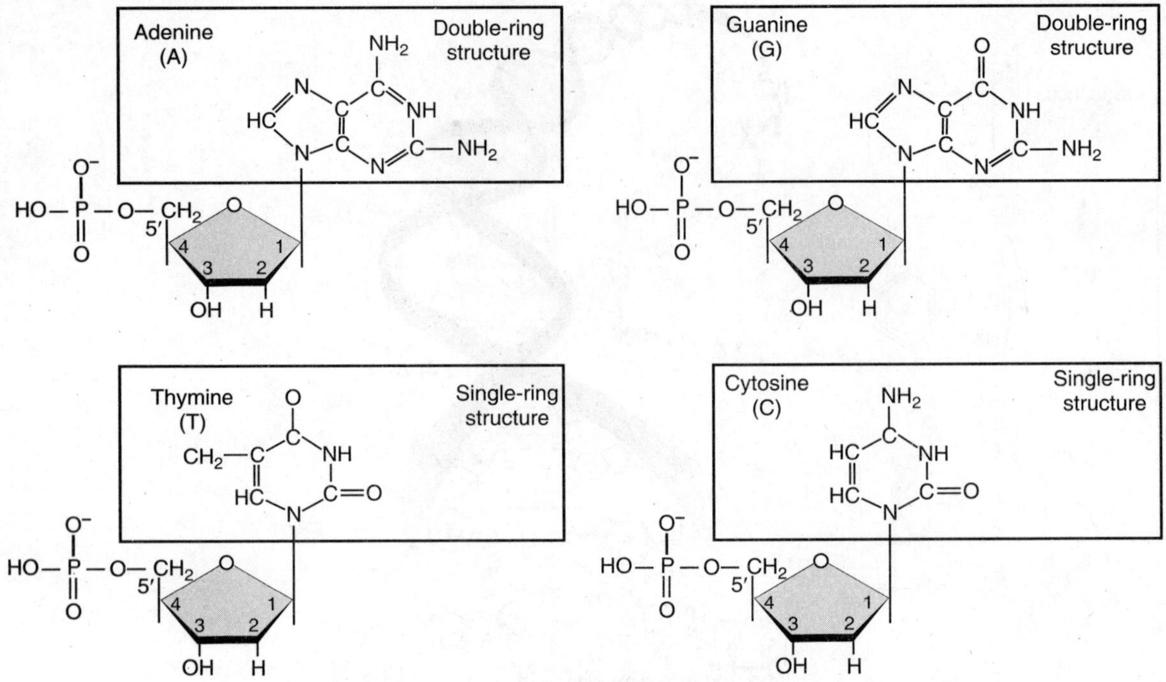

Fig. 3.2(b). The four nucleotides of DNA. The numbers 1'-5' refer to the carbon atoms on the pentose sugar to which other groups are attached.

chromosomes passed onto the child. This mutation then replicates itself when the cell multiplies inside the embryo, resulting in millions of cells containing the flawed information. These mutations can then be passed onto the next generation as the offspring receives its set of chromosomes from the parent, along with the mutated gene. Such mutations may then lead to genetic disease.

There are four different types of genetic disorders:

1. Single-gene disorder
2. Multifactorial disorder
3. Chromosomal disorder
4. Mitochondrial disorder

1. Single-Gene (Also called Mendelian or Monogenic) Disorder

This disorder is caused by changes or mutations that occur in the DNA sequence of one gene. Genes code for proteins. Proteins are the molecules responsible for carrying out various vital functions of life, and even make-up the majority of cellular structures. When a gene is mutated so that its protein product can no longer carry out the normal function, a disorder can result.

There are more than 6,000 known single-gene disorders **(Fig. 3.3)**, which occur in about 1 out of every 200 births. Some examples are cystic fibrosis, sickle cell anemia, Marfan syndrome, Huntington's disease and hereditary hemochromatosis. Single-gene disorders are inherited in recognizable patterns—autosomal dominant, autosomal recessive, and X-linked. In the case of single-gene diseases, such as cystic fibrosis and Huntington's disease, the carrier receives a gene in which the disease will manifest itself regardless of environmental factors. Hence, a single gene can cause the manifestation of the symptoms.

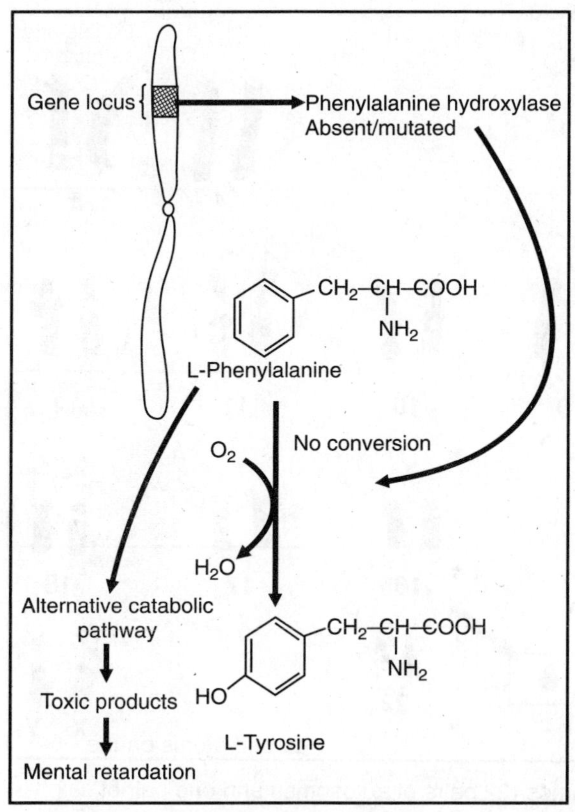

Fig. 3.3. Single-gene genetic disease, e.g., phenylketonuria.

Fig. 3.4. Multifactorial genetic diseases e.g., diabetes mellitus.

2. Multifactorial (also called Complex or Polygenic) Disorder

This disorder is caused by a combination of environmental factors and mutations in multiple genes. For example, different genes that influence breast cancer susceptibility have been found on chromosomes 6, 11, 13, 14, 15, 17 and 22. Its more complicated nature makes it much more difficult to analyze than single-gene or chromosomal disorders. Some of the most common chronic disorders are multifactorial disorders **(Fig. 3.4)**. Examples include heart disease, high blood pressure, Alzheimer's disease, arthritis, diabetes, cancer, and obesity.

Multifactorial inheritance is also associated with heritable traits such as fingerprint patterns, height, eye color and skin color. Multifactorial disorder may not manifest in the absence of certain other factors. These disorders rely on the "interaction of numerous genetic and environmental factors". For example, a person may have a gene that makes him susceptible to lung cancer. If he avoids smoking, he might not develop lung cancer. In other words, by eliminating the environmental factors, the disease may not appear.

3. Chromosomal Disorder

Chromosomes, distinct structures made up of DNA and protein, are located in the nucleus of each cell. Since chromosomes are carriers of genetic

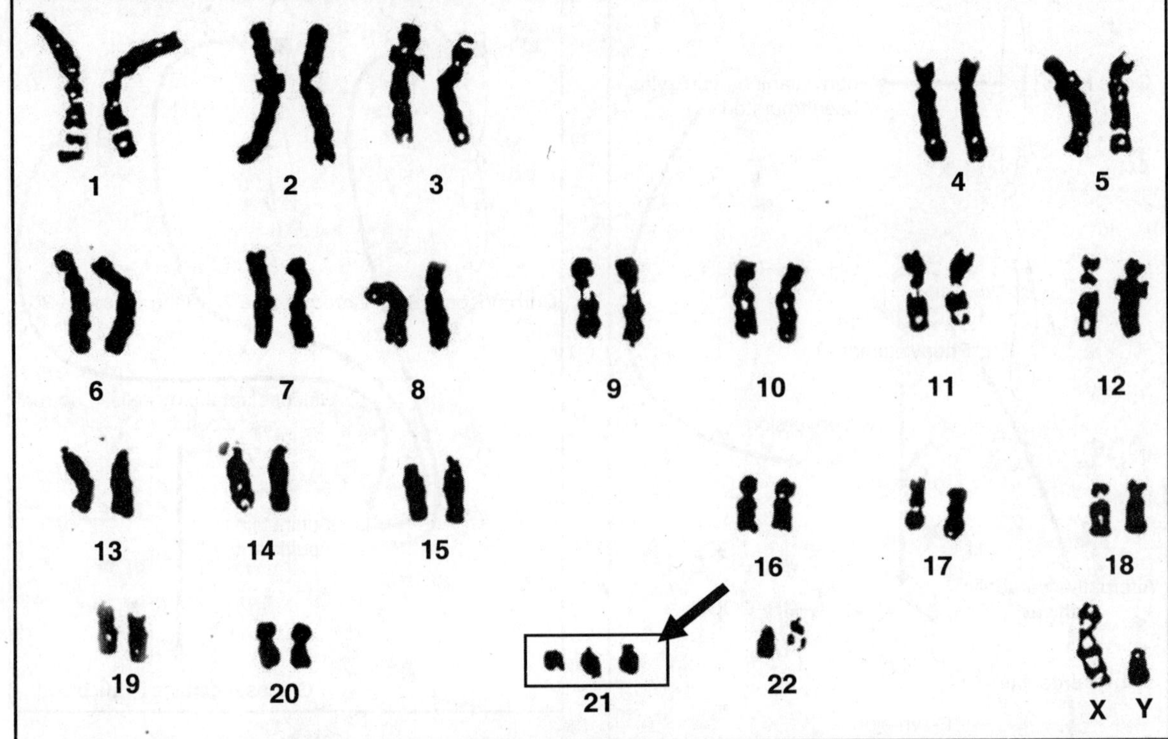

Fig. 3.5. Karyotype of an individual's 46 chromosomes (22 pairs of autosomes and one pair of sex chromosome, in this case, X and Y) can be arranged by size and banding pattern into a karyotype for the diagnosis of genetic disease. In this example, an extra copy of chromosome 21 identifies the individual as having Down's syndrome.

material, abnormalities in chromosome structure such as absence or presence of the extra copies of chromosomes, and gross breaks and rejoinings in chromosomes as missing or extra copies or gross breaks and rejoinings (trans gene locations), can result in disease. Some types of major chromosomal abnormalities can be detected by microscopic examination. Down syndrome or trisomy 21 **(Fig. 3.5)** is a common disorder that occurs when a person has three copies of chromosome 21.

4. Mitochondrial Disorder

This genetic disorder is caused by mutations in the nonchromosomal DNA of mitochondria and rarely occurs in humans. Mitochondria are small, round, or rod-like organelles that are involved in cellular respiration and found in the cytoplasm of plant and animal cells. Each mitochondrion may contain 5 to 10 circular pieces of DNA.

Mitochondria are intracellular organelles that generate energy via a series of respiratory chain of individual cells complexes. They contain a unique circular chromosome that codes for 13 proteins, various RNAs and several regulating enzymes. However, nuclear genes code for 90% of mitochondrial proteins. Each cell has several hundred mitochondria in its cytoplasm.

Mitochondrial diseases **(Fig. 3.6)** are due to the abnormalities in the mitochondrial DNA (e.g., deletions, duplications, mutations). High-

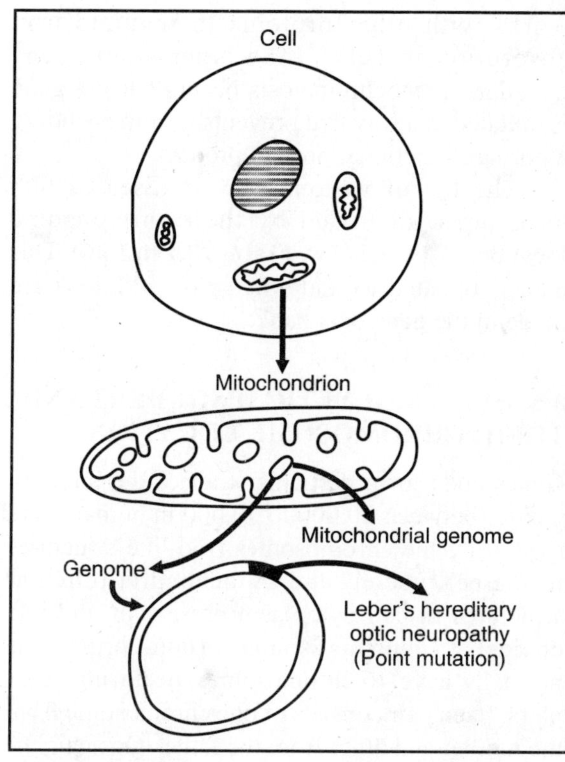

Fig. 3.6. Mitochondrial genome showing defective gene contributing to the development of visual loss due to point mutation.

energy tissues, such as muscle, heart, and brain tissues are particularly at risk, but pancreas and liver are also at risk. Patterns of tissue involvement correlate with particular mitochondrial DNA mutations, e.g., chronic progressive external ophthalmoplegia; its variant, the multisystem **Kearns-Sayre syndrome** (chronic progressive external ophthalmoplegia, heart block, retinitis pigmentosa, CNS degeneration); **Pearson syndrome** (sideroblastic anemia, pancreatic insufficiency, and progressive liver disease that begins in the first few months of life and is frequently fatal in infants); **Leber's hereditary optic neuropathy** (a variable but often devastating bilateral visual loss that often occurs in teenagers, and is due to a point mutation in mitochondrial

DNA); **MERRF** (*M*yoclonic *E*pilepsy, *R*agged *R*ed *F*ibers, dementia, ataxia, myopathy); and **MELAS** (*M*itochondrial *E*ncephalomyopathy, *L*actic *A*cidosis, *S*trokelike episodes).

Maternal inheritance characterizes abnormalities of mitochondrial DNA, because all mitochondria is inherited *via* the egg. Thus, all offspring of an affected female are at risk of inheriting the abnormality, whereas no offspring of an affected male is at risk. Variability in clinical manifestations is the rule, and may be due in part to variable mixtures of mutant and normal mitochondrial genomes (heteroplasmy) within cells and tissues.

3.4. GENETIC SCREENING AND ITS HISTORY

Genetic screening began in the 1960s with the development of the Guthrie test for phenylketonuria (PKU). PKU is easily treated by restricting certain foods in the diet, but if left untreated, the disorder causes severe mental retardation. Although hundreds of human genes have been identified and mapped, in many cases, a corresponding cure or treatment regime has not yet been developed. Moreover, although many common diseases are suspected of being genetically linked, in most cases, the risk must be augmented by other genetic or environmental factors in order for the disease to occur. Nor does the detection of a chromosomal disorder, such as Down's Syndrome, predict the severity with which the syndrome will be expressed. Advocates of new reproductive technologies, particularly prenatal technologies, argue that this technology will expand the women's reproductive choice and decrease the incidence of disability in the society. This argument is over simplistic at best, and belies the significant social and ethical implications underlying such technologies for women's reproductive autonomy and the equality rights of women and men with disabilities. It is possible that prenatal diagnostic testing can offer women more control over the occurrence of the birth of a disabled child by providing relevant information about the genetic status of the fetus during pregnancy. However, because there are no therapeutic interventions available for the majority of conditions, which prenatal testing detects, the options open to the women following a positive diagnosis consist of preparing for a life with a disabled child or terminating the pregnancy. Negative perceptions about a life with disability, combined with the difficulties in obtaining adequate social support may cause a woman to believe that her only real choice is to terminate the pregnancy.

Presymptomatic genetic diagnosis technologies will not eliminate disability from the society. The 'Council of Canadians with Disabilities' estimates that only 3 percent of genetic conditions may be affected by gene therapy. Eighty-five percent of adult disability is caused after the age of 13, and more than 90 percent of infant disability is due to social and not genetic causes. A person is far more likely to become disabled due to situations such as ageing, illness, unsafe working conditions, toxic environments, violence, poverty, lifestyle choices, poor nutrition etc. Consequently, despite the hype offered by the media and the scientific community, it is unlikely that gene identification will significantly reduce the incidence of disease, disability, or improve the social status of people with disease or disabilities.

Cause of Genetic Disorders

Many genes are named after the disorders to which they have been linked. This can be very confusing at times. For example, the gene associated with hereditary hemochromatosis is called the "hemochromatosis gene." This name implies that the gene exists for the sole purpose of causing disease, which of course is not the case. The normal function of a gene is not to cause illness, but to encode a protein. Disease occurs when genes are unable to work properly. The hemochromatosis gene actually codes for a membrane protein that works with other proteins to regulate iron absorption in cells. Like other single-gene disorders, hemochromatosis occurs when a gene is mutated in a way that prevents it from encoding a normal, functional protein product.

The list of various genetic diseases that have been identified by their chromosome locations is given below **(Tables 3.1 to 3.26)**. This data is updated as and when new additions are made in the gene data bank.

3.5. HUMAN GENETIC DISORDERS AND THEIR CHROMOSOME LOCATIONS

Genes code for a particular trait of the humans. Genes (between 20,000 to 30,000 in humans) are carried by the chromosomes (rod-like structures in the cell nuclei) and mitochondria (circular structures in cell cytoplasm present in multiple copies). In humans, somatic (non-germ) cells normally have 46 chromosomes, occurring as 23 pairs. Each pair consists of one chromosome from the mother and other from the father. One pair, the sex chromosomes, determines a person's sex. Women have two X chromosomes in every somatic cell nucleus, whereas men have one X and one Y chromosome (i.e., heterologous chromosomes). The X chromosome carries genes responsible for many hereditary traits, whereas the small, differently shaped Y chromosome carries the genes that initiate male sex determination. The remaining 22 chromosome pairs, the autosomes, are usually homologous (i.e., identical in size, shape, and position, and number of genes). Germ cells (egg and sperm) undergo meiosis, which reduces the number of chromosomes to 23—half that of somatic cells (46)—so that when an egg is fertilized by a sperm during conception, the normal number of chromosomes is reconstituted. In meiosis, the genetic information inherited from a person's mother and father is recombined through crossing over or exchange between the homologous chromosomes.

Genes, the basic units of heredity, are arranged linearly within the DNA along the chromosomes;

each gene has a specific location (locus) or position on the chromosome. The number and arrangement of loci on homologous chromosomes is usually identical. However, the structure of a specific gene may have minor variations (polymorphisms). The specific nucleotide sequences of the genes that occupy the two homologous loci along the two chromosomes of a pair are called alleles. The two alleles (i.e., the one inherited from the mother and the one inherited from the father) may have slightly different nucleotide sequences or may be the same. A person with a pair of identical alleles for a particular gene is a homozygote; a person with a pair of dissimilar alleles is a heterozygote. Large numbers of genes responsible for their characters or diseases are identified by their exact locations, and list is going to be changed at a rapid rate as new genes are discovered. The concise lists of these genes that are known to present on each chromosomes are listed in **Tables 3.1 to 3.26**. For the recent data on this, the readers are advised to see gene data based website.

3.6. GENE TESTING

Gene tests (also called DNA-based tests), the newest and most sophisticated of the techniques used for testing the genetic disorders, involve direct examination of the DNA molecule. Other genetic tests include biochemical tests for gene products such as enzymes and other proteins, and for microscopic examination of stained or fluorescent chromosomes.

Genetic testing is performed mainly for following reasons:

- Carrier screening, which involves identifying unaffected individuals who carry one copy of a gene for a disease that requires two copies for the disease to be expressed.
- Preimplantation genetic diagnosis (screening embryos for disease).
- Prenatal diagnostic testing.
- Newborn screening.

- Presymptomatic testing for predicting adult-onset disorders such as Huntington's disease.
- Presymptomatic testing for estimating the risk of developing adult-onset cancers and Alzheimer's disease.
- Confirmational diagnosis of a symptomatic individual.
- Forensic/identity testing.

Table 3.1. The list of number of genes identified in each chromosome.

Chromosome	Count
Chromosome unknown or multiple gene locations	97
Chromosome 1	131
Chromosome 2	82
Chromosome 3	75
Chromosome 4	60
Chromosome 5	58
Chromosome 6	85
Chromosome 7	70
Chromosome 8	53
Chromosome 9	68
Chromosome 10	42
Chromosome 11	98
Chromosome 12	71
Chromosome 13	26
Chromosome 14	44
Chromosome 15	36
Chromosome 16	54
Chromosome 17	99
Chromosome 18	21
Chromosome 19	71
Chromosome 20	27
Chromosome 21	18
Chromosome 22	32
Chromosome Y	10
Chromosome X	219

Table 3.2. Genes, gene locations and genetic disorders on chromosome 1.

GDB Accession	Gene Location	Disease Factor
GDB:370748	1p13-1p13 1p22-1p21	Macular Degeneration, Senile Stargardt Disease 1; STGD1 ATP Binding Cassette transporter; ABCR Retinitis Pigmentosa-19; RP19
GDB:131485	1p22-1p21	Peroxisomal Membrane Protein 1; PXMP1
GDB:118958	1p31-1p31	ACYL-CoA Dehydrogenase, Medium-Chain; ACADM
GDB:132644	1p21-1p21	Glycogen Storage Disease III
GDB:118750	1q42-1q43	Angiotensin I; AGT
GDB:9958827	1p36-1p36	Hyperprolinemia, Type II
GDB:118730	1p36.1-1p34	Phosphatase, Liver Alkaline; ALPL Hypophosphatasia, Infantile
GDB:119677	1p13-1p13	Adenosine Monophosphate Deaminase-1; AMPD1
GDB:119685	1q21-1q23 1p36-1p36	Apolipoprotein A-II; APOA2 Breast Cancer, Ductal, 2; BRCD2
GDB:119042	1p36.3-1p34.1	Complement Component 1, q Subcomponent, Alpha Polypeptide; C1QA
GDB:119043	1p36.3-1p34.1	Complement Component 1, q Subcomponent, Beta Polypeptide; C1QB
GDB:128132	1p36.3-1p34.1	Complement Component 1, q Subcomponent, Gamma Polypeptide; C1QG
GDB:119735	1p32-1p32	Complement Component-8, Deficiency of
GDB:119736	1p32-1p32	Complement Component-8, Deficiency of, Type II
GDB:126431	1q31-1q32	Calcium Channel, Voltage-dependent, L Type, Alpha 1s Subunit; CACNA1S Periodic Paralysis I Malignant Hyperthermia Susceptibility-5; MHS5
GDB:1336655	1p36-1p36 1p36.3-1p36.3	Cataract, Congenital, Volkmann Type; CCV
GDB:119766	1q22-1q23	CD3Z Antigen, Zeta Polypeptide; CD3Z
GDB:127827	1p36-1p36	Protein Kinase p58; PK58
GDB:135222	1q31-1qter 1q42-1qter	Choroideremia-like; CHML
GDB:4568202	1q42.1-1q42.2	Chediak-higashi Syndrome; CHS1
GDB:9957338	1q43-1q44	Cold Hypersensitivity Urticaria, Deafness, And Amyloidosis
GDB:698472	1p36-1p36	Chloride Channel, Kidney, B; CLCNKB
GDB:434478	1p1-1q1	Cardiomyopathy, Dilated 1A; CMD1A
GDB:137324	1q3-1q3	Cardiomyopathy, Familial Hypertrophic, 2; CMH2
GDB:119059	1p36-1p36	Melanoma, Malignant
GDB:120595	1p21-1p21	Collagen, Type Xi, Alpha-1; COL11A1

Table 3.3. Genes, gene locations and genetic disorders on chromosome 2EDT 2005.

Gene	GDB Accession ID	Gene Location	Disease Factor
ABCB11	GDB:9864786	2q24-2q24 2q24.3-2q24.3	Cholestasis, Progressive Familial IntraHepatic 2; PFIC2
ABCG5	GDB:10450298	2p21-2p21	Phytosterolemia
ABCG8	GDB:10450300	2p21-2p21	Phytosterolemia
ACADL	GDB:118745	2q34-2q35	ACYL-CoA Dehydrogenase, Long-chain, Deficiency of
ACP1	GDB:118962	2p25-2p25	Phosphatase, Acid, of Erythrocyte; ACP1
AGXT	GDB:127113	2q37.3-2q37.3	Oxalosis I
AHHR	GDB:118984	2pter-2q31	Cytochrome P450, Subfamily I, Polypeptide 1; CYP1A1
ALMS1	GDB:9865539	2p13-2p12 2p14-2p13 2p13.1-2p13.1	Alstrom Syndrome
ALPP	GDB:119672	2q37.1-2q37.1	Alkaline Phosphatase, Placental; ALPP
ALS2	GDB:135696	2q33-2q35	Amyotrophic Lateral Sclerosis 2, Juvenile; ALS2
APOB	GDB:119686	2p24-2p23 2p24-2p24	Apolipoprotein B; APOB
BDE	GDB:9955730	2q37-2q37	Brachydactyly, Type E; BDE
BDMR	GDB:533064	2q37-2q37	Brachydactyly-mental Retardation Syndrome; BDMR
BJS	GDB:9955717	2q34-2q36	Torti and Nerve Deafness
BMPR2	GDB:642243	2q33-2q33 2q33-2q34	Pulmonary Hypertension, Primary; PPH1 Bone Morphogenetic Receptor Type II; BMPR2
CHRNA1	GDB:120586	2q24-2q32	Cholinergic Receptor, Nicotinic, Alpha Polypeptide 1; CHRNA1
CMCWTD	GDB:11498919	2p22.3-2p21	Familial Chronic Mucocutaneous, Dominant Type
CNGA3	GDB:434398	2q11.2-2q11.2	Colorblindness, Total Cyclic Nucleotide Gated Channel, Olfactory, 3; CNG3
COL3A1	GDB:118729	2q31-2q32.3 2q32.2-2q32.2	Collagen, Type III; COL3A1 Ehlers-Danlos Syndrome, Type IV, Autosomal Dominant
COL4A3	GDB:128351	2q36-2q37	Collagen, Type IV, Alpha-3 Chain; COL4A3

Table 3.4. Genes, gene locations and genetic disorders on chromosome 3. Last updated: Sun Aug 21 22:00:00 EDT 2005.

Gene	GDB Accession ID	Gene Location	Disease Factor
ACAA1	GDB:119643	3p23-3p22	Peroxisomal 3-Oxoacyl-coenzyme A Thiolase Deficiency
AGS1	GDB:10795417	3p21-3p21	Encephalopathy, Familial Infantile, with Calcification of Basal Ganglia
AGTR1	GDB:132359	3q21-3q25	Angiotensin II Receptor, Vascular Type 1; AT2R1
AHSG	GDB:118985	3q27-3q27	Alpha-2-HS-Glycoprotein; AHSG
AMT	GDB:132138	3p21.3-3p21.2 3p21.2-3p21.1	Hyperglycinemia, Isolated Nonketotic, Type II; NKH2
ARMET	GDB:9959049	3p21.1-3p21.1	Arginine-rich Protein
BBS3	GDB:376501	3p-3p 3p12 3p12.3-3q11.1	Bardet-biedl Syndrome, Type 3; BBS3
BCHE	GDB:120558	3q26.1-3q26.2	Butyrylcholinesterase; BCHE
BCPM	GDB:433809	3q21-3q21	Benign Chronic Pemphigus; BCPM
BTD	GDB:309078	3p25-3p25	Biotinidase; BTD
CASR	GDB:134196	3q21-3q24	Hypocalciuric Hypercalcemia, Familial; HHC1
CCR2	GDB:337364	3p21-3p21	Chemokine (C-C) Receptor 2; CMKBR2
CCR5	GDB:1230510	3p21-3p21	Chemokine (C-C) Receptor 5; CMKBR5
CDL1	GDB:136344	3q26.3-3q26.3	De Lange Syndrome; CDL
CMT2B	GDB:604021	3q13-3q22	Charcot-marie-tooth Disease, Neuronal Type, B; CMT2B
COL7A1	GDB:128750	3p21-3p21 3p21.3-3p21.3	Collagen, Type VII, Alpha-1; COL7A1
CP	GDB:119069	3q2 3-3q25 3q21-3q24	Ceruloplasmin; CP
CPO	GDB:119070	3q12-3q12 9pter-9qter	Coproporphyria
CRV	GDB:11498333	3p21.3-3p21.1	Vasculopathy, Retinal, With Cerebral Leukodystrophy
CTNNB1	GDB:141922	3p22-3p22 3p21.3-3p21.3	Cancer, Hepatocellular Catenin, Beta 1; CTNNB1
DEM	GDB:681157	3p12-3q11	Dementia, Familial Nonspecific; DEM
ETM1	GDB:9732523	3q13-3q13	Tremor, Hereditary Essential 1; ETM1
FANCD2	GDB:698345	3p25.3-3p25.3 3pter-3p24.2	Fanconi Pancytopenia, Complementation Group D
FIH	GDB:9955790	3q13-3q13	Hypoparathyroidism, Familial Isolated; FIH
FOXL2	GDB:129025	3q23-3q23 3q22-3q23	Blepharophimosis, Epicanthus Inversus, And Ptosis; BPES
GBE1	GDB:138442	3p12-3p12	Glycogen Storage Disease IV
GLB1	GDB:119987	3p22-3p21.33 3p21.33-3p21.33	Gangliosidosis, Generalized GM1, Type I
GLC1C	GDB:3801941	3q21-3q24	Glaucoma 1, Open Angle, C; GLC1C
GNAI2	GDB:120516	3p21.3-3p21.2	Guanine Nucleotide-binding Protein, Alpha-inhibiting, Polypeptide-2;

Table 3.5. Genes, gene locations and genetic disorders on chromosome 4. Last Updated: Sun Aug 21 22:00:00 EDT 2005.

Gene	GDB Accession ID	Gene Location	Disease Factor
ADH1B	GDB:119651	4q21-4q23 4q22-4q22	Alcohol Dehydrogenase-2; ADH2
ADH1C	GDB:119652	4q21-4q23 4q22-4q22	Alcohol Dehydrogenase-3; ADH3
AFP	GDB:119660	4q11-4q13	Alpha-fetoprotein; AFP
AGA	GDB:118981	4q23-4q35 4q32-4q33	Aspartylglucosaminuria; AGU
AIH2	GDB:118751	4q11-4q13 4q13.3-4q21.2	Amelogenesis Imperfecta 2, Hypoplastic Local, Autosomal Dominant;
ALB	GDB:118990	4q11-4q13	Albumin; ALB
ASMD	GDB:119705	4q-4q 4q28-4q31	Anterior Segment Ocular Dysgenesis; ASOD
BFHD	GDB:11498907	4q34.1-4q35	Dysplasia, Beukes Type
CNGA1	GDB:127557	4p14-4q13	Cyclic Nucleotide Gated Channel, Photoreceptor, cGMP Gated, 1; CNCG1
CRBM	GDB:9958132	4p16.3-4p16.3	Cherubism
DCK	GDB:126810	4q13.3-4q21.1	Deoxycytidine Kinase; DCK
DSPP	GDB:5560457	4pter-4qter 4q21.3-4q21.3	Dentin Phosphoprotein; DPP Dentinogenesis Imperfecta; DGI1
DTDP2	GDB:9955810	4q-4q	Dentin Dysplasia, Type II
ELONG	GDB:11498700	4q24-4q24	
ENAM	GDB:9955259	4q21-4q21	Amelogenesis Imperfecta 2, Hypoplastic Local, Autosomal Dominant; Amelogenesis Imperfecta, Hypoplastic Type
ETFDH	GDB:135992	4q32-4q35	Glutaricaciduria IIC; GA IIC
EVC	GDB:555573	4p16-4p16	Ellis-van Creveld Syndrome; EVC
F11	GDB:119891	4q35-4q35	PTA Deficiency
FABP2	GDB:119127	4q28-4q31	Fatty Acid Binding Protein 2, Intestinal; FABP2
FGA	GDB:119129	4q28-4q28	Amyloidosis, Familial Visceral Fibrinogen, A Alpha Polypeptide; FGA
FGB	GDB:119130	4q28-4q28	Fibrinogen, B Beta Polypeptide; FGB
FGFR3	GDB:127526	4p16.3-4p16.3	Achondroplasia; ACH Bladder Cancer Fibroblast Growth Factor Receptor-3; FGFR3
FGG	GDB:119132	4q28-4q28	Fibrinogen, G Gamma Polypeptide; FGG

Table 3.6. Genes, gene locations and genetic disorders on chromosome 5. Last updated: Sun Aug 21 22:00:00 EDT 2005.

Gene	GDB Accession ID	Gene Location	Disease Factor
ADAMTS2	GDB:9957209	5q23-5q34 5q35.3-5q35.3	Ehlers-danlos Syndrome, Type VII, Autosoma Recessive
ADRB2	GDB:120541	5q31-5q32 5q32-5q34	Beta-2-Adrenergic Receptor; ADRB2
AMCN	GDB:9836823	5q35-5q35	Arthrogryposis Multiplex Congenita, Neurogenic Type
AP3B1	GDB:9955590	5q12.1-5q13.3	Hermansky-pudlak Syndrome; HPS
APC	GDB:119682	5q21-5q22 5q22.1-5q22.1	Adenomatous Polyposis of the Colon; APC
ARSB	GDB:119008	5p11-5q13 5q11-5q13	Mucopolysaccharidosis Type VI; MPS VI
B4GALT7	GDB:9957653	5q35.2-5q35.3	Syndrome, Progeroid Form
BHR1	GDB:9956078	5q31-5q33	Asthma
C6	GDB:119045	5p14-5p12 5p13-5p13	Complement Component-6, Deficiency of
C7	GDB:119046	5p14-5p12 5p13-5p13	Complement Component-7, Deficiency of
CCAL2	GDB:5584265	5p15.2-5p15.1	Chondrocalcinosis, Familial Articular
CKN1	GDB:128586	5pter-5qter 5q12.1-5q12.1	Cockayne Syndrome, Type I; CKN1
CMDJ	GDB:9595425	5p15.2-5p14.1	Craniometaphyseal Dysplasia, Jackson Type; CMDJ
CRHBP	GDB:127438	5q-5q 5q12-5q13.3 5q12-5q13.3	Corticotropin Releasing Hormone-binding Protein; CRHBP
CSF1R	GDB:120600	5q33.3-5q34 5q33.2-5q33.3	Colony-stimulating Factor-1 Receptor; CSF1R
DHFR	GDB:119845	5q11.2-5q13.2 5q12-5q14	Dihydrofolate Reductase; DHFR
DIAPH1	GDB:9835482	5q31-5q31	Deafness, Autosomal Dominant Nonsyndromic Sensorine-ural, 1; DFNA1 Diaphanous, Drosophila, Homolog of, 1
DTR	GDB:119853	5q23-5q23	Diphtheria Toxin Sensitivity; DTS
EOS	GDB:9956083	5q31-5q33	Eosinophilia, Familial
EPD	GDB:11509746	5q31.2-5q31.3	Pyridoxine Dependency with Seizures

cont.....

Table 3.6 contd.

Gene	GDB Accession ID	Gene Location	Disease Factor
ERVR	GDB:9835857	5q13-5q14	Hyaloideoretinal Degeneration of Wagner
F12	GDB:119892	5q33-5qter 5q34-5qter	Hageman Factor Deficiency
FBN2	GDB:128122	5q21-5q33 5q23-5q31	Contractural Arachnodactyly, Congenital; CCA
GDNF	GDB:450609	5p13.3-5p13.1	Glial Cell Line-derived Neurotrophic Factor; GDNF
GHR	GDB:119984	5p14-5p12 5p13-5p12	Growth Hormone Receptor; GHR
GLRA1	GDB:118801	5q32-5q32	Glycine Receptor, Alpha-1 Subunit; GLRA1 KOK Disease
GM2A	GDB:120000	5q32-5q33 5q31.3-5q33.1	Tay-sachs Disease, AB Variant
HEXB	GDB:119308	5q13-5q13	Sandhoff Disease

Table 3.7. Genes, gene locations and genetic disorders on chromosome 6. Last updated: Sun Aug 21 22:00:00 EDT 2005.

Gene	GDB Accession ID	Gene Location	Disease Factor
ALDH5A1	GDB:454767	6p22-6p22	Succinic Semialdehyde Dehydrogenase, NAD(+) Dependent; SSADH
ARG1	GDB:119006	6q23-6q23 6q22.3-6q23.1	Argininemia
AS ASSP2	GDB:135697	6p21.3-6p21.3 6pter-6qter	Ankylosing Spondylitis; as Citrullinemia
BCKDHB	GDB:118759	6p22-6p21	Maple Syrup Urine Disease, Type IB
BF	GDB:119726	6p21.3-6p21.3	
C2	GDB:119731	6p21.3-6p21.3	Complement Component-2, Deficiency of
C4A	GDB:119732	6p21.3-6p21.3	Complement Component 4A; C4A
CDKN1A	GDB:266550	6p21.2-6p21.2	Cyclin-dependent Kinase Inhibitor 1A; CDKN1A
COL10A1	GDB:128635	6q21-6q22	Collagen, Type X, Alpha 1; COL10A1
COL11A2	GDB:119788	6p21.3-6p21.3	Collagen, Type XI, Alpha-2; COL11A2 Stickler Syndrome, Type II; STL2 Deafness, Autosomal Dominant Nonsyndromic Sensorineural, 13; DFNA13
CYP21A2	GDB:120605	6p21.3-6p21.3	Adrenal Hyperplasia, Congenital, Due To 21-hydroxylase Deficiency

cont.....

Table 3.7 contd.

Gene	GDB Accession ID	Gene Location	Disease Factor
DYX2	GDB:437584	6p22-6p21.3 6p21.3-6p21.3	Dyslexia, Specific, 2; DYX2
EJM1	GDB:119864	6p21.1-6p11 6p21.2-6q13	Myoclonic Epilepsy, Juvenile; EJM1
ELOVL4	GDB:11499609	6q14-6q14	Stargardt Disease 3; STGD3
EPM2A	GDB:3763331	6q23-6q25 6q23.3-6q25	Epilepsy, Progressive Myoclonic 2; EPM2
ESR1	GDB:119120	6q25.1-6q25.1	Estrogen Receptor; ESR
EYA4	GDB:700062	6q22.3-6q23.2	Deafness, Autosomal Dominant Nonsyndromic Sensorineural, 10; DFNA10
F13A1	GDB:120614	6p25.1-6p24.3	Factor XIII, A1 Subunit; F13A1
FANCE	GDB:1220236	6p22-6p21	Fanconi Anemia, Complementation Group E; Face
GCLC	GDB:132915	6p12-6p12	Gamma-Glutamylcysteine Synthetase Deficiency, Hemolytic Anemia Due
GJA1	GDB:125196	6q21-6q23.2	Gap Junction Protein, Alpha-1, 43 KD; GJA1
GLYS1	GDB:136421	6p21.3-6p21.3	Glycosuria, Renal
GMPR	GDB:127058	6p23-6p23 6pter-6qter	Guanine Monophosphate Reductase
GSE	GDB:9956235	6p-6p	Disease; CD
HCR	GDB:9993306	6p21.3-6p21.3	Psoriasis, Susceptibility To
HFE	GDB:119309	6p22-6p22 6p22.1-6p21.3	Hemochromatosis; HFE
HLA-A	GDB:119310	6p21.3-6p21.3	HLA-A Histocompatibility Type; HLAA
HLA-DPB1	GDB:120636	6p21.3-6p21.3	HLA-DP Histocompatibility Type, Beta-1 Subunit
HLA-DRA	GDB:120641	6p21.3-6p21.3	HLA-DR Histocompatibility Type; HLA-DRA
HPFH	GDB:9849006	6q23-6q23	Heterocellular Hereditary Persistence Of Fetal Hemoglobin
ICS1	GDB:136433	6p21.3-6p21.3	Immotile Cilia Syndrome-1; ICS1
IDDM1	GDB:9953173	6p21.3-6p21.3	Diabetes Mellitus, Juvenile-onset Insulin-dependent; IDDM
IFNGR1	GDB:120688	6q23-6q24 6q24.1-6q24.2	Interferon, Gamma, Receptor-1; IFNGR1
IGAD1	GDB:6929077	6p21.3-6p21.3	Selective Deficiency of
IGF2R	GDB:120083	6q26-6q26 6q25.3-6q25.3	Insulin-like Growth Factor 2 Receptor; IGF2R

Table 3.8. Genes, gene locations and genetic disorders on chromosome 7. Last updated: Sun Aug 21 22:00:00 EDT 2005.

Gene	GDB Accession ID	Gene Location	Disease Factor
AASS	GDB:11502144	7q31.3-7q31.3	Hyperlysinemia
ABCB1	GDB:120712	7q21-7q21 7q21.1-7q21.1	P-Glycoprotein-1; PGY1
ABCB4	GDB:120713	7q21-7q21 7q21.1-7q21.1	P-Glycoprotein-3; PGY3
ACHE	GDB:118746	7q22-7q22	Acetylcholinesterase Blood Group—Yt System; YT
AQP1	GDB:129082	7p14-7p14	Aquaporin-1; Aqp1 Blood Group—Colton; Co
ASL	GDB:119703	7q21.3-7q22 7cen-7q11.2	Argininosuccinicaciduria
ASNS	GDB:119706	7q21-7q31	Asparagine Synthetase; ASNS; as
AUTS1	GDB:9864226	7q32-7q34	Disorder
BPGM	GDB:119039	7q31-7q34	Diphosphoglycerate Mutase Deficiency of Erythrocyte
BRAF	GDB:127513	7q34-7q34	Cancer Of Colon V-RAF Murine Sarcoma Viral Oncogene Homolog B1; BRAF
C7orf2	GDB:10794644	7q36-7q36	Acheiropody
CACNA2D1	GDB:132010	7q21-7q22	Calcium Channel, Voltage-dependent, L Type, Alpha-2/Delta Subunit; Malignant Hyperthermia Susceptibility-3
CCM1	GDB:580824	7q11.2-7q21	Cerebral Cavernous Malformations 1; CCM1
CD36	GDB:138800	7q11.2-7q11.2	CD36 Antigen; CD36
CFTR	GDB:120584	7q31.3-7q31.3 7q31.2-7q31.3 7q31.2-7q31.2	Pancreatitis, Hereditary; PCTT Cystic Fibrosis; CF Deferens, Congenital Bilateral Aplasia OF; CBAVD; CAVD
CHORDOMA	GDB:11498328	7q33-7q33	
CLCN1	GDB:134688	7q32-7qter	Chloride Channel 1, Skeletal Muscle; CLCN1
CMH6	GDB:9956392	7q3-7q3	Cardiomyopathy, Familial Hypertrophic, with Wolff-Parkinson-White
CMT2D	GDB:9953232	7p14-7p14	Charcot-marie-tooth Disease, Neuronal Type, D
COL1A2	GDB:119062	7q21.3-7q22.1	Collagen, Type I, Alpha-2 Polypeptide;
		7q22.1-7q22.1	COL1A2 Osteogenesis Imperfecta Type I, Type IV; OI4
CRS	GDB:119073	7p21-7p21	Craniosynostosis, Type 1; CRS1
CYMD	GDB:366594	7p21-7p15	Macular EDEMA, Cystoid

cont.....

Table 3.8 contd.

Gene	GDB Accession ID	Gene Location	Disease Factor
DFNA5	GDB:636174	7p15-7p15	Deafness, Autosomal Dominant Nonsyndromic Sensorine-ural, 5; DFNA5
DLD	GDB:120608	7q31-7q32	Lipoamide Dehydrogenase Deficiency, Lactic Acidosis due to
DYT11	GDB:10013754	7q21-7q31	Myoclonus, Hereditary Essential
EEC1	GDB:136338	7q11.2-7q21.3	Ectrodactyly, Ectodermal Dysplasia, and Cleft Lip/Palate; EEC
ELN	GDB:119107	7q11.23-7q11.23	Elastin; ELN Williams-Beuren Syndrome; WBS
ETV1	GDB:335229	7p22-7p22	ETS Variant Gene 1; ETV1
FKBP6	GDB:9955215	7q11.23-7q11.23	Williams-Beuren Syndrome; WBS
GCK	GDB:127550	7p-7p 7p15-7p13	Diabetes Mellitus, Autosomal Dominant, Type II Diabetes Mellitus, Type II Glucokinase; GCK
GHRHR	GDB:138465	7p14-7p14	Growth Hormone-releasing Hormone Receptor; GHRHR

Table 3.9. Genes, gene locations and genetic disorders on chromosome 8. Last updated: Sun Aug 21 22:00:00 EDT 2005.

Gene	GDB Accession ID	Gene Location	Disease Factor
ACHM3	GDB:9120558	8q21-8q22 8q21-8q21	Pingelapese Blindness
ADRB3	GDB:203869	8p12-8p11.1 8p12-8p11.2	Beta-3-Adrenergic Receptor; ADRB3
ANK1	GDB:118737	8p11.2-8p11.2 8p12-8p11.2	Spherocytosis, Hereditary; HS
CA1	GDB:119047	8q22-8q22 8q13-8q22.1	Carbonic Anhydrase I, Erythrocyte, Electrophoretic Variants of; CA1
CA2	GDB:119739	8q22-8q22 8q13-8q22.1	Osteopetrosis with Renal Tubular Acidosis
CCAL1	GDB:512892	8q-8q	Chondrocalcinosis with Early-onset Osteoarthritis; CCAL2
CLN8	GDB:252118	8p23-8p23	Epilepsy, Progressive, with Mental Retardation; EPMR
CMT4A	GDB:138755	8q-8q 8q13-8q21.1	Charcot-marie-tooth Neuropathy 4A; CMT4A
CNGB3	GDB:9993286	8q21-8q22	Pingelapese Blindness
COH1	GDB:252122	8q22-8q23	Cohen Syndrome; COH1
CPP	GDB:119798	8q21.13-8q23.1	Ceruloplasmin; CP

cont.....

Table 3.9 contd.

Gene	GDB Accession ID	Gene Location	Disease Factor
CRH	GDB:119804	8q13-8q13	Corticotropin-releasing Hormone; CRH
CYP11B1	GDB:120603	8q21-8q21 8q21-8q22	Adrenal Hyperplasia, Congenital, due to 11 @Beta-Hydroxylase Deficiency
CYP11B2	GDB:120514	8q21-8q21 8q21-8q22	Cytochrome P450, Subfamily XIB, Polypeptide 2; CYP11B2
DECR1	GDB:453934	8q21.3-8q21.3	2,4-@Dienoyl-CoA Reductase; DECR
DPYS	GDB:5885803	8q22-8q22	Dihydropyrimidinase; DPYS
DURS1	GDB:9958126	8q13-8q13 8q13.1-8q13.1	Duane Syndrome
EBS1	GDB:119856	8q24-8qter 8pter-8qter	Epidermolysis Bullosa Simplex, OGNA TYPE
ECA1	GDB:10796318	8q24-8q24	Juvenile Absence
EGI	GDB:128830	8q24-8q24	Epilepsy, Generalized, Idiopathic; EGI
EXT1	GDB:135994	8q24.11-8q24.11	Exostoses, Multiple, Type I; EXT1 Chondrosarcoma
EYA1	GDB:5215167	8q13.3-8q13.3	Branchiootorenal Dysplasia Eyes Absent 1; EYA1
FGFR1	GDB:119913	8p12-8p12 8p11.2-8p11.1	Acrocephalosyndactyly Type V Fibroblast Growth Factor Receptor-1; FGFR1
GNRH1	GDB:133746	8p21-8p11.2	Gonadotropin-releasing Hormone 1; GNRH1 Familial Hypogonadotrophic
GSR	GDB:119288	8p21.1-8p21.1	Glutathione Reductase; GSR
GULOP	GDB:128078	8p21.1-8p21.1	SCURVY

Table 3.10. Genes, gene locations and genetic disorders on chromosome 9. Last updated: Sun Aug 21 22:00:00 EDT 2005.

Gene	GDB Accession ID	Gene Location	Disease Factor
ABCA1	GDB:305294	9q22-9q31 9q31.1-9q31.1	Analphalipoproteinemia ATP-Binding Cassette 1; ABC1
ABL1	GDB:119640	9q34.1-9q34.1	Abelson Murine Leukemia Viral Oncogene Homolog 1; ABL1
ABO	GDB:118956	9q34.1-9q34.2	ABO Blood Group; ABO
ADAMTS13	GDB:9956467	9q34-9q34	Thrombocytopenic Purpura
AK1	GDB:119664	9q34.1-9q34.1	Adenylate Kinase-1; AK1
ALAD	GDB:119665	9q32-9q34	Delta-Aminolevulinate Dehydratase; ALAD
ALDH1A1	GDB:119667	9q21-9q21	Aldehyde Dehydrogenase-1; ALDH1

cont.....

Table 3.10 contd.

Gene	GDB Accession ID	Gene Location	Disease Factor
ALDOB	GDB:119669	9q22-9q22 9q22.3-9q22.3 9q22.3-9q31	Fructose Intolerance, Hereditary
AMBP	GDB:120696	9q32-9q33	Protein HC; HCP
AMCD1	GDB:437519	9p22-9q22.3	Arthrogryposis Multiplex Congenita, Distal, Type 1; AMCD1
ASS	GDB:119010	9q34.1-9q34.1	Citrullinemia
BDMF	GDB:9954424	9p22-9p21	Bone Dysplasia With Medullary Fibrosarcoma
BSCL	GDB:9957720	9q34.1-9q34.1	SEIP Syndrome
C5	GDB:119734	9q33-9q33 9q34.1-9q34.1	Complement Component-5, Deficiency of
CDKN2A	GDB:335362	9p21-9p21	Melanoma, Cutaneous Malignant, 2; CMM2 Cyclin-Dependent Kinase Inhibitor 2A; CDKN2A
CHAC	GDB:6268491	9q21-9q21 9q21.32-9q21.33	Choreoacanthocytosis; CHAC
CLA1	GDB:119781	9q34.2-9qter	Cerebelloparenchymal Disorder III
CMD1B	GDB:677147	9q13-9q22	Cardiomyopathy, Dilated 1B; CMD1B
COL5A1	GDB:131457	9q34.3-9q34.3	Collagen, Type V, Alpha-1 Polypeptide; COL5A1
CRAT	GDB:359759	9q34.1-9q34.1	Carnitine Acetyltransferase; CRAT
DBH	GDB:119836	9q34.3-9q34.3	Dopamine Beta-hydroxylase, Plasma; DBH
DNAI1	GDB:11500297	9p21-9p13	Immotile Cilia Syndrome-1; ICS1
DYS	GDB:137085	9q31-9q33 9q31-9q31 9q31.2-9q31.3	Dysautonomia, Familial; DYS
DYT1	GDB:119854	9q34-9q34	Dystonia 1, Torsion; DYT1
ENG	GDB:137193	9q34.1-9q34.1	Endoglin; ENG
FANCC	GDB:132672	9q22.3-9q22.3	Fanconi Anemia, Complementation Group C; FACC
FBP1	GDB:141539	9q22.3-9q22.3	Fructose-1,6-Bisphophatase 1; FBP1
FCMD	GDB:250412	9q31-9q31	Fukuyama-type Congenital Muscular Dystrophy; FCMD
FRDA	GDB:119951	9q13-9q21.1	Friedreich Ataxia 1; FRDA1
GALT	GDB:119971	9p13-9p13	Galactosemia
GLDC	GDB:128611	9p24-9p23 9p22-9p22	Hyperglycinemia, Isolated Nonketotic, Type I; NKH1

Table 3.11. Genes, gene locations and genetic disorders on chromosome 10. Last updated: Sun Aug 21 22:00:00 EDT 2005.

Gene	GDB Accession ID	Gene Location	Disease Factor
CACNB2	GDB:132014	10p12-10p12	Calcium Channel, Voltage-dependent, Beta-2 Subunit; CACNB2
COL17A1	GDB:131396	10q24.3-10q24.3	Collagen, Type XVII, Alpha-1 Polypeptide; COL17A1
CUBN	GDB:636049	10p12.1-10p12.1	Megaloblastic Anemia 1; MGA1
CXCL12	GDB:433267	10q11.1-10q11.1 10q11.2-10q11.2	Stromal Cell-derived Factor 1; SDF1
CYP17	GDB:119829	10q24.3-10q24.3	Adrenal Hyperplasia, Congenital, due to 17-Alpha-hydroxylase Deficiency
CYP2C19	GDB:119831	10q24.1-10q24.3	Cytochrome P450, Subfamily IIC, Polypeptide 19; CYP2C19
CYP2C9	GDB:131455	10q24.1-10q24.1	Cytochrome P450, Subfamily IIC, Polypeptide 9; CYP2C9
EGR2	GDB:120611	10q21.1-10q21.1	Early Growth Response-2; EGR2 Hypertrophic Neuropathy of Dejerine-Sottas
EMX2	GDB:277886	10q26.1-10q26.1	Empty Spiracles, Drosophila, 2, Homolog of; EMX2
ERCC6	GDB:119882	10q11-10q21 10q11-10q11 10q21.1-10q21.1	Excision-Repair Cross-Complementing Rodent Repair Deficiency, Complementation
FGFR2	GDB:127273	10q25.3-10q26 10q26-10q26	Acrocephalosyndactyly Type V Fibroblast Growth Factor Receptor-2; FGFR2
HK1	GDB:120044	10q22-10q22	Hexokinase-1; HK1

Table 3.12. Genes, gene locations and genetic disorders on chromosome 11. Last updated: Sun Aug 21 22:00:00 EDT 2005.

Gene	GDB Accession ID	Gene Location	Disease Factor
AA	GDB:568984	11p15-11p15	Atrophia Areata; AA
ABCC8	GDB:591370	11p15.1-11p15.1	Sulfonylurea Receptor; Sur Persistent Hyperinsulinemic Hypoglycemia of Infancy
ACAT1	GDB:126861	11q22.3-11q23.1	Alpha-Methylacetoaceticaciduria
ALX4	GDB:10450304	11p11.2-11p11.2	Parietal Foramina, Symmetric; PFM
AMPD3	GDB:136013	11p15-11p15 11pter-11p13	Adenosine Monophosphate Deaminase-3; AMPD3
ANC	GDB:9954484	11q22-11qter	Canal Carcinoma
APOA1	GDB:119684	11q23.3-11q23.3	Amyloidosis, Familial Visceral Apolipoprotein A-I Of High Density Lipoprotein; APOA1

cont.....

Table 3.12 contd.

Gene	GDB Accession ID	Gene Location	Disease Factor
APOA4	GDB:119000	11q23-11q23 11q23-11qter	Apolipoprotein A-IV; APOA4
APOC3	GDB:119001	11q23.1-11q23.2 11q23.2-11q23.3	Apolipoprotein C-III; APOC3
ATM	GDB:593364	11q22-11q23 11q22.3-11q22.3	Ataxia Telangiectasia; AT
BSCL2	GDB:9963996	11q13.1-11q13.5 11q13.1-11q13.1	Seip Syndrome

Table 3.13. Genes, gene locations and genetic disorders on chromosome 12. Last updated: Sun Aug 21 22:00:00 EDT 2005.

Gene	GDB Accession ID	Gene Location	Disease Factor
A2M	GDB:119639	12p13.3-12p12.3	Alpha-2-Macroglobulin; A2M
AAAS	GDB:9954498	12q13-12q13	Glucocorticoid Deficiency and Achalasia
ACADS	GDB:118959	12q22-12qter	ACYL-CoA Dehydrogenase, Short-Chain; ACADS
ACLS	GDB:136346	12p13.3-12p11.2	Acrocallosal Syndrome; ACLS
ACVRL1	GDB:230240	12q11-12q14 12cen-12qter 12q13.12-12q13.13	Osler-rendu-weber Syndrome 2; ORW2 Activin A Receptor, Type II-Like Kinase 1; ACVRL1
ALDH2	GDB:119668	12q24.2-12q24.2	Aldehyde Dehydrogenase-2; ALDH2
AMHR2	GDB:696210	12q13-12q13	Anti-mullerian Hormone Type II Receptor; AMHR2
AOM	GDB:118998	12q12-12q13.1	Stickler Syndrome, Type I; STL1
AQP2	GDB:141853	12q13-12q13	Aquaporin-2; AQP2 Diabetes Insipidus, Renal Type Diabetes Insipidus, Renal Type, Autosomal Recessive
ATD	GDB:696353	12p12.2-12p11.21	Asphyxiating Thoracic Dystrophy; ATD
ATP2A2	GDB:119717	12q23-12q24.1 12q24.1-12q24.1	ATPase, Ca(2+)-Transporting, Slow-Twitch; ATP2A2 Darier-White Disease; DAR
BDC	GDB:5584359	12q24-12q24	Brachydactyly, Type C; BDC
C1R		12p13-12p13	Complement Component-C1r, Deficiency of
CD4	GDB:119767	12p12-12p12 12pter-12p12	T-Cell Antigen T4/LEU3; CD4
CDK4	GDB:204022	12q13-12q13	Cyclin-Dependent Kinase 4; CDK4
CNA1	GDB:252119	12q21.31-12q21.32	Cornea Plana 1; CNA1
COL2A1	GDB:119063	12q12-12q13.2 12q13.11-12q13.2	Stickler Syndrome, Type I; STL1 Collagen, Type II, Alpha-1 Chain; COL2A1 Achondrogenesis, Type II; ACG2
CYP27B1	GDB:9835730	12q12-12q13 12q13.3-12q14	Pseudovitamin D Deficiency Rickets; PDDR

Table 3.14. Genes, gene locations and genetic disorders on chromosome 13. Last updated: Sun Aug 21 22:00:00 EDT 2005.

Gene	GDB Accession ID	Gene Location	Disease Factor
ATP7B	GDB:120494	13q14.3-13q21.1 13q14.3-13q14.3	Wilson Disease; WND
BRCA2	GDB:387848	13q12-13q13 13q12.3-13q12.3	Breast Cancer 2, Early-Onset; BRCA2
BRCD1	GDB:9954522	13pter-13qter	Breast Cancer, Ductal, 1; BRCD1
CLN5	GDB:230991	13q21.2-13q32 13q22-13q22	Ceroid-Lipofuscnosis, Neuronal 5; CLN5
CPB2	GDB:129546	13pter-13qter 13q14.1-13q14.1 13q14.11-13q14.11	Carboxypeptidase B2, Plasma; CPB2
ED2	GDB:9834522	13q11-13q12.1	Ectodermal Dysplasia, Hidrotic; HED
EDNRB	GDB:129075	13q22-13q22	Endothelin-b Receptor; EDNRB Hirschsprung Disease-2; HSCR2
ENUR1	GDB:594516	13q13-13q14.3	Enuresis, Nocturnal, 1; ENUR1
ERCC5	GDB:120515	13q32-13q32 13q33-13q33	Excision-repair, Complementing Defective, in Chinese Hamster, 5; ERCC5
F10	GDB:119890	13q34-13q34	X, Quantitative Variation in Factor X Deficiency; F10
F7	GDB:119897	13q34-13q34	Factor VII Deficiency
GJB2	GDB:125247	13q11-13q12.1	Gap Junction Protein, Beta-2, 26 KD; GJB2 Deafness, Neurosensory, Autosomal Recessive, 1; DFNB1 Deafness, Autosomal Dominant Non-syndromic Sensorineural, 3; DFNA3
GJB6	GDB:9958357	13q12-13q12	Ectodermal Dysplasia, Hidrotic; Hed Deafness, Autosomal Dominant Nonsyndromic Sensorineural, 3; DFNA3
IPF1	GDB:448899	13q12.1-13q12.1	Insulin Promoter Factor 1; IPF1
MBS1	GDB:128365	13q12.2-13q12.2	Moebius Syndrome; MBS
MCOR	GDB:9954520	13q31-13q32	Congenital
NYS4	GDB:11519111	13q31-13q33	Disorder with Predominant Ocular Signs
PCCA	GDB:119473	13q32-13q32	Glycinemia, Ketotic, I

cont.....

Table 3.14 contd.

Gene	GDB Accession ID	Gene Location	Disease Factor
RB1	GDB:118734	13q14.12-13q14.2 13q14.2-13q14.2 13q14.3-13q14.3	Bladder Cancer Retinoblastoma; RB1
RHOK	GDB:371598	13q34-13q34	Rhodopsin Kinase; RHOK
SCZD7	GDB:9864734	13q32-13q32	
SGCG	GDB:3763329	13q12-13q12	Muscular Dystrophy, Limb Girdle, Type 2C; LGMD2C
SLC10A2	GDB:677534	13q33-13q33	Solute Carrier Family 10, Member 2; SLC10A2
SLC25A15		GDB:120042 13q34-13q34	Hyperornithinemia-Hyperammonemia Homocitrul-linuria Syndrome
STARP1	GDB:635459	13pter-13qter	Steroidogenic Acute Regulatory Protein; Star
ZNF198	GDB:6382650	13q11-13q12	Zinc Finger Protein-198; ZNF198

Table 3.15. Genes, gene locations and genetic disorders on chromosome 14. Last updated: Sun Aug 21 22:00:00 EDT 2005.

Gene	GDB Accession ID	Gene Location	Disease Factor
ACHM1	GDB:132458	14pter-14qter	Colorblindness, Total
ARVD1	GDB:371339	14q23-14q24	Arrhythmogenic Right Ventricular Dysplasia, Familial, 1; ARVD1
BCH	GDB:118758	14q13.1-14q21.1	Chorea, Hereditary Benign; BCH Thyroid Transcription Factor 1; TITF1
CTAA1	GDB:265299	14q24-14qter	Cataract, Anterior Polar 1; CTAA1
DAD1	GDB:407505	14q11-14q12	Defender Against Cell Death; DAD1
DFNB5	GDB:636176	14q12-14q13	Deafness, Neurosensory, Autosomal Recessive, 5; DFNB5
EML1	GDB:6328385	14q32-14q32	Usher Syndrome, Type IA; USH1A
GALC	GDB:119970	14q31-14q31	Krabbe Disease
GCH1	GDB:118798	14q22.1-14q22.2	Dystonia, Progressive, with Diurnal Variation GTP Cyclohydrolase I Deficiency GTP Cyclohydrolase I; GCH1
IBGC1	GDB:10450404	14q12-14q13.1	Cerebral Calcification, Nonarteriosclerotic

Table 3.16. Genes, gene locations and genetic disorders on chromosome 15. Last updated: Sun Aug 21 22:00:00 EDT 2005.

Gene	GDB Accession ID	Gene Location	Disease Factor
ACCPN	GDB:5457725	15q13-15q15	Corpus Callosum, Agenesis of, with Neuronopathy
AHO2	GDB:9954535	15q11-15q13	Hereditary Osteodystrophy-2; AHO2
ANCR	GDB:119678	15q11-15q12	Angelman Syndrome
B2M	GDB:119028	15q21-15q22.2	Beta-2-microglobulin; B2M
BBS4	GDB:511199	15q22.3-15q23	Bardet-Biedl Syndrome, Type 4; BBS4
BLM	GDB:135698	15q26.1-15q26.1	Cancer of Colon Bloom Syndrome; BLM
CAPN3	GDB:119751	15q15-15q15 15pter-15qter 15q15.1-15q21.1	Calpain, Large Polypeptide L3; CAPN3 Muscular Dystrophy, Limb-Girdle, Type 2; LGMD2
CDAN1	GDB:9823267	15q15.1-15q15.3	Dyserythropoietic Anemia, Congenital, Type I
CDAN3	GDB:386192	15q21-15q25	Dyserythropoietic Anemia, Congenital, Type III; CDAN3
CLN6	GDB:4073043	15q21-15q21 15q21-15q23	Ceroid-Lipofuscinosis, Neuronal 6, Late Infantile, Variant; CLN6
CMH3	GDB:138299	15q2-15q2	Cardiomyopathy, Familial Hypertrophic, 3; CMH3
CYP19	GDB:119830	15q21-15q21 15q21.1-15q21.1	Cytochrome P450, Subfamily XIX; CYP19
CYP1A1	GDB:120604	15q22-15q24	Cytochrome P450, Subfamily I, Polypeptide 1; CYP1A1
CYP1A2	GDB:118780	15pter-15qter 15q22-15qter	Cytochrome P450, Subfamily I, Polypeptide 2; CYP1A2
DYX1	GDB:1391796	15q21-15q21	Dyslexia, Specific, 1; DYX1
EPB42	GDB:127385	15q15-15q15	Hereditary Hemolytic Protein 4.2, Erythrocytic; EPB42
ETFA	GDB:119121	15q23-15q25	Glutaricaciduria IIA; GA IIA
EYCL3	GDB:4590306	15q11-15q21	Eye Color-3; EYCL3
FAH	GDB:119901	15q23-15q25	Tyrosinemia, Type I

Table 3.17. Genes, gene locations and genetic disorders on chromosome 16. Last updated: Sun Aug 21 22:00:00 EDT 2005.

Gene	GDB Accession ID	Gene Location	Disease Factor
ABCC6	GDB:9315106	16p13.1-16p13.1	Pseudoxanthoma Elasticum, Autosomal Dominant; PXE Pseudoxanthoma Elasticum, Autosomal Recessive; PXE
ALDOA	GDB:118993	16q22.2-16q22.2	Aldolase A, Fructose-Bisphosphate; Aldoa
APRT	GDB:119003	16q24.2-16qter	Adenine Phosphoribosyltransferase; APRT
ATP2A1	GDB:119716	16p12.1-16p12.1	ATPase, Ca(2+)-Transporting, Fast-Twitch 1; ATP2A1 Brody Myopathy
BBS2	GDB:229992	16q21-16q21	Bardet-biedl Syndrome, Type 2; BBS2
CARD15	GDB:11026232	16q12-16q12	Synovitis, Granulomatous, with Uveitis and Cranial Neuropathiesregional Enteritis
CATM	GDB:701219	16p13.3-16p13.3	Microphthalmia-Cataract
CDH1	GDB:120484	16q22.1-16q22.1	Cadherin 1; CDH1
CETP	GDB:119773	16q21-16q21 16q13-16q13	Cholesteryl Ester Transfer Protein, Plasma; CETP
CHST6	GDB:131407	16q22-16q22 16q22.1-16q23.2	Corneal Dystrophy, Macular Type
CLN3	GDB:120593	16p12.1-16p12.1 16p12.1-16p11.2	Ceroid-lipofuscinosis, Neuronal 3, Juvenile; CLN3
CREBBP	GDB:437159	16p13.3-16p13.3	Rubinstein Syndrome Creb-Binding Protein; CREBBP
CTH	GDB:119086	16pter-16qter	Cystathioninuria
CTM	GDB:119819	16q22.1-16q23.1	Cataract, Zonular
CYBA	GDB:125238	16q24-16q24	Granulomatous Disease, Chronic, Autosomal Cytochrome-B-Negative form
CYLD	GDB:701216	16q12-16q13	Epithelioma, Hereditary Multiple Benign Cystic

Table 3.18. Genes, gene locations and genetic disorders on chromosome 17. Last updated: Sun Aug 21 22:00:00 EDT 2005.

Gene	GDB Accession ID	Gene Location	Disease Factor
ABR	GDB:119642	17p13.3-17p13.3	Active BCR-Related Gene; ABR
ACACA	GDB:120534	17q21-17q21 17q12-17q12	Acetyl-CoA Carboxylase Deficiency
ACADVL	GDB:1248185	17p13-17p13 17p11.2-17p11.1	ACYL-CoA Dehydrogenase, very-long-chain, Deficiency of
ACE	GDB:119840	17q23-17q23	Dipeptidyl Carboxypeptidase-1; DCP1
ALDH3A2	GDB:1316855	17p11.2-17p11.2	Sjogren-larsson Syndrome; SLS
APOH	GDB:118887	17q23-17qter	Apolipoprotein H; APOH
ASPA	GDB:231014	17pter-17p13	Spongy Degeneration of Central Nervous System
AXIN2	GDB:9864782	17q23-17q24	Cancer of Colon
BCL5	GDB:125178	17q22-17q22	Leukemia/Lymphoma, Chronic B-Cell, 5; BCL5
BHD	GDB:11498904	17p11.2-17p11.2	With Trichodiscomas and Acrochordons
BLMH	GDB:3801467	17q11.2-17q11.2	Bleomycin Hydrolase
BRCA1	GDB:126611	17q21-17q21 17q11-17q21	Breast Cancer, Type 1; BRCA1
CACD	GDB:5885801	17p-17p 17p13-17p13	Choroidal Dystrophy, Central Areolar; CACD
CCA1	GDB:118763	17q24-17q24	Cataract, Congenital, Cerulean Type 1; CCA1
CCZS	GDB:681973	17q11-17q12	Cataract, Congenital Zonular, with Sutural Opacities; CCZS
CHRNB1	GDB:120587	17p12-17p11	Cholinergic Receptor, Nicotinic, Beta Polypeptide 1; HRNB1

Table 3.19. Genes, gene locations and genetic disorders on chromosome 18. Last updated: Sun Aug 21 22:00:00 EDT 2005.

Gene	GDB Accession ID	Gene Location	Disease Factor
ATP8B1	GDB:453352	18q-18q 18q21-18q22 18q21-18q21	Cholestasis, Progressive Familial Intrahepatic 1; PFIC1 Intrahepatic Cholestasis Familial Intrahepatic Cholestasis-1; FIC1
BCL2	GDB:119031	18q21.33-18q21.33	B-Cell Cll/Lymphoma 2; BCL2
CNSN	GDB:9954580	18q21.3-18q21.3	Carnosinemia
CORD1	GDB:118773	18q21.1-18q21.3	Cone-Rod Dystrophy-1; CORD1
CYB5	GDB:125236	18q22.3-18q23 18q23-18q23	Methemoglobinemia due to Deficiency of Cytochrome B5
DCC	GDB:119838	18q21.1-18q21.1 18q21.3-18q21.3	Deleted in Colorectal Carcinoma; DCC
F5F8D	GDB:6919858	18q21-18q21	Factor V and Factor VIII, Combined Deficiency of; F5F8D
FECH	GDB:127282	18q21.1-18q21.31 18q21.3-18q21.3	Protoporphyria, Erythropoietic
FEO	GDB:4378120	18q21.3-18q21.3	Polyostotic Osteolytic Dysplasia, Hereditary Expansile; HEPOD
LAMA3	GDB:251818	18q11.2-18q11.2	Laminin, Alpha 3; LAMA3
LCFS2	GDB:9954578	18q11-18q12	Cancer
MADH4	GDB:4642788	18q21.1-18q21.1	Polyposis, Juvenile Intestinal Mothers Against Decapentaplegic, Drosophila, Homolog of, 4; MADH4
MAFD1	GDB:120163	18p-18p	Manic-Depressive Psychosis, Autosomal
MC2R	GDB:135163	18p11.2-18p11.2	Adrenal Unresponsiveness to ACTH
MCL	GDB:9954574	18p11.32-18p11.32	Leiomyomata, Hereditary Multiple, of Skin
MYP2	GDB:9862232	18p11.31-18p11.31	Myopia

Table 3.20. Genes, gene locations and genetic disorders on chromosome 19. Last updated: Sun Aug 21 22:00:00 EDT 2005.

Gene	GDB Accession ID	Gene Location	Disease Factor
AD2	GDB:118748	19pter-19qter	Alzheimer Disease-2; AD2
AMH	GDB:118996	19p13.3-19p13.3	Persistent Mullerian Duct Syndrome, Types I And II; PMDS Anti-Mullerian Hormone; AMH
APOC2	GDB:119689	19q13.2-19q13.2	Apolipoprotein C-II Deficiency, Type I Hyperli-poproteinemia due to
APOE	GDB:119691	19q13.2-19q13.2	Apolipoprotein E; APOE
ATHS	GDB:128803	19p13.3-19p13.2	Lipoprotein Phenotype; ALP
BAX	GDB:228082	19q13.3-19q13.4	BCL2-Associated X Protein; BAX
BCKDHA	GDB:119723	19q13.2-19q13.2	Maple Syrup Urine Disease
BCL3	GDB:120561	19q13.2-19q13.2	B-Cell Leukemia/Lymphoma-3; BCL3
BFIC	GDB:9954584	19q12-19q13.11	Benign Familial Infantile Convulsions
C3	GDB:119044	19p13.3-19p13.3	Complement Component-3; C3
CACNA1A	GDB:126432	19p13-19p13	Ataxia, Periodic Vestibulocerebellar
		19p13.1-19p13.1	Hemiplegic Migraine, Familial; MHP Spinocerebellar Ataxia 6; SCA6 Calcium Channel, Voltage-Dependent, P/Q Type, Alpha 1A Subunit; CACNA1A
CCO	GDB:119755	19q12-19q13.1	Central Core Disease of Muscle
CEACAM5	GDB:119054	19q13.2-19q13.2	Carcinoembryonic Antigen; CEA

Table 3.21. Genes, gene locations and genetic disorders on chromosome 20. Last updated: Sun Aug 21 22:00:00 EDT 2005.

Gene	GDB Accession ID	Gene Location	Disease Factor
ADA	GDB:119649	20q12-20q13.11 20q13.11-20q13.11	Adenosine Deaminase; ADA
AHCY	GDB:118983	20cen-20q13.1	S-Adenosylhomocysteine Hydrolase; AHCY
AVP	GDB:119009	20p13-20p13	Diabetes Insipidus, Neurohypophyseal Type Arginine Vasopressin; AVP
CDAN2	GDB:9823270	20q11.2-20q11.2	Dyserythropoietic Anemia, Congenital, Type II
CDPD1	GDB:11505748	20p13-20p13	Dystrophy and Perceptive Deafness
CHED1	GDB:3837719	20p11.2-20q11.2	Corneal Dystrophy, Congenital Endo-thelial; CHED
CHED2	GDB:9957389	20p13-20p13	Corneal Dystrophy, Congenital Hereditary
CHRNA4	GDB:128169	20q13.2-20q13.3	Cholinergic Receptor, Neuronal Nicotinic, Alpha Polypeptide 4; CHRNA4 Epilepsy, Benign Neo-natal; EBN1

cont.....

Table 3.21 contd.

Gene	GDB Accession ID	Gene Location	Disease Factor
CST3	GDB:119817	20p11.2-20p11.2	Amyloidosis VI
EDN3	GDB:119862	20q13.2-20q13.3	Endothelin-3; EDN3 Waardenburg-Shah Syndrome
EEGV1	GDB:127525	20q13.2-20q13.3	Electroencephalogram, Low-voltage
FTLL1	GDB:119235	20q12-20qter	Ferritin Light Chain; FTL
GDF5	GDB:433948	20q11.2-20q11.2	Chondrodysplasia, Grebe Type Cartilage-Derived Morphogenetic Protein 1
GNAS	GDB:120628	20q13.2-20q13.3 20q13.2-20q13.2	Guanine Nucleotide-Binding Protein, Alpha-Stimulating Polypeptide;
GSS	GDB:637022	20q11.2-20q11.2	Glutathione Synthetase Deficiency of Erythro-cytes, Hemolytic Anemia Pyroglu-tamicaciduria
HNF4A	GDB:393281	20q12-20q13.1	Diabetes Mellitus, Autosomal Dominant Trans-cription Factor 14, Hepatic Nuclear Factor; TCF14

Table 3.22. Genes, gene locations and genetic disorders on chromosome 21. Last updated: Sun Aug 21 22:00:00 EDT 2005.

Gene	GDB Accession ID	Gene Location	Disease Factor
AIRE	GDB:567198	21q22.3-21q22.3	Autoimmune Polyendocrinopathy-Candidiasis-Ectodermal Dystrophy; APECED
APP	GDB:119692	21q21.2-21q21.2 21q21.3-21q21.3	Alzheimer Disease; AD Amyloid Beta A4 Precursor Protein; APP
CBS	GDB:119754	21q22.3-21q22.3	Homocystinuria
COL6A1	GDB:119065	21q22.3-21q22.3	Collagen, Type VI, Alpha-1 Chain; COL6A1 Myopathy, Benign Congenital, with Contractures
COL6A2	GDB:119793	21q22.3-21q22.3	Collagen, Type VI, Alpha-2 Chain; COL6A2 Myopathy, Benign Congenital, with Contractures
CSTB	GDB:5215249	21q22.3-21q22.3	Myoclonus Epilepsy of Unverricht and Lundbor Gcystatin B; CSTB
DCR	GDB:125354	21q22.2-21q22.3	Trisomy 21
DSCR1	GDB:731000	21q22.1-21q22.2 21q22.12-21q22.12	Trisomy 21
FPDMM	GDB:9954610	21q22.1-21q22.2	Core-Binding Factor, Runt Domain, Alpha Subunit2; BFA2 Platelet Disorder, Familial, with Associated Myeloid Malignancy
HLCS	GDB:392648	21q22.1-21q22.1 21q22.2-21q22.2 21q22.13-21q22.13	Multiple Carboxylase Deficiency, Biotin-Responsive; MCD

Table 3.23. Genes, gene locations and genetic disorders on chromosome 22. Last updated: Sun Aug 21 22:00:00 EDT 2005.

Gene	GDB Accession ID	Gene Location	Disease Factor
ADSL	GDB:119655	22q13.1-22q13.1	Adenylosuccinate Lyase; ADSL
ARSA	GDB:119007	22q13.31-22qter 22q13.33-22q13.33	Metachromatic Leukodystrophy, Late-Infantile
BCR	GDB:120562	22q11.2-22q11.2 22q11.21-22q11.21 22q11-22q11	Breakpoint Cluster Region; BCR
CECR	GDB:119772	22pter-22q11	Cat Eye Syndrome; CES
CHEK2	GDB:9958730	22q11-22q11	Li-fraumeni Syndrome; Lfsosteogenic Sarcoma
COMT	GDB:119795	22q11.21-22q11.23	Catechol-o-methyltransferase; COMT
CRYBB2	GDB:119075	22q11.2-22q12.1 22q11.2-22q11.2	Crystallin, Beta B2; CRYBB2 Cataract, Congenital, Cerulean Type, 2; CCA2
CSF2RB	GDB:126838	22q12.3-22q13.1 22q13.1-22q13.1	Granulocyte-Macrophage Colony-stimulating Factor Receptor, Beta Subunit;
CTHM	GDB:439247	22q11-22q11	Heart Malformations; CTHM
CYP2D6	GDB:132127	22q13.1-22q13.1	Cytochrome P450, Subfamily IID; CYP2D
CYP2D7P1	GDB:119832	22q13-22q13 22q13.1-22q13.31	Cytochrome P450, Subfamily IID; CYP2D
DGCR	GDB:119843	22q11.21-22q11.23	Digeorge Syndrome; DGS
DIA1	GDB:119848	22q13.31-22qter 22q13.31-22q13.31	Methemoglobinemia due to deficiency of Methemoglobin Reductase
EWSR1	GDB:135984	22q12.1-22q12.3 22q12.1-22q12.1	Ewing Sarcoma; EWS
GGT1	GDB:120623	22q11.2-22q12.1	Glutathionuria

Table 3.24. Genes, gene locations and genetic disorders on chromosome Y. Last updated: Sun Aug 21 22:00:00 EDT. 2005.

Gene	GDB Accession ID	Gene Location	Disease Factor
AMELY	GDB:119676	Yp11.2-Yp11.2	Amelogenin, Y-Chromosomal; Amely
ASSP6	GDB:119020	Ypter-Yqter	Citrullinemia
AZF1	GDB:119027	Yq11-Yq11	Azoospermia Factor 1; AZF1
AZF2	GDB:456131	Ycen-Yqter	Azoospermia Factor 2; AZF2
DAZ	GDB:635890	Yq11.22-Yq11.22 Yq12-Yq12 Yq11-Yq11	Deleted in Azoospermia; DAZ
GCY	GDB:119267	Yq11-Yq11	Control, Y-Chromosome Influenced; GCY
RPS4Y	GDB:128052	Yp11.3-Yp11.3	Ribosomal Protein S4, Y-linked; RPS4Y
SMCY	GDB:5875390	Ycen-Yq11.23 Ypter-Yqter	Histocompatibility Y Antigen; HY; HYA
SRY	GDB:125556	Yp11.3-Yp11.3	Sex-determining Region Y; SRY
ZFY	GDB:120503	Yp11.3-Yp11.3	Zinc Finger Protein, Y-linked; ZFY

Table 3.25. Genes, gene locations and genetic disorders on chromosome X. Last updated: Sun Aug 21 22:00:00 EDT 2005.

Gene	GDB Accession ID	Gene Location	Disease Factor
ABCD1	GDB:118991	Xq28-Xq28	Adrenoleukodystrophy; ALD
ACTL1	GDB:119648	Xp11.22-Xp11.21	Actin-like Sequence-1; ACTL1
ADFN	GDB:118977	Xq25-Xq26	Albinism-Deafness Syndrome; ADFN; ALDS
AGMX2	GDB:119661	Xpter-Xqter	Agammaglobulinemia, X-Linked, Type 2; AGMX2; XLA2
AHDS	GDB:125899	Xq21.1-Xq21.1 Xq21-Xq21	Mental Retardation, X-linked, with Hypotonia
AIC	GDB:118986	Xp22-Xp22	Corpus Callosum, Agenesis of, with Chorio-retinal Abnormality
AIED	GDB:119663	Xp11.4-Xq21	Albinism, Ocular, Type 2; OA2
AIH3	GDB:131443	Xq22-Xq28	Amelogenesis Imperfecta-3, Hypoplastic Type; AIH3
ALAS2	GDB:119666	Xp11.21-Xp11.21	Anemia, Hypochromic
AMCD	GDB:5584286	Xp11.3-Xq11.2	Arthrogryposis Multiplex Congenita, Distal
AMELX	GDB:119675	Xp22.31-Xp22.1	Amelogenesis Imperfecta-1, Hypoplastic Type; AIH1
ANOP1	GDB:128454	Xpter-Xqter	Clinical; ANOP1
AR	GDB:120556	Xq11.2-Xq12	Androgen Insensitivity Syndrome; AIS Androgen Receptor; AR
ARAF1	GDB:119004	Xp11.3-Xp11.23	V-RAF Murine Sarcoma 3611 Viral Oncogene Homolog 1; ARAF1
ARSC2	GDB:119702	Xpter-Xqter	Arylsulfatase C, f FORM; ARSC2
ARSE	GDB:555743	Xp22.3-Xp22.3	Chondrodysplasia Punctata 1, X-Linked Recessive; CDPX1

Table 3.26. Genes, gene locations and genetic disorders on chromosome unknown or multiple gene locations last updated: Sun Aug 21,2005.

Gene	GDB Accession ID	Gene Location	Disease Factor
ABAT	GDB:581658		Gamma-aminobutyrate Transaminase
AEZ	GDB:128360		Acrodermatitis Enteropathica, Zinc-deficiency Type; AEZ
AFA	GDB:265277		Filiforme Adnatum and Cleft Palate
AFD1	GDB:265292		Dysostosis, Treacher Collins Type, with Limb Anomalies

cont.....

Table 3.26 contd.

Gene	GDB Accession ID	Gene Location	Disease Factor
ASAH1	GDB:6837715		Farber Lipogranulomatosis
ASD1	GDB:6276019		Atrial Septal Defect; ASD
ASMT	GDB:136259		Acetylserotonin Methyltransferase; ASMT Acetylserotonin Methyltransferase, Y-Chromosomal; Asmty; Hiomty
CCAT	GDB:118738		Cataract, Congenital or Juvenile
CECR9	GDB:10796163		Cat Eye Syndrome; CES
CEPA	GDB:581848		Control, Congenital Failure of
CLA3	GDB:128453		Cerebelloparenchymal Disorder I; CPD I
CLN4	GDB:125229		Ceroid-Lipofuscinosis, Neuronal 4; CLN4
CSF2RA	GDB:118777		Colony Stimulating Factor 2 Receptor, Alpha; CSF2RA Granulocyte-Macrophage Colony-Stimulating Factor Receptor, Alpha Subunit,
CTS1	GDB:118779		Carpal Tunnel Syndrome; CTS; CTS1
DF	GDB:132645		Factor D
DIH1	GDB:439243		Diaphragmatic

Testing for the presence of genes involves taking a person's cells and examining their DNA. Geneticists performing these tests focus on a particular segment of DNA in order to find missing information sequences, added sequences, or transposed sequences. Testing involves not only looking for altered genes, but also looking for gene products which could signal an aberrant gene.

Gene testing is used to determine if a person is a carrier of a particular gene, which may then lead to the conclusion whether the person is predisposed to a particular disease or presymptomatic for that disease. However, a person may neither be predisposed nor presymptomatic. In such a case, he or she is just a carrier of a particular allele, which is a different form of the same gene. Since some diseases, such as cystic fibrosis, require the presence of two same alleles in order for the diseases to manifest themselves, a person possessing only one of the requisite alleles is said to be a carrier of such disease. Some diseases, such as Huntington's disease, require only the presence of one particular allele for manifestation.

Gene testing can determine whether a person is a carrier of a dominant or recessive allele, and therefore the patient can be informed whether he is predisposed or presymptomatic for a particular disease. This information helps the patient to know, which genetic information can possibly be passed down to his next generation. This technique can also be used to determine the possibility of mutant alleles in parents and siblings. It can even be used to track the potential for a mutant gene in grandchildren.

Genetic tests can be expensive. A 1995 report found that a cystic fibrosis test costs between $125 and $150; a Huntington's disease test costs between $250 and $300; and a Tay-Sachs test costs about $150.61. Commercial tests for breast and ovarian cancer are available, and tests for Alzheimer's disease are being commercially developed. As the technology advances and more information is acquired about genes and gene products, these tests could become less expensive.

In gene tests, scientists scan a patient's DNA sample for mutated sequences. A DNA sample can be obtained from any tissue, including blood. For some types of tests, researchers design short pieces of DNA called probes, whose sequences are complementary to the mutated sequences. These probes will seek their complement among the three billion base pairs of an individual's genome. If the mutated sequence is present in the patient's genome, the probe will bind to it and flag the mutation. Another type of DNA testing involves comparing the sequence of DNA bases in a patient's gene to a normal version of the gene. Cost of testing can range from hundreds to thousands of dollars, depending on the size of the genes and the numbers of mutations tested.

3.7. MERITS AND DEMERITS OF GENE TESTING

Gene testing has dramatically improved the lives of people. Some tests are used to clarify a diagnosis and direct a physician towards appropriate treatments, while others allow families to avoid having children with devastating diseases, or identify people at high risk for conditions that may be preventable. Aggressive monitoring for the removal of colon growths in those inheriting a gene for familial adenomatous polyposis, for example, has saved many lives. On the horizon is a gene test that will provide doctors with a simple diagnostic test for a common iron-storage disease, transforming it from a usually fatal condition to a treatable one.

Commercialized gene tests for adult-onset disorders, such as Alzheimer's disease and some cancers are the subject of most of the debate over gene testing. These tests are targeted to healthy (presymptomatic) people who are identified as being at high risk because of a strong family medical history for the disorder. The tests give only a probability for developing the disorder. One of the most serious limitations of these susceptibility tests is the difficulty in interpreting a positive result, because some people who carry a disease-associated mutation do not develop the disease. Scientists believe that these mutations may work together with other unknown mutations or with environmental factors to cause the disease.

The possibility of laboratory errors is one limitation of the medical testing. This might be due to sample misidentification, contamination of the chemicals used for testing, or other factors.

In the medical establishment, many people feel that the uncertainties surrounding the test interpretation, the lack of available medical options for these diseases, the tests' potential for provoking anxiety, and the risk of discrimination and social stigmatization could outweigh the benefits of the testing.

Diseases Diagnosed by Gene Testing

Currently, more than 900 genetic tests are available in the laboratories. Some gene tests available in the clinical genetics laboratories are given below. Names of the test, and a description of the diseases or symptoms are given in parentheses. Susceptibility tests, marked by an asterisk, provide only an estimated risk for developing the disorder. More information can be obtained from GeneTests Inc. for comprehensive information on test availability and genetic testing facilities.

Some Currently Available DNA-Based Gene Tests

- *Alpha-1-antitrypsin deficiency* (AAT; emphysema and liver disease).
- *Amyotrophic lateral sclerosis* (ALS; Lou Gehrig's Disease; progressive motor function loss leading to paralysis and death).
- *Alzheimer's disease** (APOE; late-onset variety of senile dementia).
- *Ataxia telangiectasia* (AT; progressive brain disorder resulting in loss of muscle control and cancers).

- ***Gaucher disease*** (GD; enlarged liver and spleen, bone degeneration).
- ***Inherited breast and ovarian cancer**** (BRCA 1 and 2; early-onset tumors of breasts and ovaries).
- ***Hereditary nonpolyposis colon cancer**** (CA; early-onset tumors of colon and sometimes other organs).
- ***Charcot-Marie-tooth*** (CMT; loss of feeling in ends of limbs).
- ***Congenital adrenal hyperplasia*** (CAH; hormone deficiency; ambiguous genitalia and male pseudohermaphroditism).
- ***Cystic fibrosis*** (CF; disease of lung and pancreas resulting in thick mucous accumulations and chronic infections).
- ***Duchenne muscular dystrophy/Becker muscular dystrophy*** (DMD; severe to mild muscle wasting, deterioration, weakness).
- ***Dystonia*** (DYT; muscle rigidity, repetitive twisting movements).
- ***Fanconi anemia, group C*** (FA; anemia, leukemia, skeletal deformities).
- ***Factor V-Leiden*** (FVL; blood-clotting disorder).
- ***Fragile X syndrome*** (FRAX; leading cause of inherited mental retardation).
- ***Hemophilia A and B*** (HEMA and HEMB; bleeding disorders).
- ***Hereditary hemochromatosis*** (HFE; excess iron storage disorder).
- ***Huntington's disease*** (HD; usually midlife onset; progressive, lethal, degenerative neurological disease).
- ***Myotonic dystrophy*** (MD; progressive muscle weakness; most common form of adult muscular dystrophy).
- ***Neurofibromatosis type 1*** (NF1; multiple benign nervous system tumors that can be disfiguring; cancers).
- ***Phenylketonuria*** (PKU; progressive mental retardation due to missing enzyme; correctable by diet).
- ***Adult polycystic kidney disease*** (APKD; kidney failure and liver disease)
- ***Prader Willi/Angelman syndromes*** (PW/A; decreased motor skills, cognitive impairment, early death).
- ***Sickle cell disease*** (SS; blood cell disorder; chronic pain and infections).
- ***Spinocerebellar ataxia, type 1*** (SCA1; involuntary muscle movements, reflex disorders, explosive speech).
- ***Spinal muscular atrophy*** (SMA; severe, usually lethal progressive muscle-wasting disorder in children).
- ***Thalassemias*** (THAL; anemias—reduced red blood cell levels).
- ***Tay-Sachs disease*** (TS; fatal neurological disease of early childhood; seizures, paralysis).

Cost of Genetic Testing

Costs include running and maintaining the screening program, as well as follow-up visits for infants who test positive, and their parents. The combined cost for PKU and hypothyroidism tests, which all US states and many countries require, was about $6 in 1986. The cost of laboratory detection and follow-up for galactosemia and maple syrup urine disease (MSUD) was about $1.50 per specimen in 1986. However, the recent advent of tandem mass spectrometry analysis may cause an reduction in the cost. If adding a new test to the screening program exceeds a base cost level, the additional expense needs to be justified by specific benefits.

Table 3.27 gives a list of costs for genetic tests that are not included in all state or country programs. For most disorders, several laboratories can provide genetic testing. The following factors should be considered while evaluating the quoted cost at a particular laboratory:

- Is the laboratory associated with a reputable company or university?
- Is the laboratory's work published in medical literature?
- What testing method is used? For example, tests for specific mutations in a gene are less expensive than sequencing the entire gene.
- Which testing strategy is used? Some labs test for a large number of mutations all at once. Others test first for the most common mutations, and then proceed according to the results.
- How many individuals need to be tested? Several family members may be tested to obtain a meaningful result.
- Can contractual agreements lower the costs? Hospitals, insurers, and laboratories negotiate contracts to set the price of testing and amount of reimbursement. Contracts to establish large-scale screening efforts can lower the costs substantially.

Genetic Testing and its Regulation

Currently, in the United States and other countries, no regulations are in place for evaluating the accuracy and reliability of genetic testing. Most genetic tests developed by laboratories are categorized as services, which the Food and Drug Administration (FDA) does not regulate. Only a few US states have established some regulatory guidelines. This lack of government oversight is particularly troublesome in light of the fact that a handful of companies have started marketing test kits directly to the public. Some of these companies make dubious claims about how the kits not only test for disease but also serve as tools for customizing medicine, vitamins, and foods to each individual's genetic make-up. Another fear is that individuals, who purchase such kits, will not seek out genetic counseling to help them interpret results and make the best possible decisions regarding their personal welfare. More information on these questionable test kits is available from Dubious Genetic Testing, an online report provided by Quackwatch.

Genetic Testing and Insurance Cover

In most cases, an individual will have to contact his or her insurance provider to see if the genetic tests, which cost between $200 and $3000, are covered in the insurance. Usually the insurance companies do not cover genetic tests; those that do will have an access to the results. Insured people would need to decide whether they want the insurance company to have this information. States have a patchwork of genetic-information nondiscrimination laws, none of which is comprehensive. Existing state laws differ in coverage, protections afforded, and enforcement schemes. The National Conference of State Legislatures provides a listing of current legislation, regarding genetic information and health insurance. The recent marketing of genetic test kits directly to consumers may lead to an increase in demand for insurance coverage.

3.8. TESTING FOR GENETIC DISEASES AND THEIR IMPORTANCE IN THE HEALTHCARE SYSTEM

Throughout the world, human populations vary both phenotypically and genotypically. Heterogeneity results from different alleles and mutations of the multiple genes, which is a part of almost all biochemical pathways. Mutations in different parts of a gene may cause different disorders. More than 900 specific causes of congenital diseases have been identified; some are genetic (involving nuclear and mitochondrial genes), whereas others result from rubella virus or environmental agents. Concern about the teratogenic effects of drugs taken during pregnancy is growing. For instance, women who consume alcohol during pregnancy are at an increased risk of having infants with mental retardation, and behavioral disturbances, and intrauterine growth retardation.

Table 3.27. Costs of testing for selected genetic disorders.

Disorder	Laboratory	Cost
Alpha-1-antitrypsin	Baylor College of Medicine, DNA Diagnostic Lab	$285
Alpha-1-antitrypsin	University of Florida	Testing kit available free from Alpha-1 Association
Alzheimer's disorder (APOE e4 mutation)	Athena Diagnostics	$279
Breast cancer (BRCA1/BRCA2)	Baylor	$350 (only available for limited population)
Breast cancer (BRCA1/BRCA2)	University of California, San Francisco, Clinical Molecular Diagnostics Lab	$354
Breast cancer (BRCA1/BRCA2)	Myriad Genetics	$2580 (full gene sequencing)
Cystic fibrosis	Baylor	$145 (mutation only) $285 (family linkage analysis)
Cystic fibrosis	UCSF	$146 (31 mutations)
Cystic fibrosis	University of Colorado, DNA Diagnostic Lab	$150 (carrier status based on family analysis) $300 (mutation analysis)
Cystic fibrosis	Comprehensive Genetic Services (CGS)	$50 (mutation analysis)
Galactosemia	Mayo Clinic, Biochemical Genetics Lab	$107 (enzyme assay) $250 (isoelectric focusing if enzyme assay is unclear)
Galactosemia	CGS	$198
Huntington's disease	Baylor	$285
Huntington's disease	UCSF	$325
Huntington's disease	U. CO., DNA	$250
Huntington's disease	CGS	$175
Maple syrup urine disease	Baylor	$25 (As part of Supplemental Newborn Screening Program kit which also screens for 27 other disorders)
Neurofibromatosis 1	Boston University School of Medicine, Center for Human Genetics	$1250 (family linkage analysis for up to 4 people)
Neurofibromatosis 1	Children's Hospital, DNA Diagnostic Laboratory, Boston, MA	$415 (FISH analysis)

cont.....

Table 3.27 contd.

Disorder	Laboratory	Cost
Neurofibromatosis 1	Laboratory Corporation of America	Protein assay only $550 (if unknown mutation) $400 (if known mutation)
Sickle cell disorder	Baylor	$285
Sickle cell disorder	UCSF	$181
Sickle cell disorder	CGS	$98
Smith-Lemli-Opitz syndrome	Boston University School of Medicine, Center for Human Genetics	$1250 (family linkage analysis for up to four people) $225 (analysis of one mutation)
Smith-Lemli-Opitz syndrome	Kennedy Krieger Institute, Baltimore, MD	$150
Smith-Lemli-Opitz syndrome	Oregon Health Sciences University	$140

Hence, phenotypically, similar conditions may be due to mutations in different genes, non-genetic factors, or both. To establish the cause of the person's problem and to determine the risk to future offspring, the physician must know the family history properly should inquire about the possible environmental factors, keeping heterogeneity in mind. Identifying the specific causative agent is not always possible.

Now-a-days, powerful molecular genetic techniques have made it possible to study the structure of DNA and to track changes during the development of different tissues. The structure of a gene is complex and includes control elements (e.g., promoters, enhancers), expressed elements (exons), intervening elements that are not expressed (introns), and a termination signal. The configuration of the DNA of a gene in a tissue that does not express the gene is likely to be different (e.g., methylated, condensed) than that of a gene that is expressing actively.

Genetic Screening

Genetic screening may be used in populations, which are at a higher risk for a particular genetic disorder. Genetic screening is appropriate only when the natural history of the disease is understood; the screening tests are valid and reliable; sensitivity, specificity, false-negative, and false-positive rates are acceptable; and effective therapy is available. A sufficient benefit must be derived from a screening program to justify its cost.

Heterozygote Screening

The process of heterozygote screening can determine if a person is a carrier for a specific genetic disorder. If the partner is also a heterozygote, the couple is at risk of having an affected child. Screening allows the couple to make informed reproductive choices. Screening a susceptible population (e.g., Tay-Sachs disease in Ashkenazic Jews, sickle-cell anemia in blacks, thalassemia in various ethnic groups) may be appropriate because of the high frequency of heterozygotes.

Presymptomatic Genetic Screening

Presymptomatic genetic screening may be appropriate for persons with a family history of a dominantly inherited disorder (e.g., Huntington's disease, breast cancer). Identifying a definite carrier of the genetic disorder may allow the patient to make informed decisions (e.g., monitoring in case of breast cancer, reproductive choices in case of Huntington's disease or adult polycystic kidney disease).

Prenatal Diagnosis

Many techniques such as amniocentesis, chorionic villus sampling, umbilical cord blood sampling, maternal blood sampling, maternal serum screening, and fetal visualization with ultrasound and radiography are useful in prenatal diagnosis. The reasons for prenatal screening in medicine include maternal age above 35 years, a family history of a condition that can be diagnosed by prenatal techniques, abnormal maternal serum screening results, and certain complications of pregnancy.

Newborn Screening

Now-a-days newborn screening is becoming a general practice. Screening for phenylketonuria, galactosemia, and hypothyroidism in the newborn allows prophylaxis (i.e., special diet or replacement therapy) to be initiated early enough to prevent severe complications.

Construction of a Family Pedigree

The family history is often a key to determine the genetic risk among family members. It is most easily recorded in a family pedigree (family tree), which uses conventional symbols. The pedigree provides a ready view of problems or illnesses within the family, and facilitates analysis of inheritance patterns, including the range and degree of affliction and variation among people and generations. The familial disorders with identical phenotypes have several patterns of inheritance in humans. For example, cleft palate can have an autosomal dominant, autosomal recessive, or X-linked recessive pattern of inheritance, or it may be multifactorial (i.e., familial, but with no precisely predictable inheritance pattern).

Commonly generations in the family pedigree are numbered with Roman numerals; with older generations at the top and the most recent at the bottom. Within each generation, persons are numbered from left to right with Arabic numerals. Siblings are usually listed by age, with the oldest on the left. Hence, each member of the pedigree can be identified by two numbers (e.g., II, 4). A spouse is also assigned an identifying number.

The genetic study of a trait or disease begins with the affected person (the proband, propositus [male], proposita [female], or index case). When taking a family history, the physician usually draws the pedigree as the relatives are described. The inquiry begins with the siblings of the proband and proceeds to the parents; relatives of the parents, including brothers, sisters, nephews, and nieces; grandparents; and so on. The number of relatives included in the pedigree is determined by the inheritance pattern of the condition and by the extent of the informant's memory or knowledge. Usually, at least, three generations are included. Illnesses, hospitalizations, causes of death, miscarriages, abortions, congenital anomalies, and any other unusual features are recorded.

Genetic Counseling

Genetic counseling involves obtaining a thorough family history, and addressing the family's concerns and questions by gathering appropriate information from the literature and genetic specialists. Consultation with an expert in the specific condition is often important. The information provided should include the diagnosis and diagnostic methods, including identification of the carriers; the natural history of the disorder and its complications; the recurrence risk for the patient and various family members; potential therapies; and reproductive options. Communicating genetic risks and options is an involved process that often requires follow-up visits and written communication.

Genetic counseling centers exist at medical schools in the USA and Canada. Patients and families can be referred for diagnosis, counseling, management, and information. A working relationship between the genetics center and the family physician is essential for the family's best interest and follow-up.

Preventive Medicine and Customized Therapies

Studies of gene function will lead to a deeper understanding of normal biological processes and how they go awry in diseased state. These insights will allow the development of better and earlier predictive tests, and eventually usher in a field of prevention-based medicine and diagnostics.

Within the next decade, researchers will also begin to understand how DNA variations underlie our individual responses to medical treatments. Thousands of people are hospitalized each year as a result of toxic responses to medications that are beneficial to others. Some cancers respond dramatically to the current therapeutic regimens, while the same treatment has no effect on disease progression in others. Scientists in major pharmaceutical companies are trying to sort out the specific regions of DNA associated with drug responses. They are also trying to identify particular subgroups of patients, and develop drugs customized for those populations. These capabilities are expected to make drug development faster, cheaper, and more effective, while drastically reducing the number of adverse reactions.

Drug design itself will be revolutionized, as researchers use gene sequence and protein structure information to create new classes of medicines based on a reasoned approach, rather than the traditional trial-and-error methods for finding new drugs. The new drugs, targeted to specific sites in the body and particular points in the cascade of biochemical events leading to disease, will likely cause fewer side effects than many current medicines. Ideally, they would act earlier in the disease process.

Gene Therapy and Genetic Enhancement

The exploitation of genes to treat diseases has captured the imagination of the public and the biomedical community. This rapidly developing field—gene transfer or gene therapy—holds great potential for treating or even curing long genetic and acquired diseases, such as cancers and AIDS by using normal genes, to replace or supplement defective genes, or bolster a normal function like immunity.

Over 1000 clinical gene therapy trials are now in progress worldwide; most for different kinds of cancers. Performed on the patients in advanced stages of disease, most current studies aim to establish the safety of gene delivery procedures rather than determine their effectiveness. The technology still has to face many obstacles before it can become a practical approach for treating disease; however, novel experimental approaches look very promising.

Besides preventing and treating inherited and infectious diseases, gene transfer technologies will probably make possible the enhancement or replacement of genes that influence other traits, such as height, weight, strength, stamina, and even intelligence. These capabilities will generate many questions about the regulation of such technologies and the fairness of access to these expensive protocols, as well as safety and privacy issues, among others.

3.9. SOME OF THE MAJOR GENETIC DISEASES AND THEIR CHARACTERISTICS

Genetic testing is used to determine if a person is a carrier of a particular gene, This further establishes whether the person is predisposed to a particular disease or is presymptomatic for that disease. However, a person may be neither predisposed nor presymptomatic. In such case, he or she is just a carrier of a particular allele, which is a different form of the same gene. Since some diseases, such as cystic fibrosis require the presence of two same alleles in order to manifest themselves, a person possessing only one of the requisite alleles is said to be the carrier of the disease. Some diseases, such as Huntington's disease, require only the presence of one particular allele for manifestation **(Table 3.28)**.

Genetic tests can determine whether a person is a carrier of a dominant or recessive allele. This kind of information helps the patient to know, which genetic information can possibly be passed down to the next generation. This information can also be used to determine the possibility of mutant alleles in parents and siblings. It can even be used to track the possibility of a mutant gene in grandchildren.

Table 3.28. The most common genetic disorders and their symptoms in humans

Disorder	Incidence (Approximate)	Symptoms if not treated	Treatment
Alpha-1-antitrypsin	1 in 2,000	Early lung disorder (by 40 years old).	Lifestyle choice and environmental precautions(for example, no smoking, clean workplace); Prolastin injections.
Alzheimer's disorder (associated with APOE e4 allele)	1–3 percent of the population is homozygous for APOE e4. 30-50% of these people will develop Alzheimer's	Increasing dementia. Progressive loss of mental ability and changes in personality.	Alzheimer's disorder can be managed but not cured.
Breast cancer	1 in 8 women will develop breast cancer. Only 5–10% of cases have a genetic link; two thirds of those are associated with mutations in BRCA1 or BRCA2 genes	Increased risk for breast cancer.	Improved diet and exercise. Early screening. Surgical removal of all or part of a breast before tumor onset.
Cystic fibrosis	1 in 2,000 Caucasians 1 in 17,000 African-Americans 1 in 9,000 Hispanics	Obstructive lung disorder, frequent infections and heart failure, eventually resulting in death.	Physical therapy, antibiotics, oxygen therapy and organ transplants. Gene therapy is currently under trial.
Galactosemia	1 in 40,000	Severe brain and kidney damage; can cause death if untreated.	Galactose-restricted diet.
Huntington's disease	1 in 20,000 in Western countries, less frequent in Africa and Asia	Gradual deterioration of the nervous system.	None. However, genetic screening allows the individual to plan for their future and decide whether to have children.
Hypothyroidism	1 in 4,000	Mental retardation, other brain damage and growth delay.	Provide thyroid hormone.

cont.....

Table 3.28 contd.

Disorder	Incidence (Approximate)	Symptoms if not treated	Treatment
Maple syrup urine disease	1 in 90,000 to 1 in 300,000 1 in 176 in Pennsylvania Mennonites	Vomiting, seizures, severe mental retardation, coma. Death in 2 years	Lifelong strict dietary regimen and monitoring of amino acid levels.
Neuro-fibromatosis 1	1 in 4,000	Skin tumors, vision problems, learning difficulties, and skeletal abnormalities	Monitoring of tumors. Treatment of rare cancers. Treatment of skeletal abnormalities through surgery. Intervention for learning disabilities.
PKU	1 in 10,000 to 1 in 25,000	Severe mental retardation and seizures	Restricted phenylalanine diet.
Sickle cell disorder	1 in 400 African Americans 1 in 58,000 Caucasians 1 in 1,100 in Hispanics from eastern U.S. 1 in 32,000 in Hispanics from western U.S. 1 in 11,500 Asians 1 in 2,700 Native Americans	Acute pain from blocked blood vessels, tissue and organ damage, increased risk of infection, and associated complications	Early comprehensive care and penicillin treatment reduces illness and death.
Smith-Lemli-Opitz Syndrome	1 in 10,000 to 1 in 50,000	General growth retardation, developmental delay, and a variety of different malformations	Supplying missing cholesterol early in life can eliminate a large portion of the disorder symptoms.

3.10. DIAGNOSIS OF GENETIC DISEASES

A large numbers of methods are available at present for the screening of genetic testing in presymptomatic conditions. Some of them are explained below.

Preparation of the Chromosomes

Preparing the chromosomes for analysis is relatively simple. Circulating blood lymphocytes are generally used in this technique. On exception is in the case of fetus, in which amniocytes from amniotic fluid or chorionic villus cells from the placenta are used instead. The cells are cultured in the laboratory with phytohemagglutinin to stimulate cell division. Colchicine is then added to arrest mitosis during metaphase, when each chromosome has replicated into two chromatids attached at the centromere. The cells, which are spread onto microscope slides, are then stained. Chromosomes from single cells are usually photographed. Their images are cut out of the print and pasted onto a piece of paper, forming a karyotype. Computer imaging can also be used to produce a visual display of the chromosomes.

Chromosome Staining

Chromosome staining is performed by G (Giemsa) or Q (Fluorescent) banding techniques. Additional staining procedures and techniques for extending the chromosome length have greatly increased the precision of cytogenetic diagnosis.

New Molecular Techniques for Genetic Diseases

New molecular techniques use DNA probes (which can have fluorescent tags) to locate specific genes or DNA sequences on the chromosomes. *Fluorescent In Situ Hybridization* (FISH) **(Fig. 3.7)**, restriction fragment length polymorphism (RFLP) **(Fig. 3.8)**, and polymerase chain reaction (PCR) **(Fig. 3.9)** are used to identify the organization of genes, and to visualise deletions, rearrangements and duplications of the chromosomes. These molecular techniques provide highly sensitive methods for testing the defective genes (mutated alleles in an individuals or embryos) in the uterus.

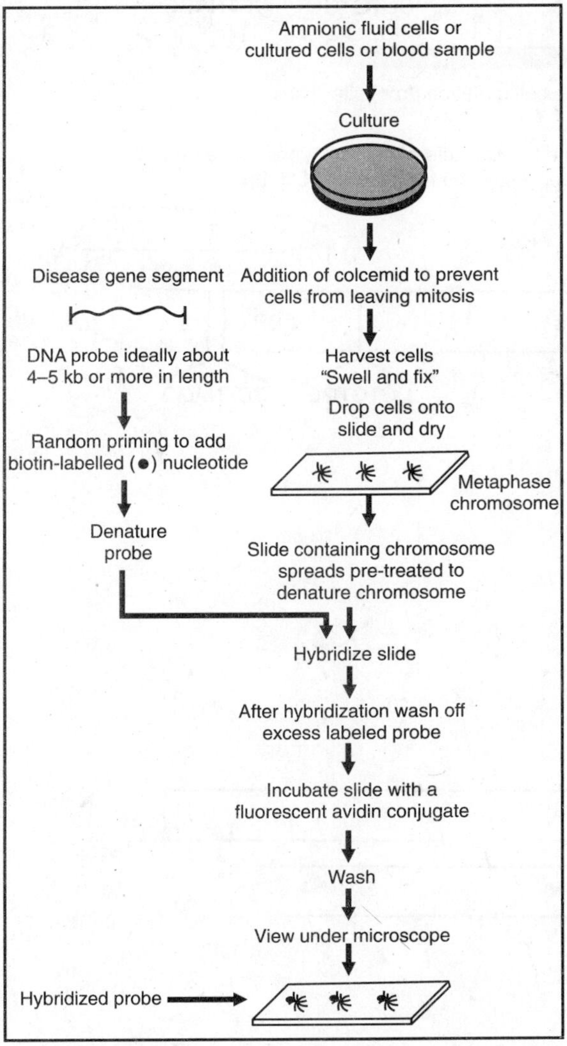

Karyotype Analysis

The normal male is designated as 46,XY, and the normal female as 46,XX. In Down syndrome, due to an extra chromosome 21 (trisomy 21), the notation is 47,XY,21+ for a male and 47,XX,21+ for a female. In Down syndrome, due to a translocation (two chromosomes that are stuck together), the typical 14/21 "balanced translocation carrier" mother is written as 45,XX, t(14q;21q). The translocation chromosome (t) is formed from 14q and 21q (in which q is the long arm); the short arms (p) are lost. For the deletion of a short arm of chromosome 5 (as in the 5p deletion syndrome), the female karyotype is 46,XX,5p.

Each arm of a chromosome is divided into four major regions. Depending on the chromosomal length, each band—positively or negatively stained—is given a number, which rises as the distance from the centromere increases. For example, 1q23 designates the chromosome (1), the long arm (q), the second region distal to the centromere (2), and the third band (3) in that region.

Fig. 3.7. Fluorescence *in situ* hybridization (FISH) is used where gene identification is done by *in situ* hybridization. It uses a non-radioactive labelling system involving a attached biotin. When viewed under special filters, the specific sites of DNA probe can be seen on a chromosome.

Genetic Diseases

Alzheimer's Disease (APOE e4)

Alzheimer's disease is characterized by a progressive loss of function and death of nerve cells in several areas of the brain. This disorder is the fourth leading cause of death in the United

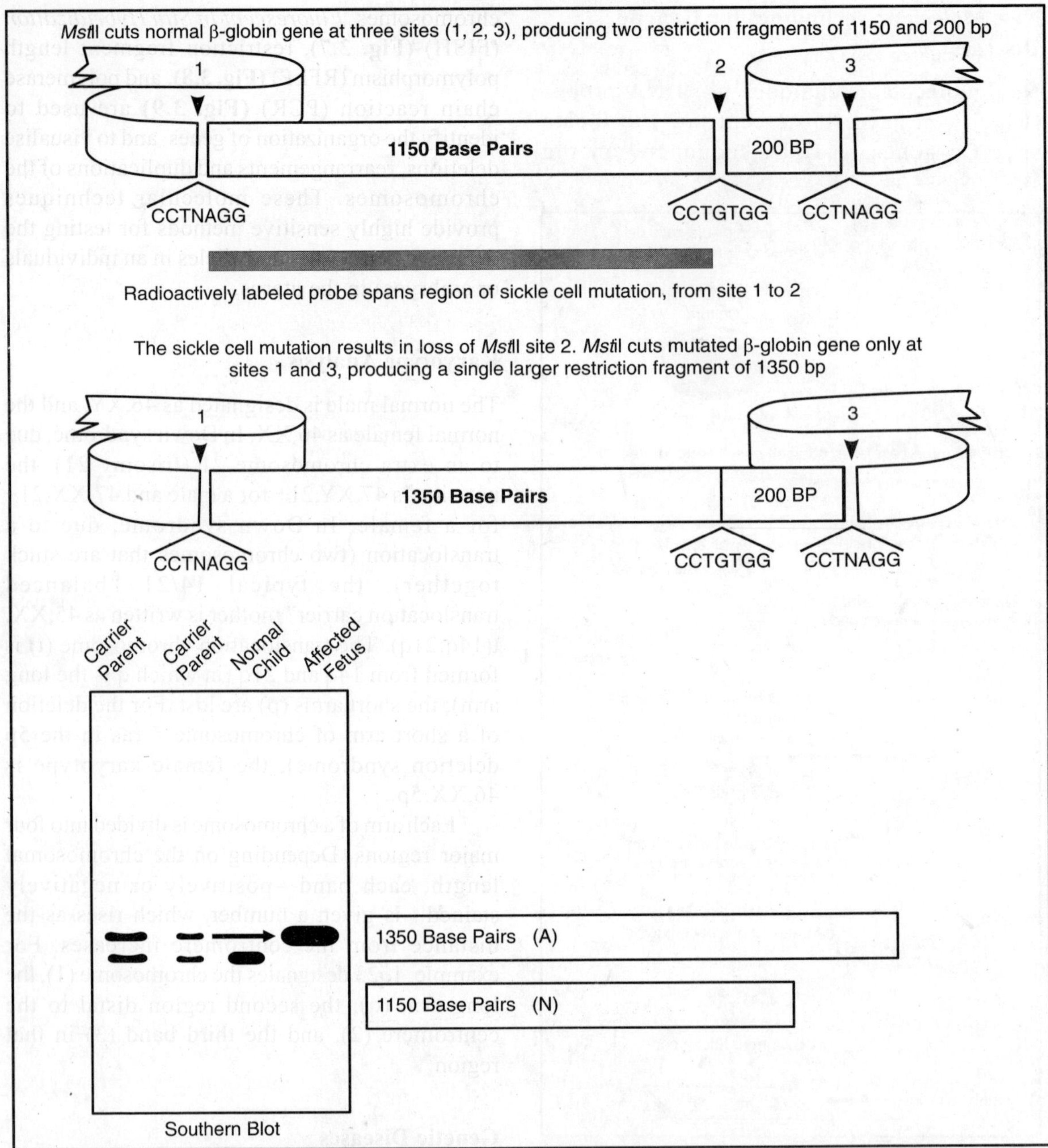

Fig. 3.8. RFLP diagnosis of Sickle Cell Anemia. The Southern blot shows the RFLP patterns of two carrier parents, a normal offspring (N), and amnionic fluid from an affected fetus (A). The carrier parents show a single copy of each RFLP: the 1350-bp fragment is associated with the defective allele, and 1150-bp fragment is associated with the normal allele; the 200-bp fragment is not detected by the probe. The normal child shows the 1150-bp fragment, and the affected fetus shows the 1350-bp fragment. Both normal and affected offspring are homozygous and thus show a single relatively thick band, denoting two chromosomal copies.

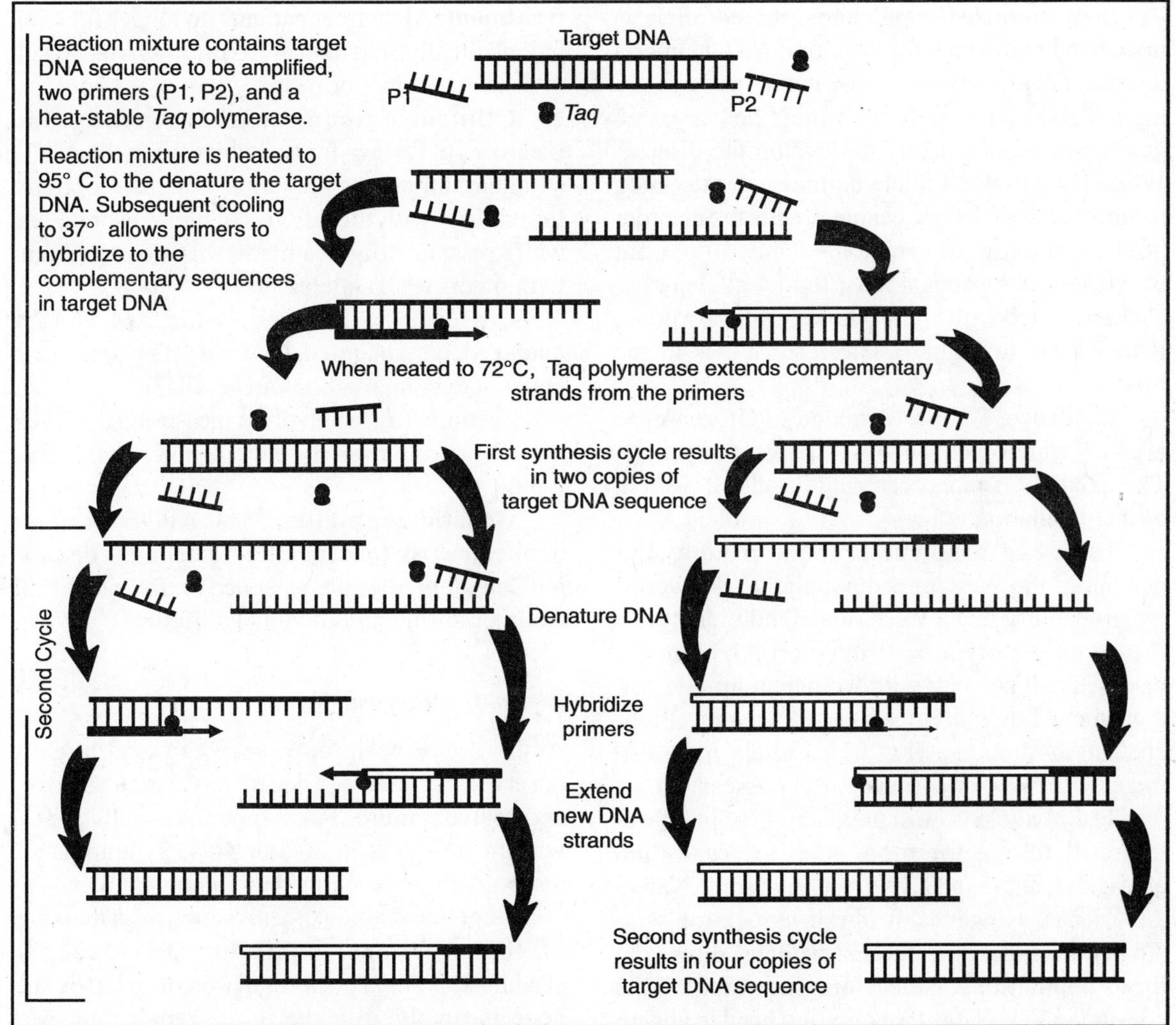

Fig. 3.9. DNA amplification using polymerase chain reaction for disease diagnosis. P1 and P2 are primers for specific disease, for example breast cancer.

States, affecting over four million Americans; most of them over the age of 70. Common symptoms include memory loss, disorientation, impaired judgement and personality changes.

Genetics: Many genes have been shown to be involved in the onset of Alzheimer's disease. One of these is Apolipoprotein E (APOE) located on chromosome 19. The APOE protein is responsible for moving fats between the cells and absorbing cholesterol from the food in the intestine.

Inheritance: The APOE gene has three alleles: e2, e3 and e4. Since everyone inherits one allele of the APOE gene from each parent, there are six possible genotypes:

- e2/e2
- e3/e3
- e4/e4
- e2/e3
- e2/e4
- e3/e4

In Westernized populations, the e4 allele is associated with a risk for developing Alzheimer's disease. People with the e4/e4 genotype have the highest risk, but people with the e2/e4 or e3/e4 genotypes are also likely to develop the disease. While the APOE e4 allele defines a greater risk, the presence of e4 alone cannot predict the disorder prior to the onset of symptoms. Only 40 percent of Alzheimer's patients have the e4 allele. e4 is also associated with higher cholesterol absorption, which leads to higher cholesterol levels in the blood.

Incidence: The most common APOE genotype is e3/e3, which occurs in 40–90 percent of people. The e2 allele is rare, occurring in only 2 percent of the population.

The e4/e4 genotype is found in only 1–3 percent of the Westernized population. However, the probability that a Westernized individual with the e4/e4 genotype will develop Alzheimer's disease is 60 percent, with women at greater risk than men. For individuals who consume high-cholesterol diets, having the e4 allele may also increase the risk of coronary artery disease.

The e4 allele is most prevalent in populations that still forage for food, such as Australian aborigines, sub-Saharan Africans and some Native Americans. However, it is not associated with Alzheimer's disease or coronary artery disease in these populations. Since foraging populations consume a low-cholesterol diet and need to absorb a higher amount of cholesterol, carrying the e4 allele may actually be advantageous.

Diagnosis without genetic screening: Standard clinical methods of diagnosing Alzheimer's disease combine physical and psychological testing. New diagnostic tools and criteria make it possible for the physicians to make a positive clinical diagnosis of Alzheimer's with around 90 percent accuracy. Many criteria are involved in diagnosis, including medical history, physical and neurological examination and psychiatric evaluation.

Clinical outcome without screening and treatment: Alzheimer patients no longer function independently, and damage to the brain may result in seizures, coma, or death.

Clinical outcome with screening and treatment: Testing for and identifying the APOE e4 allele in patients may alter treatment options. Individuals with the APOE e4 allele respond less well to certain drug treatments than do individuals with other APOE alleles.

Currently, Alzheimer's disease can be managed, but not cured. Cure might be possible if better screening methods are available.

Testing: Testing involves measuring the levels of certain proteins in the patient, and APOE genotyping.

Genetic counseling: The APOE e4 allele defines the risk for acquiring Alzheimer's disease, but cannot predict the presence or absence of the disorder during an individual's lifetime.

Alpha-1-antitrypsin

Alpha-1-antitrypsin deficiency can lead to early onset of emphysema (a degenerative lung disorder) and/or liver failure. The symptoms usually appear when a person is in 30's or 40's. Symptoms are more severe in smokers than in non-smokers.

Genetics: This disorder is caused by a mutation in the proteinase inhibitor (PI) gene on chromosome 14. The normal protein coded by this gene is involved in the tissue repair. Disorder symptoms depend on which type of mutation an individual has in the PI gene. There are more than 70 different alleles of the PI gene. The M allele is the most common non-disease causing (normal) variant. The mutant alleles S and Z are the most common disease causing variants. Individuals homozygous for the Z allele (PI ZZ) are at high risk for emphysema and liver disease. Individuals homozygous for the S allele (PI SS) do not display symptoms of the disease. However, individuals with one copy of the Z allele and one copy of the S allele (PI SZ) may develop emphysema.

Inheritance: Autosomal recessive.

Incidence: The incidence of the disorder in the United States is approximately 1 in 2,000. Only 10–20 percent of these individuals develop liver diseases. PI ZZ and PI SS are found in 1 in 2,500 Caucasians in the United States. Since the heterozygote genotype occurs in relatively high frequency (1 in 10 people of European origin are either MZ or MS), the PI SZ genotype is also relatively frequent.

Diagnosis without genetic screening: Diagnosis is made only during the onset of emphysema or other symptoms of the disorder.

Clinical outcome without screening and treatment: Early onset emphysema and/or liver failure are expected.

Clinical outcome with screening and treatment: Early identification can enable the individuals to receive the education, which is essential for them to effectively evaluate the career options, lifestyle choices and insurance needs. In some situations, onset of emphysema, liver failure and other symptoms can be delayed by altering the lifestyle. For example, parents of an affected child can choose not to smoke near him. The affected adult can choose not to smoke, and can work in a smoke-free environment.

There has been great success in using weekly injections of a drug called Prolastin to restore the level of alpha-1-antitrypsin to normal in patients with the PI ZZ genotype. This drug can delay or prevent the onset of symptoms. However, it is generally given to the patients at the first onset of symptoms, such as mild emphysema, because many people do not know that they have alpha-1-antitrypsin deficiency until the symptoms appear. The safety and effectiveness of this treatment for children has not yet been determined.

Prolastin costs about $25,000 US per year per patient, and is not available in the UK. The cost is very high because prolastin is derived from human blood plasma, which fluctuates in supply, and only one company, Bayer, manufactures it.

The price of prolastin and its limited availability has caused a great deal of controversy. In January, 2000, a home healthcare company, anticipating a shortage in the supply of the drug, stockpiled prolastin and raised the price even more for patients. PPL Therapeutics (famous for their cloned sheep, Dolly) has now developed a technique for producing the drug using transgenic sheep. They have inserted a gene for alpha-1-antitrypsin into the sheep, which excrete the gene product in their milk, from where it can be separated and purified. They have formed a partnership with Bayer, and are currently conducting clinical trials. Since it does not rely on the supply of human blood plasma, this technique may significantly reduce the cost of treatment.

Other treatment options are limited. However, gene therapy is well in a developmental stage and has shown preliminary success in mice.

Testing: This disorder can be detected by testing the level of alpha-1-antitrypsin in blood. If it is abnormally low, the next step is to identify the exact alpha-1-antitrypsin protein variants the person carries. Abnormal forms of the alpha-1-antitrypsin protein can be detected using dried blood as a sample for gel electrophoresis.

Genetic counseling: The association between alpha-1-antitrypsin and liver disease was not identified until 30 years ago. So many families are unaware that they may be transmitting alpha-1-antitrypsin deficiency to their progeny. Genetic counseling can help families find out as much as they can about the disorder, its possible outcome, and the effect it will have on their child's life. However, the effects of alpha-1-antitrypsin deficiency are not predictable. This presents a moral dilemma to the families who know that they carry the mutation, and to the healthcare professionals who may be in the position of counseling them.

Breast Cancer Genes: BRCA1 and BRCA2

Cancers cause cells in the body to change, divide out of control (forming masses of tissue called tumors), and spread to other parts of the body. They are named after the part of the body where they begin. Breast cancer, the second major cause of death in American women, is often detected when there are visible changes in the breast, such as a lump, thickening, swelling, skin irritation or nipple discharge. The risk of breast cancer increases with age.

Genetics: Only 5–10 percent of all breast cancer cases are believed to have a genetic link. Of these, an estimated two-thirds are caused by mutations in either BRCA1 or BRCA2. About 50–60 percent of individuals with certain mutations in either of these two genes will develop breast cancer by the age of 70.

Inheritance: Autosomal dominant.

Incidence: The frequency of breast cancer in women is 1 in 8. Caucasian women have a higher risk of developing breast cancer than African-American, Asian or Hispanic women. The vast majority of affected women do not have a genetic predisposition to the disease. The prevalence of cancer-predisposing mutations in BRCA1 is 1 in 500–1000. The prevalence of BRCA2 mutations is unknown. A higher frequency of mutations in both genes is seen in the Ashkenazi Jewish population.

Diagnosis without genetic screening: Mutations in BRCA1 and BRCA2 genes are associated with an early cancer onset, cancer in both breasts, and male breast cancer. Clinical breast exams and mammography can detect breast cancer during onset.

Clinical outcome without screening and treatment: Cancer can spread throughout the body if not treated, and survival rate is reduced.

Clinical outcome with screening and treatment: There is no certain way to prevent the breast cancer. However, some lifestyle risk factors have been identified, including diet and alcohol use. It is important to identify breast cancer early to optimize the treatment. If a genetic predisposition to the breast cancer is known, drugs, which have been shown to reduce the likelihood of developing the disorder, can be used. Some

high-risk women may consider a preventive mastectomy (removal of one or both breasts), though this does not guarantee that breast cancer will not develop. After such an operation, breast reconstruction may be possible and has become an important part of rehabilitation and therapy.

Testing: More than 235 mutations have been identified in BRCA1 gene and about 100 mutations have been identified in BRCA2. None of the currently available techniques can detect all of the cancer-predisposing mutations in BRCA1 or BRCA2. Myriad Genetics has developed a test that analyzes about 16,500 base pairs for the two genes. The test costs $2,580 US. However, it is estimated that about 30 percent of BRCA mutations are not detected by currently available tests.

Genetic counseling: Genetic counseling for adults is offered before and after genetic testing. It is recommended that at-risk families consider a genetic testing strategy. Individuals diagnosed with breast cancer are strongly encouraged to be tested for mutations before his unaffected family members are tested. The test results of the affected family members will identify a particular type of mutation, which is more likely to be associated with breast cancer. This information can then be used to test unaffected family members for the same mutation, thus making their results more meaningful.

Cystic Fibrosis (CF)

Cystic fibrosis (CF) is caused by changes in a protein that controls the transfer of chloride and sodium ions (salts) across cell membranes. Disruption of salt transfer results in abnormal gland secretions and dehydration due to increased loss of salt and water during sweating. CF affects almost all of the glands in the body that secrete the fluid, resulting in a variety of symptoms. Secretions may be thick and cause blockage in the pancreas, intestines and lungs. Mucus blockage also provides the place for bacteria to multiply, increasing the probability of infection. CF affected children show poor digestion, dehydration, coughing and vomiting. As the disease progresses, teenagers show slowed growth, delayed puberty and reduced physical endurance. Adults show more serious complications, such as collapsed lung, heart failure, infertility and frequent infections, which eventually leads to death.

Genetics: Cystic fibrosis is caused by mutations in the cystic fibrosis transmembrane regulator (CFTR) gene on chromosome 7, which codes for the protein that controls ion transfer across cell membranes. Molecular analysis has identified approximately 100 mutations in the CFTR gene. Different mutations determine the severity of symptoms seen in CF patients.

Inheritance: Autosomal recessive.

Incidence: CF is the most common hereditary disease leading to death among Caucasian people in the United States. The incidence of the disorder is 1 in 2,500 Caucasians; 1 in 14,000 blacks; 1 in 11,500 Hispanics; and 1 in 25,000 Asians.

Diagnosis without genetic screening: Half of the patients with CF remain undiagnosed during the first year of life, and the other 25 per cent remain undiagnosed by the end of the second year.

Clinical outcome without screening and treatment: If not diagnosed early, 13 percent of the newborns and infants will die. Most untreated CF individuals will not live past their late 20's. In general, males live longer than females.

Clinical outcome with screening and treatment: If diagnosed and treated, newborn and infant mortality can be reduced. Half of the people with CF live longer than 28 years due to availability of an increasingly wide range of treatments. These include physical therapy, enzyme replacements, supplemental salt and antibiotics to control the infection, oxygen therapy, surgery and organ transplantation. Treatment for CF costs an average of $40,000 US per year per patient in direct medical costs alone. Gene therapy is under investigation and evaluation.

Testing: Mutation in the CFTR gene results in an increase in an enzyme called trypsinogen.

The newborns are screened tests for this enzyme using a dried blood sample. However, this is not a conclusive test for CF. Measuring the salt levels in sweat can usually confirm the diagnosis.

Genetic counseling: Prenatal diagnosis is possible for most families. Carrier screening of the general population is possible using DNA mutation analysis. CF carrier testing is not recommended, unless a family history of CF is present. It is difficult to detect all the carriers. Only 80–90 percent of carriers will be identified with the available tests. Newborn screening is recommended to prevent malnutrition, as well as to improve lung condition. Embryo screening for CF is available for carrier parents prior to embryo implantation during an *in vitro* fertilization procedure.

Galactosemia

Galactosemia is caused by the lack of an enzyme in the liver necessary for the breakdown of galactose. Galactose is the product of digested lactose, and is found in milk products. Symptoms include lethargy, diarrhea, vomiting, seizures, cataracts, susceptibility to infections, jaundice, mental retardation and liver disease.

Genetics: Galactosemia is caused by disruption of the galactose-1-phosphate uridyl transferase (GALT) gene on chromosome 9. More than 130 mutations in the GALT gene have been associated with galactosemia. Some mutations occur more commonly in particular ethnic groups, and result in variable severity of the disorder.

Inheritance: Autosomal recessive.

Incidence: 1 in 50,000–80,000 for classic galactosemia, and 1 in 16,000 for other forms of galactosemia.

Diagnosis without genetic screening: The symptoms of galactosemia appear early in infants, often leading to diagnosis. Jaundice, diarrhea, vomiting and failure to gain weight begin within a few days of milk ingestion. Cataract appears within a few days of birth.

Clinical outcome without screening and treatment: If not detected early, galactosemia can result in severe liver disease and mental retardation. There is a high frequency of death during one to two weeks of age due to severe bacterial infection.

Clinical outcome with screening and treatment: Galactosemia can be treated by eliminating galactose and lactose from the diet throughout the life. With prompt treatment, survival increases to close of 100 percent. A low IQ, learning disabilities and speech delays are often seen despite seemingly adequate treatment. This is thought to be due to an overrestrictive galactose-free diet. Galactosemic women rarely become pregnant due to ovarian failure. Although some clinical care and monitoring is involved, the cost for the treatment of galactosemia is minimal as it does not require a specially-formulated replacement diet.

Testing: This disorder can be detected by testing a blood sample for the presence of high levels of galactose, or low GALT activity. Additional tests are often used to confirm the results.

Genetic counseling: Genetic counseling should be available for the parents of infants identified as having galactosemia, since it is important for these children to receive proper care and treatment. Treatment requires utilizing the services of specialized metabolic and genetic clinics that provide nutritional, psychological, nursing, biochemical and pediatric care. In addition, parents are usually unaware of this genetic disorder and require information in order to understand what needs to be done. Newborn screening programs generally have pamphlets and brochures to provide this information.

Huntington's Disease

Huntington's disease is characterized by the progressive death of certain nerve cells in the brain. Symptoms generally appear between the ages of 35–40 years and include depression, mood swings, forgetfulness, involuntary twitching and lack of coordination. As the disease progresses, involuntary movements increase, memory declines, and walking, speaking and swallowing ability gradually diminishes. Eventually, the affected persons are unable to care for themselves, and death soon follows from choking, infections or heart failure. About 10% of the patients have juvenile Huntington's disease, in which symptoms develop before the age of 20.

Genetics: Huntington's disease is caused by an excessive repeating of the DNA bases CAG (trinucleotide repeats) in the huntingtin gene on chromosome 4. The more repeats an individual has, the earlier the age of onset. The normal number of repeats is 10–35. But patients affected with Huntington's disease have repeats of 36-121. The function of the huntingtin protein is not yet fully understood.

Inheritance: Autosomal dominant.

Incidence: 1 in 20,000 Caucasians (except in Finland, where the incidence is much lower); 1 in 100,000 African Americans, 1 in 1,000,000 Africans; and 1 in 300,000 Asians.

Diagnosis without genetic screening: Symptoms are usually not observed before the onset of the disease. Individuals with a family history of this disorder may show symptoms earlier.

Clinical outcome without screening and treatment: Involuntary movements and mental disturbances continue to increase.

Clinical outcome with screening and treatment: No cure is currently available for Huntington's disease. However, involuntary movements, rigidity and psychiatric symptoms can be suppressed or reduced with certain drugs. Neural and stem cell transplantation may be a potential option for treatment in the future.

Testing: The huntingtin gene is analyzed in a blood sample to determine the number of CAG repeats.

Genetic counseling: Genetic counseling is offered before and after testing the individuals who have a family history of Huntington's disease.

Since this disorder shows dominant inheritance, a child with an affected parent has a 50 percent chance of inheriting the disorder. Some individuals may choose not to know their prognosis, rather than live with the knowledge that they have a disorder with no cure. Others would rather have a definitive diagnosis and choose to be tested. Personal life choices, such as decisions about childbearing can be affected by test results. Genetic screening allows the individuals to plan for their future as well as the future of their families. There are support groups for individuals who test positive for Huntington's disease and for their caretakers, which helps to ease the distress they face.

Hypothyroidism

Congenital hypothyroidism is one of the most common disorders detected in the newborns. Infants with this disorder do not produce adequate amounts of thyroid hormone due to the failure of the thyroid gland to develop properly. Thyroid hormone plays a vital role in normal growth, development, and mental function in children. Although infants with hypothyroidism usually appear normal for up to three months after birth, early symptoms sometimes occur, including increased birth weight, puffy face, constipation and lethargy.

Genetics: In most newborns with hypothyroidism, there is no specific reason explained why the thyroid gland did not develop normally. However, about 10–20 percent of infants have an inherited form of the disorder, which is often characterized by the visible enlargement of the thyroid gland (goiter). The most common inherited form of hypothyroidism is a defect in the thyroid peroxidase gene on chromosome 2. This gene plays an important role in the manufacture of thyroid hormone.

Inheritance: Autosomal recessive.

Incidence: 1 in 4,000 overall; 1 in 12,000 African-Americans; 1 in 1,000 Hispanics; and 1 in 700 Native Americans.

Diagnosis without genetic screening: Affected infants may not show any symptoms. This can delay the diagnosis of disorder for up to three months, and by that time, brain damage is likely to have occurred.

Clinical outcome without screening and treatment: If not treated early, thyroid hormone deficiency causes mental retardation, delayed puberty, ataxia (loss of the ability to coordinate muscular movement), and stunted growth. Fifteen percent of the individuals with untreated congenital hypothyroidism are institutionalized from five years of age till their whole life; 25 percent require foster care from the age of 5 to 20; and 40 percent require adult care from the age 20 till they are alive.

Clinical outcome with screening and treatment: Hypothyroidism is treated by replacing the missing thyroid hormone through the pills, which a person has to take daily throughout his life. The cost for one-month supply is $10–$20 US. Regular blood tests are needed to monitor hormone levels. If infants are treated early and treatment is followed properly, normal growth, intelligence and life span can be achieved.

Testing: Infants are screened 48 hours after their birth using a dried blood sample, which is obtained from a heel stick, to detect the thyroid hormone level. False positives may occur if the infant is tested prior to 48 hours. The is due to the altered hormone levels caused by the stress during birth.

Genetic counseling: The family of a patient diagnosed with an inherited form of hypothyroidism is referred to a genetic counselor. Brochures for parents are provided by state newborn screening programs, and are available before or after delivery. Additional information and counseling is available with a positive test.

Maple Syrup Urine Disease (MSUD)

Individuals with maple syrup urine disease (MSUD) are unable to properly metabolize (break down) three amino acids—leucine, isoleucine and valine. These amino acids are present in all protein foods, such as meat, eggs and milk. Smaller amounts are also found in cereals, vegetables and fruits. In individuals with MSUD, the enzymes required to process these three amino acids are absent, inactive or only partially active. Since these amino acids do not get broken down completely, their high levels accumulate in the blood, urine and sweat. The byproduct of isoleucine has a characteristic sweet smell, which gives the disorder its name. The three amino acids and their derivatives can be toxic at high levels, and can lead to brain injury, mental retardation, seizures, vomiting, coma and even death.

Genetics: Defects in three genes on different chromosomes can lead to four types of MSUD. The most common type is classic MSUD, which is caused by a defect in the BCKDHA gene on chromosome 19. In classic MSUD, enzyme activity is less than two percent of normal. The other variations are less severe.

Inheritance: Autosomal recessive.

Incidence: Classic MSUD occurs 1 in 90,000–300,000 births, and is seen in all ethnic groups worldwide. Incidence is 1 in 176 births in the Old Order Mennonites of Pennsylvania, United States.

Diagnosis without genetic screening: Without genetic screening, classic MSUD will be diagnosed reliably only after the first week of life. Three to four days after birth, these infants show poor appetite, lethargy, irritability, the characteristic urine odor, boxing and pedaling limb movements muscle rigidity, and they emit a high-pitched cry. In children with less severe MSUD variants, symptoms may not appear until several months after birth.

Clinical outcome without screening and treatment: The severity of mental retardation that develops when MSUD is left untreated depends on the level of enzyme deficiency. Death occurs within the first two years of life in classic MSUD.

Clinical outcome with screening and treatment: MSUD is treated and managed similarly to PKU or diabetes. Lifelong treatment beginning at the earliest possible age allows MSUD individuals to live a normal life span with normal intelligence. Treatment involves a carefully controlled diet to prevent the accumulation of the amino acids in the blood. The cost for special low-protein foods can be very high. For example, a box of crackers is about $16 US. An MSUD infant formula has been developed, which provides all the vitamins and minerals necessary for the proper development, without leucine, isoleucine or valine. This can be supplemented with very small amounts of regular baby formula. Cost for this special formula is about $1,200 US per month. Systems are available for frequent monitoring of amino acid levels, and diet can be adjusted as needed. Some infants can also be helped with a thiamine supplement.

Testing: Newborn screening programs that test for MSUD use the same blood sample collected for PKU and galactosemia tests. Generally, blood is analyzed for elevated levels of leucine.

Genetic counseling: Information is available from US state screening programs concerning dietary requirements. Counseling is available for the family of an identified individual. Carrier screening is also available for the Mennonites of eastern Pennsylvania.

Neurofibromatosis 1

Neurofibromatosis 1 (NF1) is a genetic disorder that primarily affects the development and growth of cells in the nervous system. The disorder can cause neurofibromas (tumors on nerves), freckling, cafe-au-lait spots (light brown spots on the skin), learning disabilities and bone deformities. The type and severity of symptoms shown by NF1 patients can vary widely, even between affected individuals from the same family.

Genetics: The NF1 gene, located on chromosome 17, produces a large and complex protein called neurofibromin. Scientists theorize that this protein acts as a switch to regulate cell growth.

Inheritance: This is an autosomal dominant disease. However, about 50 percent of the cases are caused by spontaneous new mutations in the NF1 gene that occur in the sperm or egg prior to fertilization. This mutant allele can then be passed onto the individual's offspring.

Incidence: NF1 is one of the most common dominantly inherited genetic disorders with an incidence of 1 in 3,500–4000. The incidence of spontaneous mutations in the NF1 gene is 1 in 10,000. This is one of the highest known mutation rates for a disorder-associated gene. The reason for this is currently under investigation, but may be due to the large size of the NF1 gene.

Diagnosis without genetic screening: It is often difficult to diagnose NF1, because the type and severity of symptoms vary among the patients. For example, one child might develop obvious signs of NF1 shortly after the birth, while another person might live a lifetime without even knowing that he or she has the disorder. However, early diagnosis is possible without genetic screening. The patient is examined for particular symptoms evident at an early age. Additional physical symptoms appear as the childrens older.

Clinical outcome without screening and treatment: Disfiguring, rarely cancerous skin tumors may develop, as well as bone deformation such as scoliosis (spinal curvature). Elevated blood pressure (hypertension) may develop in adults. The average life expectancy is reduced by at least 15 years.

Clinical outcome with screening and treatment: Treatment is currently designed to control the symptoms. Surgery can be used to remove the tumors in inconvenient areas, such as belt and neck lines, but tumors may grow back and often in increased numbers. Back braces or surgery can be used to treat scoliosis. Learning disabilities may require educational intervention. Cancerous tumors (occurring in 3–5 percent of all cases) can be treated by surgery, radiation or chemotherapy. Medical care should be given by a physician who is familiar with the development of the disease.

Testing: Although genetic testing is available for NF1, it is rarely used to diagnose the disorder. The large size of the NF1 gene and the variety of mutations, which can cause the disorder, make the identification of specific mutations in patients difficult. Testing is, however, more widely used for families with documented cases of NF1, where a specific mutation has been identified.

Genetic counseling: Genetic counseling may be requested by a couple affected by NF1 to assess the risk of passing on the disorder to their child. Prenatal diagnosis is available in these cases, but often is not requested because the affected child may or may not develop severe complications from NF1.

Phenylketonuria (PKU)

PKU is caused by the lack of an enzyme that processes the amino acid phenylalanine. Phenylalanine is present in all protein foods, such as meat, eggs and milk. Smaller amounts are also found in cereals, vegetables and fruits. In PKU, phenylalanine is not broken down and accumulates in the blood. Phenylalanine is toxic to the brain. Untreated individuals with PKU show progressive developmental delay in the first year of life, mental retardation, seizures, autistic-like behavior and a peculiar body odor.

Genetics: In PKU individuals, the phenylalanine hydroxylase gene on chromosome 12 is disrupted. This gene encodes the protein that processes the amino acid phenylalanine to reduce its level in the body. When the gene is mutated, phenylalanine builds up in the body.

Inheritance: Autosomal recessive.

Incidence: 1 in 10,000–25,000. Incidence in the United States is 1 in 16,000 live births.

Diagnosis without genetic screening: PKU is rarely diagnosed before 6 months of age. After this, mental retardation is apparent.

Clinical outcome without screening and treatment: Life expectancy is reduced. IQ is less than 50 in 95 percent of affected individuals. Convulsions and hyperactivity are common in untreated individuals. Sixty-four percent of individuals with untreated PKU are institutionalized since five years of age till their life; 18 percent require foster care from five years to 20; and 36 percent require adult residential support services.

Clinical outcome with screening and treatment: PKU is treated by eliminating phenylalanine from the diet for life. Normal early growth and development are expected with early treatment. The normal range of intelligence is possible with optimal dietary control. Patients who do not follow the regulated diet have memory deficits and other mental problems. The cost for special, low-protein foods can be very high. For example, a box of crackers is about $16 US.

The issue of maternal PKU (the responsibility of PKU mothers to maintain their phenylalanine-free diets while pregnant) needs to be addressed by the medical profession and the individuals with PKU. High levels of maternal phenylalanine (present if a mother does not maintain her diet) can adversely affect the development of the fetus. This issue has become a problem since the implementation of genetic testing for PKU. Previously, few PKU individuals had children of their own. Now, individuals diagnosed with PKU through newborn genetic screening can lead normal reproductive lives, and therefore must address these issues.

Testing: The blood phenylalanine level can be measured using a spot of dried blood. The PKU test (the Guthrie test) was the first genetic screening test. Automated tests are now used in some screening programs. The timing of the test is important. The test should be completed after the first day and before the seventh day of life. If done too soon, low levels in the newborn can be masked by the presence of maternal phenylalanine, yielding a false-positive result.

Genetic counseling: Information is available from US state screening programs, and is often mandated by state regulations or laws. Counseling is available for the family of an identified individual.

Sickle Cell Disorder

Sickle cell disease describes a group of inherited disorders of red blood cells. Red blood cells are responsible for delivering oxygen to different parts of the body. They are usually round and contain a molecule called hemoglobin, which carries oxygen. If the gene encoding hemoglobin is mutated, it causes a change in the shape of the molecule. When the mutated hemoglobin delivers oxygen to the tissues, the red blood cell collapses, resulting in a long, flat sickle-shaped cell. These cells clog blood flow, resulting in a variety of symptoms including pain, increased infections, lung blockage, kidney damage, delayed growth, and anemia (low blood cell count).

Genetics: The gene encoding the beta chain of the hemoglobin molecule, located on chromosome 11, can be mutated in a variety of ways, which leads to different types of sickle cell disease. Some mutations are more common than others. The three most common types of sickle cell disease in the United States are hemoglobin SS (Hb SS), hemoglobin SC (Hb SC), and hemoglobin sickle beta-thalassemia (Hb S beta-thalassemia).

Inheritance: Autosomal recessive.

Incidence: Sickle cell disease affects more than 50,000 Americans. Although the disease occurs in high frequency in individuals of Mediterranean, Caribbean, Indian, Arab and Southeast Asian descent, it exhibits the highest frequency in the people of African descent.

In African-Americans, the incidence is 1 in 375 for HbSS, 1 in 835 for HbSC, and 1 in 1,667 for HbS beta-thalassemia. In addition, 1 in 12 African-Americans are carriers for the disorder (have sickle cell trait).

In United States, the prevalence of all types of sickle cell diseases is equal to 1 in 58,000 Caucasians; 1 in 1,100 Hispanics (eastern states); 1 in 32,000 Hispanics (western states); 1 in 11,500 Asians; and 1 in 2,700 Native Americans.

Diagnosis without genetic screening: Clinical diagnosis is rarely made before 1 year of age.

Clinical outcome without screening and treatment: Complications include increased infections, kidney damage, leg ulcers, bone damage and delayed growth. Ten percent mortality occurs in early infancy and childhood. Hospitalizations for sickle cell disease cost the US government an estimated $475 million per year, at an average of $6,300 per hospitalization.

Clinical outcome with screening and treatment: With treatment, the incidence of early death is significantly reduced. After being diagnosed as suffering from sickle-cell disorder, newborns are placed on penicillin until six year of age to prevent infection. Parents are educated in particular guidelines to follow with their child, including taking folic acid to make new red blood cells, drinking lots of water, avoiding extreme temperatures, and getting regular checkups with a physician.

Testing: Sickle cell diseases can be identified using isoelectric focusing of hemoglobin from filter paper blood spots.

In the past, screening programs for sickle cell disease led to the discrimination against individuals identified as being carriers for the disorder (having sickle cell trait). These individuals did not display the disorder, but were treated as potentially ill. They were often restricted from certain jobs and barred from joining the military. Problems that developed from the early sickle cell disease screening programs have resulted in regulations that govern current screening programs.

Genetic counseling: Genetic counseling is needed in order to provide rapid access to medical care for the affected individual. In addition, it is important to provide genetic counseling for carrier individuals to prevent any feeling of stigmatization or confusion about the meaning of their genetic status, so that they may make informed decisions about future offspring. Fears about confidentiality, discrimination in the work place, and insurability need to be addressed.

Smith-Lemli-Opitz (SLO/RSH) Syndrome

The first three cases of this disorder were described in 1964 in three families; their surnames beginning with R, S and H. The researchers (Smith, Lemli and Opitz) therefore called this syndrome "RSH".

Body of the individuals suffering from SLO/ RSH syndrome is unable to make cholesterol, which is an essential nutrient required for the proper development of cell membranes and the white matter of the brain. Since cholesterol is not provided to the developing fetus by the mother during pregnancy, symptoms develop before birth. As development continues after birth, lack of cholesterol can result in a variety of symptoms that vary in severity from individual to individual. Symptoms include growth retardation, developmental delay, feeding difficulties, behavioral problems, and physical malformations such as small head, cleft palate, cataracts, drooping eyelids, extra (sixth) fingers or toes, webbing between the second and third toes, male genital malformations, and heart defects.

Genetics: SLO/RSH syndrome results from a mutation in either the DHCR7 gene on chromosome 11, or the SLOS gene on chromosome 7. These genes code for essential enzymes required for the synthesis of cholesterol.

Inheritance: Autosomal recessive.

Incidence: SLO/ RSH syndrome occurs in relatively high frequency: 1 in 20,000 to 30,000 in Europe, the United States, and Canada.

Diagnosis without genetic screening: General poor growth, developmental delay, and congenital malformations may prompt testing.

Clinical outcome without screening and treatment: The affected person shows continued failure to thrive, and developmental delays lead to a limited life span. Prognosis is related to an individual's cholesterol levels; the lower the cholesterol level, the more severe the condition and shorter the life-expectancy.

Clinical outcome with screening and treatment: Patients given cholesterol as infants or adults show improved growth, a lessening of behavioral problems, and more rapid developmental progress. Although adults benefit from this treatment, the earlier the cholesterol can be supplied, the better the eventual outcome. With correct diagnosis and treatment, many children can learn to walk, talk, and acquire basic living skills. Normal life span is possible with good care and nutrition.

Testing: Typically, testing occurs when physical abnormalities are observed. Females often go undiagnosed, because they do not show the genital malformations seen in boys. A test for low levels of cholesterol and abnormally high levels of a cholesterol precursor is usually used to diagnose SLO/RSH. However, this test can be unreliable. DNA analysis for mutations of the DHCR7 gene is available on a limited basis. Prenatal testing is also available.

Genetic counseling: Parents need to be educated about the nature of this disorder to prevent them from overfeeding their children in an attempt to make them grow faster. They also need to monitor carefully the dietary formula given to the infant.

3.11. POINTS TO BE CONSIDERED IN GENETIC TESTING

Genetic data and tools offer enormous benefits to humankind, but poses varieties of significant risks as well. As the impact of the new genetics grows, we can expect the courts to be increasingly confronted with many novel, challenging, and sometimes disturbing issues. Scientific progress continues to advance rapidly, but no one can anticipate the ways the more powerful future DNA technologies will be put to use, nor their unintended and potentially controversial or adverse effects. As we begin to realize the benefits of the new genetics, maintaining a cautious approach will help to minimize the risks. As a society, select those issues that seem most important, or for which you have information. You may also identify other issues that you want to consider.

Choice of Tests

- What are the effects of each genetic disorder considered by the task force?
- Which treatment is currently available for each genetic disorder, and at what age does the treatment begin?
- Are tests currently available for each genetic disorder? If so, which are those?
- Should tests be restricted only to those disorders that are treatable, or should they be restricted only to those in which treatment, early in life, makes a difference? Why?
- Based on demographics, how many people are likely to be affected by each genetic disorder?
- How do demographics influence the choice of tests?
- Should all newborn infants receive the same screening tests? Why or why not?

Technology

- How sensitive and reliable is each test?
- Can a test in question be performed on a large scale in laboratories? Why or why not?

Costs

- What is the cost per test and the total for screening?
- Will early identification of persons with the genetic disorder lead to cost savings in treatment or care? Why or why not?
- Who will pay for testing?
- Should citizens, who do not have children, help pay for newborn testing? Why or why not?

Rights

- Is testing fair to a newborn child who cannot speak for his or her own rights? Why or why not?
- Should parents be able to choose or refuse tests? Why or why not?
- If the symptoms of a genetic disorder are not expected to appear until a child reaches legal adulthood, should that disorder be included in a newborn testing program? What are the pros and cons involved in this?
- Who has access to the test results?
- What happens to the blood samples after testing?
- Are laws needed to protect citizens' or doctors' rights? If so, what types of laws?

Personal and Societal Impacts

- Will the test information result in prejudice against certain individuals? If so, why? How might this be expressed?
- Will the testing (or lack of testing) be a source of distress for parents or newborns? If so, why?
- Will the test information make it difficult for an individual to obtain health insurance? Will a predisposition to a genetic disorder be considered a "pre-existing condition"?
- How will early diagnosis of genetic disorders help or hurt society?
- What are current policies in different states or countries?

SUMMARY

Human development depends on genetic and environmental factors. A person's genetic composition (genome) is established at conception. The genetic information is carried in the DNA of the chromosomes and mitochondria. Most diseases probably have some genetic component, the extent of which varies. Environmental factors may alter genetic information through mutation or other structural alteration and can affect classic genetic disorders. Genes, the basic units of heredity, are arranged linearly within the DNA along the chromosomes; each gene has a specific location (locus) or position on the chromosomes. The number and arrangement of loci on homologous chromosomes are usually identical. However, the structure of a specific gene may have minor variations (polymorphisms) without causing disease. The specific nucleotide sequences of the genes that occupy the two homologous loci along the two chromosomes of a pair are called alleles. The two alleles (i.e., the one inherited from the mother and the one inherited from the father) may have slightly different nucleotide sequences or may be the same. A person with a pair of identical alleles for a particular gene is a homozygote; a person with a pair of dissimilar alleles is a heterozygote.

If a trait or disorder occurs when only one allele is abnormal, the disorder is said to be dominant. A disorder is said to be recessive if it occurs only when both alleles at the loci on both chromosomes are abnormal. A few genes are located in the mitochondrial DNA. Many copies of mitochondrial DNA are present in cytoplasm and may have the same DNA structure (homoplasmy) or different structures (heteroplasmy). There are three types of genetic disorders: Mendelian, or single-gene, mutations are inherited in recognizable patterns; multifactorial conditions involve more than one gene and environmental factors interacting in ways that are not always clearly recognizable, but have been described by observation (empirically); and chromosomal abnormalities include structural defects and deviations from the normal number. More recently, mitochondrial and nontraditional patterns of inheritance have been recognized.

Molecular gene tests involve direct examination of the DNA molecule itself in the presymptomatic condition. A DNA sample can be obtained from any tissue, including blood. To do a gene test, scientists scan the sample, looking for a specific mutation in a particular DNA region that has been linked to a disorder. Cost can range from hundreds to thousands of dollars, depending on the sizes of the genes examined and the number of mutations tested for, which can vary from a few to hundreds. Although there are several hundred DNA-based tests for different conditions, most are still offered only as research tools. Fewer than 150 gene tests are available commercially, and most are for mutations associated with rare diseases in which just a single gene is involved. Even though some current gene tests have been beneficial and their potential benefit enormous, the science is very new and dynamic. Researchers themselves are unsure how to interpret the results of some commercially available gene tests. Genetic information is the most private information. Once an individual is tested, the result is a blueprint of that individual's genetic makeup. This blueprint can be used to trace genes through past generations, predict the likelihood of the gene's presence in present generations, and can be used to predict their presence in future generations. In other words, obtaining one person's genetic information could lead to finding genetic disorders for that person's parents, siblings, and children. This could become a concern if insurers discover a genetic disorder in one person, and they use that information to classify not only the person tested, but also that person's family members.

Through testing of genetic diseases, the patients are being properly informed about the risks and limitations of genetic technology. The genetic testing also has many ethical, legal, and social implications such as how does personal genetic information affect an individual and society's perceptions of that individual? Who owns and controls genetic information? If you are tested for or diagnosed with a genetic disorder and this information becomes a part of your medical record, insurance companies, employers and other agencies may be able to access this information. Without adequate legal measures to protect individuals from misuse of their medical information, individuals diagnosed with certain genetic conditions potentially could be denied insurance coverage, employment, or other benefits.

Questions arise as to whether employers, insurance companies, and the government may access an individual's genetic information resulting from such tests. The insurers' ability to access this information has received a great deal of attention. More specifically, may an insurance company classify its applicants based on their genetic make-up? Many different views have been expressed to answer this question. On one hand, insurance companies, particularly life insurers, believe that they should be entitled to genetic information if the risk classification process is to survive. Insurers believe that widespread fraud would result from the ability to withhold genetic information, raising the premiums for all policyholders. Consumers, on the other hand, believe that allowing insurers' access to genetic information would violate rights of privacy, prevent patients from getting needed help, and lead to widespread discrimination against applicants. These debates have resulted in legislative intervention at the national level. Government legislation has also been proposed, however, no significant efforts have been passed directly relating to this issue.

Cystic fibrosis ("CF") is one of the most common genetic disorders causing death in the white population. Pulmonary disease is responsible for 90% of CF-related deaths. Liver disease, trauma, and suicide are responsible for the other 5% of CF-related deaths. Huntington's disease ("HD") causes a gradual deterioration of physical and mental capabilities around the age of 40 and lasts for about 15 years, until death. Present studies have shown that although the majority of breast and ovarian cancers develop sporadically, a small percentage (approximately 5-10%) of these cancers are genetically related. Alzheimer's disease is different from other genetically related diseases in that there are different copies of the same gene. If one of the copies of the gene is received from the parent, the process of Alzheimer's is hastened. As of 2005, tests for cystic fibrosis, Duchenne/

Becker muscular dystrophy, hemophilia, Gaucher's disease, Huntington's disease, Lou Gehrig's disease, Tay Sachs disease, and thalassemia among other genetic testing were available. Cystic fibrosis tests, which cost around $400 in 1993, cost around $125-$150 in 1995, and even less in 1997. The Michigan State University Genetics Clinic quoted a price of $52 for a cystic fibrosis carrier screening test. If this trend continues, it is predicted that these tests will cost five to ten dollars.

Presymptomatic genetic diagnosis technologies will not eliminate disability from society. The council of Canadians with disabilities estimates that only 3% of genetic conditions may be affected by gene therapy. Eighty-five percent of adult disability is caused after the age of 13, and more than ninety percent of infant disability is because of social and not genetic causes. A person is far more likely to become disabled because of situations such as ageing, illness, unsafe working conditions, toxic environments, violence, poverty, lifestyle choices, poor nutrition etc. Consequently, despite the hype offered by the media and the scientific community, it is unlikely that gene identification will significantly reduce the incidence of disease, disability or improve the social status of persons with disease or disabilities.

Powerful molecular genetic techniques have made it possible to study the structure of DNA and to track changes during development and in different tissues. More than 900 specific causes of congenital diseases have been identified; some are genetic (involving nuclear and mitochondrial genes), whereas others result from rubella virus or other environmental agents.

EXERCISE

1. What is meant by genetic disorders?

2. Name different types of genetic disorders based on their mutations in the chromosomes locations that are commonly noticed in humans.

3. Explain the merits and demerits of the screening of genetic disorders in an individual or an embryo.

4. What do you mean by presymptomatic genetic screening? Why is it gaining importance now-a-days?

5. Define single-gene disorder in humans, and name those which have been identified.

6. Explain how molecular method of genetic testing will prove beneficial for the genetic counseling and the human society.

7. Write notes on different facts that need to be addressed in the genetic testing procedures that are followed for human screening.

8. Elucidate the precautions that are essential to be followed for the conduct of molecular method of genetic testing in human beings.

9. Distinguish between the predisposed and presymptomatic genetic condition observed in the patients.

10. Briefly write about the uses of molecular method of genetic testing over the conventional symptom based diagnosis of diseases.

11. Critically comment on the chromosomal disorders in the human population and add a note on the social aspects of these diseases in the human life.

12. What is meant by multifactorial or complex or polygenic disorder? Add a note on its effects on the human population.

13. Enumerate the different gene based diseases and the molecular methods of genetic testing available for them.

14. Give a note on the mitochondrial disorders and how mitochondrial gene will be also one of the contributing factors for various diseases in an organism.

15. Describe various types of genetic diseases and their implications on the ethical, legal and social issues associated with gene testing.

17. Mention the merits and demerits associated with the gene testing for the genetic disorders in the human beings.

18. Define gene testing and its importance in the biotechnological laboratories for the number of disease diagnosis in prenatal state.

19. Explain why gene testing is followed for the population screening in the humans, and mention its impact on the medical genetics.

20. Describe how gene testing evolved and grew to the present status in the field of medical genetics. Write down some of its demerits.

21. Enumerate the need of regulations that are required to control the genetic testing in case of human medical diagnosis of the diseases.

22. Write about the importance of the gene testing of prenatal screening of the genetic disorders in the human beings in medicine.

23. Give an account of different genetic disorders in the human beings for which gene testing is done.

24. Explain how gene testing for genetic disorders in the human beings is becoming a negative factor for social issues.

25. Write about the present situation of genetic testing for the genetic disorders and their insurance cover by the insurance company.

26. Illustrate different genetic diseases that are common in human beings throughout the world and identified by the presymptomatic methods.

4

CHAPTER

Prenatal Diagnosis, Genetic Manipulations and their Ethical Issues

4.1. INTRODUCTION

The achievement made by the Human Genome Project has attracted media attention over the past few years. The public has received a steady stream of information on the latest gene to be discovered by science. Such announcements are generally greeted with enthusiasm and hope that genetic solutions will save us from diseases, disorders, discomforts, inabilities and frailties. But, this hope may be overly optimistic and ambitious. While genetic technologies may offer a few more keys to unlock the genetic puzzle, they may not necessarily produce the solutions required to solve the puzzle. For example, although hundreds of human genes have been identified and mapped, in many cases, a corresponding cure or treatment regime has not yet been developed. Moreover, although many common diseases are suspected of being genetically linked, in most cases, the risk may be augmented by other genetic or environmental factors in order for the disease to occur. Nor does the detection of a chromosomal disorder, such as Down's syndrome, predict the severity with which the syndrome will be expressed. Advocates of new reproductive technologies, particularly prenatal technology, argue that this technology will expand women's reproductive choice and decrease the incidence of disability in society.

Biomedical researchers are currently redefining human geography. These modern explorers are elaborating a new human map, based on genes, which is likely to alter our views about the world and our place in it. This argument is over simplistic at best and belies the significant social and ethical implications underlying such technologies for women's reproductive autonomy, and the equality rights of women and men with disabilities. It is possible that prenatal diagnostic testing can offer women more control over the occurrence of the birth of a disabled child by providing relevant information about the genetic status of the fetus during pregnancy.

4.2. TYPES OF PRENATAL TECHNOLOGIES

The term 'prenatal technologies' is used broadly to refer to those technologies, which are designed to assess the genetic make-up of a fetus, and do not include technologies aimed at embryo and preimplantation diagnosis. Examination of the fetus in uterus probably occurred almost 100 years ago, when an X-ray picture was taken of a dead fetus. Prenatal testing was first introduced into medical protocols in the 1970's with the emergence of amniocentesis. Since then, advancements in genetic knowledge and technology have resulted in a variety of screening and diagnostic tests being made available. However, prenatal screening and diagnosis have been available routinely to certain groups of women and selectively to others since the late 1970s. Amniocentesis and other technologies of prenatal screening and diagnosis, currently in use, provide access to the fetal cells and amniotic fluid, and enable the visualization of the developing fetus. These technologies allow the detection, before birth, of all recognizable chromosomal variations, many selected developmental malformations, over 150 biochemical disorders, and sex of the fetus. The list continues to expand, and the techniques are being applied earlier in pregnancy.

The number of conditions that can be diagnosed in uterus grows as days progress. Some of these tests can be used to detect late-onset disorders that can be identified before birth, but which do not manifest themselves until adulthood. In addition, prenatal testing can detect susceptibility genes (genes shown to increase an individual's susceptibility to certain conditions that may or may not develop later in life, such as cancer or heart disease).

Prenatal genetic screening has become a routine aspect of prenatal care for most pregnant women in the world. Physicians monitor the vast majority of pregnancies with at least one ultrasound, and maternal serum screening is fast becoming a component of standard medical prenatal management. In contrast, diagnostic testing is considered a routine aspect of responsible prenatal care only for women identified as at 'high risk' due to the detection of certain factors which are associated with fetal disability. Prenatal screening is conducted on a population-wide basis in an attempt to identify women who are at high risk for carrying a fetus with a disease or disability. It is important to note that screening methods are implemented to define women according to risk factors associated with fetal anomaly, and not to actually determine if the fetus does carry a disability. As a result, women defined through the screening process as 'high risk' are then offered the option of diagnostic testing. In essence, screening methods can identify possible high risk cases, but diagnostic tests are needed to reach a decisive diagnosis on the actual status of the fetus. The most common indication for referral to testing is advanced maternal age. The prenatal diagnosis is very essential for the eugenics. Some of the most common screening and diagnostic techniques currently being utilized in medical field are as follows:

Ultrasonography

Ultrasound is a routinely used, non-invasive technique for assessing gestational age, tracking fetal growth, and detecting major structural abnormalities. Ultrasound functions both as a screening and diagnostic method. The routine use of ultrasound as a standard part of prenatal care illustrates its importance as a screening procedure. It may uncover the fetal conditions associated with disability that would place the pregnancy at 'high risk', and thus the woman would be a candidate for diagnostic testing. However, ultrasound can also provide a decisive diagnosis on its own in certain situations, when major structural abnormalities are uncovered.

Maternal Serum Screening

These screening procedures [i.e., maternal serum alpha-fetoprotein (MSAFP) and triple marker tests] analyze the levels of certain proteins and/or hormones found in a pregnant woman's blood serum. Certain elevated protein or hormonal levels have been found to be associated with open neural tube defects in the fetus, while suppressed levels are suggestive of possible Down's syndrome.

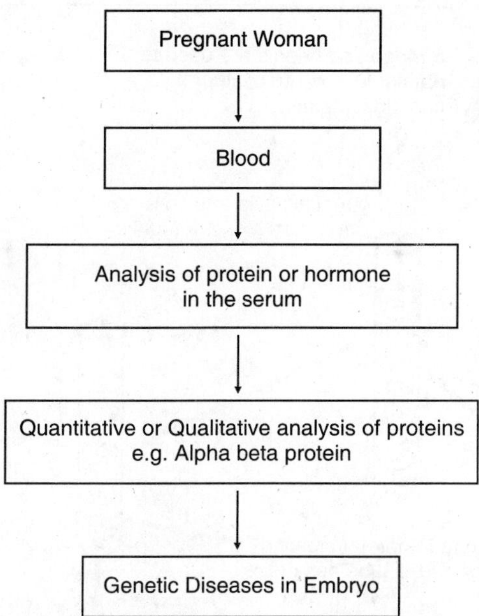

Fig. 4.1. Maternal serum analysis for the identification of genetic diseases.

Maternal serum alpha-fetoprotein analysis **(Fig. 4.1)** is another technique on its way to become a universal technology for prenatal screening. It entails the withdrawal of a sample of blood from a pregnant women. This is then further analyzed for the presence of alpha-fetoprotein, and in some jurisdictions, a limited number of other serum constituents that tend to be found in increased or decreased levels when the fetus has a malformation of the neural tube or Down's syndrome.

The high rate of false-positive and false-negative results associated with these serum tests have led medical researchers to emphasize that these procedures are to be utilized only for screening purposes. Specifically, 95 percent of women with an MSAFP value in the clinical range are carrying a normal child. Thus, more precise techniques must be implemented subsequent to an abnormal serum result in order to reach a more conclusive diagnosis. Recent experimental techniques have been able to isolate fetal cells circulating in the maternal bloodstream for the purpose of analysing them for abnormalities. Such techniques will result in a non-invasive prenatal diagnostic testing method, which poses no risk to the fetus. This is because the serum sample is obtained from the woman's arm. It is also expected that the results of such serum diagnostic procedures would be available sooner, and it would be less expensive to conduct than the current methods of prenatal diagnosis.

Amniocentesis

Amniocentesis is a procedure by which a sample of amniotic fluid surrounding the fetus is withdrawn **(Fig. 4.2)** and analyzed for chromosomal, biochemical and DNA abnormalities. It is usually performed between 14–16 weeks of gestation. Examination of amniotic fluid obtained by needle aspiration was first attempted in the late 1920s. Initially, amniocentesis was employed as a part of the effort to manage Rh group (Rh^+/Rh^-) disease, in which blood type incompatibility between fetus and mother results in a deadly condition called erythroblastosis fetalis. Management consisted of complete blood transfusion of the so-called 'blue baby' during birth. So, amniocentesis was performed in the last trimester of pregnancy. Advances during 1960s and beyond in somatic cell genetics have permitted diagnosis of genetic diseases, and diseases resulting from abnormal chromosomes through examination of fetal cells obtained from the amniotic fluid.

Depending on the quantity of cell material needed for the assay, either directly examining the fetal cells or culturing them permits identification of chromosomal abnormalities (e.g., extra, partial, or missing chromosomes), biochemical or DNA markers of various genetic disorders, and missing gene products. The time needed for some of these diagnostic techniques, as well as temporal pressures when abortion is contemplated, necessitated amniocentesis much earlier in pregnancy (early to mid-second trimester) than previously employed. The human genome project

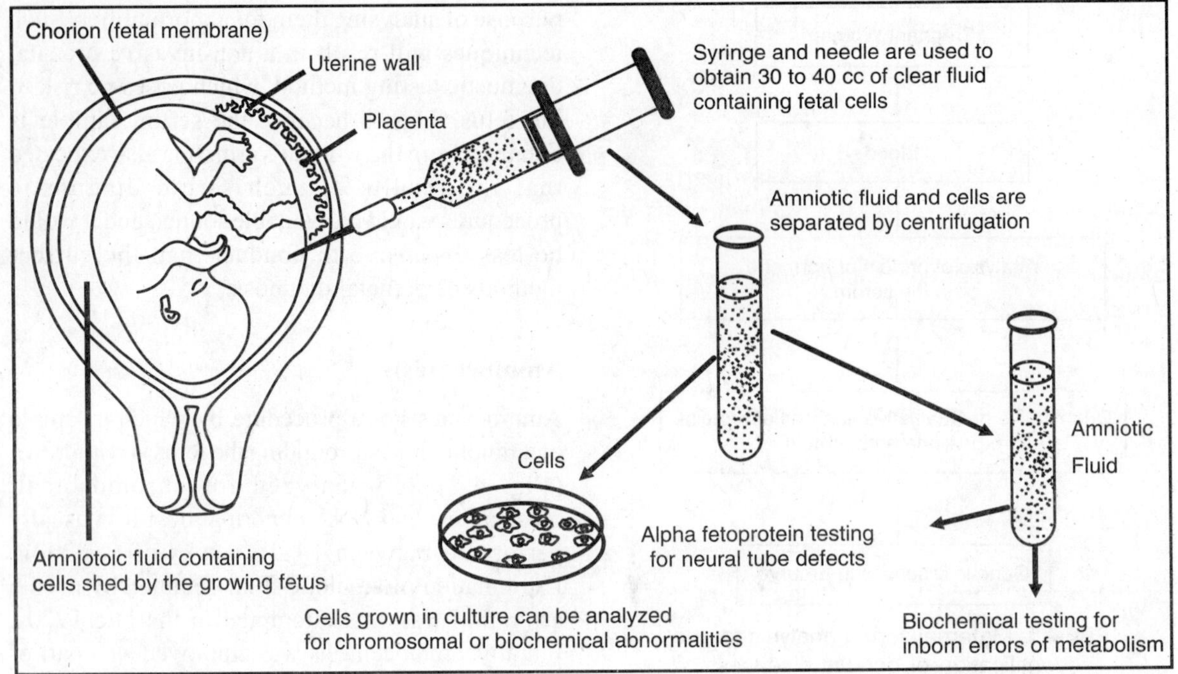

Fig. 4.2. Different steps involved in amniocentesis for a test of genetic diseases.

(the mapping of an entire human genome) promises a much larger range of possible diagnoses emerging from amniocentesis.

Under sterile conditions, a needle is inserted through the abdominal wall, uterine wall, and amniotic sac. Amniotic fluid, in which detached fetal cells float, is than aspirated in a quantity sufficient to obtain several cells.

Then various tests are performed on the uncultured cells, cultured cells, and cell-free amniotic fluid. With the development of sonography—a technique which uses low-frequency sound waves that are reflected and transformed into visual images (similar to underwater sonar)—fetuses and placentas can be visualized. This technique has decreased the incidence of bloody taps and consequent misleading information. The risk of fetal death from this procedure is less than 0.5 per cent. Sonography has also permitted the identification of twins. So, in cases where family history

indicates a risk of disease, confirmation of the status of each twin is possible.

Parents generally seek prenatal diagnosis of genetic and chromosomal abnormalities when they believe that their offsprings are at risk for serious abnormality due to family history, age of the mother, or exposure of the fetus to environmental risks (such as rubella, cytomegalovirus, irradiation, drugs, toxoplasmosis).

The most difficult dilemmas occur when amniocentesis discloses only probabilities. It is imperative that a woman seeking prenatal tests be fully informed of the inherent uncertainties and risks of any diagnostic procedures or actions that may be taken. Due to its invasive nature, the test carries an estimated risk of fetal loss between 0.5–1.0 percent. It is important to note that this test is not 100 percent accurate, and the false-positive and false-negative rates for detecting chromosomal abnormalities via amniocentesis are approximately 0.0007 and 0.004 respectively.

Chorionic Villus Sampling (CVS)

CVS involves obtaining and analyzing a sample of the chorionic sac that surrounds the fetus. Chorionic villus sampling (also chorion villus sampling, or chorion villous sampling) is a technique, which involves the removal of tissue from the chorion—the outermost embryonic membrane in mammals. Chorion is generated by the embryo, and attaches with a multitude of fingerlike villi to the wall of the uterus, eventually forming the placenta. Sampling of tissue from this structure may be done as early as the seventh week of pregnancy.

Prenatal Diagnosis

A biopsy needle is introduced either through the cervix or the abdominal wall under local anesthesia, and guided through the wall of the uterus to the site of implantation. A tissue is taken from the site, and the cells of that tissue may be subjected to various genetic and chromosomal tests for detecting the abnormalities **(Fig. 4.3)**. This sampling is possible in the seventh to eighth week, some eight to ten weeks before enough amniotic fluid has been produced to permit amniocentesis. The technique is therefore attractive to the women concerned about the possibility of abnormalities

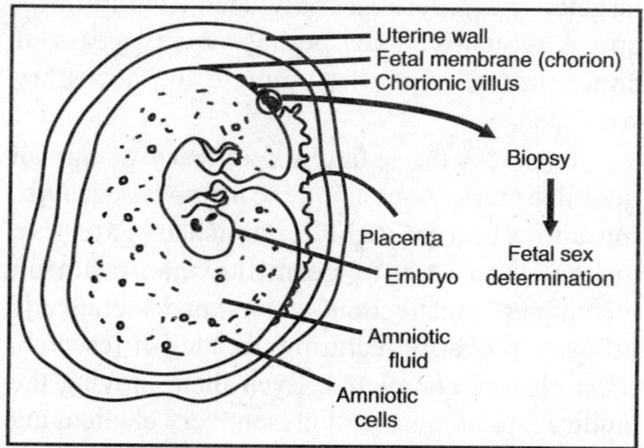

Fig. 4.3. Chorionic villi sampling for prenatal diagnosis of genetic diseases.

in their offspring. The technique provides diagnosis substantially earlier than many of the other techniques.

Early diagnosis of some conditions opens the door for those who wish to develop genetic-manipulation technologies. It allows discussion of the possibility of fetal surgery with time for the parents to weigh possible consequences. And it permits the planning for immediate medical treatment of the newborn, including delivery at a site in the presence of experienced personnel. Many of these procedures and treatments are still highly experimental, and women subjected to chorionic villus sampling may not realize that for a wide range of detectable abnormalities, the possibility of corrective treatment is remote.

Chorionic villus sampling generally results in feto-maternal transfusion, or the introduction of fetal blood cells into the mother's circulation. This may create an Rh factor and complicate future pregnancies. Fetal loss rates associated with this type of screening are generally 1 to 2 per cent, which is substantially higher than for amniocentesis (ranging between 0.2 and 1.3 per cent). There is also a failure rate that may run as high as 6 per cent, necessitating retesting. A number of congenital abnormalities, particularly limb defects, have been associated with chorionic villus sampling. It has also been associated with a significantly higher incidence of neonatal respiratory distress (as high as 8 per cent). Finally, an increased spontaneous abortion rate has been correlated with early sampling done in women above the age of thirty-five years. It is rising from just below 2 per cent to nearly 11 per cent for women forty years and older. These observations are controversial, and not all are universally regarded as risks imposed by the procedure. Nonetheless, persons considering chorionic villus sampling should carefully consider the possibility of risk to a healthy fetus. Compared to amniocentesis, CVS is accompanied by about 1.0 percent higher rate of fetal loss, a comparable false-negative rate, and an elevated false-positive rate. Its main advantage lies in the fact that it is performed earlier in pregnancy (between 9-12 weeks of gestation), and thus it provides diagnostic information to women within the first trimester. A substantial limitation, however, is that CVS does not detect neural tube defects. Thus subsequent screening and/or testing (via ultrasound and amniocentesis) is necessary to provide diagnostic information regarding these conditions.

4.3. NEW DEVELOPMENTS IN PRENATAL DIAGNOSIS

Fluorescent Dye Technique

As there is some sort of risk associated with sampling from embryo, researchers are looking for non-invasive methods. An increasing number of researchers have begun to report their ability to obtain and make diagnosis from fetal cells in the mother's circulation. Using increasingly sophisticated technologies, these cells, which have crossed the placental barrier and entered a woman's bloodstream, can be separated from the woman's blood cells. These cells are grown in the laboratory, subjected to procedures that amplify their genetic material, and stained with specific (fluorescent) dyes to locate the presence or absence of DNA sequences. If this approach does provide accurate results, it will, in principle, replace procedures such as amniocentesis or chorionic villus sampling with the much less invasive procedure of drawing blood.

Blastocyst Analysis Before Implantation (BABI) Technique

Another approach that may become more wide-spread in the future is embryo or preimplantation diagnosis. Its advocates have assisted public acceptance of this problematic technique by affixing it with the acronym BABI (blastocyst analysis before implantation). Preimplantation diagnosis is a welcome approach. It will circumvent the abortion of genetically defective embryos, because only unaffected embryos will be (re)implanted. Hence, risk of carrying a defective embryo will not appear when this method is employed.

In this approach, which is already used in programs of advanced infertility techniques, a very early embryo is removed from a pregnant woman (or created in the laboratory following *in vitro* fertilization), and tested. If it is found acceptable it is then, (re)placed in the uterus to continue its development. This technology has so far been used only as a research tool for the diagnosis of certain specific genetic disorders in embryos.

4.4. GENETIC MANIPULATION

Genetic manipulation can be done either in the somatic cells of an individual who is suffering from the disease, or in an embryo before it is going to born. Any genetic modifications in the somatic cells of an individual is referred to as **somatic genetic manipulation**, and the treatment process is called **somatic gene therapy (Fig. 4.4)**. Similar type of genetic manipulation made in the germline cells or embryonic cells is known as **germline manipulation**, and the treatment process is called **germline therapy (Fig. 4.5)**. Transgenic technology, germline gene manipulation of animals, has been successfully applied in mice, pigs, sheep, and cows. However, the more tedious gene targeting procedure has only been used in mice, and it typically takes perhaps many years of concentrated effort to complete the necessary work.

To assess the safety and efficacy of human germline therapy, somatic gene therapy procedures must first be thoroughly established. Extensive animal studies into germline manipulation techniques, transfection vectors, and therapeutic efficacy, probably requiring decades of research, must also be completed. Even then, moving the studies into humans will present very challenging and more ethical problems. Relatively simple questions like "do aspects of long-term cell culture, with its unnatural selection pressures, affect the

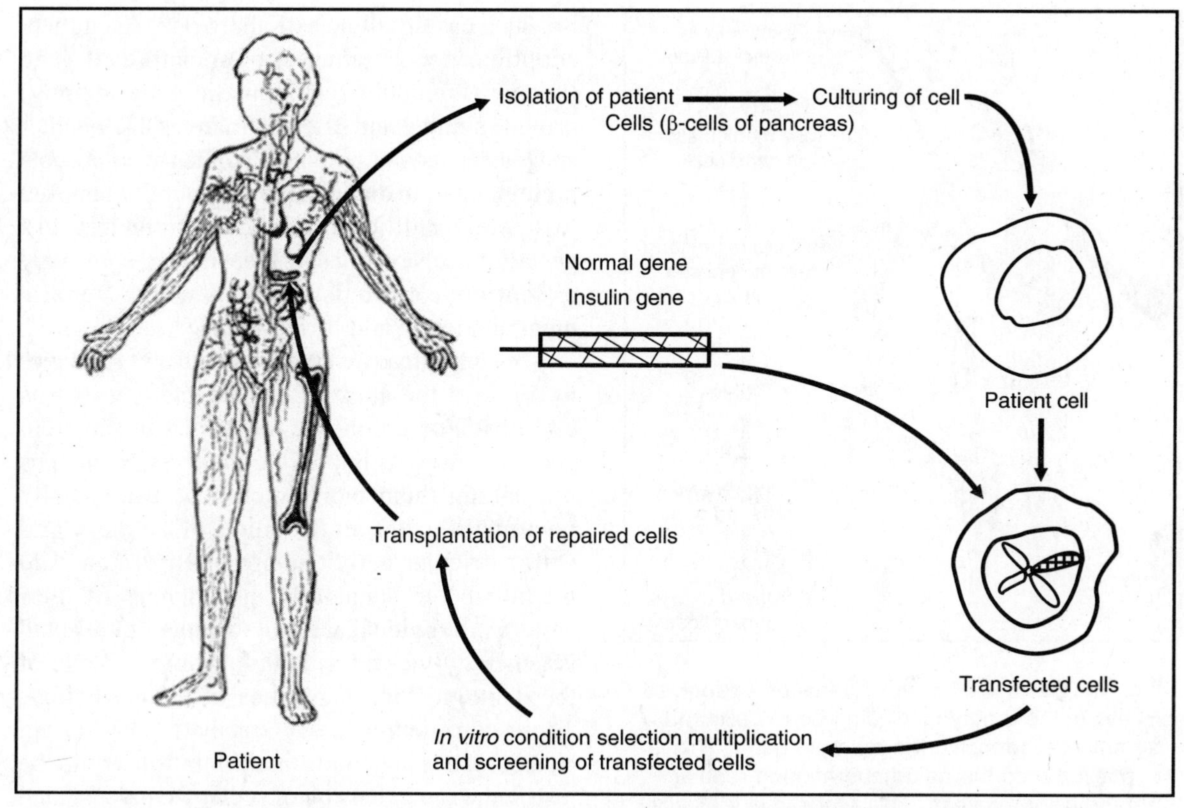

Fig. 4.4. Somatic gene therapy. In this therapy, somatic cells from the patient are isolated and normal gene segment is intoduced for normal function. No ethical issues are involved in this therapy.

person these cells may eventually become," will need to be answered specifically in humans, as animals make poor models for disorders such as mental retardation. The recent successful transgenic animal work of is impressive, but high failure rates, problems induced by culturing techniques, and the presence of genetic markers cannot be ignored.

With the increased prenatal diagnosis technologies, parents get more and more information about the embryo before it reaches maturity. In the present time, selection through abortion or discarding of preimplanted embryos is the only way to avoid the effects of an undesired gene that has been detected through prenatal diagnosis. In the future, however, genetic techniques may allow manipulation of an undesired

gene without discarding or aborting the embryo. However, while transgenic animal research provides a powerful tool for studying gene function, human germline gene therapy is unlikely to be a procedure with significant clinical utility, and is probably not worth pursuing, considering the danger posed by its comparative usefulness for eugenic enhancement.

From a medical perspective, uses of this technology in the treatment of genetic disease are limited, and more efficient alternative procedures (such as preimplantation diagnosis followed by selective reimplantation) will likely become available. Since germline gene therapy confers the same benefits as embryo selection, its risks will always be greater. But there are few situations necessitating germline gene therapy. The first is a

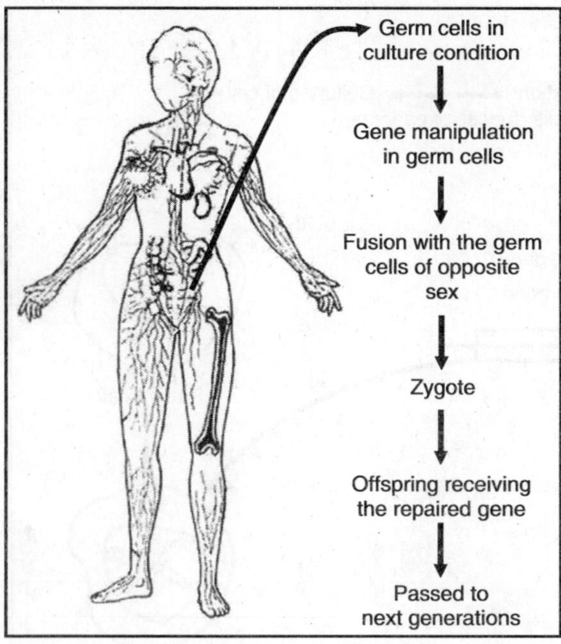

Fig. 4.5. Germ line therapy. In this procedure, all cells of the embryo including germ cells and somatic cells receive the repaired gene. This is passed onto the next generation in regular reproduction process. This process is ethically problematic in nature.

case where a couple, both homozygous for a serious genetic disease, wish to conceive a child free from the disease, but would not like to use donor gametes. Embryo screening and selection of unaffected embryos is impossible, since all zygotes conceived from this couple's gametes will carry the disease.

However, if both parents have survived to reproductive age, the disease they carry cannot be one of the most severe "maladies" untreatable by other methods, ruling out many diseases commonly cited as sufficiently severe to warrant germline gene therapy (for example, Lesch-Nyhan syndrome, Tay-Sachs disease). Even for a relatively common severe genetic disease like cystic fibrosis, the chance of two homozygotes marrying (admittedly making the assumption of independent assortment) is about 1 in four million.

So, such cases will be extremely rare. Even then, adoption, donor gametes, or somatic cell gene therapy (including fetal somatic gene therapy) provide simpler and safer alternatives. DeWachter suggests a second case of a woman who accepts preimplantation diagnosis, but is morally opposed to discarding afflicted embryos, desiring instead a genetic cure. Nevertheless, there may be some very exceptional circumstances in which genetic manipulation would be reasonable.

Society already employs a number of strategies to improve the abilities or characteristics of its citizens. For example, the education system generally aims at improving the lives of citizens by making them more intelligent and socially responsible. Parental choice of schools and extracurricular activities are often designed to maximize the capacities and chances of their children. Individual self-improvement of all kind is generally viewed as both laudable and one of the distinguishing capacities of human beings. These considerations suggest that under certain circumstances, genetic manipulation could be justified as an extension of society's, parents' and individuals' continuing efforts to enhance the capacities and improve the lives of citizens.

In the medical field, after prenatal diagnosis, the doctors prescribe the eugenics to be followed. It is difficult to say whether in fact genetic manipulations can be justified as extensions of current practices. Such manipulations are still speculative, and the mechanism by which they would be performed are unknown. Since there is still much to learn about the possibilities of genetic manipulation, a total ban on manipulation would be premature. However, because of the potentially grave dangers of manipulation, genetic interventions to enhance traits should be permissible only in severely restricted situations. The following criteria should be followed:

First, there would have to be a clear and meaningful benefit for the fetus or the child that will be born. This criterion would ensure that parents do not impose their own idiosyncratic values on the children, but engage in genetic manipulation only when there would be a scope for improvement in the child.

Second, there would have to be no trade-off with other characteristics or traits.

Genetic manipulation would be permissible only if the benefit could be gained without a change in some other characteristic or trait. Whether a gain in one area is worth a loss in another area is a choice that depends upon subjective values. Consequently, only the individual, who would experience the consequences of the trade-off, should decide whether the trade-off should be made.

There also would have to be clear evidence that no trade-off will occur. It would not be sufficient to show merely that no trade-off was known, since adverse effects are often unanticipated. Instead, the mechanisms of genetic manipulation would have to be understood in detail to ensure that trade-offs would not occur.

Third, there would have to be equal access to genetic technologies, irrespective of income or other socio-economic characteristics. If access depended on wealth, social divisions would widen and the promise of equal opportunity for all citizens would quickly become an illusion. These potential criteria should be viewed as a minimal, not an exhaustive, test of the ethical propriety of non-disease-related genetic intervention. As genetic technology and knowledge of the human genome develop further, additional restrictions or guidelines may be required.

As genetic understanding and technology continues to develop, prenatal manipulation of genetic material may become possible. As with genetic selection, the hypothetical nature of manipulation precludes precise guidelines at this time, but it is nevertheless important to begin the discussion of genetic manipulation to ensure that appropriate guidelines are developed in time. In general, it would be ethically acceptable for physicians to participate in the manipulation of genetic material in order to prevent, cure, or treat the genetic disease. There might be exceptional situations in which genetic manipulation would not be appropriate to treat or cure diseases, such as when manipulation would cause serious side effects.

Genetic Diseases in Humans and their Treatment

The use of genetic manipulation to avoid disease is consistent with ethical principles. The tools of medicine are designed to prevent functional abnormalities, such as Tay-Sachs disease, Down's syndrome, or cystic fibrosis. But the fact that genetics is used, rather than surgery or antibiotics, does not make the treatment unacceptable. Genetic manipulation in this instance would be a legitimate extension of the physician's preexisting and fundamental duty to combat disease.

Debate whether only somatic-cell gene therapy should be used to combat disease, or germ-line therapy could also be used, still continues. In somatic-cell therapy, genetic alterations affect only the fetus, but in germ-line therapy, alterations would be inherited by the fetus' future offspring as well. So, there are two explanations, which suggest that genetic manipulation should be limited only to somatic cells.

1. The profound impact of germ-line manipulations on future generations requires that we proceed with extreme caution. Until all the effects of a genetic manipulation, both long and short-term, are completely certain, germ-line manipulations should not be attempted.

2. Germ-line manipulations give parents unprecedented control over the lives of multiple generations of descendants. Such far-reaching control would curtail the autonomy of future generations, limiting their control over their own lives and restricting the decisions they can make about their own children. Parents' ability to extend the authority beyond their own immediate offspring is already restricted in property law, in which individuals are prevented from dictating the fate of their estate in perpetuity: If society has an interest in protecting future generations' autonomy in property matters, it may very well have an even stronger interest in protecting that autonomy in matters of genetic control over offspring. Initially, it would seem appropriate to limit genetic intervention to somatic cells only, though further consideration of this issue may be advisable in the future.

There may be exceptional situations in which genetic manipulation would not be appropriate to treat or cure diseases. Occasionally, genetic manipulation might harm the fetus in some way, even if it were successful in treating the targeted disease. For instance, the manipulation might damage or interfere with other genes, which in turn could cause increased susceptibility to a serious illness, impairment of mental or physical faculties, or other adverse effects. At some point, the adverse effects would be so serious that it would no longer be appropriate to assume that the future child would prefer the intervention.

Benign Trait Manipulation

When genetic technology would be used to change characteristics that are not disease-causing, the dangers of abuse discussed in this report require that extreme caution be exercised. In general, it would not be appropriate to use genetic technology to manipulate a fetus' or pre-embryo's traits or characteristics. Such manipulation risks the severe adverse consequences discussed earlier: exacerbaton of discriminatory practices, the commoditization of children, and the prospects of repugnant eugenic practices. These considerations counsel against genetic manipulation of benign characteristics.

Human Germline Gene Therapy and its Protocol

After prenatal diagnosis, if the embryo is identified having genetic disorders, it is later considered for the genetic manipulations. Preimplantation diagnosis has been used to screen embryos for cystic fibrosis and several other recessive genetic diseases, but could equally be applied to dominant disorders. Because of the inefficiencies in egg recovery and reimplantation, steps which are used in germline gene therapy, preimplantation diagnosis is costly and currently is only experimental. Available standard prenatal diagnosis by chorionic villus sampling or amniocentesis can achieve the same end, though only through second trimester abortion. Prenatal diagnosis presents a (small) risk to a potentially healthy fetus—a problem that may be circumvented by research in new diagnostic methods directed at fetal cells shed into the maternal blood-stream.

Thus, there is very little clinical need for germline gene therapy in the prevention of genetic diseases. The goal of germline gene therapy is the prevention of severe genetic diseases in newborns. The same can be achieved more easily through preimplantation diagnosis and selective reimplantation. This procedure is identical to the first and last parts of the germline therapeutic procedure, but skips entirely the *in vitro* culturing, transfection, selection, and nuclear transfer procedures. Thus, human germline gene therapy presupposes the development of efficient preimplantation diagnosis/selective reimplantation, and this alternative technique would be available as a simpler, more efficient and less dangerous option.

The steps followed in the germline genetic manipulation and therapy are summarized as follows:

Isolation of Totipotent Embryonic Cells at an Undifferentiated Stage

In the technique, a preimplanted embryo is required within four days of conception. This embryo can be obtained either by laparoscopic flushing of oviducts to recover a naturally-fertilized ovum before implantation, or by *in vitro* fertilization. Both these strategies are currently possible in humans, and lots of basic steps are clearly understood. But these techniques are highly expensive, and often require months or years of repeated attempts to get embryonic cells successfully. Although it may be morally preferable to genetically manipulate the gametes instead of embryos, this technique is not feasible to be adopted for regular genetic manipulations in the embryonic cells. Hence, this strategy does not seem to be a viable option for long-term usage.

Determination of the Genetic State of the Embryo

The isolated embryonic cells are used for the determination of genetic conditions for almost all the known genetic disorders. "Preimplantation diagnosis" is already available on an experimental basis. Note that if normal embryos can be differentiated from abnormal embryos, the simplest option by far is selective reimplantation, obviating the need for genetic manipulations. On the other hand, determination of the genetic state of the embryo can be skipped in rare cases (both parents homozygous for a recessive disorder, or one parent homozygous for a dominant disorder) where all embryos will be afflicted, and would not be a necessary step in "enhancement" procedures. The large numbers of molecular techniques such as PCR, RFLP, DNA probes or molecular kits are available for the determination of genetic status of the embryo.

Culturing of Embryonic Stem Cells

The culturing of embryonic totipotent stem cells and their standardization is already established in a number of organisms. The maximum success has already been achieved in mice, and more recently, in monkeys too. But this technique would raise several issues if applied to human embryonic cells. There is a poor success rate for establishing such cultures, and maintenance is expensive and labour-

intensive. A high technical skill is associated with the culturing of the cells and their interpretations. A normal karyotype occurrence has a limited lifespan in culture conditions, and, in particular, second X chromosomes are frequently lost during growth in *in vitro* conditions. This makes culture systems in humans that perhaps only male embryos could be successfully treated. Thus, achieving success by culturing the human embryonic stem cells without alterations or aberrations is a very difficult task.

Transfer of Genetic Material into Embryonic Cells

Generally, transfection processes are used for the genetic modifications of embryonic cells. Since millions of cells must be transfected to obtain even one targeted recombinant, mass transfection techniques will have to be employed. Available methods have limited efficiency. They kill some of the target cells, and to varying degrees cause undesired DNA integration and recombination events. Controlling such events is not possible in culture conditions.

Selection of Cells which have Stably taken up the Transfected Gene

After successful transfection, the required cells of repaired genetic constitutions are to be selected from the non-transfected cells. Because the present techniques allows only at a frequency of roughly 1 per 10,000 (depending on the transfection technique). At this end, markers, such as those coding for resistance to a cytotoxic drug, are added to the transfected DNA, and after weeks of careful culturing in the presence of drug selection, clonal transfectant embryonic cell lines can be grown out. While gene supplementation tactics, addition of a normal copy of the transfected gene, which inserts into a random position in the cell's genome by illegitimate recombination, are technically the easiest way to replace the function of a defective gene, they are inherently unpredictable in their downstream effects and can only be used for somatic gene therapy. Unfortunately, current technologies permit only the addition of a supplementary gene copy causing human germline gene therapy to adopt the more difficult targeted gene replacement. Lot of work and standardization protocols have to be worked out and success rate has to be increased.

Targeted Gene Replacement

This step represents a true genetic cure, where the deleterious DNA base pair sequence in its native chromosomal position is changed to a normal sequence that permits proper functioning of the gene product. However, mammalian cells favour illegitimate recombination events by a factor of at least 1000 over the homologous recombination events required for a targeted gene replacement. Differentiating targeted gene replacement events from illegitimate insertions is tedious and difficult, as in both cases, cells have taken up selectable marker tags attached to the transfected DNA. Molecular biologists have devised transfection vectors that increase the proportion of targeted recombinants, but only by selecting against marker-expressing illegitimate integration events. Currently, there is no way available to induce significantly higher levels of homologous recombinations, which remain at a low rate of around 1 per 10 million cells transfected. While gene targeting techniques are probably obligatory for human germline manipulation, biological limitations make targeted gene replacement inefficient and difficult.

Marker Removal

After the successful gene placement in the genome, it is always necessary to make as possible as similar to the original genome sequence of the cell. Leftover marker genes in the vicinity of the modified locus may influence neighbouring genes

in undesired ways. Hence, they should not become part of every cell in a mature embryo or a person, or passed onto the future generations. Fortunately, molecular biologists have developed strategies to permit the removal of markers, which leaves the genome completely unaltered except for the targeted sequence change. This requires re-expansion of the culture, and another round of drug selection which, in turn, makes the embryonic cell similar to the original level with modifications only in the desired gene or genes.

Confirming Genomic Integrity

With the removal of marker gene or any other segment of gene cassette, genome of the cell remains in the original genetic constitution in the future procedure. An embryonic cell line with the exact, desired genetic change has finally been established. However, genetic changes can occur during long term culture in the *in vitro* conditions, and there is evidence to suggest that subtle mutations are frequently induced in regions treated by targeted homologous recombination. This induces the cells to change their own genetic constitution. Thus, it may be necessary to recheck the genomic integrity of the cell line before attempting to reimplant an embryo into the uterus. How this step could be accomplished is uncertain; it would require tedious, extensive, expensive, yet necessarily incomplete investigation. Months or years of growth in culture would be required to reach this point, but the cells should still be viable for nuclear transfer. This is one of the major drawback of the germline therapy process.

Nuclear Transfer

With the technological achievements made in the animal nuclear transfer methods, the same can be practiced in humans. The well-publicized work of Wilmut *et al.*, who cloned sheep from the cultured cells arrested in the G_0 phase of the cell cycle, made the steps of nuclear transfer theoretically possible in humans, though it has never been tried. Unfertilized ova would need to be recovered from a superovulated woman, either the mother or a donor (who would then donate the progeny's mitochondrial DNA). The nucleus would be removed and replaced with a nucleus from the earlier step of confirming the genome integrity. From this nucleus an embryo would grow, carrying the modified human genome; again, a difficult, expensive, and tedious step. Hence, further focus has to be given to these aspects of the human embryonic stem cells culturing.

Reimplantation into the Mother

Even with the successful protocols developed for a variety of animals and humans, the real achievement is very low. Embryo loss is roughly 98 percent, and remains the limiting factor in human *in vitro* fertilizations. Subsequent spontaneous abortion is frequent; only about 15 percent of the couples attempting *in vitro* fertilization ever have a successful pregnancy. To increase the chance of success, multiple embryos are reimplanted simultaneously, which can result in identical twins or triplets. Later progress is also continuously observed for the better results.

4.5. INDIAN SCENARIO IN PRENATAL DIAGNOSIS AND THEIR APPLICATIONS

The programme was initiated in India with the main objective to develop basic capabilities; strengthen existing institutions, which have good expertise, to initiate work in molecular genetics; and to support some application oriented projects, so as to reduce the burden of genetic disorders in the country. The programme also aims at identifying, mapping and characterizing genes associated with genetic disorders and diseases prevalent in India and exploiting this knowledge, as created by the available human genome, microbial and parasitic genomic sequences, so as to usher in molecular medicine tools for better management of disorders and diseases. Since the inception of this programme, the Department of Biotechnology has implemented several projects, including genetic diagnosis cum counseling units to provide diagnosis and counseling to the affected families for the common genetic disorders prevalent in the country.

- Major programmes have been initiated in the area of functional genomics, human genome diversity, gene therapy, microbial and computational genomics, biocomputing, DNA micro array facilities, proteomics etc.

Genetic Diagnosis and Genetic Counseling Units

- In India sixteen genetic diagnosis cum counseling units were established to provide diagnosis and counseling to the affected families. So far, about 18039 affected families and more than 4,500 tribal families got benefited from these units for common genetic disorders like thalassemia, sickle cell disease, DMD, haemophilia, cystic fibrosis etc.

- The unit established for haematological disorders at CMC, Vellore has provided genetic diagnostic services to about 3,366 patients and family members suspected to have thalassemia, haemophilia, Glanzman's thrombasthenia, leukemia etc. The group has provided antinatal diagnosis to about 103 cases and the first antenatal diagnosis for a twin pregnancy.

- The unit established at Vivekananda Institute of Medical Sciences, Kolkata, provided diagnosis and counseling to the affected families. Among them were about 6800 individuals of β-thalassemia and about 3118 individuals of haemoglobionpathies.

- It extended screening services to schools, colleges, institutions and rural societies at West Bengal, and also referred cases from Tripura, Arunachal Pradesh, Bihar, Sikkim, Bhutan and Bangladesh.

- Prenatal molecular diagnosis resulted in identifying new mutants in the b globin gene.
- Under the project implemented at LVPEI, Hyderabad, on "molecular genetic analysis of retinoblastoma", analysis was carried out in 37 patients. Mutations were found in 14 patients, seven each with unilateral and bilateral disease. The mutations were largely confined to the exons 12–24, which code for the pocket domain of the RB protein.

Under the "Programme on Functional Genomics" implemented at Institute of Genomics and Integrative Biology (formerly CBT), Delhi, it established good linkages with clinicians in the three major areas, i.e., spinocerebellar ataxia (SCA), schizophrenia and biploar disorder and asthma and developed high throughput capabilities and highly skilled manpower in the field of genomics and computational biology. The significant achievements are as follows:

- A multiplex PCR has been standardized for all known spinocerebellar ataxia: diagnosis services provided to 59 families.
- Identified a novel mutation and single nucleotide polymorphism (SNP) in β-thalassaemic patient.
- Identified two novel single nucleotide polymorphisms (SNP) in the vicinity of spinocerebellar ataxia (SCA).
- A US patent was also filed in November 2000 for a method of detection for human spinocerebellar ataxia two gene variants.
- The programme has entered into knowledge alliance with genomed programme of nicholas piramal to develop genotype-based drugs.
- A major proposal on molecular genetic studies of type-II diabetes and diabetic retinopathy has been developed at Madras Diabetes Research Foundation, Chennai.

Under the Functional Genomics programme at CDFD, Hyderabad, scientists are evaluating a strategy employing dual gene knock outs on gene functions in organisms. The initial results have suggested that the strategy is used for identifying the novel "synthetic lethal combinations", such a nusG-rho, nusG-rnhA, and nusG-nusB. This suggests models of the roles of transcription termination and antitermination reactions in cell physiology and metabolism.

New Initiatives for the Tenth Plan

Large number of institutions and research organizations are planning to identify the various genetic diseases and their genetic variations in the populations to understand the genetic diseases and their inheritance. Some of the major research areas are as follows:

- Genome diversity studies—population based genotype studies to identify susceptible/resistance genes for various disorders and diseases.
- Sequencing of some specific pathogens relevant to India—*Helicobacter pylori*, *Mycobacterium, w. Leishmania donovani, mycobacterium tuberculosis, Leptospira etc.*
- Single nucleotide polymorphism (SNP) studies relevant to common diseases like hypertension, coronary heart diseases, diabetes, cancer, neurological and psychiatric disorders.
- SNP studies in ethnic population of India.
- Pharmcogenomics—drug metabolism, identification of drug target, drug design, and genotype analysis.
- Microarray based studies on gene expression, identification of candidates for vaccines, and diagnosis of infectious and non-infectious diseases.
- Studies related to proteomics, including development of protein microarrays.
- DNA banking and EBV transformed cell banking.
- Initiation of gene therapy research, especially for oral cancer, thalassemia etc.
- Development of knock-out/transgenic animals as experimental models.
- To create academic chairs in medical colleges for human genetics.

4.6. GENETIC MANIPULATIONS AND ETHICAL CONSIDERATIONS

In the 1980s, initial discussion about human gene therapy evoked strong opposition, which was based on claims that it entails playing God, or violates "natural law". A background paper submitted to the Canadian parliament stated bluntly that "there is a feeling among segments of the general public that the genetic manipulation of humans is simply not an acceptable activity". However, the public is thought to dramatically overestimate the risk of something, which is unfamiliar, hard to understand, or outside public scrutiny. Technophobia regarding human gene therapy has, to a degree, subsided, and the debate has evolved to focus more on the potential benefits of such treatment. A US government poll in the late 1980s found that 84 percent of Americans supported the genetic manipulation of human cells to cure fatal genetic diseases.

Somatic cell gene therapy is broadly considered to be acceptable in principle, gaining the approval of all international policy bodies, which have studied the issue as it "poses no new ethical problems, and is not morally different from, for instance, organ transplantation".

Since its safety and efficacy in humans is not fully established, somatic cell gene therapy is currently employed only under strict criteria. An example of a current protocol yielding encouraging results is the replacement of the adenosine deaminase gene in lymphocyte progenitor cells to cure children suffering from severe combined immunodeficiency. The acceptance of somatic cell gene therapy means that questions regarding germline gene therapy focus on its unique and crucial differences: the effects are permanent and passed onto the future generations, who cannot give their consent, and human germline interventions are extremely difficult to study experimentally.

The major argument raised in favour of developing germline gene therapy derives from the ethical principle of beneficence: up to 2 percent of newborns suffer from some kind of genetic defect, such as Lesch-Nyhan syndrome. Germline interventions may, in some cases, be the only way to prevent development of a genetic disease, furthermore, benefits would extend to future generations. A second important argument derives from the principle of liberty, which holds that the majority, who derive no benefits from this technology, should not, in the name of "human dignity" or "integrity of the gene pool", curtail the specific interests of a minority seeking to cure genetic diseases without compelling cause.

The "ecological argument" has been put forward as an argument against germline gene therapy. It claims that the human gene pool, a product of thousands and millions of years of carefully balanced evolution, will potentially be weakened in unintended and unpredictable ways by germline gene therapy.

Furthermore, the impact would be far lower than the modifications in the gene pool induced by pediatric therapy in general. Of greater concern is the claim that germline gene therapy would open the door to enhancement engineering of the human genome for eugenic purposes, which is consistently cited as the major problem. Fears of possible eugenic uses of germline gene therapy are based upon both the recent past history of the Western world, and on persisting large-scale, sometimes state-sponsored, programs in developing countries. A 1987 US government poll found that a shocking 44 percent of Americans approved of the scientists changing the make-up of human cells to improve the intelligence level and physical characteristics that children would inherit. Mean-while, human embryo work has largely fallen into the hands of private fertility clinics, which would benefit financially by offering eugenic procedures to those who are able to afford it. Ethicists and government commissions have expressed unanimous opposition to enhancement engineering, arguing that it carries huge social risks for abuse of power, discrimination, and inequality. The line between clinically required therapy and eugenic enhancement can be fuzzy, whereas the line between germline and somatic gene therapy is clear. Considering that "it may be impossible, as a practical matter, to prevent these (germline gene therapy) techniques from being used for enhancement", national commissions, suggesting policy guidelines for the research, frequently drew the clearer line, banning germline experimentation to prevent any chance of a progression to eugenic applications.

In response, it has been argued that germline genetic alterations are unlike classical eugenics. They eliminate defective genes rather than human beings, and multifactorial traits like intelligence or attractiveness may never prove amenable to manipulation. More importantly, the eugenics argument presupposes a slippery slope; however, morally relevant lines can be drawn to delineate acceptable versus unethical applications of the technology. To this end, Berger and Gert define the term "malady" as a physical disorder, which provides a direct challenge to the fundamental principles of benevolence and non-maleficence, by imposing the universally-accepted evils of death, pain, or disability, precluding the freedom and pleasure available in life to the vast majority of other human beings. Relief of morbidity and mortality fall within the acceptable line, changing sex or increasing, otherwise, normal intelligence or height—decisions not based on medical need—would lie outside. Thus, the ethical line should be drawn to make the therapy acceptable and enhancement unacceptable, rather than the current state permitting somatic manipulations but not allowing germline experiments.

Informed consent poses a special problem in germline gene therapy, because it is impossible to obtain the direct consent of the embryo, or of future generations that will be affected. The way to tackle this problem, as suggested by the *Inuyama Declaration of the Council for International Organizations of Medical Science*s, is to assume consent only in cases where genetic intervention would seem obviously acceptable

to the affected future generations. However, procedures must first be refined, and the actual risks to experimental subjects become more clear through ongoing, careful animal studies.

A last ethical issue integral to germline gene therapy arises because development of this technology necessitates experimentation on human embryos, and embryos will invariably be lost or sacrificed as part of the engineering procedure. To the extent that human embryos are assigned moral weight, they must be factored into the ethical deliberation, which therefore requires a more thorough understanding of the technical aspects of exactly how germline gene therapy would be accomplished.

4.7. FUTURE ASPECTS OF GENETIC MANIPULATION AND GERMLINE THERAPY

There is no convincing case to be made that human germline gene therapy needs development for medical applications. Even projections of technical innovation imply long years of human experimentation, with significant risks and unproven benefits for subjects who cannot consent. All these risks will be incurred to develop a procedure, which will probably remain inefficient, expensive, slow, wasteful of embryos, and risky. These drawbacks can be avoided by using readily available, cheaper, simpler, faster and less-risky alternative embryo selection procedures. Scenarios where these alternatives are not applicable are more the product of hypothetical philosophy than of any foreseeable reality.

Is human germline gene therapy even worth pursuing? Medical benefits are minimal, and considering the risks of the technology, germline manipulations may in fact have more potential for social harm than good. Many authors have set a clear ethical line between acceptable therapeutic uses and unacceptable eugenic enhancement. And it is important to ask whether this ethical line can and will be respected by human societies, should the opportunity to cross it become available. Germline genetic manipulation represents an ideal technology for eugenic applications.

Government bodies have been found to be a strong support for the establishment of national regulations to prevent misuse of genetic technology. For example, the US National Institute of Health has its Recombinant DNA Advisory Committee, which carefully scrutinizes, from an ethical and scientific perspective, all applications breaking new ground in somatic gene therapy technology. New protocols are approved only when they make relatively cautious, small extensions to the existing procedures. Such bodies may well be effective in preventing the mistakes of the past, where unapproved attempts were made on patients whose consent was not clear. But limits may not work in the private sector, or across international boundaries.

Exciting new technology tends to get used even when inappropriate, despite better alternatives being available, and risks that are not fully assessed or likely to be underestimated. If human germline gene therapy techniques ever become available, the possibility remains that their use could not be constrained within a moral line, allowing only therapeutic corrections of severe genetic maladies. While rarely exploring, in depth, the detailed technological limitations of the procedure, government commissions in many first world countries have opposed the research into human germline gene therapy. For example, the Canadian Royal Commission on New Reproductive Technologies proposed a blanket ban on both somatic cell enhancement interventions and on all germline alterations, therapeutic or otherwise. In fact, legislation is not needed at this point, because a *de facto* ban on human germline interventions is already in place, as there remain far too many technical problems to be overcome before any experimental attempts could be contemplated. If, for political reasons, a government ban is imposed, then the scientific and medical communities should not waste effort fighting it.

Transgenic animal technology should be vigorously pursued. It has proven its value in the studies of gene function, when combined with results from the human genome project. Such work could lead to spectacular advances in genetics and biology. In contrast, human germline gene therapy is not likely ever to become an efficient or cost-effective way of dealing with genetic disease, in the face of simpler options. Most other research areas are more worthy. They offer more possible benefits to the society. Other methods for preventing and treating genetic disease seem much more practical.

SUMMARY

Prenatal diagnosis includes amniocentesis, chorionic villus sampling, umbilical cord blood sampling, maternal blood sampling, maternal serum screening, and fetal visualization. Common reasons for prenatal screening include maternal age above 35 years, a family history of a condition that can be diagnosed by prenatal techniques, abnormal maternal serum screening results, and certain complications of pregnancy.

As biomedical researchers are currently redefining human geography, the usefulness of these technologies is also increasing. The public has received a steady stream of information on the latest gene to be discovered by science from the work of human genome project. Such announcements are generally greeted with enthusiasm and the hope that genetic solutions will save us from diseases, disorders, discomforts, inabilities and frailties. But, this hope may be overly optimistic and ambitious. The term prenatal technologies' are used broadly to refer to those technologies, which are designed to assess the genetic make-up of a fetus well in advance. Prenatal screening and diagnosis have been available routinely to certain groups of women and selectively to others since the late 1970s. Amniocentesis and other technologies of prenatal screening and diagnosis currently in use provide access to fetal cells and amniotic fluid, and enable the visualization of the developing fetus. These technologies allow the detection of all recognizable chromosomal variations, many selected developmental malformations, over 150 biochemical disorders, and fetal sex—the list continues to expand—before the birth of the child and the techniques are being applied earlier in pregnancy.

The prenatal diagnosis is essential for the eugenics. Some of the most common screening and diagnostic techniques currently being utilized in medical field are Ultrasonography, Maternal Serum Alpha-fetoprotein (MSAFP) screening, Amniocentesis, Chorionic Villus Sampling. Ultrasonography,which includes ultrasound, is a routinely used, non-invasive technique for assessing gestational age, tracking fetal growth, and detecting major structural abnormalities.

Ultrasound functions both as a screening and diagnostic method. The screening procedures viz., Maternal Serum Alpha-fetoprotein (MSAFP) and Triple Marker Tests analyze the levels of certain proteins and/or hormones found in a pregnant woman's blood serum. Certain elevated protein or hormonal levels have been found to be associated with open neural tube defects in the fetus, while suppressed levels are suggestive of possible Down's syndrome. Maternal serum alpha-fetoprotein analysis is another technique on its way to become a universal technology for prenatal screening. Amniocentesis is a procedure by which a sample of the amniotic fluid surrounding the fetus is withdrawn and analyzed for chromosomal, biochemical and DNA abnormalities. It is usually performed between 14–16 weeks of gestation. Chorionic villus sampling (also chorion villus sampling, or chorion villous sampling) is a technique by which tissue is removed from the chorion—the outermost embryonic membrane in mammals. Chorion is generated by the embryo and attaches with a multitude of finger-like villi to the wall of the uterus, eventually forming the placenta. Sampling of tissue from this structure can be done as early as the seventh week of pregnancy.

Any modifications of the genetic content of an individual's somatic cells is referred to as somatic genetic manipulation, and treatment process is known as somatic gene therapy. Similar type of genetic manipulation made in the germline cells or embryonic cells is known as germline manipulation, and the treatment process is called germline therapy.

The human germline gene therapy means the introduction of genetic changes in early embryos that become incorporated into all cells of the body, and, as such, are passed onto the future generations. This has elicited considerable ethical, scientific, and political controversy. Human germline interventions raise unique ethical concerns that must be addressed, such as its possible use for eugenics. The basic technology of germline gene therapy is being successfully applied in animal experiments for the investigation of gene function. However, while transgenic animal research provides a powerful tool for studying gene function, human germline gene therapy is unlikely to be a procedure with significant clinical utility. It is not worth pursuing, considering the danger posed by its comparative usefulness for eugenic enhancement. From a medical perspective, the use of this technology in the treatment of genetic diseases is limited, and more efficient alternative procedures (such as preimplantation diagnosis followed by selective reimplantation) are likely to become available. The ability to treat genetic diseases through prenatal genetic manipulation should not divert attention from efforts to find postnatal treatment or cures for those diseases.

EXERCISE

1. What do you mean by prenatal diagnosis and its importance in the society for the better future generation?

2. Briefly write about the present situation of prenatal diagnosis in the developing countries, with emphasis on India.

3. Define prenatal diagnosis and its impact on the human generations.

4. What is meant by genetic manipulations? Add a note on ethical issues associated with fetal genetic manipulations?

5. Write a note on chorion villus sampling, or chorion villous sampling for fetal sex determination in biotechnolgy.

6. Describe various types of prenatal diagnosis used in the medical field to gain information about genome.

7. Explain the newer methods of prenatal diagnosis used in the medical field.

8. Enumerate the different methods of genetic manipulations, which are used after the prenatal diagnosis is given enough knowledge of conditions of embryo.

9. Give an account of genetic diseases associated with the humans that are identified by prenatal diagnosis.

10. What is meant by benign trait genetic manipulations? Write in brief how are these manipulations achieved in the humans after prenatal diagnosis.

11. How does fetal prenatal determination lead to the genetic enhancement, and add a note on its social impact on the people?

12. Write about the importance of prenatal diagnosis and genetic manipulations in the genetic information rich world.

13. Explain briefly why the consideration of ethical issues is important before carrying out the genetic manipulations in case of humans.

14. List out different steps involved in the genetic manipulations of germline cells of the humans and its protocol.

15. Elucidate the future aspects of genetic manipulations and germline therapy adopted in the human beings.

16. Distinguish between the somatic genetic manipulations and germline genetic manipulations, and discuss their ethical issues.

17. Discuss how chorion villus sampling is helpful in the prenatal diagnosis and genetic manipulations. Add a note on its use in eugenics.

18. What is meant by BABI? Explain its applications in genetic manipulations, and ethical issues raised by this technique.

19. Explain how technological advancement in prenatal identification, sex selected and abortion prenatal diagnosis and genetic manipulations?

20. Critically comment on the ethical, social and legal issues raised by the fetal sex determination.

21. Give an account of maternal serum alpha-feta-protein (MSAFP) analysis, and ethical issues raised by this analysis.

22. Describe how amniocentesis is helpful in the prenatal diagnosis and genetic manipulations, and its use in eugenics.

5

CHAPTER

Ethical, Legal and Social Implications of Human Genome Project

5.1. INTRODUCTION

Human genome project (HGP) was completed well ahead of its scheduled time in 2003. Since its inception in 1990, it was planned meticulously and executed with the phases of development in molecular biology, genetic engineering, and other fields of biology and medicine. The HGP was coordinated by the US Department of Energy and the National Institutes of Health. During the early years of the HGP, additional contributions came from Japan, France, Germany, China, and others countries.

Upon realising the importance, many countries around the world started research on human genome. The major programs are in Australia, Brazil, Canada, China, Denmark, European Union, France, Germany, Israel, Italy, Japan, Korea, Mexico, Netherlands, Russia, Sweden, United Kingdom, and the United States. Some developing countries also participated through studies of molecular biology techniques for genome research and studies of organisms that are particularly interesting to their geographical regions. The human genome organization (HUGO) helped to coordinate international collaboration in the genome project. With the mapping of the human genome and increasing interest in genetic testing and therapies, the potential for ethical problems has increased. Nursing and medicine have the ethical responsibility to 'do no harm' and to protect the privacy of clients. However, our clients may also be family members and descendants of those we care for. The goals of human genome project were to:

- *identify* all the 20,000–25,000 genes in human DNA;
- *determine* the sequences of the 3 billion chemical base pairs that make-up human DNA;
- *store* this information in databases;
- *improve* the tools for data analysis;
- *transfer* related technologies to the private sector; and
- *address* the ethical, legal, and social issues (ELSI) that may arise from the project.

When the human genome project was being established, concerns were raised about how the new genetic information would be used, and how individuals and society could be protected from the possible harm. The planners of the human genome project recognized that the information gained from mapping and sequencing of the human genome would have profound implications for individuals, families and society. While this information would have the potential to dramatically improve the human health, they also realized that it would raise a number of complex ethical, legal and social issues. How should this new genetic information be interpreted and used? Who should have access to it? How can the people be protected from the harm that might result from its improper disclosure or use?

The ethical, legal, and social implications (ELSI) of Human Genetics Research program was developed to examine these issues and assist in the development of policy recommendations and guidelines to ensure that genetic information is used appropriately. The US Department of Energy (DOE) and the National Institutes of Health (NIH) devoted 3 percent to 5 percent of their annual human genome program budgets towards studying the ethical, legal, and social issues surrounding the availability of genetic information. This represents the world's largest bioethics program. It has become a model for ELSI programs around the world.

To accomplish this, the ELSI program was charged with:

- developing a program which could help in understanding the implications of the human genome project.
- identifying and defining the major issues of concern, and developing initial policy options to address them.

Now in its sixth year, the ELSI program at the National Center for Human Genome Research (NHGRI) has not only met these goals, but has moved beyond them to set the agenda for future work in this area. A research and education program has been established to examine the ethical, legal, and social implications of the genome research, and a number of workshops and policy fora have been held to identify the most urgent ELSI issues and develop appropriate policy recommendations.

To accomplish its goals, the NHGRI ELSI program has established a number of functional components to address specific areas, including a joint NIH/DOE working group to analyze critical issues and provide guidance to the NIH and DOE; an extramural ELSI branch responsible for supporting research studies and education projects addressing ELSI issues; an Office of Policy Coordination to address specific policy concerns raised by the HGP; and a newly formed Intramural Genome Ethics Office that will provide advice to the intramural research program.

A vital component of the ELSI program is the extramural research effort, which has funded over 125 research and education projects and related activities. These projects have resulted in the rapid increase in understanding of ELSI issues, the publication of over 150 journal articles and books, the development of education programs aimed at health professionals and the general public, and the establishment of policy recommendations on issues ranging from the use of genetic tests to preventing discrimination based on genetic information.

5.2. ETHICAL, LEGAL AND SOCIAL IMPLICATIONS (ELSI) AND ITS ORIGIN

The ELSI program was designed to provide a new approach to the scientific research by identifying, analyzing and addressing the ethical, legal and social implications of human genetics research at the same time when the basic scientific issues were being studied. In this way, problem areas can be identified and solutions can be developed before the scientific information gained is integrated into health care practice.

The ELSI program is viewed as essential to the success of the genome project in the United States and is supported by federal HGP funds. The national institutes of health's 'National Human Genome Research Institute (NHGRI)' has committed 5 percent of its annual research budget to study ELSI issues. The US Department of

Energy Office of Energy Research, NHGRI's partner in the US Human Genome Project, also reserves a portion of its funding for ELSI research and education.

5.3. ELSI ESTABLISHMENT

During its January 1989 meeting, the program advisory committee on the human genome established a working group on ethics to develop a plan for the ELSI component of the human genome program. This working group(ELSI Working Group-ELSIWG), later named as the 'National Institutes of Health-Department of Energy Joint Working Group on Ethical, Legal, and Social Implications of Human Genome Research, held its first meeting in September 1989. In January, 1990, the working group issued its first report. In the report, the group agreed that the purpose of the ELSI program should be to:

- anticipate and address the implications of mapping and sequencing the human genome for the individuals and society;
- examine the ethical, legal, and social consequences of mapping and sequencing the human genome;
- stimulate public discussion of the issues; and
- develop policy options that would assure that the information is used for the benefit of individuals and society.

The working group envisioned a program that would anticipate potential problems before they actually occurred, and identify possible solutions for the problems. It suggested a number of means for accomplishing these goals. Specifically, it encouraged the research community to explore and gather the data on a wide range of issues pertinent to the human genome program that could be used to develop educational programs, policy recommendations or possible legislative solutions. A number of areas for focus were identified, including fairness in the use of genetic information; the impact of the knowledge of genetic variation on individuals; and privacy and confidentiality of genetic information, to name a few.

In 1990, in response to the working group's report, the NHGRI established the ELSI branch (later renamed the **ELSI Research Program**) in its Division of Extramural Research and the DOE established an **ELSI program** in its Office of Energy Research. Since the beginning, these two programs have collaborated closely, including the joint support of the ELSI working group, the development of complementary research priority areas, and the cofunding of ELSI activities of mutual interest.

ELSI Program at NHGRI

The NHGRI uses a variety of mechanisms to carry out its ELSI research, policy and education activities. It supports the acquisition of knowledge through the funding of basic and applied research grants. It disseminates information through professional and public education activities, and assists in the development of policy recommendations through the sponsorship of workshops and policy conferences. These activities are organized and implemented by three organizational units.

The ELSI Research Program in the Division of Extramural Research (ELSI/DER) was established in 1990, and is responsible for funding and managing research grants and education projects that examine ELSI issues at institutions throughout the United States. It also supports workshops, research consortia, and policy conferences related to funded research and education projects.

Established in 1995, the Office of Policy Coordination (OPC) in the office of the director provides information and analysis on ELSI policy and legislative issues, and sponsors workshops and conferences, which assist in the development of policy options and recommendations related to ELSI issues.

The Office of Genome Ethics (OGE) in the Division of Intramural Research was established in 1996 to assist genome researchers involved in the NHGRI intramural program in identifying and addressing the ethical issues arising from the genome research.

Each of these offices focus on a different aspect of the ethical, legal and social implications of the human genome project, and contribute to the overall strength of the NHGRI ELSI program.

The ELSI research program is essential for the success of HGP in the United States, and is supported with federal funds. The National Human Genome Research Institute (NHGRI) commits more than $18 million annually from its HGP budget to ELSI research, making it the largest supporter of research in the ethical, legal and social implications of genetics research. The United States Department of Energy (DOE) 'Office of Energy Research', which partners with NHGRI in the HGP, also reserves a portion of its funding for ELSI research and education.

The human genome project has generated widespread interest in a large spectrum of questions regarding the ethical, legal, and social implications of the existence and use of human genetic sequences. Many groups concur that sequencing of the human genome does not raise ethical questions. Questions center on how the information is used, and its implications for individuals, society and the human species.

It is generally accepted that the questions raised by the accessibility of the information generated by the human genome project are not new. Rather, the volume, variety, and the ease with which such information can be obtained, and the vast number of people affected, make it imperative that the possible use or misuse of the project's findings be anticipated and addressed.

The ability to generate any individual's genetic profile raises important questions of privacy, confidentiality, ownership and autonomy. How should information be protected? Who should have access to the information and under what circumstances? What rights, if any, do employers, insurers, and family members have to an individual's genetic information?

In addition, the potential trend towards understanding the diseases or traits in genetic terms may raise questions about human nature and our fundamental beliefs about ourselves and our species. Will we view an individual as merely a product of interacting genes? How will we define normalcy, abnormalcy, or disability?

The recognition of the importance of such questions, and the increased dialogue surrounding them has resulted in an international commitment, which addresses issues of ethical, legal, and social issues. Many nations and organizations participating in the human genome project have allocated funding and/or assembled committees to address such concerns. It is likely that these initial efforts will increase, as the national programs and the number of countries involved expands.

Below is a sample of the literature on social, political, and ethical aspects of the human genome project, and a brief listing of references on human genome mapping.

Societal Concerns Arising from the Human Genome Project

• Fairness in the use of genetic information by insurers, employers, courts, schools, adoption agencies, and the military, among others.

• *Who should have access to personal genetic information, and how will it be used?*

Privacy and confidentiality of the genetic information.

• *Who would own and control the genetic information?*

Psychological impact and stigmatization due to an individual's genetic differences.

• *How does personal genetic information affect an individual, and society's perceptions of that individual?*

• *How does genomic information affect members of minority communities?*

Reproductive issues including adequate informed consent for complex and potentially controversial procedures, use of genetic information in reproductive decision-making, and reproductive rights.

- *Do healthcare personnel properly counsel parents about the risks and limitations of genetic technology?*
- *How reliable and useful is fetal genetic testing?*
- *What are the larger societal issues raised by new reproductive technologies?*

Clinical issues including the education of doctors and other health service providers, patients, and the general public in genetic capabilities; scientific limitations, and social risks; and implementation of standards and quality-control measures in testing procedures.

- *How will genetic tests be evaluated and regulated for accuracy, reliability, and utility? (Currently, there is little regulation at the federal level).*
- *How do we prepare healthcare professionals for the new genetics?*
- *How do we prepare the public to make informed choices?*
- *How do we, as a society, balance current scientific limitations and social risk with long-term benefits?*

Uncertainties associated with gene tests for susceptibilities and complex conditions (e.g., heart disease) linked to multiple genes and gene-environment interactions.

- *Should testing be performed when no treatment is available?*
- *Should parents have the right to have their minor children tested for adult-onset diseases? Are genetic tests reliable and interpretable by the medical community?*

Conceptual and philosophical implications regarding human responsibility, free will vs genetic determinism, and concepts of health and disease.

- *Do people's genes make them behave in a particular way?*
- *Can people always control their behavior?*

- *What is considered acceptable diversity?*
- *Where is the line between the medical treatment and enhancement?*

Health and environmental issues concerning genetically modified foods (GM) and microbes.

- *Are GM foods and other products safe for the humans and the environment?*
- *How will these technologies affect developing nations' dependence on the West?*

Commercialization of products including property rights (patents, copyrights, and trade secrets), and accessibility of data and materials.

- *Who owns genes and other pieces of DNA?*
- *Will patenting DNA sequences limit their accessibility and development?*

Rapid advances in the science of genetics and its applications have presented new and complex ethical and policy issues for individuals and society. ELSI programs that identify and address these implications have been an integral part of the US HGP since its inception. These programs have resulted in a body that promotes education, and guides the conduct of genetic research and the development of related medical and public policies.

A continuing challenge is to safeguard the privacy of individuals and groups who contribute DNA samples for large-scale sequence-variation studies. Other concerns have been to anticipate how the resulting data may affect concepts of race and ethnicity; identify potential uses (or misuses) of genetic data in workplaces, schools, and courts; identify commercial uses; and foresee impacts of genetic advances on the concepts of humanity and personal responsibility.

5.4. ELSI RESEARCH GOALS

- Examine issues surrounding the completion of the human DNA sequence, and the study of human genetic variation.

- Examine issues raised by the integration of genetic technologies and information into health care and public health activities.
- Examine issues raised by the integration of knowledge about genomics and gene-environment interactions in non-clinical settings.
- Explore how new genetic knowledge may interact with a variety of philosophical, theological, and ethical perspectives.
- Explore how racial, ethnic, and socioeconomic factors affect the use, understanding, and interpretation of genetic information; the use of genetic services; and the development of policy.

Genetics Privacy and Legislation

Policy History

No federal legislation has been passed relating to genetic discrimination in individual insurance coverage, or genetic discrimination in the work-place. Several bills were introduced during the last decade. Some of these bills attempted to amend the existing civil rights and labor laws, while others stood alone. The primary public concerns are that:

- insurers will use genetic information to deny, limit, or cancel the insurance policies; and
- employers will use genetic information against existing workers or to screen potential employees. Since DNA samples can be held indefinitely, there is the added threat that the samples will be used for purposes other than those for which they were collected.

Executive Order Protecting Federal Employe

On February 8, 2000, US President, Clinton signed an *executive order* prohibiting every federal department and agency from using genetic information in any hiring or promotion action. This executive order, endorsed by the American Medical Association, the American College of Medical Genetics, the National Society of Genetic Counselors, and the Genetic Alliance:

- *prohibits* federal employers from requiring genetic tests as a condition of being hired. Employers cannot request or require the employees to undergo genetic tests in order to evaluate an employee's ability to perform his or her job;
- *prohibits* federal employers from using protected genetic information to classify employees in a manner that deprives them of future opportunities. Employers cannot deny employees promotions or overseas posts to the employees because of their genetic predisposition for certain illnesses; and
- *provides* strong privacy protections to any genetic information used for medical treatment and research. Under the EO, obtaining or disclosing genetic information about employees is prohibited, except when it is necessary to provide medical treatment for the employees to ensure workplace health and safety, or provide occupational and health researchers access to data. In every case where genetic information about employees is obtained, it will be subjected to all Federal and state privacy protections.

Bills Introduced to US Congress

The following bills have been referred to Congressional committees:

- H.R. 3636—Genetic Privacy and Nondiscrimination Act of 2003. Introduced to the House of Representatives, November 21, 2003.
- S. 1053—Genetic Information Nondiscrimination Act of 2003—Introduced to the Senate, May 13, 2003.
- H.R. 1910—To prohibit discrimination on the basis of genetic information with respect to health insurance. Introduced to the House of Representatives, May 1, 2003.
- S. 16—Equal Rights and Equal Dignity for Americans Act of 2003—To protect the civil rights of all Americans, and for other purposes. Introduced to the Senate, January 7, 2003.

- Previous Bills (No longer candidates for law).
  - — S.1995: Genetic Information Nondiscrimination Act of 2002.
  - — S.318: Genetic Nondiscrimination in Health Insurance and Employment Act 2001.
  - — S. 382: Genetic Information Nondiscrimination in Health Insurance Act of 2001.
  - — H.R. 602: Genetic Nondiscrimination in Health Insurance and Employment Act 2001.

Existing Federal Anti-Discrimination Laws and how they Apply to Genetics

Although no specific federal genetic nondiscrimination legislation has been enacted, some believe that parts of existing nondiscrimination laws could be interpreted to include genetic discrimination. Here is a brief overview of these laws and how they apply to genetics.

Americans with Disabilities Act of 1990 (ADA)

The most likely current source of protection against genetic discrimination in the workplace is provided by laws prohibiting discrimination based on disability. Title I of the Americans with Disabilities Act (ADA), enforced by the Equal Employment Opportunity Commission (EEOC) and similar disability-based antidiscrimination laws such as the Rehabilitation Act of 1973, does not explicitly address genetic information, but provides some protection against disability-related genetic discrimination in the workplace. These are as follows:

- Prohibits discrimination against a person who is regarded as having a disability.
- Protects individuals with symptomatic genetic disabilities same as the individuals with other disabilities.

- Does not protect against discrimination based on unexpressed genetic conditions.
- Does not protect potential workers from requirements or requests to provide genetic information to their employers after a conditional offer of employment has been extended, but before they begin work (Note: this is a heightened concern because genetic samples can be stored).
- Does not protect workers from requirements to provide medical information that is job related and consistent with business necessity.

In March 1995, the EEOC issued an interpretation of the ADA. The guidance, however, is limited in scope and has legal effect. It is policy guidance that does not have the same legal binding effect on a court as a statute or regulation and has not been tested in court. According to the interpretation:

- Entities that discriminate on the basis of genetic predisposition are regarding the individuals as having impairments, and such individuals are covered by the ADA.
- Unaffected carriers of recessive and X-linked disorders, and individuals with late-onset of genetic disorders, who may be identified through genetic testing or family history as being at high risk of developing the disease, are not covered by the ADA.

Health Insurance Portability and Accountability Act of 1996 (HIPAA)

The Health Insurance Portability and Accountability Act (HIPAA) applies only to employer-based and commercially issued group health insurance only. It is the only federal law that directly addresses the issue of genetic discrimination. There is no similar law applying to private individuals seeking health insurance in the individual market. The law:

- prohibits group health plans from using any health status-related factor, including genetic information, as a basis for denying or limiting eligibility for coverage or for charging an individual more for coverage;

- limits exclusions for preexisting conditions in group health plans to 12 months, and prohibits such exclusions if the individual has been covered previously for that condition for 12 months or more;

- states explicitly that genetic information in the absence of a current diagnosis of illness shall not be considered a preexisting condition; and

- doesn't prohibit the employers from refusing to offer health coverage as a part of their benefits packages.

HIPAA National Standards to Protect Patients' Personal Medical Records

This regulation would protect medical records and other personal health information of the patients maintained by the health care providers, hospitals, health plans and health insurers, and health care clearinghouses. The regulation was mandated when Congress failed to pass comprehensive privacy legislation (as required by HIPAA) by 1999. The new standards limit the nonconsensual use and release of private health information; give rights to the patients to access their medical records and to know who else has accessed them; restrict the disclosure of health information to the minimum needed for the intended purpose; establish new criminal and civil sanctions for improper use or disclosure; and establish new requirements for access to records by researchers and others. They are not specific to genetics, rather they are sweeping regulations governing all personal health information.

Title VII of the Civil Rights Act of 1964

An argument could be made that genetic discrimination based on racially or ethnically linked genetic disorders constitutes unlawful race or ethnicity discrimination.

- Protection is available only when an employer engages in discrimination based on a genetic trait that is substantially related to a particular race or ethnic group.
- A strong relationship between race or national origin has been established only for a few diseases.

Minorities, Race and Genomics

In 2004, DOE sponsored a *Nature Genetics* supplement called 'Genetics for the Human Race'. This supplement originated from a May 2003 workshkop held by the National Human Genome Center at Howard University in Washington DC. The workshop, 'Human Genome Variation and Race', and this special issue of Nature Genetics were proposed by scientists at Howard University, and financially supported by the genome programs of the US Department of Energy through its Office of Science; the Irving Harris Foundation; the National Institutes of Health, through the National Human Genome Research Institute; and Howard University. The supplement contains articles based on the presentations at this workshop.

In 2002, DOE sponsored the production of a 24-page genomics insert in the Journal for Minority Medical Students. Distributed in all volumes of its spring issue, the supplement covers basic genetics, Zeta Phi Beta's genetics education program, the genetics of sickle-cell anemia, medical genetics, and ELSI issues.

Another ELSI project was with the National Educational Foundation of Zeta Phi Beta Sorority, Inc. The sorority planned and conducted major informational conferences on the Human Genome Project and its impact on minority communities. The conferences covered a variety of topics from basic genetics and HGP history to gene testing and careers in genetics. Held in Chicago in 2003, Atlanta and Washington, DC in 2001, Pennsylvania Legislative Black Caucus in 2001, Philadelphia in 2000, and New Orleans in 1999, the conferences have sparked numerous follow-up meetings and training sessions led by members of the educational foundation.

About 150 leaders of minority communities came together in June, 1997, at the University of Maryland, Baltimore (UMB) to learn about the human genome project. The goals of the meetings sponsored by DOE HGP were to:

- inform minority communities about the human genome project by explaining its potential benefits and clarifying its possible ELSI implications; and
- make the aspirations and interests of these communities known to the scientists involved in genome project and policymakers.

This program grew from organizers' concerns about an information vacuum among minorities regarding the genome project and the possibility that suspicions would arise about the project's intent. Proceedings from this conference were combined with an additional essay to form the book, *The Human Genome Project and Minority Communities: Ethical, Social and Political Dilemmas*, edited by Raymond Zilinskas (Monterey Institute of International Studies) and Peter Balint (University of Maryland).

The ELSI Research Program is essential to the success of the HGP in the United States and is supported with federal funds. The National Human Genome Research Institute (NHGRI) commits more than $14 million annually from its HGP budget to ELSI research, making it the largest supporter nationwide of research into the ethical, legal and social implications of genetic research. The United States Department of Energy Office of Energy Research, who partners with NHGRI in the HGP, also reserves a portion of its funding for ELSI research and education.

ELSI Working Group

The ELSI Research Planning and Evaluation Group (ERPEG) Final Report.

In July 1997, NHGRI's National Advisory Council for Human Genome Research (NACHGR) and DOE's Biological and Environmental Research Advisory Committee (BERAC) established the ELSI Research Planning and Evaluation Group (ERPEG). ERPEG's mission was to review and analyze the portfolio of ELSI research grants at both NHGRI and DOE, to participate in the development of the ELSI component of the new five-year plan for the HGP and to prepare a report for submission to NACHGR and BERAC summarizing its findings and making recommendations.

After performing an initial analysis of the ELSI research grant portfolio, ERPEG developed five goals to guide ELSI research programs over the next five years. These goals were designed both to emphasize areas of research of ongoing importance and identify emerging issues requiring additional attention. The goals were seen by ERPEG as a work in progress that should be regularly updated and expanded. Following the publication of these goals (as part of the HGP five-year plan) in the October 1998 issue of *Science*, ERPEG continued its analysis of the ELSI portfolio.

In February 2000, ERPEG released its final report, "A Review and Analysis of the ELSI Research Programs at NIH and DOE." This document summarizes the findings of ERPEG's extended analysis and provides recommendations for strengthening the ELSI research programs at NHGRI and DOE. With publication of its final report in 2000, ERPEG concluded its mission.

ELSI Research Advisors Group

ELSI research activities at NHGRI are now reviewed by the ELSI Research Advisors Group (ERA), established in 2000 as a working group of the National Advisory Council for Human Genome Research (NACHGR). The ERA was created to advise NHGRI on the progress and goals of the ELSI research program and to help identify emerging issues and implications of genomic research.

The NHGRI ELSI Research Program is organized around four program areas as mentioned earlier.

The first five-year goals of the ELSI program have also been met. A research and education program has been established which is designed to examine the ethical, legal and social implications surrounding the Human Genome Project and the most urgent issues of concern have been identified. However, while a number of policy options have been developed and implemented, it has become clear that some of the issues that have been identified are highly complex and controversial. Reaching consensus about such issues is not easy. Some of the issues are of such magnitude as to require substantial changes in our society's institutions (including health care and insurance systems) in order for this new information to be put to its optimal use.

5.5. DETAILS OF ELSI PROGRAM

Goal 1: *Develop programs addressed at understanding the ethical, legal and social implications of the human genome project.*

Over the past five years, the NHGRI has utilized a variety of mechanisms to identify, analyze, and address the ethical, legal, and social issues surrounding the HGP. It has accomplished this by the acquisition of new knowledge through the funding of basic and applied research grants and education programs and the sponsorship of workshops and policy fora. NHGRI supports several components to accomplish its identified ELSI goals. They include the NIH/DOE Joint Working Group on Ethical, Legal, and Social Implications (ELSI Working Group), the Extramural ELSI Branch, the Office of Policy Coordination (OPC), and the Intramural Genome Ethics Office.

Formed in 1989, the ELSI Working Group has provided overall guidance to the NHGRI and OHER ELSI programs; facilitated a number of early policy discussions; and participated in the development of a number of policy options and recommendations related to these issues. The working group has also spun off two task forces aimed at analyzing and developing recommendations about:

* genetic information and health insurance; and
* genetic testing.

The task force on genetic information and insurance, whose report was published in 1994, was charged with examining genetic information and health insurance. The Task Force's recommendations have been utilized in a number of legislative proposals related to health care reform. The Task Force on Genetic Testing, whose report is due in early 1997, is examining the issues surrounding the quality and marketing of genetic tests. A document summarizing a set of interim principles has recently been released for public comment (March, 1996).

* The extramural ELSI branch, which was established in 1990, is responsible for supporting studies designed to examine ELSI issues. It funds and manages research grants and education projects at institutions throughout the United States, and supports workshops, research consortia, and policy fora related to funded research and education projects. Established in 1995, the Office of Policy Coordination (OPC) provides information and analysis on ELSI policy issues, directly sponsors policy fora, and supports the work of the ELSI Working Group. The NHGRI's Intramural Ethics program is now being established to assist intramural genome researchers to identify and address ethical issues arising from genome research.

* The ELSI program budget has increased from 3 percent in fiscal year (FY) 1990 ($1.5 million) to an average of 5.1 percent in fiscal years 92–95 ($6.3 million in FY 1995). This contribution currently represents the largest single investment in bioethical research from any one source.

- The ELSI program has also established collaborative and co-funding relationships with a number of other NIH Institutes, such as the National Cancer Institute (NCI), the National Institute of Child Health and Human Development (NICHD), the National Institute of Mental Health (NIMH), and the National Institute of Nursing Research (NINR). It has also formed strong cooperative relationships with a number of other federal agencies, such as the Department of Energy (DOE), the National Centers for Disease Control and Prevention (CDC), the Health Resources Services Administration (HRSA), the National Science Foundation (NSF), and the Food and Drug Administration (FDA), to name a few. Further signs of the ELSI program's acceptance and success are the establishment of similar programs in other countries participating in the human genome project (e.g., Canada, UK, Japan); its cooperation with many other NIH components and agencies who have sought guidance from the ELSI program (e.g., National Institute of General Medical Sciences (NIGMS), NCI, NIMH, CDC); and the recent co-sponsorship of a number of ELSI program initiatives, including ELSI's revised Program Announcement (NIMH and NINR).

Goal 2: *Identify and define the major issues, and develop initial policy options to address them.*

- The ELSI program has developed a vital research program at the NIH to identify and define the ethical, legal, and social implications surrounding the HGP. The initial Program announcement, published in 1990, casted a very broad net to accomplish this goal. The original issues identified in the ELSI program announcement were: questions of fairness in the use of genetic information; the impact of genetic information on individuals; privacy and confidentiality of genetic information; the impact of the HGP on genetic counseling; the impact of genetic information on reproductive decision-making; the impact of genetic information on medical practice; the uses and misuses of genetics in the past; questions of commercialization; and conceptual and philosophical implications raised by the HGP.

- Since its existence, the number and type of research and education applications being submitted to the ELSI branch has grown significantly (from just over 20 applications in the first year to an average of 75 applications per year after that). Since the beginning of the program, of those applications accepted, the ELSI branch has been able to fund between 20 and 25 percent. The investigators seeking funding through the ELSI program come from a wide range of disciplines—from bioethics and philosophy to psychology, sociology, medicine, and law. Forty-six per cent of principal investigators currently funded by ELSI are women, and a substantial number are young investigators or new applicants to the National Institutes of Health.

- As a result of the ELSI branch's initial program announcement, a number of research and education projects designed to address a set of highly diverse issues were funded. The diversity and sophistication of the extramural research and education program has increased significantly over the past five years; 126 research and education projects and related activities have been funded. These efforts have resulted in the accrual of empiric information about ethical, legal, and social issues surrounding the genetics research, and the discovery and use of genetic information; an assessment of the impact of knowledge and attitudes about genetic information; and the impact of the increasing clinical integration of these new technologies. During this period, four high priority areas have emerged. These include:

- Privacy and fairness in the use and interpretation of genetic information.
- Clinical integration of new genetic technologies.
- Issues surrounding genetics research.
- Public and professional education.

As a result of the evolution and maturation of these four high priority areas, the ELSI program has now more clearly articulated its highest priority research and policy areas for the next several years, and has recently released a revised program announcement. Although much work has already occurred in the above-identified areas, it is anticipated that the revised statement of priorities will serve to focus future program activities even further.

High Priority Program Areas

Privacy and Fairness in the use and Interpretation of Genetic Information

Genetic information is being discovered at an increasingly remarkable pace. In some cases, however, not enough is known about the meaning of genetic information or how it is being interpreted and used. The risk of genetic discrimination grows as new disease susceptibility genes are identified. Genetic research results are attracting a great deal of attention, yet often, information that is presented about discoveries is confusing, misleading, or factually incorrect. As more genetic factors are identified as cause and predictors of human diseases, a new framework for defining the meaning of health, disease and predisposition to disease may need to be developed.

Clinical Integration of New Genetic Technologies

As a result of the human genome project, genetic technologies have developed at a rapid rate. Defective genes are being located and their functions are beginning to be understood at an increasingly extraordinary pace. Scientists are searching for genes that cause diseases; identifying people, who are at increased risk for having children with a genetic disorder; and helping distinguish those individuals, who are at increased risk for developing a certain disease, from those who are not.

As new genetic technologies are moving rapidly from research into the clinical practice arena, concerns have been raised that in some cases, much is not known about the impact of these findings on people's lives and health. Furthermore, questions have arisen as to whether health professionals have been adequately educated about genetics, genetic technologies and the ethical, legal, and social implications surrounding their use, in order to optimally provide genetic services to their patients. The NHGRI, along with the National Institute of Child Health and Human Development and the National Institute of Nursing Research released an RFA "Studies of Testing and Counseling for Cystic Fibrosis Mutations". Its objective was to solicit applications to examine alternative approaches for genetics education, testing, and counseling in order to help establish practices that improve professional interpretation and patient understanding of CF testing.

As a result of this initiative, eight research projects were funded to examine these issues and to gain actual experience in providing genetic testing for cystic fibrosis mutations in a research environment. The findings of these studies suggested that interest in testing for cystic fibrosis mutations was much lower than had been expected in the general population. Investigators also discovered that although there was limited knowledge about genetics and cystic fibrosis in the population, it was possible to develop a variety of satisfactory alternative education strategies (e.g., videos and brochures) for testing. Furthermore, the investigators saw no evidence of undue anxiety in most individuals tested. The results of these projects have been published in the literature and presented at several professional and scientific meetings. An article summarizing the themes, which emerged from the studies, has been submitted for publication. NHGRI is now exploring whether the empiric information gained by these projects can be used to inform policy development regarding genetic testing and reproductive risk counseling through an NIH consensus development conference.

A second major initiative was undertaken in 1994 by the NHGRI in anticipation of the discovery of a number of cancer predisposing genes. This initiative (also an RFA) was co-sponsored by the National Cancer Institute, the National Institute of Mental Health and the National Institute of Nursing Research. It solicited applications for studies designed to examine the psychosocial and clinical impact of using gene-based diagnostic tests in families with heritable forms of breast, ovarian, and colon cancer to identify those individuals who have an increased risk of developing cancer, and those who do not. Knowledge and attitudes about genetic testing for cancer risks are also being assessed, and information is being gathered to establish clinical protocols for the optimum use of these risk assessment technologies in the future. These projects are now in their second year, and preliminary results are becoming available. Once completed, these projects will provide valuable experience-based guidance for genetic testing of cancer susceptibility genes.

Issues Surrounding Genetics Research

Genetics research may result in the discovery of information that is powerful and potentially predictive. In addition, such information may have familial implications. While, in some cases, such information may be beneficial to research subjects and their families, there is also potential for misinterpretation or misuse. Special concerns have arisen about the process of informed consent, particularly when the risks and benefits of research participation may not be fully known. Concerns have also arisen about how to prevent the preliminary or premature release of research results, and protect the privacy of individuals who choose to participate in genetics research. As a result, a third priority area identified for the ELSI program pertains to issues surrounding the conduct of genetics research. Examination of existing research guidelines and recommendations over the past five years has revealed that current guidance and protections need to be enhanced in order to deal with the special considerations related to genetics research.

Public and Professional Education

It has become clear that general population and most health professionals are not knowledgeable about genetics, genetic technologies, and the possible ethical, legal, and social implications of having genetic information. This has become more apparent, as the results of a number of ELSI-funded surveys have become available. The new information generated by the human genome project and human genetics research is changing biomedical research, the practice of medicine, and public perceptions about genetic information and technologies. It is imperative that public have an adequate knowledge of the meaning of newly discovered genetic information. It is also essential that our nation's health professionals have the knowledge, skills, and resources to effectively utilize this new knowledge and technology for the diagnosis, prevention, and treatment of disease. Without well-informed health professionals and general public, the advances in genetics research will not be fully realized. Thus, a fourth high priority area that has been identified for the ELSI program is public and professional education. In genetics, program (U of Virginia-Fletcher) is designed to educate appellate judges and journalists about the Human Genome Project and its implications for the future. During the course of this project, an integrated textbook, a casebook, and a teaching manual appropriate for each group is developed, and educational workshops are being provided.

A fourth project funded (MCET-Texter) is designed to develop and field-test a semester-long high school course to be disseminated through a public broadcasting network using multiple telecommunications networks, including satellite, computer, audio, and print materials.

One last example of an educational resource, which the ELSI program has funded, is a grant (George-town University-Walters) that supports the further development of two databases of the ELSI literature. The first database, bioethics line, which is a part of medline, is an annotated bibliographic listing of the ethics literature. The second, ETHX, is a broader bibliography containing all of the bioethics line citations, in addition to citations from the popular literature. The number of Medline articles published annually on ethics and genetics topics has nearly tripled over the past ten years. The number of genetics citations in the ETHX database has nearly doubled since its inception in 1988. These databases are now easily accessible by calling (800)633-ETHX, or by visiting the website.

Ethical, legal and social implications of human genome project research can be summarised as follows:

- Fairness in the use of genetic information
- Privacy and confidentiality
- Psychological impact and stigmatization
- Genetic testing
- Reproductive issues
- Education, standards, and quality control
- Commercialization
- Conceptual and philosophical implications

ELSI: Fairness and Privacy

- How much informations should be disclosed?
- Who should have access to one's genetic information?
- To whom should be the responsibilities given—government agencies or private organizations?
- How should it be used by researchers and medical professionals?
- Who owns and controls it?
- Consider these would-be owners:
  - Insurers, employers, courts, schools, adoption agencies, the military.

ELSI: Psychological Stigmatization

- How does knowing one's predisposition to disease affect an individual?
- What are the reactions of adults to their age old parents or children?
- Is it good to know the disease, for which the treatment is not available?
- How does it affect friends' and family's perceptions of that individual?
- How does it affect society's perceptions of that individual?

ELSI: Genetic Testing

- Should testing be done only when there is a family history to make informed prenatal/diagnosis decisions?
- Should population screening be done (new-born, premarital, and occupational)?
- Should testing be done when no treatment is available?
- Confidentiality levels.
- Exploitation by the insurer, employer, army, social organizations, high profile jobs.
- Should testing be done for susceptibility genes?
- Role of environmental factors in disease development.
- Should parents have the right to have their minor children tested for adult-onset diseases?
- Are current genetic tests reliable and interpretable by the medical community?

ELSI: Reproductive Issues

- Informed consent for procedures.
- Use of genetic information in decision-making.
- Reproductive rights.
- Increased abortions.
- Unethical social trends.

ELSI: Conceptual and Philosophical Implications

- Human responsibility.
- Free will versus genetic determinism.
- Concepts of disease and health.

ELSI: Commercialization

- Property rights.
- Patents, copyrights and trade secrets.
- Accessibility of data and materials.

ELSI: Clinical Issues

- Education of health service providers, patients, and the general public.
- Implementation of standards and quality control measures in testing procedures.
- Change in the patient care, expenditures and other costs involved activities.

SUMMARY

The ethical, legal and social implications of Human Genetics Research (ELSI) program was developed to examine these issues and assist in the development of policy recommendations and guidelines to ensure that genetic information is used appropriately. The U.S. Department of Energy (DOE) and the National Institutes of Health (NIH) devoted 3% to 5% of their annual Human Genome Program budgets toward studying the ethical, legal, and social issues (ELSI) surrounding availability of genetic information. This represents the world's largest bioethics program. It has become a model for ELSI programs around the world. A vital component of the ELSI program is the extramural research effort, which has funded over 125 research and education projects and related activities. These projects have resulted in the rapid increase in understanding of ELSI issues, the publication of over 150 journal articles and books, the development of education programs aimed at health professionals and the general public and the establishment of policy recommendations on issues ranging from the use of genetic tests to preventing discrimination based on genetic information.

The U.S. Human Genome Project (HGP) recognized that the information gained from mapping and sequencing the human genome would have profound implications for individuals, families, and society. While this information would have the potential to dramatically improve human health, they realized that it would also raise a number of complex ethical, legal and social issues. How should this new genetic information be interpreted and used? Who should have access to it? How can people be protected from the harm that might result from its improper disclosure or use?

As the ELSI Program has evolved, four high priority areas of focus for research and policy activities have emerged. They are privacy and fairness in the use and interpretation of genetic information; activities in this area examine the meaning of genetic information and how to prevent its misinterpretation or misuse, clinical integration of new genetic technologies. These activities examine the impact of genetic testing on individuals, families and society and inform clinical policies related to genetic testing and counseling. Issues surrounding genetics research and activities in this area focus on informed consent and other research ethics review issues related to the design, conduct, participation in and reporting of genetics research; and public and professional education. This area includes activities that provide education on genetics and related ELSI issues to health professionals, policy makers and the general public.

From the inception of ELSI from January 1989 up to the February 2000 a huge areas are covered under these researches encompassing almost all issues raised as the progress of the human genome project achieved. This has considered and to explore the ways in which new genetic knowledge may interact with a variety of philosophical, theological and ethical perspectives. It also involved in how to explore how socioeconomic factors and concepts of race and ethnicity influence the race. And to interpretations of genetic informations, the utilization of genetic services to the development of the society in the larger extent. ELSI also examined the issues raised by the integration of knowledge about genomics and gene environment interactions into non-clinical settings. These issues will be ever ending in nature as new and newer informations generated the newer ELSI issues will be created and that should be addressed with greater scientific and ethical considerations.

EXERCISE

1. Explain the importance of ethical, legal and social issues raised by the completion of human genome project.

2. Mention how the funds are initially planned for the ethical, legal and social issues studies in the various periods of human genome studies.

3. Describe the goals the ethical, legal and social issues addressed in the inception of the research programme by the human genome project.

4. Elucidate the societal concerns arising from the human genome project and its importance to the world new genetic information era.

5. Write a note on high priority area addressed by the ethical, legal and social issues in the present genomic information situations.

6. How work place discrimination starts with the human genome project completion and its understanding in fullest details?

7. Discuss in what way the insurance discrimination became a major problem after the human genome project work was completed.

8. Name different anti-discrimination acts that are present or passed to protect the human rights issues after the better understanding of the genomes of the human beings.

9. Critically comment on the advantages and disadvantages of knowing the detail informations of the human genome and its impact on the human society.

10. List out the ethical, legal and social issues of the eugenics that affect the future generations of human beings on the genetic information world.

11. Briefly explain the progress of ethical, legal and social issues in human genome project and its development made after the ELSI research programme.

12. Enumerate different organizations and programmes developed under the ethical, legal and social issues research programme.

13. Depict how ethical, legal and social issues address the genetic privacy in the human genome era.

14. Define genome and its importance to the organisms in the heredity of individual species in the ecosystem.

15. Explain how developed countries are affected by the genomic informations and their use in the genetic counseling?

16. Give a note on policies adopted for the use of genetic informations in the medical field and the education of patients for the use of their genetic private data.

17. Write notes on how the ethical, legal and social issues programme successful addressed the issues after completion of human genome project.

18. What are the issues raised by the integrations of the clinical medicine and new genetic informations?

19. Comment on the necessity of genetic information of an individual in the present day advanced biological information trends.

20. Mention the fields of biology that are benefited from the advancement made in the genetics and genomics.

21. Elaborate the need of genomics, proteomics and bioinformatics in the fast developing world of biotechnology.

22. Name the institutions that are contributed to study of ethical, legal and social issues concerned with the advanced biotechnological inventions.

Genetics Studies on Ethnic Races

6.1. INTRODUCTION

Humans are much more than simply the product of a genome, but in a sense we are, both collectively and individually, defined within the genome. The mapping, sequencing and analysis of the human genome is therefore a fundamental advance in self-knowledge; it will strike a personal chord with many people. The application of this knowledge will, in time, materially benefit almost everyone in the world.

After the successful completion of human genome sequencing, new genetic data enabled the researchers to re-examine the relationship between human genetic variations and races. With the genomic era in full swing, scientists are interested in cataloging genetic variation among indigenous groups and studying their susceptibility to chronic diseases for several reasons. Indian population or any other tribes in the world with remote access are a convenient sample in terms of geographic area, because they tend to be concentrated in areas such as reservations. American Indian populations are also very homogenous. Since there has been no influx of genes from other races and ethnic groups, scientists can study ancestral gene sequences in modern tribe people. They can also easily control genetic variation and study other factors, such as the role of environmental exposure in the American Indian tribes, as their genes are closely conserved.

The cytogenetic map, described on plots landmarks across the genome and anchors the physical map to the underlying chromosomal positions. A comparison of the genetic and physical map charts the rate of recombination, or exchange between each pair of 46 chromosomes, which occurs as the genome is passed on through generations. The International SNP (single nucleotide polymorphism) Map Working Group documents 1.42 million polymorphic sites in the genome, providing for the first time a variant in virtually every gene and in each genomic region. This map of human variation will facilitate efforts to reconstruct our evolutionary history and dissect the genetic basis of human traits and disease. According to the data published in nature, each pair of human differs 1 in 1000 nucleotides, and thus, on an average, one to three million base pair differences can be noticed.

Hence, for genetic variations, we have to look in only 0.1% of the human DNA. Analysis of the draft genome sequence provides the first panoramic view of the landscape of our genome. We learn the extent to which it has been colonized by parasitic DNA elements. During past, these elements underwent massive proliferation, playing an important role in shaping the evolution of the human genome. Another notable feature is the much lower gene tally than anticipated, which indicates that human complexity does not arise solely from the number of genes.

Although computational predictions can provide 'best guesses' on gene number and structure, definitive answers require experimental data on gene expression, in terms of both temporal and positional analyses of gene products. As a step towards this, microarray technology has been used to validate gene predictions, and more accurately define gene structures. One principle at the heart of the human genome project, reflected in the Universal Declaration on the Human Genome and Human Rights, is free and unlimited access to the sequence.

In its original usage, to be ethnic was to be heathen. Deriving from the Greek work *'ethnosnation'*, contemporary usage has often preserved the exclusionary sentiments implicit in this conflation of religion and nationality. To be ethnic is to be different, foreign and marginal; not "one of us". More recently, social scientists have extended the concept of ethnicity to encompass all the ways in which people seek to differentiate themselves from others. In this sense, ethnicity has largely replaced "race"—differentiation according to physical appearance—as an appropriate way to think about human difference.

The precise meaning of the term 'ethnicity' as a variable in health research, however, is hotly debated. A number of definitions abound for this complex social phenomenon, for which there still appears to be little consensus. The debate indicates caution in categorizing populations and points to a consideration of their social significance as well as precision in the terminology used. It appears that each field needs to examine its use of the term "ethnic" or "ethnicity" to define its precise meaning in the context in which it is being used, rather than an unthinking use of existing terminologies. So, in the field of genetics, precisely, which variable is being noted while recording ethnicity, and what purpose does this form of data collection serve? This is an unresearched area that requires further consideration to reach a consensus of opinion that can inform practice. The following offers a preliminary introduction.

Genealogical studies always have advantages in the biomedical issues, such as identification of disease in the ethnic races, increased percentage of diseases and resistance to pathogens, and other inciting agents of diseases. Categories presently used by researchers and service providers for monitoring ethnicity vary considerably; the lack of consensus itself diminishes the usefulness of data. The categories are also criticized for being too broad to be useful. For example, the term "white" indicates English, Irish, Turkish Cypriot and Jewish people originally from Eastern and Central Europe. The term "Indian" encompasses Sikh Punjabis, Hindu Gujaratis and Bengali Muslims, who differ at a number of levels from one another. Thus, the usage of broad terms render invisible whole communities leading to political invisibility and prevention from being included in the development of a policy agenda. Cultural assumptions are also made on the basis of ethnicity related data rather than accurate information seeking. For example, the practice of consanguineous marriage is generally thought to have significance in relation to Muslims of Pakistani origin, whereas it is also practiced amongst Irish travelers. Attentions to the genetic and service delivery implications for this group, Irish travelers in this case, are subsumed if they are ethnically defined as white.

Simply to record ethnic origin may not be enough in the context of genetic service delivery and genetic variations identifications. Are the variables being inferred from this term largely related to geographical origin and popular notions of cultural differences? In which case would it not be more useful to ask questions related to specific variables? Which other variables impinge on genetic service delivery? Most importantly, in what way will the collected data have an impact on service delivery? In other words, once collected, how will the information about minority ethnic families be used to improve their experience of genetic conditions? These questions, amongst others, need to be addressed as an integral part of developing an appropriate genetic service for minority ethnic populations.

At present, ethnicity related data is essential for gathering accurate information on genetic epidemiology, gene functions, gene flow and gene mixing, finding the ancestry, planning appropriate services for an ethnically diverse population, and conducting an effective audit. Genetics service providers need to examine the precise relevance of ethnicity-based data in genetics, and practice caution in using terminologies borrowed from other disciplines.

6.2. ETHNICITY

Health professionals collecting data on ethnicity related issues do so against a historical backdrop of racism against minority ethnic populations, and of ethnocentrism, wherein minority cultures are perceived as less desirable than the white norm. Questions relating to place of birth, marriage patterns, religion, language and so on are reminiscent of the interrogations carried out by immigration authorities. Even if such interrogations have not been the direct experience of each individual, they, nevertheless, are a part of the historical cultural experience of the groups as a whole, of which there is a general awareness.

In the context of genetics, the background of racism is supplemented by the history of eugenics and its powerful legacy. This backdrop affects the perceptions of people regardless of which ethnic group they belong to. The difference lies in the dynamics that operate when a white genetics professional is the one asking the questions from a minority ethnic population, bringing into play the power relations that already exist between the two groups in the wider society. The result is a reticence and discomfort on the part of white professionals in asking ethnicity related questions and for respondents to regard the questions and the person asking them with suspicion. But the questions need to be asked. How to ask them with sensitivity, allaying any fears, is a training and education issue that needs to be addressed.

Consultation with health professionals reveals that lack of knowledge and understanding about why ethnicity related data is being collected and how it will feed back into service delivery, undermines their confidence in asking pertinent, ethnicity related questions. It is an issue that has also been noted in fields other than genetics. Clear guidelines and training of health professionals on the collection of ethnicity related data are an essential component of operationalizing a policy on ethnic monitoring.

Genetics and Minority Ethnic Culture—
The Case of Consanguineous Marriage

Consanguineous marriage forms part of the cultural practices of a number of minority ethnic populations in Britain, including people originally from Pakistan, the Middle East, and Irish travelers. The issue of consanguineous marriage and its genetic implications has assumed an exaggerated importance beyond its clinical relevance to rare recessively inherited conditions, causing confusion amongst health professionals and communities practicing the marriage pattern. As Bundey and Roberts note:

"Once a consanguineous couple has produced a child with a recessive defect, the risk of its recurrence is virtually the same (1 in 4) as for any other parents of a child with this condition, though there is an increased risk of some other defect. It is doubtful whether in counseling the fact that the parents are consanguineous need even be mentioned. If one is asked whether this tragedy is due to their relationship, one can only say that this has contributed, but point out that if they had both chosen other partners, they might have produced children with some other condition, for everybody carries bad genes for some disorder or other".

However, health professionals citing consanguineous marriage as the "cause" of genetic conditions is not a rare occurrence. Darr, in a study of thirty-one British Pakistani families at risk for thalassaemia, reported that virtually every family in a consanguineous marriage had been told by a health professional that the condition was caused by their marriage pattern. A more recent study encompassing the eight District Health Authorities in the Northern and Yorkshire and North Western regions by Atkin shows that Pakistani families are still subjected to the same misinformation. Consultation with Asian genetic specialist workers also reveals the same situation in other parts of the country.

The development of novel diagnostics, therapeutics and health services is increasingly predicated on the search for significant biological differences within and between populations. This has led to the creation of large-scale scientific projects that map genetic variation, including human genetic databases focused on populations defined according to geography, nationality, race/ ethnicity or disease. However, comparisons of genetic differences between populations raise important scientific, social, ethical and policy questions. On one hand, there is the potential to provide improved research, treatment and health services that are targeted for specific populations.

However, by linking genetic information to socially defined categories, there is the potential to reinforce crude biological notions of race/ ethnicity and for the research to be misused in political, scientific or clinical contexts. If these latter issues are not addressed, scientific research in this area may be misconceived and lack public support.

Genetics Studies on Ethnic Races and their Uses

Genetics studies on ethnic races help to investigate how the categories of race/ethnicity are used in applied population genetics research, and the implications this may have for scientific practice, public health and access to healthcare. The research is focused on a detailed analysis of two case studies of population genetic databases and pharmacogenetics. Operationally, it is designed around four discreet, but intersecting 'workpackages' that explore the conceptual, practical and policy implications in commercial, scientific and health policy settings.

Firstly, extensive search of scientific, medical, social scientific literatures, national health policy, and healthcare planning documents will be undertaken, with the purpose of analyzing the referencing, application and validity of categories of race and ethnicity across these areas. Secondly, the project will involve 60–70 qualitative interviews with a range of individuals including:

- managers and research staff of genetic databases;
- academic researchers, scientists, and clinical researchers in the area of pharmacogenetics; and
- policy makers, public health professionals, and health services researchers involved in planning and assessing healthcare delivery.

The purpose of these tests is to understand the ways by which race and ethnicity are conceptualized and applied in these specific environments, and their impact on practice.

6.3. MAIN ADVANTAGE OF THE GENETICS STUDIES

- Describe the use of race/ethnicity in the scientific literature covering two case study areas (genetic databases and pharmacogenetics), and debate about how these categories are used more generally in genetics and biomedical research.

- Examine how concepts of race/ethnicity are deployed in population genetics research, and analyze their influence on research agendas, study design and scientific findings, and their social, ethical and policy implications.
- Assess the potential implications of applying the findings of population genetics research, based on the use of racial/ethnic categories, in developing health policy and the provision of healthcare services.
- Identify 'good practice' for the use of racial/ ethnic categories in population genetics research. Recommendations can be made about how notions of racial/ ethnic differences might be used to enhance scientific knowledge and promote equity in health and healthcare.

Applications

The projects done on ethnic genetic populations aims to seek ways to improve the scientific acuity of population genetics, and explore how social experiences of race/ethnicity might be used to develop categories that are biologically meaningful, and aetiologically useful. Therefore, one application of the study findings will be to suggest that information derived from population genetics research can be used in a manner that is scientifically meaningful, socially sensitive and will help improve equity in health and healthcare. It also has the following applications:

- Identifying the defective gene movement in the families.
- Nature of gene interaction.
- Helpful in recognizing the pattern of variabilities of the genes.
- Predicting the pattern of diseases and its variations.
- Studying the influence of environment on humans and their development.
- To know the drug responses by different human populations.
- To trace the evolution of a particular human population.

- Used in different types of clinical trials for vaccines, antibodies, interferons-alpha, etc.
- Safety and effectiveness evaluation.
- To study disease incidence and its inheritance.

At present, human populations around the world to adequately characterize race and ethnicity for studies conducted indicates the following categories described below. For ethnicity,

- Hispanic or Latino.
- Not Hispanic or Latino.

For racial designations in clinical studies conducted in foreign countries, it is recommended that the categories be modified to reflect the following as appropriate:

- American Indian or Alaska Native.
- Asian.
- Black, of African Heritage.
- Native Hawaiian or other Pacific Islander.
- White.

6.4. CATEGORIES AND DEFINITIONS

The Department of Health and Human Services (HHS), USA, program administrative reporting, and civil rights compliance reporting have defined each category as follows:

American Indian or Alaska Native

A person having origin in any of the original peoples of North and South America (including Central America), and who maintains tribal affiliation or community attachment.

Asian

A person having origin in any of the original peoples of the Far East, Southeast Asia, or the Indian subcontinent including, for example, Cambodia, China, India, Japan, Korea, Malaysia, Pakistan, the Philippine Islands, Thailand, and Vietnam.

Black or African American

A person having origins in any of the black racial groups of Africa. Terms such as "Haitian" or "Negro" can be used in addition to "Black or African American".

Hispanic or Latino

A person of Cuban, Mexican, Puerto Rican, South or Central American, or other Spanish culture or origin, regardless of race. The term, "Spanish origin", can be used in addition to "Hispanic or Latino".

Native Hawaiian or other Pacific Islander

A person having origins in any of the original peoples of Hawaii, Guam, Samoa, or other Pacific Islands.

White

A person having origins in any of the original peoples of Europe, the Middle East or North Africa.

The advantages of such studies can be seen in the following example.

6.5. ROLE OF RACE AND ETHNICITY IN HEART DISEASE

Differences in heart disease can be seen across all ethnic groups. For example, one study showed that African American women with coronary artery disease (CAD) were two times more likely to have a heart attack than white women. The study also showed that they were twice as likely to die from CAD.

- As a group, African Americans have high rates of heart disease, which includes CAD, stroke, high blood pressure, and heart failure.
- More than one quarter of Mexican Americans have some form of heart disease.

- People of South Asian descent are at high risk for heart disease, but people of Japanese descent are at low risk.
- Historically, Native Americans had very low rates of heart disease, but today it is the leading cause of death in this group.

Many people are surprised to learn that race and ethnicity may also play a role in heart disease. Some groups of people may be more at risk for heart disease, and other groups may not respond as well to certain treatments.

The study brought up some disturbing issues. African American women had higher rates of high blood pressure, high cholesterol, and diabetes. These women were also less likely than white women to control these risk factors. Despite being at greater risk, the African American women were less likely to take aspirin to prevent heart attacks, or drugs to lower cholesterol. Although the researchers did not know why aspirin was taken less, they did say that African American women need to have their risk factors for heart disease controlled better.

The American Heart Association (AHA) says that about 40% of both African American men and women have some form of heart disease. In contrast, among white people, the numbers are about 30% for men and 24% for women. Heart disease affects more than one quarter of Mexican American men and women. (The data did not include all Latino or Hispanic groups.)

While racial groups may be useful, it's important to remember that they give only a general picture. For example, Asian Pacific Islanders are very diverse. People of South Asian descent are at high risk for developing heart disease, but people of Japanese descent are not. Scientists have found that over the past few decades, some groups of people seem to be getting more heart disease. In the past, Native Americans had very low rates of heart disease. However, it is now the leading cause of death in this group, ahead of cancer, accidents, chronic liver disease, and diabetes.

Heart Disease is a Product of the Environment and Genes

Scientists have long blamed environmental (or external) factors, such as diet, income, education, social or economic factors (racism and poverty), and access to healthcare or quality of treatment as causes of disease. In the past, ethnic minorities have made up a large number of the poor. In addition, lower socioeconomic status has been linked to many diseases, such as high blood pressure, or atherosclerosis—the buildup of fatty material inside the arteries.

Although environment has an important role in this, scientists are also studying how genes may affect heart disease within specific racial or ethnic groups.

The National Institutes of Health (NIH) is funding a large project which is involved in search of genes that may cause heart disease. This task is part of the larger Multi-Ethnic Study of Atherosclerosis (MESA), which began in 2000. The eight-year study will track how heart disease develops in four major US ethnic groups: White people, people of Hispanic descent, African Americans, and Chinese Americans. The MESA study, which includes more than 6,000 people, may help to answer many questions related to heart disease.

Other scientists are taking a more specific approach. The medical school at Howard University, mostly a African American campus in Washington DC, has started a large gene bank called *Genomic Research in the African Diaspora* (or GRAD) *Biobank*. DNA will be collected from about 25,000 African American people. The goal is to find out whether genes can be blamed for the higher rates of high blood pressure, stroke, heart disease, obesity, and other diseases in this group.

Other scientists have already found some genetic differences among races. For example, researchers have discovered that 13% of African American people carry a gene change that puts them at greater risk for having a rare kind of abnormal heart rhythm. This gene occurred at a much lower rate in White and Asian people than in Latin subjects.

New information about the role of genes and race may help find better ways to prevent and treat diseases. Some studies have found differences in the way groups of people respond to drugs. For example, research suggests that African American people may not respond as well to some blood pressure drugs, such as beta-blockers and angiotensin-converting enzyme inhibitors. As scientists learn more about how certain drugs affect different races, patients may be able to get treatments that work better in their particular group.

As researchers discover genes that can lead to heart conditions, they may find that not all people who have these genes will get the disease. The result may depend on a complex relationship between the genes and the environment. What's more, many genetic trends can be modified by lifestyle changes and medical treatment. For example, people may have genes that make them more susceptible to high cholesterol. However, they can take steps that will go a long way towards preventing heart disease by eating well and taking cholesterol-lowering drugs as prescribed by their doctor.

6.6. RESEARCH AND ADVANCEMENT IN HUMAN RACES

The field of genetics has made rapid advances in the past ten years. It has been enthusiastically embraced and supported by the researchers, public, and world governments as a means to understand, prevent, or treat human diseases in the future. While the possibilities for ameliorating human suffering are exciting and the benefit of scientific advances are great, so are the potential risks involved. A formerly secret or unknown code for human development is now becoming accessible. The question of how accessible it should be and who will have access, remains. Without an understanding and respect for the issues involved, the scientific advances have the potential to harm people through lack of truly informed consent or breaches in privacy. After the successful completion of the human genome project, it is clearly understood that races which are present today are only the cultural formations. But there is no evidence to prove that different races are genetically entirely different. The HGP clearly indicated that approximately 99.9% is same in almost all individuals. It is also observed that the greater individual genetic variations within each race than the difference between the races. Dr. Harold Freeman, professor, North General Hospital, Manhattan, mentioned that only about 0.01% of our genes are responsible for our external appearance, on which racial categorizations are done. The human brain is very well adjusted to recognize differences in appearance between the two individuals. Hence, great importance will always be given to the external appearances. However, these external differences in appearances are having least consequences on the genetic informations each one is having.

Dr. Craig Venter, a cofounder of private HGP organization, Celera Genomics, which made the great contribution to the human genome sequencing, said that race is a social concept and is not based on the scientific principles. Concepts of race and ethnicity are among the most controversial, contentious, and misunderstood classifications in our social and scientific landscapes. Among geneticists, there are two dominant historical perspectives on race and ethnicity. On the one hand, there is a perspective rooted in the eugenics movement and early population genetics that treats racial and ethnic categories as biological classifications (Kevles, 1995), attempting to use scientific analysis to specify the precise nature of presumed biological differences between socially labeled populations (Huxley, 1951).

On the other hand, there is a perspective with its roots in the work of physical anthropologists and social scientists who argue that race and ethnicity are primarily cultural and historical constructions with little biological significance (Boas 1942). Although there has been much debate among geneticists regarding these two competing perspectives throughout the 20th century, geneticists at the beginning of the 21st century have yet to reconcile these divergent views on race and ethnicity.

In this regard, a mid-century celebration of the first 50 years of modern genetics by Laurence Snyder (1951), one of the first presidents of the American Society of Human Genetics, bears a noteworthy resemblance to the present state of affairs:

"Human populations differ from one another almost entirely in the varying *proportions* of the allelic genes of the various sets of hereditary factors, and not in the *kinds* of genes they contain. The extreme positions held by those who on the one hand maintain that there are no significant genetic differences between human races, and those who on the other hand hold that certain races are 'superior' and others 'inferior,' require drastic modification in the light of the accumulated data on the gene frequency dynamics of human populations".

Now researchers have accumulated a far more extensive collection of data on allelic distributions within and between human populations. Despite this ever increasing pool of information, geneticists have yet to resolve the basic tension between social and biological conceptions of race and ethnicity.

Debates about race and ethnicity have changed in one important respect today: nearly all geneticists reject the idea that biological differences belie racial and ethnic distinctions. Geneticists have abandoned the search for "Indian" or "African" genes, for example, and few, if any, accept racial typologies. Even so, although simplistic biological interpretations of race and ethnicity have been discredited for decades, studies in clinical and population genetics continue to associate biological findings with the social identities of research participants. In some cases, scientists use genetic features to reconstruct population histories, applying biological criteria to define and redefine commonalities among research participants recruited on the basis of shared linguistic, racial, ethnic, or geographical identities. In these and other cases, researchers name the racial or ethnic communities being studied, thereby implicitly indicating that genetic features can be used to characterize contemporary social populations. Thus, although the simplistic biological understanding of race and ethnicity associated with the eugenics movement may be dead, the far more subtle presumption that racial and ethnic distinctions nonetheless capture "some" meaningful biological differences is alive and flourishing (Kaufman and Cooper, 2001).

It was hoped by some that the sequencing of the human genome would undermine the view that racial and ethnic classifications have biological significance (Gilbert, 1992). This position was based on the prospect that by showing that there are numerous genetic similarities across all social classifications, and no genetic features, which are unique to any particular racial or ethnic population genomics, would provide definitive evidence that race and ethnicity are social, not biological, classifications. Ironically, the sequencing of the human genome has instead renewed and strengthened interest in biological differences between racial and ethnic populations, as genetic variants associated with disease susceptibility (Collins and McKusick, 2001), environmental response (Olden and Guthrie, 2001), and drug metabolism (Nebert and Menon, 2001) are identified, and frequencies of these variants in different populations are reported.

This renewed interest in identifying inter-individual, and interpopulation differences have prompted the construction of several types of genomic resources designed to facilitate the study of human genetic variation. Existing genomic resources include:

- biological materials for use in coordinated sequencing projects;
- electronic databases of sequence information; and
- databases describing genotype-phenotype associations.

These resources seek to describe, as best as possible with existing technologies, the broad range of genetic variation that exists across human populations. For example, anthropologists and evolutionary biologists are interested in creating a collection of human cell lines for the study of genetic diversity among culturally defined populations (http://www.nsf.gov/pubs/2001/nsf01120/nsf01120.html). Other projects include a cell line collection for the discovery of single-nucleotide polymorphisms (SNPs; Collins *et al.,* 1998); a publicly funded electronic database for clinical investigations of genetic influences on drug response (http://www.pharmgkb.org); a research database for investigators examining how genetic variation influences response to adverse environmental agents (Olden and Wilson, 2000); and more recently, a federally supported collection of cell lines for use in assembling a comprehensive human haplotype map (Helmuth, 2001). But Dr. Craig Ventor and large number of human geneticists still argue and declare that there is only one race—human race—and nothing else.

Proponents of the HGDP were interested in sampling from small, geographically isolated populations thought to have experienced relatively little or recent admixture. Their hope was to correlate specific genetic features with particular language families, cultural affiliations, and racial and ethnic identities (Kidd *et al.,* 1993). Members of some of those populations to be sampled, however, argued that the HGDP effort amounted to reducing distinctive histories of social populations to analyses of shared biological ancestry, in effect giving priority to evidence for genetic homogeneity in the face of compelling records of considerable linguistic, cultural, and social interaction and heterogeneity.

The historical processes, which have produced that heterogeneity include intermarriage; cultural preferences for claiming only one's father's or mother's identity; fictive and adoptive kinship; instrumental and situational choices in asserting alternative identities; colonialist, racist, and nationalist ideologies that impose new identities on subjugated populations; voluntary and forced migration; and economic, religious, and other barriers to reproduction within larger social categories. These processes (and others) undermine the view that small "isolated" populations with distinctive racial, ethnic, linguistic, or cultural identities are better proxies of biological relatedness or homogeneity than are larger populations comprised of members with multiple such identities.

Despite these problems, social categories cannot be completely dismissed as irrelevant to genomic analysis. The lines that exist between social categories can constitute partial barriers to interaction, reproduction, and migration. Consequently, some members of a particular social population may share similar single nucleotide polymorphisms, similar conserved or ancestral haplotypes, or similar variants of disease-susceptibility, drug metabolism, and environmental- response genes in frequencies and patterns that are not found in other social populations. These differences owe due to distinctive demographic structures that result from historical events, such as population bottlenecks and founder effects that affect the ways in which genetic features are distributed in subsequent generations. These historical events also influence the techniques, which geneticists use to discover the genetic features.

For example, some of the techniques such as, linkage disequilibrium, which are successful in identifying defective genes, have been developed explicitly for use in the context of discrete, socially identifiable populations with unique demographic histories (Jorde, 2000). Haplotype mapping techniques also depend on the analysis of discrete populations with differing demographic histories and structures (Goldstein and Weale, 2001).

Clearly, deciphering the relationships that may exist between social classifications and biological categories is not a simple matter. The biological significance that a social distinction may have for one purpose can dissolve when the same social categories are used to answer other biological questions. Thus, it may be appropriate to use social categories as a proxy for biological relatedness (or unrelatedness) in some circumstances, but not in others. Since social categories like race and ethnicity are problematic as proxies for genomic features, the burden for showing their scientific utility in any specific resource or study should fall on researchers.

Efforts to meet this burden of establishing that social identities should be used in the construction of genomic resources can be divided into two general strategies. First, it has been claimed that the use of social identities helps to increase the likelihood of representing a more complete range of variation in human populations, particularly if the social identities used correspond to relatively disparate geographical localities. For example, genomic resources developed using biological samples from members of social groups from different continents (and from different regions within each continent) can serve as a proxy for human genetic variation in general, even though they may not include samples from all racial or ethnic groups within each continent.

Although imperfect because of subsequent population migrations and intermarriages, this proxy strategy remains powerful because of the initial but still relatively recent ancestral movement of human populations out of Africa into other continents. Second, the use of social identities in genomic research has been defended on the grounds that recruiting individuals with shared social identities may increase the likelihood that the collected samples reflect a common demographic history (such as a population bottleneck) that may be useful in identifying discrete health-related genomic structures or features (such as specific polymorphisms, genes, or haplotypes).

Notably, both of these two strategies use social categories as loose proxies of genetic features. In the first instance, differing social identities are used to approximate genetic differences in the human population as a whole, whereas in the second strategy, shared social identities are used to approximate common genealogical relationships. In contrast to these two applications, the use of social identity is most problematic, both scientifically and ethically, when used to associate all members of a population with a particular genetic feature or features. When social identities are used in this manner in the construction of genomic resources, the complex history of social populations are reduced to naive biological characterizations.

An important genomic resource for the discovery of polymorphic variation is the Polymorphism Discovery Resource (PDR). The PDR was assembled to approximate genetic variation among all human populations, and consisted of samples from socially diverse populations within the United States (Collins *et al.*, 1998). The creation of this resource caused considerable disagreement regarding the number of human cell lines needed to establish a standard reference set—which populations should be included in the collection, and what proportion of samples should come from each population. Much of this debate focused on scientific questions of inter- and intrapopulation variation. An additional consideration, however, was the potential social risk posed by the association of genetic features, with particular racial or ethnic groups represented in the resource. To address these concerns, the PDR removed all social identifiers once the collection was established, thereby removing the possibility of using the PDR to estimate allelic distributions in racial or ethnic populations. As a result of this decision, the PDR has been described as less useful to researchers than it might otherwise have been were this information retained (Davidson, 2000).

Biological collections, such as the PDR illustrate a difficult challenge in the construction of genomic resources, namely, how to decide whether the social identities of sample donors ought to be made available to users of the resource. Naming the socially defined populations of which individual donors are members means that all members of those populations may be affected by research findings, including those who did not consent to or take part in the resource. Some genomic resources, such as the CEPH (Centre du Etude Polymorphisme Humain) family samples and the proposed haplotype map repository, will be used by many more researchers and may take on greater significance than other resources (such as disease-specific DNA repositories) will. This higher profile may entail greater risks for both participating individuals (whose entire genomes may become publicly available on the Internet) and the social populations of which they are members (about which many more genomic findings may be generated than for other populations).

Certainly, research that does not explicitly link socially defined populations with genetic findings has fewer ethical risks. At the same time, the distribution of benefits resulting from genomic resources may depend on identifying populations from which study participants are recruited. For common alleles that cross social boundaries, identification is less crucial. However, for less-common alleles, or for rare variants of common alleles that tend to occur most frequently in the social populations in which they originated ancestrally, identification may be more important for clinical diagnosis, targeted delivery of appropriate health services, and public health education.

Choosing how to socially identify donors as members of social populations is not a trivial or self-evident matter. An individual donor, for instance, may be known simultaneously as a resident of a particular Indian village in Arizona; a member of the Hopi tribe; a descendant of a Laguna family (through a paternal ancestor who is not explicitly noted in matrilineal Hopi society);

a Native American; and as someone of Spanish ancestry (owing to 18th-century intermarriages between Lagunas and Spaniards) in addition to being a member of the general US population. Each of these identities, or several in combination, would have different social implications (as well as resulting in somewhat different scientific findings about genetic variation in relation to specific social categories), depending on which label or labels are placed on an individually anonymous DNA sample.

This potential for scientific inaccuracy in defining social categories and in recording them for specific donors could reduce the value of their association with specific genetic findings, while creating additional risks for members of the populations that happen to be named. Particularly, in a larger socially defined population in which there is considerable difference in members' haplotypes and SNPs, a small number of donor participants may not represent the full range of genetic variation present in the population. Ascertaining the intrapopulation range of variation will thus be important for any specific downstream clinical benefits that basic genetic research may have for persons with that social categorization (apart from any general benefits from the discovery of common genes that occur across social boundaries).

Many genetic studies rely solely or primarily on participant self-reporting to assess membership in a socially defined population. Although this standard is appropriate for studying race and ethnicity as social phenomena, it does not allow for the critical scientific investigation of their biological accuracy. Moreover, genetic studies often use social labels, such as "Chinese", "Nigerian", and "African American" that conceal a great deal of cultural, linguistic, and biological variation (Cooper *et al.*, 2000). Even a more specific ethnic label, such as "Yoruba", which refers to a population of >10 million people distributed across a large multinational region of West Africa, can mask significant intragroup variation (Reich *et al.*, 2001). Consequently, the

accuracy (or lack thereof) in the choice of social label for a group of study participants may affect the scientific validity of genetic findings for that larger social category.

Among members of smaller populations that are, in effect, a small number of extended families, there will be fewer differences in conserved haplotypes and SNPs than those among members of larger social categories as discovered by Tarazona-Santos *et al.*, 2001. These smaller population studies will, essentially, be pedigree studies and may require more stringent human subjects protections to preserve family and individual privacy (Botkin *et al.*, 2001).

Moreover, many smaller populations already experience the disadvantages of minority status within larger polities and, for that reason as well as their size, may be more vulnerable to social stigmatization. These observations indicate that smaller social populations (particularly those that already are economically or politically disadvantaged) should not be identified in genomic resources or publications unless there is the potential for direct benefits to those populations, such as identifying genetic variants that predispose members to disease and that are less common in other populations.

In the case of larger social populations, identifying donors' racial or ethnic identities frequently involves a trade-off between gaining additional information that may be useful in designing genetic studies and interpreting their results (a process that may produce downstream benefits to members of donor populations), while exposing all members of those populations to greater risks of discrimination and stigmatization. The evaluation of this risk-benefit calculus can be complex and should include members of the populations in question (Sharp and Foster, 2000). When faced with the concerns of members of populations, many of which are informed by historical experiences of mistreatment and exploitation by outside researchers, the goal of minimizing the risks of social identifiability can become a central element in the scientific design of the resource or project, as in the case of the PDR.

The need for diversity in genomic resources is compelling, but the use of social classifications as a proxy for biological heterogeneity can be problematic. A biologically diverse resource will identify more genetic variants and increases the generalizability of findings. A resource that draws donors from multiple populations also recognizes that genetic variants may be more or less common among individuals with particular social identities. Without studying different socially defined populations, we cannot assume that all will benefit equally. Moreover, a basic genomic resource (such as a haplotype map) may provide important "spillover" benefit to members of sampled populations if it increases the probability that the genetic structures of greatest relevance to their health will be found expeditiously. Correspondingly, the decision to "exclude" certain groups could have the effect of "depriving" them of these potential (albeit indirect) benefits. Thus, choices made today about the ways in which genomic research resources are established, and how the social identities of sample donors are viewed by researchers will have significant implications.

Inclusion in a genomic resource is not just a question of potential benefits. It is also a question of perceived risks. This point is clear when one recalls the many ethical controversies surrounding the HGDP (National Research Council, 1997). In addition to its problematic scientific assumption that many social populations are equivalent to biological demes, the project also floundered because of an inability to adequately address concerns from members of indigenous communities (Lone Dog, 1999). These included concerns about the commodification of indigenous persons and their bodies; worries about potential uses that might conflict with cultural beliefs and practices; harmful legal and political implications resulting from genetic interpretations of population histories and memberships; and lack of provision of both short-term and long-term benefit for participating communities.

Many anthropologists and geneticists argue that the genome diversity information challenges the legitimacy of racial categorizations, and clearly indicate that race categorization, at present, is meaningless. Hence, an American anthropological association (AAA) in 1997 made a clear statement regarding the race concept, and directed the government to cease the present concept of racial categorization in the scientific work. The implications of population-specific genomic research are not always self-evident, as both risk and benefit tend to be culturally defined (Douglas and Wildavsky, 1982). In case of many nonmajority communities, researchers and institutional review boards will be unable to anticipate all culturally specific risks (e.g., those associated with the creation of immortalized cell lines and public databases (Foster and Sharp, 2000). Outsiders are often unable to fully appreciate community-specific risks of these kinds once they have been identified for them, in part because minority community members' perceptions of these risks may have been heightened by their historical experiences of being economically and politically disadvantaged with respect to the majority of society (of which biomedical research is a tangible manifestation).

The difference in power and privilege between researchers and socially defined populations lacking in significant economic and political resources may affect the ability of the latter to fully conceptualize and negotiate the conditions for research participation, and to take effective action on any subsequent concerns about sample misuse or adverse interpretations of genetic findings. Moreover, differences in economic power may prevent donor populations from taking advantage of genetic tests and therapies developed from genomic research using members' samples. For these reasons, community involvement and consultation are essential in planning genomic resources that include and/or identify socially defined populations (Sharp and Foster, 2000).

Categories such as race and ethnicity can be useful as heuristic starting-points for the investigation of genetic variation across populations. As such, social categories used to recruit participants should themselves be investigated in the course of assembling genomic resources, with the goals of discovering smaller population subgroups that are more predictive of biological relatedness and better understanding of the social factors that confound the relationship between socially defined populations and human genetic variation. These factors will also affect how genomic findings and technologies will be perceived and used by the general public. In this way, we will be able to improve our understanding of genetic variation within social populations, while simultaneously minimizing the risks of identifiability reification and maximizing the potential benefits of genomics. Thus, genomic research with socially defined populations requires as much attention to the social organization of populations as it does to the genetic organization of chromosomes.

6.7. EXAMPLES OF STUDIES ON ETHNIC RACES

Nuu-Chah-Nulth Indians of Vancouver Island

The Nuu-chah-nulth Indians of Vancouver Island in British Columbia have one of the highest rates of arthritis in the world. This ethnic population suffered with nearly two-thirds of the population afflicted. Often passed down family lines, the genetics of arthritis in this unique population was studied extensively during the 1980s by Richard Ward, who later became head of the Institute of Biological Anthropology at the University of Oxford in England. As a part of his study, Ward traveled through Nuu-chah-nulth territories in 1985 and collected 833 vials of blood from subjects who signed a consent form, allowing him to screen the samples for arthritis biomarkers. But, Ward was not able to find the gene he was looking for. He kept the blood samples, and unbeknownst to his donors, used them in genetic anthropology studies

that identified the Nuu-chah-nulth as a distinct indigenous population dating back nearly 70,000 years.

Although he insists that his actions were neither wrong nor unethical, the Nuu-chah-nulth say that Ward should have given the samples back once his arthritis studies were finished. That Ward shifted his research into areas not previously approved by the tribe, while also allowing other scientists' access to the samples without the subjects' consent, has angered the tribe greatly. This incident, covered by the Canadian media in September, 2000, is perhaps the best known example of a broader clash between genetic researchers and indigenous groups that is taking place throughout North America.

American Indians

Blood and tissue samples from indigenous peoples are getting harder to come by. As in other countries, many US tribes are suspicious of genomics and believe that geneticists and biotechnologies they spawn are out to exploit their genes and other sacred life-forms for commercial gain, usually to be enjoyed by the people outside the tribe. Already wary of bioengineered transformations of sacred foods, such as corn and fish, many American Indians see human genomics as a science that challenges the spiritual basis of their existence.

Genetic researchers have also developed a reputation among Indian tribes as being culturally ignorant and arrogant. Some American Indians call such researchers "helicopter scientists" to describe how they fly into isolated communities claiming to be researching important health issues, and then, after collecting the necessary information and samples, fly back out, never to be heard from again. "They don't want to help us preserve the culture or language we evolved in," says Judy Gobert, dean of math and sciences at Salish Kootenai College on Montana's Flathead Indian Reservation. "They just want our DNA."

By examining the opposition of American Indians to genomics, a number of broad themes began to emerge. One item of concern is the indefinite storage of tissue samples for unknown future studies. Judith Greenberg, a scientist at the National Institute of General Medical Sciences in Bethesda, Maryland, is the project officer of the Human Genetic Cell Repository at the Coriell Institute for Medical Research in Camden, New Jersey. She acknowledges that the repository currently has no samples from indigenous peoples in its collection. She also says that the repository "recognizes and respects the concern and resistance among the tribes" to provide such samples.

Recently, these indigenous samples held at the repository were relinquished when the anonymous tribal donors demanded their return. Says Clifton Poodry, director of the Division of Minority Opportunities in Research at the National Institute of General Medical Sciences, who is himself an American Indian from the Seneca Nation in western New York, "Native Americans will give informed consent to have their samples analyzed for a specific study. But they object to having those samples amplified and reproduced for future studies of which they have no knowledge."

According to Gobert, opposition to this practice is based partially on the spiritual attachment American Indians place on human tissue. "For me, even plucking a hair from my grandmother's head is abhorrent," she says. Since the samples are sacred, American Indians vehemently oppose their being used in ways that are inconsistent with tribal spiritual beliefs. Gene patenting, for example, is seen as a violation of nature by all of the American Indian community. To ensure that blood and tissue samples aren't used in ways that contribute to such practices, American Indians insist on retaining control over where the samples go and how they are studied.

The study of genomics is also worrisome to American Indians from several legal perspectives. For example, many American Indians believe that genetic confirmation of the Bering Strait theory (which describes how early cultures migrated to North America by crossing the Bering land bridge) could be used to challenge aboriginal rights to territory, resources, and self-determination.

Also at issue, suggests Poodry, is whether genomic studies might ultimately be used to challenge blood quantum measurements, which define membership in a given tribe. Tribal membership is an increasingly controversial issue for many American Indians, as more and more people seek to identify themselves with tribes, in part to share the revenues generated by tribal casinos. This potentially destabilizing variable is something that American Indians, would prefer not to deal with. Some tribes also worry that if genomic studies show that genes for diseases, such as alcoholism are highly expressed among their populations, American Indians will be further stigmatized in the eyes of the public. "What we may attribute to genetics, (laity) might be simply labeled as race," says William Freeman, director of research at the federal Indian Health Service in Rockville, Maryland. "This kind of information could be used to suggest that Native Americans are genetically inferior." He adds "We've heard these kinds of arguments many times before."

Finally, there is a growing belief among some American Indians that high-priced genetic research is diverting funds that could be used to promote public health on the reservations. On 26 June, 2000, when the completion of a rough draft of the mapping of the human genome was announced, the Indigenous Peoples Council on Biocolonialism (IPCB), an activist organization based in Wadsworth, Nevada, released a statement saying that "genetic research of this scale hurts, rather than benefits, indigenous peoples because it diverts public funds away from direct healthcare and prevention programs."

Gobert, who has close ties with the IPCB, is emphatic in her view that genomics does nothing to improve the health of people afflicted with diabetes and other diseases common to indigenous populations. "We haven't seen any benefits from genetic studies," she says. "These diseases don't have a genetic fix; they are lifestyle diseases. Genomics give Native Americans false hope."

Indian Caste Origins and Genomic Studies

A caste is a collection of people who share similar cultural and religious values and practices. Members within a caste generally marry among themselves; intercaste marriages are a cultural taboo. These social regulations governing the institution of marriage have resulted in a substructuring of the Indian gene pool. There are also elaborate social regulations of avoidance of marriages within castes, and thus there is genomic substructuring even within a caste. The origins of the castes in India remain an enigma.

Many castes are known to have tribal origins, as evidenced from various totemic features that manifest themselves in these caste groups (Kosambi 1964). The caste system in northern India may have developed as a class structure from within the tribes. As agriculture spread from the Indus River valley to the Gangetic basin, knowledge and ownership about the means of food production may have created hierarchical divisions within tribal societies (Kosambi, 1964).

The Aryan world comprised three classes (varnas): priests, nobles, and commoners. Aryans, as the conquering people, possibly placed their three classes in the indigenous Indian society. The varna organization is hierarchical. Initially, the system had names for two ranks, Brahma (Brahmin) and Kshatra (Kshatriya); Brahmin being of a socially higher rank than Kshatriya. The third rank was made up of Vis, i.e., all the subjects. To this society, a fourth rank, Shudra was added, who had no rights to Aryan ritual. In southern India, the menial workers, the so-called "untouchables", were placed in a new varna, Panchama (meaning fifth). It is conceivable that the Aryan speakers had greater contact, including genetic admixture, with the Brahmins, who were professionally the

torchbearers and promoters of Aryan rituals. The Aryan contact should have been progressively less as one descended the varna ladder. The genetic expectation, therefore, is that the proportions of these genes (or genomic features, such as haplotypes or haplogroups), which "characterized" the Aryan speakers, should progressively decline from the highest varna to the lowest, and a reverse trend should be observed with respect to these genes that "characterized" the indigenous Indians. Although some previous studies have sought to test this expectation, the observed trends were equivocal. The primary reason was the lack of data on a large uniform set of markers from populations of India and central/west Asia (the region from which the Aryans speakers, who entered India, originated).

The study by Bamshad *et al.* (2001), who have also sought to test the above expectation, is clearly a landmark. Using a very large battery of genomic markers and DNA sequences, spanning autosomal, mitochondrial, and Y-chromosomal genomic regions, they have shown that the observed trend of genetic admixture estimated from castes belonging to different varnas is congruent with expectations. This trend was observed in each of the three data subsets. The only exception was in respect of mtDNA restriction site haplotypes, which was also noted in a recent study conducted by Roychoudhury *et al.,* 2000. However, after combining these haplotype data with DNA sequence data, Bamshad and colleagues were able to capture the expected trend. Thus, this study not only provides a wonderful genomic view of the castes and of their origins, but also underscores the need for careful statistical analysis of genomic data for drawing appropriate inferences. The use of "upper", "middle", and "lower" to designate caste hierarchy is much more recent than the use of varna. Whereas varnas are traditionally defined, anthropologists have used different definitions of upper, middle, and lower castes, in terms of the castes that they included in each of these clusters. These differences in definitions sometimes stemmed from socio-cultural similarities or differences as noted or perceived by different anthropologists, and sometimes ranked caste-cluster compositions were altered for convenience, such as pooling to adjust for small sample sizes. As noted earlier, in studies such as Bamshad *et al.,* the most appropriate classification is by varna. As the reader will note, the authors have analyzed their data using different compositions of hierarchical caste-clusters and have obtained homologous results. However, it needs to be emphasized that traditional varna system is the only unequivocally accepted hierarchical system. In studies pertaining to the origin of castes, one is liable to draw incorrect inferences by including castes belonging to different varnas in the same ranked cluster. Bamshad *et al.,* have chosen to study caste populations drawn from a restricted geographical region of India. They have rightly emphasized the need to replicate their findings. This is absolutely essential because, as Karve (1961) has noted, "It is not generally realized that the caste society in a sense was a very elastic society."

Indeed, a caste bearing the same name may have very different origins in different geographical regions. There are examples in which a tribe dispersed over a large geographical region, took up different occupations in different subregions, and "fitted" itself into the caste hierarchy on different rungs.

6.8. HUMAN GENOME PROJECT AND HUMAN RIGHTS

With the mapping of the human genome and increasing interest in genetic testing and therapies, the potential for ethical problems has increased. Nursing and medicine have the ethical responsibility to "do no harm" and to protect the privacy of clients. However, the clients may also be family members and descendants of those who are cared in medical health center. The purpose of this chapter is to discuss issues related to genetic privacy using a deontological approach, and to outline methods for protecting the clients and research participants.

Although genetics has been in news and medical literature in greater amounts in the past fifteen years, it is not a new scientific topic. O. Avery, C. MacLeod, and M. McCarty identified deoxyribonucleic acid (DNA) in 1944. In 1953, F. Crick and J. Watson described the three-dimensional structure of DNA. M. Nirenberg and H.G. Khorana broke the genetic code in 1966, when they found that triplet base pairs of messenger ribonucleic acid (mRNA) specified each of the twenty amino acids. Recombinant DNA molecules, or the combination of genetic materials from different sources through gene splicing, were first produced in 1972.

During the rapid increase of genetic knowledge, the development of potential ethical problems was recognized. By 1975, scientists from around the world adopted guidelines for recombinant DNA experiments. In 1989, the Department of Energy (DOE) and the National Institutes of Health (NIH) formed an Ethical, Legal, and Social Issues (ELSI) working group to study the implications of genome research. The working group became a branch of the National Center for Human Genome Research (NCHGR) in 1990 [ELSI, 2000; National Human Genome Research Institute (NHGRI)]. Three to five percent of the NHGRIs and the DOEs budgets are dedicated to support ELSI (ELSI, 2000; NHGRI, 2000). The ELSI program provides the largest amount of federal funding for bioethics research with a budget of over $10 million a year (NHGRI, 2000).

6.9. ETHICAL CONCERNS INVOLVED IN GENETIC RESEARCH

Potential ethical concerns are for privacy, confidentiality, and informed consent. Areas of concern in privacy and confidentiality are breaches that may lead to discrimination and stigmatization of racial groups or individuals in employment or health insurance, or whether to inform family members of genetic markers for disease, especially if the original participant donor (proband) refuses to release the information. Respect for persons issues include:

- the use of DNA stored samples without consent;
- autonomy issues, such as who owns the data after it is collected and stored, and if participants and families have the right to choose whether to know about genetic risks or not; and
- the use of genetic information by other researchers, especially for profit ventures.

SUMMARY

After the human genome sequencing was completed, new genetic data enabled the researchers to reexamine the relationship between human genetic variations and races. With the genomic era in full swing, scientists are interested in cataloging genetic variation among indigenous groups and studying their susceptibility to chronic diseases for several reasons. Indian populations or any other tribes in the world with remote access are a convenient sample in terms of geographic area because they tend to be concentrated in areas such as reservations. American Indian populations are also very homogenous. Since there tends not to have been an influx of genes from other races and ethnic groups, scientists can study ancestral gene sequences in modern tribes people. Finally, an enormous benefit for genetic research is that because the genes of American Indian tribes are so closely conserved, it allows researchers to more easily control for genetic variation and study other factors such as the role of environmental exposures.

Genetics studies on ethnic races helps to investigate how the categories of race/ethnicity are used in applied population genetics research and the implications this may have for scientific practice, public health and access to healthcare.

The research is focused on a detailed analysis of two case studies of population genetic databases and pharmacogenetics. Operationally, it is designed around four discreet, but intersecting, 'workpackages' that explore the conceptual, practical and policy implications in commercial, scientific and health policy settings.

It was hoped by some that the sequencing of the human genome would undermine the view that racial and ethnic classifications have biological significance. Thus it helps in the better understanding of various problems that are associated with the human biology that cannot be researcher in any other work. This position was based on the prospect that by showing that there are numerous genetic similarities across all social classifications and no genetic features that are entirely unique to any particular racial or ethnic population, genomics would provide definitive evidence that race and ethnicity are social, not biological, classifications. Ironically, the sequencing of the human genome has instead renewed and strengthened interest in biological differences between racial and ethnic populations, as genetic variants associated with disease susceptibility, environmental response, and drug metabolism are identified and frequencies of these variants in different populations are reported.

EXERCISE

1. Define genetic studies on ethnic races, and its impact on the ethical, legal and social issues raised in the present day scientific world.
2. How are the ethnic races important for genetic studies? Mention its goals and expectations in the field of medical genetics and biotechnology.
3. What do you mean by ethnic races, and how are they associated with the human genome project studies and understanding?
4. List out various ethnic groups considered for the genetic studies, and their importance in understanding difficulties associated with human genetic studies.
5. Write a note on the advantages of ethnic genetic studies, and its importance in the field of human biology and biotechnology, in particular.

6. Distinguish between the American Indian or Alaska Native and Asian. Add a note on their importance in genetic studies.

7. Describe how race and ethnicity play a role in heart disease, and mention how genetic studies on the ethnic races are helpful for the better understanding of the disease.

8. Critically comment on the ethical, social, and legal issues raised by the human genetic studies on ethnic races.

9. Write about the involvement and methodologies followed by the national institutes of health and department of energy for genetic studies on ethnic races.

10. Discuss how human genetic studies on ethnic races are helpful for understanding the genetic diseases, and explain its applications and risk involved in the medicine?

11. Enumerate the importance of human genome project, ethnic populations genetic studies and its adaptations in the understanding the different biological processes of human beings.

12. Write a note on the role played by the single nucleotide polymorphisms, and its applications in the human genome project.

13. Briefly write about the advantages of the human genome project and their applications in the ethnic genomic studies.

14. Explain different types of limitations associated with the genetic studies on ethnic races and storing of the ethnic samples.

15. What is meant by ethnic populations? Explain its importance in the patent rich gene technology world.

16. Give an account of the arguments for and against human genetic studies on ethnic races and their applications.

17. Elucidate the future plans of human genetic studies on ethnic races and their usage by private companies patents, and its impact on the human research and development.

18. What are the ethical issues associated with the different human populations of ethnic groups? Add a note describing its impact on the understanding of human genome.

19. Explain briefly how scientists are supporting the continued studies on human genetic studies on ethnic races and technological need of the human biology.

20. What is meant by arthritis? Add a note on Nuu-chah-nulth Indians of Vancouver Island in British Columbia for use in genetic studies?

21. Describe how different countries are benefiting from the information of human genetic studies on ethnic races, and its impact on their society.

7

CHAPTER

Biosafety Guidelines and Regulations

7.1. INTRODUCTION

Biotechnology falls within the tradition of improving crops and livestock to meet the human needs in a better way. It also greatly expands the ability to move genes within and across species, and creates a new ability to move genes across distantly related species and biological kingdoms. It is this attribute of biotechnology that makes it a potentially powerful tool for modifying nature, but which also raises safety, ethical, health, and environmental issues. Biotechnology can offer some solutions in the fight for sustainable, clean production, and significant mitigation of some of the negative side effects of other, currently non-sustainable technologies that are used to provide the necessities of everyday life in an industrialized society. For instance, cereal carbohydrates, cellulose waste streams and other biological materials can be converted using existing technologies into industrial feed stocks, such as plastics, polymers and fine chemicals, while clean, renewable energy sources such as alcohols and methyl esters can be obtained by relatively simple processing technologies. Biotechnology, especially through the application of genetic modification, can enhance the efficiency of these processes, speeding them up and increasing yield-reducing net emissions of greenhouse gases while at the same time, often sequestering carbon.

Living organisms such as microbes, plants and animals can be genetically enhanced so as to produce many of the sophisticated protein and carbohydrate molecules currently being manufactured in energy intensive, non-sustainable ways. Molecules, which are being produced right now in genetically enhanced plants and animals include enzymes for laundry detergent and industrial processes, human pharmaceuticals, blood products, biological pesticides and fertilizers, synthetic textile fibers, and a host of other raw materials. Many other possibilities exist for the application of the tools of biotechnology to clean, sustainable development. Two applications of biotechnology in the forest industry include enhancing the rate of growth of trees, and reducing the lignin content of pulpwoods such as poplar. Trees that reach harvestable size faster, and trees for pulping that require less harsh chemicals to break down lignin in the pulping process, are in field trial right now. These novel trees will help reduce the use of heavy industrial chemicals and pesticides that can end up as environmental pollutants and which require greenhouse gas-producing energy for their manufacture. These and other technologies are in existence today, they are available now, to help the industrialized world move towards cleaner, sustainable production.

The convention on biological diversity (CBD) on biosafety entered into effect on December 29, 1993. It establishes important principles regarding all aspects of biological diversity. Under this convention, the cartagena protocol was adopted as a guiding framework for activities on safety in biotechnology. However, it recognizes the ecological and economic importance of an effective biosafety regulation in biotechnology development. It also recognizes the need to have in place an appropriate safety regulation before large-scale field trials of GMOs are conducted and released to the environment. At present, large number of third world countries are in a process of designing, developing safety regulation, and have brought into the government consideration for rectification and approval of cartagena protocol.

7.2. DEFINITION OF BIOSAFETY

In general, "biosafety" refers to the efforts that ensure safety in using, transporting, transferring, handling, releasing, and disposing of biological organisms, including genetically modified organisms (GMOs), which are capable of harming humans, animals, plants, or the environment **(Fig. 7.1)**.

GMOs are used in a variety of fields ranging from agriculture to medicine. Most developed countries (industrialized countries) have laws and regulations that ensure the biosafety of potentially hazardous organisms in domestic use and trade. However, most developing countries lack similar legal biosafety safeguards. The "cartagena biosafety protocol", completed in January 2000, is the international agreement that addresses

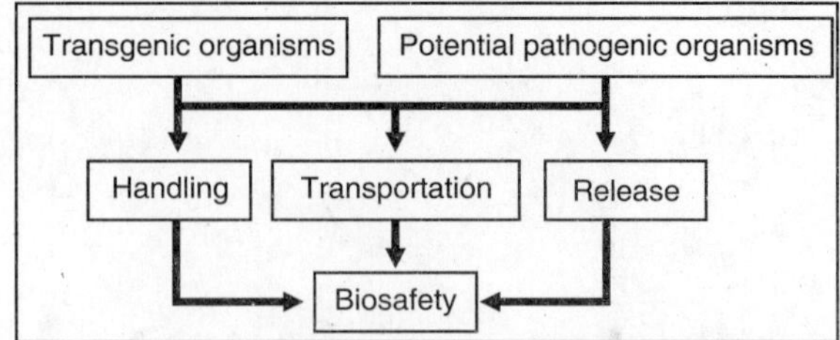

Fig. 7.1. Process involved in biosafety.

safety concerns related to trade in genetically modified organisms. The protocol applies to the transboundary movement, handling and use of GMOs that may have impacts on human health, environment, or biological diversity. It is viewed by many as the main mechanism for dealing with biosafety concerns when GMOs cross international borders. This biosafety agreement is a protocol for the 1992 convention on biological diversity (popularly known as the biodiversity convention)— a global agreement that addresses major aspects of biological diversity, including species conservation, genetic resources, and ecosystem protection, with a general goal of reducing extinction of the species. The CBD entered into force in 1993, and at present, 179 nations are parties to this treaty.

Even with its biotechnologically developed status, the United States is not a party to the CBD (the Senate has declined to approve the treaty), and thus could not participate directly in the official proceedings that drafted the biosafety protocol. However, the US observer delegations have attended all CBD meetings since 1992, and have played an active role in making its positions known and influencing treaty deliberations. Drafting of the biosafety protocol started in 1996, and in January 2000, the parties to the CBD agreed upon a final draft protocol in Montreal, Canada. Many, especially within the Clinton Administration, thought that US actions were key in shaping the draft protocol to protect US interests, yet they acknowledged the US capacity to influence CBD's deliberative bodies was affected by its non-party status. Other analysts think that the protocol will change international trade rules for biotechnology in ways contrary to US interests. They have argued that provisions in the protocol are mostly based on controversial concepts, such as the precautionary approach (popularly known as the "precautionary principle"), which they see as veiled threat to fair trade. Other potentially controversial protocol provisions, still subject to further negotiation, will require mandatory labeling of bulk commodities or could call for environmental studies, and may establish compulsory product information and disclosure procedures. In general, provisions in the biosafety protocol are likely to shape market rules and impact US interests in biotechnology trade.

Convention on Biological Diversity and its Importance

As mentioned earlier, the biosafety protocol is an outgrowth of the convention on biological diversity. Only parties to the parent treaty, the CBD, are part of the official negotiations on protocols, which are effectively adjunct treaties on specific issues that, the parties believe, require elaboration or treatment. The United States played an active role in the negotiations that produced the CBD, even though the Congress has declined to approve its ratification. Similarly, because of the importance of the biosafety protocol in US agriculture and biotechnology, the United States has remained actively involved, sending delegations for its negotiations. These delegations actively participated in substantive discussions and formulation of positions that were made part of the official proceedings of the protocol negotiations. Since, the United States is not a party to the CBD, it is not in a position to sign the protocol. Thus, the history and status of the biodiversity convention is a key element in considering the role of US in the biosafety protocol. The convention on biological diversity (CBD) dates back to intensive negotiations in the

early 1990s. In 1992, world leaders came together in Rio de Janeiro, Brazil, to attend the world's largest summit—the United Nations Conference on Environment and Development (UNCED—popularly known as the "Earth Summit")—to take up the wide array of issues related to environmentally and economically sustainable use of the world's resources. One major area of concern was biological diversity—the concern that rapid loss of species is occurring around the world, and international cooperation was needed to address the problem. The CBD was one of two major treaties opened for signature at UNCED (the other was on climate change).

7.3. AIMS OF THE NATIONAL INSTITUTES OF HEALTH (NIH) GUIDELINES

The purpose of the NIH guidelines is to specify practices for constructing and handling recombinant deoxyribonucleic acid (DNA) molecules, and organisms and viruses containing recombinant DNA molecules.

- Any recombinant DNA experiment, which requires approval by NIH, must be submitted to NIH. If the approvals or other applicable clearances have been obtained from a government agency other than NIH (whether the experiment is referred to that agency by NIH, or sent directly there by the submitter), the experiment may proceed without the necessity for NIH review or approval.

- For experiments involving the deliberate transfer of recombinant DNA, or DNA or RNA derived from recombinant DNA, into human research participants (human gene transfer), no research participant shall be enrolled until the RAC review process has been completed.

- No research participant shall be enrolled at the clinical trial site until the following documentation has been submitted to NIH Office of biotechnology activities (OBA):

— IBC approval (from the clinical trial site)
— Institutional review board approval
— Institutional review board approved informed consent document
— Curriculum vitae of the principal investigator(s)
— NIH grant number(s) if applicable

Definition of Recombinant DNA Molecules

In the context of the NIH guidelines, recombinant DNA molecules are defined either as:

* molecules constructed outside the living cells by joining natural or synthetic DNA segments to DNA molecules that can replicate in a living cell; or
* molecules that result from the replication of those described above.

Synthetic DNA segments, which are likely to yield a potentially harmful polynucleotide or polypeptide (e.g., a toxin or a pharmacologically active agent) are considered as equivalent to their natural DNA counterpart. If the synthetic DNA segment is not expressed *in vivo* as a biologically active polynucleotide or polypeptide product, it is exempted from the NIH guidelines. Genomic DNA of plants and bacteria that have acquired a transposable element, even if the latter was donated from a recombinant vector no longer present, are not subject to the NIH guidelines unless the transposon itself contains recombinant DNA.

Risk Groups

Risk group is ultimately a subjective process. The investigator must make an initial risk assessment based on the Risk Group (RG) of an agent. Classification of human etiologic agents on the basis of hazard includes the biological agents, which are known to infect humans as well as selected animal agents that may pose theoretical risks if inoculated into humans. List of representative genera and species known to be pathogenic is included; mutated, recombined, and non-pathogenic species and strains are not considered. Non-infectious life cycle stages of parasites are excluded.

Agents are classified into four Risk Groups (RGs) according to their relative pathogenicity for healthy adult humans as follows:

* Risk Group 1 (RG1) agents are not associated with disease in healthy adult humans.
* Risk Group 2 (RG2) agents are associated with human diseases, which are rarely serious and for which preventive or therapeutic interventions are often available.
* Risk Group 3 (RG3) agents are associated with serious or lethal human diseases for which preventive or therapeutic interventions may be available.
* Risk Group 4 (RG4) agents are likely to cause serious or lethal human disease for which preventive or therapeutic interventions are not usually available.

Comprehensive Risk Assessment

While deciding on the appropriate containment for an experiment, the initial risk assessment from Classification of Human Etiologic Agents on the basis of hazard should be followed by a thorough consideration of the agent itself and how it is to be manipulated.

Factors to be considered while determining the level of containment include agent factors or biotic factors, and types of manipulation.

Biotic factors include virulence, pathogenicity, infectious dose, environmental stability, route of spread, communicability, operations, quantity, availability of vaccine or treatment, and gene product effects such as toxicity, physiological activity, and allergenicity. Any strain that is known to be more hazardous than the parent (wild-type) strain should be considered for handling at a higher containment level. Certain attenuated strains or strains that have been demonstrated to have irreversibly lost known virulence factors may qualify for a reduction of the containment level compared to the risk group assigned to the parent strain or organism. A final assessment of risk based on these considerations is then used to set the appropriate containment conditions for the experiment. The containment level required may be equivalent to the risk group classification of the agent, or it may be raised or lowered as a result of the above considerations. The institutional biosafety committee must approve the risk assessment and biosafety containment level for recombinant DNA experiments.

Types of manipulation

Careful consideration should be given to the types of manipulation planned for some higher risk group agents. For example, the RG2 dengue viruses may be cultured under the Biosafety Level (BL) 2 containment; however, when such agents are used for animal inoculation or transmission studies, a higher containment level is recommended. Similarly, RG3 agents such as Venezuelan equine encephalomyelitis and yellow fever viruses should be handled at a higher containment level for animal inoculation and transmission experiments.

Classification of Genetic Manipulation Research

NIH guidelines points out the six types of rDNA experiments that are mentioned below.

- Those that require Institutional Biosafety Committee (IBC) approval, RAC review, and NIH Director approval before initiation.
- Those that require NIH/OBA and Institutional Biosafety Committee approval before initiation.
- Those that require Institutional Biosafety Committee and institutional review board approvals and RAC review before research participant enrollment.
- Those that require Institutional Biosafety Committee approval before initiation.
- Those that require Institutional Biosafety Committee notification simultaneous with initiation.
- Those that are exempt from the NIH guidelines.

The details of each classification types are as follows:

- The Institutional Biosafety Committee must approve the risk assessment and the biosafety containment level for recombinant DNA experiments. Experiments require Institutional Biosafety Committee Approval, RAC review, and NIH Director approval before initiation. During the deliberate transfer of a drug resistance trait to microorganisms that are not known to acquire the trait naturally, if such acquisition could compromise the use of the drug to control disease in humans, veterinary medicine, or agriculture, it will be reviewed by the RAC.
- Experiments that require NIH/OBA and Institutional Biosafety Committee approval before initiation experiments in this category cannot be initiated without submission of relevant information on the proposed experiment to NIH/OBA. The containment conditions for such experiments will be determined by NIH/OBA in consultation with ad hoc experts. Such experiments require Institutional Biosafety Committee approval before initiation. Experiments involving the cloning of toxin molecules with LD50 of Less than 100 Nanograms per kilogram body weight deliberate formation of recombinant DNA containing genes for the biosynthesis of toxin molecules lethal for vertebrates at an LD50 of less than 100 nanograms per kilogram body weight (e.g., microbial toxins such as the botulinum toxins, tetanus toxin, diphtheria toxin, and *Shigella dysenteriae* neurotoxin). Specific approval has been given for the cloning in *Escherichia coli* K-12 of DNA containing genes coding for the biosynthesis of toxic molecules that are lethal to vertebrates at 100 nanograms to 100 micrograms per kilogram body weight.

- Experiments that require Institutional Biosafety Committee and Institutional Review Board Approvals and RAC Review include an experiment involving the deliberate transfer of recombinant DNA, or DNA or RNA derived from recombinant DNA, into human research participants (human gene transfer). In its evaluation of human gene transfer proposals, the RAC will consider whether a proposed human gene transfer experiment presents characteristics that warrant public RAC review and discussion.

- Experiments that require Institutional Biosafety Committee approval before initiation: Prior to the initiation of an experiment that falls into this category, the Principal Investigator must submit a registration document to the Institutional Biosafety Committee, which contains the following information:
  — The source(s) of DNA.
  — The nature of the inserted DNA sequences.
  — The host(s) and vector(s) to be used.
  — An attempt will be made to obtain expression of a foreign gene, and if so, indicate the protein that will be produced.
  — The containment conditions that will be implemented as specified in the NIH guidelines. For experiments in this category, the registration document should be dated, signed by the Principal Investigator, and filed with the Institutional Biosafety Committee. This committee shall review and approve all experiments in this category prior to their initiation. Requests to decrease the level of containment specified for experiments in this category will be considered by NIH.

- Experiments involving the formation of recombinant DNA molecules containing no more than two-thirds of the genome of any eukaryotic virus (all viruses from a single family being considered identical) may be propagated and maintained in cells in tissue culture using BL1 containment. For such experiments, it must be demonstrated that the cells lack helper virus for the specific families of defective viruses being used. If helper virus is present, procedures are specified. Experiments involving the use of infectious animal or plant DNA or RNA viruses, or defective animal or plant DNA or RNA viruses in the presence of helper virus in tissue culture systems, should be used. The DNA may contain fragments of the genome of viruses from more than one family, but each fragment should be less than two-thirds of a genome.

- The following recombinant DNA molecules are exempt from the NIH guidelines, and their registration with the Institutional Biosafety Committee is not required:
  — Those that are not in organisms or viruses.
  — Those that consist entirely of DNA segments from a single nonchromosomal or viral DNA source, though one or more of these segments may be a synthetic equivalent.
  — Those that consist entirely of DNA from a prokaryotic host, including its indigenous plasmids or viruses when propagated only in that host (or a closely related strain of the same species), or when transferred to another host by well established physiological means.
  — Those that consist entirely of DNA from an eukaryotic host, including its chloroplasts, mitochondria, or plasmids (but excluding viruses) when propagated only in that host (or a closely related strain of the same species).
  — Those that consist entirely of DNA segments from different species, which exchange DNA by known physiological processes, though one or more of the segments may be a synthetic equivalent. A list of such exchangers will be prepared and periodically revised by the NIH Director with advice of the RAC after appropriate notice and opportunity for public comment.

7.4. REGULATIONS SPECIFIC TO BIOTECHNOLOGY COMPANIES AND RESEARCH INSTITUTIONS

In the present scenario, biotechnology has led to broad changes in medicine, industrial processes, engineering and agriculture. In agriculture, biotechnology is being used to enhance the nutritional content of the crops, and to increase their resistance to pests and environmental stresses. Controversies have developed among various nations over whether certain types of GMOs constitute a potential hazard to human health or to the environment. GMOs have long been considered the subject of a number of biosafety concerns. Many genetically modified plants, for example, are produced by techniques that include the DNA sequences from common plant pathogens. Other biosafety concerns commonly cited are: the possibility of adverse impacts on non-target species or ecosystems; the potential for enhanced weediness of GM crops (i.e., a plant becomes invasive, or transfers resistance genes to its wild relatives); and the instability of inserted genes (i.e., the possibility that a gene will lose its effectiveness). To date, transfer of genes to weedy relatives has been observed, and a controversial laboratory study has shown potential damage to a non-target species. The types of research described in the following categories are conducted at biotechnology companies as well as research institutions, such as universities and hospitals.

Drugs, Medical Devices, Biological Products, and Diagnostic Products

Worldwide, many biotechnology companies are involved in the development of one or more therapeutics, medical devices, biological products or diagnostic products. The US Food and Drug Administration (FDA) regulates all of these products in the USA. Other countries in the world are almost following the same guidelines for the usage of these products. The materials, facilities, equipment and processes used to make these products require companies to submit an Environmental Impact Statement to the FDA. This statement evaluates not only the safety and usefulness of the product, but also the environmental and health impacts of processes, wastes and by-products involved in its manufacture. There are very few manufacturers in other industrial sectors that are regulated in this way. And majority of these companies has followed these regulations.

Animal Biologics, Plants, Pests, and Pesticides

The living organisms are sources of the large number of essential products. Any biotechnology company engaged in working with animal biologics, plants, pests, or pesticides is regulated by the US Department of Agriculture and the Department of Public Health. These materials have to be regulated when used as raw materials or as products themselves.

Recombinant DNA Technology

The advancement made in the fields of molecular biology and genetic engineering contributed immensely to the development of biotechnology production and utilization. Most biotechnology companies and research institutions utilize some form of recombinant DNA (rDNA) technology. This technology was developed in the research labs in 1970s, and became the basis of many new biotechnology companies in the 1980s. rDNA technology involves the manipulation of genes to produce a variety of products.

Although, there are no laws or regulations at either the federal or state level governing the use of genetically modified organisms, rDNA technology is the subject of the National Institutes of Health Guidelines (NIHG), and these guidelines are followed by most biotechnology companies in a large number of countries as if they were law. In order for a biotechnology company to receive its license for the manufacturing of pharmaceutical therapeutics and diagnostics, the FDA requires certification of compliance by NIHG as well as all other governmental and state regulations. The manufacturing of plants and pesticides is governed by the Environmental Protection Agency and the US Department of Agriculture in the USA.

National Institutes of Health Guidelines

As concern over the biosphere is increasing, stricter guidelines are established all over the world. In response to the concern in 1970s about the risks in conducting rDNA research, the National Institutes of Health (NIH) examined the known risks, and in 1976 published a guide of recommended safe practices in rDNA technology for research scientists who received NIH funding. In 1986, the NIH formalized their recommendations in the form of specific NIH Guidelines (NIHG). These guidelines are regularly updated as the history of safe use and the body of knowledge of recombinant organisms, especially those used in large-scale production in the industry, develops. These guidelines are the generally accepted standard for working with recombinant organisms. They outline how projects are classified, how and by whom they are reviewed, and what type of controls should be in place based upon the project and types of organisms used.

Community Regulations

The NIH guidelines form the basis of regulations adopted by several countries worldwide. The key element in this regulation is the provision for the establishment of an Institutional Biosafety Committee (IBC). This committee oversees the rDNA projects of an institution, with focus on biological safety. The committee has at least two members not affiliated with the institution, who represent the community-at-large and who should have backgrounds in science or public health. By having an Institutional Biosafety Committee, the community has direct review and knowledge of the research being conducted at the company. This type of community oversight is unique to the biotechnology industry.

This has given wider scope of controlling the rDNA work in the research/academic field. What should a community do to regulate rDNA work? What many communities have done, usually in cooperation with a company coming into the community, is to develop a by-law, regulation, or ordinance based on current NIH guidelines. The simplest form is a regulation under the local Board of Health. This basic regulation requires companies to register with the Board, giving assurance that its operation will follow NIH guidelines. These regulations allow the community and the commercial entity to understand what types of activities will be allowed in the locality. The community may choose to adopt the NIH guidelines as a whole as a local ordinance, or adopt specific regulations, which include the elements outlined in the guidelines including the formation of an Institutional Biosafety Committee.

Overall, the trend in community regulations has been to have less complex regulations over time. Some communities that have had regulations for ten years or more have revised their regulations to be less restrictive and complicated. The reason for this is based on history. More than 15 years of experience in Massachusetts and in the larger scientific community has shown rDNA technology to be of very low risk for both the employees as well as the community. Additional requirements, such as special permits, ordinances, or regulatory committees likely create an atmosphere, which discourages biotechnology companies from settling in a community. The NIH guidelines, together with other safety and environmental regulations, have already and will continue to serve and protect the community from any potential risks from biotechnology, while creating an atmosphere that encourages biotechnology companies to locate in a community.

Risk Assessment

Risk assessment involves determination of the potential and anticipated adverse effects of the recombinant DNA research on the concerned workers. It also involves the determination of adverse effects of the products of such research on human health and the environment, consequent to their accidental or deliberate release (in case of living organisms) or as a result of their consumption. Risk assessment should be carried out in a scientifically sound and transparent manner, and should be in accordance with recognized risk assessment techniques. Risk assessment consists of hazard identification, hazard analysis, consequence analysis, risk determination, and risk evaluation.

Sample Recombinant DNA Technology Use Regulation for Communities

Applicability

These regulations apply to all the recombinant DNA (rDNA) technology activities. All institutions performing rDNA activities must conform to the "Regulations" as defined, and other such health regulations as the Board may promulgate.

Permit requirements

- All institutions ("Institution" means any single individual, group of individuals, or organization, whether public or private) proposing any use of rDNA technology ("Recombinant DNA molecules" (rDNA) and "organisms and viruses containing rDNA" are those defined in the "Guidelines,"), as defined in and not exempted by the NIH Guidelines, must obtain an rDNA use permit from the Board before engaging in any rDNA activity, including construction or renovation of facilities for rDNA use(s).
- Permit requirements should, at minimum, include written agreement and evidence to demonstrate the capacity to:
  - follow the guidelines as defined in these regulations;
  - adhere to any other conditions as set forth in these regulations;
  - allow inspections, at reasonable times, of both the institution's facilities and records, as related to these regulations;
  - adhere to a health and safety manual prepared by the institution, which contains all procedures relevant to the use of rDNA at all levels of containment at the institution. The manual should also include a plan for waste disposal in compliance with all applicable federal, state, and local regulations;
  - establish and implement a training program of safeguards and procedures for personnel using rDNA; and
  - establish an Institutional Biosafety Committee (IBC). It will be a committee established by an institution in accordance with the "guidelines" and the terms set forth in these regulations. in accordance with the guidelines, including two members not associated with the institution.

- Applications for permits should be acted upon within sixty (60) days. Failure to act on complete applications within this time period will constitute an approval unless the applicant has been notified of a delay.
- The Board may establish requirements for regular reports from institutions, and may issue permits which contain restrictions or conditions relative to public health interests.

Large scale use

All institutions intending to perform large scale rDNA (as defined in the guidelines) require specific approval as part of the permit or as a modification of an existing permit. The Board will act within thirty (30) days from the receipt of a request for the modification of an existing permit to conduct large scale work. Failure to act within the allotted time period will constitute an approval unless the institution has been notified of a delay.

The Board may request additional information about the proposed large scale activity during its review of the request. All large scale activity must be clearly noted in minutes of the IBC.

Miscellaneous

* Institutions should provide appropriate medical surveillance for their employees, as determined by their IBC and consistent with the guidelines.
* Institutions will report any significant problems with or violations of the guidelines, and any significant accidents or illnesses related to rDNA activities to the Board within thirty days. Any such incidents, which may have an impact on public health, should be reported immediately.

Restrictions

* Biosafety level 3 work may warrant further review.
* rDNA use requiring biosafety level 4 shall not be permitted.
* Any deliberate release of surviving rDNA organisms into the environment should be done only in compliance with the guidelines and the rules, regulations and laws of the commonwealth.

Severability

If any portion of this regulation is for any reason held invalid or unconstitutional by any court of competent jurisdiction, such part should be deemed a separate and independent part, and such holding should not affect the validity of the remaining parts of the regulation.

7.5. BIOSAFETY GUIDELINES IN INDIA

The Department of Biotechnology (DBT) implements the research and development experiments utilizing GMOs and recombinant DNA products, while Ministry of Environment and Forests (MOEF) implements the large scale commercial use of these. The salient features of these guidelines and regulations are as follows:

* Every organization involved in research and development using recombinant DNA technology is required to set up an Institutional Biosafety Committee (IBC), which has a DBT nominee. IBC will provide half-yearly reports on the ongoing projects to Review Committee for Genetic Manipulation (RCGM).
* The DBT has a RCGM, which reviews all the approvals of ongoing projects on GMOs and several other issues related to recombinant DNA research and development.
* Each state has, in addition, a state Biotechnology Coordination Committee and a District Level Committee, which are also involved along with RCGM in the inspection and monitoring of experiments at field sites.
* The MOEF Committee (also called the Genetic Engineering Approval Committee (GEAC)), has subject specialists as members and is the competent authority to decide on the large-scale use of GMOs.
* The guidelines recognize 4 levels of risk in case of experiments with microorganisms, based on the pathogenicity of the micro-organisms, local prevalence of the concerned disease, and epidemic causing strains in India.
* Experiments with microorganisms, plants, and animals are grouped into the following three categories:
  — Exempt category (for self-cloning experiments).
  — Category requiring intimation of initiation to competent authority (e.g., experiments involving non pathogenic DNA vector systems).

— Category requiring review and approval by the competent authority (e.g., cloning of genes for toxins, antibiotics, resistance, etc.).

- Four different biosafety levels are recognized and containment facilities for each level are recommended for necessary safeguards.
- Physical containment envisages to limit the spread of dangerous microorganisms by:
  — good laboratory practices;
  — safety equipment; and
  — laboratory design and facilities.
- Biological containment consists of the use of vectors and hosts in such a way so that it:
  — can limit the infectivity of vector to specific hosts; and
  — control the host-vector survival in the environment.
- Experiments with plants are to be done under two types of special environment that maybe achieved by using glasshouse containment. **Glasshouse containment A** is recommended when nonpathogenic vector systems and regeneration from single cells are used. It is not used in case of plant pathogens. **Glasshouse containment B** is recommended for experiments involving:
  — genetically manipulated plant pathogens, including plant viruses.
  — the plants regenerated from the cells transformed by genetically manipulated pathogen vector system, which contains the pathogen.
- Application for recognition of research facility to carry out genetic manipulation should be forwarded to the Department of Environment before the commencement of work.
- The controlled release of GMOs should be done under appropriate containment facilities to ensure safety.

- Prerelease tests of GMOs in agriculture should include elucidation of requirements for vegetative growth, persistence, and stability in small plots and experimental fields.
- Planned field experiments with transgenic plants are permitted only after a step-wise evaluation either in India or elsewhere, to generate data on the following:
  — Promoter sequence.
  — Target gene sequence.
  — Regulatory mechanism utilized in the expression cassette.
  — Diagram of the expression cassette to describe the marker genes used in detail.
  — Cell lines used for shuttling and amplification of the cassette, and several other details of the procedure, and the consequences thereof, used for the production of the transgenic plants.

In addition, the following is also insisted upon:
- Laboratory data to show that protein products of transgenes are safe for the environment and human beings.
- Isolation distance applicable to foundation seed of the crop be provided all round the field having transgenic crops.
- A few rows of the same crop as the transgenic one should be planted beyond the isolation distance to act as pollen trap.
- Non-transgenic plants should be grown within the isolation distance at 1 or 5 m intervals to determine the distance of pollen escape.
- All the vegetative plants and leftover seeds must be destroyed by burning after the completion of the experiment.
- After the experiment, the land may be left fallow, and all plants that emerge must be destroyed.
- The experimental field may be visited by the company authorized personal only, and all records of visits should be maintained.
- Full account of transgenic seeds should be kept, and no part of this seed should be transacted or propagated further without authorization.

7.6. GUIDELINES FOR RESEARCH IN TRANSGENIC ORGANISMS

New companies are continuing to form around the developing technologies, in agricultural biotechnology, pharmacology, gene sequencing, genetic testing, DNA profiling, and human gene therapy or human gene transfer. And universities are enthusiastically setting up companies and cutting multimillion-dollar deals with industry. This made large number of organizations are ultimately focusing on the various types of transgenic organisms and their usage. As finite natural resources are rapidly being exhausted, and the global environment becomes increasingly stressed as a result of industrial and human pollution, there is mounting awareness of the need for clean, renewable, industrial and agricultural strategies that will help ensure the future supply of raw materials for the manufacture of products and services we require, while at the same time reducing stress on the planetary environment from pollution and other negative side effects, such as global climate change, that currently mark man's material progress.

Biohazards are defined as biological agents or substances present in or arising from the work environment of transgenic organisms, which may be a hazard to the health or well-being of the worker or community. Examples of biohazards are infectious organisms such as bacteria and viruses, some recombinant DNA, zoonoses (diseases which occur primarily in animals but may be transmitted to humans), and primary cell cultures that are initiated with tissues from infected humans or animals. Recombinant DNA molecules are defined as either: (i) molecules that are constructed outside living cells by joining natural or synthetic DNA segments to DNA molecules that can replicate in a living cell, or (ii) molecules that result from the replication of those described in (i) above.

The questions arise at present that transgenic organisms or genetically modified organisms (GMOs) being directed to the right research priorities. Despite some hopeful signs, the inventory suggests that this is not yet the case. There are specific technical challenges that are difficult to handle through traditional crop breeding, where transgenics may offer great possibilities for crops and forestry: these include drought and heat-tolerance; improved nutrient uptake and rooting; biological nitrogen fixation; responses to carbon dioxide; and tolerance to key abiotic stresses, such as salinity and drought. As part of a strategy of diet diversification, genetic modification may help us to address protein, vitamin and iron deficiencies as seen in the famous example of the golden rice.

To address the varieties of the human problems researchers are focusing on the different aspects of biotechnology applications. Biotechnology companies are involved in agriculture; human diseases such as AIDS, Alzheimer's, cystic fibrosis and cancer; biodegradable materials; environmental management; food and fuel production; marine science; and waste management. Biotechnology companies operate under strict federal, state and local guidelines in handling and usage of hazardous materials. Any pathogenic, infectious or otherwise hazardous microorganisms or genetic material is rendered harmless before being released into the environment.

In such biotechnologically advanced situations in addition to the government agencies, local community representatives involved in the industry review projects and visit facilities. Protective clothing is provided to protect the products and biological processes from foreign substances, such as human contamination, within the controlled environment. Naturally occurring microorganisms when introduced into that environment could interfere, delay or ruin the biological processes that are not repairable or unknown consequences.

The higher animals, primarily rodents such as mice and rats, are a necessary component of scientific research. The FDA requires animal research and testing on all drugs and medical devices before they are tested in humans. In handling animals, all companies are under the jurisdiction of the US Department of Agriculture and are required to conform to the standards of the Animal Welfare Act. In addition, the local Institutional Biosafety Committee formed for those companies involved in recombinant DNA technology has local representatives, often from the Board of Health, that visit companies on a regular basis.

At present unfortunately, multiple genes govern most of the traits needed, and it will probably be some time before farmers and consumers benefit from such research. The perceived profit potential of GMOs has already changed the direction of investment in research and development in both the public and private sectors away from systems-based approaches to pest management, and towards a greater reliance on monocultures with potentially adverse effects on landscape biodiversity. The possible long-term environmental costs of such strategies may be overlooked. If research is to address the challenges of agriculture in the future, we need to put genetic modification in context, and realize that it is but one of the many elements of agricultural change.

In advanced biotechnological thinking scientists must not be blinded by the glamour of cutting-edge molecular science for its own sake. I am also worried about the disconnection between the laboratory and the field and the decline of agronomy as an integrative science. Governments need to be vigilant and not let this glamour, or the perception by private industry of major profit opportunities, draw investment away from research in other, more traditional fields such as water and soil management or ecology, and from public sector research. At the same time, the best science is developed in a climate of intellectual freedom without much direct government interference. It is a difficult balance to strike! The cost of molecular technologies in plant breeding and the size of markets that are required to recoup the profits lead to the growing use of "hard" intellectual property rights over seeds and planting material and the tools of genetic engineering. This changes the relationship between the public and private sectors, to the detriment of the public sector.

At present in the era of biotechnology the policy question that governments must take up, in both the national and international contexts, is how to ensure that public research is not the "poor relation. In developing countries in particular, it is important for the public sector to retain enough capacity, resources and freedom of action to provide the services on which their national private sectors can build. They will also need to build their policy and regulatory capacities with regard to transgenic crops that originate elsewhere, and in this FAO has an "honest broker" role to play. Thus we need a science that, in a globalized world, addresses local agriculture and local food security and helps to keep the rural areas livable and attractive to young people in search of employment. The constraints to doing so are more economic than technical.

Globalization is not only a question of size, but also of kind: it is inextricably linked to privatization, and major economic restructuring in both developed and developing countries has changed the balance of public and private sectors, and the privatization of knowledge through intellectual property rights. This results in concentration, and there is often little economic incentive for global companies to address local needs. It has been estimated that the world's top ten companies account for about 85% of the global market for the seed and agrochemical industries.

In 1998, just four companies controlled 69% of the seed market in the USA. Of course, globalization also promises to create opportunities for millions of farmers throughout the world, can offer opportunities to the poorer countries, and can drive rapid development and local capital accumulation. But it may also exacerbate existing differences among countries. Developing transgenic crops implies massive investments, and the need for massive returns.

The small number of GM technologies currently in use suggests that there is a real danger that the scale of the investment may lead to selective concentration on species and problems of global importance, and concomitant capital inertia. In this context, I must also sound a note of warning about the rising costs of regulation. In the health industry, where regulation is heaviest, bringing a pharmaceutical to market now costs about $500 million; bringing a pesticide to market can cost about $200 million; and it has recently been reported that a GM crop costs about $30 million to produce, to which regulatory costs can add up to a further $5–6 million. The question is that these crops can bear such costs in the commercial market. The methods that are to serve local needs, small farmers and poor consumers, and promote genetic and dietary diversity are to be addressed.

Developing Agreed National and International Instruments of Governance

Most policy and regulatory activities regarding GMOs are national. International agreements have a role to play in harmonizing national regulatory systems, for example on transboundary questions, and when trade standards are involved. International agreements can also help governments, particularly the smaller governments, to rationalize their national capacities and investments through common action. In relation to GM crops, the major instruments concern risk-assessment and management in two areas: food and health; and the environment. GMOs in food and agriculture have been around long enough for us to have some background against which to assess possible risks. There has been as yet no proven effect on human life. Moreover, a recent study of GMO releases into the environment in Britain showed that they had not survived in nature, as, in fact, they only would if they had an evolutionary advantage, which remains an important possibility. But let me say that state-of-the-art knowledge may not always be adequate to assess the risk of GMOs to health or the environment. Of course, lack of evidence of adverse effects is not the same as knowledge that genetic modification is safe.

In FAO we deal with environmental and food risk-assessment questions under a more general concept of "biosecurity", namely the application of sanitary and phytosanitary measures (SPS) and a methodology of risk analysis for food and agriculture, including fisheries and forestry. They include:

- The protection of animals or plants from pests, diseases, or disease-causing organisms.
- The protection of human or animal life from risks arising from additives, contaminants, residues toxins or disease-causing organisms in foods, beverages or foodstuffs.
- The protection of human life or health from risks arising from diseases carried by animals, plants or plant products, or from pests.
- The prevention or limitation of other damage from pests.

Moreover, FAO has specific standard-setting responsibilities that relate directly to GMOs under the Technical Barriers to Trade (TBT) and the Sanitary and Phytosanitary (SPS) Agreements of the World Trade Organization. There are several important international mechanisms in the context of GMOs. Firstly, the cartagena protocol to the Convention on Biological Diversity, which was adopted in January 2000. It will come into force after fifty ratifications and is the main international agreement in relation to GMOs ("LMOs", or "Living Modified Organisms", in the language of the protocol).

It covers the transboundary movement, transit, handling and use of all LMOs (except pharmaceuticals) that may have adverse effects on the conservation and sustainable use of biological diversity, taking also into account risks to human health. It allows for standard-setting in relation to the handling, transport, packaging and identification of LMOs.

Secondly, food safety aspects of Genetically Modified Organisms are, at international level, dealt with by the FAO/WHO *Codex Alimentarius*, which covers all aspects of food safety. The *Codex* is currently working on standards for risk assessment for labeling, and for several other food safety aspects of GMOs. The *Codex* standards are recognized by the SPS and TLC agreements.

Thirdly, the International Plant Protection Convention (IPPC) has the objective of preventing the spread and introduction of plant and plant product pests, including weeds and other species that have indirect effects on both wild and cultivated plants, and to promote appropriate control measures. This also applies to risks associated with LMOs. The IPPC sets International Standards for Phytosanitary Measures (ISPMs), which are also recognized in the SPS agreement.

It is also in the process of establishing practical cooperation with the CBD and its biosafety protocol. An IPPC expert working group met in September 2001, in coordination with experts from the CBD, to develop a detailed standard specification for an ISPM that identifies the plant pest risks associated with LMOs, and ways of assessing these risks. It is highly desirable to harmonise the guidelines and regulations of various countries otherwise research workers and organizations, especially private companies, may shift their research experiments to such countries that have more lax regulations or no regulations at all. In the past, several such shifts were made, some of which became scandals. One such case occurred in 1980.

Researches of University of California working a gene therapy for thalassemia were advised by the university to continue work on animals before testing it on human subjects. Instead, the researchers treated two patients in Italy and Israel. It has been advocated that:

- Appropriate legislation should be promoted in those countries that lack them.
- Public consciousness of biosafety issue should be increased.
- A licensing may be imposed on the expert of genetically manipulated organisms.
- A voluntary code of conduct. That is an international agreement between industry and governments on various biotechnological products and processes may be promoted.

In 1991, the Informal Working Group on Biosafety (IWGB), the main body of UN on biosafety issued the draft voluntary code of conduct for release of organisms in to the environment.

The consultative group on International Agricultural Research (CGIAR) created a task force on biotechnology (BIOTASK) in 1989. BIOTASK has on its agenda, along with other things, regulatory issues and environmental release of GMOs related to agriculture.

7.7. INTRODUCTION OF GENETICALLY MODIFIED ORGANISMS INTO THE ENVIRONMENT

Recent concern over the introduction of genetically engineered organisms into the environment has stimulated biologists to look for the analogues of the invasion process; and the consequences of the introduction of species into the open environment.

A project of scientific committee on problems of the environment (SCOPE) has recently summarized the effect of biological invasions across numerous taxonomic groups. This comprehensive review of species introductions provides the basis for analysis of potential risks of introducing GEOs in the environment.

The SCOPE program was initiated in 1982 with the objectives of:

- determining the properties of invading species which contribute to their success as invaders; and
- identifying the characteristics that make certain ecosystems more vulnerable to invasions than others.

Predicting the Success of Invading Species

The reasons behind our lack of the kind of precision in predictability rest both in the complexity of the problem and the nature of the data. As noted by Ehrlich, other factors must be taken into account. These include the particular physiological and behavioral status of the founding population, as well as the unique physical and biological status of the habitat being invaded. A species could be a good invader in one ecological system, but a poor invader in another.

Population features

- Number of individuals
- Sex and sex ratio
- Physiological status maturity, state in breeding cycle, health, and acclimation status
- Genetic composition, genetic variability found in population, and ecotype of origin

Habitat and community features

- Season
- Weather
- Size and structure of the population of resource organisms
- Size and structure of the population of competitors
- Size and structure of the population of predators
- Size and structure of the population of parasites and pathogens

Hence, it is concluded that with the detailed knowledge of the biology of the potential invader and the characteristics of the target environment, relatively good predictions can be made about the potential success and impact of an invading species.

7.8. ENVIRONMENTAL BENEFITS

- A February 2003 report, published in the prestigious proceedings of the royal society, detailed how a field trial of herbicide tolerant sugar beet in the United Kingdom helped wild life thrive.
- Creative use of GM crops could bring back increasing numbers of endangered wild life and birds such as skylarks and finches.
- The benefits range from improved habitat (cleaner drinking water) for birds such as Pheasants, bobwhite quail and doves, and a reduction in harmful greenhouse gases and fuel use.
- The results clearly show that soil, air and water quality is enhanced through the responsible use of biotechnology derived soybean, corn and cotton crops.

Specifically, the study found that genetically modified crops can:

- **prevent soil erosion:** GM soybeans and cotton have led to a significant increase in the adoption of environment friendly no-till farming practices, which conserve top soil, preserve soil moisture, and reduce run off.
- **Improve water quality:** GM soybeans and cotton enable farmers to use more benign herbicides that rapidly dissipate in soil and water.
- **improve air quality:** The adoption of no-till farming practices significantly reduce the release of greenhouse emissions, which may help slow global warming.
- **increase biodiversity:** Biotech cotton has been documented to have a positive effect on the number and diversity of beneficial insects in US and Australian cotton fields. In addition, the adoption of no-till farming practices creates additional habitat for birds and other wildlife.

New biotech crop varieties are also being tested, and they will do even more to help preserve the environment. Several different crops are being developed that can thrive in difficult conditions such as areas plagued by drought; soil contaminated by salt from too much irrigation; or soil laced with aluminium, which is toxic to plants. Some examples include:

- Rice that can better withstand droughts and thrive in marginal soil.
- Tomatoes designed to grow in overly irrigated saline soil.
- Rice and corn that are more tolerant to aluminium—a common soil toxin.

- Rice that can survive long periods under water in extremely wet climate.

7.9. FIELD TRIALS WITH GENETICALLY-MODIFIED PLANTS

The risk of genetic engineering products lies in the possible unexpected effects of the exotic genes in the new cell environment and the new ecological surroundings. The type of regulation necessary during field trials depends on the ability of modified plant to survive, disperse, reproduce and hybridize with crops and wild plants. In cross-pollinated species, isolation by distance is rarely sufficient to prevent inter-population mating. Suggestions have been made to introduce male-sterility to prevent cross-pollination. Another approach to eliminate transgene transmission through pollen is to integrate it into the chloroplast genome.

Transgenic varieties of crops expressing "cry" genes are in commercial cultivation. Cry proteins are rapidly degraded by stomach juices of vertebrates, but they could have harmful effects on non target insect species, e.g., honeybees. Oilseed rape (*B. napus*) expressing CpT I (cow pea trypsin inhibitor), chitinase, and B-1,3-glucanase has been used for assaying the impact of these genes on honeybees. Chitinase did not effect learning performance, while CpT I and B-1,3-glucanase had detrimental effects.

In general, the effects of transgenic crop varieties on the incidence and dynamics of natural enemies of insect pests appear to be negligible and are often comparable to those of non transgenic varieties of the same crop, e.g., transgenic tobacco, cotton [expressing cry IA(C)], potatoes etc. But in some cases, cry gene expressing transgenic varieties may lead to adverse effects on predaters e.g., chrysopid larvae predating on caterpillars fed on cry gene-expressing maize (reduced fitness), lady bird beetles predating on peach-potato aphids fed on potatoes expressing a lectin gene (reduced reproduction rate), etc.

It has been reported that cry proteins leak into the soil from transgenic maize roots, where they persist for weeks. The cry proteins become associated with soil clay/humus, which makes them non available for microbial degradation. It has been suggested that this protein could protect the plants from insect pests that damage roots of crops, but it could also harm benign soil organisms or select resistant insect pests.

7.10. CONTAINMENT LEVELS

Effective biological safety programs have been operative in a variety of laboratories for many years. Considerable information already exists about the design of physical containment facilities and selection of laboratory procedures applicable to organisms carrying recombinant DNA. The existing programs rely upon mechanisms that can be divided into two categories:

- A set of standard practices that are generally used in microbiological laboratories.
- Special procedures, equipment, and laboratory installations that provide physical barriers, which are applied in varying degrees according to the estimated biohazard.

Four biosafety levels are described in physical containment. These biosafety levels consist of combinations of laboratory practices and techniques, safety equipment, and laboratory facilities appropriate for the operations performed, and are based on the potential hazards imposed by the agents used and for the laboratory function and activity. Biosafety level 4 provides the most stringent containment conditions, whereas biosafety level 1 is the least stringent.

Experiments involving recombinant DNA lead themselves to a third containment mechanism, namely, the application of specific biological barriers. Natural barriers exist that limit either:

- the infectivity of a vector or vehicle (plasmid or virus) for specific hosts or

- its dissemination and survival in the environment.

Vectors, which provide the means for recombinant DNA and/or host cell replication, can be genetically designed to decrease, by many orders of magnitude, the probability of dissemination of recombinant DNA outside the laboratory. Since these three means of containment are complementary, different levels of containment can be established that apply various combinations of the physical and biological barriers along with a constant use of standard practices. Categories of containment are considered separately so that such combinations can be conveniently expressed in the NIH guidelines.

Physical containment may not always be appropriate for all organisms due to their physical size, the number of organisms needed for an experiment, or the particular growth requirements of the organism. Likewise, biological containment may not be appropriate for all organisms, particularly higher eukaryotic organisms. However, significant information exists about the design of research facilities and experimental procedures that are applicable to the organisms containing recombinant DNA, which is either integrated into the genome, or into microorganisms associated with higher organisms as a symbiont, pathogen, or other relationship. This information describes facilities for physical containment of organisms used in non-traditional laboratory settings, and special practices for limiting or excluding the unwanted establishment, transfer of genetic information, and dissemination of organisms beyond the intended location, based on both physical and biological containment principles.

Research conducted in accordance with these conditions effectively confines the organism. For research involving plants, four biosafety levels (BL1-P through BL4-P) are described in Appendix P (physical and biological containment for recombinant DNA research involving plants).

BL1-P is designed to provide a moderate level of containment for experiments, in which there is convincing biological evidence that precludes the possibility of survival, transfer, or dissemination of recombinant DNA into the environment, or in which there is no recognizable and predictable risk to the environment in the event of accidental release. BL2-P is designed to provide a greater level of containment for experiments involving plants and certain associated organisms in which there is a recognized possibility of survival, transmission, or dissemination of recombinant DNA containing organisms. But the consequence of such an inadvertent release has a predictably minimal biological impact. BL3-P and BL4-P describe additional containment conditions for research in plants, certain pathogens, and other organisms that require special containment because of their detrimental impact on man made or natural ecosystems.

BL1-P relies upon accepted scientific practices for conducting research in most ordinary greenhouse or growth chamber facilities, and incorporates accepted procedures for good pest control and cultural practices. BL1-P facilities and procedures provide a modified and protected environment for the propagation of plants and microorganisms associated with the plants, and a degree of containment that adequately controls the potential for release of biologically viable plants, plant parts, and microorganisms associated with them. BL2-P and BL3-P rely upon accepted scientific practices for conducting research in greenhouse with organisms infecting or infesting plants in a manner that minimizes or prevents inadvertent contamination of plants within or surrounding the greenhouse. BL4-P describes facilities and practices known to provide containment of certain exotic plant pathogens. For research involving animals, which are of a size or have growth requirements that precludes the use of conventional primary containment systems used for small laboratory animals, four biosafety levels (BL1-N through BL4-N).

BL1-N describes containment for animals that have been modified by stable introduction of recombinant DNA, or DNA derived therefrom, into the germ-line (transgenic animals), and experiments involving viable recombinant DNA-modified microorganisms and is designed to eliminate the possibility of sexual transmission of the modified genome or transmission of recombinant DNA-derived viruses known to be transmitted from animal parent to offspring only by sexual reproduction. Procedures, practices, and facilities follow classical methods of avoiding genetic exchange between animals. BL2-N describes containment which is used for transgenic animals associated with recombinant DNA-derived organisms and is designed to eliminate the possibility of vertical or horizontal transmission. Procedures, practices and facilities follow classical methods of avoiding genetic exchange between animals or controlling arthropod transmission. BL3-N and BL4-N describe higher levels of containment for research with certain transgenic animals involving agents which pose recognized hazard.

For constructing the NIH guidelines, it was necessary to define boundary conditions for the different levels of physical and biological containment, and for the classes of experiments to which they apply. These definitions do not take into account all existing and anticipated information on special procedures that will allow particular experiments to be conducted under different conditions than indicated here without affecting risk. Individual investigators and Institutional Biosafety Committees are urged to devise simple and more effective containment procedures, and to submit recommended changes in the NIH guidelines to permit the use of these procedures.

Thus, containment may be defined as a combination of laboratory procedures, equipment and installations, and host-vector systems designed to minimize accidental release of organisms during laboratory operations, their dissemination and survival in the environment, and accidental infection of laboratory workers and persons outside the laboratory. The containment may be either:

- physical containment or;
- biological containment.

Physical Containment

Physical containment consists of the use of special laboratory design, containment equipment, and special operational procedures to restrict the number of organisms released accidentally during normal laboratory operations, and to prevent infection of laboratory workers. Physical containment is grouped into four categories, referred to as BL1 (biosafety level 1), BL2, BL3, and BL4. BL1 is applicable to non pathogenic organisms, while BL4 represents the most stringent containment level. As the biosafety level increases, the laboratory design, facilities, and the operational procedures become more and more rigid and stringent, so as to minimize the risk of accidental release of GMOs as well as infection of laboratory workers.

Physical Containment Levels

The objective of physical containment is to confine the organisms containing recombinant DNA molecules, and to reduce the potential for the exposure of laboratory workers, persons outside the laboratory, and the environment to the organisms containing recombinant DNA molecules. Physical containment is achieved through the use of laboratory practices, containment equipment, and special laboratory design. Emphasis is placed on primary means of physical containment provided by laboratory practices and containment equipment. Special laboratory design provides a secondary means of protection against the accidental release of organisms outside the laboratory or into the environment.

Special laboratory design is used primarily in facilities in which experiments of moderate to high potential hazard are performed. Four levels of physical containment, which are designated as BL1, BL2, BL3, and BL4 are described. It should be emphasized that the descriptions and assignments of physical containment detailed below are based on existing approaches to containment of pathogenic organisms. For example, the National Cancer Institute describes three levels for research on oncogenic viruses that roughly correspond to BL2, BL3, and BL4. It is recognized that several different combinations of laboratory practices, containment equipment, and special laboratory design may be appropriate for containment of specific research activities.

The NIH guidelines, therefore, allow alternative selections of primary containment equipment within facilities that have been designed to provide BL3 and BL4 levels of physical containment. The selection of alternative methods of primary containment is dependent on the level of biological containment provided by the host-vector system used in the experiment. Consideration will be given to other combinations that achieve an equivalent level of containment.

Biosafety level 1 is suitable for work involving agents of unknown or minimal potential hazard to laboratory personnel and the environment. The laboratory is not separated from the general traffic patterns in the building. Work is generally conducted on open bench tops. Special containment equipment is not required or generally used. Laboratory personnel have specific training in the procedures conducted in the laboratory, and are supervised by a scientist having a general training in microbiology or a related science.

Standard microbiological practices have to be followed by all the persons involved in the work. Access to the laboratory is limited or restricted at the discretion of the principal investigator when experiments are in progress. Work surface is decontaminated once a day and after any spill of viable material. All contaminated liquid or solid waste is decontaminated before disposal. Mechanical pipetting devices are used; mouth pipetting is prohibited. Eating, drinking, smoking, and applying cosmetics is not permitted in the work area. Food may be stored in cabinets or refrigerators designated for this purpose only. Lab personnel should always wash their hands after handling materials involving organisms containing recombinant DNA molecules and animals before leaving the laboratory. All procedures should be performed carefully to minimize the creation of aerosols. In the interest of good personal hygiene, facilities (e.g., hand washing sink, shower, changing room) and protective clothing (e.g., uniforms, laboratory coats) should be provided that are appropriate in case of exposure to viable organisms containing recombinant DNA molecules.

Special practices should always be followed. Contaminated material, which is to be decontaminated at a site away from the laboratory, is placed in a durable leak-proof container that is closed before being removed from the laboratory. An insect and rodent control program should be followed regularly. Special containment equipment is generally not required for manipulation of agents assigned to this category. The laboratory is designed so that it can be easily cleaned. Bench tops are impervious to water and resistant to acids, alkalis, organic solvents, and moderate heat. Laboratory furniture is sturdy. Space between benches, cabinets, and equipment is accessible for cleaning. Each laboratory should contain a sink for washing hands. If the laboratory has windows that open, they should be fitted with fly screens.

Biosafety level 2 is similar to level 1 and is suitable for work involving agents of moderate potential hazard to personnel and the environment.

- Laboratory personnel have specific training in handling pathogenic agents and are directed by competent scientists.
- Access to the laboratory is limited when work is being conducted.
- Certain procedures in which infectious aerosols are created are conducted in biological

safety cabinets or other physical containment equipment.

Access to the laboratory is limited or restricted by the principal investigator when work with organisms containing recombinant DNA molecules is in progress. Work surfaces are decontaminated at least once a day and after any spill of viable material. All contaminated liquid or solid wastes are decontaminated before disposal. Mechanical pipetting devices are used; mouth pipetting is prohibited. Eating, drinking, smoking, and applying cosmetics is not permitted in the work area. Food may be stored in cabinets or refrigerators designated for this purpose only. People wash their hands after handling materials involving organisms containing recombinant DNA molecules and animals, and when leaving the laboratory. All procedures are performed carefully to minimize the creation of aerosols. Experiments of lesser biohazard potential can be conducted concurrently in carefully demarcated areas of the same laboratory.

When the organisms containing recombinant DNA molecules require special provisions for entry into the laboratory (e.g., vaccination), a hazard warning sign incorporating the universal biosafety symbol is posted on the access door of the laboratory. The sign identifies the agent, lists the name and telephone number of the principal investigator or other responsible person(s), and indicates the special requirement(s) for entering the laboratory. Laboratory coats, gowns, socks, or uniforms are worn while in the laboratory. Before leaving the laboratory for non-laboratory areas (e.g., cafeteria, library, administrative offices), this protective clothing is removed and left in the laboratory, or covered with a clean coat not used in the laboratory. Special care is taken to avoid skin contamination with organisms containing recombinant DNA molecules.

Gloves should be worn while handling experimental animals, and when skin contact with the agent is unavoidable. All wastes from laboratories and animal rooms are appropriately decontaminated before disposal. When appropriate, considering the agent(s) handled, baseline serum samples of laboratory personnel and those at-risk are collected and stored. Additional serum specimens may be collected periodically, depending on the agents handled or the function of the facility.

Each laboratory contains a sink for washing hands. If the laboratory has windows that can be opened, they are fitted with fly screens. An autoclave is available for decontaminating laboratory wastes.

Biosafety level 3 is applicable to clinical, diagnostic, teaching, research, or production facilities in which work is conducted with indigenous or exotic agents that may cause serious or potentially lethal diseases as a result of exposure by inhalation. Laboratory personnel have specific training in handling pathogenic and potentially lethal agents, and are supervised by competent scientists who have experience in working with these agents. All procedures involving the manipulation of infectious material are conducted within biological safety cabinets or other physical containment devices, or by personnel wearing appropriate protective clothing and devices. The laboratory has special engineering and design features. It is recognized, however, that many existing facilities may not have all the facility safeguards recommended for BL3 (e.g., access zone, sealed penetrations, and directional airflow, etc.).

In these circumstances, acceptable safety may be achieved for routine or repetitive operations (e.g., diagnostic procedures involving the propagation of an agent for identification, typing, and susceptibility testing) in laboratories where facility features satisfy BL2 recommendations, provided the recommended "Standard Microbiological Practices", "Special Practices", and "Containment Equipment" for BL3 are rigorously followed. The decision to implement this modification of BL3 recommendations should be made only by the principal investigator.

If experiments involving other organisms that require lower levels of containment are to be conducted in the same laboratory concurrently with experiments requiring BL3 level physical containment, they should be conducted in accordance with all BL3 level laboratory practices. When organisms containing recombinant DNA molecules, or experimental animals are present in the laboratory or containment module, a hazard warning sign incorporating the universal biosafety symbol is posted on all laboratory and animal room access doors. This sign identifies the agent, lists the name and telephone number of the principal investigator or other responsible person(s), and indicates the special requirements for entering the laboratory, such as the need for immunization, respirator, or other personal protective measures.

All activities involving organisms containing recombinant DNA molecules are conducted in biological safety cabinets or other physical containment devices within the containment module. No work in open vessels is conducted on the open bench. Laboratory animals held in a BL3 area shall be housed in partial-containment caging systems, such as Horsfall units, open cages placed in ventilated enclosures, solid-wall and bottom cages covered by filter bonnets or solid-wall, and bottom cages placed on holding racks equipped with ultraviolet in radiation lamps and reflectors. Conventional caging systems may be used, provided that all personnel wear appropriate protective devices. These protective devices include at a minimum, wrap-around gowns, head covers, gloves, shoe covers, and respirators. All personnel shall shower on exit from areas where these devices are required. All wastes from laboratories and animal rooms are appropriately decontaminated before disposal. Vacuum lines are protected with high efficiency particulate air/HEPA filters and liquid disinfectant traps. A biosafety manual is prepared or adopted. Personnel are advised of special hazards, and are required to read and follow the instructions on practices and procedures.

Biosafety level 4 All other precautions that are followed in the earlier levels are must in this level also. Standard microbiological practices are followed. Work surfaces are decontaminated at least once a day, and immediately after any spill of viable material. Only mechanical pipetting devices are used. Eating, drinking, smoking, storing food, and applying cosmetics are not permitted in the laboratory similar to earlier levels. All procedures are performed carefully to minimize the creation of aerosols.

Biological materials are removed from the Class III cabinets or the maximum containment laboratory in a viable or intact state. These are then transferred to a non-breakable, sealed primary container, and enclosed in a non-breakable, sealed secondary container, which is removed from the facility through a disinfectant dunk tank, fumigation chamber, or an airlock designed for this purpose. No material, except biological material that is to remain in a viable or intact state, is removed from the maximum containment laboratory, unless it have been autoclaved or decontaminated before leaving the facility.

Equipment or material, which might be damaged by high temperatures or steam, is decontaminated by gaseous or vapor methods in an airlock or chamber designed for this purpose. Access to the facility is limited by means of secure, locked doors, Accessibility is managed by the principal investigator, biological safety officer, or other person responsible for the physical security of the facility. Before entering, persons are advised of the potential biohazards and instructed as to appropriate safeguards for ensuring their safety. Authorized persons comply with the instructions and all other applicable entry and exit procedures. A logbook signed by all personnel indicates the date and time of each entry and exit. Practical and effective protocols for emergency situations are established. All procedures within the facility with agents assigned to biosafety level 4 are conducted in the Class III biological safety cabinet, or in Class I or II biological safety cabinets used in conjunction with one-piece positive pressure personnel suits ventilated by a life-support system.

The maximum containment facility consists of either a separate building or a clearly demarcated and isolated zone within a building. Outer and inner change rooms separated by a shower are provided for personnel entering and leaving the facility. A double-doored autoclave, fumigation chamber, or ventilated airlock is provided for the passage of those materials, supplies, or equipment, which are not brought into the facility through the change room. Walls, floors, and ceilings of the facility are constructed to form a sealed internal shell, which facilitates fumigation and is animal and insect proof. The internal surface of this shell is resistant to liquids and chemicals, thus facilitating cleaning and decontamination of the area. All penetrations in these structures and surfaces are sealed. Any drains in the floors contain traps filled with a chemical disinfectant of demonstrated efficacy against the target agent, and they are connected directly to the liquid waste decontamination system. Sewer and other ventilation lines contain high efficiency particulate air/HEPA filters.

Biological Containment

Biological containment specifically aims at making genetic changes in GMOs that reduce the hazard of these organisms when they are accidentally or deliberately released into the environment.

Initially, NIH guidelines required the use of *E. coli* strains and vectors that were severely debilitated, so that they could not infect humans, survive and spread outside the laboratory, and transfer the introduced foreign genes readily into other organisms. These objectives can be achieved by using the following:

- Auxotropic mutants of *E. coli* (limits survival following escape of bacteria).
- Rec. A strains (they lack recombination function).
- Plasmid vectors that are non-self transmissible and non-mobilizable (such vectors cannot be transmitted to other bacteria).

- *E. coli* strains not carrying a transposon or antibiotic resistance gene.

Thus, biological containment is based on the vector (plasmid, organelle or virus) used for the construction of recombinant DNA, and the host (bacterial, plant or animal cell) in which the vector is propagated in the laboratory. Therefore, it is also called HV system (host vector system). Any combination of vector and host that is able to provide biological containment should be chosen or constructed so that the following types of 'escapes' are minimized:

- Survival of the vector in its host outside the laboratory.
- Transmission of the vector from the propagation host to other non laboratory hosts.

The following levels of biological containment (host-vector systems) for prokaryotes are established. Specific criteria will depend on the organisms to be used.

For example, the host *Escherichia coli* K-12 or a derivative thereof, and the vectors include non-conjugative plasmids (e.g., pSC101, Co1E1, or derivatives thereof and variants of bacteriophage, such as lambda. The *Escherichia coli* K-12 hosts shall not contain conjugation-proficient plasmids, whether autonomous or integrated, or generalized transducing phages. At a minimum, hosts and vectors shall be comparable in containment to *Escherichia coli* K-12 with a nonconjugative plasmid or bacteriophage vector. The containment systems provide a high level of biological containment as demonstrated by data from suitable tests performed in the laboratory. Escape of the recombinant DNA either via survival of the organisms or via transmission of recombinant DNA to other organisms should be < 1/108 under specified conditions. For *Escherichia coli* K-12 in which the vector is a plasmid, no more than 1/108 host cells shall perpetuate a cloned DNA fragment under the specified non-permissive laboratory conditions designed to represent the natural environment, either by survival of the original host or as a consequence of transmission of the cloned DNA fragment.

For *Escherichia coli* K-12, if the vector is a phage, no more than 1/108 phage particles shall perpetuate a cloned DNA fragment under the specified non-permissive laboratory conditions designed to represent the natural environment, either as a prophage (in the inserted or plasmid form) in the laboratory host used for phage propagation, or survival in natural environments and transferring a cloned DNA fragment to other hosts (or their resident prophages). Initial review will be based on the construction, properties, and testing of the proposed host-vector system by a subcommittee composed of one or more RAC members and/or ad hoc experts. The RAC will evaluate the subcommittee's report and any other available information at the next scheduled RAC meeting. The NIH Director is responsible for the certification of host-vector systems, following the advice of the RAC. Minor modifications in existing host-vector systems (i.e., those that are of minimal or no consequence to the properties relevant to containment) may be certified by the NIH Director without prior RAC review. Once certified, a notice of certification will be sent by NIH/OBA to the applicant and the Institutional Biosafety Committee Chairs.

A list of all currently certified host-vector systems is available from the office of biotechnology activities, National Institutes of Health; the NIH Director may rescind the certification of a host-vector system. If certification is rescinded, NIH will instruct investigators to transfer cloned DNA into a different system, or use the clones at a higher level of physical containment level unless NIH determines that already constructed clones incorporate adequate biological containment. Certification of an host-vector system does not extend to modifications of either the host or vector component of that system. Such modified systems shall be independently certified by the NIH Director. If modifications are minor, it may be necessary for the investigator to submit data showing that the modifications have either improved or not impaired the major phenotypic traits on which the containment of the system depends. A substantial modification in a certified host vector system requires submission of complete testing data.

In systems other than *Escherichia coli* K-12, the following types of data shall be submitted, modified as appropriate for the particular system under consideration:

- A description of the organism and vector; the strain's natural habitat and growth requirements; its physiological properties, particularly those related to its reproduction, survival, and the mechanisms by which it exchanges genetic information; the range of organisms with which this organism normally exchanges genetic information, and the type of information exchanged; and any relevant information about its pathogenicity or toxicity.
- A description of the history of the particular strain and vector to be used, including data on any mutation, which renders this organism less able to survive or transmit genetic information.
- A general description of the range of experiments contemplated, with emphasis on the need for developing such an host-vector. Investigators planning to request host-vector systems certification may obtain instructions from NIH/OBA concerning data to be submitted.

In general, the following data is required:

- Description of construction steps, with indication of source, properties, and manner of introduction of genetic traits.
- Quantitative data on the stability of genetic traits that contribute to the containment of the system.
- Data on the survival of host-vector system under nonpermissive laboratory conditions designed to represent the relevant natural environment.
- Data on transmissibility of the vector and/or a cloned DNA fragment under both permissive and non-permissive conditions.

- Data on all other properties of the system, which affect containment and utility, including information on yields of phage or plasmid molecules, ease of DNA isolation, and ease of transfection or transformation.
- In some cases, the investigator may be asked to submit data on survival and vector transmissibility from experiments in which the host-vector is fed to laboratory animals or one or more human subjects. Such *in vivo* data may be required to confirm the validity of predicting *in vivo* survival on the basis of *in vitro* experiments. Data shall be submitted 12 weeks prior to the RAC meeting.

7.11. EXPERIMENTS WITH MICROORGANISMS

The biosafety levels for experiments with microorganisms are designated as BL1, BL2, BL3 and BL4. A brief description of the physical containment need for these is given below.

Biosafety Level 1 (BL1)

BL1 is suitable for work with microorganisms which are an unknown or minimal potential hazard to laboratory workers and the environment. The laboratory is not separated from the general traffic patterns in the building. Work is generally done on open bench tops, and special containment equipment is not required. Laboratory personnel are trained in the laboratory procedures. They are supervised by a scientist, who is an expert in microbiology or a related science. The details of containment equipment and practices are given below.

Standard microbiological practices

- Access to the laboratory is restricted at the discretion of principal investigator (PI), when the experiments are in progress.
- Work surfaces are decontaminated once a day and after any spill of viable material.
- All contaminated liquid or solid wastes are decontaminated before disposal.
- Mouth pipetting is prohibited.
- Eating, drinking, smoking and applying cosmetics are not permitted in the laboratory premises.
- Persons wash their hands after handling materials and animals involving organisms containing recombinant DNA molecules, before leaving the laboratory.
- All procedures must be performed carefully, so that a minimum of aerosols are produced.
- Facilities for good personal hygiene and protective clothing, e.g., uniforms and laboratory coats, must be provided.

Special laboratory practices

- Contaminated materials that are to be decontaminated at a site away from the laboratory are placed in a durable leak-proof container, which is closed before being removed from the laboratory.
- An effective insect and rodent control programme is implemented.

Containment equipment

Special containment equipment is generally not required.

Laboratory facilities

- Laboratory design must facilitate easy cleaning.
- Bench tops must be impervious to water and resistant to acids, alkalies, organic solvents and moderate heat.
- Laboratory furniture must be sturdy, and spaces between benches, cabinets and equipment must be accessible for cleaning.

- Each laboratory must have a sink for washing hand.
- Laboratory windows, if they open, must be fitted with fly screens.

Biosafety Level 2 (BL2)

BL2 is similar to BL1. It is suitable for working with organisms, which are moderate potential hazard to personnel and the environment. It differs from BL1 in the following respects:

- Laboratory personnel must be specifically trained in handling pathogenic agents, and are directed by competent scientists.
- Access to the laboratory is limited when the experiments are in progress.
- Certain procedures, in which infectious aerosols are produced, are conducted in biological safety cabinets or other containment equipment.
- The PI has the responsibility of determining who may enter the laboratory or animal rooms, and also the conditions for such entry. Where necessary, a hazard warning sign incorporating the universal biosafety symbol is posted on the access door of the laboratory area.
- Gloves should be worn while handling experimental animals, and when skin contact with the microorganisms is unavoidable.
- A biosafety manual is prepared, and personnel are required to read and follow the manual.
- Spills and accidents, which result in overt exposures to organisms containing recombinant DNA molecules, are immediately reported to the Institutional Biosafety Committee (IBC) and higher bodies concerned with biosafety.
- An autoclave for decontaminating laboratory waste must be available.

Biosafety Level 3 (BL3)

BL3 is applicable to clinical, diagnostic, teaching, research or production facilities in which work is done on microorganisms—indigenous or exotic—which may cause serious or potentially lethal diseases as a result of exposure during inhalation. BL3 has the following additional chief safety measures over BL2:

- Laboratory personnel have specific training in handling pathogenic and potentially lethal agents, and are supervised by competent scientists who are experienced in working with these agents.
- All procedures involving the manipulation of infectious material are conducted within biological safety cabinets or other physical containment devices by personnel wearing protective clothing.
- The laboratory has special engineering and design features, e.g., access zone, sealed penetrations, directional air flow, etc. Passage through two sets of doors is the basic requirement for entry into the laboratory. This may be provided by a double-doored clothes change room. Diagnostic procedures may be done in laboratories with BL2 facilities provided BL3 standard microbiological practices, special practices and containment equipment are rigorously followed. The decision to implement this modification of BL3 recommendations can be made by the PI. Windows in the laboratory are closed and sealed; access doors to the laboratory are self-closing.
- Persons under 16 years of age should not enter the laboratory, and laboratory doors are kept closed when work is in progress.
- Laboratory clothing is not worn outside the laboratory, and it is decontaminated prior to laundering or disposal.
- Molded surgical masks or respirators are worn in rooms containing experimental animals.
- Vacuum lines are protected with high efficiency particulate air (HEPA) filters and liquid disinfectant traps.

Biosafety Level 4 (BL4)

BL4 is appropriate for microorganisms that are likely to cause serious or lethal human diseases for which preventive or therapeutic interventions are not usually available.

- Limited only to those who work in the area; the access is limited by means of secure, locked doors. A logbook signed by all personnel indicates the date and lime of each entry and exit.
- Personnel enter and exit the facility only through the clothing change and shower rooms; personnel shower each time they exit the facility.
- All procedures with microorganisms are done in Class III or Class II safety cabinets used in conjunction with one-piece positive pressure personnel suits ventilated by a life support system.
- Biological materials to be removed from Class III cabinets or from the maximum containment laboratory in a viable state are transferred into a non-breakable sealed primary container, which is then enclosed in a non-breakable, sealed secondary container. These are removed from the laboratory through a disinfectant dunk tank, fumigation chamber, or an air-lock designed for this purpose.
- All other biological materials, equipment, and other materials are removed from the maximum containment laboratory only after decontamination by autoclaving or some other effective method.
- The laboratory is designed for the maximum containment facility. It is either a separate building or a clearly demarcated and isolated zone within a building. Outer and inner change rooms separated by a shower are provided for personnel entering and exiting the facility.
- A double-doored autoclave, fumigation chamber, or ventilator air-lock is provided for passage of those materials, supplies or equipment, which are not brought into the facility through the change room. Walls, floor and ceilings of the facility are constructed to form a sealed internal shell, which facilitates fumigation and is animal and insect proof. Sewer and other ventilation lines contain HEPA filters. In addition, the laboratory design must have several other specific features as detailed in the guidelines.

7.12. RESEARCH INVOLVING PLANTS

When the size, number, or growth requirements of plants, plant-associated microorganisms, and small animals are such that the containment conditions described for microorganisms cannot be used, the conditions and practices suitable for greenhouse conduct of experiments are used. The principal purpose of plant containment is to avoid the unintentional transmission of a recombinant DNA containing plant genome, including nuclear or organelle hereditary material, or release of recombinant DNA-derived organisms associated with plants.

The containment principles are based on the recognition that the organisms used pose no health hazard to humans or higher animals, and containment conditions minimize the possibility of an accidental release of the organisms, e.g., a plant pathogen, from the greenhouse into the ecosystem, outside the experimental facility. Four biosafety levels, (BL1-P, BL2-P, BL3-P, and BL4-P) have been established. When plants are grown in a laboratory, the containment practices of BL1 to BL4, as appropriate for the experiment, are followed. These practices include the use of plant tissue culture rooms, growth chambers within laboratory facilities, etc. But in case botanical reproductive structures are produced on these plants, additional biological containment practices should be added by the IBC.

Biosafety Level 1-Plants (BL1-P)

Greenhouse access is limited at the discretion of greenhouse incharge when experiments are in progress. Prior to entering the greenhouse, personnel must read and follow instructions on BL1-P greenhouse practices and procedures. A record is maintained of the experiments currently in progress in the greenhouse facility. A programme to control undesired species, e.g., weeds, rodents, arthropod pests and pathogens, should be in effect. Experimental arthropods and other microorganisms are housed in appropriate cages. If microorganisms, e.g., flying arthropod, nematodes, are released into the greenhouse, precautions should be taken to minimize their escape from the greenhouse. Experimental organisms are rendered biologically inactive before their disposal outside the greenhouse. Experiments with organisms requiring containment lower than BL1-P can be done concurrently, only if BL1-P containment practices are followed.

The 'greenhouse' has walls and roof constructed of transparent or translucent material, and a floor made up of gravel or other porous material. It is designed for growing plants in controlled and protected environment. The material of the walkways in the greenhouse must be, at a minimum of impervious material, e.g., concrete.

Biosafety Level 2-Plants (BL2-P)

BL2-P differs from BL1-P in the following respects:

- When work is in progress, access is limited to personnel directly involved with the experiments.
- A record is kept of experimental plants, microorganisms, or small animals that are brought into or removed from the greenhouse facility.
- The PI should report any inadvertent release or spill of microorganisms to greenhouse incharge, IBC, and higher bodies concerned with biosafety.
- If a part of the greenhouse is composed of gravel or similar material, periodical treatments should be done to inactivate organisms present in the gravel.

- A sign is posted indicating that a restricted experiment is in progress. A sign is also posted if an experimental organism has a potential for causing serious damage to the ecosystem/agriculture.
- If there is a risk to human health, a sign is be posted incorporating the universal biosafety symbol.
- Materials containing experimental microorganisms are brought into or removed from the greenhouse in a closed non-breakable container.
- A greenhouse practice manual appropriate for BL2-P is prepared.
- The greenhouse floor is made of an impervious material, e.g., concrete.
- Windows and other openings in the greenhouse must have screen to exclude small flying animals, i.e., arthropods and birds.
- An autoclave is available for the treatment of contaminated greenhouse materials.

Biosafety Level 3-Plants (BL3-P)

BL3-P has the features of BL2-P in addition to the following measures:

- The entry into greenhouse is restricted to those who are needed for the programme or support purposes.
- All experimental (except those that are to remain in a viable or intact state for experimental purposes) materials are sterilized in an autoclave or inactivated by some other effective means before disposal.
- Experimental materials brought into or taken out of the greenhouse facility in a viable or intact state, are kept in a non-breakable sealed secondary container as well, and the surface of the secondary container is decontaminated.
- Disposable clothing is worn in the greenhouse if considered necessary by the greenhouse incharge. This can be removed before leaving the greenhouse, and decontaminated prior to laundering or disposal.

- Personnel are required to thoroughly wash their hands upon leaving the greenhouse.
- All procedures are performed carefully to minimize production of aerosols.
- The greenhouse is a closed self-contained structure.
- It has facility to maintain negative pressure.
- It has a provision for collection and decontamination of liquid run-off.
- Windows are closed and sealed.
- The greenhouse facility is surrounded by a security fence.
- The internal wall, ceilings, and floors are resistant to penetration by liquids and chemicals, and bench tops and other work surfaces are seamless.
- A foot, elbow or automatically operated sink is located near the exit door.
- An individual supply and exhaust air ventilation is provided.
- The exhaust air from the greenhouse facility is filtered through HEPA filters.

Biosafety Level 4-Plants (BL4-P)

BL4-P represents a greater stringency in containment over BL3-P in the following respects:

- Access to the greenhouse is limited by the means of secure, locked doors.
- Individuals are informed about the potential environmental hazards and appropriate safeguards to be followed, which they comply with.
- Personnel enter and exit the greenhouse facility only through the clothing change and shower rooms, and take shower each time they exit the facility.
- A record is kept of all personnel entering and leaving the greenhouse facility.
- Standard microbiological procedures are followed for the decontamination of equipment and materials.
- Experimental material brought into or removed from the greenhouse in a viable or intact state is transferred to a non-breakable, sealed primary container. After that, it is enclosed in a non-breakable, sealed secondary container, which is removed from the facility through a chemical disinfectant, fumigation chamber, or airlock designed for this purpose.
- Supplies and materials are brought into the facility through a double-door autoclave, fumigation chamber, or airlock that is appropriately decontaminated after each use.
- Street clothing is removed in the outer clothing change room. Complete laboratory clothing is worn by all personnel entering the facility. It is removed before personnel enter the shower area during their exit from the facility and shall be autoclaved before laundering.
- The greenhouse facility has inner change rooms separated by a shower.
- The access doors to the greenhouse are self-closing and locking.
- The walls, floors and ceilings of the greenhouse are constructed to form a sealed internal shell, which facilitates fumigation and is animal and arthropod-proof.
- A double-door autoclave, fumigation chamber, or ventilated airlock is provided for the passage of all materials, supplies or equipment.
- Liquid effluents from sink, floor and autoclave chambers are decontaminated by heat or chemical treatment before their release.
- If there is a central vacuum system, it will not serve areas outside the greenhouse facility.

7.13. RESEARCH INVOLVING ANIMALS

These guidelines cover experiments into which whole transgenic animals are involved or recombinant DNA-modified microorganisms are tested on whole animals. The whole animals should be of a size or have growth requirements that preclude the use of containment for laboratory animals. Examples of such animals are cattle, swine, sheep, goats, horses and poultry. The containment levels are based on the classification of risk. Four levels of containment, via., BL1-N, BL2-N, BL3-N and BL4-N, are recognized. The following procedure is followed at all biosafety levels.

Biosafety Level 1-Animals (BL1-N)

In case of BL1-N:
- Access to the containment area is limited or restricted when experimental animals are being held.
- The containment area is monitored at frequent intervals.
- All transgenic newborns are permanently marked within 72 hours of their birth or, at least, kept in marked containers. In addition, they should contain easily assayable DNA sequences for unambiguous identification from non transgenic animals.
- A double barrier separates male and female animals unless reproductive studies are a part of the experiment.
- The animals are confined to security-fenced areas or in enclosed structures (animal rooms) to minimize the possibility of theft or unintentional release.

Biosafety Level 2-Animals (BL2-N)

It differs from BL1-N in the following respects:
- The containment area is locked.
- The containment building is controlled and have a locking system.
- Only such persons can enter the laboratory who have been advised of potential hazards and who meet other requirements like vaccination, etc.
- Animals not involved in the work are not allowed in the area.
- Materials to be decontaminated at a site away from the laboratory are placed in a closed durable leak-proof container prior to removal from the laboratory.
- Needles and syringes, etc. are placed in a puncture-resistant container and decontaminated.

- When special provisions, e.g., vaccination, etc., are needed for entry, a warning sign incorporating the universal biosafety symbol is posted on all access doors to the animal work area.
- Laboratory coats, gowns, smocks or uniforms are worn while in the work area. These are removed before entering areas such as cafeteria, library, etc.
- Protective gloves are worn while handling experimental animals.
- Any incident involving spills and accidents that result in environmental release or exposure of animals or laboratory workers to organisms containing recombinant DNA are reported immediately to IBC and the higher bodies concerned.
- Base-line serum samples are collected and stored for animal care and at-risk personnel.
- Viable biological materials are removed from the containment area as in case of BL4-P.
- Appropriate steps are taken to prevent horizontal transmission or exposure of laboratory personnel.
- Eating, drinking, smoking and applying cosmetics is prohibited.
- Individuals who handle materials/animals containing recombinant DNA wash their hands before exit.
- A biosafety manual is prepared or adapted.
- Animals are contained in animal rooms or equivalent structures with provision to avoid entry or escape of arthropods (only when the agent in use is transmissible by arthropods).
- Surfaces should be impervious to water and resistant to acids, alkali, organic solvents and moderate heat.
- The design should permit easy cleaning and windows shall have fly screens.
- An autoclave shall be available for decontamination of laboratory wastes.
- If arthropods are used in the study, or if the agent used is transmitted by arthropods, interior work areas shall be appropriately screened.

Biosafety Level 3-Animals (BL3-N)

BL3-N differs from BL2-N in the following respect:

- Animal room doors, gates or other closures are kept closed when experiments are in progress.
- The work surfaces of containment equipment are decontaminated when work with organisms containing recombinant DNA is finished.
- All animals are euthanized at the end of their experimental usefulness.
- Special safety testing, decontamination procedures, and IBC approval is required to transfer agent or tissue/organ specimens to a low biosafety level area.
- Liquid effluent is decontaminated by heating before being released into the sanitary system.
- Appropriate respiratory protection is worn in rooms containing experimental animals.
- Removal of biological material in a viable or intact state is done as in BL4-P, but only after approval of animal facility incharge.
- Animal holding area is cleaned at least once a day, and decontaminated immediately following any spill of viable materials.
- Creation of aerosols must be minimized.
- The animal rooms or equivalent are as in case of BL2-N but their windows are closed, sealed and breakage-resistant.
- The access doors to the containment area shall be self-closing.
- The animal area is separated from all other areas. Passage through a double-door clothes change room, equipped with integral showers and airlock is the basic requirement.
- An exhaust air ventilation system is provided so that it draws air into the animal rooms through the entry area. If the agent is transmitted through aerosol, the exhaust air must pass through HEPA filter.
- Vacuum lines are protected with HEPA filters and liquid disinfectant traps.
- There shall be a wash basin (foot, elbow or automatically operated) near the exit door.

Biosafety Level 4-Animals (BL4-N)

BL4-N represents greater stringency than BL3-N in the following main respects.

- Individuals under 16 years of age are not allowed entry.
- Individuals enter and exit the animal facility only through the clothing change and shower rooms. Complete street clothing is removed and kept in the outer clothing change room.
- All waste from the animal rooms and laboratories is decontaminated before disposal in an approved manner.
- When ventilated suits are required, the animal/personnel shower entrance/exit area is equipped with a chemical disinfectant shower to decontaminate the surface of the suit before exiting the area.
- Supplies and materials are brought in through a double-door autoclave, fumigation chamber or airlock that is appropriately decontaminated between each use.
- A ventilated head-hood or a one-piece positive pressure suit, which is ventilated by a life-support system, is worn by all personnel entering rooms that contain experimental animals when appropriate.
- Reports are made to IBC and other relevant authorities regarding exposure to organisms containing recombinant DNA, employee absenteeism and medical surveillance of potential laboratory associated illnesses.
- A permanent record book indicating the date and time of each entry is signed by all personnel.

- An essential adjunct to the reporting surveillance system, is the availability of a facility for quarantine, isolation and medical care of personnel with potential or known laboratory-associated illnesses.
- All perimeter joints and openings are sealed to form an arthropod-proof structure.
- The BL4-N laboratory has a double barrier to prevent the escape of recombinant DNA-containing micro-organisms designed, so that if the barrier of the inner facility is breached, the outer barrier will prevent release into the environment.
- A necropsy room is provided.
- The exhaust air is double HEPA-filtered or passed through a certified HEPA filter and an air incinerator. All HEPA filter's frames and housings are certified to have no detectable smoke leaks.
- All equipment and floor drains are equipped with deep (minimum 5″) traps.
- The supply water distribution system is fitted with a back-flow preventer or break-tank.
- All utilities, liquid and gas services are protected with devices that avoid back-flow.
- Sewer and other atmospheric ventilation lines are equipped with atleast one HEPA filter.

Future Plans

It is very important to develop national projects, which focus on the preparation of a national biosafety framework including regulatory, administration and decision-making systems, and mechanisms for public participation and information. Here are some of the activities:

- Gathering of necessary information concerning the use of modern biotechnology, existing legislation on biotechnology and/or biosafety.
- Establishing Committees such as National/ Institutional Biosafety Committees, GM Advisory Committee, Review Committee, and GMO Approval Committee to enforce the regulation.
- Training biosafety committees at both institutional and national levels to conduct scientifically sound biosafety reviews, etc.
- Developing National Biosafety Database and linkages to the Biosafety Clearing House.
- Drafting legal instruments as guidelines for implementing the regulation.
- Sharing experience on the organization and risk analysis with other countries, which are in a more advanced stage of drafting and enforcing biosafety regulations.
- Cooperating with other countries and international organizations in developing relevant educational, public awareness programs and intensive workshops on biosafety.

7.14. BIOSAFETY PROTOCOL [UN CARTAGENA BIOSAFETY PROTOCOL (CBP)]

The introduction of genetic engineering to agriculture has produced new challenges in the fields of environmental safety, human health, trade, and development. In the last decade, a growing web of global rules and institutions has been created to govern agricultural biotechnology. The most recent of these is the "cartagena protocol on biosafety" negotiated under the auspices of the convention on biological diversity, which regulates the transboundary transfer and use of genetically modified organisms (GMOs). The protocol seeks to facilitate informed decision-making by GMO importing countries about whether or not to permit entry of particular GM seeds and food. It entered into force in September 2003, and had been ratified by 129 countries and the european community as of December 2005. We find ourselves at a critical moment in global biotechnology governance. With the cartagena protocol now in force, its implementation will reveal whether international governance of biosafety can keep up with the rapid pace of technological change and globalization in genetic engineering. Some important challenges lie ahead, especially in the developing world: will the biosafety protocol be implemented on the ground, despite severe capacity constraints? And will implementation lead to an internationally harmonized science-based approach to GMO regulation (as hoped for by GMO producer countries and industry), or will it allow for a diversity of regulatory models and practices to co-exist? How might the ongoing transatlantic GMO trade conflict between the US and the EU affect developing countrys' regulatory choices? We address these questions through comparative analysis of biosafety policy in selected developing/emerging economies, Mexico, China and South Africa, which are currently both producers and importers of transgenic crops. In doing so, we explore how global biosafety governance and global trade-safety conflicts influence domestic choices about powerful new technologies such as genetic engineering.

The use of genetic engineering techniques is rapidly expanding in key sectors of food production, particularly in globally traded commodity crops such as maize, canola, soybean and cotton. Producers of these transgenic crops, which have been genetically engineered largely to be herbicide-tolerant or insect-resistant, offer a range of benefits to farmers.

Although data about the spread of GM crops worldwide is difficult to come by, statistics compiled by the International Service for the Acquisition of Agri-biotech Applications (ISAAA) claim that since 1996, when genetically modified varieties were first grown commercially, the global area planted to such crops has increased over 50-fold—1.7 million hectares in 1996 to 90.0 million hectares in 2005. Biotech crops are now grown in 21 countries, of which the United States is leading with 49.8 million hectares, followed by Argentina (17.1 million ha), Brazil (9.4 million ha), Canada (5.8 million ha), China (3.3 million ha), Paraguay (1.8 million ha), India (1.3 million ha), and South Africa (0.5 million ha). Mexico and twelve other countries make-up the rest, with less than 0.3 million ha each. It is important to note

that although both developed and developing countries are growing transgenic crops, the United States alone accounts for over half of the total area devoted to such crops.

Apart from the few developing countries growing transgenic crops in commercial quantities, the rest are still carrying out field testing and experimental research, if they participate in the process at all. However, irrespective of whether the countries grow transgenic crops, most have to contend with an increasing global trade in agricultural commodities and food containing genetically modified material. The growth of a globalized biotechnology industry and trade in biotech crops requires the countries to develop regulatory systems, forcing them to consider the impact that the spread of biotech seed and crops to their countries might have on the sustainability of their agricultural systems, prospects for biosafety and food security, and their current and future position in global agricultural trade.

The cartagena protocol on biosafety (often referred to simply as the biosafety protocol) is an international agreement on trade in GMOs that was negotiated between 1996 and 2000, under the auspices of the United Nations Convention on Biological Diversity. The final negotiating session took place in Montreal in January, 2000, under intense media scrutiny. At that time, GMOs were high on the public's radar screen, and Europe had recently enacted a de-facto moratorium on the release of new GMOs into the environment. The combination of strong leadership by the representatives of developing countries, excellent mobilization by Canadian and international NGOs, and the European shift to a more critical stance on GMOs, meant that the Cartagena Protocol was much stronger than GMO-exporting countries would have liked. These countries, which included Canada, the United States, Australia, Chile, Argentina, and Uraguay in a negotiating block called the 'Miami Group' were forced to make a number of concessions in an effort to show that they are not simply trying to force GMOs down other countrys' throats.

The resultant draft protocol states that countries are expected to provide "advance informed agreement" for the import of modified organisms (called living modified organisms, or LMOs, in the protocol) intended to be introduced into the environment (such as modified seeds). Decisions on whether or not to import an LMO must be based on the results of risk assessments of those organisms. In cases where there is insufficient scientific information to make a decision, or when that information is uncertain, state's can still decide to keep an LMO out based on the precautionary principle. In April, 2002, the Intergovernmental Committee on the Cartagena Protocol on Biosafety (ICCP) met in The Hague, Netherlands, to work out more details on how to implement this international agreement governing trade in genetically modified organisms. The cartagena protocol is the centerpiece of an emerging global architecture designed to govern uptake of genetic engineering in agriculture. Other key elements of this architecture include the World Trade Organization's (WTO) agreement on the application of sanitary and phytosanitary measures, (SPS Agreement) and Agreement on Technical Barriers to Trade (TBT Agreement). The Codex Alimentarius Commission, a global food safety standard-setting body, is also debating global safety standards for food produced via use of genetic engineering. This emerging governance framework has to contend with a wide range of concerns (including ecological, human health, social and ethical) associated with the use of genetic engineering in agriculture. The challenge is made more complex by the fact that the nature and manageability of risks associated with modern biotechnology remain contested. Moreover, the emerging system of rules and institutions is far from coherent or consistent. Instead, it remains unclear how components of this rapidly expanding set of global rules interact with and influence one another. This is partly because these regimes are still evolving, and their obligations are still being interpreted or expanded within global fora as well as via national implementation. It is also, however, because of the potential for conflict between norms and rules contained in the cartagena protocol and WTO agreements.

Cartagena protocol and WTO rules were the key stumbling blocks on the way to agreement on the biosafety protocol. Even today, it remains one of the most controversial aspects of implementing the protocol. The potential for conflict centres around how countries will interpret the protocol's provisions on precautionary and socioeconomic factors in making GMO trade decisions, and whether these interpretations will conflict with the WTO-SPS agreement's more narrowly circumscribed 'scientifically sound' approach to domestic decision-making. Developing countries and the EU have been concerned about the possibility of WTO disciplines trumping biosafety measures based on the cartagena protocol.

Although the earlier meetings of the ICCP were characterized as being fairly amicable, the meeting in 'The Hague' showed that two key issues, each of which was hotly debated in the years leading to the cartagena protocol 'consensus', are still highly contentious and far from resolved. First, the question of the detail of documentation to accompany bulk commodity shipments of modified seeds; and second, the issue of liability and redress in the event that GMOs cause harm to the environment (including impacts on human health) in an importer country. Both of these issues are extremely important in the struggle for strong regulation over GMOs, and they deserve careful attention by activists in various countries around the world. On 30 January, 2000, the *Cartagena Protocol on Biosafety* (CPB) was adopted in Montreal by the delegates of 128 Parties, in the *Convention on Biological Diversity* (CBD). This must be seen as a landmark in the development of multinational environmental agreements. With it, the *Precautionary Approach* (PA) became the guiding principle for the import of genetically modified organisms (GMOs). It subjected this import to an *Advance Informed Agreement* (AIA), and secured its legal status in relation to the *World*

Trade Organization (WTO). By 15 September, 2000, 75 governments including those of 44 developing countries had signed this environmental agreement.

Transboundary Movement of Living Modified Organisms

The CPB covers only "living modified organisms" (LMOs), and not "genetically modified organisms". This was due to the opposition by the US and other big agriculture exporters, who opposed the wording 'GMO' in order to exclude non-living modified organisms and limit the effect of the law to those living modified organisms destined to grow in the fields, e.g., seeds. At the same time, the use of the definition 'LMO' excludes products derived from GMOs, e.g., processed food derived from a GM plant such as maize flour that might be used in a tortilla.

The CPB focuses primarily on regulating the transboundary movement of LMOs between the parties of the protocol, but there are some provisions on transit, handling and use of LMOs as well. The general scope applies to all LMOs, but is very limited to LMO pharmaceuticals, LMOs in transit, and LMOs for contained use.

Advance Informed Agreement

Advance Informed Agreement (AIA) is the key mechanism of the protocol, in which the importing party must give a prior informed consent before an LMO enters into its territory. It applies only to the first transboundary movement for intentional introduction into the field trials or planting. The decision whether to give the prior informed consent will be taken by the importing party based on risk assessment and the precautionary principle. Unfortunately, the AIA mechanism will not apply to LMOs intended for direct use as food, feed or processing, which constitutes the bulk of the LMOs traded today. This category will be subjected only to an internet system of information called the "biosafety clearing house". This system requests that minimal information related to any LMO, approved at national level by any party of the protocol, should be posted on the biosafety clearing house.

Labeling and Segregation

An earlier version of the biosafety protocol required that living modified organisms used as food, feed or for processing (LMO-FFPs) would be accompanied by a documentation that "*clearly identifies them as living modified organisms*", but in the final hours of negotiation, Argentina claimed that as a developing country, it would not be able to implement an adequate identification system. The final text (Article 18.2) now states that LMO-FFPs have to be accompanied by documentation, which "*clearly identifies that they 'may contain' (LMOs) and are not intended for introduction into the environment, as well as a contact point for further information.*"

This text does not fulfil the provisions of many existing national labeling laws, and fails to provide clear information.

The parties to this protocol are supposed to decide on a more detailed procedure, such as specifying identity and unique identification within two years of the protocol coming into force. As a result, the CPB requires the type of mandatory labeling already demanded by the EU and Japan, who are currently the main importers of GMOs. The majority of emerging biosafety laws in developing countries also adhere to the European concept of mandatory labeling, and the international markets are about to accept and implement such labeling and segregation conditions in response to consumer demands.

Capacity Building

Many developing countries do not have a regulatory framework governing GMOs, nor the resources to enforce such legislation. Capacity is, therefore, key in the areas of administration, legislation, science, enforcement, and monitoring.

Liability

The development of a system for liability and redress is foreseen in the protocol, and has to be established within four years after the entry into the protocol. There is an urgent need for the establishment of an effective liability mechanism under the protocol to ensure that corporations that harm the environment, for instance, through contamination by GM crops, pay for the pollution they create.

Other Provisions of the Protocol

The protocol also requires all exporters of LMOs take measures for the prevention of contamination of GM seed products by implementing an identity preservation system. Living modified organisms that are intended for intentional introduction into the environment have to be clearly identified as living modified organisms. On the other hand, LMOs destined for food, feed or processing will have to be identified only as "may contain" living modified organisms.

The inclusion of the precautionary principle is another important accomplishment of the CBP. This principle requires that where there are threats of serious or irreversible damage, despite lack of scientific certainty, it is better to act now to be safe than wait to be sorry.

Solving the Deficiencies

The biosafety protocol is a landmark agreement, which for the first time set out regulations on GMOs, particularly on transboundary movements. Despite its significance, the biosafety protocol is still a young agreement and not robust enough. The protocol contains several deficiencies, like rather limited nature of its scope and advance informed agreement.

This is due to the fact that it was a heavily negotiated agreement, which was always opposed by the US and the so-called Miami group (main agriculture exporting countries). In spite of not being a robust agreement, the biosafety protocol was an important step since it set the pillars for international regulation on the issue—something that industry and some governments had rejected. International instruments often start weak, but continue growing progressively. In the long-term, the aim should be to improve the protocol by correcting the deficiencies it has at present.

Article 23 of the Cartagena Protocol on Biosafety

*   Parties (to the protocol) should:
    — promote and facilitate public awareness, education, and participation concerning the safe transfer, handling and use of living modified organisms in relation to the conservation and sustainable use of biological diversity, taking also into account risks to human health. In doing so parties should cooperate, as appropriate, with other states and international bodies.
    — endeavour to ensure that public awareness and education encompass access to the information on living modified organisms identified in accordance with this protocol.
*   The parties should, in accordance with their respective laws and regulations, consult the public in the decision-making process regarding the living modified organisms, and should make the results of such decisions available to the public, while respecting confidential information in accordance with article 21.
*   Each party should endeavour to inform the public about the means of public access to the biosafety clearing house.

A Minimum Standard

Urgent need for building comprehensive national biosafety regimes

The protocol establishes only a set of minimum standards, and it is necessary that countries build their own biosafety laws to improve the provisions of the current text of the protocol. The international notification system under the protocol does not replace national biosafety legislation, so enacting stricter national legislation on biosafety is still needed at the domestic level. This is allowed by the protocol, which recognizes the right of a party to take action that is more protective of the conservation and sustainable use of biological diversity than that called for in the protocol if consistent with its objective and the provision. One of the main challenges with regulating GMOs via the CPB is the US government. The US is the main grower of GM crops. On the other side, it is not a party to the CPB and therefore not bound by its rules. This shifts the burden to recipient countries, and that is why it is so important that strict national legislation on biosafety is implemented at the national level by recipient countries, and adequate capacity is built. Countries that would like to establish protection against unregulated imports of GMOs should impose national legislation immediately. In their relationship with the United States, developing countries cannot rely on the biosafety protocol.

Way Forward

Governments that wish to establish a high level of protection for the environment, human health, and socioeconomic concerns against the potential risks derived from GMOs are urged to:

- sign and ratify the biosafety protocol as soon as possible. The biosafety protocol membership must be strengthened; the more countries that join it, the stronger the protocol will be.
- implement comprehensive biosafety regimes at the national level that go further than the biosafety protocol, in order to properly regulate GMOs and products thereof. National biosafety legislation is needed to compensate for the deficiencies of the protocol, and to protect the recipient countries against exports

of GMOs from the countries, which are not parties to the protocol.

Scope of the CPB

The CPB regulates the transboundary movement of GMOs, but there are several exceptions to the modified regulations.

Commodities

During the biosafety negotiations a new category of GMOs was created: *living modified organisms intended for direct use as a food or feed, or for processing* (LMO-FFPs). It replaces the expression "commodities" and makes a clear distinction between GMOs as seeds, and GMOs that are intended for food or feed and which are therefore not supposed to be released into the environment. This distinction does not take into account that during crisis in developing countries, GMOs intended for food might also be used as seeds.

While the draft text totally excluded LMO-FFPs from the AIA procedure, the final negotiations in Montreal led to the development of a weaker form of AIA for LMO-FFPs. This 'AIA-light' is closely connected to the development of an internet-based *Biosafety Clearing House* (BCH)—an information system that is part of the clearing house mechanism of the CBD. The BCH will be a central portal with basic information, and links to other relevant web pages. Meanwhile, there are 144 national CBD focal points; 122 of them with e-mail, but only 52 with a CBD clearing-house mechanism website. The CBD has started an international project that will try to improve this situation within the next four years, paying particular attention to the development of the BCH structure.

Every party to the protocol must be informed about any approval given for the market introduction of an LMO by another Party either in writing or through the BCH. However, the actual import itself does not have to be announced. The potential importing party "may request additional information" and "may take a decision on the import [...] under its domestic regulatory framework which is consistent with the objective of this protocol" (Article 11). Most developing countries expressed their concerns about this procedure, which imposes the burden of monitoring the worldwide registration of LMO-FFPs on the potential importing countries, and relies entirely on highly developed telecommunication systems that are often not accessible to developing countries.

Under the regular AIA procedure, explicit consent is needed for the import of an LMO. However, Article 11.7 states that the "failure by a party to communicate its decision [...] shall not imply its consent or refusal to the import" of LMO-FFPs. The party's decision has to be announced through the BCH. In practice, this might lead to a matrix that shows which country has allowed the import of which LMO-FFP. This matrix can help exporters decide whether a particular consignment of agricultural commodities that "may contain" LMO-FFPs can be exported legally into a given country or not.

Exporting Countries

How do the three main GMO growing countries and other potential GMO growers react on these CPB requirements? The USA as the main producer of GMOs cannot be a party to the CPB until they ratify the CBD, which at present seems very unlikely. It is also unclear whether Canada as the third biggest exporter of GMOs, and Australia as a potential GMO growing country will sign the CPB. On the other hand, the developing countries Argentina—the second biggest GMO exporter, and Chile, both members of the Miami-Group, signed the CPB very early.

It will be extremely difficult if major exporting countries refuse to enter the CPB and accept its obligations. In the draft text, the transfer of LMOs from non-parties to parties was forbidden, but this provision could not be maintained. The CPB now states that the movement of LMOs between parties and non-parties "shall be consistent with the objective of this protocol", and might be performed under bilateral agreements. Non-parties shall be encouraged "to adhere to this protocol and to contribute appropriate information to the (BCH)".

Transit

LMOs in transit or intended for contained use are included in the CPB, but they are neither subject to the AIA procedure nor to a risk assessment. The exclusion of LMOs for contained use from risk assessment does not reflect the position of the majority of the parties to the CBD or 'green' NGOs, and has the potential to counteract the spirit of the CPB. This provision invites exporters to evade risk assessment by declaring that the imported LMOs are meant for contained use, for example, by postponing field releases for a year. In such cases, the responsibility for risk assessments would be shifted from the exporter to the importing country; if national legislation exists in that country. Furthermore, the definition of contained use as "*any operation, undertaken within a facility, installation or other physical structures, which involve living modified organisms controlled by specific measures that effectively limit their contact with, and impact on, the external environment*" is extremely vague. The introduction of the term "or other physical structures" by the USA may encourage exporters to claim, for instance, that field trials surrounded by hemp plants or other pollen traps can be regarded as contained use.

Pharmaceutical LMOs

The CPB "*shall not apply to the transboundary movement of living modified organisms, which are pharmaceuticals for humans, addressed by other relevant international agreements or organizations*" (Article 5). To date, the only pharmaceutical LMOs are living, genetically modified viruses serving as vaccines. Currently, no other agreement explicitly addresses their trans-boundary movement and effects on biodiversity. They have to be approved after a medical risk assessment is conducted according to national and international pharmaceutical legislation. It is still unclear whether and how this open issue of the CPB will be solved.

The exclusion of pharmaceutical LMOs from the protocol reflects the common interest of countries with powerful pharmaceutical industries. Many developing countries fear that genetically engineered plants that produce vaccines or drugs might fall under this article and will be able to enter their territory, unannounced. Such plants will exhibit substantial variability in vaccine or drug concentration due to variations between each organism, climate, soils etc. Taking into account the unpredictable concentration of the active ingredient in a particular plant, the plants will not fulfil the criteria that would enable being classified as drugs. They will provide the raw material for industrial processes, in contained use, necessary to extract the active ingredient and produce pure drugs. The importation of this type of plants is, therefore, most likely to fall under article 6 (Transit and Contained Use), since they cannot be used as medicine themselves. This still implies import without the application of AIA and a biodiversity risk assessment. However, labeling requirements will have to be fulfilled and this will enable LMOs to be monitored and/or become subject to national legislations.

Relation to the WTO

During negotiations of the biosafety protocol, the Miami Group tried to subordinate the CPB to WTO regulations despite the strong objection of the other countries involved in this. A subordination such as that formulated in the so-called saving clause article 31 of the draft text would have devalued the PA as a political guideline. The Miami Group regarded the WTO as a bulwark against attempts to impose "unjustified" import restrictions on GMOs. According to WTO regulations, such restrictions can only be imposed after the damage has occurred and its cause has been identified. Without any statement about its relation with other international treaties, and in accordance with the *Vienna Convention on the Law of Treaties*, the CPB, as the more recent and more specific agreement, would prevail over the WTO. As a compromise, the CPB preamble reads that the parties *"recogniz(e) that trade and environment agreements should be mutually supportive with a view to achieve sustainable development, emphasiz(e) that this protocol shall not be interpreted as implying a change in the rights and obligations of a party under any existing international agreements; and understand that the above recital is not intended to subordinate this protocol to other international agreements"*.

Nonetheless, it is unclear how the formal equalization of the CPB and WTO agreements would fare in a lawsuit arising from an import restriction. The WTO has its own jurisdiction, and is able to impose fines worth millions of dollars. This makes it structurally much stronger than the Rio documents, which cannot be enforced in court.

Technical Assistance and Capacity Building

The CPB calls for cooperation between parties with regard to financial and technical assistance, and capacity-building. Canada and the USA, and some EU countries have been funding biosafety meetings for many years. These mainly serve to promote modern biotechnology and the *Technical Guidelines* of the *United Nations Environment Programme* (UNEP), which are not legally binding, are not based on the precautionary principle, and do not anticipate labeling. The promotion of these guidelines can be seen as an extension of the USA regulatory system. It would be fatal for the biosafety process if the implementation of the CPB were to be led by countries that are opposed to the precautionary principle and the spirit of the CPB.

Questions about the governance of risk are critical to the disputes over GMOs, and these issues are compounded by the international character of the disputes, which include questions about:

- the authority, role, and credibility of national and international regulatory authorities.
- the reconciliation of divergent national views of GMO technologies and international free trade.
- the special circumstances and needs of developing countries in relation to GMO technologies.
- the social, cultural, and economic dimensions of GMO agricultural technologies, as well as health and safety risks in context of an emerging precautionary approach.
- the influence of non-governmental organizations (NGOs) on global GMO politics and legal regimes.
- the role and influence of consumers.
- the role of scientists and the acceptability of technical expertise.
- the authority and credibility of international organizations in addressing GMO regulatory controversies.

A number of developments compound the difficulties of developing a system for governance of GMO risks that is both effective and enjoys widespread public confidence in the international context. First, the agricultural biotechnologies are new. While their proponents present substantial arguments in favor of the safety and benefits of GMOs, opponents claim that products have not been adequately tested and present many uncertainties and potential risks. At the same time, there are no well-developed or widely accepted regulatory principles or strategies for addressing the uncertainties posed by the new agricultural biotechnologies. The precautionary principle invoked to address uncertainty has so many different formulations and interpretations that its

practical utility is uncertain, and will remain so until its meaning has been clarified by authoritative international interpretation, whether legislative, or more likely, judicial.

7.15. MECHANISM OF IMPLEMENTATION OF BIOSAFETY GUIDELINES

In principle, all parties to an international treaty should have an interest in strong compliance mechanisms to ensure full implementation, and to prevent free-riding by the few at the cost of the many. In practice, multilateral environmental agreements contain only 'soft' mechanisms that seek to facilitate implementation through creating transparency and providing assistance. The possibility of taking stronger, even punitive, measures remains the exception in multilateral environmental agreements (MEAs). Environmental treaty-making has thus developed a practice of creating compliance mechanisms that are non-judicial, participatory and of a facilitative nature. These procedures aim at preventing disputes arising from instances of non-compliance, and at clarifying the application of MEA rules and provisions. Products arising from modern biotechnology provide new opportunities to achieve sustainable productivity gains in agriculture. Concerns over their possible environmental and health implications stimulated regulatory mechanisms for food safety and environmental risk assessment. Over the past two decades, national biosafety frameworks, guidelines, and regulatory systems have often been implemented on a "piece-by-piece" basis in response to the demands or urgent needs of the moment. Ideally, a biosafety system would be developed from a comprehensive plan **(Fig. 7.2)**. However, building such a system and making it operational is complicated by the fact that there is no single best approach or standard that reflects national, environmental, cultural, political, financial, and scientific heterogeneity. At the Meeting of the Parties (MOP) on the cartagena

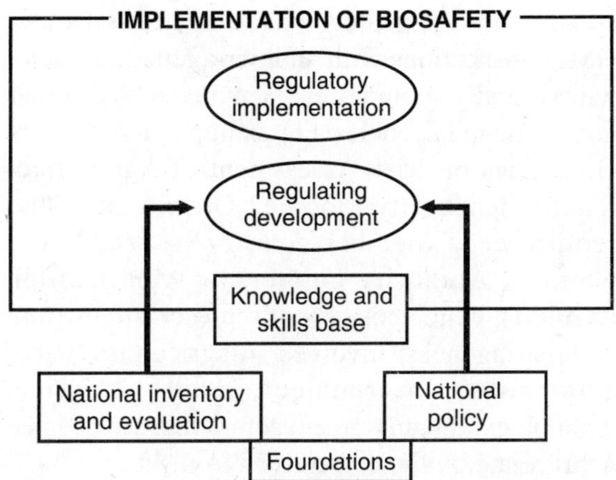

Fig. 7.2: Basic steps essential for implementation of biosafety regulations

protocol on biosafety (COP/MOP-1) in Kuala Lumpur, developing countries expressed concerns about proposed language that sought to strengthen the biosafety protocol's compliance mechanism.

The most outspoken opposition to strong compliance rules came from the group of GMO-exporting nations, most of which, however, as non-parties were not able to influence the negotiations. The draft negotiation text prepared by ICCP was littered with unresolved issues in square brackets. Among them were the questions of who would be entitled to initiate the compliance procedure; in what capacity the members of the compliance body—the Compliance Committee—would be serving; what information they would be able to consider, and from whom; and what measures could be taken against non-compliant countries. All countries with functional biosafety regulatory systems have gradually developed these systems, beginning with voluntary guidelines and standards that were developed cooperatively by stakeholders in academia, industry, and government. Over time these were incorporated in laws, either under existing legislation covering food and agricultural products or new legislation dealing specifically with biotechnology. Even in countries with long-established systems, biosafety policy and its implementation continues to evolve, and it is not

unusual to have a mix of voluntary and mandatory measures. For countries seeking to develop a national biosafety system, it must be emphasized that there is no model for a single best approach.

The central issues around the implementation of biosafety regulations involve the establishment of appropriate mechanisms for risk assessment, risk management, and risk communication within existing financial, technical, and human resource constraints. Decisions made during the implementation phase directly affect the costs associated with assessing and managing risks, ensuring compliance with regulations.

Mechanisms of implementations are the primary tools that countries use to examine the merits of a GMO or other biohazard in the areas of biosafety, food safety or consumer protection. The oversight mechanisms that have been established around the world are generally premised on a GMO's "first time" use in a particular context—importation, in-country research or commerce/marketing and, sometimes, export. Legal and non-legal instruments describe the oversight process and various institutions that may be involved with implementation and oversight. Requirements of submitting to oversight are either mandatory and typically described in legislation, or they are voluntary and described in guidelines. Common components of mechanisms are:

- The designation or establishment of institutions to undertake the review and/or provide advice.
- Safety assessment.
- Decision-making. In the systems examined, stakeholder participation is only a common element of mandatory oversight mechanisms promulgated by law.

At the international level, only the biosafety-related review bodies mention competent national authorities. For example, the International Plant Protection Convention requires its contracting parties "to make provision for" an official national plant protection organization. A list of responsibilities is enumerated including *inter alia* surveillance of growing plants, wild flora, plants and products in storage or transportation, inspection of international consignments for plant pests, disinfection, and the conduct of pest risk analyses.

The cartagena biosafety protocol to the CBD requires each of its contracting parties to designate one or more competent national authorities. These are to be authorized to be responsible for performing the administrative functions required by the protocol. The FAO Code of Conduct on Biological Control Agents lists some of the responsibilities of competent authorities in situations before and upon release, including *inter alia* "critical assessment", encouraging monitoring and ensuring corrective action where necessary. In its chapter on "biosafety and environmental concerns", the FAO preliminary International Code of Conduct on Plant Biotechnology suggests that governments should designate "competent national authorities to review, assess, implement and monitor biosafety and other concerns such as genetic erosion and agro-ecological disruption" from the introduction of biotechnological products (Article 11). Multidisciplinary and multiinterest "national committee(s) on biosafety and other environmental concerns" should contribute to the competent national authority's work. National instruments dealing with biosafety address institutional issues in far greater detail than international instruments. For example, the instruments examined either establish new institutions or designate existing institutions and give them new responsibilities related to GMOs.

Whenever the existing line ministries or their agencies are tasked with regulatory oversight, they do so within their traditional areas of competence. In Indonesia, for example, the category of organism determines the agency within the Ministry of Agriculture that reviews the application. Where new national level institutions are created they may be interdisciplinary or inter-agency in nature and either have an oversight function or an advisory function to the competent

authority that ultimately makes the decisions on a GMO. Institutions with an inter-agency character will typically include representatives from other governmental agencies. For example, the ASEAN Guidelines on Risk Assessment of Agriculture-related Genetically Modified Organisms (1999) require each member country to establish a National Authority on Genetic Modification (NAGM), which consists of representatives from national agencies involved in agriculture, trade, economics, environment, health, science, technology and any other sectors that are deemed appropriate by the respective NAGMs.

In some countries, representatives may also be from the academic and scientific communities and other major stakeholder groups. Bringing an interdisciplinary and, ideally, an independent perspective to the oversight review process could strengthen the determination of where the benefits and risks of the particular GMO lie.

For example, in the Philippines, government created a national committee on biosafety (NCBP) that is attached to the Department of Science and Technology. The NCBP has a multidisciplinary membership including various scientists, a social scientist, citizens and representatives from various governmental agencies. The NCBP has a number of functions. These include *inter alia*:

- identifying and evaluating potential hazards related to genetic engineering experiments, introduction of new species and GMOs, and recommending risk minimization measures;
- formulating and reviewing national biosafety policies and guidelines;
- formulating and reviewing national policies and guidelines on risk assessment;
- publishing the results of internal deliberations, and holding public deliberations on proposed national policies, guidelines and other biosafety issues; and
- assisting in the formulation of laws.

The department of science and technology provides the NCBP's secretariat.

The NCBP created the Philippine National Biosafety Guidelines in 1991. The NCBP must review and approve any work covered by the guidelines. However, institutions and involved scientists have the primary responsibility to enforce biosafety rules and regulations that is accomplished through institutional biosafety committees and biosafety officers. The NCBP has the power to impose sanctions on erring personnel and institutions. Other countries establish advisory bodies to focus on particular issue areas. Australia offers an example where a new competent national authority has been created, and is advised by three newly created committees.

These regulatory bodies perform functions related to issuing GMO licences, developing draft policy principles and codes of practice, and providing advice to the public, other regulatory agencies, and the ministerial council. They also establish a scientific committee (Gene Technology Technical Advisory Committee); a community committee (Gene Technology Community Consultative Committee); and an ethics committee (Gene Technology Ethics Committee).

The committees are interdisciplinary and share cross membership. On matters within their competence, the committees provide advice, upon request, to the regulator and the ministerial council.

They also provide advice on the need for policy principles and codes of practice is a function common to all three committees.

The ethics committee advices on ethical issues related to gene technology; the need for and content of codes of practice in relation to ethics and conducting dealings with GMOs; and the need for a content of policy principles in relation to dealings with GMOs that should not be conducted for ethical reasons. All committee members are subject to disclosure and conflict of interest rules.

Some mechanisms of implementation also require all institutions working with GMOs to create institutional biosafety committees (IBC). IBCs are typically given the ultimate responsibility to ensure the safety of any GMO-related work within the institution. When used effectively, IBCs could have a particularly important role in maximizing the benefits and minimizing the risks of GMOs. This is because projects could be screened early at the level of the researcher or institution before the government oversight is more formally applied.

A cornerstone of all mechanisms of implementation systems examined—whether voluntary or mandatory—is to assess the GMO's for safety. Biosafety regimes attempt to identify the risks posed by the GMO's to the environment and human health. Food safety regimes attempt to identify the risks posed by the GMO to human health. Safety assessment generally consists of the following:
* Hazard identification.
* Risk assessment.
* Risk management (UNEP, 1995).

Only risk assessment and risk management will be discussed here.

Risk Assessment

Any biohazard is dangerous for the natural environment, hence, the risk assessment is very essential. The underlying principle of risk assessment is to prevent the harm by identifying the probability of the occurence of the particular hazard. Since case-by-case risk assessment is quite burdensome, other principles such as "familiarity" and "substantial equivalence" have evolved with which certain assumptions can be made about the GMO under scrutiny, in order to facilitate the review. The principle of familiarity is used primarily in the biosafety area to determine the level of oversight applied to a particular GMO. It is premised on knowledge and experience with the host and recipient organisms. This can be used to extrapolate the potential risks of the modified organism. The UNEP biosafety guidelines note that familiarity does not imply that an organism is safe, while unfamiliarity does not imply that an organism is unsafe. Unfamiliarity means, however, that an organism should be assessed on a case-by-case basis. With experience and knowledge, risk assessment may apply to a group of organisms having characteristics, which are functionally equivalent on a physiological level. The development of generic risk assessment approaches or exemptions in one country does not necessarily mean that other countries will apply similar approaches. Monitoring can provide knowledge and experience on the use of organisms with novel traits. The principle of substantial equivalence is used primarily in the food safety area, where because of the complex nature of food and the inadequacy of traditional risk assessment techniques, there is a need for targeted approach. Substantial equivalence is primarily applied to food derived from genetically modified plants, and it attempts to take into account both intended and unintended changes in the plant or food derived from it (WHO, 2000).

The codex proposed draft guideline for the conduct of food safety assessment of food derived from recombinant DNA plants. It points out that substantial equivalence is not a safety assessment *per se*, rather, it is a way to structure food safety assessments relative to a conventional counterpart. Substantial equivalence is used to identify similarities and differences between the new food and the conventional counterpart, acknowledging that, for the foreseeable future, food derived from modern biotechnology will not be used as conventional counterparts. The safety assessment then assesses the safety of identified differences, taking into consideration unintended effects due to genetic modification. Risk managers subsequently judge this and design risk management measures as appropriate. "Risk communication" is one final principle related to risk assessment that may soon be introduced into the international instruments. Risk communication is an area related to public participation and access to information. Within the food safety area, the risk communication principles found in the Codex proposed draft principles for risk analysis of food

derived from modern biotechnology are premised on the belief that effective communication is essential in all phases of risk assessment and management.

Risk Management

The underlying principle of risk management is to identify and take steps to eliminate or minimize, to an acceptable level, the risks identified in the risk assessment. Risk management is typically practised at the level of the regulatory decision maker, who must process risk assessment data along with other factors that may be required to determine whether the approval should be granted or denied. The decision maker must determine the acceptable risk for society in relation to other possible benefits and costs. This is an inherently political decision. Risk management strategies vary with circumstances, and can embrace a number of techniques ranging from an outright ban to softer approaches that might include educating users of the proper application of an end product. In particular, post-approval monitoring, labeling and traceability can be used within risk management strategies, and are mentioned below.

A cornerstone of risk management practice, at least in toxicity studies related to human health, has been to build a safety factor to ensure that risks are truly minimized, if not eliminated. The evolution of this practice to a wider number of application, such as GMOs, may be reflected in part now by the precautionary principle, which should be applied by decision-makers where there is scientific uncertainty. The recognition of the need for a precautionary approach is greatest at the international level.

In the biosafety area, the cartagena protocol on biosafety is, at the moment, the foremost international instrument referring to a precautionary approach. In its preamble, the protocol "reaffirms" the "precautionary approach contained in Principle 15 of the Rio Declaration on Environment and Development". This approach is also referred to in Article 1. According to the protocol, decisions of the contracting party importing a GMO destined for first-time release into the environment (and where necessary, for GMOs intended for direct use as food or feed, or for processing) must be according to a risk assessment. However, lack of scientific certainty due to insufficient relevant scientific information and knowledge regarding the extent of potential adverse effects should prevent the contracting party from taking a decision, as appropriate, in order to avoid or minimize potential adverse effects.

In the food safety area, it appears that Codex Commission is embracing a precautionary approach, even if the term is not explicitly referred to in the Codex itself. For example, the Codex proposed draft principles for the risk analysis of food derived from modern biotechnology state that risk managers are to account for the uncertainties identified in the risk assessment, and manage the uncertainties.

In the area of trade, the WTO SPS Agreement provides some flexibility for member states to provisionally adopt sanitary and phytosanitary measures (SPMs), when scientific evidence for the measures is insufficient. Provisional SPMs can be adopted on the basis of "available pertinent information" derived from a variety of sources. However, member states must subsequently seek additional information to more objectively assess the risk and to review the SPM within a reasonable period of time.

Biosafety and Post-Approval Monitoring

Biosafety does not permit the release of GMOs. Post-approval monitoring is very essential part in the biosafety implementation. It is a mechanism to ensure compliance, after a permit is issued, to gather general information, and to identify unexpected consequences resulting from an approval. Post-approval monitoring, therefore, may be an important way to minimize the risks of modern biotechnology.

After receiving the consent, the authorization holder may be required to comply with certain conditions related to the release or marketing of a GMO that contributes to risk management. Monitoring may be one such condition. Another condition may be for the authorization holder to notify authorities when a problem occurs, and to take corrective measures. Monitoring may also take place for all the releases within a country, over a period of time. The FAO preliminary draft International Code of Conduct on Plant Biotechnology addresses both strategic and post-approval monitoring. For example, governments and international organizations should monitor and assess socio-impacts of biotechnologies as a part of their technology assessment programmes. Technology assessment procedures should include monitoring and long-term assessment of environmental impact. Finally, a proposer must ensure adequate and proportional monitoring of the actual effects that the organisms had on the environment, as a part of technology assessment procedures.

A number of international instruments in the biosafety and food safety areas refer to post-approval monitoring in a very general way. These instruments include the Codex Proposed Draft Principles for the Risk Analysis of Food Derived from Modern Biotechnology (risk management measures could include post-marketing monitoring); the Pest Risk Analysis Guidelines of the Plant Protection Convention (the effectiveness of phytosanitary measures should be monitored and risk management options should be reviewed if necessary; the convention on biological diversity (identify processes and categories of activities, which have or are likely to have significant adverse impacts on biodiversity, and monitor their effects; and the UNIDO Voluntary Code of Conduct for the Release of Organisms into the Environment (researchers/proposers have the general responsibility to notify unexpected or adverse public health or environmental impacts to the appropriate national authorities).

Labeling has a dual role—as a mechanism to provide access to information and as a means to manage risks. Traceability—the ability to track a GMO is an emerging issue within the biosafety and food safety areas. The concept behind traceability is to create a system to ensure that information is available on the origin of a genetically modified product as it moves from its point of manufacture to the end user. The system established would enable authorities to trace the organism back to those responsible for the import and export, as well as those responsible for the GMO's original development.

Implementation of the Cartagena Protocol

Implementation of the protocol is now a key issue, and there is an enormous amount of work to be done to ensure that developing countries can properly benefit from the rules which the cartagena protocol on biosafety establishes for the transboundary movements of GMOs. In particular, the international organizations urge the countries to ratify the cartagena protocol to strengthen or adopt national measures on biosafety, and to adopt interim measures for the implementation of the protocol's provisions in advance to its entry into force.

To ensure that the cartagena protocol on biosafety is implemented effectively as an international instrument for biosafety, the international organizations believe that the ICCP should undertake the following as a matter of priority:

- Set up a working group to make recommendations of international rules and procedures in the field of liability, and redress for damage resulting from transboundary movement of GMOs, for adoption by the first meeting of the parties to the protocol. The international organizations advocate that in the event of harm arising as a consequence of GMOs, including GMOs intended for food, feed or processing (GMO-FFPs), the exporter or the party of origin of those GMOs should be strictly liable for that harm and for providing compensation.

- Ensure that the procedures and mechanisms developed by the ICCP to facilitate decision-making by parties of import concerning imports of GMOs and GMO-FFPs:
  - are fully operational and apply the precautionary principle.
  - take into account risks to human health, socio-economic and cultural factors, as well as scientific information.
  - provide for the imposition of limitations and ceilings on the import of GMOs and GMO-FFPs, and conditions on their use. For example, where and how they are to be grown and used.
  - make it clear to States that it is legitimate for them to base their decisions on these factors.

- Set out detailed requirements on handling, transport, packaging and identification (article 18 of the cartagena protocol) of GMOs. These should include provisions for the traceability of all GMOs, including GMO-FFPs. Furthermore, the international organizations call for the clear labeling of all GMOs, GMO-FFPs, and their products, and for segregation of GMOs, including GMO-FFPs, from non-GMOs in their handling, storage and transport.

- Set up a working group to make recommendations on effective procedures and mechanisms for monitoring, compliance and dispute resolution in the context of the cartagena protocol for adoption by the first meeting of the parties to the protocol. The UKFG advocates that such procedures should ensure that disputes regarding the protocol and its implementation are resolved within the framework of the protocol rather than being transferred to other international fora.

- Establish, as a part of the biosafety clearing house and bring into operation as quickly as possible, an adverse impact reporting mechanism for reporting and sharing of information on any adverse effects of GMOs, including GMO-FFPs.

- Set guidelines for biosafety capacity building to ensure that all relevant technical, ecological and socio-economic aspects are covered, as well as the development and strengthening of the overall regulatory frameworks for biosafety implemented by States.

The international organizations propose that in order to ensure the credibility of capacity building, these guidelines should exclude the involvement in capacity building for the biosafety of private sector organizations, including parent companies and their subsidiaries, that may potentially engage in the supply or distribution of the GMOs including GMO-FFPs on a commercial basis. In addition, countries should also:

- establish, if they have not already done so, national focal points and competent authorities for dealing with GMOs and biosafety.
- establish and strengthen domestic regulatory frameworks to ensure that these fully cover GMOs and GMO-FFPs. The UKFG recommends that States should, *inter alia,* take the following further steps:
  — Implement seed testing and approval requirements for all imported seeds as well as domestically produced seeds, including GMO seeds.
  — Develop and strengthen appropriate domestic regimes for liability and compensation, so that these include coverage of any failures of GMOs to perform in accordance with claims made by the seed originator; adverse effects of GMOs arising from intentional or unintentional release; or sale or distribution by seed companies of contaminated seed.
- require companies to provide adequate and intelligible information at point-of-sale to farmers, especially poor farmers, concerning GMOs; husbandry requirements; risk assessment and risk management measures:

names and addresses of patent owner(s), seed originator, exporter, and importer.

- put in place measures, including national legislation, to implement exporter-based responsibilities for compliance with all relevant provisions of the biosafety protocol in relation to GMOs and GMO-FFPs that may be exported from their territories.
- develop public consultation under the decision- making procedures on AIA for all GMOs, including GMO-FFPs, in accordance with their national laws and regulations.

7.16. BIOSAFETY AND POLITICS

Genetically modified (GM) agricultural crops such as cotton, maize, or soybeans have been planted widely by farmers since 1995 in three western hemisphere countries—the United States, Argentina, and Canada. The GM crops have performed as advertised, allowing the farmers to reduce costs by controlling insects and weeds with fewer, less toxic, and less persistent chemicals. Yet the planting of GM crops has not spread significantly beyond these three countries, which still accounts for roughly 98% of the total GM crop acreage worldwide. It is perhaps understandable that GM crops have not made inroads into Europe or Japan, where consumers are well fed, farmers are prosperous without GM crops, and green parties and anti-globalization nongovernmental organizations (NGOs) promote a highly precautionary approach to food safety and environmental protection.

But why haven't more developing countries begun planting GM crops? In poor countries, many farmers are far from being fully productive, and many citizens worry more about food availability or food cost than unconfirmed food safety dangers. In poor countries, the development trumps the issues of environmental precaution. So, why has a pattern of policy resistance to GM crops emerged instead? Varieties of Bt cotton received only formal approval for commercialization in India in 2002.

However, it was discovered that an unauthorized variety had been marketed and planted during two growing seasons on 10,000 hectares in Gujarat and else-where, without being detected.

The rapid adoption of illegal Bt varieties indicates a high level of demand for GM cotton among some farmers. National experiences, whether in Africa, Asia or Latin America, must be considered in the wider context of international trade politics. This is highlighted by the ongoing World Trade Organization dispute between the United States (US) and the European Union (EU). This presents many risks and opportunities for developing countries. Some developing countries are vulnerable to the risks of losing markets in the EU through GM contamination. In Namibia, where approximately 80% of the country's meat exports go to the EU, livestock farmers are concerned that GM animal feed entering the country unofficially could undermine the confidence of European consumers.

There are, however, dilemmas with this, and this is illustrated by the attempts of those scientifically or bureaucratically marginalized within the regulatory process to push for more wide-ranging consideration of the environmental impacts of insect-resistant cotton, or potential commercialization of GM food crops. These deliberations bring questions of uncertainty, precaution and the social nature of risk centre stage. Such an analysis of regulation shows that science-policy cultures are not only central to the politics of GM crops, but they also open up far wider questions about how China negotiates the unchartered waters of constructing appropriate institutions to effectively manage the risks associated with the new technologies it is so rapidly embracing in its pursuit of modernization and economic growth. The export of US GM food aid to famine-affected countries in southern Africa has also provoked the suspicion that the US government is attempting to achieve acceptance of GM crops gradually. Senior US officials argue that it is more important to feed starving people than worry about the concerns of EU countries. However, African governments are concerned about both the safety of biotechnology and protecting their future trading relations with key export markets in the EU. These concerns contributed to Zambia's decision to refuse US food aid altogether, while Malawi, Mozambique and Zimbabwe agreed to accept shipments on the condition that they were milled to prevent planting.

Larger countries may be relatively insulated from the EU-US tussle. China and India have large domestic markets, which may enable them to commercialize certain GM crops without threatening exports. A recent analysis of GM commercialization in China shows that the country could realize significant gains domestically from commercializing some GM crops, regardless of the policies adopted by potential export markets. Brazil could have taken advantage of the difficulties faced by US producers and shippers in meeting the EU demand for non-GM supplies of crops such as soybeans. However, the Brazilian government has recently accepted GM soya production and export, preventing them from taking advantage of this opportunity.

In some cases (Kenya), the caution grows partly out of a weak national capacity to test GM crops for all possible local biosafety risks. Despite this weak capacity, the international donor community—pushed by environmental NGOs— has insisted that GM crops should not be introduced into poor countries, until regulations, facilities, and trained personnel are in place to permit thorough case-by-case biosafety screening. Many poor countries now have the regulations, but few have the facilities, trained personnel, or budgetary resources. In such countries, rather than be caught failing to measure up to their own announced biosafety standards, regulators often decide to err on the side of not granting any GM crop approvals. In other cases (Brazil and India), the non-approval of GM crops has been a direct result of the media campaigns and actions by local and international NGOs (Green-peace) opposed to GM crops. National regulatory authorities have been blocked from giving commercial release to GM crops by court actions, challenging the constitutional authority of national biosafety committees, or they have been deterred from moving ahead by NGO acts, such as attacks on crop field trials, designed to create social and media resistance to GM crops. In still other cases (China), the release of GM food and feed crops has recently been slowed in part by anxieties regarding consumer acceptance abroad. Regulators working in China's more tightly controlled political system have had less to fear from domestic media criticism or actions by NGOs and partly, as a result they have been able to approve and promote the production of an important GM crop—Bt cotton. Yet, partly for the fear of losing access to export markets in Asia or Europe, Chinese regulators have decided not to approve the commercial planting of GM feed or food crops such as maize or rice. Anxiety about the future acceptance of GM crops in export markets is also a growing factor in Brazil's policy making process. The governments of the developing countries are using biosafety regulations to go slow on GM crops, even in cases where biological safety is not the true concern.

SUMMARY

The biosafety refers to "efforts to ensure safety in using, transporting, transferring, handling, releasing and disposing of biological organisms, including genetically modified organisms (GMOs), when they are considered potentially capable of harming human, animal or plant health, or the environment". The use of biotechnology has led to broad changes in medicine, industrial processes, engineering and agriculture. In agriculture, biotechnology is being used to improve crops' resistance to pests and environmental stresses and to enhance their nutritional content.

Controversies have developed among various nations over whether certain types of GMOs constitute a potential hazard to human health or to the environment. GMOs have long been considered the subject of a number of biosafety concerns. Many genetically modified plants, for example, are produced with techniques that include the DNA sequences from common plant pathogens. Other biosafety concerns commonly cited are:

* The possibility of adverse impacts on non-target species or ecosystems.
* the potential for enhanced weediness of GM crops i.e. a plant becomes invasive, or transfers resistance genes to its wild relatives.
* The instability of inserted genes, i.e. the possibility that a gene will lose its effectiveness.

To date, transfer of genes to weedy relatives has been observed and a controversial laboratory study has shown potential damage to a non-target species. Whether such effects prove to be widespread or manageable has not been fully established.

Containment may be defined as a combination of laboratory procedures, laboratory equipments and installations, and host-vector systems designed to minimize accidental release of organisms during laboratory operations, their dissemination and survival in the environment and accidental infection of laboratory workers and of persons outside the laboratory. The containment may be either physical containment or biological containment methods. Physical containment consists of the use of special laboratory design, containment equipments and special operational procedures to restrict the number of organisms accidentally released during normal laboratory operations and to prevent infection of laboratory workers. Physical containment is grouped into four categories, referred to as BL1 (Biosafety level 1), BL2, BL3 and BL4. Biological containments specifically aim at making such genetic changes in GMOs that reduce the hazard from these organisms when they are accidentally or deliberately released in to the environment.

On 30 January 2000, the *Cartagena Protocol on Biosafety* (CPB) was adopted in Montreal by delegates of 128 Parties to the *Convention on Biological Diversity* (CBD). This must be seen as a landmark in the development of multinational environmental agreements. With it the *Precautionary Approach* (PA) became the guiding principle for the import *of Genetically Modified Organisms* (GMOs). It subjects this import to an *Advance Informed Agreement* (AIA) and secured its legal status in relation to the *World Trade Organization* (WTO). By 15 September 2000, 75 governments including those of 44 developing countries had signed this environmental agreement. The CPB is to set internationally binding safety standards for the import of *genetically modified organisms* (GMOs), and defaults for the risk assessment of these GMOs for biological diversity taking also into account risks to human health. It is also intended to guarantee the right of member states to make a decision about the import of GMOs after being informed in advance. These GMOs can, amongst other things, be food. However, such food that does not contain living organisms is not regulated by the CPB but by national or regional regulations like the *Regulation (EC) No 258/97 of the European Parliament and of the Council of 27 January 1997 concerning novel foods and novel food ingredients*. Minimal standards for risk assessment and safety measures for the transboundary movement of GMOs can be set using the CPB. This is particularly significant for importing countries that do not yet have GMO regulations. The CPB can thus be seen as an instrument that protects biological diversity in developing countries and this explains why it has received such strong support from these quarters.

Analysts agree that the future impact of the Protocol on trade is difficult to gauge because so many key aspects are still to be decided. Also, the current climate of controversy surrounding trade in agricultural GMOs in many countries, especially in Europe, further complicates assessments of trade implications. The Clinton Administration and the US biotechnology industry have stated that the new rules will make it easier to harness the promise of biotechnology without unduly disrupting world food trade. Others, especially critics in Congress, have viewed the accord with skepticism, and have charged that many of its provisions will harm biotechnology trade by opening a potential "flood gate" to restrictive and costly labeling and documentation requirements for US goods.

What does the adoption of international environmental safety standards for the transboundary shipment of GMOs mean for the export of such organisms as seed and/or food to developing countries? Will the commodity markets of these countries be flooded with GMOs because of the weaknesses in the CPB as mentioned above, the absence of appropriate national regulations and a lack of local capacity to deal with the problem? As far as LMO-FFPs are concerned this seems unlikely because, in the first place, commodity flows follow market capacities. The big, prosperous commodity markets are in Europe and in some Asian-Pacific countries not in developing countries. Those markets are the target of GMO exports from the three GMO growing countries—USA, Canada and Argentina, even though in most of these importing countries strong biosafety legislation exists.

The GMO market entered a critical stage when European and Japanese retailers reacted to the wide spread critical consumer attitude towards *genetically modified* (GM) food and started to import GMO free raw products.

The USA reacted to this by stabilizing the market for GMOs by indirect agricultural subsidies. Non-marketable harvests could be sold to the government and then distributed in food aid programmes. The amount of GMOs entering developing countries in this way and not following the provisions of the CPB—may well be grossly in excess of regular commodity flows. NGOs from developing countries have been critically analyzing this issue recently. The import of GM seeds into developing countries will be subject to the full provisions of the CPB. The analysis of the ongoing development of modern biotechnology in developing countries show that the major efforts of private and public development programs focus on establishing research facilities that can produce GMOs on national territory. This investment is closely connected to efforts to convince the countries concerned to establish the legal framework for exclusive intellectual property rights, e.g., through patents on plants or through the adoption of the *International Union for the Protection of New Varieties of Plants* (UPOV) agreement from 1991. Given this situation, the necessary biosafety evaluation depends exclusively on the legal framework of the individual developing country. Recognizing the efforts being made by developing countries to elaborate a CPB with high protective standards, attempts must now focus on implementing the CPB and on drafting of appropriate national biosafety legislation. They must be supported by the industrialized countries that signed the CPB.

In the face of continued controversy over GMO regulation, the first meeting of the parties to the cartagena protocol was a significant step forward. The fact that most GMO exporting nations, including the United States, Canada and Argentina, have yet to ratify the agreement gave the existing parties an opportunity to take some strong decisions on implementing the biosafety treaty. Whether the cartagena protocol can provide a working system of biosafety governance is to be seen. Future meetings of the parties will have to add more components to the treaty—including more specific rules on GMO identification in trade, compliance and liability—in order to make it fully operational. But in taking further decisions over the next few years, the parties will have to balance the desire to strengthen the protocol with the need to encourage ratification by some of the world's largest agricultural trading nations.

EXERCISE

1. Explain the need of biosafety in improved conditions of biotechnology in the worldwide environment.
2. Define biosafety.
3. Name different types of biosafety levels followed for the handling and maintenance of transgenic organisms.
4. Discuss how Convention on Biological Diversity (CBD) on biosafety entered into effect, and its impact on biotechnology.
5. Describe how widespread use of hazardous and genetically manipulated organisms affect the natural ecosystem.
6. Write a note on the advantages of adopting types of biosafety levels followed for the handling and maintenance of transgenic organisms.
7. Give an account of National Institutes of Health guidelines in transgenic research and their applications in science.
8. What do you mean by risk assessment? Mention its types with advantages in determination of risk assessment.
9. Name different types of animals biosafety levels in the biotechnology. What are their applications in the research?
10. Define containment and list different types of containments in biotechnological work.
11. Critically comment on the physical containment and biological containment used in the research and testing in biotechnology.
12. Differentiate between the different levels of biosafety levels used in the biotechnology research in the laboratory.
13. List out the national guidelines used for the recombinant research work in the biotechnological applications.
14. What do you mean by cartagena biosafety protocol (CBP)?
15. Elucidate different principles that are necessary to follow for genetically modified organisms business.
16. Mention different types of biosafety levels followed for the microbiological research.
17. What do you mean by Advance Informed Agreement (AIA)? Give its importance in the research and testing in biotechnology.
18. Distinguish between physical containment and biological containment.
19. Briefly explain the biosafety protocol or UN Cartagena biosafety protocol (CBP) and its key issues.

20. What are the experiments that require the approval of Institutional Biosafety Committee before initiation.

21. Explain the applications of biosafety level 4 and their importance in the studies of animals.

22. Mention the examples of biosafety equipment and their relevance to the biotechnology laboratories.

23. Illustrate how future plans are necessary for better biosafety measurement in the field of biotechnology.

24. How does biosafety level 1 work in ensuring the safety in the working conditions in the biotechnological research laboratories?

25. Define LMOs or "Living Modified Organisms".

26. Explain the problems that are to be addressed at biosafety level 2 and biosafety level 3, with their importance in the transgenic work.

27. Give an account of the mechanisms of implementation of biosafety guidelines for effective protection of the environment and human health.

28. Write a note on ecological consequences of genetically modified organisms and genetically modified foods in the present world.

Legal and Socio-economic Impacts of Biotechnology

8.1. INTRODUCTION

Kark Ereky, the famous Hungarian economist, coined the term 'biotechnology' in 1919. Since then it has become spot filed in the science, and now even common man has started speaking about the biotechnology. We are living in a wondrous world of inventions, which have made human lives better. These inventions are derived from the materials found in the nature, and through the precise understanding of the laws of physics. This lead to the technology, by adopting physical sciences (law of nature) to develop devices. This technology has been blended with living organisms to develop products for the betterment of human society. The necessity to feed the ever increasing population, and the reduced production rates made the biotechnology **(Fig. 8.1)** reach a greater heights in different fields of human life such as:

- plant sciences;
- animal sciences;
- medicine; and
- the environments.

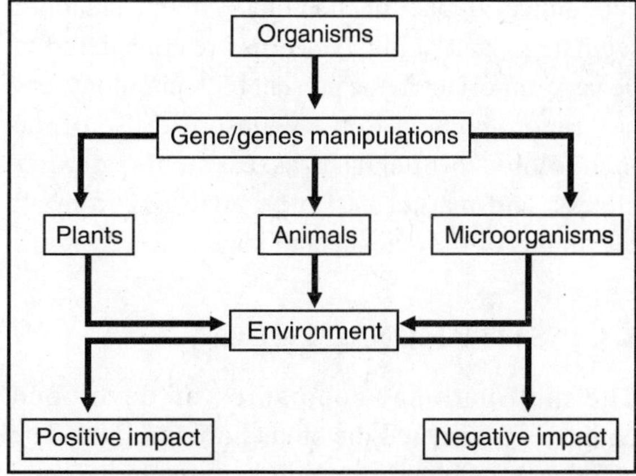

Fig. 8.1: Impact of biotechnologically produced organisms on environment

Large number of development and inventions made in the above fields by the biotechnology has paved the way to the present day situations that perfect human life and health. In doing so, there are some negative impacts, which are not considered in the fields of agriculture and food sector issues, business and consumer issues, institutional issues and social issues.

8.2. AGRICULTURE AND FOOD SECTOR ISSUES

Humans have changed the nature of plants. The plants that are available know are dramatically different from those that are found in the wild. This is due to both conventional and biotechnological processes.

Today, tremendous developments are made for resistance characters, improvements of nutritional values, pharmaceutical productions. Thus, impact of biotechnology on the organization, structure and behavior of agricultural industry, the coexistence of conventional organic agriculture and biotechnology oriented (industrialized) agriculture, impacts on competitions involved, segregation of genetically modified commodity from normal commodity and their specific market shares, trade methods on these products, economical impact of the both systems, standard regulations and public concerns are considered to be very important in the present biotechnology era.

Before coming to the widespread use of the technology, proper risk assessments, quality checks, and market has to be analyzed from all the angles of the science and society needs.

8.3. INSTITUTIONAL ISSUES

The multinational companies of developed countries dominated the biotechnology field. The efficiency, economic strength, knowledge, and market strengths are far greater when compared to the other developing and under developing countries local institutions.

Thus, impacts of biotechnology are on local individual small scale forms or groups of forms about buying or selling biotechnology products and processes, changes in business practices, alliances, domestic and international markets including the markets in the third world countries. The situations are not same in the developed and the ill-equipped third world poorer countries.

8.4. SOCIAL ISSUES

Different countries around the world have different patterns of social activities and functions. With the biotechnology developments, they need meaningful information for their involvement in taking decision on use of biotechnology products and their development.

The roles of civic participation at the local, state, national and international levels are the urgent need in the present scenario. The perceived and actual risks-benefits to the consumers and the general environmental protection, agroterrorism, research vandalism and their impacts on the poorer nations are the issues that are socially very important, as the poorer nations are not having the required facilities to cope with the impacts of the developments.

8.5. CULTURAL ISSUES

Many countries are culturally varied in their food, religious belief, agriculture and industrial matter. The biotechnology products and processes are not considering any of these cultural issues. This is creating a great impact on large number of cultural issues that has to be considered before this could become a major problem for the market of biotechnology products and processes.

8.6. BUSINESS AND CONSUMER ISSUES

With the aggressive development in the field of biotechnology, various products reached the market at tremendous speed. The influence on culture, scientific information availability, advertising methodology, product labeling and new events on consumer decision making the use of GM products and other biotechnology products, effectively increasing the awareness for biotechnology products and decision making capacity of consumers are very important in this category.

Biotechnology is an industry that commands higher and higher R&D outlays at the same time the relatively inexpensive copying technologies are available. Since biotechnology is so complex and unusual, and because the industry that deploys them is largely international in scope, the politics of adjusting domestic intellectual property rights (IPR) rules invariably coexists with political pressure for foreign IPR reform. This duality suggests that external IPR negotiations are influenced by international competitive pressures as well as by domestic political bargaining, concerning the creation and regulation of biotechnology products and processes.

In terms of market structure, the biotechnology industry is relatively new and based on rapidly changing, information-intensive technology. In addition, the market for biotechnology is global in the sense that products are developed, produced and sold on an international scale. However, neither condition is uniform. Technology varies in terms of its level of development, and markets vary in terms of the product cycle. Because of this variability, IPR reform should vary across different technological and market circumstances. In a new market involving developing technologies (like biotechnology), IPR reform has more long-term implications for economic competitiveness. IPR reforms primarily affect the potential for developing new products and markets. Consequently, IPR reform does not provoke immediate trade conflicts.

IPR policy in biotechnology reflects the industry's relatively low level of commercialization, and generally cooperative intra-industry relations. Although biotechnology is highly international in terms of R&D and marketing requirements, it is still a developing technology and consequently confers a relatively low degree of trade leverage. Since international markets are underdeveloped and expanding, IPR disputes have not provoked trade tensions, except in the secondary markets of pharmaceutical sales. In drug sales, countries having lower economic standards of living are pressured to either allow for sales of "generic" copies that replace the more expensive foreign (mostly US) patented drugs, or negotiate (either singly or through an international organization) for markedly lower prices from the drug companies for these much needed drugs.

Biotechnology is characterized by a symbiotic and generally cooperative relationship between small, highly innovative R&D firms, and large multinational enterprises, both of which stand of gain from stronger patent protection for biotechnology products. Their ability to pursue IPR reform has been aided by the lack of a contrary industry coalition, either from the biotechnology field or form unrelated holders of industrial patent rights. The primary resistance to increased IPR protection for biotechnology products has come from consumer advocates, and some public sector agencies who are wary of both the ethical implications of patenting life and the uncertain health and safety. In addition, although the human genome project (to identify all the genetic composition of the human body) has been funded and undertaken as an international research project by various governments, several well-financed private commercial companies have competed to identify and patent gene sequences. Public concern with potential restriction to these gene forms and their potential for medicines has caused several well-known biotechnology researchers to speak out against patenting of these markers.

Some of the legal, social and economic points are listed below.

8.7. LEGAL IMPACTS OF BIOTECHNOLOGY

- Uncertainty with performance of GMO's in field condition (transgenic fish, m.o.'s, plants, animals), and their implications on health, plants, and environment.
- No long term study on GMO's in field, thus releasing without unexplained problems.
- Safe in what context, quality and quantity of scientific data not mentioned.
- Ethics in animal, plant and environmental biotechnology is not regulated/controlled in third world countries.

- .No legal binding in GM farm animals for xenotransplantation and pharmaceuticals
- Animal and/or human health impact of animal biotechnology based on the nature and intended use.
- Different agencies operates on same product, thus complicate the issues.
- Lack of any established regulatory framework for research and commercial exploitation by multinational companies.
- No proper uniform guidance of food, agriculture, medicine, pharmaceuticals, cloning, and transplantation biotechnology.
- Potential lack of clarity about regulatory responsibilities and data collection requirements and its potential environmental hazard by different regulatory commitee.
- Absence of legal aspect for control of biotechnology, which affects social, economic, religious, cultural and ethical values.

8.8. SOCIAL IMPACTS OF BIOTECHNOLOGY

- Shift from cultural practice in food, agriculture, health.
- Dependence heavily on developed countries and companies.
- Loss of rural jobs and industries, thus migration to urban-creations of slums.
- Thick populated urban area.
- Poverty alleviations.
- Changed land use practices, social system, and resultant health impact.
- Non-performance of biotechnology crops and animal products ruined the land owners upto suicides.
- No market for conventional products, thus loss of labor, money, land, wealth, health.
- Monoculture, thus depends on others (outsiders) for basic need of food and life.
- Industrialized agriculture affected natural environment and local communities.
- The norms or religious traditions violated by GM crops and animals.

- Lack of information on biotechnology in different consumers goods.
- Signing for not to save GM seeds.
- Reproductive and health effects of biotechnology chemicals (glyphosates).
- Antibiotic resistance.
- Breaking all types of barriers in genetic mixing, i.e., spp., genera, phyla/kingdom.
- Unintended, unexplained and unexpected effects on biotic and nonbiotic components.
- Conflicting and variable opinion by scientists/experts on biotechnology.

8.9. ECONOMIC IMPACTS OF BIOTECHNOLOGY

- Due to industrial agriculture, small scale farmers and producers loose job and money-earning capacity.
- Reduced labor results in large population will be under below poverty line.
- Gap between richer and poorer is still wider.
- Loss of export by third world countries due to GM, new varieties/species/component.
- Food insecurity.
- Increased cost of environmental damage treatment processes.
- Switching of production centre from third world countries to developed countries (better IPR and biotechnology protection law), thus export income loss.
- Paying royalities on the products of locals or own knowledge.
- Penalties for hired pinkerton investigators (hired police).
- No confidentialities.

SUMMARY

Biotechnology can offer some solutions in the fight for sustainable, clean production, and significant mitigation of some of the negative side effects of other, currently non-sustainable technologies that are used to provide the necessities of everyday life in an industrialized society. For instance, cereal carbohydrates, cellulose waste streams and other biological materials can be converted using existing technologies into industrial feed stocks, such as plastics, polymers and fine chemicals, while clean, renewable energy sources such as alcohols and methyl esters can be obtained by relatively simple processing technologies.

Biotechnology, especially through the application of genetic modification, can enhance the efficiency of these processes, speeding them up and increasing yield, reducing net emissions of greenhouse gasses while at the same time, often sequestering carbon. But in doing so the legal, social and economic implications are very important to be considered for the prospective growth of any country. These issues should be addressed before any technology to become available for the people, since these issues are not easily understood and analyzed by the ordinary people of the country. The people were particularly interested in knowing if the research labs were open to the public. They also wanted to know if scientists talk to the public about biotechnology, especially genetic engineering and cloning. Several of the people felt there is a need for a "body or organizations or company" in country to deliver accurate and non-biased information on biotechnology in a manner that is easily understood. They felt that people need to understand before they accept and people fear what they do not understand. Hence, it is the responsibility of the scientists, academicians, policy makers and governmental organizations to give the right informations at the right time.

EXERCISE

1. What do you mean by "biotechnology"? Who coined this word for the first time?
2. Name different types of fields that made to feed the ever-increasing population and reduced production rates in the biological world.
3. Discuss the agricultural issues that are essential to be addressed for the biotechnological developments.
4. What do you mean by social, ethical and legal issues?
5. Elucidate different social issues that are affected by the developments made in the biotechnological fields.
6. Mention large number of legal issues raised by the improvements made in the biology and biotechnology in particular.
7. What do you mean by 'agroterrorism'? What is its impact on the research and development?
8. Distinguish genetic barrier and gene mixing.
9. Briefly explain which business and consumer issues should be addressed in the biotechnology, and their impact on third world countries.
10. How does intellectual property rights affect the poorer nationals in the protection of their identity?
11. Critically comment on the economic issues raised in the biotechnology fields of developing countries and their importance.
12. Mention how the development in the field of biotechnology made the developing and under developed countries suffer?
13. Differentiate between the convention methods and biotechnology methods of xenotransplantation in the animal biotechnology.
14. List out the ethical issues that are to be addressed in the animal biotechnology and agricultural biotechnology.
15. Illustrate how religious practices are affected by the use of biotechnology, and its developments made in the recent past.

Use of Genetically Modified Organisms and their Release in the Environment

9.1. INTRODUCTION

Genetic engineering processes are becoming common, and are being applied to a wide variety of organisms. Genetic modification involves identifying genes, and inserting them into foreign organisms which would express the desired traits. These genes can be taken from species of animals, plants, bacteria, and even humans. There are several processes used to insert "new" DNA into the foreign organisms that range from inserting genetic material directly into the eggs by subjecting the eggs of the fish to electrical pulses, which form pores and allow foreign DNA to access the eggs. The precise location where the new genetic material has attached to the original DNA is unknown and may vary among individual organisms, so scientists need to check if the inserted gene is present and functioning as expected.

These increasingly common practices, coupled with a lack of safeguards and regulations, make governmental action crucial for the protection of public and environmental health. Due to the commercialization of the products of genetic engineering, regulation to ensure safety became a key issue. European regulations on this issue, aspire to being precautionary, have been put in place before the evidence of environmental damage. It is recognized that scientific understanding of the implications of the release of genetically modified organisms (GMOs) into the environment is limited, but agreement says that the risks are potentially serious and irreversible. During such efforts, genetic engineering also raises ethical concerns.

Genetic engineering is altering the genome of a variety of species for the purpose of increasing aquaculture production, medical and other research, cleaning up water pollutants, and ornamental reasons. Since the development of the first transgenic plants, health and environment issues concerning the safety of genetically modified food and feed have been raised. Genetically modified organisms (GMOs) are a fact of modern biotechnology, and are here to stay for longer **(Fig. 9.1)**.

In the present situation, great deal of unease about genetic engineering was expressed throughout the world. The concerns focused on

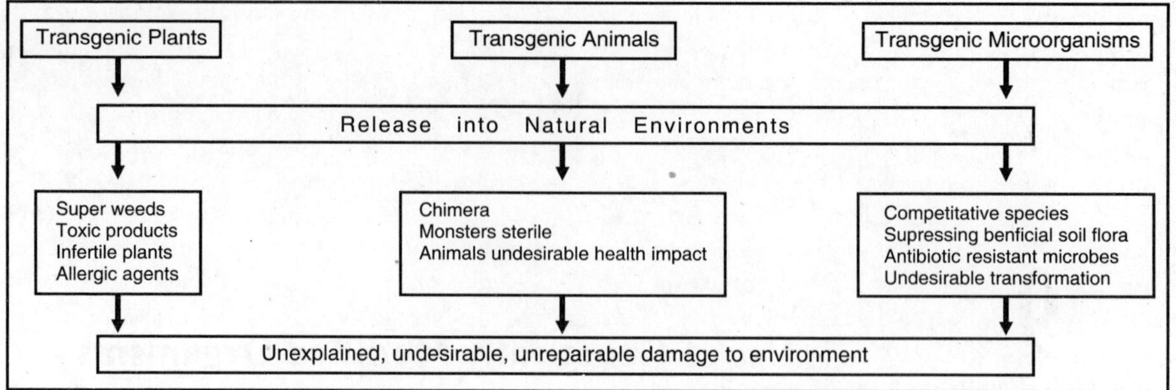

Fig. 9.1: Effect of transgenic organisms on the environment

the potentially irreversible nature of the risks; the ethical implications of producing transgenic species; uncertainty over the ability to predict accurately the effects of the release of GMOs; and the artificially restricted definition of risk used as the basis of policy. These concerns were interwoven with broader concerns about the environment such as the use of animals, the future shape of the countryside, and the overall trajectory of both the agriculture and the food industry.

Animal biotechnology has given new opportunities to the scientists. They are developing transgenic animals for a variety of purposes, which includes treating human disease, easing the shortage of organs for transplant patients, improving the efficiency of food production, and providing more nutritious foods. Conventional risk assessment was thought to be inadequate not only because extrapolation from small scale trials to large scale applications might be unreliable as the unknowns were so great, but also because the context of use was not fully considered. An example of this is the possibility of genetically engineered herbicide resistant plants becoming troublesome weeds. At the same time, as pointed out in "Animal Biotechnology: Science-Based Concerns", a report recently released by the National Research Council (NRC), both transgenic

and cloned animals, raise new concerns about food safety, animal welfare, human health, and the environment. A number of the researchers had participated in research and writing the NRC report, and gave the conclusions as well as their own perceptions of the discussion.

Whether genetic engineering was needed in agriculture or not, and whether the claimed benefits could justify the nature and scale of the potential risks, were some of the important questions. There was a desire to see alternative approach, such as organic agriculture, which was taken more seriously. Whether the improvement in quality of the product from GMOs was really needed, and whether GMOs were the answer to food shortages were also questioned. However, some expressed the opinion that GMOs could be used as an alternative to some of the most damaging conventional agricultural practices.

9.2. PRESENT SCENARIO OF GMOs

Presently, it is not possible to talk about GMOs if we remain at the level of generalities. Any statement must refer explicitly to specific crops, genes, ecologies, and production systems. For this reason, FAO has been conducting a worldwide inventory of agricultural biotechnology applications and products (Table 9.1), with special reference to developing countries. They wished to test the hypothesis that biotechnology has been uneven, and that most current biotechnology applications in crop science, particularly GM crops, have not addressed the special needs of developing countries. They have looked specifically at current as well as pipeline research on GM crops. The sources of information have been varied, informal discussions with scientists as well as formal reports. The inventory will be available electronically and will be updated regularly. Some of the preliminary results are as follows:

Table 9.1: Different types of organisms used for recombinant protein expression and their advantages

Organism	Amount expressed	Amount available for purification	Post translation modifications
Bacteria	++++	++	+
Yeast	++++	+++	++
Fungi	++++	+++	++
Transgenic plants	++++	++?	++
Baculo virus	++++	+++	+++
Mammalian cells	+	++++	++++
Transgenic animals	++++	++++	++++

The total area cultivated with GMO crops in the world currently stands at about 44.2 million hectares. This is in comparison to 11 million hectares only three years ago. About 75% of the area planted with GM crops is in industrialized countries. Substantial plantings, so far, largely concern only four crops—soybean, maize, cotton, and canola, and about 16% of the total area planted with these major crops is now under GM varieties. Two traits—insect resistance (mainly based on Bt) and herbicide tolerance—dominate. There are also small areas of potato and papaya, with genes for delayed ripening and virus-resistance inserted.

Only ten developing countries commercially cultivate GMO crops, and with the exception of Argentina and China, the areas involved are less than 100,000 hectares. Here again, the dominant crops are soybean and cotton, and the traits are herbicide tolerance and insect resistance. Only in China, one of the GM cotton was locally developed and commercialized. Other countries obtained genetic constructs or varieties from industrialized countries. Although several tree species, such as conifers, poplar, sweet gum and eucalyptus, have been transformed using recombinant DNA technology, they have not so far been released for commercial purposes. Tropical fruit tree species seem to have been largely neglected. Current releases are still very narrow in terms of crops and traits, and countries involved. So, what is in the pipeline? Throughout the world, several thousand GMO field tests have been conducted or are underway, mostly in industrialized countries. Some 200 crops are currently under field testing in developing countries. The vast majority of these are in Latin America (152), followed by Africa (33) and Asia (19). Many more countries than the seven that have already released GMOs are involved, and many more crop-trait combinations are being investigated, with greater focus on virus resistance, quality, and in some cases, tolerance to abiotic stresses. It can, therefore, be expected that the number of GMOs ready for commercial release in these countries will expand considerably in the next few years, though many important crops such as pulses, vegetables, fodder, and industrial crops, and certain traits such as drought and aluminum tolerance are still almost entirely neglected.

Today, animal transgenic technology offers a number of commercial and medical applications. Biotech firms are developing transgenic animals to provide organs for human transplantation; proteins for pharmaceutical and industrial production; limit environmental harm from agricultural practices; and improve production traits such as disease resistance. The benefits of transgenic animals to humans are great, in particular, the use of transgenic animals to produce blood proteins, pharmaceuticals and vaccine components in their milk. According to the researchers, blood products produced through transgenics would be much safer than current blood transfusions from human donors. Products of transgenic animals impose less threat to human health than older products. Researchers are pointing out that tainted blood transfusions have been responsible for spreading diseases such as leukemia and AIDS. Industrial uses are also on the way. For example, one company is producing spider silk in the milk of transgenic goats. It is the finest and the strongest. It is so strong that the military is proposing to use it in body armor.

9.3. FUTURE OF GMOs

So, while the portfolio of GM applications is increasing day by day, we need to deal rather rapidly with some overall questions to ensure that GM crops make an optimal contribution to world food security, safety, quality, sustainability, and is available to the public at large. There are three main problems here. Firstly, directing genomics and related research to meet these key challenges. Secondly, implementing international instruments and developing international standards of governance, which ensure health and environmental safety, and so command public support. Thirdly, facilitating the access to research and new technologies, for developing countries, poor producers and consumers. A related issue is that if developing countries are to be full partners in the international economy, they need assistance to build up their national legislation and regulations, as well as the relevant skills and infrastructures.

Assessing the Risk of Genetically Manipulated Organisms

United Kingdoms regulation over the release of GMOs implements a European Union Directive in which a precautionary approach is taken. Potential environmental risks are evaluated by an expert committee, the Advisory Committee on Releases to the Environment (ACRE), which advises the secretary of state for the environment.

The evaluation of risk by ACRE addresses only the immediate technical dimensions of risk, and relies on expert judgement to determine acceptability of risk as defined in this framework. Although this approach has an air of impartial assessment, recommendations also involve implicit value judgements. The most central one is that the important parameters of risk in the GMO domain are those in ACRE's essentially technical and single-issue remit.

ACRE also has to make value judgements about identified risks. For example, if genes flow, then the transfer of genes between organisms is possible, does it matter? ACRE's guidelines state that it matters if populations of plants or animals are likely to be seriously affected. What defines a 'serious' impact inevitably involves a set of value judgements about how much harm society may be willing to tolerate. A further, rather concealed value stance is the fact that ACRE operates under a 'taken-for-granted' assumption that genetic engineering is desirable.

These issues were the subject of three seminars held in 1995, organized by the Green Alliance and the Centre for the Study of Environmental Change, Lancaster University, and supported by the ESRC Global Environmental Change Programme. Each seminar was tailored for a different group: non-governmental organizations and conservation agencies; regulators and research councils; and the biotechnology industry. Key outcomes were as follows:

- Current approaches to environmental risk assessment exclude wider dimensions of risk, such as justification, need, benefit, the context of use, and public confidence in the trustworthiness of regulatory provisions and institutions themselves.

- Environmental and public interest groups felt that their broader concerns about GMOs lay outside the regulatory framework of risk assessment. These concerns include the implications of GMOs for agricultural strategies (such as pesticide use), and the extent to which GMOs are an appropriate technology or needed at all. Since these concerns have no clear route of expression, the groups in question have feelings of frustration and disempowerment, and their lack of engagement in the issues may give a false impression of the lack of interest or concern.

- This frustration is exacerbated by a general lack of trust in those creating and managing the risks—the biotechnology industry and the regulators—and should be viewed as a part of the diminishing trust in established institutions.

- Representatives of the industry expressed the frustration that the benefits of the new GMO technologies were not explicitly articulated in the risk assessment process. GMOs could bring benefits through progress in agriculture and the development of a new wealth generating industry. These factors may be denied through excessive emphasis on risks.

- Since the regulatory system, which addresses the risks of release of the GMOs, is the only focus for public involvement in the debate, much attention is directed at the Advisory Committee on Releases to the Environment (ACRE). However, both non-governmental organizations' concerns about wider issues, and industry's concerns about the potential benefits of GMOs were seen to lie outside ACRE's remit. ACRE and other regulators thus find that neither public interest groups nor industry are entirely happy with their role.

- There is an immense pressure on regulatory mechanisms, which are legally defined in narrow technical terms, to take responsibility for wider issues too. The mechanisms in question are not appropriate for addressing such issues adequately, and there are no other fora for addressing these dimensions.

Regulator Perspectives

There were some aspects of the risk of releasing GMOs, which were of legitimate public concern and lay beyond the current risk assessment system. However, for the regulators, factors such as agricultural practices were seen by many as 'non-scientific' or demanding a different area of expertise. They were, therefore, largely seen as not relevant to the risk assessment processes governing GMOs but there might need to be criteria established by others with appropriate expertise which could then be applied by ACRE.

Despite recognizing the importance of promoting confidence among the public that risks were being properly considered, there was a sense of puzzlement and frustration that inappropriate concerns were brought to the regulators. For example, whether herbicide resistant crop plants were a good thing in agricultural terms could not be a matter for ACRE. Similarly, ACRE was unable to consider benefits of any particular application.

New modes of communication between the scientific community and the public would be welcomed. The presses were seen by some as an unhelpful intermediary, and the legitimacy of pressure groups for representing the public's views was questioned.

Directing Research Towards the Key Challenges

Are GMOs being directed to the right research priorities? Despite some hopeful signs, the inventory suggests that this is not yet the case. There are specific technical challenges, which are difficult to handle through traditional crop-breeding. In that case, transgenics may offer great possibilities for crops and forestry. These include drought and heat-tolerance; improved nutrient uptake and rooting; biological nitrogen fixation; responses to carbon dioxide; and tolerance to key abiotic stresses, such as salinity and drought. As a part of the strategy of diet diversification, genetic modification may help us address protein, vitamin and iron deficiencies—see the famous example of the golden rice. Unfortunately, most of the traits needed are governed by multiple genes, and it will probably be some time before farmers and consumers benefit from such research. The perceived profit of GMOs has already changed the direction of investment in research and development in both the public and private sectors away from systems-based approaches to pest management, and towards a greater reliance on monocultures, with potentially adverse effects on landscape biodiversity. The possible long-term environmental costs of such strategies may be overlooked. If research is to address the challenges of agriculture in the future, we need to put genetic modification in context, and realize that it is but one of the many elements of agricultural change. Scientists must not be blinded by the glamour of cutting-edge molecular science for its own sake. One worry is also about the disconnection between the laboratory and the field, and the decline of agronomy as an integrative science. Governments need to be vigilant and not let this glamour, or the perception by private industry of major profit draw investment away from research in other, more traditional fields such as water and soil management or ecology, and from public sector research. At the same time, the best science is developed in a climate of intellectual freedom without much direct government interference. It is a difficult balance to strike! The cost of molecular technologies in plant breeding, and the size of markets that are required to recoup the profits lead to the growing use of "hard" intellectual property rights over seeds and planting material and the tools of genetic engineering. This changes the relationship between the public and private sectors, to the detriment of the public sector. A policy question that governments must take up, in both the national and international contexts, is how to ensure that public research is not the poor relation. In developing countries, in particular, it is important for the public sector to retain enough capacity, resources, and freedom of action to

provide the services on which their national private sectors can build. They will also need to build their policy and regulatory capacities with regard to transgenic crops that originate elsewhere, and in this, FAO has an "honest broker" role to play. Thus, we need a science that, in a globalized world, addresses agriculture and food security, and helps to keep the rural areas liveable and attractive to young people in search of employment. The constraints on doing so are more economic than technical. Globalization is not only the question of size, but also of kind. It is inextricably linked to privatisation, and major economic restructuring in both developed and developing countries has changed the balance of public and private sectors, and the privatisation of knowledge through intellectual property rights.

There is often little economic incentive for global companies to address local needs. It has been estimated that the world's top ten companies account for about 85% of the global market for the seed and agro-chemical industries. In 1998, just four companies controlled 69% of the seed market in the USA. Of course, globalization also promises to create opportunities for millions of farmers throughout the world. It also offers opportunities to the poorer countries, and can drive rapid development and local capital accumulation. But it may also exacerbate existing differences among countries. Developing transgenic crops imply massive investments and the need for massive returns.

The small number of GM technologies, currently in use, suggest that there is a real danger that the scale of the investment may lead to selective concentration on species and problems of global importance, and concomitant capital inertia. In this context, they must also sound a note of warning about the rising costs of regulation. In the health industry, where regulation is heaviest, bringing a pharmaceutical to market now costs about $500 million. Bringing a pesticide to market can cost about $200 million; and it has recently been reported that a GM crop costs about $30 million, to which regulatory costs can add up to a further \$5–6 million. What crops can bear such costs in the commercial market? How are we to serve local needs, small farmers and poor consumers, and promote genetic and dietary diversity?

9.4. DEVELOPING AGREED NATIONAL AND INTERNATIONAL INSTRUMENTS OF GOVERNANCE

Most policy and regulatory activities regarding GMOs are national. International agreements have a role to play in harmonizing national regulatory systems, for example, on transboundary questions, and when trade standards are involved. International agreements can also help governments, particularly the smaller governments, to rationalize their national capacities and investments through a common action. In relation to GM crops, the major instruments concern risk-assessment and management in two areas—food and health—and the environment. GMOs in food and agriculture have been around us for long to have some background against which to assess possible risks. There has been, as yet, no proven effect on human life. Moreover, a recent study of GMO released into the environment in Britain showed that they had not survived in nature. They would, in fact, if they had an evolutionary advantage, which remains an important possibility. But the state-of-the-art knowledge may not always be adequate to assess the risk of GMOs on the health and environment. Of course, lack of evidence of adverse effects is not the same as knowledge that genetic modification is safe.

In FAO we deal with environmental and food risk-assessment questions under a more general concept of "biosecurity", namely the application of sanitary and phytosanitary measures (SPS), and a methodology of risk analysis for food and agriculture, including fisheries and forestry. These include:

- the protection of animals or plants from pests, diseases, or disease-causing organisms.
- the protection of human or animal life from additives, contaminants, or residual toxins, or disease-causing organisms in foods, beverages or feedstuffs.
- the protection of human life from the diseases carried by animals, plants or plant products, or pests.
- the prevention or limitation of other damage from pests.

Moreover, FAO has specific standard-setting responsibilities that relate directly to the GMOs under the Technical Barriers to Trade (TBT) and the Sanitary and Phytosanitary (SPS) Agreements of the World Trade Organization. There are several important international mechanisms in the context of GMOs. Firstly, the cartagena protocol to the Convention on Biological Diversity, which was adopted in January 2000. It will come into force after fifty ratifications and is the main international agreement in relation to GMOs ("LMOs", or "Living Modified Organisms", in the language of the protocol). It covers the transboundary movement, transit, handling and use of all LMOs (except pharmaceuticals) that may have adverse effects on the conservation and sustainable use of biological diversity, taking also into account risks to human health. It allows for standard-setting in relation to the handling, transport, packaging and identification of LMOs.

Secondly, food safety aspects of genetically modified organisms are, at international level, dealt by the FAO/WHO *Codex Alimentarius*, which covers all aspects of food safety. The *Codex* is currently working on standards for risk assessment for labeling, and for several other food safety aspects of GMOs. The *Codex* standards are recognized by the SPS and TLC agreements.

Thirdly, the International Plant Protection Convention (IPPC) has the objective of preventing the introduction and spread of plant and plant pests, including weeds and other species that have indirect effects on both wild and cultivated plants, and to promote appropriate control measures. This also applies to risks associated with LMOs. The IPPC sets International Standards for Phytosanitary Measures (ISPMs), which are also recognized in the SPS agreement. It is also in the process of establishing practical cooperation with the CBD and its biosafety protocol. An IPPC expert working group met in September 2001, in coordination with experts from the CBD, to develop a detailed standard specification for an ISPM that identifies the plant pest risks associated with LMOs, and ways of assessing these risks.

Access and Benefit-Sharing

In a world governed by the intellectual property rights and concentrated research investments, genetic resources are the raw material to which biotechnology is applied. In agriculture, these are not the wild products of nature, but the fruit selection and development by farmers throughout the world since the Neolithic. In contrast to other areas of industry, this poses the immediate question of how to guarantee continued access to those who have been involved before—the farmers and the breeders. Governments are now completing the negotiation, within FAO's Commission on Genetic Resources for Food and Agriculture, of a new, binding convention—the International Undertaking on Plant Genetic Resources—which was expected to be approved by the conference in November. The undertaking will create a multilateral system of facilitated access and benefit-sharing for the world's key crops. This is a major step, because dealing bilaterally with such widespread resources would involve unacceptably high transaction costs, and could impede agricultural progress. Multilateral access provides multilateral benefit-sharing, which includes the sharing of the benefits arising from the commercialization of materials from the multilateral system through a mandatory payment. The access of breeders to genetic material for further breeding, which becomes ever more difficult with GM crops under patents, is a

public good that needs to be protected. FAO is involved in several instances, such as in the World Intellectual Property Organization, where concerns for food, agriculture and IPRs are being discussed.

9.5. GMOs AND SCIENTIFIC RESPONSIBILITY

Science never took place in a vacuum, but now more than ever, it needs to stand up to public scrutiny, to engage with public opinion, and address technical and ethical issues for the application of science to the needs of the poor. The current debate over transgenics and the labeling of foods containing ingredients from GMOs has highlighted the need for transparency. There is a growing awareness of the public's ambivalence towards genetic engineering and its aims. The scrupulous honesty of the peer review process, under the watchful eye of scientific institutions, including scientific academies, the great journals, is a guarantee for quality, if not always for relevance. The food and agriculture sector could well take the example of the declaration made last week by twelve of the world's most prominent medical journals of a policy to prevent the corporate sponsors of research exercising control over the analysis, interpretation and reporting of research results. However, the general public is largely unaware of peer reviews and standard setting (which takes place through similar processes), and public trust remains low.

The increasing use of modern biotechnology in agriculture has generated significant debate, much of which centers on the rapidly growing food crops that have been genetically modified to make them more resistant to pests or chemical herbicides. As a result, the debate has not addressed the potential products of agricultural biotechnology that are on the horizon. While technology developers believe that these new products will offer benefits in the form of food, fuel and fiber, as well as novel industrial and pharmaceutical uses, some of these future products are also likely to raise environmental and other concerns that will need to be addressed by the regulatory system.

It is intended to enrich both the knowledge and dialogue surrounding agricultural biotechnology by profiling some of the genetically engineered products being developed by the industry and scientists. The report reviews some of the current research on large-scale crops like corn and soybeans, but it also outlines ongoing research on a much broader range of plants, trees, grasses, animals, insects and fish. While not a comprehensive inventory, harvest on the horizon reveals the breadth and scope of current research activities and gives a snapshot of how industry and university researchers are thinking about potential future agricultural biotechnology products.

The report should not be viewed as an endorsement of biotechnology or any of the potential future applications discussed. Indeed, many of the applications in the report are likely to generate strong public debate over their relative risks and benefits. The report does not address these policy issues deliberately in order to provide a clear picture of the current agricultural biotechnology research and development efforts, the understanding of which is also critical to the public debate. Nor should the report be construed as a forecast of which products will be successful and brought to market. Much of the research cited is at an early stage, and many of the applications face significant technical, economic, marketing and regulatory challenges before they can be commercialized. By demonstrating what could be the "next generation" of products created by agricultural biotechnology, the Pew Initiative on Food and Biotechnology hopes to facilitate and enhance the ongoing public debate about this technology.

Context of Biotechnology

Biotechnology falls within the tradition of improving crops and livestock to meet the human needs in a better way. It also greatly expands the ability to move genes within and across the species, and creates a new ability to move genes across distantly related species and biological kingdoms. It is this attribute of biotechnology that makes it a potentially powerful tool for modifying nature, but it also raises ethical, health, and environmental issues.

Although the public is not widely knowledge-able about biotechnology and genetically modified food, consumer awareness appears to be growing. When the Pew Initiative on food and biotechnology compared the findings of two polls conducted in January 2001 and June 2001, respectively, it found that consumer awareness of genetically modified foods sold in grocery stores had increased by 11% between the polls. At the same time, use of the technology is also increasing in American fields, with farmers planting more genetically modified soybeans and cotton this year, according to a recent US Department of Agriculture survey. The government's report also notes that currently 68% of soybean acreage, 69% of cotton acreage, and 26% of total corn acreage planted is genetically modified.

9.6. FOOD CROPS

Making Food Production Easier

To date, much of the focus of industrial research has been on the development of products to reduce crop losses due to pests and disease, while maintaining yields. The first applications of biotechnology, now in the market since several years, include staple crops such as corn and cotton bioengineered to make toxins capable of killing insect pests. Other crops such as squash, potatoes, wheat, papaya and raspberries, engineered to resist common plant diseases, have recently become commercially available or may soon reach the market. Genetically modified soybeans, corn, canola and cotton resistant to the herbicide-glyphosate have been developed and are widely used in the United States. Scientists are continuing their work on other methods to directly improve the crop yields, and to help crops grow in difficult environments, such as those having limited water supplies or soils lacking nutrients.

Food Quality and Nutrition

Improving the crop yields and reducing the losses are important to farmers, but they do not necessarily deliver obvious benefits to consumers. Research is going on to develop products with benefits to consumers, such as improved food quality and nutrition. In particular, scientists are using biotechnology to add nutrients to the food traditionally lacking in these nutrients. For example, research is on ways to add Vitamin A to rice, a staple part of the diet in developing countries, to increase the nutritional quality of diets in these regions. Similar research is being undertaken to add vitamin E, an anti-oxidant thought to prevent cancer, to vegetable oils; to reduce the harmful saturated fats of cooking oils; to increase protein quality in vegetable staples; and to reduce the allergenic properties of milk, wheat, and other products to make them available to those who are ordinarily sensitive to them. The report does not address the likelihood that these products will be as beneficial as some proponents believe—an issue of current policy debate.

Future developments in this area will focus on the following:

Establishing scientific safety assessment frameworks based on sound science

There is a need to facilitate the establishment of safety and risk assessment frameworks for food derived from modern biotechnology. In addition to providing input to the Codex Inter-governmental Task Force on Food Derived from Biotechnology, WHO will develop the principles and guidelines for addressing emerging issues based on sound science.

Standardizing methods for nutritional aspects in safety assessment of food derived from modern biotechnology

There is a need for an increased focus on nutritional assessment of new foods derived from modern biotechnology and novel foods, in general, with specific nutritional traits. These types of considerations are already a part of the FAO/WHO Expert Consultations on food derived from modern biotechnology, but a strengthened effort is needed. This effort should be coordinated with efforts in other intergovernmental fora, such as the OECD Task Force on Novel Foods.

Linking risk assessment to risk management and communication

Effective mechanisms and approaches are urgently needed both at national and international level to bridge the outcomes of risk assessment into risk management and risk communication efforts. New thinking towards involving consumers and other interested parties, already at the planning stages of risk communication efforts, need to be developed. And the experience gained in this area could be used in other aspects of food safety.

Trees, Turf, and Decorative Plants

The applications of biotechnology in agriculture are not limited to food and feed crops. Some research on forestry is currently underway. Researchers are exploring ways to make trees grow more quickly and disease and stress resistant. For example, scientists are working on changing the amount of a tree's lignin—a substance that provides rigidity. Reducing the lignin could improve the efficiency of processing trees into paper. Alternatively, increasing the lignin content could be desirable for lumber, and possibly provide energy production when wood is used as a, fuel. Again, the report does not discuss any of the ethical and environmental issues that have been raised about some potential applications of this research.

Genetic engineering is also being applied to the recovery of tree species threatened by disease. The American chestnut is a hardwood tree formerly common in Appalachia and the East Coast. It was destroyed by a blight that killed 3.5 billion trees—all during the first 40 years of the 20th century. The discovery of genes, in a few surviving trees, responsible for resistance to the fungus may make it possible to develop resistance in genetically modified American chestnut trees.

Turf, commonly found on golf courses and also used in commercial landscaping, requires enormous quantities of water, herbicides and pesticides to preserve its appearance and functional quality. Industry researchers are genetically engineering grass to resist herbicides, which may allow for better weed control. In the more aesthetic arena, researchers are looking forward to engineer flowers and potted plants to have different shades of color and intensity than would normally be found in nature, and are investigating ways to change their size and shape (e.g., dwarf plants). Biotechnology techniques may also be able to extend the shelf life of cut flowers.

Biotech Industrial Products, Pharmaceuticals, and Environmental Clean-up

Researchers are modifying plants to produce industrial products and pharmaceuticals. Examples of products currently under development include, proteins and enzymes for diagnostic, therapeutic and manufacturing purposes; modified fatty acids and oils for paints and manufacturing; and biopolymers as substitutes for plastics. One potential product that is currently being explored is the creation of fibers for clothing and other materials. Using bacteria, engineered to make polymers that closely resemble natural fibers (such as silk, elastin, collagen and keratin), it may be possible to produce unique fibers similar to some of the most popular textiles used in the production of latest fashions.

Plants and animals can also be modified to act as biosensors—detecting or monitoring hazardous materials in the environment. For example, researchers have modified bacteria to be sensitive to TNT—an ingredient in many explosives—making it potentially useful in tasks like landmine detection. There is also work going underway to engineer zebra fish, so that it can act as biosensor for pollutants such as dioxin or PCBs.

Scientists are also working on a number of plant species to enhance their natural ability to absorb and store toxic and hazardous substances, which could assist in the cleanup of oil spills and chemical leaks.

It has been found that genetically modified plants can produce vaccines for human and animal illnesses, ranging from colon cancer, traveler's diarrhea to tooth decay. Technology developers believe that using food to deliver vaccines could have the benefit of permitting the vaccine to be consumed directly by humans or animals as food or feed, and eliminating the need for purification of the vaccine strain and the hazards associated with injections. Some of the plants used to develop the vaccines include corn, spinach, tobacco, lettuce, tomato, soybeans and potatoes. Plants may also be used to produce medical compounds for use in humans, such as monoclonal antibodies, hormones, or blood proteins.

Animals

In addition to research on plants, scientists are also investigating ways to bioengineer animals so as to provide treatment for disorders or diseases such as cystic fibrosis. Efforts are underway to modify sheep so that it could be able to produce fibrinogen—a major component in blood clotting—used for wound treatment. Scientists are also engineering animal tissue and organs for transplantation into humans. Researchers may be able to genetically modify pigs by suppressing a natural protein that causes the animal's tissue or organs to be rejected in humans. Harvest on the Horizon does not address animal welfare, ethical or any other policy issue that has been raised with respect to biotechnology research on animals.

Another area of animal research is the production of spider silk from genetically modified goats. Spider silk, an ultra-strong material that can be used in a variety of industrial products, including sutures and bulletproof vests, is difficult to cultivate commercially. Scientists are now able to produce spider silk by harvesting a protein from the milk of goats engineered with a spider gene.

Fish and Aquatic Organisms

According to the UN's Food and Agriculture Organization (FAO), approximately 100 billion tons of fish are consumed worldwide annually. It is, therefore, not surprising that researchers have focused on biotechnology to increase yields for the production of salmon and other fish. For example, the US Food and Drug Administration (FDA) is currently reviewing an application for a genetically modified salmon that grows three to five times the rate of its unmodified counterpart. At least nine species of fish, none of which are commercially available yet, have been genetically modified for enhanced growth including, the common carp, crucian carp, channel catfish, loach, tilapia, pike, rainbow trout, atlantic salmon, pacific salmon, and sockeye salmon. The possible ecological issues raised by the unintended introduction of genetically modified fish into the environment are beyond the scope of this report.

Since, diseases can spread rapidly in aquaculture with potentially devastating economic consequences, scientists are investigating techniques to protect fish and seafood from such diseases by improving disease resistance. In addition, by modifying salmon for tolerance to sub-freezing temperatures, the geographic range, where caged salmon could be farmed, could be extended.

Insects

Nowadays, researchers are investigating alternatives to chemical pesticides. They are modifying insects to make them more effective predators of unwanted insects, less virulent pests themselves, or to make them unable to carry diseases that are harmful to humans.

One area of research has centered on reducing the ability of mosquitoes to spread disease, such as malaria. The ability to make mosquitoes resistant to the malaria parasites they host could reduce the insects' ability to transmit devastating diseases in places like Africa and other developing countries, where some 300 to 500 million cases of the disease occur each year.

Current biotechnology research is addressing a wide range of possible future uses, from enhancing the nutritional value of food and increasing the food supply to controlling or preventing diseases the in humans and animals. While some applications are already in the market and used widely, others are a decade or more away.

Whether today's research projects become tomorrow's products depends not only on continued scientific progress, but also on addressing concerns about environmental impacts and other risks, meeting regulatory requirements, and dealing with marketplace realities. In this context, understanding the potential uses of this technology is one critical part of the process by which the public's decisions about biotechnology products will be made.

Broader perspective of health and development policy

The development of new foods through modern biotechnology has not always been perceived to be guided by a perspective of "public good". New products with potential health or production benefits in developing countries could change this. Such products will need a more holistic assessment, looking into all aspects of such production changes like health benefit, nutrition, safety, development and socio-economical issues, etc. To accomplish this, WHO should strengthen its collaboration with inter-sectoral partners. Such efforts would entail not only technical issues related to areas such as production efficiency and safety assessments, but also more general issues of how food derived from modern biotechnology could be made useful by addressing the needs of developing countries.

GM crops are being grown commercially or during field trials in over 40 countries and 6 continents. In 2000, about 109.2 million acres land was planted with transgenic crops; the principal ones being herbicide and insecticide resistant soybeans, corn, cotton and canola. Among crops grown commercially or field-tested are a sweet potato resistant to a virus that could decimate most of the African harvest, rice with increased iron and vitamins that may alleviate chronic malnutrition in Asian countries; and a variety of plants able to survive extremes of weather.

On the horizon are bananas that can be used to produce vaccines against infectious diseases such as hepatitis B; fish that mature more quickly; fruit and nut trees that yield years earlier; and plants that produce new plastic with unique properties.

In 2000, countries that grew 99% of the global transgenic crops were the United States (68%), Argentina (23%), Canada (7%), and China (1%). Although growth is expected to plateau in industrialized countries, it is increasing in developing countries. The next decade will see exponential progress in GM product development, as researchers are gaining increasing and unprecedented access to genomic resources that are applicable to organisms beyond the scope of individual projects.

Genetically modified (GM) food promises to meet some of challenging areas in the 21st century. But, like all new technologies, they also pose some risks—known and unknown. Controversies surrounding GM foods and crops commonly focus on human and environmental safety, labeling and consumer choice, intellectual property rights, ethics, food security, poverty reduction, and environmental conservation.

9.7. GM PRODUCTS: BENEFITS AND CONTROVERSIES

Benefits

Crops

- Enhanced taste and quality
- Reduced maturation time
- Increased nutrients, yields and stress tolerance
- Improved resistance to disease, pests and herbicides
- New products and growing techniques

Animals

- Increased resistance, productivity, hardiness, and feed efficiency
- Better yields of meat, eggs and milk
- Improved animal health and diagnostic methods

Environment

- "Friendly" bioherbicides and bioinsecticides
- Conservation of soil, water and energy
- Bioprocessing for forestry products
- Better natural waste management
- More efficient processing

Society

- Increased food security for growing populations

Controversies

Safety

- Potential human health impact—allergens, transfer of antibiotic resistance markers, unknown effects potential environmental impact—unintended transfer of transgenes through cross-pollination, unknown effects on other organisms (e.g., soil microbes) and loss of flora and fauna biodiversity.

Access and intellectual property

- Domination of world food production by a few companies
- Increasing dependence of developing countries on industrialized nations
- Biopiracy—foreign exploitation of natural resources

Ethics

- Violation of natural organisms' intrinsic values
- Tampering with nature by mixing genes among species
- Objections to consuming animal genes in plants and *vice versa*
- Stress for animals

Labeling

- Not mandatory in some countries (e.g., United States)
- Mixing the GM crops with non-GM crops confounds labeling attempts

Society

- New advances may be skewed to interests of rich countries

The report's findings have had a significant impact on policy debates, in particular, the Royal Commission on Environmental Pollution's highly regarded report on setting environmental standards. This in turn has enlightened several inquiries in both the House of Commons and Lords that have put ministers under pressure to explain the adequacy of the existing regulatory arrangements for GMOs.

9.8. PRESENT PROBLEMS OF GM TECHNOLOGY

In an increasingly technological society, the ability to manage the introduction of new technologies and practices is an essential skill for governments and firms. In our society, a range of risks and uncertainties are associated with new technologies that bring with them worries about human health, industrial competitiveness, ecological disruption, and trade disputes.

Governments need to reconcile a number of objectives for managing GM technology. We know that the Prime Minister is committed to a vision of the country becoming a 'knowledge economy', and sees biotechnology as an important part of this. So, powerful parts of the government consider that their first objective is to support a modern, knowledge—an intensive industry that has the potential to generate high-skill jobs, economic growth and a higher quality of life.

Government also has the job to ensure the safety of the consumers and general public, and protection of the environment. This involves making decisions about potential effects on human health, ecology and the countryside, more generally. The central question is how these decisions can be made when the technology and, therefore, the information on which decisions must be based, are subject to such profound and open-ended uncertainties.

This is the first of a range of problems with the way in which government is grappling with the GM issue and beyond. These problems are explored next, and include:

- the role of science;
- official characterizations of public perceptions; and
- defining the task for regulators.

Science Cannot Answer all the Questions

Science seeks to provide explanations for natural processes based on theory and evidence. But the way the scientific advice is used is heavily influenced by the way the official advisory system is put together. Until recently, this has been determined by a precise yet narrow interpretation of the sort of knowledge required to form judgements about GM technology. For example, no ecologist has been included in the various advisory committees.

Research also questions the widespread assumption that the safety and acceptability of commercial GM agriculture can be settled by science alone (Wynne 1999). Summarising recent research, Brian Wynne challenges this assumption on the following grounds:

- Genetic specimens used in test may not be replicable when produced industrially, since existing production is more a hit-and-miss than admitted.
- Crucial factors are excluded from the tests, such as the influence of birds, due to the fact that they circulate wider than the single-farm 'laboratory'.
- While demanding rules are imposed on crop management for tests, no systematic assessment is made of the extent and consequences of variation from these artificial conditions in real world uses of the technology.

What is more, while assessing the benefits and costs of new technologies, science faces a fundamental difficulty: not much is yet known, and there may be scientific disputes about what it is. Beyond this, there are all sorts of things we don't know that we don't know: remember CFCs and their unexpected damage to the ozone layer. In the face of uncertainty, there are two options—to conduct experiments to generate new information, or to make informed guesses about the potential effects of the technology. Either way, scientists are under real pressure to reach conclusions that can inform commercial and political decisions.

Science not only has problems in producing conclusive evidence about the impacts of new technologies, it is also ill-equipped to tackle the diffuse effects of existing technologies. In a global review of the evidence, Dr. Chris Williams attempted to build a comprehensive picture of the evidence concerning the health effects of industrial hazards, and in particular how these are damaging people's brains (Williams 1997). He found that science has great difficulty in rising to the challenge, due to the following reasons:

- Science generally analyses the effects of single substances, while most of the serious problems involve interactions between different industrial hazards, such as lead and fluoride (when added to water, fluoride increases the absorption of lead from pipes). Few comprehensive studies of multi-substance hazards have been undertaken due to lack of funding.
- The complexity of the issue means that scientists, who undertake such studies, risk arriving at inconclusive results.
- Clear results in a more narrow field of study bring more recognition.
- There are often long time lags and large distances between cause and effect.
- Scientific evidence of hazard or harm is frequently ignored, suppressed or attacked.
- Scientists themselves are unwilling to undertake such studies due to the above factors, and the effect that these might have on their careers.

Crisis management consultants were soon brought in. They advised companies like Tesco and Monsanto to avoid public discussion, and Tesco refused to answer further research questions. One important lesson of this example is that, as with Brent Spar, almost irrespective of the science or the law, it is the dynamics of public opinion around the world that has dictated the behavior of otherwise powerful companies (Williams 1998).

It has been found that particular difficulties have been posed by the one-at-a-time, product-by-product character of the assessments used by the key Committees. The limited basis for the existing scientific advice and the way it is framed has left such committees under-equipped to provide the desired comprehensive, scientifically robust under-pinning for regulatory decision-making in this area (Stirling and Grove-White 1999). Some of the issues left unaddressed include:

- the need for GM foods and the social benefits envisaged from GM strategies;
- the potential for indirect, cumulative and synergistic ecological and/or health effects arising from GM crops and food;
- the wider effects on the agricultural industry and the countryside;
- how to compare the significance of the risks and uncertainties, such as for human health, biodiversity, pesticide use, etc., that are attached to different agricultural strategies,
- the degree of public control and international pluralism that might be desirable and possible in a global food system dominated by a small number of large companies the contribution to global food production and the elimination of hunger; and
- systematic and transparent ways for regulatory appraisal to take account of different values and interests in society.

Feast of the Assumptions

By aspiring to a single definitive 'sound scientific' result, orthodox risk assessments fail to address the degree to which the resulting picture may be sensitive to crucial subjective 'framing assumptions'. Andy Stirling, working with Sue Mayer, is analysing the effects of people's assumptions and values on how they judge GM strategies against other options. By using a technique called multicriteria mapping, he has shown that people's different but equally reasonable starting assumptions can over-whelmingly affect the outcome of their analysis.

Dr. Stirling asked twelve leading players from all sides of the GM debate to list all of the criteria against which they would wish to judge GM crops and other alternatives. The 117 criteria that

emerged included factors such as the use of chemicals, impacts on wildlife, human health and cost to consumers. The participants were then able to assign scores to all of their criteria, and weights to reflect the relative importance of the different criteria. They were also asked to give a measure of how uncertain or variable they felt matters to be by assigning both an optimistic and pessimistic score.

Participants in this 'multi-criteria mapping' process are entirely able to express what is important to them and to assign weightings accordingly. They are also able to go back and adjust these in the light of results—they are in the driving seat at all times. However, when confronted with the outcomes of their analysis, remarkably, no one wanted to seriously amend the criteria or scores or weights they initially selected, so the technique seemed to be producing a robust reflection of people's evaluations.

The full results are described in a recent publication (Stirling and Mayer 1999), but the most significant finding was that the final results are most strongly influenced by each participant's early framing of the debate—their selection of criteria—rather than the scores or weightings assigned later. This finding stresses the importance of ensuring that the entire spectrum of values and interests are represented. Yet many criteria chosen by the participants in this study lie outside the scope of official risk assessments. For no participant is their whole range of criteria explicitly considered in the formal evaluation of GM crops in the UK.

9.9. TRANSGENIC ANIMALS AND THEIR ESCAPE

William Muir, director of the high definition genomics center at Purdue University, and a member of the committee that authored the NRC's animal biotechnology report, identified two principle environmental risk issues associated with transgenic animals—an invasion risk and an extinction risk.

In terms of both risks, Muir likened transgenic organisms to exotic species, which invade new areas, disrupting ecosystems and displacing other species. He referred to the gypsy moth as an example of such an exotic species. It was introduced on the East Coast to increase silk production. It got loose and is moving westward, causing extensive destruction of the plant life wherever as it goes.

He also said that like an exotic species, a transgenic escapee could pose an invasion risk if the new gene present in it increased its ecological adaptation, allowing it to extend its range into different environments. For instance, if a freshwater catfish, which was genetically engineered to tolerate salt-water, escaped into the ocean, it could displace species already present there.

Transgenics in new environments could displace other species through increased competition for limited food, or through changing predatorprey relationships. Fish that are genetically modified to grow more quickly or larger than normal could displace other fish by eating larger prey that other species had previously relied upon. At the same time, the transgenics would less likely become prey themselves.

Escaped transgenic animals engineered in ways that remove natural biological barriers, which limit their numbers, could also displace other species. One example is a transgenic catfish that uses a moth gene to excrete its own antibiotic. No longer subject to diseases that formerly culled its numbers, the transgenic catfish could multiply, putting pressure on other species.

Finally, Muir explained the workings of "Trojan Genes". Based on this hypothesis, if an escaped transgenic animal introduces a maladapted new gene—one that lowers the net fitness of the animal—into wild populations through interbreeding, the end result could be the extinction of both—the escaped transgenic population and its wild relatives. Trojan genes give transgenic animals an initial advantage over wild relatives, but lower their long-term fitness. For instance, an escaped male transgenic fish engineered to grow larger than normal would enjoy an outstanding mating, because female fish see large males as more "fit" than small males. Thus, it would rapidly spread the transgene through the wild population. In this example, however, the large transgenic male also carries a maladapted gene, one which results in decreased offspring viability. Since it is rapidly spreading a maladapted gene, in the long run, both the transgenic fish population and the wild species that carries its gene may face extinction.

Muir said that not all transgenic species present the same ecological risks. Those that are highly mobile, able to escape captivity, and can easily return to a wild state would be more worrisome. Mice, insects, and fish, therefore, create the greatest concern. Chickens, cows, and pigs have low mobility and are generally of lower concern, but pigs can readily return to the wild state.

Scientists can estimate the risk posed by a particular transgenic animal prior to release by analyzing the likelihood that it can spread into the environment. Muir described a new technology, the net fitness component methodology, to do this. To assess risk, it evaluates six fitness components—juvenile viability, age of sexual maturity, mating success, male fertility, female fecundity, and adult viability.

Since this risk assessment methodology provides a rigorous scientific test that regulatory agencies can easily adopt, this will, in turn, result in a greater acceptance of the technology because we know that we have sound scientific principles governing it. More importantly, we can debate those scientific principles for long enough.

The technology used to create the transgenic animals is one factor to consider. For example, microinjection—whereby the new gene is integrated at random spots in the genome—could, in theory, activate a host gene that produces a novel toxin or allergen, or even a prion-like element (similar to that responsible for "mad cow" disease). The risk that the transgene itself could cause allergenicity or toxicity depends on whether its protein product is already part of the human diet. If so, the risk would be very small. Earlier humans used to consume more animal body parts than today, but in fact, we are already eating—in some form or another—the proteins of most genes that are expressed in animals. If the transgene's protein is not a part of the human diet, it must be carefully assessed beforehand to make sure that it is not a potential allergen or toxin. This could be done either experimentally or by using databases to compare the protein structure with that of known allergens and toxins. Another safety issue is leaky expression of the transgene. This is especially important for bioactive molecules such as hormones. For instance, if a transgene targeted for expression in the mammary gland gets expressed in other animal tissues as well, this could affect both the physiology of the animal and the food safety of its meat.

Localization of transgene is an important issue in the transgenic organisms. Similar problems could be caused if the hormone leaked out of the mammary gland into the animal's blood circulation.

Multi-criteria mapping provides a systematic and transparent tool for exploring assumptions in judgements about issues such as GM crops and the alternatives. It does so in a way which directly addresses the practical business of choosing between policy options. By being simple and open, it builds trust among participants. It fully acknowledges uncertainties. It can include a wider range of issues than orthodox risk assessments. And it can be applied in a wide range of situations and social groups.

Government needs to be a more 'intelligent customer' for a wider range of advice. This involves specifying the criteria used to assess the quality of that advice. A provisional set of simple quality criteria for the regulatory appraisal of GMO's are emerging from the ESRC's research. As formulated in a recent report by Andy Stirling to the EU Forward Studies Unit, these criteria suggest that government should, among other things:

- broaden the scope of regulatory appraisal to address social benefits, and include synergistic, cumulative, and indirect effects.
- conduct assessments on a comparative rather than a case-by-case basis, including, for instance, an account of a variety of agricultural strategies.
- maintain a culture of humility and pluralism in the face of the many sources of uncertainty and ignorance in the appraisal of GM foods and other agricultural strategies.
- provide for iteration and open-enddeness in the interactions between sustained scientific monitoring, continued analysis, and inclusive deliberation in appraisal.
- uphold the primacy of institutional legitimacy and political accountability in the final justification of regulatory decisions.

9.10. FUTURE GOVERNANCE OF NEW TECHNOLOGIES

The development of GMOs is just one of the many technological innovations that will require careful handling by governments in the coming years. Governments and others need to learn a lesson from the current difficulties encountered with GM foods. They need to adopt a precautionary approach, some of the implications of which are now becoming clear.

Science is an essential engine of social progress. It produces much of the knowledge and technologies on which we all rely. Not only does it identify social and environmental problems, but help them solve too. As suggested, scientific monitoring of the effects of new technologies needs to be maintained and enhanced. But even with thorough monitoring, science is a necessary but not sufficient basis for government decisions. The issue is the use of science and scientific advice.

As advisors to Ministers, scientists often come under great pressure to reach unambiguous judgements about the safety of new technologies such as GM foods. This means that they find it difficult to justify precautionary advice due to the lack of evidence to back it up. The onus of proof tends to fall on those that fear negative effects. By contrast, justice systems around the world are beginning to develop various precautionary ways of dealing with uncertainty.

Six Rules for a Precautionary World

Tim O'Riordan has identified the following rules for precautionary action:

1. Where unambiguous scientific proof of cause and effect is not available, it is necessary to act with care.
2. Where the benefits of early action are judged to be greater than the likely costs of delay, it is appropriate to take a lead and inform society why such action is being taken.
3. Where there is the possibility of irreversible damage to natural life support functions, precautionary action should be taken irrespective of the foregone benefits.
4. Always listen to calls for a change of course. Incorporate representatives of such calls into deliberative forums, and maintain transparency throughout.
5. Never shy away from publicity, and never try to suppress information, however unpalatable. In the age of the internet, someone is bound to find out if information is being distorted or hidden.
6. Where there is public unease, act decisively to respond to that unease by introducing extensive discussions and deliberative techniques.

Risk of New Diseases

Do transgenic animals pose a significant risk of infectious disease among humans? Not according to John Coffin, a retrovirologist and professor at the Tufts University School of Medicine. Coffin described several possible diseases linked to transgenics, each one highly improbable and requiring a number of steps of a sort of increasing cumulative improbability. Coffin said that the risks were technology-dependent, and it was possible to mitigate or eliminate them by changing the technology. He outlined several possible routes by which transgenic animals could cause infectious disease in humans.

For instance, antibiotic resistant genes used in some transgenic technologies could enter the environment and taken up by other bacteria. Given the right chain of events, this could lead to an increase in antibiotic resistant pathogens. Coffin pointed out that one simple way to eliminate this possibility would be to stop the use of antibiotic resistant markers. In another scenario, a retroviral vector used to deliver the transgene could recombine with a latent virus buried in the animal's genome, or with another virus that infects the animal. This new recombinant virus could potentially spread into the food chain and then to humans. Coffin said that minimizing the transgene's viral sequences would lessen this risk, while stopping the use of viral vectors would eliminate it.

Similarly, a retroviral vector, such as murine leukemia virus could recombine with a related virus, leading to rescue of the vector and the restoration of pathogenicity. The rescued virus could cause leukemia, immunodeficiency, or any of the other diseases that retroviruses are famous for. Coffin emphasized that not using retroviral vectors would eliminate the risk.

Finally, a transplantation organ from a transgenic animal infected with an undetected animal virus could spread that virus to the human who receives the donor organ. The same genetic modification that would prevent rejection of the organ by the patient's immune system could also destroy the patient's immunity to the foreign virus. Once the organ is transplanted into the patient, the undetected virus could adapt itself to the new human host and cause infection. In a worse scenario, the virus would not only infect the transplant recipient, but would also spread to others and cause a new disease epidemic.

Presently, four requirements for the regulatory system in transgenic analysis are essential and are as follows:

- It needs the authority and expertise to ask the right questions.
- It should be able to generate the right data.
- It needs scientifically sound processes that are, by definition, open to new information and diverse points of view to evaluate that data.
- It needs a process that is publicly transparent and participatory. In addition, the process needs to place the burden of proof and the consequences of uncertainty on the sponsor of the technology.

The researcher pointed out that the FDA or other regulatory bodies has not had to deal with other disciplines that will be relevant to assessing ecological risk, such as population diversity, marine ecology and evolutionary biology. It is considered throughout the world that the regulatory process lacks transparency. As an example, Degnan pointed out that the licensing process under INAD is totally privileged information. The public would never know when an INAD has been filed, or for what product. Furthermore, the private license process wasn't designed to answer some of the broad scientific questions and novel issues raised by the technology.

Regulations in the World Biotechnology

Globally, a large number of national and multilateral organizations often hold conflicting policies on issues that affect transgenic animals, including policies on animal welfare, environmental impact, intellectual property rights, food safety and labeling. Lonzell (Bud) Locklear, an economic specialist at the United States Embassy in Ottawa, Canada, gave some examples of significant disputes in biotech products that are as follows:

- The European Union (EU) has imposed a moratorium on approvals of biotech applications. This has had a direct effect on the United States' biotech industry by factoring into decisions US farmers make regarding the crops they intend to plant and their willingness to embrace new biotech applications.
- The lack of an internationally agreed standard for labeling biotech food has led to a wide and often incompatible array of national labeling policies, which could possibly function as a barrier to trade.
- The European Union recognizes an overriding precautionary principle regarding transgenic animals. The US government approaches precaution as a flexible context-specific policy tool.

These three examples portray that at this juncture, there is a lack of agreed-upon rules, which, in turn, is having significant impact on trade and biotech products. Likewise, commerce and biotech animal products, whether for food, feed, or medicinal purposes, will be affected by the rules agreed in the international multilateral organizations.

Locklear pointed towards additional contradictory actions and policies that have been implemented around the globe. While some countries take steps to curtail the development of transgenics, others take steps to encourage it. In the United Kingdom, for example, The Royal Society has recommended that the UK government fund research into the welfare implications of transgenic plants and animals for agriculture. Similarly, The Royal Society of Canada recommends a moratorium on ocean pens for transgenic aquaculture. On the other hand, in October 2001, President Bush met with the leaders of the Asian Pacific Economic Cooperation organization to confirm their support for the development of biotechnology to help feed growing populations and its safe use based on sound science.

Internationally, the main organization involved with regulating transgenics is the World Trade Organization (WTO). The WTO describes itself as "the only global international organization dealing with the rules of trade between nations". It has two major agreements that affect biotechnology. The Agreement on the Applications of Sanitary and Phytosanitary Measures sets out basic rules to ensure that a country's consumers are supplied with food that is safe to eat, while at the same time ensuring that regulations are not used as an excuse to protect domestic producers. The second major WTO agreement, the Agreement on Technical Barriers to Trade, tried to ensure that regulations, standards, testing and certification procedures do not create unnecessary obstacles to trade.

In addition to these agreements, Locklear described an "alphabet soup" of organizations that affects issues of transgenic animals. Researchers stressed that international law and policy is evolving with many players in the game. As a consequence, there's not yet a comprehensive, well defined, globally agreed upon set of rules to manage commerce in biotech animals and animal products. However, he cautioned that international organizations and rules not only affect commerce, but also help to define public perception and acceptance of the technology.

Although many researchers agreed that society needs a way to identify transgenic plants and animals in the population, questions were raised about how long traceability needs to be maintained —just through the second or third generation or in perpetuity—as transgenic animals go through the breeding process.

There may be, one day, a national identification system for plants and livestock, perhaps in the form of a computerized mini-chip, that would allow people to know where the plants and animals came from and whether they were genetically engineered.

At present, views on labeling in biotechnology differ widely, with some saying it is crucial to give consumers the opportunity to make an informed choice in their purchases. Labeling, however, raises many questions: What would be its impact on agricultural production, particularly on producers? Would labeling be useful without widespread public education? How would consumers interpret the labels? Would labels be equally applicable to cloned and transgenic animals?

9.11. ENVIRONMENTAL EFFECTS

From a scientific perspective, the committee saw a need to place potential impacts of transgenic crops within the context of environmental effects caused by other agricultural practices and technologies. There is a substantial evidence that ecological effects of farming practices exert simplifying and destabilizing effects on neighboring natural ecosystems.

These effects are of concern because they appear to weaken or destroy the ecosystems' capacity for resilience, which is the ability of ecosystem to return to its initial state despite disturbance. Ecological effects of transgenic crops and other crops bearing novel traits may be heightened in this destabilized ecological milieu. This argues for a cautious approach to the release of any crop that bears a novel trait, or any extensive change in agricultural practices.

The two plant pest statutes (Federal Plant Pest Act and the Federal Plant Quarantine Act), which are used by APHIS to regulate transgenic plants, were originally developed to regulate the introduction of non-indigenous plant species. Since the amount of novel genetic information added to an ecosystem by the introduction of a new species is much greater than that added by a single transgene, the use of these plant pest statutes to regulate transgenic plants has been criticized. On the other hand, the use of these statutes for regulating transgenic but not conventionally improved plants has been defended, because the genes introduced in transgenic plants can have a much more distant taxonomic origin (for example, bacterial genes transferred to plants). These arguments are based on a general assumption that the risk associated with the introduction of genetic novelty is related to the number of genetic changes and the origin of the novel genes.

The committee compared empirical evidence of environmental impacts involving small to large amounts of genetic novelty to the taxonomically related and unrelated sources, and found no general support for this assumption. More specifically, it was found that:

- small and large genetic changes have had substantial environmental consequences.
- the consequences of biological novelty depend strongly on the specific environment, including the genomic, physical and biological environments into which they are introduced.
- the significance of the consequences of biological novelty depend on societal values.
- introduction of biological novelty can have unintended and unpredicted effects on the recipient community and ecosystem.
- *a priori* there is no strict dichotomy between the possibility of environmental hazard associated with releases of cultivated plants with novel traits, and the introduction of nonindigenous plant species. However, the highly domesticated characteristics of many cultivated plants decrease the potential of certain hazards.

The conventional development of semi-dwarf, short-season varieties of rice and wheat that propelled the Green Revolution of the twentieth century clearly exemplifies how a small number of genetic changes in a crop can impact the environment. In the case of rice, a single gene for short stature made rice much more responsive to fertilizer, while a few genes for more rapid maturation allowed farmers to grow one or two extra crops per year. These small genetic changes enabled massive changes in agricultural practices that increased production. However, these changes increased soil salinity, lowered water tables, and altered wetlands in some regions. Thus, there are tradeoffs between the long-term positive and negative effects of these varieties. The

environmental impact of the genes introduced in Green Revolution varieties and other crops is often indirect, which makes their assessment more complex but not less important.

In comparing conventional and transgenic approaches to crop improvement, it is found that both transgenic and conventional approaches (for example, hybridization, mutagenesis) for adding genetic variation to crops can cause changes in the plant genome that could result in unintended effects on crop traits. Genetic improvement of crops by both approaches typically involves the addition of genetic variation to the existing varieties, followed by screening for individuals that have only desirable traits. The screening component will remove many but not all of the unanticipated physical and ecological traits that could adversely affect the environment.

Based on a detailed evaluation of the intended and unintended traits produced by the two approaches to crop improvement, the committee found that the transgenic process does not present any new category of risk compared to the conventional methods of crop improvement. But specific traits introduced by both approaches could pose unique risks. There is currently no formal environmental regulation of most conventionally improved crops, so it is clear that the standards being set for transgenic crops are much higher than those in their conventional counterparts. The committee found that the scientific justification for regulation of transgenic plants does not depend on historically set precedents for not regulating conventionally modified plants. While there is a need to reevaluate the environmental effects of conventionally improved crops, for practical reasons, the committee does not recommend immediate regulation of conventional crops. Transgenic and conventional approaches are in a period of rapid change. This makes it difficult to assess the potential risks of specific traits that each approach will be able to alter in the future.

While it is not possible to assess the risks of any genetically modified plant without empirical examination, the committee found that it should be possible to screen quickly the modified plants for potential environmental risk, and then conduct detailed tests on only the subset of plants for which preliminary screening indicates potential risk.

To increase the knowledge in relevant areas, the international committee recommends substantial increase in public-sector investment in the following research areas:

- Improvement in precommercialization testing methods.
- Improvement in transgenic methods that will minimize risks.
- Research to identify transgenic micro-organism, animal and plant traits that would provide environmental benefits.
- Research to develop transgenic micro-organism, animal and plants with such traits.
- Research to improve the environmental risk characterization processes.
- Research on the social, economic, and value-based issues affecting environmental impacts of transgenic microorganisms, animals and crops.

Five Pillars of Risk Analysis

Not every environmental problem can be addressed, and priorities need to be set by the agencies and land managers. In recognition of this, risk analysis has forced itself onto the agenda for governments around the world over the last 20 years. **Risk** is the likelihood that damage can be caused by some behavior or action (including no action). Whereas **Hazard** is the agent that causes damage. While describing risk-based disciplines, risk analysis is the most general term and consists of the following four components, which are referred to as the "four pillars of risk analysis" by Davies (1996):

1. *Comparative risk analysis* entails comparing two or more types of risk, and is principally a tool for policymakers to decide on resource allocation.

2. *Risk assessment* is a set of analytical techniques for estimating the frequency of undesired events and their consequences, (damage or injury) and is accompanied by a description of uncertainty in the assessment process.
3. *Risk management*, considers social, economic, and political factors to determine the acceptability of damage, and the action that could be taken to mitigate it.
4. *Risk communication* entails conveying information about risk.
5. *Monitoring* is used to detect the impact of hazards at an early stage (although purists might include such monitoring under the heading of risk management), or to provide data to refine future risk assessments.

SUMMARY

Genetic modification has increased production in some crops. But the evidence we have suggests that the technology has so far addressed too few challenges, in few crops of relevance to production systems in many developing countries. Even in developed countries, a lack of perceived benefits for consumers, and uncertainty about their safety, have limited their adoption. The scale of investment involved, and the attraction of advanced science, may distort research priorities and investment.

Genetic modification is not a good in itself, but a tool integrated into a wider research agenda, where public and private science can balance each other. Most of the short term successes may be derived from marker assisted breeding and diagnostics rather than from transgenic crops per se. Steering research in the right direction and developing adequate and international agreements on safety and access is a difficult and responsible task. While we are more aware than ever of the need to manage international public goods responsibly, the political tools to do so are weak, and, in a globalised economy, the voices of small countries and poor producers and consumers often go unheard.

It is believe that scientists have moral responsibilities to speak for the weaker segments of society, because they sometimes best understand the likely results of not doing so. Biotechnology is a tool. It does not provide the only approach to addressing a particular set of problems, and other approaches or applications may offer better solutions for particular needs. The report does not attempt to weigh the costs and benefits of any particular biotechnology approach or to compare the relative merits of potential alternatives. The society is just beginning a long process to figure out the best regulatory system for transgenics. In order to do so, however, it need clarity about what we want the system to do and how we want it to do it.

For future novel products of biotechnology, adequate risk analysis for decision support and maintenance of authority will depend on a regulatory culture that reinforces the seriousness with which environmental risks are addressed. Public confidence in biotechnology will require that socioeconomic impacts are evaluated along with environmental risks and that people representing diverse values have an opportunity to participate in judgments about the impact of the technology. Currently, in USA, the APHIS environmental assessments focus on the simplest ecological scales, even though the history of environmental impacts associated with conventional breeding points to the importance of large-scale effects. The committee recommends that in the future APHIS should include any potential impacts of transgenic plants on regional farming practices or systems in its deregulation assessments. APHIS has been constrained in its risk analysis and decision-making process by the statutes through which it may regulate transgenic plants. In May 2000, the US Senate and House of Representatives agreed in a conference report on a new Plant Protection Act (PPA). The new regulations that will be used to enforce the PPA have not yet been developed. This has to be practiced throughout the world for better environmental protection.

The top priorities as expressed by the many national and international regulatory representatives are: improved laboratory methods involving relevant organisms that can be used to assess potential impact on non-target organisms (particularly for aquatic species, ability to predict and monitor distribution post release (including marker and detection systems, what are the safety issues and precautions necessary with DNA vaccines, containment methods development, understanding the behavior of the organism in the environment. (including survival, transport, host range, genetic transfer, vectors, virulence, etc.), use of genetic resources to distinguish pathogens from closely related taxa.

Assessing and managing environmental risks from transgenic microorganisms formed the focus of our discussion. Risk is defined as a function of hazard and exposure. Research activities aimed at assessing potential hazards from the use of transgenic microorganisms and assessing levels and routes of exposure were deemed high priorities. The group acknowledged that our ability to measure differences resulting from the use of transgenic organisms and to interpret the ecological meaning of observed change(s) was fundamental to risk assessment. Hence, priority should be given to improving the capacity to detect differences in behavior and ecology between parent and transgenic organisms that also allow us to better interpret changes at the ecosystem level. In addition to defining and prioritizing research needs for current technologies, the group recommends that the BRARG program take a proactive role in anticipating regulatory needs for technologies in the pipeline and providing funding for associated risk assessment.

EXERCISE

1. Define genetically modified organisms. Write a note on their applications in the society.
2. What are the advantages of the genetic modifications?
3. What do you mean by genetic modifications? How do genetic modifications proceed to the development of better food varieties and medicinal products?
4. What are the techniques used for genetic modifications of living organisms for the purpose of human society development.
5. Distinguish between genetically modified organisms and non-genetically modified organisms, with respect to their threat to the environment.
6. Mention the risks and benefits associated with the genetically modified agricultural biotechnology and medicinal biotechnology.
7. Define social, ethical and legal issues involved in the production of genetically modified organisms and their release into the environment.
8. Elucidate the negative effects of genetically modified organisms and their role in natural environment.
9. Depict the controversies associated with the production of genetically modified organisms and food.
10. Critically comment on the comprehensive and effective systems needed to find the solutions of the risks associated with the release of genetically modified organisms in the natural environment.
11. What do you mean by 'genetically modified food'? Write a note on the importance of the safety assessment of genetically modified organisms in the natural environment.
12. Write down briefly about different organisms produced by the genetic modification technology, and their present status with respect to their release into the environment.
13. How will you identify the potential risks associated with the release of genetically modified organisms in the natural environment?
14. List out the various fields that are getting benefits form the genetically modified organisms in the near future.
15. Mention the importance of education, certification, and regulations in the release of genetically modified organisms into the environment.
16. Differentiate between the plants and animals produced by the genetic modification technology.
17. Give an account of the environmental impact of genetic modification technology, organisms and their products that are used in the human society.
18. How is the quality of products affected by the genetic modification technology?
19. Explain the future of genetically modified organisms, and mention their merits in the research and development of biotechnology.
20. Name important criteria necessary for regulating a genetic modification technology in plants at the present applied situation of biotechnology.
21. Illustrate how regulations of genetic modification technologies in animals are used for the development and improvement of the animal health.

<table><tr><td>

10

CHAPTER

</td><td>

Hazardous Materials used in Biotechnology—their Handling and Disposal

</td></tr></table>

10.1. INTRODUCTION

The "dictionary" meaning of the hazardous material means any material that poses a danger to the living organisms. A large number of such hazardous materials are used in biotechnological research and development, industrial processes, product purification, packing, and usage.

The hazardous material could be of physical, chemical or biological origin. If may be toxic, corrosive, irritant, etc. The toxic material may not only have poisonous effects, but it also may have a detrimental effect on embryo, could cause cancer, or other heritable genetic damage. It could also be harmful if inhaled or while in contact with the skin. It could cause irritation to the eyes and skin, and may be toxic to humans and animals if they swallow it. Possible risks of impaired fertility can also result in some cases.

In the mid 1970s, when the new recombinant DNA techniques were being developed, there was a great excitement among the molecular biologists who were involved in it. But amid that excitement, there was a concern that the techniques might prove hazardous for the laboratory workers and people outside the laboratory. In some cases there was environmental concern as well.

Biohazards are defined as biological agents or substances present in or arising from the work environment that may be a hazard to the health or well-being of the worker or community. Examples of biohazards are infectious organisms such as bacteria and viruses, some recombinant DNA, zoonoses (diseases which occur primarily in animals but may be transmitted to humans), and primary cell cultures that are initiated with tissues from infected humans or animals. Recombinant DNA molecules are defined as either:

- Molecules that are constructed outside living cells by joining natural or synthetic DNA segments to DNA molecules that can replicate in a living cell.
- Molecules that result from the replication of those described above.

Containment of biohazards refers to safe methods for managing infectious agents in the laboratory environment to reduce or eliminate exposure of laboratory workers, other persons, and the outside environment. Factors to be considered while determining the level of containment include agent factors such as virulence, pathogenicity, infectious dose, environmental stability, route of spread, communicability, operations, quantity, availability of vaccine or treatment, and gene product effects.

The most important element of containment is strict adherence to standard microbiological practices and techniques. Appropriate facility design and engineering features, safety equipment, and management practices supplement this. Four biosafety levels (BSL's) define those conditions under which the agent can be handled safely.

There are six categories of chemical waste in the biotechnology laboratories, and these are commonly grouped as following:

- Unchlorinated organic waste
- Chlorinated organic waste
- Unused laboratory chemicals
- Hazardous waste—known components
- Hazardous waste—unknown components
- Dilute aqueous non-toxic waste

The biotechnology institution should be able to:

- provide resources to maintain safety and health;
- carry out risk assessments and review them when necessary;
- provide and maintain system of work that is safe;
- establish arrangements for the use, handling, storage and transport of articles and substances provided for use at work, which are safe and without risk to health;
- provide employees with information such as, instruction, training and supervision, as is necessary to secure their safety and health at work as well as of others who may be affected by their actions;
- carry out health surveillance, when required;
- ensure that all machinery, plant and equipment are maintained in a safe condition;
- make adequate provision for welfare facilities at work;
- keep the workplace safe and ensure that access and egress are safe and without risk; and
- monitor safety performance to maintain agreed standards.

10.2. DUTIES OF EMPLOYEES

- Take reasonable care of their own safety and health, as well as of others who may be affected by their acts or omissions at work.
- Cooperate with others in the university to fulfil the statutory duties.
- Not interfere with, misuse, or cause willful damage, to anything provided in the interest of safety and health.

To ensure that this policy is effective, it is necessary to:

- review it annually, or during any significant changes in the organization;
- make any such changes known to employees; and
- maintain procedures for communication and consultation between all levels of staff on matters of safety, health, and welfare.

Management Responsibility

The person with overall legal responsibility for safety and health is the president of the biotechnology institution. The academic staff are accountable for their research laboratory and ancillary areas.

Safety Structure

The head of the biotechnology institution has the ultimate responsibility for safety in the organization.

However, day-to-day implementation of the safety policy is the responsibility of the following:

- Radiation safety officer
- Radiation safety/radioactive room/disposal advisor
- Users group
- Safety advisor
- Mutagen/carcinogen handling
- Bioethics

- Pilot plant safety
- Safety advisor (Academic)

Responsibilities of the Organization Safety Advisor

- Monitoring all aspects of safety policy, and implementation and reporting to the head of organization on all these items.
- Ensuring, through the research supervisors and safety representatives that the system is implemented. The role of safety advisor is an advisory role.
- Representing the organization on the institutes safety committee, and liaison with the building's office, university's safety officer etc.
- Review of procedures.
- Ensuring that warning notices are placed at appropriate points.
- Ensure through the laboratory supervisors and safety representatives that personal protective equipment (PPE) is worn.
- Student safety.
- Major policy decisions should be processed through a staff meeting, for comment, and for approval by the Head of Organization. The safety advisor will have the full responsibility for preparing proposals and documents and implementing safety policy.

10.3. DISPOSAL OF CHEMICAL WASTE

The environmental committees and local authorities have laid down strict guidelines for the safe disposal of waste. These regulations are being constantly updated, and laboratory personnel should pay particular attention during the disposal of waste. Chemicals should be kept only for five years from the date of purchase. Chemicals should be bought in small quantities, as disposal of unused or unknown waste chemicals is extremely costly. Containers of correctly treated and labeled waste should be given to the technician, who is incharge of the waste, in the preparation room. The waste will then be stored in the outside waste solvent store to await collection and disposal.

All waste should be labeled correctly and stored in appropriate safety cabinets or safety cans.

The disposal of laboratory waste should only be done in the following manner:

Unchlorinated Organic Waste

This must be neutralized (pH 6–8 approx.) and labeled clearly. The label must include the category of waste, the neutralized pH value, the person's name and lab identification. Chlorinated and unchlorinated waste should not be mixed.

Chlorinated Organic Waste

Same for unchlorinated wasted.

Unused Laboratory Chemicals

Surplus chemicals can be donated to other laboratories along with the MSDS. If chemicals are not hazardous they can be thrown out provided the label and the cap is removed from the chemical container.

Hazardous Waste—Known Components

This category can be added to either chlorinated or unchlorinated waste respectively. If a component is incompatible, it should be clearly labeled and kept separate. If there is a mixture of chemicals, then each component should be named and the concentration indicated on the label. Some chemicals such as ethidium bromide can be neutralized or treated. Powdered chemical waste in their original containers should be catalogued with the name of the chemical, amount, and lab number. These waste chemicals should be given to the technician, who is in incharge of the waste, along with the catalogue.

Hazardous Waste—Unknown Components

The container should be labeled clearly with the lab number and person's name. The estimated amount left in the container, and any other information which might be relevant, should be written on the label.

Dilute Aqueous Waste

Dilute solutions of aqueous non-toxic/non hazardous waste may be disposed down the drain in the fume cupboard sink.

Disposal of Containers with Waste

Ring the appropriate technician in the preparation laboratory, and they will advise where and how to manage the disposal of the waste.

Disposal of Empty Containers

Small contaminated glass containers can be disposed in the cytotoxic sharps bins. Winchesters should be emptied and rinsed five times with water. Then the label can be removed and the container can be reused. Plastic containers should be emptied and rinsed five times in water, and the label removed. The container can then be safely disposed in the bin.

Chemical Storage and Inventory Control

Adverse hazardous chemical reactions

Adverse hazardous chemical reactions can occur when incompatible materials mix because of accidental breakage, fire, container failure, and mixing of gases or vapors from the containers having loose lids.

Reaction products

Reaction products from adverse hazardous chemical reactions can include generation of heat or fire, evolution of toxic or flammable gases, pressurization of containers and possible dispersal of materials or violent polymerization.

Precautions for Storing Dangerous and Reactive Substances

- Keep the laboratory chemical inventory to a minimum, and do not store excess quantities of any hazardous substance. Chemicals should be kept and stored as recommended by the manufacturer, and should not be kept beyond five years. Hazardous chemicals must have a risk assessment completed, signed and given to the safety advisor before ordering as per safe work practice. An inventory of hazardous chemicals must be kept; a copy of this inventory must be given to the safety advisor.
- All chemicals should be separated according to the guidelines for storing dangerous and reactive substances. Hazardous chemicals should be stored separately in ventilated, labeled, secure cabinets/rooms.
- Chemicals that are radioactive must be stored and controlled according to "Local Regulations for working with unsealed radioisotopes, DCU".
- Container must be in good condition and compatible with its contents. Degraded, missing label, spilled or leaking containers must be disposed off in a safe manner (see safe work practice).
- All containers must be legibly labeled with the name of the chemical, concentration, and a hazard warning symbol.
- Peroxide formers and other chemicals, which degrade over time, must be dated when received, and opened and disposed off within one year or tested for the presence of organic peroxides (see safe work practice sheet).
- Store hazardous chemicals/materials as low down as possible in case they fall.
- Flammable material should not be kept in a domestic refrigerator. It should be kept in a flammable storage refrigerator.
- Material Safety Data Sheets (MSDS) should be available for each chemical in the laboratory (see safe work practice sheet). The MSDS will give the information to the user to work with the chemical safely. MSDS's should be acquired and kept by the user.
- Excess chemicals should be donated to a laboratory, which requires that chemical, or

Table 10.1: List of incompatible chemicals (reactive hazards)

Substances	Corresponding substances
Acetic acid	• Chromic acid, nitric acid, hydroxyl-containing compounds, ethylene glycol, perchloric acid, peroxides and permanganates.
Acetone Acetylene	• Concentrated nitric and sulphuric acid mixtures. • Chlorine, bromine, copper, silver, fluorine, and mercury.
Alkali and alkaline earth metals, such as sodium, potassium, lithium, magnesium, calcium, powdered aluminium	• Carbon dioxide, carbon tetrachloride and other chlorinated hydrocarbons (Also prohibit water, foam and dry chemical on fires involving these metals—dry sand should be available)
Ammonia (anhyd.)	• Mercury, chlorine, calcium hypochlorite, iodine, bromine, and hydrogen fluoride.
Ammonium nitrate	• Acids, metal powders, flammable liquids, chlorates, nitrites, sulphur, finely divided organics or combustibles.
Aniline	• Nitric acid, hydrogen peroxide.
Bromine	• Ammonia, acetylene, butadiene, butane and other petroleum gases, sodium carbide, turpentine, benzene, and finely divided metals.
Calcium oxide	• Water.
Carbon, activated	• Calcium hypochlorite.
Chlorates	• Ammonium salts, acids, metal powders, sulphur, finely divided organics or combustibles.
Chromic acid and chromium trioxide	• Acetic acid, naphthalene, camphor, glycerol, turpentine, alcohol, and other flammable liquids
Chlorine	• Ammonia, acetylene, butadiene, butane and other petroleum gases, hydrogen, sodium carbide, turpentine, benzene and finely divided metals.
Chlorine dioxide	• Ammonia, methane, phosphine and hydrogen sulphide.
Copper	• Acetylene, hydrogen peroxide.
Fluorine	• Isolate from everything.
Hydrazine	• Hydrogen peroxide, nitric acid, any other oxidant.
Hydrocarbons (benzene, butane, propane, gasoline, turpentine etc.)	• Fluorine, chlorine, bromine, chromic acid, peroxide
Hydrocyanic acid	• Nitric acid, alkalies.
Hydrofluoric acid, anhyd. (Hydrogen fluoride)	• Ammonia, aqueous or anhydrous.
Hydrogen peroxide	• Copper, chromium, iron, most metals or their salts, any flammable liquid, combustible materials, aniline, nitromethane.
Hydrogen sulphide	• Fuming nitric acid, oxidizing gases.

Cont....

Table 10.1. contd.

Substances	Corresponding substances
Iodine	• Acetylene, ammonia (anhydr. or aqueous).
Mercury	• Acetylene, fulminic acid*, ammonia.
Nitric acid (conc.)	• Acetic acid, acetone, alcohol, aniline, chromic acid, hydrocyanic acid, hydrogen sulphide, flammable liquids, flammable gases and nitratable substances
Nitroparaffins	• Inorganic bases, amines.
Oxalic acid	• Silver, mercury.
Oxygen	• Oils, grease, hydrogen, flammable liquids, solids or gases.
Perchloric acid	• Acetic anhydride, bismuth and its alloys, alcohol, paper, wood, grease, oils.
Peroxides, organic	• Acids (organic or mineral), avoid frictions, store cold.
Phosphorus (white)	• Air, oxygen.
Potassium chlorate	• Acids (see also chlorates).
Potassium perchlorate	• Acids (see also perchloric acid).
Potassium permanganate	• Glycerol, ethylene glycol, benzaldehyde, sulphuric acid
Silver	Acetylene, oxalic acid, tartaric acid, fulminic acid*, ammonium compounds.
Sodium	See alkali metals (above).
Sodium nitrite	Ammonium nitrate and other ammonium salts.
Sodium peroxide	• Any oxidizable substance, such as ethanol, methanol, glacial acetic acid, acetic anhydride, benzaldehyde, carbon disulfide, glycerol, ethylene glycol, ethyl acetate, methyl acetate and furfural.
Sulphuric acid	• Chlorates, perchlorates, permanganates.

*Produced in nitric acid-ethanol mixtures.

disposed off according to safe work practice sheet.

List of Incompatible Chemicals (Reactive Hazards)

Substances in the left hand column should be stored and handled carefully so that they do not accidentally contact the corresponding substances in the right hand column under uncontrolled conditions, where violent reactions may occur **(Table 10.1)**.

Obtaining a Material Safety Data Sheet (MSDS)

The manufacturer or distributor of a chemical product must send material safety data sheets to the purchaser (DCU), with the initial shipment of a product/hazardous chemical, and each update of the material safety data sheet. These sheets will either accompany or precede the shipment. However, the manufacturer or distributor is not required to send material safety data sheets with subsequent orders of the same product to DCU.

The users of biological agents or chemicals are responsible for keeping their own updated MSDS's.

Purpose of MSDS

The purpose of MSDS is to ensure that the hazards of all chemicals produced or imported are evaluated, and the information is transmitted to the employers and employees. An MSDS must provide a variety of safety information related to the chemical or biological agent that will allow the user to recognize and prepare for potential hazards and emergency situations. A glossary of terms appearing in an MSDS can be obtained at www.ilpi.com/msds/ref/index.html.

Online Sources of MSDS

Some online sources of MSDS are:
- www.sigma-aldrich.com
- www.piercenet.com
- www.lennox.ie/chemicals/index.htm
- Ensure that the MSDS obtained is the exact product you are using.

Procedure for the Reduction and Disposal of Hazardous Chemical Waste

Biotechnology institution can be inspected at any time by the relevant authorities, e.g., Health and Safety Authority, Corporation and Environmental Protection Agency for compliance with the health and safety guideline and other relevant legislation, and license/s to discharge to the sewer and the Waste Management Act.

Amount of chemical and solvent waste should be minimized by the miniaturization of the experiments, and ordering the quantities of chemicals that will be used completely by the laboratory within five years. There are exceptions to this (see peroxide forming materials).

Labeling

All hazardous waste containers must have a label, which states hazardous waste and lists its constituents. The labeling must be done when waste is first placed into the container. Be specific while naming the waste, (e.g., xylene instead of non-halogenated solvents, and ethanol instead of alcohol). Do not use abbreviations or chemical formulas. Waste should be divided into two classes—chlorinated and unchlorinated—and labeled as such. Labels for hazardous waste are available from the technical staff (**Figs. 10.1 and 10.2** for copies of waste labeling).

Containers and Storage

The person who generates the waste must provide proper containers for accumulating and storing hazardous waste. Generally, the best containers for hazardous waste are the ones in which the materials come in e.g., Winchesters. All containers must have tight-fitting lids. Corks, ground glass stoppers, or parafilm are not proper substitutes for a tight-fitting lid. Unacceptable containers or containers without tight-fitting lids will not be disposed of by the technical personnel, and it will be the responsibility of the generator to transfer the material to another container or to provide a proper lid for the container.

At any time, not more than 25 litres of waste would be allowed to accumulate in any lab or storage area, prior to its removal to the outside solvent storage area. When the waste generator has a container full of waste, a waste form should be filled and the container given to the technical staff incharge of the waste.

Waste should be stored in a solvent cabinet or ventilated cabinets under the fume hoods, and the location should be secure. Incompatible materials, either waste or unused chemicals, should never be mixed. Incompatible materials when mixed together may cause explosions, fires, or may generate flammable or toxic gases resulting in serious health hazards.

Caution must be exercised in any area where chemicals or wastes are stored to ensure that incompatible materials are segregated appropriately. Segregate chemical/solvent waste by hazard class. Do not store waste with fresh/new chemicals. Flammable waste should be kept away from heating sources, and stored in accordance with the University's safety manual. The organization safety advisor can provide information on the proper storage of flammable and combustible materials. Examples of chemicals that should not be stored together, as they are incompatible, are:

- Oxidizers with flammables
- Elemental metals with hydrides
- Acids with cyanides
- Acids with sulfides
- Acids with bases
- Acids with flammables
- Acids with chlorine compounds
- Acids with alcohols
- Acids with elemental metals
- Amines with chlorine compounds
- Water or air reactive chemicals are incompatible with everything
- Organic peroxides with anything
- Phenols with formaldehyde
- Sodium azide with aqueous lead

This is not an exhaustive list. For a more detailed list, see respective chemical guide. You should always consult a Material Safety Data Sheet (MSDS) or other chemical information sources such as Bretherick's Handbook, or the Merck Index for compatibility information.

Halogenated waste materials (those containing halogen compounds such as chlorine or fluorine) should be separated from non-halogenated compounds, unless unavoidable. This is for the safety and economic reasons. The halogenated wastes, though much less flammable, are generally more toxic than non-halogenated waste materials.

Where possible, mercury compounds should be eliminated from the laboratory. It is very important not to mix mercury with other materials due to difficulty (and cost) of disposing off mercury compounds. A mercury spill kit should be available where mercury is used.

If there are queries regarding dangerous or incompatible materials, these will be handled on a case-by-case basis by the laboratory supervisor and the safety advisor.

10.4. DISPOSAL OF HAZARDOUS WASTE

To dispose off any hazardous waste, a hazardous waste form for each container should be filled. **(Figs. 10.1 and 10.2)**. These are obtained from the preparation area.

- The person who generates the waste should fill the hazardous waste form and attach it to the container.
- Hazardous waste must be divided into chlorinated or unchlorinated waste.
- Chemical formulas or abbreviations should not be used.
- Chemical names of waste constituents should be used.
- All liquids should be brought to pH 7.
- Information should be printed and it should be legible.
- One form should be filled for each container.
- Containers must also be labeled with a hazard-warning symbol.
- For animal carcasses, one form is needed for each bag or box.
- No waste material should be disposed off without properly filling a hazardous waste form.

Pharmaceuticals and Controlled Medications

Pharmaceuticals, chemotherapy agents and other controlled medications should be managed in the

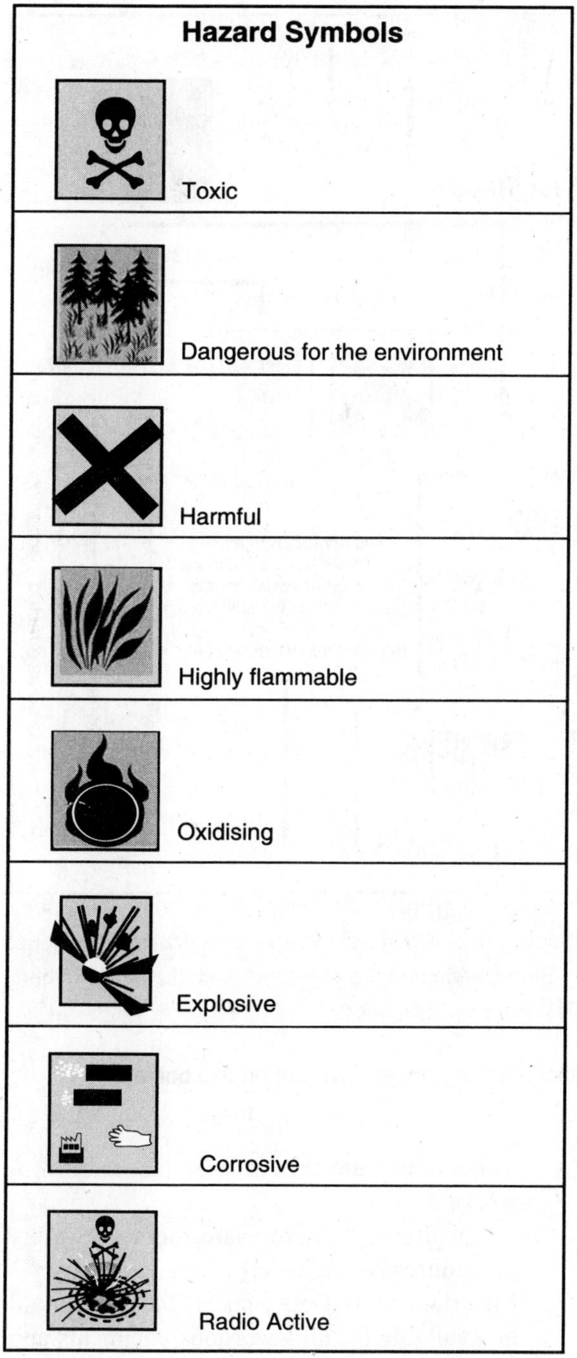

Hazard Symbols

Toxic

Dangerous for the environment

Harmful

Highly flammable

Oxidising

Explosive

Corrosive

Radio Active

Fig. 10.1: Different hazard symbols generally used for indicating various materials used in biotechnology

same manner as any other hazardous waste. While filling the hazardous waste disposal form, both the common trade name and the more definitive chemical name should be listed. Any additional information available about the substance should be written on the waste disposal form.

Ethers and Peroxide-Forming Materials

Some chemicals such as old ethers (not petroleum ether), picric acid, and organic peroxides tend to form unstable (explosive) compounds. Over time they can become extremely unstable. If peroxide formation is suspected, the containers should be isolated and the safety advisor notified. Under no circumstances should the staff attempt to open containers if crystal formations are visible in the container. It is recommended that researchers mark their containers with the date when the container was received and opened. Peroxide-forming materials, which are not needed, should be disposed of six months prior to the expiration date on the containers.

Dilute Solutions of Non-Toxic Aqueous Chemicals

Dilute solutions of non-toxic aqueous chemicals may be disposed off down the fume hood sinks only. (For guidelines on disposal see the discharge license.)

Animal Carcasses

Animals that have been contaminated with carcinogens or other highly toxic materials are considered to be hazardous waste and must be disposed of. Animals should be frozen and kept at $-20°C$. Disposal of the frozen carcass will be done by a licensed contractor.

10.5. SAFETY STATEMENT

Summary of Duties

- To ensure that laboratory practices are in accordance with the safety document.

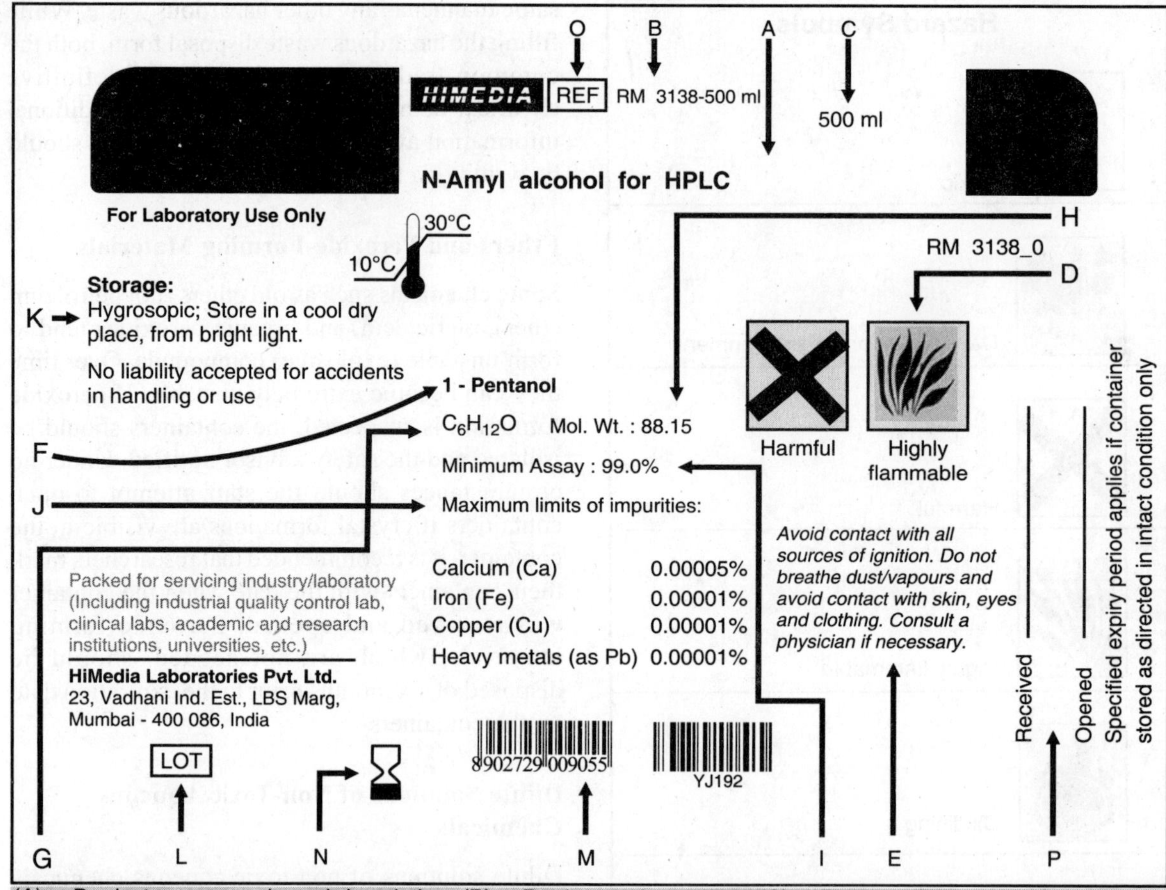

Fig. 10.2: The chemical label of N-amyl alcohol showing hazard symbol on the bottle.

- To ensure good laboratory practice/house-keeping.
- To be fully familiar with first aid and fire extinguisher procedures or know who is trained in these procedures.
- To ensure that all safety equipment is in good working condition.
- To know the university safety regulations regarding fire safety, and fire warden.
- To know the assembly point for your area, and to communicate this to your laboratory co-workers.
- To ensure that appropriate radiation safety procedures are followed where necessary.
- Material Safety Data Sheets (MSDS) should be available for all hazardous chemicals and updated regularly.
- To compile a list of hazardous chemicals and biohazards used in your laboratories, and to have the appropriate disposal procedure

written and available for the chemical/biohazard. A copy of this list should be given to the safety advisor.

- To ensure that the occupants of your laboratory are aware of any unsafe practices. Report any unsafe practices to your supervisor.
- To complete the safety inspection report/laboratory risk assessment every six months.
- To inform supervisors of safety issues, accidents or near misses affecting their laboratories. To complete an accident/incident report form and have it signed by the supervisor. To advise the first aid person about the items used from the first aid box and request replacement items. A copy of the accident/incident report form should be sent to the safety advisor and the university safety officer.
- To know the location of gas and electricity shut-off points.
- To ensure that all the occupants of the laboratory, including visitors, wear a white coat, safety glasses, and all necessary Personal Protective Equipment (PPE).

Laboratory Safety Regulations

General safety precautions

- Do not enter the laboratory without supervision.
- White coats and safety glasses must be worn in laboratories.
- Sandals or open shoes must not be worn in the laboratory.
- Read the practical manuals and pay attention to the warnings about dangerous chemicals and biohazards.
- Listen and follow the safety advice given by lecturers and demonstrators.
- Eating, drinking, smoking and applying cosmetics is prohibited in laboratories.
- Know the locations of the emergency exits, first aid boxes, fire extinguishers, fire blankets, safety showers and emergency contacts.
- Evacuate the building once the fire alarm has sounded.
- Entrances and exits to laboratories must be kept free. Bags, coats etc., should be stored in lockers.
- Tie long hair back.
- Use scissors to cut tape. Never put tape, pen etc., into your mouth.
- Protective gloves should be worn when necessary (anyone with allergies to latex etc., please inform your lecturer or demonstrator).
- Do not handle door knobs, laboratory equipment, telephone etc., with contaminated hands or gloves.
- Contact lenses should not be worn in laboratory areas.
- Report all accidents to the lecturer or demonstrator incharge.
- Avoid rubbing your hands against your eyes or face while in the laboratory.
- Pipette pumps or automatic pipettes must be used while pipetting.
- Do not leave pipettes sticking out of bottles, flasks, or beakers.
- Solvents and harmful chemicals should be disposed off.
- Broken glass or sharp objects should be placed in a sharps bin.
- Never leave experiments unattended.
- Working areas and equipment must be kept clean and tidy.
- Mobile phones should be switched off while in the laboratory.
- Always wash your hands when leaving the laboratory.
- Turn off lights, equipment, water and electricity when leaving the laboratory.

Safe handling of cultures

- The general safety precautions apply.
- Appropriate immunizations should be

obtained when handling mammalian tissue, blood, cells etc.

- Treat all cultures (cell and microbial) as potential pathogens.
- Familiarize yourself with the hazardous properties of the organism you are using.
- Perfect your aseptic technique.

Risk phrases: R45 (may cause cancer); R46 (may cause heritable genetic damage); R20/21/22 (Harmful by inhalation and in contact with the skin and if swallowed); R25 (toxic if swallowed); R36/38 (irritating to the eyes and skin); R43 (may cause sensitization by skin contact); R48/23/24 (toxic: danger of serious damage to health by prolonged exposure through inhalation, and in contact with skin; and R62 (possible risk of impaired fertility).

Storage: Fridge, locked.

Disposal: Solutions can be mixed with combustible solvent and incinerated.

Residual hazards: (i.e., hazards still remaining), operator should work in a safe manner. Nitrile gloves should be used instead of latex.

Existing controls (query their adequacy/suitability):

- A fumehood should be used when preparing solutions.
- Wear safety goggles, white coat, gloves at all times.

Latex gloves are only splash resistant and should be discarded immediately upon splash contamination. Nitrile gloves offer a better level of chemical protection. Hands should be washed after working with a chemical.

State the amount of time spent working with this agent under controlled conditions— Actual time working with chemical—10–15 mins.

State occupational exposure limit value (OELV)—OELV = 0.03 mg/m^3 for an 8 hour period.

Safety phrases: S53 (Avoid exposure— Obtain special instructions before use); S26 (in case of contact with eyes, rinse immediately with plenty of water and seek medical advice); S36/37 (wear suitable protective clothing and glasses, wear

suitable gloves); and S45 (In case of accident seek medical advice and show label where necessary).

Elimination of the Risk Altogether, or Substantial Reduction

Do not use acrylamide powder. The protocol can be altered to reduce exposure by purchasing it pre-made at a concentration of 30%. This has eliminated the necessity to weigh out and dissolve the chemical. At present, it is not known how to eliminate or reduce the chemical further except by non-use of the chemical.

10.6. CONTROLLING THE EXPOSURE TO HAZARDOUS SUBSTANCES

Wherever possible, it is necessary to eliminate all exposure to hazardous substances. You should ask:

- Can you replace the substance?
- Is there a substance that will carry out the same function that isn't hazardous to health or to the environment?
- Can you change your processes to avoid the need for the substance completely?

If there is no way to eliminate the use of hazardous substances, you must control exposure to them. Following are the three key steps in the order of priority:

Use Appropriate Work Processes

Review how you use the hazardous substances, looking particularly at how you can minimize the amount of material used and equipment that can totally enclose the process. Find out more about how to reduce your use of hazardous substances.

Control Exposure at Source

Reduce to a minimum the number of people— including contractors and customers—exposed, the duration of their exposure, and the number of hazardous substances used. Use suitable exhaust ventilation to keep substances in the air away from where they could be breathed in.

Provide Personal Protective Equipment

If you cannot control the exposure using the first two steps, you must make protective clothing available to all the employees, who are at risk. You can read more information in the guide on how to protect yourself. Remember this must be the last resort, and never a replacement for the previous two steps.

Cryogenic Liquids

Due to the inherent danger, only fully trained personnel should handle cryogenic materials, fluid-piping systems, and related equipment. A variety of physical hazards are associated with this class of material.

Physical Hazards

- Serious burns to the skin can result from direct contact with a cryogen or related equipment.
- Permanent damage to the eyes can result from contact with liquid cryogens.
- Liquid cryogens warmed above their critical temperature will generate high pressures that can cause a confining vessel to rupture or even explode.
- Cryogens have significant potential for creating oxygen deficiency, because they have a large liquid-to-gas expansion ratios; therefore, they should not be used in confined spaces or areas that are unventilated.

Storage

The main issue to consider in the safe storage of cryogenic gases is ventilation. Due to the potential risk of asphyxiation, liquid nitrogen and solid carbon dioxide should not be stored in cold or growth rooms, i.e., confined spaces.

Liquid nitrogen

It is stored only in a purpose-designed vessel. "Dewar" is the generic term used for such a vacuum insulated non-pressurized vessel. They vary in volume, and the types used here for decanting are 25 litre containers known as 'onions'. Culture storage vessels often have a volume of 50 litre. Liquid nitrogen is delivered weekly and decanted directly into the 'onions' and pressurized containers. Temporary storage can be in small Dewar vessels of 1 or 2 litre capacities. However, any lid must be vented to avoid the build-up of pressure, which could happen in a sealed vessel. Pressurized vessels are used where dispensing via a flexible hose is needed.

Solid carbon dioxide (Dry ice)

Pellets of solid carbon dioxide should be stored in insulated containers. It should not be stored in an airtight container, as the ice will sublimate and possibly cause an explosion. Dry ice should not be stored in cold rooms or other confined spaces, as it will replace the air in the room and cause suffocation. Solid carbon dioxide can usually be disposed off by giving it to another laboratory that requires it e.g., organic chemistry.

Dry ice is heavier than air, and precautions should be taken to ensure that there is no accumulation of carbon dioxide gas. Windows or a door should be left open.

Transport

Pressurized containers of liquid nitrogen should not be accompanied by anyone while being transported in the lift, due to the potential risk of sudden release of gas if a bursting disc failed. The 25 litre unpressurized containers can be moved either with the decanting trolley (if the 'onion' is fitted with side lugs, or trunions) or with a set of small wheels. Unpressurized containers of liquid nitrogen should not be carried up, or downstairs.

Safe Handling of Cryogens

- Do not decant from liquid nitrogen storage vessels if you are not confident in handling it. Seek help from trained personnel.

- Decant only into the insulated liquid nitrogen containers (Dewar or vacuum flasks).
- Do not seal the lids.
- Do not use other types of containers (such as polystyrene 'igloos' or plastic ice buckets).
- Liquid nitrogen is extremely cold, and there is a risk of freezing your fingers if you handle ultra-cold items. Liquid nitrogen boils at $-196°C$, and the liquid is even colder.
- Always handle dry ice with care and wear protective gloves, glasses and a white coat, as it is extremely cold. Prolonged contact with the skin will freeze the cells, as dry ice is extremely cold with a temperature of $-78°C$.

Personal Protective Equipment

Visor

A full-face visor is a minimum requirement while decanting liquid nitrogen, as vigorous boiling and consequent splashing can be anticipated.

Gloves

Non-absorbent, insulated gloves should be worn while handling liquid nitrogen. Avoid splashes, which could be trapped against the body and make burns more severe. Avoid gauntlet style gloves.

Cryogloves should be worn if one is handing ultra-cold items.

Lab coat

Lab coat is advisable to minimize skin contact. Wear trousers over safety shoe/boot tops to prevent shoes filling in the event of a spillage. The sleeves of the coat should have a tight cuff so that they can seal over gloves and wrist.

10.7. BIOLOGICAL AGENTS (BIOHAZARDS)

Biological Agents

A microorganism, including those that have been genetically modified, a cell culture and a human endoparasite, which may be able to provoke an infection, allergy or toxicity.

Classification

There are four groups of biological agents. They are classified, in ascending order, according to the increasing hazard they present to the laboratory workers and should display biohazard sign (Fig. 11.3).

Group 1 Biological agent

These are unlikely to cause human diseases. These rely on standard microbiological practices, with no special primary or secondary barriers, other than a sink for hand washing.

Group 2 Biological agent

These can cause human diseases and might be a hazard to employees.

These rely on standard microbiological practices, and access is restricted to nominated workers only. Design of laboratories and containment level is according to safety, health and welfare at work (biological agents) regulations.

Group 3 Biological agent

These can cause severe human diseases and present a serious hazard to the employees. These agents can spread to the community.

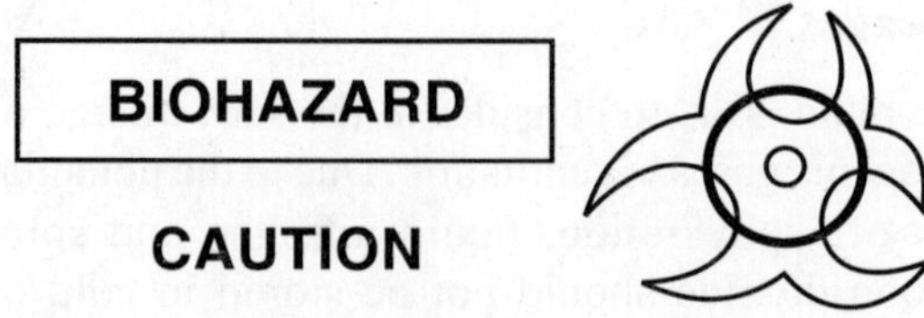

Fig. 10.3: Biohazard Symbol.

These rely on standard microbiological practices, and access is restricted to nominated workers only. Design of laboratories and containment level is according to safety, health and welfare at work (biological agents) regulations.

Group 4 Biological agent

These cause severe human diseases and are a serious hazard to employees. These may present a high risk of spreading to the community; there is usually no treatment available.

These rely on standard microbiological practices, and access is restricted to nominated workers only. Design of laboratories and containment level is according to safety, health and welfare at work (biological agents) regulations.

It is necessary that the Health and Safety Authority (HSA) should be notified thirty days prior to the commencement of work involving the use, for the first time, of group 2, 3 or 4 biological agents. A licence should be obtained from the Environmental Protection Agency (EPA) prior to commencing work with GMM's (genetically modified microorganisms) and GMO's (genetically modified organisms).

Procedure for the Reduction and Disposal of Biohazardous Waste

The organizations can be inspected at any time by the relevant authorities (Health and Safety Authority, Dublin Corporation, and Environmental Protection Agency) for compliance with hazardous waste regulations. Under the waste management act, the generator of the waste is responsible for ensuring that the waste is disposed off properly. Biohazardous waste, for disposal, will be assessed for risk before being added to the system.

Minimization of Waste

Biohazardous waste is minimized by ensuring that only contaminated waste is put in a biohazard bag, sharps bin etc. The waste, after autoclaving, is shifted to a store in the basement, where it awaits removal. The waste management company removes the waste from the biotechnology institution. A completed C1 form, which can be obtained from the local authority, must accompany the waste. A copy of the form is kept in the biotechnology institution (pink copy).

On completion of research projects or while leaving the organization, undergraduate students, postgraduate students, postdocs, visiting researchers, and staff must complete the relevant clearance form (safe work practice sheet). This will help to ensure that waste etc., is disposed off correctly before individuals leave the university. These forms can be obtained from the secretary of the biotechnology institution.

Categories of Biohazardous Waste

- Solid waste e.g., contaminated gloves, paper towel, cotton wool, disposable loops, tissue culture bottles, petri-dishes, flasks, and disposable pipettes etc.
- Liquid waste e.g., tissue culture medium, microbiological media, cultures, broths, media from fermentors etc.
- Sharps e.g., broken glass, needles, pipettes, scalpel blades, small glass vials, tips and ampoules.
- Carcasses etc., should be frozen to await disposal.
- Mixed wastes e.g., biohazard waste containing solvents.

Labeling

All biohazard waste containers must have a biohazard sign. Individual containers, bags etc., should be labeled with lab number and operator researcher.

Storage and Storage Containers

The generator of solid biohazardous waste should store the waste in a suitable container marked with a biohazard symbol. Biohazardous waste should be segregated from the general waste.

Do not put sharp objects into an autoclave bag; use a sharps container.

Autoclave bag

If an autoclave bag is used, it should be double bagged. Liquid waste should not be put in these bags, as leakage occurs. The autoclave bag should be filled only two-third, and the neck should be tied loosely with autoclave tape. A 5 cm gap should be left at the top of the bag to ensure proper autoclaving. A laboratory name tag should be attached to the bag for identification. Sharp items should not be placed in an autoclave bag. Large yellow bins marked with a biohazard sign can be obtained for storing semi-filled autoclave bags in your laboratory.

Sharps container

Sharps container should be filled only to the mark, and then it should be sealed. It can be placed in the biohazardous waste wheelie bins for disposal. Sharps containers do not need to be autoclaved. The technician incharge of the biohazardous waste should be informed when the waste is left for disposal.

Storage of Liquid Waste

Liquid waste should be put into pyrex bottles, and filled only upto two-thirds. The bottles should be marked with autoclave tape. The lids should be loosened at least a ½ turn before autoclaving.

Storage of Biohazardous Waste

In the laboratory, solid biohazardous waste, petri-dishes, gloves etc., should be stored in two autoclave bags suitably marked with a biohazard symbol, and then locked. The bags should be stored in a yellow solid bin with a biohazard symbol, which will ensure that the bag will not spill out its contents. Biohazardous waste should be segregated from the general waste, and should be removed regularly from laboratories. Large quantities should not be allowed to build up. Ensure that the cleaning staff does not remove this waste.

Storage of Carcass Waste

Carcass waste should be stored in a freezer to await disposal. It should be labeled with a date, laboratory name, and amount of chemicals added.

Autoclaving

All biohazardous waste should be sterilized in an autoclave, which is capable of reaching 121°C at 15 psi for 30 min.; higher temperatures can be used if necessary. Spore strips should ideally be used to ensure complete sterilization. Chemical indicators are also used within a batch to be autoclaved. Sterilized solid waste should be stored in the biohazard wheelie bins in XG29. The bins will be shifted to the external biohazardous store, and will be taken from the university by a licensed waste contractor.

Agar (waste) should be autoclaved in your lab, if possible, and disposed off as mentioned above. The Anstell Scientific autoclave (in XG29c) should only be used for large amounts of waste. Please follow the rota for your laboratory. XG29c cannot be used as a storage area for your waste. Once the waste has been brought from your laboratory, it should be auto-claved immediately. While loading the autoclave, it should be remembered that adequate space should be left between the items for sterilization. Do not overfill the autoclave.

Using the guidelines for use of the Anstell Scientific Autoclave (posted in XG29c), run the preset biohazard autoclave cycle. After the sterilization process is complete, the autoclave can only be opened when the temperature has fallen to below 70°C. The autoclave tape should have black stripes. The autoclaved items can then be removed.

Sterilized waste bags should be placed in the plastic bag within the wheelie bin for collection. Sterilized liquid media from Class 1/Group 1 biological agents, with the exception of GMO's, can be washed down the drain with plenty of water, according to the discharge license from corporation.

Removal of Waste

A licensed (biohazard) waste disposal contractor will remove the biohazardous waste, sharps and frozen carcasses etc. A completed local authority C1 form should accompany the waste (a completed pink copy should be kept in the organization and filed). It is necessary to inform the technician incharge of the biohazardous waste if the waste has been left for disposal.

Identification of Sharps

Sharps are materials such as hypodermic needles, lancets, scalpel blades, broken glass, tips, pipettes, or any material that is likely to cause inadvertent injury to the staff, students or the general public. Sharp objects should be carried, handled and used in a safe manner. Sharps must be segregated from other waste, and placed in a clearly identifiable containers, which are yellow and contain a label clearly stating that the boxes are for sharps and contain the UN number. There are two types of sharps boxes. One is for cytotoxic sharps (purple lid), i.e., sharps contaminated with cytotoxic drugs and chemicals. These include carcinogens, cytostatic, cytotoxic, mutagens, and teratogens. The second box is for the sharps contaminated with cell culture or microbiological reagents, and has a blue lid.

Disposal

Sharp and pointed items should be placed in sharps box for safe disposal. The sharps boxes should be sealed when they are ¾ full, and discarded as solid hazardous waste. At present, a licensed disposal company is collecting the sharps boxes. The sharps box should be left with the technician incharge of the biohazards waste disposal. The fully sealed box should be placed in the biological hazardous store for collection. Do not attempt to break the glass further to ensure a better packed box. This will result in an injury or a punctured box, which defeats the purpose of the sharps bin.

Biohazard Spill

If there has been a spill, immediately notify other individuals in the area. If there is a hazard associated with the release of the aerosol from a spill, everyone should be advised to leave the area (applicable to class 2, class 3, and class 4 organisms). If necessary, cordon off the area.

Individuals involved in the spill should check for contaminated clothing, footwear and skin before attempting to clean up. Spare coveralls can be obtained from the preparation room. Contaminated skin should be washed immediately with soap and water.

Basic Clean Up Procedure

While cleaning the spill, a white coat, safety glasses, and gloves should be worn. Identify the area of the spill allowing sufficient area around the spill for splashing or drying, which may have occurred. Decontamination of the area with an appropriate disinfectant e.g., 10% sodium hypochlorite should be done the appropriate disinfectant solution can be around the outside of the spill and not on the spill. This will avoid the creation of aerosols. Use absorbent materials e.g., paper towel to work the decontaminant into the area of the spill. Always cover the entire spill area with absorbent paper soaked in disinfectant and allow it to remain in contact with the spill for minimum of 30 min. Place the used absorbent material into a biohazard bag for sterilization. Place any broken glass in a sharps bin. Ensure your hands are washed thoroughly with soap and water. Immediately complete an accident/incident report form.

10.8. IMMUNIZATION AND FIRST AID

Immunization of the laboratory staff may be required in situations, where it is known that the staff will come in contact with blood or human tissue on a regular basis. The vaccine given is either Hep A or Hep B, or a combined vaccine.

Tetanus vaccine can also be obtained. (Please read biohazards).

Obtaining a Vaccine

The biotechnology institution doctor will administer the vaccine or arrange serological monitoring as appropriate. The biotechnology institution advisor will contact HR if the laboratory worker is a member of staff, postgraduate student etc., and an appointment will be arranged.

Emergency Response or Laboratory First Aid

First aid trained staff should be available in the Organization of Biotechnology, a list of qualified first aid staff and current telephone numbers are on display in the corridors and beside phones.

First aid boxes are available in each laboratory, as are sterile bottles of saline for eye irrigation and safety showers. If first aid supplies are required, an accident/incident report form must be completed and given to the safety advisor.

First aid courses are available on a yearly basis, and refresher courses are ongoing through the Health and Safety Office. The biotechnology institution abides by the biotechnology institution policy on first aid safety.

Fire Fighting Equipment

Fire extinguishers and fire blankets are available at the safety station in the laboratories. There are a number of trained fire wardens in the biotechnology institution. Fire warden courses are run on a regular basis and are available through the Health and Safety Office. The biotechnology institution abides by the biotechnology institution policy on fire safety.

General Safety

Safety showers

Single head emergency showers should be located close to all areas where chemicals are being used or where clothing fires could occur. To ensure adequate operation of the units, all individuals working in laboratories should be instructed in the proper use of the shower. If an individual becomes contaminated with chemicals etc., the affected area should be rinsed immediately for at least 15 min., and a doctor or nurse should be contacted.

Emergency shower units are not a substitute for the use of proper primary personal protective equipment. To protect against flying solid particles and splashing hazardous liquids, staff and students should wear eye and/or face protectors, protective clothing, and appropriate gloves. Fume hoods should be used as primary protection, as they contain fumes, splashing, and a spillage.

Testing

Safety showers should be tested once a week using the safety shower test kit. This can be obtained from the safety advisor. The shower should be run for at least 10 minutes to ensure that there are no blockages in the mains, and to check that the water is clean. A minimum of 30 liter of water per minute should be the flow rate. A testing tag is attached to each unit and is used to record the date of the test and the initials of the tester. The safety representative for each laboratory should ensure that the safety showers are tested each week. If there is a fault in the eyewashers, it should be reported to the safety advisor of the organization.

Chemical burns

Remove contaminated clothing, which shows no sign of sticking to the skin, and flush all affected parts of the body with plenty of clean, cool water for 15 minutes, ensuring that the chemical is diluted as to be rendered harmless. Apply a sterilized dressing to the damaged skin and clean towels to the damaged areas where the clothing cannot be removed. Take care while treating the casualty to avoid contamination. A doctor or nurse should be contacted immediately.

Eye/Eye-Face Wash Fountains

Warning

Eye/eye-face wash fountains should not be used if it is known that the eye contamination is due to the metal or some other rigid solid fragment. In such an event, both the victim's eyes should be gently immobilized in accordance with the current standard first aid manual, and medical attention immediately sought.

Emergency eye/eye-face wash units are not a substitute for proper primary personal protective equipment. To protect against flying solid particles and splashing hazardous liquids, all individuals working in lab areas should wear eye and/or face protectors.

Emergency eye/eye-face wash units should be in all areas where chemicals are being used. To ensure adequate operation of the units, all individuals working in lab areas should be instructed in the proper use of the eye/eye-face wash fountain. If an individual becomes contaminated with chemicals etc., the affected area should be rinsed immediately for at least 15 minutes, and a doctor or nurse should be contacted immediately.

Testing

Eye/eye-face wash fountains should be tested once a week, and can be checked inconjunction with the shower unit. The kit for testing can be obtained from the safety advisor. The eye-wash should be run for at least 10 minutes to ensure that there are no blockages in the mains and to check that the water is clean. A jet of water from each nozzle should meet in the middle. The safety representative for each laboratory should ensure that the eyewashes are tested each week. If there is a fault in the eyewashes, it should be reported to the safety advisor.

SUMMARY

The hazardous material means any materials that causing danger or posing the risk to the living organisms including humans. In biotechnology large number of such hazardous materials are used in research and development, industrial processes, product purification, packing and usage.

The hazardous materials are from physical, chemical or biological origin. The type of hazard or danger is that it may be toxic, corrosive, irritant etc. The toxic materials may cause poisonous effect, may cause cancer, may cause heritable genetic damage. Also harmful by inhalation and in contact with the skin. Toxic if swallowed. Irritating to the eyes and skin. May cause sensitization by skin contact. There is a danger of serious damage to health by prolonged exposure through inhalation, in contact with skin and if swallowed. Possible risks of impaired fertility can also results.

There are six categories of chemical waste in the biotechnology laboratories and these are commonly grouped as:

- Unchlorinated organic waste.
- Chlorinated organic waste.
- Unused laboratory chemicals.
- Hazardous waste.
  — Known components.
- Hazardous waste.
  — Unknown components.
- Dilute aqueous non-toxic waste.

Along with these there are hazardous materials derived from the physical and biological sources. Hence proper handling and storage is very essential in the biotechnology as there is a great danger of causing ill effects on humans and environment.

The environmental committees and local authorities have laid down strict guidelines for the safe disposal of waste. These regulations are being constantly updated and laboratory personnel should pay particular attention during the disposal of waste. Chemicals should only be kept for five years from the date of purchase. Only buy chemicals in small quantities, as disposal of unused or unknown waste chemicals is extremely costly. Full containers of correctly treated and labeled waste should be given to the technician in charge of waste in the preparation room. The waste will then be stored in the outside waste solvent store, to await collection and disposal. Equal importance should be given to the proper and effective disposal of the physical and biological hazardous materials for the work environments of the biotechnology.

EXERCISE

1. Define hazardous materials. What is their importance in the biotechnological laboratories?

2. Write notes on different types of chemical waste generated in the biotechnology laboratories and their negative effects.

3. What is meant by physical, chemical or biological hazardous materials? Add a note on the use of such hazardous materials by the researchers in biotechnology.

4. How biohazards are generated in the recombinant DNA work? Explain their safe management which is essential to be practiced in the biotechnology laboratory.

5. Elucidate the precautions for storing dangerous and reactive substances that are regularly used in the various types of biotechnological work.

6. Distinguish between the toxic materials and pathogenic organisms used in the biotechnological research and development for the product manufacture.

7. Briefly write about how hazardous materials are handled, prepared and disposed in the biotechnology laboratory.

8. Enumerate the responsibilities of the organization safety advisor in the handling and safe disposal of hazardous materials.

9. Briefly write about the disposal of chemical wastes that are generated during the biotechnological laboratory work.

10. Explain general laboratory safety regulations that are followed for the safety of the persons in the work environment.

11. Give an account of the practical list of incompatible chemicals (reactive hazards) and their impact on the researchers working in the laboratory.

12. Describe various types of responsibilities that are vested with the biotechnological research and development units followed for safe and effective disposal of hazardous materials.

13. Name the different duties performed by the employers in managing the hazardous materials that are used in the biotechnology laboratories.

14. What do you mean by Material Safety Data Sheets (MSDS)? Why now-a-days the material safety data sheets is very essential in the laboratory work environment?

15. Mention how labeling helps in better handling, usage and disposal of the hazardous materials in biotechnological laboratory and manufacturing units.

16. List out how animal carcasses are considered to be potential biohazardous materials, and explain their management in biotechnological laboratory.

17. Critically comment on the personal protective equipment that are essential in the biotechnological laboratory and manufacturing units for the safety purposes.

18. Write about the importance of autoclaving in the management of biohazards, and their affect on the potential pathogenic organisms.

19. Give an account of the classification of biohazards used in the biotechnological laboratory and manufacturing companies.

20. Explain how technological advancement is becoming a negative factor for the management of safety in the biotechnological laboratory and manufacturing units.

11

CHAPTER

Good Laboratory Practice (GLP) and Good Manufacturing Practice (GMP)

11.1. INTRODUCTION

Worldwide researchers are using biotechnology to modify microorganisms, plants and animals to produce industrial products, pharmaceuticals, and other useful products. Examples of products currently under development include proteins and enzymes for diagnostic, therapeutic, and manufacturing purposes; modified fatty acids and oils for paints and manufacturing; biopolymers as substitutes for plastics; and specialty substances. One potential product that is currently being explored is the creation of fibers for clothing and other materials. Using the bacteria engineered to make polymers that closely resemble natural fibers (such as silk, elastin, collagen and keratin), it may be possible to produce unique fibers similar to some of the most popular textiles used in the production of the latest fashions at faster rate.

Plants and animals can also be modified to act as biosensors for detecting or monitoring hazardous materials in the environment. For example, researchers have modified bacteria to be sensitive to trinitrotoluene (TNT)—an ingredient in many explosives—making it potentially useful in landmine detection. Scientists are also working

to engineer zebra fish, which can act as biosensors for pollutants such as dioxin or PCBs present in almost all environment.

Scientists are working on ways to use biotechnology for environmental preservation and remediation. For example, they are working with a number of plant species to enhance their natural ability to absorb and store toxic and hazardous substances that could assist in the clean up of oil spills and chemical leaks.

Genetically modified microbes, plants, and animals can produce vaccines for human and animal illnesses, ranging from colon cancer to tooth decay. Technology developers believe that using food to deliver vaccines could have the benefit of permitting the vaccine to be consumed directly by humans or animals as food or feed, and eliminating the need for purification of the vaccine strain and the hazards associated with injections. Some of the plants used to develop the vaccines include corn, spinach, tobacco, lettuce, tomato, soybeans, and potatoes. Plants may also be used to produce compounds essential for humans such as monoclonal antibodies, hormones, or blood proteins.

For varieties of such studies and developments, researchers have to adopt the laboratory methods that are reproducible and of high standard. To achieve these goals, different laboratories should not follow different standards. A number of organizations for economic cooperation and development (OECD), member countries have recently passed legislation to control chemical substances and others are about to do so. This legislation usually requires the manufacturer to perform laboratory studies and to submit the results of these studies to a governmental authority for assessment of the potential hazard to human health and the environment. Government and industry are increasingly concerned with the quality of studies upon which hazard assessments are based. As a consequence, several OECD member countries have, or plan to establish, criteria for the performance of these studies in a uniform manner.

To avoid different schemes of implementation that could impede international trade in chemicals, OECD member countries have recognized the unique opportunity for international harmonization of test methods and good laboratory practices. During 1979–1980, an international group of experts established under the special programme on the control of chemicals developed this document concerning the "Principles of Good Laboratory Practice (GLP)", utilizing common managerial and scientific practices, and experience from various national and international sources. This makes the test practice a standard throughout the world.

The main purpose of the principles of good laboratory practice is to promote the development of quality test data. Comparable quality of test data forms the basis for the mutual acceptance of test data among countries.

If individual countries can confidently rely on test data developed in other countries, duplicative testing can be avoided, thereby introducing economics in test costs and time. The application of these principles should help avoid the creation of technical barriers to trade, and further improve the protection of human health and the environment in a sustainable manner.

11.2. DEFINITION OF GOOD LABORATORY PRACTICE (GLP)

Good laboratory practice (GLP) has become a major criteria in all the biotechnology research and development laboratories around the world. The principles of good laboratory practice (GLP) define a set of rules and criteria for a quality system concerned with the organizational process, and the conditions under which non-clinical health and environmental safety studies are planned, performed, monitored, recorded, reported, and archived. Hence, uniformity can be achieved in all the testing procedures developed in the biotechnology **(Fig. 11.1)**.

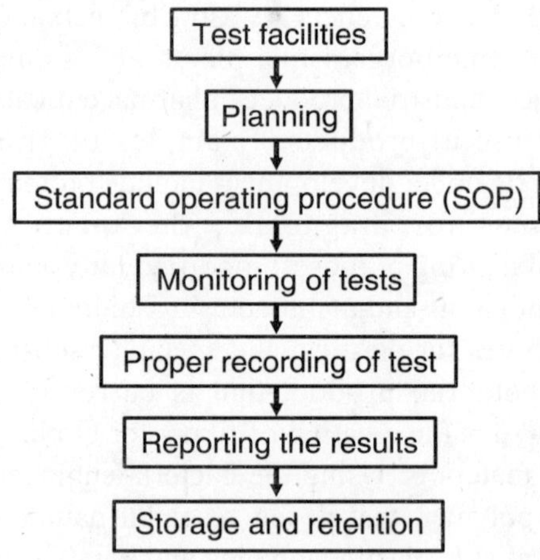

Fig. 11.1. Good laboratory practice for the organization

Scope

In biotechnology, different types of materials and organisms are used for a variety of purpose. The principles of good laboratory practice should be applied to the testing of chemicals and living organisms for obtaining data on their properties and/or their safety with respect to human health or the environment.

Biotechnology studies covered by good laboratory practice also include work conducted in field. These data would be developed for the purpose of meeting regulatory requirements. This makes the laboratory practice more environmental friendly.

Definition of Terms

Good laboratory practice

Good laboratory practice (GLP) is concerned with the organizational process and the conditions under which laboratory studies are planned, performed, monitored, recorded, and reported.

Test facility means the persons, premises, and operational unit(s) that are necessary for conducting the study.

Study director means the individual responsible for the overall conduct of the study.

Quality assurance programme means an internal control system designed to ascertain that the study is in compliance with the principles of good laboratory practice.

Standard operating procedures (SOPs) means written procedures, which describe how to perform certain routine laboratory tests or activities normally not specified in detail in study plans or test guidelines.

Sponsor means a person(s) or entity who commissions and/or supports a study.

Terms concerning the study

Study means an experiment or set of experiments in which a test substance is examined to obtain data on its properties and/or its safety with respect to human health and the environment.

Study plan means a document that defines the entire scope of study.

OECD test guideline means a test guideline, which the OECD has recommended for use in its member countries.

Test system means any animal, plant, microbial, as well as other cellular, subcellular, chemical, or physical system or a combination used in a study.

Raw data means all original laboratory records and documentation, or verified copies, which are the result of the original observations and activities in a study.

Specimen means any material derived from a test system for examination, analysis, or storage.

Terms concerning the test substance

Test substance means a chemical substance or a mixture that is under investigation.

Reference substance (control substance) means any well defined chemical substance or mixture other than the test substance used to provide a basis for comparison with the test substance.

Batch means a specific quantity or a lot of test or reference substance produced during a defined cycle of manufacture in such a way that it could be expected to be of a uniform character and should be designated as such.

Vehicle (carrier) means any agent which serves as a carrier to mix, disperse, or solubilize the test or reference substance to facilitate its administration to the test system.

Sample means any quantity of the test or reference substance.

11.3. TEST FACILITY ORGANIZATION AND PERSONNEL

Management's Responsibilities

Test facility management should ensure that the principles of good laboratory practice are complied within the test facility. At minimum it should:

* ensure that qualified personnel, appropriate facilities, equipment, and materials are available;

- maintain a record of the qualifications, training, experience, and job description of each professional and technical individual;
- ensure that personnel clearly understand the functions they have to perform, and, where necessary, provide training for these functions;
- ensure that health and safety precautions are applied according to national and/or international regulations;
- ensure that appropriate standard operating procedures are established and followed;
- ensure that there is a quality assurance programme with designated personnel;
- where appropriate, agree to the study plan in conjunction with the sponsor;
- ensure that amendments to the study plan are agreed upon and documented;
- maintain copies of all study plans;
- maintain a historical file of all standard operating procedures;
- ensure that a sufficient number of personnel is available for the studies timely and proper conduct;
- designate an individual with appropriate qualifications, training, and experience as the study director before the study is initiated. If it is necessary to replace a study director during a study, this should be documented; and
- ensure that an individual is identified as responsible for the management of the archives.

Study Director's Responsibilities

- The study director has the responsibility for the overall conduct of the study and its report.
- These responsibilities should include, but not be limited to, the following functions:
  — Should agree to the study plan.
  — Ensure that the procedures specified in the study plan are followed, and authorization for any modification is obtained and documented together with the reasons for them.
  — Ensure that data generated is fully documented and recorded.
  — Sign and date the final report to indicate acceptance of responsibility for the validity of the data, and to confirm compliance with the principles of good laboratory practice.
  — Ensure that after the termination of the study, the study plan, the final report, raw data, and supporting material are transferred to the archives.

Personnel Responsibilities

- Personnel should exercise safe working practice. Chemicals should be handled with suitable caution until their hazards have been established.
- Personnel should exercise health precautions to minimize risk to themselves, and to ensure the integrity of the study.
- Personnel known to have a health problem, which is likely to have an adverse effect on the study, should be excluded from operations that may affect the study.

11.4. QUALITY ASSURANCE PROGRAMME

General

- The test facility should have a documented quality assurance programme to ensure that the studies performed are in compliance with the principles of good laboratory practice.
- The quality assurance programme should be carried out by an individuals designated by and directly responsible to the management, and who are familiar with the test procedures.
- This individual(s) should not be involved in the conduct of study being assured.
- This individual(s) should report any findings, in writing, directly to the management and the study director.

Responsibilities of the Quality Assurance Personnel

- The responsibilities of the quality assurance personnel should include, but not be limited to, the following functions:
  — Ascertain that the study plan and standard operating procedures are available to the personnel conducting the study.
  — Ensure that the study plan and standard operating procedures are followed by the periodic inspections of the test facility and/or auditing the study in progress. Records of such procedures should be retained.
  — Promptly report to the management and the study director the unauthorized deviations from the study plan and standard operating procedures.
  — Review the final reports to confirm that the methods, procedures, and observations are accurately described, and that the reported results accurately reflect the raw data of the study.
  — Prepare and sign a statement, to be included in the final report, which specifies the dates when inspections were held and any findings reported to the management and the study director.

11.5. FACILITIES

General

- The test facility should be of suitable size, construction and location to meet the requirements of the study and minimize disturbances that would interfere with the validity of the study.
- The design of the test facility should provide an adequate degree of separation of different activities to assure the proper conduct of each study.

Test System Facilities

- The test facility should have a sufficient number of rooms or areas to assure the isolation of test systems and individual projects involving substances known or suspected to be biohazardous.
- Suitable facilities should be available for the diagnosis, treatment, and control of diseases, in order to ensure that there is no unacceptable degree of deterioration of test systems.
- There should be storage areas for supplies and equipment. Storage areas should be separated from the areas housing the test systems, and should be adequately protected against infestation and contamination. Refrigeration should be provided for perishable commodities.

Facilities for Handling Test and Reference Substances

- To prevent contamination or mix-ups, there should be separate areas for receipt and storage of the test and reference substances, and mixing of the test substances with a vehicle.
- The storage areas for the test substances should be separated from the areas housing the test systems, and should be adequate to preserve the identity, concentration, purity, and stability, and ensure safe storage for hazardous substances.

Archive Facilities

- Space should be provided for archives for the storage and retrieval of raw data, reports, samples, and specimens.

Waste Disposal

- Handling and disposal of wastes should be carried out in such a way as not to jeopardise the integrity of studies in progress.

- The handling and disposal of wastes generated during the performance of a study should be carried out in a manner, which is consistent with pertinent regulatory requirements. This would include provision for appropriate collection, storage, and disposal facilities, de-contamination and transportation procedures, and the maintenance of records related to the preceding activities.

11.6. APPARATUS, MATERIAL AND REAGENTS

Apparatus

- Apparatus used for the generation of data and controlling environmental factors relevant to the study should have appropriate design and adequate capacity. It should be suitably located.
- Apparatus used in the study should be periodically inspected, cleaned, maintained, and calibrated according to the standard operating procedures. Records of procedures should be maintained.

Material

- Apparatus and materials used in the studies should not interfere with the test systems.

Reagents

- Reagents should be labeled, as appropriate, to indicate source, identity, concentration, and stability information. It should also include the preparation date, earliest expiration date, and specific storage instructions.

11.7. TEST SYSTEMS

Physical/Chemical

- Apparatus used for the generation of physical/chemical data should be suitably located and of appropriate design and adequate capacity.

- Reference substances should be used to assist in ensuring the integrity of the physical/chemical test systems.

Biological

- Proper conditions should be established and maintained for the housing, handling and care of animals, plants, microbial, as well as other cellular and sub-cellular systems, in order to ensure the quality of the data.
- In addition, conditions should comply with appropriate national regulatory requirements for the import, collection, care and use of animals, plants, microbial, as well as other cellular and sub-cellular systems.
- Newly received animal and plant test systems should be isolated until their health status has been evaluated. If any unusual mortality or morbidity occurs, the lot should not be used in studies and humanely destroyed.
- Records of source, date of arrival, and arrival condition should be maintained.
- Animal, plant, microbial, and cellular test systems should be acclimatised to the test environment for an adequate period before a study is initiated.
- All information needed to properly identify the test systems should appear on their housing or containers.
- The diagnosis and treatment of any disease before or during a study should be recorded.

11.8. TEST AND REFERENCE SUBSTANCES

Receipt, Handling, Sampling, and Storage

- Records including substance characterization, date of receipt, quantities received, and used in studies should be maintained.
- Handling, sampling and storage procedures should be identified so that the homogeneity and stability is assured to the degree possible, and contamination or mix-up are precluded.

- Storage container(s) should carry identification information, earliest expiration date, and specific storage instructions.

Characterization

- Each test and reference substance should be appropriately identified (e.g., code, chemical abstract number (CAS), name).
- For each study, the identity, including batch number, purity, composition, concentrations, or other characteristics to appropriately define each batch of the test or reference substances should be known.
- The stability of test and reference substances under conditions of storage should be known for all studies.
- The stability of test and reference substances under the test conditions should be known for all studies.
- If the test substance is administered in a vehicle, standard operating procedures should be established for testing the homogeneity and stability of the test substance in that vehicle.
- A sample for analytical purpose from each batch of test substance should be retained for studies in which the test substance is tested for more than four weeks.

11.9. STANDARD OPERATING PROCEDURES

General

- A test facility should have written standard operating procedures, approved by management, which are intended to ensure the quality and integrity of the data generated in the course of the study.
- Each laboratory unit should have standard operating procedures relevant to the activities being performed therein. Published text books, articles, and manuals may be used as supplements to the standard operating procedures.

Application

Standard operating procedures should be available for, but not limited to, the following categories of laboratory activities. The details given under each heading should be considered as illustrative examples.

- Test and reference substance receipt, identification, labeling, handling, sampling, and storage.
- Apparatus and reagents, maintenance, cleaning, calibration of measuring apparatus and environmental control equipment; preparation of reagents.
  Record keeping, reporting, storage and retrieval coding of studies, data collection, preparation of reports, indexing systems, handling of data, including the use of computerised data systems.
- Test system (where appropriate):
  — Room preparation and environmental room conditions for the test system.
  — Procedures for receipt, transfer, proper placement, characterization, identification and care of test system.
  — Test system preparation, observations, examinations, before, during and at termination of the study.
  — Handling of test system, individuals found moribund or dead during the study.
  — Collection, identification and handling of specimens, including necropsy and histopathology.
- Quality assurance procedures:
  — Include operation of quality assurance personnel in performing and reporting study audits, inspections and final study report reviews.
- Health and safety precautions (as required by national and/or international legislation or guidelines).

11.10. PERFORMANCE OF THE STUDY

Study Plan

- For each study, a plan should exist in a written form prior to initiation of the study.
- The study plan should be retained as raw data.
- All changes, modifications, or revisions of the study plan, as agreed by the study director, including justification(s), should be documented, signed and dated by the study director and maintained with the study plan.

Content of the Study Plan

The study plan should contain, but not be limited to, the following information:

- Identification of the study, test, and reference substances:
  - A descriptive title.
  - A statement which reveals the nature and purpose of study.
  - Identification of the test substance by code or name (IUPAC; CAS number, etc.).
  - The reference substance used.
- Information concerning the sponsor and the test facility:
  - Name and address of the sponsor.
  - Name and address of the test facility.
  - Name and address of the study director.
- Dates:
  - The date of agreement to the study plan by signature of the study director, and when appropriate, of the sponsor and/or the test facility management.
  - The proposed starting and completion dates.
- Test methods:
  - Reference to OECD test guideline or other test guideline to be used.
- Issues (where applicable):
  - The justification for selection of the test system.

- Characterization of the test system, such as the species, strain, substrain, source of supply, number, body weight range, sex, age and other pertinent information.
  - The method of administration and the reason for its choice.
  - The dose levels and/or concentration(s), frequency, duration of administration.
  - Detailed information on the experimental design, including a description of the chronological procedure of the study, all methods, materials and conditions, type and frequency of analysis, measurements, observations and examinations to be performed.
- Records:
  - A list of records to be retained.

Conduct of the Study

- A unique identification should be given to each study. All items concerning this study should carry this identification.
- The study should be conducted in accordance with the study plan.
- All data generated during the conduct of the study should be recorded directly, promptly, accurately, and legibly by the individual entering the data. These entries should be signed or initialled, and dated.
- Change in the raw data if any, should be made so as not to obscure the previous entry, and should indicate the reason, if necessary, for change made. It should be accompanied by date, and signed by the individual making the change.
- Data generated as a direct computer input should be identified at the time of data input by the individual(s) responsible for direct data entries. Corrections should be entered separately, with the date and identity of the individual making the change. It should also mention the reason for change.

11.11. REPORTING OF STUDY RESULTS

General

- A final report should be prepared for the study.
- The use of the international system of units (SI) is recommended.
- The final report should be signed and dated by the study director.
- If reports of principal scientists from co-operating disciplines are included in the final report, it should include their sign and date.
- Corrections and additions to a final report should be in the form of an amendment. The amendment should clearly specify the reason for the corrections or additions, and should be signed and dated by the study director and the principal scientist from each discipline.

Content of the Final Report

The final report should include, but not be limited to, the following information:

- Identification of the study, test, and reference substance:
  - A descriptive title.
  - Identification of the test substance by code or name (IUPAC; CAS number, etc.).
  - Identification of the reference substance by chemical name.
  - Characterization of the test substance including purity, stability, and homogeneity.
- Information concerning the test facility:
  - Name and address.
  - Name of the study director.
  - Name of other principal personnel who contributed reports to the final report.
- Dates:
  - Dates on which the study was initiated and completed.
- Statement:
  - A quality assurance statement certifying the dates of inspections and any findings reported to the management and the study director.

- Description of materials and test methods:
  - Description of methods and materials used.
  - Reference to OECD or other test guidelines.
- Results:
  - A summary of results.
  - All information and data required in the study plan.
  - A presentation of the results, including calculations and statistical methods.
  - An evaluation and discussion of the results, and where appropriate, conclusions.
- Storage:
  - The location where all samples, specimens, raw data, and the final report are to be stored.

11.12. STORAGE AND RETENTION OF RECORDS AND MATERIAL

Storage and Retrieval

- Archives should be designed and equipped for the secure storage of:
  - the study plans.
  - the raw data.
  - the final reports.
  - the reports of laboratory inspections and study audits performed according to the quality assurance programme.
  - samples and specimens.
- Material retained in the archives should be indexed, so as to facilitate orderly storage and rapid retrieval.
- Only the personnel authorized by the management should have access to the archives.

Movement of material in and out of the archives should be properly recorded.

Retention

- The following should be retained for the period specified by the appropriate authorities:

— The study plan, raw data, samples, specimens and the final report of each study.

— Records of all inspections and audits performed by the quality assurance programme.

— Summary of qualifications, training, experience and job descriptions of personnel.

— Records and reports of the maintenance and calibration of equipment.

— The historical file of standard operating procedures.

- Samples and specimens should be retained only as long as the quality of the preparation permits evaluation.

- If a test facility or an archive contracting facility goes out of business and has no legal successor, the archive should be transferred to the archives of the sponsor(s) of the study(s).

- Results:

— A summary of results.

— All information and data required in the study plan.

— A presentation of the results, including calculations and statistical methods.

— An evaluation and discussion of the results, and where appropriate, conclusions.

- Storage:

— The location where all samples, specimens, raw data, and the final report are to be stored.

The complete adaptations and following these guidelines and practices is helpful for obtaining complete control over all the activities done in the laboratory. The repetitive and accurate results can be obtained that are comparable to the international standards. Thus, these activities provide maximum reliabilities and dependability of the results without any demerits.

11.13. GOOD MANUFACTURING PRACTICES (GMP)

Introduction

The increasing use of modern biotechnology has generated significant debate, much of which centers on the rapidly growing use of food crops that have been genetically modified to make them more resistant to pests or chemical herbicides. The debate has not addressed the potential products of agricultural biotechnology that are on the horizon. While technology developers believe that these new products will offer benefits in meeting needs for food, fuel and fiber, as well as for novel industrial and pharmaceutical uses, some of these future products are also likely to raise environmental and other concerns that will need to be addressed by the regulatory system and safety issues.

The current good manufacturing practice (GMP) requirements set forth in the quality system (QS) regulation are promulgated under section 520 of the food, drug, and cosmetic (FD & C) Act of USA. They require that domestic or foreign manufacturers have a quality system for the design, manufacture, packaging, labeling, storage, installation, and servicing of finished medical devices intended for commercial distribution in the United States. The regulation requires that various specifications and controls be established for devices; devices be designed under a quality system to meet these specifications; devices be manufactured under a quality system; finished devices meet these specifications; devices be correctly installed, checked, and serviced; the quality data be analyzed to identify and correct quality problems; and complaints be processed. Hence, biotechnological products can achieve large-scale acceptance worldwide.

Good manufacturing practice (GMP) also helps the QS regulation to assure that medical devices are safe and effective for their intended use. The food and drug administration (FDA) monitors device problem data and inspects the operations and records of device developers and manufacturers to determine compliance with the GMP requirements in the QS regulation. Thus, tractability of any negative impact can be identified in the production phase.

Good manufacturing practice (GMP) in the quality system regulation is contained in Title 21 Part 820 of the code of federal regulations of USA. This regulation covers quality management and organization, device design, buildings, equipment, purchase and handling of components, production and process controls, packaging and labeling control, device evaluation, distribution, installation, complaint handling, servicing, and records **(Fig. 11.2)**. The preamble describes the public comments received during the development of the QS regulation, and the FDA Commissioner's resolution of the comments. Thus, the preamble contains valuable insight into the meaning and intent of the QS regulation. This is now generally followed throughout the world for proper production of biotechnology products.

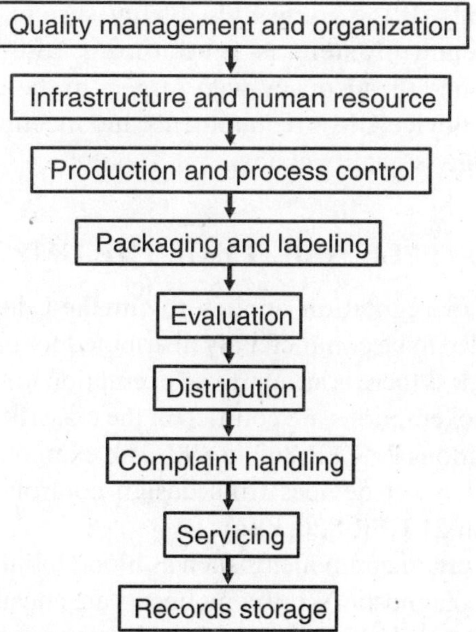

Fig. 11.2. Good manufacturing practice followed by the biotechnology industry for quality products

The good manufacturing practice (GMP)/ quality system regulation can also obtain informations and details from the "medical device quality systems manual": A small entity compliance guide, first edition of which details the requirements of the new QS regulation and provides detailed care needed and guidance in the following areas:

- Obtaining information on GMP requirements.
- Determining the appropriate quality system needed to control the design, production, and distribution of the proposed device.
- Designing products and processes.
- Training employees.
- Acquiring adequate facilities
- Purchasing and installing processing equipment.
- Drafting the device master record.
- Noting how to change the device master records.
- Procuring components and materials.
- Producing devices.
- Labeling devices.
- Evaluating finished devices.
- Packaging devices.
- Distributing devices.
- Processing complaints and analyzing service and repair data.
- Servicing devices.
- Auditing and correcting deficiencies in the quality system.
- Preparing for an FDA inspection.

Flexibility of the GMP

Manufacturers should use good judgement while developing their quality system, and apply those sections of the QS regulation that are applicable to their specific products and operations—21 CFR 820.5 of the QS regulation. Operating within this flexibility, it is the responsibility of each manufacturer to establish requirements for each type or family of devices, which will result in devices that are safe and effective, and to establish methods and procedures to design, produce, and distribute devices that meet the quality system requirements. FDA has identified in the QS regulation the essential elements, which a quality system should embody for design, production, and distribution, without prescribing specific ways to establish them.

Since the QS regulation covers a broad spectrum of devices and production processes, it allows some leeway in the details of quality system elements. It is left to the manufacturers to determine the necessity for, or extent of, some quality elements, and to develop and implement specific procedures tailored to their particular processes and devices. For example, if it is impossible to mix up labels at a manufacturer because there is only one label or one product, then there is no necessity for the manufacturer to comply with all of the GMP requirements under device labeling. This ensures a product of better quality.

The biotechnology medical device QS regulation requires an "umbrella" quality system intended to cover the design, production, and distribution of all medical devices from simple surgical hand tools to very complex computerized axial tomography (CAT) scanners. It is not practical for a regulation to specify details of quality system elements for such a wide range of products. Rather, the QS regulation specifies general objectives such as the use of trained employees, design reviews, design validation, calibrated equipment, process controls, etc., rather than methods, because a specific method would not be appropriate to all operations. The risk is tremendously reduced to nil.

In majority of the cases, it is left to the manufacturer to determine the best methods to attain quality objectives. In some cases, however, the QS regulation does specify the particular type of method to be used, such as written procedures or written instructions. This does not mean that manufacturers cannot vary from the method specified if the intent of the GMP requirement can be met by another method, such as using an engineering drawing plus a model device as manufacturing instructions. Written procedures are not restricted to paper copies. Written procedures may be filed and distributed by automated data processing equipment.

In biotechnology, large manufacturers have a quality system that exceeds the medical device QS regulation. Small manufacturers have a proportionally simpler system. FDA recognizes that a small manufacturer may not need the same amount of documentation as that of a large manufacturer, to achieve a state of control, and that some of the records maintained to fulfill the GMP requirements for written procedures may not be as long and complex for a small manufacturer.

After a manufacturer establishes a quality system, it should be maintained. Each manufacturer should assure that with growth, process, or product changes, their quality system would still be adequate. This assurance is obtained through change control, day-to-day observance of operations, and by periodic audits of the quality system. The auditor should first identify the elements of the company's quality system, how well each element is functioning, and then determine its adequacy with respect to the intent of the device GMP requirements and meeting the company's quality claims.

11.14. APPLICABILITY OF THE GMP

The QS regulation applies to finished devices intended to be commercially distributed for human use unless there is an approved exemption in effect. GMP exemptions are codified in the classification regulations 21 CFR 862 to 892. The exemption of most Class I devices from design controls is in section 21 CFR 820.30(a).

Certain components such as blood tubing and major diagnostic X-ray components are considered by FDA to be finished devices, because they are accessories to finished devices. The manufacturer of such accessories is subject to the QS regulation when the accessory device is labeled and sold separately from the primary device for a health related purpose to a hospital, physician, or other user.

The designation of a device as a "custom" or "customized" device does not confer a GMP exemption.

Contract manufacturers and specification developers should comply with the sections of the QS regulation that apply to the functions they perform.

Contract test laboratories are considered an extension of a manufacturer's quality system, and presently, they are not routinely scheduled for GMP inspections. The finished device manufacturer should meet the requirement of the QS regulation, particularly 21 CFR 820.50, purchasing, when they obtain products or services. Internal test laboratories, which are part of a corporate manufacturer that provides services to individual corporation factories, should meet GMP requirements. Internal laboratories are inspected as part of the FDA GMP inspection of the member factories.

Situations are discussed in the remainder of this chapter, where various manufacturers are exempt from the QS regulation or are not routinely inspected. However, these manufacturers are still subject to the FD & C Act. If these manufacturers or any manufacturer render devices, which are unsafe or ineffective, adulterated, and/or misbranded, they will be subject to the penalties of the FD & C Act.

GMP Exemptions

FDA has determined that certain types of establishments are exempt from GMP requirements; and FDA has defined GMP responsibilities for others. Exemption from the GMP requirements does not exempt manufacturers of finished devices from keeping complaint files (21 CFR 820.198) or from general requirements concerning records 21 CFR 820.180. Sterile

devices are never exempted from GMP requirements. Medical devices manufactured under an investigational device exemption (IDE) are not exempt from design control requirements under 21 CFR 820.30. A device that normally would be subject to GMP requirements may be exempt under the following conditions:

- When FDA has issued an exemption order in response to a citizen's petition for exemption.
- When FDA, in the absence of a petition, has exempted the device and published the exemption in the federal register.
- When the device is exempted by FDA classification regulations published in the federal register and codified in 21 CFR 862 to 892.
- When the device is an intraocular lens (IOL) under an IDE, and meets the requirements of the IDE regulation for IOL's (except for design controls 21 CFR 820.30).
- Through a policy statement, FDA may decide not to apply GMP requirements to some types of devices and processes, though the devices may not have been exempted from GMP requirements.

Manufacturers should be aware of the GMP exemption status of their devices. In addition, manufacturers should keep file records of any specific GMP exemption granted to them by FDA. Upon request, during a factory visit, the exemption records need to be shown during normal business hours to the FDA investigator in order to verify that an exemption has been granted.

Types of Establishments Exempt from GMP

Component manufacturers

A "component" is defined by 21 CFR 820.3(c) as "any raw material, substance, piece, part, software, firmware, labeling, or assembly that is intended to be included as part of the finished, packaged, and labeled device". Component manufacturers are excluded from the QS regulation by 21 CFR

820.1(a)(1). Current FDA policy is to rely upon the finished device manufacturers to assure that components are acceptable for use. Component manufacturers are not routinely scheduled for GMP inspections; however, FDA encourages them to use the QS regulation as a guidance for their quality system.

When finished device manufacturers produce components specifically for use in medical devices, whether in the same building or another location, such production of components is considered part of the device manufacturing operations, and it should comply with the QS regulation as detailed under manufacturers of accessories. Types of establishments subject to the GMP:

- Remanufacturers
- Custom device manufacturers
- Contract manufacturers
- Contract testing laboratories
- Repackagers, relabelers and specification developers
- Manufacturers of accessories
- Initial distributors of imported devices

Remanufacturers

A remanufacturer, as defined in 21 CFR 820.3(w), is any person who processes, conditions, renovates, repackages, restores, or does any other act to a finished device, which changes the finished device's performance or safety specifications, or intended use. Remanufacturers are considered manufacturers. As such, these manufacturers are subject to inspection by FDA and should meet the applicable requirements of the medical device QS regulation. These manufacturers should establish and implement quality systems to assure the safety and effectiveness of the devices that are distributed. Such activities include drafting of device master records, rebuilding per the device master records, inspection and testing, calibration of measurement equipment, control of components, updating of labeling, processing of complaints, and any other GMP requirement applicable to the activities being performed.

Remanufacturers are also required to comply with the labeling requirements of 21 CFR 801.1(c). This labeling regulation requires that where the person or manufacturer named on the label of the device is not the original manufacturer, the name should be qualified by an appropriate phrase, which reveals the connection that person has with the device, e.g., remanufactured by XYZ Company.

Custom device manufacturers

Section 520(b) of the FD & C Act and the IDE regulation [21 CFR 812.3(b)] defines a custom device. Custom devices are exempt from certain statutory requirements. For example, manufacturers of custom devices are not required to comply with premarket approval requirements (Section 515), and are exempt from premarket notification requirements [Section 510(k)]. Custom devices are NOT exempt from the GMP requirements. Manufacturers of custom devices should comply with the GMP requirements while considering the flexibility allowed.

Contract manufacturers

A person(s) who manufactures a finished device under the terms of a contract with another manufacturer is a contract manufacturer. The agreement between the manufacturers should be documented in a written contract. Contract manufacturers should comply with applicable requirements of the quality system, and register their establishment with FDA. Depending on the circumstances, both the contractor and manufacturer may be held jointly responsible by FDA for the activities performed.

Contract testing laboratories

Contract laboratories that design or test components or finished devices for a manufacturer according to his specifications are considered an extension of the manufacturer's quality system.

These laboratories may provide services to a number of customers, many of which are not medical device manufacturers. These contract laboratories are not subject to routine GMP inspections. Through the conduct of purchasing assessment, the finished device manufacturer is responsible for assuring that equipment and procedures used by a lab are adequate and appropriate (21 CFR 820.50). However, an internal test laboratory, if part of a manufacturer that does testing for various facilities within the corporation, is subject to inspection when FDA GMP inspections are conducted at the individual manufacturing facilities. That is, the test laboratory is simply a part of medical device manufacturer, the device-related divisions of which should comply with the QS regulation.

Repackagers, relabelers, and specification developers

Repackaging and relabeling of a device and specification development are defined as manufacturing in 21 CFR 820.3(o) and 21 CFR Part 807, establishment registration and device listing for manufacturers of devices. Some definitions from 21 CFR 807.3(d) are reprinted below because they affect the applications of the QS regulation.

(d) "Manufacture, preparation, propagating, compounding, assembly, or processing" of a device means the making by chemical, physical, biological, or other procedures of any article that meets the definition of a device in section 201(h) of the Act. These terms include the following activities:

- Repackaging or otherwise changing the container, wrapper, or labeling of any device package in furtherance of the distribution of the device from the original place of manufacture to the person who makes final delivery or sale to the ultimate consumer.
- Distribution of domestic or imported devices.

- Initiation of specifications for devices that are manufactured by a second party for subsequent commercial distribution by the person initiating specifications".

As defined above, repackaging and relabeling are manufacturing operations. Further, a repacker, repackager or relabeler is a manufacturer per 21 CFR 820.3(o) and subject to the applicable requirements of the QS regulation. Individuals are repackers or relabelers if they:

- package and/or label previously manufactured finished devices or accessories.
- receive finished devices in bulk (e.g., surgical tubing, syringes, media, etc.), repack them into individual packages, and label them.
- receive previously manufactured devices that have been packaged and labeled by another manufacturer, and combine them into a kit with other unpackaged devices which are received in bulk.

Individuals are not considered repackers, or relabelers, or a manufacturer, for the purpose of applying the QS regulation, if they pack only previously packaged and labeled individual devices into packages for the convenience of the user (Note that this activity is essentially the same as a drug store employee placing packaged items into a bag labeled with the name of the drug store).

A distributor, who only adds a label bearing their name and address, is exempt from the GMP requirements. A manufacturer simply affixing a sticker label bearing the distributor's name and address would not require record keeping demonstrating compliance with labeling control requirements.

Specification developers provide specifications to contract manufacturers, who produce devices to meet the specifications. The contract manufacturer may package and label the device, or the finished device may be shipped to the specification developer for packaging and labeling.

Specification developers are manufacturers and are subject to the GMP requirements that apply to the activities they conduct, such as various design controls including correct transfer of the design information to a contract manufacturer [21 CFR 820.30 (h)]. This activity, in turn, requires an adequate device master record (21 CFR 820.181) and adequate document change control [21 CFR 820.40 (b)]. Further, if the product carries the specification developer's label, the developer is responsible for maintaining a complaint file and processing complaints, in addition to maintaining the device specifications and other appropriate documents in the device master record.

Manufacturers of accessories

When the finished device manufacturers produce components specifically for use in medical devices they produce, whether in the same building or another location, such production of components is considered part of the device manufacturing operations, and the production should comply with the QS regulation.

Accessory devices are discussed in 21 CFR Part 807, establishment registration and device listing for manufacturers of devices [21 CFR 807.20(a)(5)]. Devices such as hemodialysis tubing or major diagnostic X-ray components which are packaged, labeled and distributed separately to a hospital, physician, etc., for health-related purpose are sometimes inappropriately referred to as components. However, FDA considers them finished devices because they are suitable for use and are distributed for health-related purposes; the QS regulation applies to their manufacture. Similarly, a device or component including software that is sold as an addition to a finished medical device to augment or supplement its performance is also termed an accessory. An accessory to a medical device is considered a finished device and, therefore, is subject to the QS regulation.

Additional Quality System Information

- Quality system 21 CFR 820.
- The good manufacturing practice (GMP—quality system regulation) final rule (*Federal Register*).
  http://www.fda.gov/cdrh/humfac/frqsr.html
  http://www.fda.gov/cdrh/fr1007ap.pdf
- Guideline on general principles of process validation.
  http://www.fda.gov/cdrh/ode/425.pdf
- Quality management systems process validation guidance, Global harmonization task force.
  http://www.accessdata.fda.gov/scripts/cdrh/cfdocs/linkwarning/linkwarning.cfm?link
- ISO 9001 comparison to the quality system.
- Comparison chart: 1996 quality system reg. vs. 1978 good manufacturing practice reg. *vs.* ANSI/ISO/ASQC Q9001 and ISO/DI 13485:1996.
- ISO 9001:2000 and FDA quality system.
- GMP guidance documents—CDRH office of compliance.
  http://www.fda.gov/cdrh/comp/gmp.html

The GMP institute (GMPI), founded in 1977 to help companies comply with FDA's GMP regulations, was acquired by ISPE in January 2000. The philosophy of the GMPI is simple—GMP makes good business sense because it helps a company to implement practices, assuring the quality of their products.

For 25 years, the GMP Institute has offered public training sessions, in-house training sessions, and a series of training materials available to companies who wish to include these materials in their own training programs. ISPE continues this tradition of public and in-house training sessions along with the training products available for purchase.

SUMMARY

For studies and developments, researchers have

to adopt the laboratory methods that are reproducible and of high standard. To achieve these goals, organizations for economic cooperation and development (OECD) member countries have recently passed legislation. This legislation usually requires the manufacturer to perform laboratory studies and to submit the results of these studies to a governmental authority for assessment of the potential hazard to human health and the environment. Government and industry are increasingly concerned with the quality of studies upon which hazard assessments are based. As a consequence, several OECD member countries have, or plan to establish, criteria for the performance of these studies in a uniform manner.

In biotechnology, to avoid different schemes of implementation that could impede international trade in chemicals, OECD member countries have recognized the unique opportunity for international harmonization of test methods and good laboratory practices. During 1979–80, an international group of experts established under the special programme on the control of chemicals developed this document concerning the "Principles of good laboratory practice (GLP)" utilizing common managerial and scientific practices and experience from various national and international sources. This makes the test practice a standard throughout the world.

The principles of good laboratory practice (GLP) define a set of rules and criteria for a quality system concerned with the organizational process and the conditions under which non-clinical health and environmental safety studies are planned, performed, monitored, recorded, reported and archived. Hence, uniformity can be achieved in all the testing procedures that are developed in the biotechnology.

Good laboratory practice principles include different criteria in using and handling of various items in the predesigned manner and they are test facility organization and personnel, quality assurance programme; facilities for test system; facilities for handling test and reference substances; apparatus, materials, reagents; physical, chemical and biological test systems; test and reference substances receipt, handling and their storage; standard operating procedures for the test procedures used in the laboratory; performance of the study; reporting the results; storage and retention of records and materials. Thus, from the beginning of test to the end point, everything can be known in an international standards and the reproducibility, tracability along with consistency can be achieved for all types of tests performed in the laboratory. This makes the reliability of the test results without any doubt.

With the biotechnology developments, the debate has not usually addressed the potential products of different biotechnology segments that are on the horizon. While technology developers believe that these new products will offer benefits in meeting needs for food, fuel and fiber, as well as for novel industrial and pharmaceutical uses, some of these future products are also likely to raise environmental and other concerns that will need to be addressed by the regulatory system and safety issues.

The current good manufacturing practice (GMP) requirements set forth in the quality system (QS) regulation are ear marked. They require that domestic or foreign manufacturers have a quality system for the design, manufacture, packaging, labeling, storage, installation, and servicing of finished medical devices intended for commercial distribution in the any country. The regulation requires that various specifications and controls be established for devices; devices be designed under a quality system to meet these specifications; devices be manufactured under a quality system; finished devices meet these specifications; devices be correctly installed, checked and serviced; quality data be analyzed to identify and correct quality problems; and complaints be processed. Hence, the biotechnology product can achieve large-scale acceptance worldwide.

Good manufacturing practice (GMP) also helps the QS regulation helps assure that medical devices are safe and effective for their intended use. The international body inspects the operations and records of device developers and manufacturers to determine compliance with the GMP requirements in the QS regulation. Thus, tractability of any negative impact can be done in the production phase.

The good manufacturing practice (GMP)/ quality system regulation can also obtain informations and details from any suitable manuals or set procedures for achieving maximum quality of the products. The good manufacturing practice regulations enforce the law for manufacturers, processors and packagers of biotechnology products (food, drugs, therapeutic proteins, medical devices, blood products) to take proactive steps to ensure that their products are safe, pure and effective for the consumption. The good manufacturing practice regulations also make a quality approach to manufacturing, enabling organization to minimize or totally eliminate chances of contamination, mixing-ups and errors in the final products. Thus, it ensures that the consumers are purchasing a product that is totally effective and free from hazardous effects. Any organizations that are not following the good manufacturing practice regulations are liable for serious consequences such as recall, seizure, penalties and imprisonment for long terms.

EXERCISE

1. What do you mean by good laboratory practice (GLP)? Give its importance in the society for the better reliability of tests.
2. Describe various types of terms that are commonly used in the good laboratory practice in the laboratory.
3. Write a note on the scope of good laboratory practice (GLP) in biotechnology and mention its merits.
4. Explain the organizations for economic cooperation and development (OECD) processes necessary for the good laboratory practice.
5. Briefly write about the standard operating procedures that are essential in the good laboratory practice.
6. Define test facility system and its impact on the good laboratory practice. Mention different methods of facilities required for handling test and reference materials in the laboratory.
7. What is meant by management responsibilities in the good laboratory practice? Add a note on the issues associated with the quality assurance programme.
8. How physical, chemical and biological systems should be maintained for good laboratory practice for consistent results and accuracy?
9. What is meant by the quality assurance programme associated with the good laboratory practice? Brief its importance to achieve quality in the test procedures.
10. What care should be taken while handling of various test and reference materials in the laboratory for good laboratory practice.
11. Enumerate the importance of standard operating procedures in the laboratory under the guidelines of good laboratory practice.
12. Distinguish between the test substances and references substances.
13. Explain different types of limitations that are associated with the good laboratory practice.
14. What is meant by "study plan"? Mention different types of study plans required in the good laboratory practice.
15. Describe how storage and retention of test results and materials are done in the good laboratory practice.
16. List out the importance of reporting the test result in the standard manner, and name different contents of final results reporting.
17. Define good manufacturing practice (GMP).
18. Mention the applications of GLP for the quality achievement in the products.
19. Briefly explain the applicability of good manufacturing practice in the biotechnology industries and production systems.
20. Elucidate the future of good manufacturing practice exemptions that are operated in the good manufacturing practice.
21. Write a note on the types of organizations that are exempted from the good manufacturing practice.
22. Why are contact testing laboratories not exempted from the guidelines of good manufacturing practice?
23. What is meant by manufacturers of accessories? Is it necessary to include it under the guidelines of good manufacturing practice?

12

CHAPTER

Public Education of Producing Transgenic Organisms

12.1. INTRODUCTION

The genes are made up of genetic material—either DNA or RNA. Friedrich Miescher first discovered the DNA in 1869. But the structure of DNA was not determined for another 84 years. James Watson and Francis Crick made the breakthrough in 1953. They came up with their model after analyzing X-ray crystallographs produced by Rosalind Franklin and Maurice Wilkins. It is fair to say that Watson and Crick made an intellectual leap that Franklin and Wilkins had failed for long to make. Later Watson, Crick, and Wilkins were awarded nobel prize. The structure of DNA is so elegant that it has become iconic in the science and paved the way for identification of genes in the organisms.

As finite natural resources are rapidly being exhausted, and the global environment becomes increasingly stressed as a result of industrial and human pollution, there is an awareness for the need of clean, renewable, industrial, and agricultural strategies which will help ensure the future supply of raw materials for the manufacture of products, while, at the same time, protecting the environment from pollution and other negative side effects, such as global climate change, which currently marks man's material progress. Man is interested in improving all these, without affecting the nature.

In order to achieve this desired state, it will be necessary to change the sources of the raw materials, and the way in which they are processed into products and services—food, energy, industrial feed stocks, plastics, transportation, consumer goods, and all the rest. Current methods such as distillation of fossil fuels to provide energy sources and petrochemicals are both non-sustainable and environmentally damaging. The manufacture of chemical fertilizers to increase crop yield falls into the same category. Despite relatively stringent environmental controls, greenhouse gas emissions have increased as a result of the extraction and subsequent processing of these non-renewable resources into products.

Since the development of the first transgenic plants, health and environment issues concerning the safety of genetically modified foods and feeds have been raised. The main concerns regarding genetically modified (GM) foods include toxin or allergen production, changes in nutrient levels, and development of antibiotic resistance. Gene flow and effect on non-target organism of GM crops are among environmental issues. As an effort to manage globally-debated issue of genetically modified organisms (GMOs), the Law on transboundary Movement of Living Modified Organisms has in hot discussion among the public.

There has been a considerable upsurge of interest in biotechnology in various countries including India in recent months. Discussions on biotechnology are no longer confined to scientific seminars, conferences and symposia. People from various sections of the society are curious about the role biotechnology will play in the next millennium. Newspapers are replete with articles on genetic cloning and cloning ambitions. Biotechnology has entered numerous fields ranging from development of varieties of plants to production of vaccines. The techniques of biotechnology namely, tissue culture, recombinant DNA technology and cloning have naturally attracted the attention of people in India. Farmers' organizations, non-Governmental Organizations (NGO's), people's representatives, Industrial Houses, business enterprises, policy makers and academicians are curious about the crucial role biotechnology will play in future.

These sector people know that this is the technology of the next century. In the face of continuous population growth, the land holdings will be further differentiated in future. The current technology, based on the Green Revolution, will not be able to support the population in future. New technologies will have to be evolved and introduced. Biotechnology is, therefore, looked upon for augmenting food grains production in future. Concomitant with the success of biotechnology many apprehensions have crept in the minds of people about the efficacy of biotechnology for increasing agricultural production. People are inquisitive about the deleterious effects that transgenic varieties may produce on biodiversity. People are also inquisitive about environmental safety, as the effects of transgenic varieties on the environment cannot be easily gauged. The doubts cannot simply be dismissed and due attention will have to be paid to these doubts.

An appropriate strategy for harnessing biotechnology research will ensure enhanced food grains production as well as production of meat, fish, eggs and milk. The techniques of biotechnology will also enable the growers of fruits and vegetables to produce items of international standards with long shelf life.

In the present situation of ever increasing populations of the world merger of green revolution with gene revolution, what we want to do is to have a blend of the conventional technologies along with new biotechnological tools, the whole understanding of genetic engineering, and their applications to get something better. So the merger is a very important phase as we move into the next millennium. There has been a 15-fold increase in transgenics world over from 1996 to 1998. Eight countries are basically the key players. From 16.8 million hectare to 27.8 million hectare 1998, it reached to 81 million hectares in 2004 in 17 countries. We have soybean, corn, cotton, tomato, potato, canola and so on. Many of these crops are already in the field.

The traits that are considered are herbicide resistance, insect/viral resistance, and nutritional quality improvement. Are these not required? The whole areas of molecular genetics/breeding program are picking up very fast not only in developed countries but even in developing countries. But to come back to the issues to develop the field of biotechnology, risks must be addressed. Now it is necessary to work very closely on all aspects of biosafety, environmental safety, toxicity, and side effects on human kind everything has to be looked at very carefully. Agricultural universities should have positive roles in these aspects. They are located all over the world in every agroclimatic region and they must assume the responsibility of regularly getting a feedback from the farmer educating them and giving them new agritechnology packages.

12.2. NEED OF THE TRANSGENIC ORGANISMS

Biotechnology can offer some solutions in the fight for sustainable, clean production, and significant mitigation of some of the negative side effects of other, currently non-sustainable technologies that are used to provide the necessities of everyday life in an industrialized society. For instance, cereal carbohydrates, cellulose waste streams and other biological materials can be converted using existing technologies into industrial feed stocks, such as plastics, polymers and fine chemicals. Clean, renewable energy sources such as alcohols and methyl esters can be obtained by relatively simple processing technologies. Biotechnology, especially through the application of genetic modification, can enhance the efficiency of these processes, speeding them up and increasing the yield, reducing net emissions of greenhouse gases, and at the same time, often sequestering carbon. Plants and animals can be genetically enhanced so as to produce many of the sophisticated protein and carbohydrate molecules currently being manufactured in energy intensive, non-sustainable ways. Molecules which are being produced right now in genetically enhanced plants and animals include enzymes for laundry detergent and industrial processes, human pharmaceuticals, blood products, biological pesticides and fertilizers, synthetic textile fibers, and a host of other raw materials.

"Biotechnology" and "genetic modification" commonly are used interchangeably. GM is a special set of technologies that alters the genetic make-up of living organisms such as animals, plants, or bacteria while, biotechnology, a more general term, refers to the use of living organisms or their components, such as enzymes, to make products that include wine, cheese, beer, and yoghurt.

Combining genes from different organisms is known as recombinant DNA technology, and the resulting organism is said to be "genetically modified", "genetically engineered", or "transgenic". GM products (current or in the pipeline) include medicines and vaccines, food and food ingredients, feeds, and fibers.

Locating genes for important traits, such as those conferring insect resistance or desired nutrients is one of the most limiting steps in the process. However, genome sequencing and discovery programs for hundreds of different organisms are generating detailed maps along with data-analyzing technologies to understand and use them.

GM crops are grown commercially or in field trials in over 40 countries and 6 continents. In 2000, about 109.2 million acres were planted with transgenic crops; the principal ones being herbicide and insecticide-resistant soybeans, corn, cotton, and canola. Among other crops, grown commercially or field-tested, are sweet potato resistant to a virus that could decimate most of the African harvest; rice with increased iron and vitamins that may alleviate chronic malnutrition in Asian countries; and a variety of plants able to survive weather extremes.

The use of genetically modified organisms for studies in taxonomy, ecology and insect pathology is essential for advancement of fundamental knowledge in ecology and biology. Genetically marked microbes allow studies to be carried out to a level of detail that was not previously possible, with a consequent increase in knowledge of ecosystem function. There is little uncertainty about such an outcome and it is likely that appropriate use of molecular biology and marked organisms will become an increasingly common part of laboratory practice. Such work provides a pathway to understanding that would be impossible without using genomic techniques. Similar to any other new technology, this transgenic organism technology also raised several issues with merits and demerits. These are as follows:

Benefits

Crops

- Enhanced taste and quality.
- Reduced maturation time.
- Increased nutrients, yields and stress tolerance.
- Improved resistance to disease, pests, and herbicides.
- New products and growing techniques.

Animals

- Increased resistance, productivity, hardiness, and feed efficiency.
- Better yields of meat, eggs, and milk.
- Improved animal health and diagnostic methods.

Environment

- "Friendly" bioherbicides and bioinsecticides.
- Conservation of soil, water, and energy.
- Bioprocessing for forestry products.
- Better natural waste management.
- More efficient processing.

Society

- Increased food security for growing populations.
- Reach all types of people in the society.

Controversies

Safety

- Potential human health impact: allergens, transfer of antibiotic resistance markers, unknown effects potential environmental impact: unintended transfer of transgenes through cross-pollination, unknown effects on other organisms (e.g., soil microbes), and loss of flora and fauna biodiversity.

Access and Intellectual Property

- Domination of world food production by a few companies.
- Increasing dependence on industrialized nations by developing countries.
- Biopiracy—foreign exploitation of natural resources.

Ethics

- Violation of natural organisms' intrinsic values.
- Tampering with nature by mixing genes among species.
- Objections to consuming animal genes in plants and *vice versa*.
- Stress for animal.

Labeling

- Not mandatory in some countries (e.g., United States).
- Mixing GM crops with non-GM confounds labeling attempts.

Society

- New advances may be skewed to interests of rich countries.
- Not immediately access to the third world countries.

12.3. PUBLIC EDUCATION AND THEIR INVOLVEMENT

The technological achievements made in the various fields ultimately have to reach the general public, who are the beneficiaries of the development. They are the people who have to decide whether to accept or reject the product or processes. In the early 1990's, members of US and Australian public and private research organizations, including universities, began proactive efforts to educate the public about genetic engineering and transgenic organisms. The target audience included members of the media and public opinion leaders. These play a pivotal role in determining exactly what information people hear and read regarding an issue; in what way that information is presented; and in what manner this information is used to shape public policy.

The society should be able to decide some things new in an:
* unbiased;
* unidirectional manner;
* logical way;
* non-influential way;
* based on scientific principles;
* ethically correct;
* unambiguously;
* clearly acceptable manner; and
* truthful manner.

Trusted governmental and professional agencies need to be actively involved in information dissemination and education, for e.g., General C. Everett Koop issued a statement that milk from rBST-injected cows was safe. In addition, several other governmental and public sector agencies, e.g., American Medical Association, Food and Drug Administration, and American Dietetic Association, released information in the popular press, peer reviewed scientific journals, as well as made a hot line available to answer questions on the safety of the product, thereby increasing the openness of information exchange. If this kind of engagement does not occur, an informational void can result.

Hence, a proper information should be made available for all people irrespective of their status, education, religious strata, and country. Scientific information should be in a simple language, **(Fig. 12.1)** so that it can reach the entire population around the world. This can be achieved through the following:
* Newspapers

* Television
* Radio
* Internet
* Magazines
* Journals
* Periodicals
* Popular articles
* Booklets
* Pamphlets
* Handouts
* Advertising panels
* Conferences
* Workshops

Information Dissemination Tools

Awareness-raising workshops at national and sub-national level, and/or targeted at particular groups such as journalists (as in India, Japan, Estonia, Namibia and Zimbabwe).

Training workshops targeted at particular groups that need to know about biosafety risk assessment, risk management and the regulatory framework (as in India, Japan, Estonia, Namibia and Zimbabwe). **Printed information** such as:
* Technical fact sheets (Scotland, UK).
* Non-technical publications in the form of leaflets, magazines and newsletters (as in India, Canada, Estonia, Norway and Zimbabwe).
* Editions translated into local languages (as in India, Estonia, Kenya and New Zealand).

Electronic dissemination using web-sites, sometimes for two-way communication (as in India, Canada, Denmark, Estonia, New Zealand, Norway, the UK and the US).

Use of national and local media and working with journalists to encourage balanced and informative coverage of biosafety issues (as in Estonia, the UK, Namibia and Norway).

Making use of NGOs or other intermediaries for outreach to their networks (as in Estonia (REC-Estonia), Kenya (African Centre for Technology Studies, ACTS), Malaysia (Third World Network,

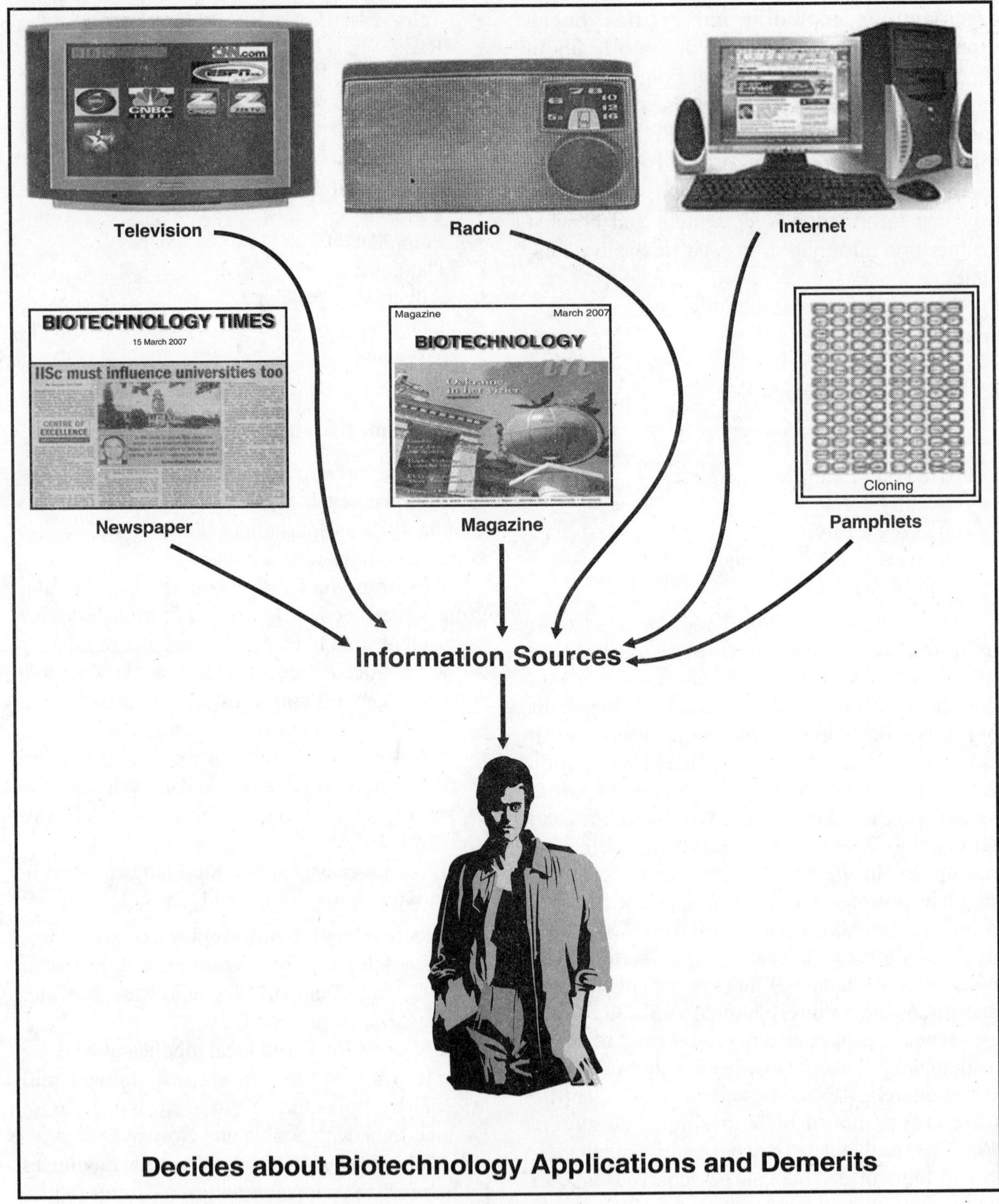

Fig. 12.1. Different source of information to the public regarding biotechnology

TWN) and Zimbabwe (Intermediate Technology, ITDG). (In Brazil and India, among other places, NGOs were able to initiate information and awareness campaigns on their own initiative).

Establishing independent intermediary organizations with a mandate to disseminate information and promote public awareness (as in Namibia (Namibian Biotechnology Alliance, NABA), Norway (Norwegian Biotechnology Advisory Board, NBAB) and the UK [Agriculture and Environment Biotechnology Commission, AEBC/Steering Committee for the Public Debate)].

Public open days/information days (as in Estonia and Ethiopia).

12.4 DIFFERENT EXAMPLES OF PUBLIC EDUCATION ABOUT TRANSGENIC TECHNOLOGY

Danish consensus conference on genetically modified foods, march 1999

The consensus conference was intended to promote a dialogue between experts and layman, (non-experts) and to convey a picture of citizens' views to politicians. The exercise lasted three days and was open to the public. The resulting report was submitted to the Folketing (Parliament).

The consensus conference is regarded as being suitable for addressing reasonably discrete issues having current social relevance, where public attitudes are unclear and contributions from experts appear to be necessary for the resolution of the public debate. The experts contribute by providing knowledge about the technology and its implications to a panel of citizens. Afterwards, the citizens' panel drafts a final document clarifying the issue and expressing a position on it.

The panel of layman is selected by sending out invitations to a random sample of around 1,000 adult citizens. Among those who wish to participate, 14 are selected to reflect a representative mix of age, gender, education, profession, and geographical spread. The panel

receives a thorough briefing on the subject consisting of information material on the topic and two weekend courses, during which the citizens get to know each other. The panel formulates the questions the conference will deal with, and participates in selecting the experts who will give 'evidence'.

The three days of the conference proceed as follows:

- On day one, the experts make presentations addressing the questions posed in advance by the layman panel. The presentations cover a diverse range of aspects on the issue, e.g., financial, biological, legal, social and ethical and perspectives.

- On day two, the panel and the audience questions the individual experts for elaboration and clarification of their presentation. The panel then begins drafting its final document. The first draft is then discussed and refined in smaller groups. The panel strives to find unanimous formulations.

- On day three, the panel presents the final document to the experts, audience, and media. The experts have the opportunity to challenge misunderstandings and factual errors. The final document and the written contributions of the experts are then submitted to the Folketing.

- The Danish Board of Technology claims that consensus conferences in Denmark have stimulated public debate about technology, and succeeded in making politicians aware of public attitudes, aspirations, and concerns. On several occasions such conferences have helped to initiate new regulation.

Participatory biotechnology policy processes in Zimbabwe and Sri Lanka

The UK-based NGO ITDG has developed a participatory methodology designed to assess the potential impacts of modern agricultural biotechnology on poor people. To date their work has focused on Sri Lanka and Zimbabwe. The exercise consisted of the following elements:

- National **scoping studies** were undertaken to discover what biotechnologies were currently being applied or would potentially be applied in the countries involved, and to gain an initial understanding of the actual or potential impacts and risks associated with those technologies.
- The scoping studies were discussed in national **stakeholder workshops**.
- The discussions in the stakeholder workshops were incorporated into a **communications package** for use in community-level studies.
- The communications package was used as input into community level **participatory sessions**, including **focus groups** and **brainstorming exercise**s. Participants were encouraged to critically review their experiences and knowledge of the biotechnologies identified in the scoping studies, and to consider their potential impacts at community level. Brainstorming was used to enable people to assess the feasibility of risk management strategies in their own context, and to identify the roles of different actors (at farmer level, local level and national level, including the private sector) in applying such risk management requirements.

The ITDG experience in Zimbabwe and Sri Lanka demonstrates the importance of carrying out preliminary but rigorous research work in order to identify which preconditions would be necessary for designing and implementing a risk assessment and risk management strategy (ITDG 2000). ITDG's ability to perform the scoping studies effectively depended on the existence of a central inventory of biotechnology work being undertaken. This resource existed in Zimbabwe because of the work of the Biotechnology Trust of Zimbabwe (BTZ), but was absent in Sri Lanka. Other problems inhibited the studies, including the unwillingness of some researchers to divulge the details of the biotechnology work underway in their laboratories. Another complaint was that virtually no information was available concerning risk assessments and risk management relating to transgenic crops and other LMOs in either Sri Lanka or Zimbabwe. The lack of local literature tended to drive the discussions towards a more theoretical debate about personal preferences rather than factual assessments of local impacts. ITDG concluded that 'it would be difficult to implement this Impact Assessment Methodology unless the lack of good information based on local empirical data is remedied'. A key challenge was to find ways of communicating technological and risk concepts to the stakeholders in an accessible and intelligible manner. This was especially important, given that participants lacked direct experience of transgenic crops. At first, many participants were reluctant to speak because they lacked familiarity with the issues and concepts, and many found it difficult to absorb the information being conveyed to them. This underscores the benefits of organizing a gradual and iterative process that enabled understanding to grow over time, rather than conducting one-off events. Effective communication relied on the contributions of a multi-disciplinary team, which included biotechnologists and social scientists, as well as people with experience of communicating information at the local level. 'Experts' were required who would be able to translate the technical language into a language, which is understood by and relevant to the public.

Present Situations of GMOs

At present we cannot talk intelligently about GMOs if we remain at the level of generalities. Any statement must refer explicitly to specific crops, genes, ecologies and production systems. For this reason, FAO has been conducting a worldwide inventory of agricultural biotechnology applications and products, with special reference to developing countries. We wished to test the hypothesis that biotechnology has been uneven, and that most current biotechnology applications in crop science—particularly GM crops—have not addressed the special needs of developing countries. We have looked specifically at current as well as pipeline research on GM crops.

Our sources of information have been varied, informal discussions with scientists as well as formal reports. Our inventory will be available electronically and will be updated regularly. Some of the preliminary results are as follows. The total area cultivated with GMO crops in the world currently stands at about 44.2 million hectares. This compares with 11 million hectares only three years ago. About 75% of the area planted to GM crops is in industrialized countries. Substantial plantings so far largely concern only four crops: soybean, maize, cotton and canola, and about 16% of the total area planted to these major crops is now under GM varieties. Two traits—insect resistance (mainly based on Bt) and herbicide tolerance—dominate. There are also small areas of potato and papaya, with genes for delayed ripening and virus-resistance inserted.

SUMMARY

After discovery of DNA by Friedrich Miescher in 1869 and its structure determination by James Watson and Francis Crick in the year 1953, its applications in the public domain grown enormously and helped the society immensely. It is not only scientists and researchers speaking the importance of DNA, the public also started speaking about DNA and its applications. At the early period the people came to the realizations for importance of the transgenic organisms. The work on the Human Genome Project, Rice Genome Project and other transgenic researches made a huge impact on the society and common public are getting interested in these activities. The media reports it has attracted over the past several years, the public has received a steady stream of information on the latest gene to be discovered by science and their uses. Such announcements are

generally greeted with enthusiasm, and the hope that genetic solutions will save us from our human frailties, hunger and environmental problems.

The transgenic organisms productions seen a greater jump after thorough basic studies and it moved to the greater dimensions. Especially in agricultural biotechnology the GM crops are grown commercially or in field trials in over 40 countries and on 6 continents. In 2000, the huge extent of land i.e. about 109.2 million acres were planted with transgenic crops, the principal ones being herbicide and insecticide-resistant soybeans, corn, cotton, and canola. Other crops grown commercially or field-tested are a sweet potato resistant to a virus that could decimate most of the African harvest, rice with increased iron and vitamins that may alleviate chronic malnutrition in Asian countries, and a variety of plants able to survive weather extremes.

In present day population growth biotechnology, especially through the application of genetic modification, can enhance the efficiency of these processes, speeding them up and increasing yield— reducing net emissions of greenhouse gasses while at the same time, often sequestering carbon. Plants and animals can be genetically enhanced so as to produce many of the sophisticated protein and carbohydrate molecules currently being manufactured in energy intensive, non-sustainable ways. Molecules that are being produced right now in genetically enhanced plants and animals include enzymes for laundry detergent and industrial processes, human pharmaceuticals, blood products, biological pesticides and fertilizers, synthetic textile fibers and a host of other raw materials.

Hence, the public populations should be educated in the proper manner for their right information in the transgenics technology. Because these are the group who has to decide for accepting or rejecting the product or processes. Similar to any other new technology, this transgenic organisms technology also raised the several issues with merits and demerits. In the early 1990's, members of US and Australian public and private research organizations including universities began pro-active efforts to educate the public about genetic engineering and transgenic organisms production. The target audience included members of the media and public opinion leaders, who play a pivotal role in determining exactly what information people hear and read regarding an issue, in what way that information is presented and in what manner this information is used to shape public policy.

The proper information should be made available for all peoples irrespective of their status, education, religious strata and country. Public should get the scientifically based informations in a simple language to reach the entire populations around the world. This can be done through the following media such as; news papers, television, radio, internet, magazines, journals, periodicals, popular articles, booklets, pamphlets, handouts, advertising panels, conferences, workshops. Through these media the details of producing transgenics are given in a laymen's language to reach the large number of publics. Thus, the society at a large scale clearly knows about the procedure and facts of producing the transgenics and they can decide in the right way with out affecting any of their requirements.

Beyond choices about which tools, for which purpose and at which stage of the process, the key issue is political commitment. The reports show quite clearly that where there is genuine receptivity to the experimental and flexible use of an imaginative array of tools for participation and consultation and education and awareness-raising, change can occur. Likewise, we have seen examples where lack of government commitment to a process has resulted in a breakdown of the process amid distrust and suspicions that people are being asked to endorse a policy that, in effect, has already been made. It is very much necessary to have repeatedly drawn attention to the importance of context to understanding which tools are likely to work in which settings. But we should not allow inevitable differences in political culture and available resources, for example, to become an excuse for inaction. These factors present challenges that require us to think creatively about what can be done in the face of often pressing constraints, but they should not be seen as justifications for a lack of concerted effort to meet the obligations contained in article 23 of the CBD protocol.

Ultimately, while there are many cases of governments being criticised for not consulting widely and sincerely enough, there are no cases we have come across where a government has been accused of promoting too much public participation or of expending too much effort in public education and awareness-raising. What governments are embarking on is a journey without a clear end and there is much reflection that has to take place along the way to ensure that we learn from oneanothers' experiences.

EXERCISE

1. What do you mean by transgenic organisms?
2. Name the scientists responsible for the discovery of DNA. What is its importance in different fields of biotechnology?
3. Explain the merits and demerits of transgenic organisms.
4. Describe how India requires transgenic technology in the near future, and its impact on the rural agricultural sector that has already adjusted to the conventional agriculture.
5. Write a note on genetic modifications in agricultural field, and the spread of the transgenic plants in the world.
6. Give an account of how developed countries claimed the dominance in agricultural biotechnology, and their interest in the developing countries market.
7. Define biopiracy.
8. What do you mean by bioprospecting?
9. Discuss the present situation of transgenic organisms or genetically modified organisms in the biotechnology all over the world.

10. Elucidate the importance of public education regarding the production of genetically modified organisms in biotechnology, and comment on its methods.

11. Briefly explain labeling of genetically modified organisms, and their advantages and disadvantages to the agricultural biotechnology.

12. Enumerate the risks and benefits of genetically modified organisms that are commonly used to develop the agriculture and medicine.

13. How are genetically modified organisms helping the poor farmers in the developed countries?

14. Critically comment on the developed countries' control over the genetically modified organisms.

15. Mention the necessary regulations of modern genetically modified organisms in biotechnology that are essential in the developed countries, with an emphasis on India.

16. Differentiate between the applications of genetically modified organisms in developed countries' and third world countries agriculture.

17. List out different bioresources of third world countries that are sources for the exploitment by the developed countries in the agricultural biotechnology and their companies.

18. Illustrate the impact of genetically modified organisms on the socioeconomic conditions of the developing countries.

19. Define public education. Add a note on "if the biotechnology companies or government organizations do not adopt the public educations."

13

Intellectual Property Rights (IPRs)

13.1. INTRODUCTION

Intellectual property means slightly different things to different people, but in the legal sense it is thought of as a material that may be protected under the patent, trademark, and/or copyright laws. It is very important for the survival and development of the biotechnology company.

Intellectual property (IP) is a class of property emanating primarily from the activities of the human intellect or intangible personal property i.e., it cannot be perceptible by touching. The property, movable or immovable, that are tangible (perceptible by touch) physical matter, is legally protected to prevent it from being stolen by others. The legal rights accrued on the intellectual property are termed as intellectual property rights (IPRs). Intellectual property is a generic term for intangible personal property that includes patents, trademarks, copyrights, and trade secrets.

For an innovative business, intellectual property may be the most valuable asset—more valuable than the plant and equipment. Intellectual property has always been important to agriculture. Much of the new agricultural equipment of the nineteenth century that made the expansion of the agriculture possible was subject to patent protection.

Many commentators have seen intellectual property as a key to innovation and economic prosperity. Without the protection of intellectual property, companies could not profit from their research and development, and they risk by having the good name that they have spent years developing plundered by imitators selling inferior products. This chapter will provide a general introduction to the intellectual property, and issues specific to means biotechnology.

The legal rights secured for an intellectual property under appropriate legislation can be enforced only within the boundaries of the country, which grants such rights. For example, the rights secured by the grant of a patent in USA can be enforced only in the USA. In other words, there is nothing like a world patent or an international patent. Hence, if one desires protection for an IP created for his invention in different countries, he has to give separate applications in each country according to give law and practice prevailing there. This situation will prevail under the WTO regime too. In other words, the trade-related intellectual property right (TRIPS) provisions of the WTO agreement do not envisage one patent that will be valid in all member countries.

13.2. FORMS OF INTELLECTUAL PROPERTY

The convention establishing the world intellectual property organization (WIPO)—one of the specialized agencies of the United Nations System—provided that IPRs should include rights related to:
- literary, artistic, and scientific work;
- performance of artists, phonograms, and broadcasts;
- inventions in all fields of human endeavour;
- scientific discoveries;
- industrial designs;
- trademarks, service marks, commercial names, and designations; and
- protection against unfair competition.

On the basis of the WIPO guidelines, the TRIPS, in the agreement of the WTO, recognizes seven forms of IPR:
1. Patents
2. Industrial designs
3. Trademarks
4. Copyrights
5. Geographical indication
6. Integrated circuits
7. Trade secrets or know-how

Patents

These are legal rights granted for new inventions employing scientific and technical knowledge. For example, a new drug for the treatment of AIDS; new technology for the production of antibiotics; and a new cell phone.

Patent is available only for a limited period of time. Patent, in the law, is a property right and hence can be gifted, inherited, assigned, sold, or licensed. As the right is conferred by the particular country, it can also be revoked by the country, though under very special circumstances.

An invention must satisfy the following three conditions to be patentable:
- Novelty (i.e., it does not form a part of the global state of art).
- Inventiveness (non-obviousness).
- Usefulness (industrially applicable).

Industrial Designs

A design is an idea or conception as to the features of shape, configuration, pattern, ornament or composition of lines or colors applied to any article, two or three-dimensional or both, by any industrial process or means, which in the finished article appeal to and are judged solely by the eye or product, e.g., designs as applied to shoes, TV, textiles.

Design means the features of shape etc., applied to an article, and not the article itself. The features are conceived in the author's intellect. He/she gives those ideas as a pictorial illustration, or as a specimen, prototype, or model. By registration under the designs act, the features are protected as design.

Trademarks

A trademark is an identification symbol, which is used in the course of trade to enable the public to distinguish one trader's goods from the similar goods of other traders. Trademark is represented graphically **(Fig. 13.1)**, and may include shape of goods, their packaging and combination of colors. When a trademark is used in connection with services, it may be called 'service mark'.

Copyrights

A copyright is basically the right to copy and make use of literary **(Fig. 13.2)**, dramatic, musical, artistic works, cinematographic films, records, and broadcasts. It is a proprietary right and comes into existence as soon as the work is created.

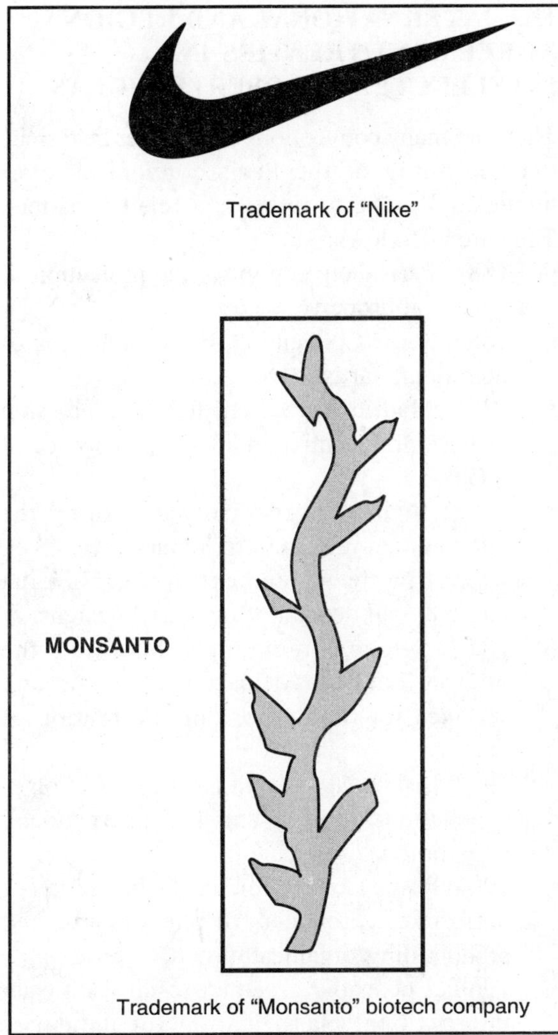

Fig. 13.1. Trademark—a type of intellectual property rights

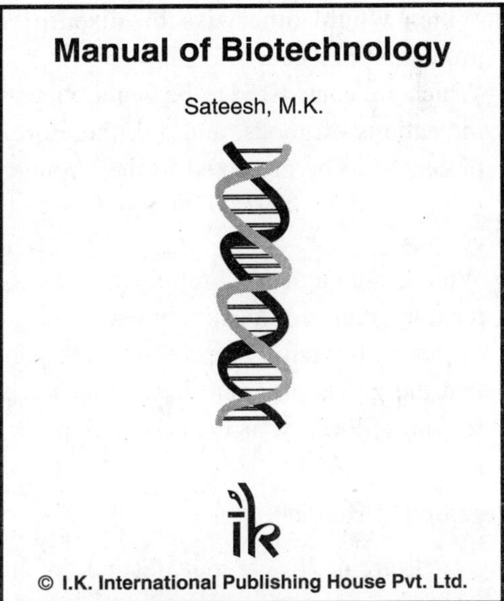

Fig. 13.2. Copyright—a type of intellectual property rights

Types Included in Copyright

- Literary, dramatic and musical work (computer programmes/software are covered within the definition of literary work).
- Artistic work.
- Cinematographic films which include sound track and video films.
- Recording disc, tape, perforated roll or other device.

Geographical Indications

These are indications that identify goods as originating in the territory of a country, region, or a locality in that territory, where a specific quality, reputation, or other characteristics of the goods are attributed to their geographical origin. For example, Mysore silk, Darjeeling tea, Kancheepuram sari, Champagne, Scotch, etc.

Any association of persons or producers, or any organization or authority established by or under any law for the timebeing in force, representing the interest of the producers of the concerned goods, who are desirous of registering geographical indication in relation to such goods, may apply for such registration in a prescribed manner. However, the following geographical indications cannot be registered:

- The use of which would deceive, cause confusion, or is contrary to any law.
- Which comprises or contains scandalous or obscene matter, or any matter likely to hurt the religion susceptibility of any class or section of the citizens of India.

- Which would otherwise be disentitled to protection in a court.
- Which are considered to be generic names or indications of goods, and are, therefore, not or ceased to be protected in their country of origin or which have fallen into disuse in that country.
- Which, although are literally true as to the territory, region or locality in which the goods originate, but falsely represent to the persons that the goods originate in another territory, region, or locality, as the case may be.

Integrated Circuits

Integrated circuit (IC) layout designs include a layout of transistor and other circuitry elements, lead wires connecting such elements, which are expressed in any manner in a semiconductor IC. Semiconductor IC is a product having transistors and other circuitry elements, which are inseparably formed on a semiconductor or an insulating material, or inside the semiconductor material and designed to perform an electronic circuitry function.

Trade Secret or Know-How

Know-how is another important form of intellectual property, generated by R&D institutions, that does not have the benefit of patent or copyright protection. This could be in the form of an aggregation of known processes/procedures, accumulation of data, a secret formulation, or a combination of any of these. Such know-how is kept undisclosed as trade secrets. Know-how is often transferred together with licensing of patents and under transfer of technology arrangements. Know-how, being an undisclosed/secret thing, need not be registered. Hence, the term of the trade secrets is as long as its owner wishes to keep it secret.

13.3. INTERNATIONAL AND REGIONAL AGREEMENT/TREATIES IN INTELLECTUAL PROPERTY RIGHTS

There are many conventions and treaties held since the beginning of the Paris convention over intellectual property rights and related matters. They are as follows:

- 1883: Paris Convention for the protection of industrial property.
- 1886: Berne Convention for the protection of literary and artistic works.
- 1891: Madrid Agreement for the repression of false or deceptive indications of source of goods.
- 1891: Madrid Agreement concerning the international registration of marks.
- 1925: Hague Agreement concerning the international deposit of industrial designs.
- 1957: Nice Agreement concerning the international classification of goods and services for the purpose of registration of marks.
- 1958: Lisbon Agreement for the protection of appellations of origin and their international registration.
- 1961: Rome Convention for the protection of performers, producers of phonograms, and broadcasting organizations.
- 1968: Locarno Agreement establishing an international classification for industrial designs.
- 1970: Patent Cooperation Treaty (PCT).
- 1971: Strasbourg Agreement concerning the international patent classification.
- 1971: Geneva Convention for the protection of producers of phonograms against unauthorized duplication of their phonograms.
- 1973: Vienna Agreement establishing an international classification of the figurative elements of marks.
- 1974: Brussels Convention related to the distribution of programme-carrying signals transmitted by satellite.

- 1977: Budapest Treaty on the international recognition of the deposit of microorganisms for the purpose of patent procedure.
- 1981: Nairobi Treaty on the protection of the olympic symbol.
- 1989: Washington Treaty on intellectual property in respect to integrated circuits.
- 1989: Protocol related to the Madrid Agreement concerning the international registration of marks.
- 1994: Trademark Law Treaty (TLT).
- 1994: Trade Related Intellectual Property Rights (TRIPs).
- 1996: Community Trademarks.
- 1996: Documents of the diplomatic conference on certain copyright and neighboring rights questions (Geneva, December 2–20).
- 1996: WIPO Copyright Treaty (WCT).
- 1996: WIPO Performance and Phonograms Treaty (WPPT).

13.4. INTELLECTUAL PROPERTY RIGHTS RELATED LEGISLATIONS IN INDIA

Similar to the international changes and modifications that are made to the intellectual property rights, India also made several modifications to meet the requirements of the international standards and requirements of the various bodies:

- 1856: The Act on Protection of Invention based on the British Patent Act of 1852.
- 1859: The Act modified as Act XV; patent monopolies called exclusive privileges (making selling and using invention in India and authorizing others to do so for 14 years from date of filing specification).
- 1872: The Patents and Designs Protection Act.
- 1883: The Protection of Inventions Act.
- 1888: Consolidated as the Inventions and Designs Act.
- 1911: The Indian Patents and Designs Act (under the management of the Controller of Patents and Designs). This came into force from August 15, 1947.

- 1914: Copy of the British Copyright Act of 1911, with suitable modifications for British India.
- 1940: Legislation for Protection of the Trademark Act, brought into force from June 1, 1942.
- 1957: Adopted many principles of the British Copyright Act of 1956. This came into force to cope with new problems in law of copyright created by technological advances in the fields of communication, broadcasting, microfilming, photolithography, movies, cinemas, and talkies.
- 1959: On November 25, 1959, the Act, Trademark of 1940 amended as the Indian Trade and Merchandise Marks Act, 1958 (Replaced by new Act in 1999).
- 1967: The Patent Bill introduced in Parliament.
- 1970: The Indian Patents Act, 1970. The Design Act of 1911 retained without changes.
- 1972: The Patents Act (Act 39 of 1970) came into force on April 20, 1972.
- 1983: Amendments for availing the benefits arising from the revision of the Berne Convention and the Universal Copyright Convention to which India adheres.
- 1984: Amendments to discourage and prevent piracy prevailing in the video film and records.
- 1992: Amendment to increase protection time to author's lifetime and extended later to 60 years.
- 1992: Amendment proposed for a "New Act" in line with internationally accepted norms. Debate continuing.
- 1994: Amendment to give effect to the obligation arising from the General Agreement on Trade and Tariff (GATT). Copyright protection extended to new areas of creative work, including the computer industry. Special rights introduced to cover the performing arts.
- 1999: The Trademarks Act replaces the four-decade-old Trade and Merchandise Act. It seeks to make trademark offences cognisable, and incorporates provisions in tune with

developments around the world. It provides registration of trademarks for the services; registration of collective marks owned by the associations; setting up of an appellate board; appointment of a registrar of trademarks, who shall be known as the Controller General of patents, designs, and trademarks, to meet the requirement of the postulates of the World Trade Organizations and the emerging globalization of Indian economy.

- 1999: The Copyright (Amendment) Act amends the Copyright Act, 1957. It seeks to extend the protection of performers' rights from 25 to 50 years, and provide powers to the government to extend the provisions of the Act to broadcasts and performances made in India. This is intended to benefit Indian broadcasting organizations and performers to get reciprocal protection for their rights in other countries, which are signatories to the agreement on trade related aspects for intellectual property rights (TRIPS).
- 1999: The Geographical Indications of Goods (Registration and Protection) Act seeks to appoint the Controller-General of patents, designs, and trademarks to function as the registrar of geographical indications.
- **1999: First Amendment:** Under the exclusive marketing right (EMR) system, it is possible to make a patent to application for a substance that is intended for use or is capable of being used as a medicine or drug if certain conditions are fulfilled. India had a 10-year-transition period (starting from the end of 1994) before the application for product claims would be processed.
- **2002: Second Amendment:** This amendment to the 1970 Patent Act sought to bring India's patent laws in line with WTO obligations:
  - New definition of 'invention' defined as a new product or process involving an inventive step and capable of industrial application.
  - Method or process of testing during the process of manufacture made patentable.
  - Software patents excluded specifically.
  - Provided for patenting of microorganisms and microbiological processes, thus enabling patent protection for biotechnological inventions, both in the agricultural sector and pharmaceuticals produced through biotechnology.
- **2003: NDA Ordinance:** It puts in place the product patent regime:
  - Computer programmes with technical applications patentable.
  - Microorganisms remain patentable.
- **2004: UPA Ordinance:** The term 'new use' substituted for 'mere new use' in the list of what is not patentable, thus expanding the scope of what can be patented.
  - Computer programmes with technical applications patentable.
  - Microorganisms remained patentable.
- **2005: Third Patent** (Amendment) Bill 2005 to replace the December 2004 ordinance promulgated to meet the January 1, 2005, WTO deadlines to move to new patents regime.

SUMMARY

In a world governed by intellectual property rights and concentrated research investments, genetic resources are the raw material to which biotechnology is applied. In agriculture, these are not wild products of nature, but the fruit selection and development by farmers throughout the world since the Neolithic. In contrast to other areas of industry, this poses the immediate question of how to guarantee continued access to those who have been involved before: the farmers and the breeders. Governments are now completing the negotiation within FAO's commission on genetic resources for food and agriculture of a new binding convention, the international undertaking on plant genetic resources, which we expect to be approved by our conference in November.

The undertaking will create a multilateral system of facilitated access and benefit-sharing for the world's key crops. This is a major step, because dealing bilaterally with such widespread resources would involve unacceptably high transaction costs, and could impede agricultural progress. Multilateral access provides multilateral benefit-sharing, which includes the sharing of the benefits arising from the commercialization of materials from the multilateral system through a mandatory payment. The access of breeders to genetic material for further breeding, which becomes ever more difficult with GM crops under patents, is a public good that needs to be protected. FAO is involved in several instances, such as in the world intellectual property organization, where concerns of food and agriculture and IPRs are being discussed. While technology developers believe that these new products will offer benefits in meeting needs for food, fuel and fiber, as well as for novel industrial and pharmaceutical uses, some of these future products are also likely to raise environmental and other concerns that will need to be addressed by the regulatory system.

Historically, systems for the protection of intellectual property were applied principally to mechanical inventions of one kind or another, or to artistic creations. The assignment of IPRs to living things is of relatively recent origin in developed countries. Vegetatively propagated plants were first made patentable in the US only in 1930. And the protection of plant varieties (or plant breeder's rights—PBRs), a new form of intellectual property, only became widespread in the second half of the 20th century. Hence, an intellectual property rights of biotechnology products and processes results in better protection methods for research and development units. This also results in the encouragement of industries to allocate more labor, funds, setting of more and more R&D units, boost up the production, stimulate the innovative ideas and research, to gain an advantage over illegal firms and to attract lucrative investments from foreign countries.

EXERCISE

1. Define intellectual property rights.
2. Explain the importance of intellectual property rights in the present day biotechnological developments.
3. What do you mean by patents? Write down the criteria used for obtaining the patent from the government.
4. Give an account of trademark, and the need of trademark in the business of biotechnological products in the market.
5. Enumerate different intellectual property rights that are used for the protection of the inventions and ideas.
6. Briefly write about copyrights. What forms of intellectual property rights are covered under the copyrights issues.
7. Elucidate the applications of intellectual property rights and its need for the research and development.
8. Describe various types of intellectual property rights and its relevance to research communities in the biotechnology.
9. Write a note on international and regional agreement /treaties in intellectual property rights, which are significant to the biotechnology.
10. Mention how geographical indications are useful for the biotechnology business, and their importance in the biotechnology.
11. What is meant by trade secret and how is it important in biotechnology?
12. List out the intellectual property rights related legislature, which were passed in the India during recent past.
13. Distinguish between the use of trade secret and patents in the biotechnological research and development for the inventions.
14. How are intellectual property rights helpful for the researcher for the gain of his inventions in the competition oriented field?
15. Give a note on the ethical issues raised by the intellectual property rights in the field of biotechnology.
16. Name the important intellectual property rights in biotechnology that made the news in the recent past.
17. Briefly explain how intellectual property rights framed by the developed countries have an impact on the third world countries.
18. Critically comment on the intellectual property rights and their improvements in the Indian scenario in biotechnology research and developments.

<table><tr><td>

14
CHAPTER

</td><td>

Intellectual Property Rights and Agricultural Technology, and their Implications for India and other Developing Countries

</td></tr></table>

14.1. INTRODUCTION

Intellectual property, in the legal sense, is thought of as material that may be protected under the patent, trademark and/or copyright laws. Intellectual property (IP) is a class of property emanating primarily from the activities of the human intellect or intangible personal property. Any property, movable or immovable, is legally protected to prevent it from being stolen. The legal rights accrued on the intellectual property created are termed intellectual property rights (IPRs). Biotech industry claims that biotechnology will help to increase crop yields and, therefore, the availability of food to feed a growing population. They point to the projections of population growth, increase in food demand, and declining rates of growth in yields from Green Revolution technologies to support their claims. They claim that modification of crop plants can contribute to sustainability—a 'Doubly Green Revolution'—by reducing the amount of fertilizers, herbicides and pesticides used in farming, and improving the efficiency of water and fertilizer use. The revolutions in biotechnology and intellectual property protection began in the developed countries. The USA led the global transformation of intellectual property protection, and has been the leader in commercialization of biotechnology in agriculture.

At present, all members of the World Trade Organization are committed to offer intellectual property protections for agriculture. Will the benefits of agricultural biotechnology proliferate globally as a result? Can we now rely on the dynamism and focus of the private sector to exploit the potential of biotechnology to meet the needs of developing nations? In this chapter we look forward, drawing some inferences from the record thus far, to the future relevance of agricultural biotechnology for developing countries.

Critics point to the consolidation and vertical integration of the agriculture biotechnology (ag-biotech) industry, which they fear will entrench the dominance of powerful trans-national corporations and undermine the interests of poor farmers and consumers. They claim the technology will make poor farmers dependent on expensive external inputs, and pose a risk to developing world biodiversity. They blame the seductive allure of biotechnology and the interests of business for distracting attention and resources away from the promotion of appropriate technologies and sustainable agricultural practices. They point out that food insecurity is about 'entitlements' rather than 'availability'.

Worldwide, 30 million acres were planted with genetically modified (GM) crops in 1997. Almost 15% of the 1997 US soya harvest was grown from GM seed, and China is thought to be growing over four million acres of GM tobacco and tomatoes. In the US, twenty-three GM crop varieties have reached the stage, where strict regulations are no longer required for field testing. Until last year, most commercial transgenic crops were engineered for single gene traits, mostly herbicide tolerance and pest resistance. But in 1997, for the first time, crops were marketed with "stacked gene traits", i.e., more than one engineered trait in a single variety. In 2004, the whole 81 million hectares of land was covered with genetically modified (GM) crops. Thus, more and more IPR protected agricultural varieties are coming to the agricultural technology, and their use has become very important for the benefits of the people.

The large and overwhelmingly public agricultural research effort of India and other developing countries has, with a few notable exceptions, made relatively little progress in developing and commercializing agricultural biotechnology innovations. Some applications, such as virus elimination used *in vitro* propagation, marker-assisted breeding, and use of genetic cultivars identification to improve the efficiency of germplasm conservation, have been used for years in many countries. But here we shall concentrate on the technology that has captured the interest of farmers, breeders and private firms—the development of new cultivars with transgenic traits. As in North America, transgenic traits offering tolerance to broad-spectrum herbicides or resistance to insects involving *Bt* genes, have dominated agricultural biotechnology innovations adopted by farmers in other countries. Although contributions from China are becoming

increasingly significant in terms of variety of traits and the range of species of transgenic plants and animals under development, achievements in development of new transgenic plant varieties elsewhere have been very modest.

In the agricultural sector, the overwhelming majority of the value and area of transgenic cultivars adopted globally involve traits developed in North America. Indeed, most, if not all of the transgenic cultivars used widely by farmers in developing countries other than China appear to have been either developed in the USA, or bred from transgenic parents developed there (FAO, 2005; Pardey *et al.*, 2006). The ability to obtain, and if necessary adapt, transgenic cultivars developed in the USA has been crucial for the diffusion of this technology to farmers' fields that has occurred in the developing countries, where the largest adopters by area are Argentina, Brazil, China, Paraguay, India, and South Africa (James, 2004). Argentina, China and Brazil dominate biotech crop value outside North America, accounting for about one third of the world total of $44 billion in 2003–2004.

The three important questions in the present context are:

First, have intellectual property claims held by the developed countries been a major hindrance to biotechnological innovation in the developing world thus far?

Second, how are the stronger intellectual property regimes, currently being adopted by developing countries, likely to affect exploitation of the opportunities offered by agricultural biotechnology in the region?

Third, what types of institutional initiatives might improve the effectiveness of agricultural innovation for the developing world in the new global intellectual property environment?

The UN Food and Agriculture Organization (FAO) estimates that food output must increase by 60% over the next 25 years to keep up with demand. In a report on the bioengineering of crops, written for the World Bank and the Consultative Group on International Agricultural Research (CGIAR) in October last year, a group led by Henry Kendall, chair of the Washington DC-based Union of Concerned Scientists, said that transgenic crops could improve food yields by upto 25% in developing countries, and could help to feed an estimated additional three billion people over the next 30 years. To meet this demand, all the developing countries (third world countries) including India has kept up the relations with the developed countries, especially in the biotechnology sector.

Future Developments

Future developments in the agricultural biotechnology:

- Continued development of herbicide-tolerant, virus, and pest-resistant crops.
- Methods to speed up traditional plant breeding.
- Further development of fruit and vegetables in which the production of ethylene is suppressed, so that they take longer to ripen.
- Modification of oils, fats, and starches to improve processing or dietary characteristics.
- Improvement of flavor, texture, bioabsorbability, nutritional content, and elimination of genes for toxic substances and allergens.
- Identification of genes controlling salt tolerance, resistance to drought, flood and extreme temperatures, and response to day length.
- GM crops that fix nitrogen with greater efficiency, thereby reducing the need for fertilizers.
- GM plants that produce vaccines or therapeutic agents.
- GM biodegradable plastics grown in plants, such as oilseed rape, could begin to replace plastics from fossil fuels within a decade.
- GM plants for bioremediation, i.e., removing toxic chemicals and agrochemical residues from the soil.

14.2. INTELLECTUAL PROPERTY RIGHT PROBLEMS AND ITS HINDRANCE TO DIFFUSION OF AGRICULTURAL BIOTECHNOLOGY

Intellectual property rights have a great impact on the development, investment, and business in the agricultural technology worldwide. In the international agricultural research community, the belief has been widespread that patents have been hindering access of developing countries to important plant biotechnologies. For example, in a survey by Taylor and Cayford (2002), about two-thirds of 'stakeholders' with interests in plant breeding, intellectual property, and agricultural development reported that US patents adversely affect the ability of researchers to access and use specific gene traits, transformation tools, transformation marker systems, and genetically modified germplasm for developing countries. The survey included the information that US patents were valid only in that country. Many of the respondents were located in developing countries. It may be that some were confused on this issue, and perhaps others perceived that the USA and other donors use their funding of international aid as a means of de facto extension of patent rights. This affects the interest of third world countries.

Nevertheless, it affects the international research and their implementation for the poorer nations. Referencing a communication involving the leadership of an international agricultural research centre, Taylor and Cayford (2002) report the view that "National agricultural research systems and CGIAR institutions could jeopardize their funding if they systematically violated US patents to develop useful applications of biotechnology". This talk of 'violation' indicates that the reach of US patents rights in the non-profit sector can extend well beyond the geographic bounds of their legal, if not their political, reality, and certainly beyond the scope of protection recognized by well-informed private firms.

This has a greater impact on the technically difficult and socially important issues. Such confusion is encouraged by well-publicized news stories of donations of intellectual property rights for the technologies such as 'Golden Rice', and virus-resistant potatoes, sweet potatoes, and yams for use by poor farmers in developing countries patented in the Europe, the USA and other rich countries. The reports often imply that IPRs would otherwise constrain such research and innovation in those developing countries. Thus, the Nuffield Council on Bioethics (2004) commented that the recent example of Golden Rice shows that patented technologies need not necessarily be a barrier. In fact, few or no relevant IPRs existed in most of the developing countries among the top 15 importers or producers of rice (Kryder *et al.*, 2000; Nottenburg *et al.*, 2002).

In the present situation, almost all third world countries are not giving much importance to the IPRs issues. Innovators have generally been either unable or unwilling to file for patent protection in many developing countries. Moreover, the main staples for the poor in such countries are largely consumed domestically; the portion exported to rich countries, where imports are subject to relevant domestic patents, is typically small (Binenbaum *et al.*, 2003). Finally, modern biotechnology has been applied predominantly at the pre-commercialization stages of research. Patent holders typically have little or no incentive to constrain this type of activity. Prior to commercialization, little or no recoverable damages are generated. In fact, patent-holders are often happy to see their technology locked into an innovation, because the investment committed, in time as well as resources, improves the patentee's bargaining position is the innovation proceeds to commercialization.

It is clear that intellectual property claims have not greatly hindered international diffusion of agricultural biotechnology thus far. Some recent papers even suggest that the reverse is true. Kanwar

and Evenson (2003), and Lesser (2005) find a positive empirical relationship in the aggregate between intellectual property protection and foreign direct investment, respectively. Although it is difficult to be sure that such aggregate relationships are not unduly influenced by other confounding influences, it is possible that applications of biotechnology will increase in developing countries in response to the strengthening of IPR protection under the trade related aspects of intellectual property rights (TRIPS) agreement (TRIPS, 2002). On the other hand, as noted in the companion paper, Lanjouw (2005) finds that lengthening product patent protection can, in fact, delay the diffusion of patented pharmaceuticals. Whether a similar effect could occur in agriculture is an important unresolved question.

Agricultural Research

Agricultural research support: volumes, intensities and sources

From an economic development perspective, the pre-eminent concern with intellectual property rights is the degree to which their incentive effects impede or enhance investments in agricultural innovation over the long haul. With respect to global research effort, agricultural research expenditures are a special case. In developed countries, the private-sector share of total agricultural R&D is about 55%, comparable to other areas of rich-country R&D, where the private share typically (but not always) accounts for substantially more than half (often more than 60%) of the overall science spending. Moreover, the expenditure share of the developing countries now slightly exceeds the public or private shares of the rich-country group. Agricultural research is globally dispersed, more so than science spending generally, and public and private involvement are both important.

The public and private roles in agricultural science have been evolving, reflecting changing economic conditions in the broader economy as well as in agriculture, along with trends in the institutional environment, including intellectual property rights, and in public attitudes and perceptions. While many elements of the changes have been common among countries, reflecting common influences at work, there have been some important differences between the countries as well—especially between the richest and the poorest countries.

Over the past several decades, worldwide public investments in agricultural research increased by 51%, from an estimated 15.2 billion international dollars (year 2000 prices; Pardey *et al.*, 1992) in 1981 to $23 billion in 2000. Over this period, the developing-country proportion increased, so much so that during the 1990s, the developing countries, as a group undertook more of the world's public agricultural research than either the public or private sectors in the developed countries. Notably, only 6.5% of the estimated global total of $13.4 billion of private agricultural R&D took place in developing countries, and the lion's share of that private research was done in the Asia and Pacific region. Agricultural R&D in rich countries is dominated by just four countries, and likewise agricultural R&D in the poorer parts of the world is dominated by a small set of countries. Five countries—China, India, Brazil, Thailand, and South Africa—undertook 53.3% of the developing world's public agricultural research in 2000, up from 40% two decades earlier. Meanwhile, only 6.3% of the agricultural R&D worldwide was conducted in 80 other countries, which was home to some 625 million people in 2000.

Spending by low-income countries, dominated in terms of share of population and spending by India and China, grew rapidly during the 1990s, so that by 2000, these two countries alone accounted for 39% of the developing world's public agricultural R&D, significantly higher than their 22.9% share in 1980. Notably, spending on crop biotech research in China shot up dramatically, from an estimated total of just $17 million in 1986 to $112 million in 1999 (Huang *et al.*, 2002). In stark contrast, Sub-Saharan Africa has continued losing market share, falling from a 17.3% to 11.4% share of the global agricultural R&D between 1981 and 2000. In interpreting research expenditure shares, it warrants noting that in total value of agricultural production, the developing countries are dominant. For developing countries, public agricultural research was only about 0.53% of agricultural GDP in 2000, (and still lower in China), compared with about 2.36% in developed countries as a group. Expenditure per person engaged in agriculture was vastly higher ($11.92 per person in 2000) in developed than developing countries ($2.72 per person). By these 'research intensity' measures, research was a less important contributor to agricultural production in developing countries. There is an ample opportunity for conventional agricultural research to contribute more in these countries. A serious concern is that the research intensity is actually falling in Africa (Pardey *et al.*, 2006; Bientema and Stads, 2004)—a region in desperate need of more agricultural research, despite evidence of high returns on agricultural research projects (Oehmke and Crawford, 1996; Maredia *et al.*, 2000).

Furthermore, research intensities understate the scientific knowledge gaps between different regions. As Pardey and Beintema (2001) emphasised, the total cumulated agricultural research resource stock, at reasonable rates of interest and depreciation, is at least 12 times higher in proportion to value of agricultural output in the USA than in Africa (compared with a 3.4-fold difference in their respective research intensity ratios in 2000).

In sum, agricultural research throughout large parts of the developing world is hindered largely by inadequate expenditure intensity. Further, there is no near-term prospect that the private sector will fulfil the biotech innovation role, at least with respect to most staple food crops. Absent support from rich countries—either in kind as a part of publicly performed research, or in cash through public or private philanthropy—and effective public-private innovation partnerships are unlikely to materialise. (Doering, 2005). In acquiring applications of a new technology pioneered elsewhere, a natural initial role of the researhers of developing country is the adaptation of the technology for local conditions. As has long been recognised, this adaptive effort is crucial for the efficient dissemination of promising technologies globally (Alston, 2002). For specific well-targeted projects, it is feasible to transfer the application of biotechnology to a country with advanced innovative capacity. For example, yams for introduction in Africa have been transformed by Monsanto in the USA, and the investments necessary for application in Africa have been supported through the Agricultural Technology Foundation and the Danforth Foundation (Kent, 2004). But even with this unusual foreign support, commercialization of local adaptations of biotechnological innovations has been blocked or at least delayed by domestic biosafety regulations in Africa.

14.3. BIOSAFETY REGULATIONS

Genetically modified organisms including agricultural plants and animals are subject to the cartagena protocol on biosafety adopted in 2000 by parties to the convention on biodiversity, and to other domestic regulations in developed and developing countries. Even in the USA, the regulation of genetically modified plants involves insufficiently coordinated responsibilities of different public agencies, and has significant gaps in coverage (Pew Initiative on Food and Biotechnology, 2004). In developing nations, the regulatory structure is typically less well established, and the regulations in force might themselves be viewed, in particular cases, as institutional experiments evolving towards feasible implementation.

Compliance, if achievable at all, can be very costly. Redenbaugh and McHughen (2004) report these costs to be at least $1 million per allele for both horticultural and field crops, if approval is sought solely for the USA. Cohen (2005, p.30) reports estimates of compliance costs of $700,000, $4 million and $2.25 million for transgenic papaya and soybeans in Brazil, and rice in Costa Rica, respectively. He observes that paradoxically, because they are novel, locally developed products pose unique challenges for institutes seeking regulatory approval… In contrast, GM crops pre-approved in Western markets are more successful in gaining approvals in developing countries. Kent (2004), drawing on his experience with biotechnology transfer to Africa, notes that biosafety regulations risk stifling the emergence of locally adapted technologies by making it too expensive to take even the first step in moving them out of the lab and into field conditions. This problem affects public researchers even more seriously than the companies. Even if the developing countries do not share European concerns with the safety of transgenic foods, and adopt technology already tested and approved in North America, fears of trade disruption by importers citing risk of contamination are another deterrent to the approval of adoption of transgenics (Cohen and Paarlberg, 2004).

In the long run, this could hinder biotech adoption in export crops. As discussed below, the major food staples are in general likely to be largely unaffected by such concerns. Regulation of large firms may be easier to implement than regulation of a multitude of domestic farmers. Recent experience with soy in Brazil and cotton in India shows that regulations and laws that delay or prevent importation or domestic commercialization of transgenic germplasm by large, highly visible plant breeding firms can be ineffectual in preventing smaller breeders or farmers from introducing and adopting transgenic cultivars without any formal approval or regulation. The latter occurred on such a massive scale in both the Brazilian and the Indian cases that there appeared to be a loss of effective public regulatory control. Both examples illustrate that dissemination of transgenic technologies, if sufficiently attractive, can occur rapidly without intellectual property protection, and without sponsorship by public authorities or large firms. As familiarity with transgenics and their management increases in developing countries, it is possible that barring the emergence of new, adverse information, the domestic influence of groups with concerns about consumption of transgenics will subside. If it happens, formal regulations could be relaxed somewhat and administered more efficiently and predictably, at least with respect to staple food and fibre crops, which are not intended for export to countries with more entrenched opponents of genetic modification. In that case, under the most optimistic assumptions about research funding in the public sector, the rapidly changing IPR environment might have a prominent influence on development and commercialization of appropriate agricultural biotechnology. In this context, managing intellectual property rights and related policies will be important for developing countries.

Even under optimistic assumptions, biosafety regulations, intellectual property protection and some components of adaptive research will impose fixed minimum costs on the applications of biotechnology, such as genetic transformation. Public involvement in these innovations will not be justified in cases where the added social value of the innovative effort is below the fixed costs incurred. Absence of private interest is not per se justification for public investment. The larger the value of the crop, the more the value of a given improvement. Before the biotechnology revolution, breeders of crops with small markets relied on a narrow germplasm base, and made little use of gene banks to explore sources of genetic improvements. The costs of doing more exploration of genetic resources could not be justified by the prospective returns. On the other

hand, absence of private interest, even when institutions are well-developed and private investment is highly competitive, is not proof positive that research is not justified. In the USA, wheat breeding has not attracted sustained attention of major private crop breeders. Yet the social returns to wheat breeding accruing domestically in the USA, (including prominently domestic adaptation of innovations generated in developing countries) have been high in the USA on a sustained basis.

Freedom to Operate for Public Non-Profit Researchers

If ownership of the multiple rights necessary to develop new agricultural biotechnologies is diffuse and uncertain, the multilateral bargaining needed to access all of these rights at arms length, and bring a product to market, can become difficult if not impossible. In the private sector, this problem has been solved by bringing key technology 'in-house' within oligopolies *via* merger and acquisition. On the other hand, the facts regarding global agricultural research, reviewed above, indicate that development of agricultural biotechnologies for use by farmers will be largely a public enterprise (particularly for staple foods) over the next few decades at least, especially in developing economies. The same will continue to be true in the developed countries for all but the handful of the most lucrative crops, and for upstream research that is still largely performed by universities and other nonprofits with public funding.

For the public and non-profit sectors, merger with or acquisition by the private sector is not generally an option for obtaining freedom to operate, or to secure crucial complementary assets, though mixing of public and private forms of organization appears to be a feature of the Chinese environment currently (Fan *et al.*, 2006). Public-private partnerships and research consortia are other options that have been somewhat successful in some cases.

In one controversial case, a partnership between a life-science corporation, Novartis, and the Department of Plant and Molecular Biology in the College of Natural Resources at the University of California, Berkeley, provided substantial funding to the latter while offering no intellectual property rights to Novartis beyond rights to first negotiation of patents on a subset of the department's research output. But this relationship proved problematic for the university, and apparently unrewarding for the private partner, who opted not to continue the arrangement beyond the initial five years of the contract. It is unlikely to be used as a model for future university-industry collaboration.

But do non-profits and public-sector research organisations have offsetting advantages in obtaining freedom to operate in agricultural biotech? A priori, it would seem that they should. After all, the applied research interests in cultivar development of non-profits are largely focused on products of no great interest to the private-sector agricultural biotechnology firms, which continue to emphasise soybean, hybrid corn, cotton, canola and some specialty horticultural crops.

Furthermore, the more basic research advances, which are likely to be generated as complements to applied public research in this area (Stokes, 1997), can be very valuable to the private sector. Finally, the producers of transgenics currently have a formidable public image problem. Obviously, they wish to be seen as 'helping feed the world's poor' by offering key enabling technology to the relevant public/non-profit research agencies working on behalf of the world's poor people, whether that research is done in developed or developing countries.

Do these positive factors ensure that worries about the 'anticommons' are unfounded, and freedom to operate is not and will not become a significant problem for non-profit agricultural biotech research? Many well-informed observers appear, in casual conversation, to believe that there is, in fact, no problem worthy of serious attention

by policy makers. Evidence on this question from this relatively new, dynamic sector is not surprisingly scant and hardly definitive, especially at a time when public acceptance and regulation are overwhelming obstacles in the generation and diffusion of new cultivars produced using transgenic methods. The available sources relevant to the implications of IPR for freedom to operate in public/non-profit agricultural biotechnology include surveys, econometric analyses of citation behavior, and case studies. We briefly consider each of these.

In a survey by Zhen *et al.* (2006), which focused on researchers in plant biology and related fields at four public land grant institutions (UC Berkeley, UC Davis, UC Riverside and the University of Arizona), with 93 respondents, researchers indicate that when they send research tools to colleagues, whether at other universities or in industry, the interaction is more frequently informal than formal (Receipts of tools from industry are predominantly formal). Researchers rarely check the IPR status of the tools they use. On the other hand, more than one-third of the survey respondents had made invention disclosures, implying more than casual acquaintance with the patent system.

Most of the respondents act as if they had a bonafide research exemption. For them, concern with freedom to operate is focused on the lack of easy and quick access to materials held by others.

Over the past five years, more than one-third of the respondents reported delays in obtaining access to research tools, with a mean of delays, and mean duration of over eight months. More than one-quarter of the respondents reported one or more cases in which difficulties in obtaining research tools affected their research projects in ways other than causing delays. In one-third of these cases, alternative tools of equivalent effectiveness were used, but in about 40% of cases, researchers resorted to less effective tools.

More seriously, in about one-quarter of the problematic cases, a project or line of research that was a part of the project had to be abandoned due to the lack of access to research tools. There does not appear to be any strong evidence distinguishing academia from industry as the major source of these problems. The majorities of all respondents, and of the subset who have made invention disclosures, believe that IPR on research tools is, overall, having a negative effect on research in their area. The attitudes of the agricultural biotechnologists surveyed are largely consistent with those of scientists actively engaged in genomics, proteomics and related fields surveyed in a study by Walsh *et al.* (2005), commissioned by the National Academy of Sciences Committee on Intellectual Property Rights in Genomic and Protein Related Innovations, formed at the request of the National Institutes of Health. They are also in line with a key finding of an earlier survey (Walsh *et al.*, 2003). Public sector and non-profit scientists usually ignore intellectual property claims in accessing, making and using their research tools. Scientists continue to consider they to have a de facto research exemption. From their viewpoint, they encounter 'freedom to operate' as a problem when they cannot get their hands on a research tool, held by another, as quickly and easily as usual. In such cases, the problem arises not only from patent-related intellectual property claims, but also from more traditional academic concerns, with priority of publication and competition for support.

The reported perceptions of scientists regarding freedom to operate appear quite rational. First, the expected penalties for infringement are the financial damages suffered by the patentee from infringement. When the infringement relates to research usage rather than commercialization, these will typically be too small to cover the cost of litigation, even if, as is possible in the USA, there is a prospect of an award of triple damages. The use of patented processes to produce innovations marketed much later may be difficult to verify, and, if not embodied in the innovation itself, or revealed in publications or in the course of meeting regulatory requirements, may become

subject to a statute of limitations (six years in the USA). On the other hand, the cost of a full freedom-to-operate analysis on each and every technology and material used in the lab would be prohibitive for a public or non-profit researcher. The only feasible way to avoid any risk of infringing others' patents might well be to cease cutting-edge research. Patentees rationally wait until a program has succeeded in bringing an innovation close to commercialization before warning that the technology infringes. Freedom to operate problems materialises in the subset of projects that have survived to this point and show the greatest promise, and they are experienced as constraints on commercialization. Usually, it is not feasible for the researchers to substitute away from the infringing embodied technology at this late stage.

Comparison of the two surveys conducted in 2005 suggests that the effects on research of lack of access to needed technology have been more serious on average for biotechnologists working on agriculture than for those focused on human health. This might reflect the smaller set of promising technologies in agriculture and the lower level of resources available to help scientists surmount or invent around roadblocks.

Econometric Citation Studies

Until recently, many economists and lawyers viewed claims of the negative effects of IPR on research as purely 'anecdotal'. One reason for this is that each case of research problem has its own special features. The standardisation that facilitates quantification is a real challenge. The metrics that have been used are at best only indirectly related to the productivity of the research enterprise, which by its very nature is not standardised.

Two recent papers have used forward citations as an indicator of the effects of patenting on subsequent research in biotechnology. Murray and Stern (2005) considered the 169 peer-reviewed scientific papers published in

Nature Biotechnology between 1997 and 1999 that could be linked to a US patent. Sampat (2004) analysed the genomic patents assigned to the top 15 recipients of NIH funding over the period 1990–2000, and identified 499 patent-paper matches. Using different approaches, they found similar overall post-grant declines in citations (between 9% and 17%, and 14%, respectively). These papers are important, because they find the first quantifiable, econometrically estimated effect of patenting on a relatively standardised metric of the flow of research information. However, the causal links between patent issue, changes in citation behaviour and changes in research require further examination and explanation. Nevertheless, the citation studies do at a minimum indicate that prior patenting appears to affect a quantifiable aspect of the behavior of other scientists.

Case Studies of Public/Non-Profit Agricultural Innovation Blocked by Domestic IPR

As discussed above, in the agricultural research sector, public research institutions have responsibility to see research through to commercialization in all but the few lucrative markets that attract the bulk of private-sector attention. Public biotechnology research related to health, on the other hand, is (with prominent exceptions such as research complementary with clinical use of diagnostic tools) directed at providing upstream input to private-sector development. Negative effects of IPR on non-profit 'commercialization' of innovations have until now been most apparent in the agricultural sector and in non-profit institutions such as medical centres, where clinical researchers wish to use patented diagnostic tools to treat patients, for a fee, as a part of their research program. It is the effect of IPR on the mission of these integrated enterprises, rather than on the environment perceived by the bench scientist, which is the key issue for the prospects for biotechnology innovations for agriculture in

developing countries. For the public and non-profit agricultural research sector, any effects of IPRs on the overall research are difficult to measure objectively. Appeal might be made to the law in analysing samples of thousands of citations or hundreds of researcher responses. But thousands of citations might relate, in the end, to only a handful of commercialised products. Moreover, as noted above, surveys of scientists, even those located within an agricultural experiment station, reveal that they generally perceive the effects of IPR in relation to their own laboratory or area of investigation. This, no doubt, reflects the specialisation necessary for progress in modern research. Most scientists are not directly responsible for the development of a commercialised product that could be a target of an infringement suit.

The number of biotechnology innovations that have been developed anywhere near to the point of commercialised actions by public and non-profit agriculture is small. Further, had more been developed, political and social opposition to transgenic products in key export markets would very likely have prevented commercial adoption in many cases. Thus, a direct statistical analysis of the effects of IPR on agricultural biotechnology innovation through adoption on farmers' fields is impossible at this time.

In the USA and some other developed countries, there is some evidence that university research projects designed to produce new crops with modern biotechnology have been shut down because of the refusal of IPR-holders to permit commercialization of varieties incorporating their intellectual property. In one instructive example, University of California researchers, with financial support from producers, successfully created a tomato variety genetically engineered to express the University's endoglucanase gene to retard softening and improve shelf-life characteristics (Wright, 1998). However, the promoter they used was one for which a patent application surfaced during the development of the new cultivar. The patentee, a life science corporation, refused to negotiate terms for the use of its embodied technology for the commercialization of cultivar. The research and development effort came to naught, shattering the confidence of the producers, who helped finance the project, in the capacity of the university to successfully breed and commercialize new transgenic cultivars.

In another example, development of a fungus-resistant strawberry at the University of California was blocked by lack of access to the necessary *Agrobacterium* transformation technologies (National Research Council, 1997). Funding of the research by the Strawberry Commission was discontinued (National Research Council 1997). In another case, a dead end was encountered when transgenic barley with good herbicide tolerance was developed during a research project. The owner of the relevant herbicide tolerance patent refused to negotiate commercialization rights, and indeed refused to discuss developing the germplasm itself (Wright, 1998). Reports indicate that similar impediments to commercialization, in the form of refusal of freedom to operate, have been encountered in the development of herbicide tolerant turf grass at the University of Michigan (Erbisch, 2000) and of a herbicide tolerant lupin in Australia (Lindner, 1999). The point of these examples is not that they would all have been commercially successful, given the freedom to operate, but that freedom to operate was in these cases a serious barrier to the system of non-profit innovation that had responsibility for development to the point where they were made available to farmers in the field.

Why do these roadblocks occur? Economists expect that when there are gains to be made from a trade, the trade will occur. If the parties failed to find a mutually satisfactory solution in the above examples, by an economic tautology, the 'transaction costs' must have been too high. Perhaps the public-sector negotiators had unrealistic expectations regarding private sector largesse. The owner of the key IPR might have been concerned with protecting himself from liability or damage to his reputation due to misuse beyond his control. In some cases, the expected financial gains, given the size of the market, might have been less than the cost in time and money to the IPR owner (public or private) of making and enforcing an agreement. Or perhaps, the patent holder saw no reason to help out a potential competitor, for little financial return, in a market that could one day be of financial interest to the patentee.

But in more recent conversation, scientists who have experience with startups and public-sector development in this area, strongly support the view that IPR thickets, along with the testing costs to meet registration and regulatory requirements, are two serious impediments to development of transgenic cultivars for agriculture and horticulture in the USA, for all but the most lucrative markets. Like-minded research managers and their potential financial supporters will avoid the lines of research that could furnish new examples to add to the list—a serious reaction that would likely be even more difficult to quantify.

At present, there is more optimism with respect to genetic engineering for health-related markets. But here too, there have been problems related to the freedom to operate. At the University of California, Berkeley, researchers in 2003 developed transgenic wheat lines that were shown to be less allergenic in an animal model. Patents to some of the methods and materials used to develop the new transgenic lines of the California wheat variety, *Yecora Rojo*, were licensed to a firm that did not pursue the development of hypoallergenic wheat. Sublicensing of rights back to the university, to allow product development of the hypoallergenic wheat to proceed, was delayed, hampering negotiations for further development with other firms that were potentially interested in this opportunity at that time. Since then, the investment climate has changed, and market and consumer concerns have become the dominant obstacles to further commercial development of hypoallergenic wheat.

Definitive evidence of the effects of IPR on agricultural research will not be available soon, if ever. But the evidence from surveys, citations, and case studies constitutes, in aggregate, a strong prima facie case for a significant blocking effect of intellectual property claims in public/non-profit agricultural research that yields commercially attractive results. One commentator has complained that there is no evidence of such a blocking effect on the commercialization of an innovation in a staple in a developing country. This is not surprising, as strong patent protection of agricultural biotechnology innovations is just now being implemented in most developing countries. However, we already have cases in which US patents, later invalidated, have been used to hold up the commercialization of products from developing countries.

Where developing country producers grow crops for export to the developed-country markets, a US patent of dubious validity can be, and has been, used to disrupt exporters to the USA. For example, a Colorado firm patented a yellow bean, which it named the 'Enola bean', bred from beans purchased in Mexico. The firm had reproduced and selected the yellow beans over several self-pollinated generations. After receiving the patent, the firm proceeded to demand licenses from importers of similar Mexican beans. This patent was challenged in the USA, and after several years was surrendered in 1995 for reissue/re-examination. The patent's long-run status is unclear, although plant variety protection still covers the Enola bean (Nottenburg, 2005).

In another case, an improperly obtained patent on a new and superior variety of pineapple was referenced in a letter from the Vice President of Del Monte Fresh Produce to a Central American researcher. The writer warned him against working on pineapple plant material developed by Del Monte Fresh Produce, which, he added, owned a US Plant Patent.18 Similar warnings allegedly were sent to potential US competitors, including Dole. It later became clear that the variety in question was not patented, and indeed had been refused a US patent in 1992. The patented variety was another improved variety—a hybrid that was a sibling to the Del Monte Gold variety. That patent was later withdrawn by Del Monte Fresh Produce, which acknowledged pre-patent sales of the patented variety by a competitor.

These examples show that, even before TRIPS has had its full effect, confused perceptions of the geographic scope of patents, combined with the confusing use of US patents of dubious validity, have had a plausibly discouraging effect on agricultural research and production in the developing world. After TRIPS has been fully implemented, the potential geographic scope of protection will cover all members of the WTO, and thus could be almost global if applications are submitted in all relevant countries. Furthermore, restrictive actions or threats thereof, involving dubious patents could be made much more effective by the changes mandated by TRIPS. A dubious US patent could well be used to persuade local patent authorities in the country of the potential competitor to approve a local patent application. The cost of challenging the validity of such patents is likely to be such a heavy burden on less-developed country researchers and producers that competition in major markets might be eliminated, or delayed for years.

These examples are instructive regarding the types of potential threats associated with intellectual property claims that can confront developing-country producers and innovators. They are obviously too few to establish the seriousness of such threats. In any event, it is important to stress that, thus far, the overwhelming reason for the lack of adequate agricultural research in developing economies, and in particular in sub-Saharan Africa, has been the lack of sustained, adequate flows of investment in research in this area, noted above. For public and non-profit researchers everywhere, the lack of capacity to make the heavy investment in the testing required for regulatory approval is another major limitation on agricultural biotechnology innovation involving plant transformation. For the future, the World Intellectual Property Organization is exploring the implications for standardizing and centralizing certain aspects of patenting worldwide, consistent with at least European levels of protection, via the proposed Substantive Patent Law Treaty. The treaty could not only offer very important economies in administering patents worldwide, with evident advantages for small and less developed countries, but it could also result in a further strengthening of global IPR, which might not be in the interests of such countries.

The potential influence of patents (and IPR more generally) on agricultural research in developing countries is changing rapidly, to the degree that countries are achieving effective implementation of TRIPS and subsequent agreements. Public and non-profit researchers in these poorer parts of the world, who will continue to be responsible for the bulk of agricultural research not directed at the most lucrative cash crops, are now increasingly confronting the challenges of obtaining the necessary freedom to operate for bringing new crops, processes and products to market that have already been faced by their counterparts in the developed countries. Presently, they are very poorly equipped, in terms of funding, information and expertise, to meet this challenge.

Institutional Initiatives to Encourage Agricultural Biotechnology Innovations

For policy makers, the challenges are to get the science done, while avoiding IPR problems that might materialise when commercialising or distributing technology to the farmers. Several institutional innovations that have been designed to address these challenges deserve attention. The Public Intellectual Property Resource for Agriculture (PIPRA) initiative has several aims (Atkinson *et al.*, 2003). One is to construct data-bases with complementary informational services to facilitate, for research planners, preliminary scoping of the freedom to operate problems along a research trajectory. Another is to educate public sector institutions on best-practice licensing that reserves rights for humanitarian purposes. PIPRA also supports the development of key enabling technologies as alternatives to tightly-controlled proprietary technologies, offering freedom to operate for public and non-profit researchers. It also seeks to overcome public-sector fragmentation, encouraging collaboration in bundling technologies and reducing costs of licensing.

The Biological Information for Open Society (BIOS) initiative (Nature, 2004) is a bold attempt, extending achievements originating with CAMBIA in Australia, fostering collaborative open-source development of sets of key enabling technologies for agricultural biotechnology. It intends to develop licensing strategies inspired by the open source movement in software. It also offers a database with information from over 70 patent offices, which is used in supporting IP land-scape analyses and other informational services for researchers in agricultural biology as well as other health-related fields.

Another relevant initiative is the African Agricultural Technology Foundation. It is a public/private attempt, supported by the Rockefeller Foundation, to facilitate negotiated, case-by case royalty-free licensing of proprietary technologies owned by firms, to be developed by African institutions for small farmers in a way that reduces risks and transaction costs faced by the donors.

The Specialty Crops Regulatory Initiative launched in November 2004 is a collaborative effort to establish an organisation to facilitate and reduce the cost of the regulatory approval of biotechnology-derived specialty crops. This initiative seeks to play a role in this area similar to that of the IR-4 Project of the USDA to facilitate approval of pesticides for small crops, and the Orphan Drug Act to encourage the development of new drugs for diseases with small markets. Economists have developed other proposals that merit serious attention. Masters (2003) promoted a system of prizes to encourage innovations that offer social value to rural Africans, to reduce the gap between perceived private returns and social value (the 'consumer surplus'), and provide more incentive for profit-oriented initiatives in the public interest. Kremer and Zwane (2005) proposed a related program, with more specific goals. Lanjouw (2003) advocated a change in domestic patent law that encourages holders of key patents that have their largest potential market in developed countries to grant effective freedom to operate in a list of less developed countries, provided that the production is not exported back to other markets.

Finally, developing countries should be aware of the need for access to expertise in the law, economics and administration of intellectual property protection, in international negotiations and in formulation of domestic policies. Some informed observers believe, for example, that Australian negotiators, expert on trade questions, were outmatched on key intellectual property issues in the recent bilateral negotiations with the USA. Expertise is available on the world market, at a price. Developing countries need to make the educational investments required to establish the domestic capacity to identify the appropriate international sources of expertise, and to use their advice effectively. This is a significant challenge.

14.4. ETHICAL IMPLICATIONS

Until recently, agricultural research in developing countries, which is predominantly public, has been less constrained by valid intellectual property rights than many international researchers and other observers appear to have believed. As intellectual property rights proliferate in the most important export and consumer markets, the problems of freedom to operate in developing economies will become similar to those encountered by the public

and non-profit researchers who account for almost half of agricultural research effort in developed economies. Experience thus far in countries that are leaders in applications of biotechnology offers important insights for the dozen or so developing countries with substantial near-term potential for the development and application of agricultural biotechnology. First, scientists in the public laboratories have been relatively unconstrained by most intellectual property claims not embodied in materials, because they ignore them. For them, freedom to operate tends to mean ability to get the tools needed to do the project. Conversely, lack of freedom to operate is perceived as lack of timely access to materials physically controlled by others. Material transfers are the intellectual property claims that dominate their attention.

For research policy makers, the situation is quite different. The private firms who tend to acquire most claims on key technologies originating in the public and private sectors are focused on a handful of crops, but their technologies are much more broadly applicable. Although they cannot, and often have no incentive to, control infringers in the lab, they can prevent the commercialization of new cultivars and other innovations achieved using their proprietary technology. This is the major freedom to operate problem for the stronger public agricultural researcher programs (as distinct from individual scientists in these programs) in rich and poor countries alike. It can become evident only in that tiny but important subset of lines of research that are successful enough to have been commercialized. In the survey by Cohen 2005, only seven cases, of 201, had reached the commercialization stage.

In the years of time and effort that must be invested in a successful innovation, it is likely that one or more patented technologies will become embodied in the product by scientists oblivious to IPR claims. When this has happened in the course of public-sector transgenic innovations for use in a developing country, failure of the patentee(s) to consent to negotiate seems to have been a more prevalent roadblock than demands for excessive royalties. Concerns with control, reputation and the cost of negotiation seem to outweigh the modest licensing revenues obtainable from crops of no direct interest to the private sector. Even a single 'train wreck' in public sector development can mean frustration, withdrawal of funding sources, and indeed the failure of public research programs.

Nations with strong agricultural research capacity, like China, Brazil, India and Argentina, have shown that they need little or no overseas investment to support rapid diffusion of the most profitable transgenic technologies originating in the USA and elsewhere. Strong domestic intellectual property protection is probably not in their interests, absent the threat of trade sanctions. This might change, if they succeed in developing and patenting new transgenic rice technology, for example, that is attractive to other countries. It may be that they will demonstrate a comparative advantage in transgenic cultivar development for some crops in the medium term. In any event, these countries should give serious attention to initiatives such as PIPRA and BIOS, which also serve the needs of public agricultural research in the rich countries. Beyond the effects on the choice of programs and their successful completion, the new IPR regime is also changing the perceived mission and funding structure of the stronger public research institutions in developing countries. For example, Koo *et al.* (2006) report that EMBRAPA, the large Brazilian federal agricultural research organisation, has negotiated contracts with private seed producers that restrict the latter from pursuing further research on the germplasm they receive. This is in sharp contrast with the traditional notion of the role of public research institutions in fostering private applied research based on their own innovations.

Finally, it is important to recognise that the dominant constraint on agricultural innovation for the majority of developing countries is not at present IPR. It is a lack of sufficient and sustained funding in the face of well-documented high social returns on research using non-transgenic technologies—a problem also observed in developed economies with strong IPR, for all but a handful of crops. Especially in poor countries, another serious constraint at present is strict and costly biosafety regulation. For many, if not most, developing countries, these problems will not be solved in the near future. For such countries, applications of biotechnology are likely to be concentrated in non-food cash crops, and in particular in corn, soybean, and cotton, often using cultivars developed from transgenic events produced in the rich countries. Where substantial adaptive research or market development investment is required, strong intellectual property rights might be necessary to encourage rapid development and diffusion of imported technology by private seed producers. Where local public sector biotech research capacity continues to be weak, local development of biotech for staple food crops is unlikely to occur, regardless of intellectual property protection.

Of 15 GMOs reported as released varieties in FAO (2005), all but one of the non-Chinese varieties appear to be imported, if one includes Bulgarian maize and Indonesian cotton, for which the origin of the transgenic variety is not reported. Pardey *et al.* (2004) report that all the officially approved Monsanto/DeltaPine bioengineered cotton varieties grown in China are the same varieties grown in the USA. Likewise, the transgenic cotton varieties grown in Mexico are from the USA, while in South Africa, *NuCotn 37-B*, a US variety is widely used. The donations more typically take the form of grants of physical access to constructs and other crop improvement materials and expertise rather than intellectual property per se. All these data involved conversions from local currency units to US dollar equivalents using purchasing power parities to account for cross-country price differentials, rather than using market exchange rates. These agricultural R&D spending estimates are from Pardey *et al.* (2005). The USA, Japan, France, and Germany accounted for two-thirds of the public research done by rich countries in 2000—about the same share as two decades earlier. As Doering (2005) puts it: "Despite much talk, public-private partnerships in which both sectors make contributions and have a stake in the outcomes are rare in crop development for developing countries." He goes on to illustrate the unrealistic conditions often imposed by international donors on potential partnerships: "Project structure must guarantee the benefits to the poorest farmers and guarantee that benefits to the private sector also create public goods such as rural development, improved agricultural productivity, and seed markets". Assessing the formal impact evaluation literature, Alston (2002) contends that up to half the local productivity gains in agriculture over the past several decades are attributable to the effects of spill-in technologies developed elsewhere.

Recently, Zambia, during a famine in part of the country, refused food aid in the form of maize from the USA, on the grounds that it contained genetically modified seed. It was reported that the government was concerned with the response of the European Union to Zambian exports, in case they become contaminated with transgenic seeds. Some informal reports imply that domestic inter-regional politics influenced policy in this case. This phenomenon is characteristic of the power of farmers to trump public regulation withholding innovations when the apparent returns are sufficiently attractive. A similar loss of control occurred on two separate occasions with respect to new technologies for biological control of rabbits in Australia. DuPont Corporation negotiated much stronger claims in 1981, when Dr. Philip Leder was recruited from the corporation by Harvard University Medical School. In return for $6 million to support Leder's research, Harvard gave DuPont exclusive license to any patents that might result from that research. This arrangement led to the development of the 'OncoMouse',

in 1988—the subject of the first patent on a living animal. DuPont's license terms for use of the oncomouse as a research tool subsequently generated much controversy.

About 400 surveys were distributed. The response rate reflects the fact that at each institution the entire department was included in the sample. No attempt was made to pre-select those scientists, who were engaged in applications of biotechnology to agriculture. Follow-up interviews made it clear that reports that a project was affected referred to one line of investigation in a larger research enterprise, rather than an entire independently funded project. The exception is diagnostic tools that are used by universities in revenue-generating treatment of patients, which jointly contributes to ongoing research. Concerns with commercial application have led patentees to vigorously pursue infringers including clinicians involved in research, generating a great deal of controversy. The 14% refers to the effects found by Sampat of patents on gene sequences. Non-sequence patents ('techniques') showed no such effect. Since techniques, once published, can usually be copied, the latter result is consistent with bench scientists' perception that access means freedom to operate. Such a chilling effect has been claimed with respect to commodity boards interested in research on minor crops at the University of California (National Research Council, 1997).

The letter included the following, as reported in the Consolidate Class Action Complaint (RMB), US District Court, Southern District of New York, In Re Fresh Del Monte Pineapples Antitrust Litigation, 4th August 2004, at Fresh Del Monte Produce Company is aware that company has acquired pineapple plant material and is researching the growth and production of pineapple plants. Del Monte has also learned of an organised effort to steal this planting material from the Del Monte plantation for propagation. Be advised that Del Monte is the developer of this plant material and intends to protect its interests as necessary. In addition, be advised that Del Monte owns US Patent No. Plant 8,863, dated 16th August 1994. Many more reports of problems with intellectual property claims by the North are listed on websites of non-governmental organisations, but these tend to be unreliable, or difficult to verify. The commercial seed market has already undergone a rapid increase in concentration via acquisitions by multinational corporations (Commission on Intellectual Property Rights, 2002) in Brazil, where Monsanto received permission (currently under legal challenge) to collect royalties on soybeans via a point-of-sale payment, and in Mexico.

Benefits for the Poor?

Some independent voices argue that agricultural biotechnology, including transgenics, may benefit poor people in the South. Transgenic crops may require fewer expensive inputs, boost incomes and provide better livelihoods for people dependent on agriculture. Production efficiencies and bigger yields may bring prices down for poor urban consumers also. However, these outcomes will require that farmers are not made dependent on the seed companies or committed to use proprietary agrochemicals as a part of a licensing agreement. It will also require public sector research, or market incentives for private sector research, into so-called 'orphan crops' (Nuffield 1999).

There is concern from all sides of the biotech debate that crops of importance to the South are not being developed. The biotech firms admit they are seeking a profit, and focus on crops and technologies designed for the industrialised farmers and sophisticated markets of the North. Staple crops and the production constraints of Southern farmers are neglected. The focus on GM technology (by both biotechnologists and critics) distracts attention and diverts resources from technologies, which are more appropriate to developing country farming. There are calls for public investment in 'public good' R&D, but precious few initiatives that seem likely to bring a boost in funding.

How is the Agri-Food Sector Changing?

- Consolidation of the biotechnology industry, reflected in a wave of recent mergers, acquisitions, and strategic alliances.
- Vertical integration of the agri-food production chain from the laboratory through the seed companies and farms to food processors, with the potential in future to encompass even retailers. This is reflected in:
- The acquisition of seed companies by biotech firms.
- Enhancement of agbiotech R&D capacity by seed firms and biotech companies.
- The expansion of contract farming, technology fees and product licensing.
- Strategic alliances between biotech/seed companies and food processors.

Why is this Happening?

Driven by the high costs of biotechnology research and product development; the rapid pace of scientific discovery; financial pressure from shareholders and venture capitalists for a return on their investment; and competitive pressures from other firms, corporations need to control or gain access to a broad patent portfolio; gain access to scientific and product-development expertise and markets; and achieve economies of scale.

Toenniessen of the Rockefeller Foundation has proposed that public-sector intellectual property should be 'pooled' in a commercially-managed fund mandated to serve the needs of the farmers of developing country (Toenniessen 2001). For example, the recent merger between Novartis Agribusiness and AstraZeneca Agrochemicals to form Syngenta.

For example, Monsanto's purchase of Brazilian seed company Agroceres in November 1997 and the Cargill international seed business in June 1998, including both in-house facilities and contract research at universities. (Kloppenburg 1988; BASF 1998, 1999). For example the alliance between the chemicals and biotechnology business DuPont and food processor General Mills, 'to develop and market soy foods' (DuPont: Holliday: 1999). Meanwhile, the public international agricultural research system faces intense financial pressure and is encouraged to harness the commercial value of its germplasm collections and R&D capacity.

Regulation of Modern Agricultural Biotechnology

At the international level, a variety of institutions and agreements are implicated in the governance of biotechnology. These include the Consultative Group on International Agricultural Research (CGIAR); the Biosafety Working Group; the WTO's Agreement on Trade-Related Intellectual Property Rights (TRIPs); Agreement on Agriculture; the Convention on Biodiversity (CBD) and its Biosafety Protocol (BSP); and the FAO's International Undertaking on Plant Genetic Resources for Food and Agriculture. Little coordination exists between the various mechanisms and regimes. Two key regulatory issues are **biosafety** and **intellectual property rights (IPRs)**.

Biosafety

The key agreement in relation to agricultural trade is the BSP. At the core of the protocol is a mechanism for prior notification and consent for transborder movements of genetically modified living organisms, known as the Advance Informed Agreement procedure (AIA). It incorporates the so-called 'precautionary principle' and requires a scientifically sound risk assessment to be carried out in advance of consent. Implementation of the protocol in poor countries will probably require a significant investment in capacity building, including training of personnel to carry out risk

assessments, and establishing a national office to coordinate the AIA procedure. The protocol contains provisions on capacity building for developing countries.

Intellectual property rights

Plant variety protection (PVP) allows farmers to save seeds for their own use and permits plant breeders to use the protected variety as a basis for developing a new cultivar without having to obtain the consent of the owner of the original protected variety. PVP is governed at the international level by the International Union for the Protection of New Varieties of Plants (UPOV) Convention of 1978/1991. **Patents** grant the holder the right for a limited period to prevent others from reaping the benefit from an invention, in exchange for the publication of the details of the new technology. Farmers do not have the right to save seed from a patented variety for use in the following season. Patents pose certain restrictions on the ability of researchers to carry out further development on a patented technology. Patents are governed at the international level by the World Intellectual Property Organisation (WIPO) and the TRIPs Agreement. Poor countries have until 2002 to implement TRIPs. Again, this process will require investment in capacity-building.

Capacity Building

Some observers are concerned that the obligation to implement IPR and biosafety regimes, imposed by and designed for the North, is diverting poor-country resources away from other national priorities. Currently, many developing countries have no biotechnology to regulate. Nevertheless, where biotechnology is likely to be salient in the future, timely initiatives to build capacity seem likely to pay off. Capacity building is currently being undertaken by organisations such as the International Service for the Acquisition of Agri-biotech Applications (ISAAA); the Rockefeller Foundation and USAID's Agricultural

Biotechnology Support Program (ABSP). Corporations also protect their IP using contractual arrangements (licensing) and Genetic Use Restriction Technologies (GURTs), which have been labelled 'terminator' and 'traitor' by critics.

- **Terminator:** GE crops that produce sterile seeds which cannot be saved by farmers to plant the following year.
- **Traitor:** GE crops designed to grow or to express specific desired traits only in conjunction with external chemical inputs. Following public outcry, several of the big biotech transnationals have pledged not to commercialise GURTs. However, R&D into GURTs is still being undertaken and patents are being granted (Warwick 2000).

Impact of IPRs

Those in favour of IPRs, notably the large biotech companies, argue that systems of intellectual property rights (IPRs), particularly patents, enable innovators to reap a just reward for their investment in R&D. They argue that IPRs are necessary to encourage innovation. They point out that only the new innovation is patented, and that the existing knowledge and plant material, which were used as a basis for the innovation, remain in the public domain.

Critics retort that IPRs, particularly patents, restrict the flow of knowledge and information and therefore inhibit R&D by others. Patents have also been granted for quite trivial inventions and developments which fall short of the legal test for patentability (Barton 2000). Patents are a key mechanism for the commodification of agriculture for the benefit of northern transnational corporations and to the detriment of poor Southern farmers. The benefits of the invention are not distributed equitably because the providers of the landraces and traditional knowledge used as a foundation for R&D are not rewarded; and the price of proprietary biotechnologies puts them beyond the reach of poor farmers.

Biopiracy *vs* Bioprospecting—Indigenous Knowledge and Benefit-Sharing

In the world, richest sources of commercially useful germplasm are in the South or in the third world countries. Traditionally, this material has been publicly available and free of charge. During the Green Revolution, poor farmers provided plant material to the IARCs, which returned improved varieties as public goods to the South. Today, germplasm is still taken free from the South or from public collections held by the CGIAR, by private corporations. However, modified varieties, which are then developed by the firms are marketed commercially for profit. Critics argue that private firms are therefore profiting from 'indigenous knowledge' and farmers' innovations without compensating them ('biopiracy'). Issues of benefit-sharing need to be addressed, but these face the challenge of defining the scope of collective ownership, valuing indigenous knowledge, and identifying the appropriate recipients of proceeds from innovation.

In a report prepared by RAFI for the United Nations Development Program in 1994, it is estimated that the biopiracy of plant genetic resources and knowledge costs developing countries $4.5 billion per annum. Most bioprospecting to date has been carried out for the pharmaceutical industry, but in the future it is expected to play a greater role in research for GM crop plants. The following are some examples to date:

- In April 1997, two Australian government agencies applied for patents on two strains of chickpea, samples of which they had borrowed from the International Crops Research Institute for the Semi-Arid Tropics (ICRISAT) at Hyderabad. ICRISAT subsequently told the Australian agencies that it would be wrong to patent plants that should be owned universally, and on 13 January, 1998, they withdrew their applications.

- W.G. Grace & Co. was granted a series of patents on extracts from the neem tree, whose seeds and bark have been used for centuries in India as natural pesticides. Grace has estimated that the global market for the pesticide could reach US $50 million a year by 2000, and while there are no hard statistics on the impact on Indian farmers to date, but farmers in the south of India, where neem is harvested, are already losing out because the processing and exporting of the seeds and oil is no longer available to them.

- In 1995, two Mississippi doctors were granted a patent for the traditional use of turmeric as a healing powder. India's Council of Scientific and Industrial Research (CSIR) petitioned the US Patent and Trademark Office (PTO) on the grounds that the 'discovery' was not original, but had been chronicled in traditional Indian texts. In August, last year, the PTO rejected the patent holders' claims.

14.5. SOCIOECONOMIC IMPACT

At a meeting of the UK Food Group in April,… Indira Jaising of the Supreme Court of India noted a trend in TNCs (transnational companies)—pharmaceuticals moving from the developing world to the US, with the contingent drain of capital, because the US allows patenting of life forms. Whether or not the rest of the world falls in line with the US in accepting life patents, researchers predict that with advances in biotechnology there will be a switch in centers of production away from the developing world, accompanied by loss of export income. For example:

- Rapeseed engineered to produce lauric acid, which is used in soap and cosmetics, is already being grown commercially, and the lauric acid is sold to Procter and Gamble—one of the world's largest buyers of the substance. Lauric acid is traditionally derived from coconut and palm kernel oils produced in the tropics. The Philippines, the world's largest exporter of coconut oil, accounts for around two-thirds of global exports. The coconut industry also provides employment for about 30% of the countly population.

- Farmers in the US are expected to plant twice as much GM soybean in 1998 as in 1997, and with resistance to GM soya in Europe, there are concerns that it will be dumped in countries like India (which has a good harvest of its own).

- In April, the Africa News Service reported that tomato growers in Kenya will soon have access to the longer-ripening FlavrSavr tomato, provided it meets Kenya's safety measures and guidelines on biotechnology transfer.

Property Rights—Control of Knowledge

There are various issues to be considered here, but the central, over-arching debate (or lack of debate) concerns ownership of resources and how to reconcile the rigid, individualistic patenting system of the developed world with the community-held knowledge systems of poorer countries.

Food Security

If foreign researchers and TNCs can patent indigenous crop plants without making recompense to the communities who provided them, there are fears that farmers will end up paying royalties on the products of their own knowledge, products on which they rely for survival. The following are examples of such patents:

- In 1994, two researchers from the University of Colorado received US patent number 5, 304, 718 on male sterile plants of the traditional Bolivian 'Apelawa' quinoa variety. They claim they were the first to identify and use a reliable system of cytoplasmic male sterility in quinoa for the production of hybrids, although Andean farmers have long known that the male flower of the Apelawa variety is sterile. Quinoa is a high protein cereal and an important dietary component in Andean countries. The value of Bolivia's export market is estimated at about US $1 million per annum. The US patent claim covers any quinoa hybrid that is derived from 'Apelawa' male sterile cytoplasm, including, but not limited to, some 36 traditional varieties cited in the patent application.

- In September 1997, the US company Ricetec, Inc., was granted a patent on Basmati rice. The patent is for a variety achieved by the crossing of Indian Basmati with semi-dwarf varieties, and it covers Basmati grown anywhere in the Western hemisphere. Ricetec can also put its brand on any breeding crosses involving 22 farmer-bred Basmati varieties from Pakistan and, according to Rural Advancement Foundation International (RAFI), on any blending of Pakistani or Indian Basmati strains with the company's other proprietary seeds. Ricetec also claims the right to use the Basmati name. The Indian government has challenged Ricetec's claim, arguing that the patent jeopardizes India's annual Basmati export market of around US $277 million, and threatens the livelihood of thousands of Punjabi farmers.

The ability of farmers to save seed is seen as crucial to food security. According to RAFI, up to 1.4 billion poor farmers in the developing world depend on saved seed and seeds exchanged with farm neighbours, and up to 50% of soybean in the developing world is planted with farmer-saved seed.

- TNCs such as Monsanto require farmers who buy their GM seeds to sign contracts agreeing not to save seed. In March 1998 RAFI reported that Monsanto had taken legal action against more than 100 soybean growers in the US and had hired Pinkerton investigators (hired police) to identify those saving seeds.

- In 1997, Farmers' Weekly reported that the Biotechnology Working Party of the National Farmers' Union (NFU) in the UK was seeking legal advice on the status of contracts between farmers and agrochemical firms under UK law. British farmers were reported to be unhappy with the US-style technology fees, licenses and contracts for GM crops.

In 1998 the US Department of Agriculture and the Mississippi-based Delta and Pine Land seed company were granted a patent on the so-called "terminator technology", which involves engineering seeds so that they do not germinate if planted for a second time.

Conservation—Sustainable Agriculture

International genebanks for seed samples were originally set up in response to concerns about the loss of genetic diversity. Essentially they are vast refrigerators where seed samples are stored under controlled humidity and temperature conditions. But apart from issues of biopiracy there are practical problems with these genebanks:

- Many of the seeds lose viability because they are not grown out regularly.
- They are poorly catalogued and costly to run.
- Their material may not be accessible to farmers.

In recent years, there has been a push towards on-farm conservation and farmer-scientist partnerships are being encouraged in countries including Ethiopia, the Philippines and Chile. Over the last few years research has brought to light highly innovative, informal systems of agricultural research being carried out by indigenous farmers in developing countries. But landraces (primitive cultivar of crops) are not fixed. They shift with evolution to suit local conditions, and although many TNCs claim that the genebanks will provide them with all the basic genetic material they need to feed the world for the next 200 years, over a longer, evolutionary timeframe they may need to draw on landraces.

14.6. ENVIRONMENTAL IMPACT

Examples of the impact on the environment to date: Bt (genetic manipulation through biotechnology).

- Research at the University of Hawaii has shown that insects which survive Bt transmit genetic resistance to their immediate offspring. If Bt becomes useless as an implanted pest control strategy within one insect generation, as this suggests, then organic farmers, who have been using it effectively since the 1940s, will be robbed off a valuable biopesticide. Regional cases of Bt resistance have already been reported.
- In April, Greenpeace International demanded an urgent European ban on Novartis' Bt-maize (which was approved by the EC in 1996) after Swiss research showed that it kills insects other than the European corn borer, which it is meant to target.
- George Stuber of Greenpeace in Sweden reported that in June 1996, Monsanto planted Bt-potatoes (containing a marker gene for the antibiotic "canamicin") in three regions of Georgia, in the former Soviet Union. Georgia lacks any proper legislation controlling the release of Genetically Modified Organisms (GMOs). The crops were devastated by phytophthora disease because, says Stuber, the potatoes were not adapted to the local climate.

Similarly, GM FlavrSavr tomatoes were grown in Guatemala, which has hundreds if not thousands of indigenous varieties of tomato, without the consent or knowledge of the authorities. Although the tomatoes were grown in greenhouses, Stuber says there is no way of knowing how rigorously they were contained, and there is no information as to whether the transgenic tomatoes spread beyond those sites.

Chemical Herbicides

Glyphosate is the world's best-selling "total" herbicide. It was developed by Monsanto in the early 1970s, and its major formulation is Roundup. Largely due to the introduction of Roundup Ready crops, human and environmental exposure to the herbicide is expected to increase.

- In health terms, say Friends of the Earth, experiments have shown damaging and long lasting reproductive effects of glyphosate on laboratory animals. In 1995, a study in the Journal of Environmental Science and Health showed that exposure to glyphosate was associated with reduced semen quality, volume, and concentration in rabbits.
- Herbicide-tolerant members of the bean family produce higher levels of plant oestrogens when grown in the presence of the herbicide, and there is a risk that these could disrupt the developing reproductive systems of children who consume them.

In 1994, a report by the World Health Organisation's (WHO) International Program on Chemical Safety found that glyphosate residues in animal feeds arising from pre-harvest glyphosate treatment of cereals may result in low residues in meat, milk and eggs.

Antibiotic Resistance

Novartis' Bt-maize contains a marker gene, which codes for antibiotic resistance in *E. coli*. There is a risk that if animals or humans consume Bt-maize-based products such as cattle feed or starch, some antibiotics would be rendered useless.

Nutritionally enhanced crops… when crops such as rice and rapeseed with high Vitamin A concentrations are planted, there will be no way to distinguish them from normal rice, with the contingent risk of liver damage if too much Vitamin A is consumed.

Unpredictable crops—Monsanto is facing an increasing number of lawsuits as their GM plants are not behaving as intended. In 1997 farmers who grew Monsanto's herbicide-tolerant cotton saw the cotton balls fall off their crops… In 1996, Monsanto's pest-resistant Bt-cotton succumbed to a heat wave in the southern US and was destroyed by bollworms and other pests.

Monocultures—Most researchers agree that biotechnology in crops will exacerbate the trend towards monocropping set in train by the Green Revolution, with all the problems that that entails, e.g., higher risk of disease and pest damage.

• Jonathan Rigg, a researcher in agricultural development and rural change at Durham University, cites one example from the Green Revolution, in the late 1970s, vast areas of Indonesia planted with a single variety of rice were devastated by the brown plant hopper, an event which has been linked to a subsequent mini famine on Lombok.

• In 1996 the FAO reported that the world depends on too few crops and that many thousands of genetic varieties (landraces) have been lost, mainly due to the spread of modern commercial agriculture. The report lists the main causes of plant genetic erosion in 154 countries and in over 80 of them, "replacement of local varieties" came top. In maize, for instance, Costa Rica, Chile, Malaysia, Philippines and Thailand have documented widespread genetic erosion due to monocropping.

"What has been almost entirely overlooked is that throughout that vast continent (of Africa) can be found more than 2000 native grains, roots, fruits and other food plants. These have been feeding people for thousands of years but most are being given no attention whatever today".

SUMMARY

Intellectual property right (IPR) is a class of property emanating primarily from the activities of the human intellect or intangible personal property. Any property, movable or immovable, is legally protected to prevent it from being stolen. The legal rights accrued on the intellectual property created are termed intellectual property rights (IPRs). Hence, intellectual property, in the legal sense, is thought of as material that may be protected under the patent, trademark and/or copyright laws. For biotechnology companies, intellectual property is often essential to their financing and survival. Intellectual property has always been important to agriculture. Much of the new agricultural equipment of the nineteenth century that made possible agricultural expansion into the Great Plains was subject to patent protection.

Patents on plant varieties, as such, are only allowed in the US, Japan and Australia, and are most frequent in the US. The 1930 US Act introduced a special kind of Plant Patent for vegetatively propagated materials, but in the US standard, utility patents can also now be granted on plant varieties. Patents are the strongest form of intellectual property protection in the sense that they normally allow the rights holder to exert the greatest control over the use of patented material by limiting the rights of farmers to sell, or reuse seed they have grown, or other breeders to use the seed (or patented intermediate technologies) for further research and breeding purposes.

However, in contrast to medicines, there is the potential for companies to become attracted to crops that are widely grown in developing countries. The investment costs are correspondingly lower than for medical research, and the potential markets correspondingly larger. For instance, rice, where the value of production in India alone exceeds that of the US maize market, has hitherto been a crop where breeding has been the preserve of the national or international public sector (principally the CGIAR). Since then the private sector has become increasingly interested in rice research. Monsanto and Syngenta have worked on sequencing the rice genome of two major rice varieties. The number of patents relating to rice issued annually in the US has risen from less than 100 in 1995 to over 600 in 2000.

The development of genetic traits such as herbicide tolerance has been determined principally by the search for commercial advantage, rather than for characteristics useful to poor farmers in developing countries. But companies are introducing GM varieties that, although controversial both in developed and developing countries, are considered by some developing countries to be of potential benefit to them (for example, the Bt gene which confers insect resistance). Bt cotton or Bt maize is now grown in at least five developing countries, and other countries may be interested, if they can resolve environmental concerns. For instance, India has recently approved the planting of Bt cotton. Companies have also donated technologies of relevance to developing countries (for example, through royalty free licences), including those related to vitamin A enriched rice (Golden Rice) and cassava. Some companies have published scientific articles based on their genomic research, but have aroused controversy by not depositing the raw data in public databanks.

The potential influence of patents (and IPR more generally) on agricultural research in developing countries is changing rapidly to the degree that countries are achieving effective implementation of TRIPS and subsequent agreements. Public and non-profit researchers in these poorer parts of the world, who will continue to be responsible for the bulk of agricultural research not directed at the most lucrative cash crops, are now increasingly confronting the challenges of obtaining the necessary freedom to operate for bringing new crops, processes and products to market that have already been faced by their counterparts in the developed countries.

EXERCISE

1. Name different types of Intellectual property rights in agricultural biotechnology.
2. Explain the merits and demerits of Intellectual property rights in agricultural biotechnology.
3. Describe how India suffers from the developed countries' patents in agriculture sector.
4. Write a note on genetic modifications in agricultural field, and the spread of the transgenic plants in the world scenario.
5. Discuss how intellectual, property problems pose a major hindrance to the diffusion of agricultural biotechnology.
6. Elucidate different institutional initiatives to encourage agricultural biotechnology innovations in the worldwide arena.
7. What do you mean by Plant Variety Protection (PVP)? Write about its impact on the developing countries agricultural system in the conventional way.
8. Briefly explain genetic use restriction technologies, and their advantages and disadvantages in the agricultural biotechnology.
9. Distinguish between "terminator" and "traitor" that are commonly used to save the technology by the multinational companies for their crops.
10. How agricultural biotechnology is helping the poor farmers in the developed countries and agriculture systems?
11. Critically comment on the developed countries control over the agricultural biotechnology and its companies in the developing or third world countries.
12. Mention the necessary regulations of modern agricultural biotechnology that are very essential in the developed countries with the emphasis on the India.
13. Differentiate between the applications of intellectual property rights in developed countries agricultural biotechnology and third world countries agricultural biotechnology.
14. List out the different bioresources of third world countries that are exploited by the developed countries in the agricultural biotechnology.
15. Illustrate how intellectual property rights impact on socioeconomic conditions of the developed countries in the agricultural biotechnology.
16. Define monoculture and add a note on its effects on agriculture.

15

CHAPTER

International Organizations and Intellectual Property Rights

15.1. INTRODUCTION

The production of transgenic microorganisms, animals, and plants seems to have much to offer to the productivity and quality of products for human consumption. However, farmers fear that allowing patents on transgenic animals and plants will push them out of the market. Since altered animals, plants, and seeds may be more expensive, farmers fear that large corporations will corner the market on genetically engineered animals, etc., and thereby deprive the small farms of their livelihood. In addition, farmers fear that the initial acquisition price of genetically altered plants, seeds, and animals as well as subsequent royalties will increase rather than decrease the costs for farmers and consumers. On the other hand, transgenic farm capital assets could be stronger and more disease resistant and might balance the cost of the initial investment. Some environmental groups are concerned with what effect the release of transgenic organisms would have on the environment. The National Wildlife Foundation opposes patenting for transgenic animals because of the lack of legislation in the area concerning their release into the wild. The National Wildlife Foundation fears that allowing patents will cause a greater number of transgenic animals to be created, thus increasing the risk to the environment. These issues are becoming not only at national levels and at an international levels. Hence, many numbers of issues raised in biotechnology, market concerned, intellectual property rights, ethical aspects and socioeconomic are discussed in depth and solutions were found at the many international organizations such as World Trade Organization (WTO), General Agreement on Tariffs and Trade (GATT), Trade Related Aspects of Intellectual Property Rights (TRIPs) and World Intellectual Property Organization (WIPO).

15.2. WORLD TRADE ORGANIZATION (WTO)

The World Trade Organization (WTO) is the successor organization to the multilateral General Agreement on Tariffs and Trade (GATT). The headquarters of WTO is in Geneva, Switzerland. The World Trade Organization (WTO) was created by the Uruguay Round of negotiations for revision of the General Agreement on Tariffs and Trade

(GATT). The largest-ever agreement in history was signed on April 15, 1994, in Marrakesh, Morocco, and the WTO came into being on January 1, 1995.

The WTO, the legal and institutional basis of the multilateral trading system, aims at liberalization of world trade. Member-countries are required to apply fair trade rules covering commodities, services and intellectual property. The Uruguay Round also commits the members to lowering of tariffs on industrial goods; abolition of import duties on a variety of items; progressive abolition of quotas on garments and textiles; reduction of trade-distorting subsidies and import barriers; agreements on intellectual property and rules for civil aviation; telecommunications; financial services; and the movement of labor.

The ultimate decision-making body of the WTO is the two-yearly **ministerial meeting**. In the interregnum, the General Council consisting of representatives of all member-countries carries out its functions. The General Council is assisted by the Council for Trade in Goods; the Council for Trade in Services; the Council for Trade Related Aspects of Intellectual Property Rights (TRIPs); and by a large number of committees, working groups etc. A seven-member Standing Appellate Body under the Dispute Settlement Body oversees the process of settlement of disputes between member-countries. The WTO had 146 members in April 2003. Most of the WTO agreements are the result of the 1986-94 Uruguay Round negotiations, signed at the Marrakesh ministerial meeting in April 1994. There are about 60 agreements and decisions totaling 550 pages. Negotiations since then have produced additional legal texts such as the Information Technology Agreement, services and accession protocols. New negotiations were launched at the Doha Ministerial Conference in November 2001.

Recognizing that their relations in the field of trade and economic endeavour should be conducted with a view to raising standards of living, ensuring full employment and a large and steadily growing volume of real income and effective demand, developing the full use of the resources of the world and expanding the production and exchange of goods.

Being desirous of contributing to these objectives by entering into reciprocal and mutually advantageous arrangements directed to the substantial reduction of tariffs and other barriers to trade and to the elimination of discriminatory treatment in international commerce.

In all the Ministerial Conferences, the highest decision-making body of the WTO that have held so far: Singapore conference (1996), Geneva conference (1998), Seattle conference (1999), Doha conference (2001), Cancun conference (2003), and Hong Kong conference (2005).

Singapore Conference

The Singapore conference saw the interested WTO members negotiating an Information Technology Agreement, and the launch of work program on the four, so-called, Singapore issues:

- Relationship between trade arid investment
- Interaction between trade and competition policy
- Transparency in government procurement
- Trade facilitation

This conference identifies these four areas on which WTO could initiate negotiations between the member countries.

Geneva Conference

It gave the go-ahead signal for the preparations to negotiate the "built-in" agenda covering global trade on agriculture and services. It was agreed that WTO should discuss the difficulties that developing countries have been facing in the Marrakesh agreement (Uruguay Round). The Marrakesh agreement among GATT members in 1994, also known as the Final Act Embodying the Results of the Uruguay Round of Multilateral Trade Negotiations, called for the creation of the WTO, with annexes stating terms of the understanding reached in agriculture, intellectual property rights, goods, services, market access and technical issues such as subsidies, and dispute settlement.

Seattle Conference

The Seattle conference collapsed without being able to reach a consensus on the declaration to be adopted by the members.

Doha Conference

The fourth ministerial meeting of the WTO concluded on November 14, 2001 at Doha (Qatar) with the Doha declaring a varying trade-issues. The Doha declaring—comprising a main declaration— a declaration on TRIPS agreement and public health, and a decision on implementation related issues and concerns launches the future work programme of the WTO, and includes elaboration and timetables for the current negotiations in agriculture, and services and negotiations/possible negotiations in a range of other issues. The major functions of WTO are as follows:

- It implements and administers all WTO's trade agreements.
- It acts as a forum for trade negotiations between member countries.
- It monitors the national trade policies.
- It provides the technical assistance and training for developing countries.
- It helps in handling and solving the trade disputes between the member countries.
- It cooperates with other international organizations.
- It acts as smooth, easy conducting of the trade at international levels.
- It mediates the transmission of the required information.
- It deals with the ethical issues of the business.
- It provides the detailed information on biotechnology, GM crops, GM food, and their business.

The activities of the World Trade Organization (WTO) have important ramifications for economic and social development throughout the world. Officially, the WTO is a member-driven 'one-country one-vote organization'. In practice, however, there is a long-standing custom of decision-making "by consensus" and rich countries exert disproportionate influence within the organization. While some of the world's least developed countries are not even represented in Geneva, and have minimal capacity to participate in negotiating sessions, the richer countries have large staffs of trade specialists, lawyers and expert negotiating teams. Moreover, powerful transnational corporations have been successful in swaying trade policy to suit their interests.

Mainstream economic theory teaches that international trade is beneficial to all countries and their citizens. This belief is based on the idea of **"comparative advantage"**—each country should focus on what it does best and trade for other products in order to reach the most efficient allocation of resources in the global economy and the highest levels of output and growth in all countries. It is assumed that trade leads to growth, which in turn promotes national development and reduces poverty. While it is recognized that trade produces both "winners" and "losers" in every economy, the theory is that the "losers" can be compensated from the net gains. The evidence, however, does not support these simplistic assumptions. Furthermore, the "comparative advantage" of some countries is their low wages, poor environmental regulations and lax labor standards.

The stated objective of the WTO is to help trade flow smoothly, freely, fairly and predictably. The original GATT was limited to facilitating trade in goods by eliminating so-called 'trade barriers' (e.g., quotas, tariffs) and articulating the basic principles of free trade (i.e., nondiscrimination, fiscal and regulatory independence of states). The scope and power of the organization, however, has greatly increased. Today, the WTO's reach includes issues of service provision, intellectual property, health and safety standards, and a vast array of products. It is now the primary actor in international trade, administering multilateral agreements, hosting negotiating sessions, handling disputes, monitoring national trade policies, and providing technical assistance and training for developing countries. With good reason, the WTO has been described as "the institutional face of globalization" and has been the focus of many so-called "anti-globalization" protests in recent years. The organization's structure has been non-transparent, unaccountable, non-participatory, undemocratic, and imperialistic, and trade liberalization has become an ever-expanding end in itself. Furthermore, the WTO's contribution to human rights protection and sustainable development is questionable.

Hormone-Treated Beef Case

[*Complaints by: US and Canada. Complaints against: European Union. WTO Cases WT/DS26 (US) and WT/DS48 (Canada)*]

"In 1980, as a result of consumer concern over reports of harm caused by eating hormone-treated meat, the EU (European Union) instituted a series of bans on the use of growth hormones in meat production, and subsequently on the import of meat from animals treated with such hormones. In 1996, the United States and Canada challenged the European ban as a violation of the WTO rules".

"A WTO dispute panel found (the bans) to violate the WTO Agreements (food-safety rules), because the EU had not definitively demonstrated that the beef would cause harm to consumers. While the EU argued that it had the right to protect its citizens against uncertain risks from the hormones, the panel concluded that the WTO rules require proof of such harm before trade can be restricted".

"Despite the Appellate Body's determination that the European hormone ban violated the WTO rules, the EU refused to rescind the ban. As a result, the WTO granted the United States permission to impose $116.8 million in retaliatory trade sanctions each year that the EU maintains its ban".

Pesticide Residues Case

[*Complaint by: US Complaint against: Japan. WTO case WT/DS76*]

In this case, the US challenged Japan's public-health standards requiring testing for pesticide residues in certain imports of fruits and nuts. The testing was required when a poisonous chemical, methyl bromide, was used to fumigate these products against infestation by coddling moths. Because Japan's safety standards were higher than relevant WTO standards, the WTO found that they violated WTO agreements.

A dispute-settlement panel ruled in October 1998 that Japan's requirements for testing of agricultural imports was not based on "sufficient scientific evidence", as required by article 2.2 of the agreement on the application of sanitary and phytosanitary measures. An appellate body upheld the ruling in February 1999. The two countries finally issued a joint letter settling this dispute in August 2001.

Uncooked Salmon Case

[*Complaint by: Canada. Complaint against: Australia. WTO Cases WT/DS18 and WT/DS21*]

In June 1998, a WTO dispute-settlement panel ruled in favor of Canada against Australia's quarantine against imports of uncooked wild salmon, and in October of that year an appellate body upheld the ruling. Australia had imposed the limitation after a 1994 risk assessment found that such imports posed a threat to Australian wild salmon. Roughly 20 bacteria found in Canadian and US salmon but not in Australia, it concluded, posed a risk of spreading to and infecting wild Australian salmon.

The WTO ruling found that Australia's risk assessment was inadequate and not based on sound science. As a result, it said, the ban was "more trade-restrictive than required to achieve its appropriate level of sanitary protection", and exceeded international standards. Since it was arbitrarily and unjustifiably discriminatory, it violated the Agreement on the Application of Sanitary and Phytosanitary Measures (SPS).

This ruling shifted the burden further onto importing countries that want to exclude a product to prove that the product is unsafe, rather than requiring exporting countries to prove that the products they send abroad are safe. It required expensive risk assessment measures to determine precise levels of risk, though health authorities must often adopt protections because the exact level of risk cannot be known. With it, the WTO added yet another precedent that under SPS countries cannot err on the side of caution.

In May 1999, the US filed a related complaint alleging that the Australian prohibition on imports of fresh, chilled, or frozen salmonids appeared to be inconsistent with:

- the agreement on the application of sanitary and phytosanitary measures—Articles 2, 5, 7 and 8.
- the general agreement on tariffs and trade 1994—Article XI.

The US complaint was settled in a joint communique with Australia in October 2000. It stated that a mutually satisfactory solution had been reached involving amendments to Australia's quarantining policies on uncooked salmonids introduced in May 2000.

Automobile Fuel-Efficiency Case

[*Complaint by: EU Complaint against: US GATT Case DS31*]

[*GATT, US—Taxes on Automobiles (DS31/R), Report of the Panel, Oct. 11, 1994. (Pre-WTO, not available on WTO Web site)*]

In the wake of 1970s energy crisis, the United States passed Corporate Average Fuel Efficiency (CAFE) standards that were successful in doubling the average fuel economy of passenger cars operating in the US by the early 1990s. These rules contributed to substantially cutting US emissions of greenhouse gases, and were consistent with United Nations standards on reducing global warming.

The CAFE standards applied equally to all US and foreign automobile manufacturers. They aimed to reduce emissions by all passenger vehicles sold in the US by requiring that the average fuel-efficiency of all the cars sold by each firm fall below a given figure. For US companies, which also imported foreign-made vehicles, it required that both the foreign and the domestic fleets meet the same fuel-efficiency standards.

United States car manufacturers complied and began to produce smaller, more fuel-efficient models. Although many European manufacturers had met CAFE standards in the 70s and early 80s, they later shifted to a strategy of exporting more profitable, less fuel-efficient luxury cars to the US market. They voluntarily chose not to comply with the CAFE standards, and as a result paid substantial penalties under the law.

In 1993, Europe challenged the CAFE standards under the general agreement on Tariffs and Trade (GATT, now enforced by the WTO), arguing that their effect was to discriminate against European automobile manufacturers. The GATT panel upheld the challenge, ruling that regardless of its intent, the requirement that each separate fleet meet the standards had a negative impact on European manufacturers and thus violated GATT nondiscrimination rules.

This ruling established an apparent precedent that even if a rule is not discriminatory, a party that chooses not to comply with it and then suffers the consequences can claim discriminatory effect. This perverse logic is an invitation to foreign firms in any country to violate standards designed to protect the environment or people and then claim injury under the WTO.

Plant and Animal Patents Case

[*Complaint by: US Complaint against: India WTO Case WT/DS50*]

[*India—Patent Protection for Pharmaceutical and Agricultural Chemical Products. Report of the Panel. (WT/DS50/R, 5 September 1997) http:// www.wto.org/english/tratop_e/dispu_e/trips.wp5*]

[*India—Patent Protection for Pharmaceutical and Agricultural Chemical Products. Report of the Appellate Body. (WT/DS50/AB/R, 19 December 1997) http://www.wto.org/english/tratop_e/ dispu_e/tripab.pdf*]

In this case, the US challenged an Indian law, which, in an effort to keep prices down on pharmaceuticals and other products, excluded plant and animal varieties from patenting. The WTO agreement on Trade-Related Aspects of Intellectual Property (TRIPS) requires that by 2005, developing countries change their patent laws to allow foreign companies to patent local plant varieties.

Even though the deadline had not yet been reached, the WTO dispute-settlement panel ruled that India was not moving fast enough towards compliance. As a result, India must grant monopolies to foreign corporations on plant and animal varieties based on patents granted by any other WTO member.

Months before July 1997, decision of a dispute settlement panel of the WTO, which agreed with the US that India had failed to implement the so-called "mail-box" provisions of article 70 of TRIPs, the US Ambassador to India had announced that "certain areas of research and training will be closed to cooperation" if India failed to amend its patent laws, threatening some 130 scientific projects supported by the US-India Fund. (Despite the five-year transition period for developing countries, article 70 requires them to establish legal procedures for receiving applications for patents on pharmaceutical and agricultural chemicals immediately, so that, upon their full implementation of TRIPs, patents can be back-dated to the date of filing. India had developed a legal administrative procedure, which the dispute panel deemed insufficient)"—Kristin Dawkins, Director of the Trade and Agriculture Program, Institute for Agriculture and Trade Policy, US.

15.3. WTO TREATIES

The following treaties are enforced by the World Trade Organization (WTO).

* General Agreement on Tariffs and Trade (GATT).
* General Agreement on Trade in Services (GATS).
* Agreement on Technical Barriers to Trade (TBT).
* Agreement on Government Procurement (AGP).
* Agreement on the Application of Sanitary and Phytosanitary Measures (SPS).
* Agreement on Trade-Related Aspects of Intellectual Property Rights (TRIPS).
* Agreement on Trade-Related Investment Measures (TRIMS).
* Agreement on Agriculture (AoA).

15.4. GENERAL AGREEMENT ON TARIFFS AND TRADE (GATT)

Negotiated in 1947, GATT came into force as an interim arrangement on January 1, 1948. Originally, it had only 23 signatories—members of the preparatory committee set up to detail guidelines for the then intended International Trade Organization, which, however, never came to be. GATT remained the only world body laying down trade rules. Eight rounds of negotiations were conducted under GATT to liberalize world trade and evolve a common code of conduct in world trade and trade relations between nations. In December 1993, there were 111 contracting parties and 22 other countries applying GATT rules on a *de facto* basis. The eighth round, the last, called the Uruguay Round, was concluded on December 15, 1993 by 117 countries accounting for about 90% of international trade.

The headquarters of GATT, in Geneva, Switzerland, now serves the WTO.

The new version of GATT after completion of the Uruguay Round of negotiations is still really an arrangement of the first category listed above, but has started to take on additional features. The Bretton Woods Conference after World War II, which established the World Bank and the International Monetary Fund, also proposed the creation of an International Trade Organization. As a result of resistance from the United States Senate, this never came into existence and the GATT was adopted on a provisional basis until such time as the issue could be readdressed. This occurred during the Uruguay Round of negotiations that started in 1986 and finally ended in December 1993. In this round of negotiations, the task of the negotiators was extended from dealing simply with trade in goods to covering other matters such as trade in services, trade-related aspects of intellectual property rights, and a new disputes settlement mechanism. This has led to the decision to establish a World Trade Organization to oversee the implementation of the new.

Many of the changes in intellectual property laws that have taken place over the last few years have been made with a view to meeting the obligations that were expected to be imposed by the GATT TRIPS Agreement (the form of which was effectively finalized over two years before agreement was reached on the main GATT agreement itself). Thus, for example, Poland, Hungary, the Andean Pact countries, and Thailand have all amended their laws over the past few years in light of the GATT requirements. Even countries that are not yet part of GATT, such as China, Russia and Taiwan, have amended their laws, at least partially, as required. The final adoption of GATT even seems to have brought an end to the parliamentary squabbling in Argentina, and it seems likely that a new patent law that will include protection for pharmaceutical products will be enacted shortly. Parliaments in Australia and New Zealand are considering legislation to extend the present 16-year patent terms in these countries to 20 years. Even in Brazil, the legislature approved ratification in December 1994, not even taking advantage of the possibility of arguing that Brazil is a developing country and entitled to delay implementation of some provisions. Only in countries such as India, where the local pharmaceutical industries persuaded the government to try to exclude intellectual property from the Uruguay round of negotiations, progress towards implementation seems slow. And that slowness seems fairly well confined to the issue of patent protection for pharmaceutical products. Even in these countries, however, the prospects of action seem brighter than at any time in recent memory, and even if amending legislation is slow to come on to the statute books, the black box provisions requiring at least the ability to file patent applications relating to excluded subject matter should be in effect shortly.

Consideration of GATT's relationship to environmental policy is an emerging concern in trade and environmental policy circles. Until the recently concluded Uruguay Round of GATT negotiations, the word environment did not appear in the GATT text. Several provisions and sections of GATT may be relevant to environmental issues, however. The following sections of GATT are often referenced in the examination of trade-environment issues. The excerpts are from GATT, as amended through 1966, originally digitized by the multilaterals project of the Fletcher School of Law and Diplomacy, Tufts University:

- Article I: General Most-Favored-Nation Treatment.
- Article III: National Treatment on Internal Taxation and Regulation.
- Article XI: General Elimination of Quantitative Restrictions.
- Article XIII: Non-discriminatory Administration of Quantitative Restrictions.
- Article XVI: Subsidies.
- Article XX: General Exceptions.

The GATT final act embodying the results of the Uruguay Round contains several other relevant items:

- The trade-related aspects of intellectual property rights.
- An agreement on subsidies and countervailing measures that permits some environmental subsidies in section 8.2.
- The agreement establishing the multilateral trade organization.

This gives WTO information on the agenda set by the Doha Ministerial Conference. It covers these main areas: The main Doha declaration, and the resulting negotiations and other work; The decision on implementation, and resulting work. The November 2001 declaration of the Fourth Ministerial Conference in Doha, Qatar, provides the mandate for negotiations on a range of subjects and other work, including issues concerning the implementation of the present agreements.

GATT Member Countries

1. Angola	32. Dominican Republic	63. Liechtenstein	94. St. Kitts & Nevis
2. Antigua and Barbuda	33. Egypt	64. Luxembourg	95. St. Lucia
3. Argentina	34. El Salvador	65. Macau	96. St. Vincent & Grenadines
4. Australia	35. Fiji	66. Madagascar	97. Senegal
5. Austria	36. Finland	67. Malawi	98. Sierra Leone
6. Bahrain	37. France	68. Malaysia	99. Singapore
7. Bangladesh	38. Gabon	69. Maldives	100. Slovak Republic
8. Barbados	39. Gambia	70. Mali	101. Slovenia
9. Belgium	40. Germany	71. Malta	102. Rep. of South Africa
10. Belize	41. Ghana	72. Mauritania	103. Spain
11. Benin	42. Greece	73. Mauritius	104. Sri Lanka
12. Bolivia	43. Grenada	74. Mexico	105. Surinam
13. Botswana	44. Guatemala	75. Morocco	106. Swaziland
14. Brazil	45. Guinea-Bissau	76. Mozambique	107. Sweden
15. Brunei	46. Guyana	77. Myanmar	108. Switzerland
16. Burkina Faso	47. Haiti	78. Namibia	109. Tanzania
17. Burundi	48. Honduras	79. Netherlands	110. Thailand
18. Cameroon	49. Hong Kong	80. New Zealand	111. Togo
19. Canada	50. Hungary	81. Nicaragua	112. Trinidad and Tobago
20. Central African Rep.	51. Iceland	82. Niger	113. Tunisia
21. Chad	52. India	83. Nigeria	114. Turkey
22. Chile	53. Indonesia	84. Norway	115. Uganda
23. Colombia	54. Ireland	85. Pakistan	116. United Arab Emirates
24. Congo	55. Israel	86. Paraguay	117. United Kingdom
25. Costa Rica	56. Italy	87. Peru	118. Uruguay
26. Cote d'Ivoire	57. Jamaica	88. Philippines	119. USA
27. Cuba	58. Japan	89. Poland	120. Venezuela
28. Cyprus	59. Kenya	90. Portugal	121. Yugoslavia
29. Czech Republic	60. Korea	91. Qatar	122. Zaire
30. Denmark	61. Kuwait	92. Romania	123. Zambia
31. Dominica	62. Lesotho	93. Rwanda	124. Zimbabwe

Ideas and knowledge are an increasingly important part of trade. Most of the value of new medicines and other high technology products lies in the amount of invention, innovation, research, design and testing involved. Films, music recordings, books, computer software and on-line services are bought and sold because of the information and creativity they contain, not usually because of the plastic, metal or paper used to make them. Many products that used to be traded as low-technology goods or commodities now contain a higher proportion of invention and design in their value. For example, brand named clothing or new varieties of plants.

Creators can be given the right to prevent others from using their inventions, designs or other creations, and to use that right to negotiate payment in return for others using them. These are "intellectual property rights". They take a number of forms. For example, books, paintings and films come under copyright; inventions can be patented; brandnames and product logos can be registered as trademarks; and so on. Governments and parliaments have given creators these rights as an incentive to produce ideas that will benefit society as a whole.

The extent of protection and enforcement of these rights varied widely around the world, and as intellectual property became more important in trade, these differences became a source of tension in international economic relations. New internationally-agreed trade rules for intellectual property rights were seen as a way to introduce more order and predictability, and for disputes to be settled more systematically.

The WTO's TRIPS Agreement is an attempt to narrow the gaps in the way these rights are protected around the world, and to bring them under common international rules. It establishes minimum levels of protection that each government has to give to the intellectual property of fellow WTO members. In doing so, it strikes a balance between the long term benefits and possible short term costs to society. Society benefits in the long term when intellectual property protection encourages creation and invention, especially when the period of protection expires and the creations and inventions enter the public domain. Governments are allowed to reduce any short term costs through various exceptions, for example to tackle public health problems. And, when there are trade disputes over intellectual property rights, the WTO's dispute settlement system is now available.

The agreement covers four broad issues: how basic principles of the trading system and other international intellectual property agreements should be applied; how to give adequate protection to intellectual property rights; how countries should enforce those rights adequately in their own territories; how to settle disputes on intellectual property between members of the WTO special transitional arrangements during the period when the new system is being introduced.

Basic Principles

- National treatment
- Most favored nation treatment
- Balanced protection

As in GATT and GATS, the starting point of the intellectual property agreement is basic principles. And as in the two other agreements, non-discrimination features prominently the national treatment (treating one's own nationals and foreigners equally), and most-favored-nation treatment (equal treatment for nationals of all trading partners in the WTO). National treatment is also a key principle in other intellectual property agreements outside the WTO.

The TRIPS agreement has an additional important principle: intellectual property protection should contribute to technical innovation and the transfer of technology. Both producers and users should benefit, and economic and social welfare should be enhanced, the agreement says.

How to Protect Intellectual Property?

Common ground-rules

The second part of the TRIPS agreement looks at different kinds of intellectual property rights and how to protect them. The purpose is to ensure that adequate standards of protection exist in all member countries. Here the starting point is the obligations of the main international agreements of the World Intellectual Property Organization (WIPO) that already existed before the WTO was created—the Paris Convention for the protection of industrial property (patents, industrial designs, etc.); and the Berne convention for the protection of literary and artistic works (copyright).

Some areas are not covered by these conventions. In some cases, the standards of protection prescribed were thought inadequate. So, the TRIPS agreement adds a significant number of new or higher standards.

15.5. DIFFERENT ARTICLES OF THE GAAT

Part I: General Provisions and Basic Principles

- Article 1: Nature and scope of obligations
- Article 2: Intellectual property conventions
- Article 3: National treatment
- Article 4: Most-favored-nation treatment
- Article 5: Multilateral agreements on acquisition or maintenance of protection
- Article 6: Exhaustion
- Article 7: Objectives
- Article 8: Principles

Part II: Standards Concerning the Availability, Scope and Use of Intellectual Property Rights

Copyright and related rights

- Article 9: Relation to berne convention
- Article 10: Computer programs and compilations of data

- Article 11: Rental rights
- Article 12: Term of protection
- Article 13: Limitations and exceptions
- Article 14: Protection of performers, producers of phonograms (sound recordings), and broadcast organizations

Trademarks

- Article 15: Protectable subject matter
- Article 16: Rights conferred
- Article 17: Exceptions
- Article 18: Term of protection
- Article 19: Requirement of use
- Article 20: Other requirements
- Article 21: Licensing and assignment

Geographical indications

- Article 22: Protection of geographical indications
- Article 23: Additional protection for geographical indications of wines and spirits
- Article 24: International negotiations; exceptions

Industrial designs

- Article 25: Requirements for protection
- Article 26: Protection

Patents

- Article 27: Patentable subject matter
- Article 28: Rights conferred
- Article 29: Conditions on patent applicants
- Article 30: Exceptions to rights conferred
- Article 31: Other use without authorization of the right holder
- Article 32: Revocation/forfeiture
- Article 33: Term of protection
- Article 34: Process patents: Burden of proof

Layout-designs (topographies) of integrated circuits

- Article 35: Relation to IPIC treaty
- Article 36: Scope of the protection
- Article 37: Acts not requiring the authorization of the right holder
- Article 38: Term of protection

Protection of undisclosed information

- Article 39

Control of anti-competitive practices in contractual licences

- Article 40

Part III: Enforcement of Intellectual Property rights

General obligations

- Article 41

Civil and administrative procedures and remedies

- Article 42: Fair and equitable procedures
- Article 43: Evidence of proof
- Article 44: Injunctions
- Article 45: Damages
- Article 46: Other remedies
- Article 47: Right of information
- Article 48: Indemnification of the defendant
- Article 49: Administrative procedures

Provisional measures

- Article 50

Special requirements related to border measures

- Article 51: Suspension of release by customs authorities
- Article 52: Application
- Article 53: Security or equivalent assurance
- Article 54: Notice of suspension
- Article 55: Duration of suspension
- Article 56: Indemnification of the importer and the owner of the goods
- Article 57: Right of inspection and information
- Article 58: *Ex-officio* action
- Article 59: Remedies
- Article 60: *De Minimus* imports

Criminal procedures

- Article 61

Part IV: Acquisition and Maintenance of Intellectual Property Rights and Related Inter-Partes Procedures

- Article 62

Part V: Dispute Prevention and Settlement

- Article 63: Transparency
- Article 64: Dispute settlement

Part VI: Transitional Arrangements

- Article 65: Transitional arrangements
- Article 66: Least-developed country members
- Article 67: Technical cooperation

Part VII Institutional Arrangements—Final Provisions

- Article 68: Council for trade-related aspects of intellectual property rights
- Article 69: International cooperation
- Article 70: Protection of existing subject matter
- Article 71: Review and amendment
- Article 72: Reservations
- Article 73: Security exceptions

15.6. TRADE-RELATED ASPECTS OF INTELLECTUAL PROPERTY RIGHTS (TRIPS)

The agreement on Trade-Related Aspects of Intellectual Property Rights (TRIPS) was concluded at the end of 1994 as a part of the Uruguay Round multilateral trade agreement. It provides for the establishment of standards for protection of a full range of intellectual property rights and the enforcement of those standards both internally and at the border through legal and administrative actions. The intellectual property rights covered by the TRIPS agreement are:

- Copyrights
- Patents
- Trademarks
- Industrial designs
- Trade secrets (undisclosed information)
- Integrated circuits (semiconductors)
- Geographical indications

Following are the highlights of the provisions

Copyrights

In the area of copyrights, TRIPS obligates signatories to comply with provisions of the Berne convention, except for the convention's requirements on moral rights.

It protects computer programs as literary works and databases as compilations under copyright. It imposes an immediate obligation on signatories to grant owners of computer programs and sound recordings the right to authorize or prohibit the rental of their products. It establishes a 50-year term for protection of sound recordings, as well as requiring signatories to provide protection for existing sound recordings. It sets a minimum of 50-year term for the protection of motion pictures and other works, where companies may be the author.

Patents

In the area of patents, TRIPS obligates signatories to protect product and process patents for virtually all types of inventions, including pharmaceuticals and agricultural chemicals.

It establishes a patent term of 20 years from the date the application is filed. It requires prompt implementation of procedures to permit the filing of patent applications covering pharmaceuticals and agricultural chemicals upon entry into force of the agreement, even in members that lack product patent protection. It sets limits on the use of compulsory licensing—a method by which the government can compel a patent holder to license the patent to someone else.

Trademarks

In the area of trademark, TRIPS requires signatories to register service marks as well as trademarks. It provides protection for internationally well-known marks. It prohibits the mandatory linking of trademarks, and compulsory licensing of marks.

Other Areas

Trade secrets enable owners to prevent unauthorized use or disclosure of confidential information.

Integrated circuits eliminate the deficiencies of the 1989 Washington treaty, by lengthening the term of protection from a minimum of eight years to a minimum of 10 years.

Industrial designs consistent with the existing US laws. Nongeneric geographical indications used to identify wines and spirits.

Enforcement Requirements

In general, members are required to provide expeditious, fair, and equitable remedies that create a deterrent to further infringements. Written decisions on the merits should be provided to parties, and there should be an opportunity for judicial review of final administrative decisions.

Injunctive relief and other effective provisional measures should be available through the judiciary. Adequate damages should be available, which may include expenses to rightholder (including possible attorneys' fees); recovery of profits or payment of statutory damages is also permissible. The judiciary should have authority to provide other remedies, such as disposal of infringing goods, and should also have the authority to order the applicant to indemnify the accused party to make up any losses suffered due to abuse of enforcement procedures. Border measures, including suspension of release by customs authorities, indemnification of the importer and owner of the goods, and a right of inspection, shall be provided by members.

Criminal procedures and penalties must be available for use at least in cases of willful trademark and copyright infringement committed on a commercial scale.

Industrialized countries were given one year for implementation from the entry into force of the TRIPS accord, July 1, 1995. Developing countries and those shifting from centrally planned to market economies generally were given four additional years for implementation. However, these countries have another five years (for a total of nine additional years) for implementation in the pharmaceutical and agricultural chemicals sectors if such protection is currently unavailable. Least-developed countries have until 2006 to comply.

Although developing countries were given a long transition period, they were required, at the end of the one-year transition period afforded all countries, to provide for national treatment (equal treatment for domestic and foreign parties) and most-favored-nation treatment (equal treatment for all participating trading partners).

The Uruguay Round agreement created the World Trade Organization (WTO) to replace the General Agreement on Tariffs and Trade (GATT). The WTO is facilitating the implementation and administration of the agreement.

Unlike the GATT system, the WTO integrates all the dispute settlement procedures established under individual agreements (goods, services, TRIPS). Disputes involving TRIPS are handled by the WTO General Council, acting as the dispute settlement body.

If a dispute settlement panel finds inadequate IPR protection or enforcement in a member country, the signatory bringing the complaint has the right to retaliate in other sectors.

15.7. MAIN FEATURES OF THE TRIPS AGREEMENT

The three main features of the agreement are:
1. Standards
2. Enforcement
3. Dispute settlement

Standards

In respect of each of the main areas of intellectual property covered by the TRIPS agreement, the agreement sets out the minimum standards of protection to be provided by each member. Each of the main elements of protection is defined, namely the subject-matter to be protected, the rights to be conferred and permissible exceptions to those rights, and the minimum duration of protection. The agreement sets these standards by requiring, first, that the substantive obligations of the main conventions of the WIPO, the Paris convention for the protection of industrial property (Paris convention) and the Berne convention for the protection of literary and artistic works (Berne convention) in their most recent versions, must be complied with. With the exception of the provisions of the Berne convention on moral rights, all the main substantive provisions of these conventions are incorporated by reference and thus become obligations under the TRIPS agreement between TRIPS member countries. The relevant provisions are to be found in articles 2.1 and 9.1 of the TRIPS agreement, which relate, respectively, to the Paris convention and the Berne convention.

Secondly, the TRIPS agreement adds a substantial number of additional obligations on matters where the pre-existing conventions are silent or were seen as being inadequate. The TRIPS agreement is thus sometimes referred to as a Berne and Paris-plus agreement.

Enforcement

The second main set of provisions deals with the domestic procedures and remedies for the enforcement of intellectual property rights. The agreement lays down certain general principles applicable to all IPR enforcement procedures. In addition, it contains provisions on civil and administrative procedures and remedies, provisional measures, special requirements related to border measures and criminal procedures, which specify, in a certain amount of detail, the procedures and remedies that must be available so that right holders can effectively enforce their rights.

Dispute Settlement

The agreement makes disputes between WTO members about the respect of the TRIPS obligations subject to the WTO's dispute settlement procedures.

In addition the agreement provides for certain basic principles, such as national and most-favored-nation treatment, and some general rules to ensure that procedural difficulties in acquiring or maintaining IPRs do not nullify the substantive benefits that should flow from the agreement. The obligations under the agreement will apply equally to all member countries, but developing countries will have a longer period to phase them in. Special transition arrangements operate in the situation where a developing country does not presently provide product patent protection in the area of pharmaceuticals.

The TRIPS agreement is a minimum standards agreement, which allows members to provide more extensive protection of intellectual property if they so wish. Members are left free to determine the appropriate method of implementing the provisions of the agreement within their own legal system and practice.

15.8. ROLE OF THE TRIPS COUNCIL

The TRIPS council comprises all WTO members. It is responsible for monitoring the operation of the agreement, and, in particular, how members comply with their obligations under it.

Monitoring

The member countries review each others' laws. The reviews are central to the TRIPS Council's task of monitoring what is happening under the agreement. Each country has to make sure its laws comply with the obligations of the agreement, according to the timetable spelt out in the agreement. Most have to enact laws implementing the obligations. These laws are notified to the TRIPS council, allowing members to review each others' legislation, and promoting the transparency of members' policies on intellectual property protection. The requirement to notify comes under article 63.2 of the TRIPS agreement. Members have to supply the TRIPS council with copies of their laws and regulations that deal with the TRIPS agreements' provisions. These notifications are then used as the basis the council's reviews of members' legislation. In these reviews, countries supply written questions about each others' laws before the review meetings. The answers are also in writing. Follow-up questions and replies are made orally during the course of the meeting, and further follow-up is possible at subsequent meetings.

Consultations

The TRIPS council is also a forum that countries can use to consult each other on problems they may have with each other to do with the TRIPS agreement. It can also clarify or interpret provisions of the agreement.

Technical Cooperation

A work programme—The council follows a work programme on technical cooperation with a view to monitoring how developed countries fulfil their obligations under article 67 of the TRIPS agreement. This article sets out the developed countries' commitments on technical cooperation. The work programme ensures that developing countries can have adequate information on the assistance on offer. It also ensures that any of their unfulfilled needs are identified and responded to.

Reviews and Negotiations on Specific Subject

The WTO is a forum for further negotiations aimed at enhanced commitments in the area of intellectual property, as in other areas covered by the WTO agreements. The TRIPS agreement calls for further work in specified areas, including the negotiation of a multilateral system of notification and registration for geographical indications of wines (article 23.4); the review of the application of provisions on protecting geographical indications (article 24.2); the review, after four years, of the option to exclude from patentability certain plant and animal inventions [article 27.3(b)]; and the examination of the applicability to TRIPS of non-violation complaints under the dispute settlement process (article 64).

Review of Trips Agreements

The TRIPS council holds a general review of the agreement after five years, but it is also empowered to review it at any time in the light of any relevant new developments, which might warrant modification and amendment (article 71). According to article 2.1 of the agreement, the WTO members shall, in respect of Parts II, III and IV of the agreement, comply with articles 1 through 12, and article 19, of the Paris convention (1967) (the Stockholm Act of 14 July 1967 of the Paris convention for the protection of industrial property). Article 9.1 of the agreement requires members to comply with articles 1 through 21 of the Berne convention (1971) and the appendix thereto (the Paris Act of 24 July 1971 of the Berne

convention for the Protection of Literary and Artistic Works). However, members do not have rights or obligations under the TRIPS agreement in respect of the rights conferred under article 6bis of that convention, i.e., the moral rights, or of the rights derived therefrom. As regards protection of the layout-designs of integrated circuits, article 35 of the agreement requires members to comply with articles 2 through 7 (other than article 6.3), article 12 and article 16.3 of the Treaty on Intellectual Property in Respect of Integrated Circuits, adopted at the agreement contains some references to certain provision of the Rome convention (the International Convention for the Protection of Performers, Producers of Phonograms and Broadcasting Organizations, adopted at Rome on 26 October 1961). However, there is no general obligation to comply with the substantive provisions of that convention.

Below is a list of these agreements, which can be found at the website of the World Intellectual Property Organization:
- Paris Convention for the Protection of Industrial Property (1967).
- Berne Convention for the Protection of Literary and Artistic Works (1971).
- International Convention for the Protection of Performers, Producers of Phonograms and Broadcasting Organizations (the Rome Convention) (1961).
- Treaty on Intellectual Property in Respect of Integrated Circuits (1989).

Certain general provisions

As in the main pre-existing intellectual property conventions, the basic obligation on each member country is to accord the treatment in regard to the protection of intellectual property provided for under the agreement to the persons of other members. Article 1.3 defines who these persons are. These persons are referred to as "nationals", but include persons, natural or legal, who have a close attachment to other members without necessarily being nationals. The criteria for determining, which persons must thus benefit from the treatment provided for under the agreement, are those laid down for this purpose in the main pre-existing intellectual property conventions of WIPO, applied of course with respect to all WTO members whether or not they are party to those conventions. These conventions are the Paris convention, the Berne convention, International Convention for the protection of performers, Producers of Phonograms and Broadcasting Organizations (Rome convention), and the Treaty on Intellectual Property in Respect of Integrated Circuits (IPIC Treaty).

Articles 3, 4 and 5 include the fundamental rules on national and most-favoured-nation treatment of foreign nationals, which are common to all categories of intellectual property covered by the agreement. These obligations cover not only the substantive standards of protection, but also affect the availability, acquisition, scope, maintenance and enforcement of intellectual property rights as well as those matters affecting the use of intellectual property rights specifically addressed in the agreement. While the national treatment clause forbids discrimination between a member's own nationals and the nationals of other members, the most-favoured-nation treatment clause forbids discrimination between the nationals of other members. In respect of the national treatment obligation, the exceptions allowed under the pre-existing intellectual property conventions of WIPO are also allowed under TRIPS. Where these exceptions allow material reciprocity; a consequential exception to MFN treatment is also permitted (e.g., comparison of terms for copyright protection in excess of the minimum term required by the TRIPS Agreement as provided under article 7(8) of the Berne Convention as incorporated into the TRIPS Agreement). Certain other limited exceptions to the MFN obligation are also provided for.

The general goals of the TRIPS Agreement are contained in the Preamble of the Agreement, which reproduces the basic Uruguay Round negotiating objectives established in the TRIPS area by the 1986 Punta del Este Declaration and the 1988/89 Mid-Term Review. These objectives include the reduction of distortions and impediments to international trade, promotion of effective and adequate protection of intellectual property rights, and ensuring that measures and procedures to enforce intellectual property rights do not themselves become barriers to legitimate trade. These objectives should be read in conjunction with article 7, entitled "Objectives", according to which the protection and enforcement of intellectual property rights should contribute to the promotion of technological innovation and to the transfer and dissemination of technology, to the mutual advantage of producers and users of technological knowledge and in a manner conducive to social and economic welfare, and to a balance of rights and obligations. Article 8, entitled "Principles", recognizes the rights of members to adopt measures for public health and other public interest reasons and to prevent the abuse of intellectual property rights, provided that such measures are consistent with the provisions of the TRIPS Agreement.

Substantive Standards of Protection

Copyright

During the Uruguay Round negotiations, it was recognized that the Berne Convention already, for the most part, provided adequate basic standards of copyright protection. Thus it was agreed that the point of departure should be the existing level of protection under the latest Act, the Paris Act of 1971, of that Convention. The point of departure is expressed in article 9.1, under which members are obliged to comply with the substantive provisions of the Paris Act of 1971 of the Berne Convention, i.e., articles 1 through 21 of the Berne Convention (1971) and the Appendix thereto. However, members do not have rights or obligations under the TRIPS Agreement in respect of the rights conferred under article 6b of that Convention, i.e., the moral rights (the right to claim authorship and to object to any derogatory action in relation to a work, which would be prejudicial to the author's honour or reputation), or of the rights derived there from. The provisions of the Berne Convention referred to deal with questions such as subject-matter to be protected, minimum term of protection, and rights to be conferred and permissible limitations to those rights. The Appendix allows developing countries, under certain conditions, to make some limitations to the right of translation and the right of reproduction.

In addition to requiring compliance with the basic standards of the Berne convention, the TRIPS agreement clarifies and adds certain specific points. Article 9.2 confirms that copyright protection shall extend to expressions and not to ideas, procedures, methods of operation or mathematical concepts as such.

Article 10.1 provides that computer programs, whether in source or object code, shall be protected as literary works under the Berne convention (1971). This provision confirms that computer programs must be protected under copyright, and that those provisions of the Berne convention which apply to literary works, shall be applied also to them. It confirms further that the form in which a program is, whether in source or object code, does not affect the protection. The obligation to protect computer programs as literary works means e.g., that only those limitations that are applicable to literary works may be applied to computer programs. It also confirms that the general term of protection of 50 years applies to computer programs.

Possible shorter terms applicable to photographic works and works of applied art may not be applied.

Article 10.2 clarifies that databases and other compilations of data or other material shall be protected as such under copyright, even where the databases include data that as such are not protected under copyright. Databases are eligible for copyright protection, provided that they by reason of the selection or arrangement of their contents constitute intellectual creations. The provision also confirms that databases have to be protected regardless of which form they are in—machine readable or other form. Furthermore, the provision clarifies that such protection shall not extend to the data or material itself, and that it shall be without prejudice to any copyright subsisting in the data or material itself.

Article 11 provides that authors shall have in respect of at least computer programs and, in certain circumstances, of cinematographic works the right to authorize or to prohibit the commercial rental to the public of originals or copies of their copyright works. With respect to cinematographic works, the exclusive rental right is subject to the so-called impairment test. A member is exempted from the obligation unless such rental has led to widespread copying of such works which is materially impairing the exclusive right of reproduction conferred in that member on authors and their successors in title. In respect of computer programs, the obligation does not apply to rentals where the program itself is not the essential object of the rental.

According to the general rule contained in article 7(1) of the Berne convention, as incorporated into the TRIPS agreement, the term of protection shall be the life of the author and 50 years after his death. Paragraphs 2 through 4 of that article specifically allow shorter terms in certain cases. These provisions are supplemented by article 12 of the TRIPS agreement, which provides that whenever the term of protection of a work, other than a photographic work or a work of applied art, is calculated on a basis other than the life of a natural person, such term shall be no less than 50 years from the end of the calendar year of authorized publication, or, failing such authorized publication within 50 years from the making of the work, 50 years from the end of the calendar year of making.

Article 13 requires members to confine limitations or exceptions to exclusive rights to certain special cases, which do not conflict with a normal exploitation of the work and do not unreasonably prejudice the legitimate interests of the right holder. This is a horizontal provision that applies to all limitations and exceptions permitted under the provisions of the Berne convention and the appendix thereto as incorporated into the TRIPS agreement. The application of these limitations is permitted also under the TRIPS agreement, but the provision makes it clear that they must be applied in a manner that does not prejudice the legitimate interests of the right holder.

Related rights

The provisions on protection of performers, producers of phonograms and broadcasting organizations are included in article 14. According to article 14.1, performers shall have the possibility of preventing the unauthorized fixation of their performance on a phonogram (e.g., the recording of a live musical performance). The fixation right covers only aural, not audiovisual fixations. Performers must also be in position to prevent the reproduction of such fixations. They shall also have the possibility of preventing the unauthorized broadcasting by wireless means and the communication to the public of their live performance.

In accordance with article 14.2, members have to grant producers of phonograms an exclusive reproduction right. In addition to this, they have to grant, in accordance with article 14.4, an exclusive rental right at least to producers of phonograms. The provisions on rental rights apply also to any other right holders in phonograms as determined in national law. This right has the same scope as the rental right in respect of computer programs. Therefore, it is not subject to the impairment test as in respect of cinematographic works. However, it is limited by a so-called grand-fathe-ring clause, according to which a member,

which on 15 April 1994, i.e., the date of the signature of the Marrakesh agreement, had in force a system of equitable remuneration of right holders in respect of the rental of phonograms, may maintain such system provided that the commercial rental of phonograms is not giving rise to the material impairment of the exclusive rights of reproduction of right holders.

Broadcasting organizations shall have, in accordance with article 14.3, the right to prohibit the unauthorized fixation, the reproduction of fixations, and the rebroadcasting by wireless means of broadcasts, as well as the communication to the public of their television broadcasts. However, it is not necessary to grant such rights to broadcasting organizations, if owners of copyright in the subject-matter of broadcasts are provided with the possibility of preventing these acts, subject to the provisions of the Berne convention. The term of protection is at least 50 years for performers and producers of phonograms, and 20 years for broadcasting organizations (article 14.5).

Article 14.6 provides that any member may, in relation to the protection of performers, producers of phonograms and broadcasting organizations, provide for conditions, limitations, exceptions and reservations to the extent permitted by the Rome Convention.

Trademarks

The basic rule contained in article 15 is that any sign, or any combination of signs, capable of distinguishing the goods and services of one undertaking from those of other undertakings, must be eligible for registration as a trademark, provided that it is visually perceptible. Such signs, in particular words including personal names, letters, numerals, figurative elements and combinations of colors, as well as any combination of such signs, must be eligible for registration as trademarks.

Where signs are not inherently capable of distinguishing the relevant goods or services, member countries are allowed to require, as an additional condition for eligibility for registration as a trademark, that distinctiveness has been acquired through use. Members are free to determine whether to allow the registration of signs that are not visually perceptible (e.g., sound or smell marks). Members may make registrability depending on use. However, actual use of a trademark shall not be permitted as a condition for filing an application for registration, and at least three years must have passed after that filing date before failure to realize an intent to use is allowed as the ground for refusing the application (article 14.3).

The agreement requires service marks to be protected in the same way as marks distinguishing goods (e.g., articles 15.1, 16.2 and 62.3). The owner of a registered trademark must be granted the exclusive right to prevent all third parties not having the owner's consent from using in the course of trade identical or similar signs for goods or services, which are identical or similar to those in respect of which the trademark is registered, where such use would result in a likelihood of confusion. In case of the use of an identical sign for identical goods or services, a likelihood of confusion must be presumed (article 16.1).

The TRIPS agreement contains certain provisions on well-known marks, which supplement the protection required by article 6 bis of the Paris convention, as incorporated by reference into the TRIPS agreement that obliges members to refuse or to cancel the registration, and to prohibit the use of a mark conflicting with a mark which is well known. First, the provisions of that article must also be applied to services. Second, it is required that knowledge in the relevant sector of the public acquired not only as a result of the use of the mark, but also by other means, including as a result of its promotion, be taken into account. Furthermore, the protection of registered well-known marks must extend to goods or services, which are not similar to those in respect of which the trademark has been registered, provided that its use would indicate a connection between those goods or services and the owner of the registered trademark, and the interests of the owner are likely to be damaged by such use (articles 16.2 and 3). Members may provide limited exceptions to the rights conferred by a trademark, such as fair use of descriptive terms, provided that such exceptions take account of the legitimate interests of the owner of the trademark and of third parties (article 17).

Initial registration, and each renewal of registration, of a trademark shall be for a term of no less than seven years. The registration of a trademark shall be renewable indefinitely (article 18). Cancellation of a mark on the grounds of non-use cannot take place before three years of uninterrupted non-use has elapsed unless valid reasons based on the existence of obstacles to such use are shown by the trademark owner. Circumstances arising independently of the will of the owner of the trademark, such as import restrictions or other government restrictions, shall be recognized as valid reasons of non-use. Use of a trademark by another person, when subject to the control of its owner, must be recognized as use of the trademark for the purpose of maintaining the registration (article 19). It is further required that use of the trademark in the course of trade shall not be unjustifiably encumbered by special requirements, such as use with another trademark, use in a special form, or use in a manner detrimental to its capability to distinguish the goods or services (article 20).

Geographical indications

Geographical indications are defined, for the purposes of the agreement, as indications which identify a good as originating in the territory of a member, or a region or locality in that territory, where a given quality, reputation or other characteristic of the good is essentially attributable to its geographical origin (article 22.1). Thus, this definition specifies that the quality, reputation or other characteristics of a good can each be a sufficient basis for eligibility as a geographical indication, where they are essentially attributable to the geographical origin of the good. In respect of all geographical indications, interested parties must have legal means to prevent use of indications, which mislead the public as to the geographical origin of the good, and use that constitutes an act of unfair competition within the meaning of article 10 bis of the Paris convention (article 22.2).

The registration of a trademark, which uses a geographical indication in a way that misleads the public as to the true place of origin, must be refused or invalidated *ex officio* if the legislation so permits or at the request of an interested party (article 22.3). Article 23 provides that interested parties must have the legal means to prevent the use of a geographical indication identifying wines for wines not originating in the place indicated by the geographical indication. This applies even where the public is not being misled, there is no unfair competition, and the true origin of the good is indicated or the geographical indication is accompanied by expressions such as "kind", "type", "style", "imitation" or the like. Similar protection must be given to geographical indications identifying spirits when used on spirits. Protection against registration of a trademark must be provided accordingly.

Article 24 contains a number of exceptions to the protection of geographical indications. These exceptions are of particular relevance in respect of the additional protection for geographical indications for wines and spirits. For example, members are not obliged to bring a geographical indication under protection, where it has become a generic term for describing the product in question (paragraph 6). Measures to implement these provisions shall not prejudice prior trademark rights that have been acquired in good faith (paragraph 5). Under certain circumstances, continued use of a geographical indication for wines or spirits may be allowed on a scale and nature as before (paragraph 4). Members availing

themselves of the use of these exceptions must be willing to enter into negotiations about their continued application to individual geographical indications (paragraph 1). The exceptions cannot be used to diminish the protection of geographical indications that existed prior to the entry into force of the TRIPS Agreement (paragraph 3). The TRIPS Council shall keep under review the application of the provisions on the protection of geographical indications (paragraph 2).

Industrial designs

Article 25.1 of the TRIPS agreement obliges members to provide for the protection of independently created industrial designs that are new or original. Members may provide that designs are not new or original if they do not significantly differ from known designs or combinations of known design features. Members may provide that such protection shall not extend to designs dictated essentially by technical or functional considerations. Article 25.2 contains a special provision aimed at taking into account the short life cycle and sheer number of new designs in the textile sector: requirements for securing protection of such designs, in particular in regard to any cost; examination or publication; must not unreasonably impair the opportunity to seek and obtain such protection. Members are free to meet this obligation through industrial design law or copyright law. Article 26.1 requires members to grant the owner of a protected industrial design the right to prevent third parties not having the owner's consent from making, selling or importing articles bearing or embodying a design, which is a copy, or substantially a copy, of the protected design, when such acts are undertaken for commercial purposes. Article 26.2 allows members to provide limited exceptions to the protection of industrial designs, provided that such exceptions do not unreasonably conflict with the normal exploitation of protected industrial designs and do not unreasonably prejudice the legitimate interests of the owner of the protected design, taking account of the legitimate interests of third parties. The duration of protection available shall amount to at least 10 years (article 26.3). The wording "amount to" allows the term to be divided into, for example, two periods of five years.

Patents

The TRIPS agreement requires member countries to make patents available for any inventions, whether products or processes, in all fields of technology without discrimination, subject to the normal tests of novelty, inventiveness and industrial applicability. It is also required that patents be available and patent rights enjoyable without discrimination as to the place of invention and whether products are imported or locally produced (article 27.1). There are three permissible exceptions to the basic rule on patentability. One is for inventions contrary to ordre public or morality. This explicitly includes inventions dangerous to human, animal or plant life or seriously prejudicial to the environment. The use of this exception is subject to the condition that the commercial exploitation of the invention must also be prevented and this prevention must be necessary for the protection of ordre public or morality (article 27.2). The second exception is that members may exclude from patentability diagnostic, therapeutic and surgical methods for the treatment of humans or animals [article 27.3(a)]. The third is that members may exclude plants and animals other than microorganisms and essential biological processes for the production of plants or animals other than non-biological and microbiological processes.

However, any country excluding plant varieties from patent protection must provide an effective sui generis system of protection. Moreover, the whole provision is subject to review four years after entry into force of the agreement [article 27.3(b)]. The exclusive rights that must be conferred by a product patent are the ones of making, using, offering for sale, selling, and importing for these purposes. Process patent protection must give rights not only over use of the process, but also over products obtained directly by the process. Patent owners shall also have the right to assign, or transfer by succession, the patent and to conclude licensing contracts (article 28). Members may provide limited exceptions to the exclusive rights conferred by a patent, provided that such exceptions do not unreasonably conflict with a normal exploitation of the patent and do not unreasonably prejudice the legitimate interests of the patent owner, taking account of the legitimate interests of third parties (article 30).

The term of protection available shall not end before the expiration of a period of 20 years, counted from the filing date (article 33). Members shall require that an applicant for a patent shall disclose the invention in a manner sufficiently clear and complete for the invention to be carried out by a person skilled in the art, and may require the applicant to indicate the best mode for carrying out the invention known to the inventor at the filing date or, where priority is claimed, at the priority date of the application (article 29.1). If the subject-matter of a patent is a process for obtaining a product, the judicial authorities shall have the authority to order the defendant to prove that the process to obtain an identical product is different from the patented process.

Compulsory licensing and government use without the authorization of the right holder are allowed, but are made subject to conditions aimed at protecting the legitimate interests of the right holder. The conditions are mainly contained in article 31. These include the obligation, as a general rule, to grant such licences only if an unsuccessful attempt has been made to acquire a voluntary licence on reasonable terms and conditions within a reasonable period of time; the requirement to pay adequate remuneration in the circumstances of each case, taking into account the economic value of the licence; and a requirement that decisions be subject to judicial or other independent review by a distinct higher authority. Certain of these conditions are relaxed where compulsory licences are employed to remedy practices that have been established as anticompetitive by a legal process. These conditions should be read together with the related provisions of article 27.1, which require that patent rights shall be enjoyable without discrimination as to the field of technology, and whether products are imported or locally produced.

Layout-designs of integrated circuits

Article 35 of the TRIPS agreement requires member countries to protect the layout-designs of integrated circuits in accordance with the provisions of the IPIC Treaty (the Treaty on Intellectual Property in Respect of Integrated Circuits), negotiated under the auspices of WIPO in 1989. These provisions deal with, *inter alia*, the definitions of "integrated circuit" and "layout-design (topography)", requirements for protection, exclusive rights and limitations, as well as exploitation, registration and disclosure. An "integrated circuit" means a product, in its final form or an intermediate form, in which the elements, at least one of which is an active element, and some or all of the interconnections are integrally formed in and/or on a piece of material and which is intended to perform an electronic function. A "layout-design (topography)" is defined as the three-dimensional disposition, however expressed, of the elements, at least one of which is an active element, and of some or all of the interconnections of an integrated circuit, or such a three-dimensional disposition prepared for an integrated circuit intended for manufacture. The obligation to protect layout-designs applies to such layout-designs that are original in the sense that they are the result of their creators' own intellectual effort and are not commonplace among creators of layout-designs and manufacturers of integrated circuits at the time of their creation. The exclusive rights include the right of reproduction and importation, sale and other distribution for commercial purposes. In addition to requiring member countries to protect the layout-designs of integrated circuits in accordance with the provisions of the IPIC treaty, the TRIPS agreement clarifies and/or builds on four points. These points relate to the term of protection (ten years instead of eight, article 38), the applicability of the protection to articles containing infringing integrated circuits (last sub clause of article 36), and the treatment of innocent infringers (article 37.1). The conditions in article 31 of the TRIPS agreement apply mutatis mutandis to compulsory or non-voluntary licensing of a layout-design or to its use by or for the government, without the authorization of the right holder, instead of the provisions of the IPIC treaty on compulsory licensing (article 37.2).

Protection of undisclosed information

The TRIPS agreement requires undisclosed information—trade secrets or know-how—to benefit from protection. According to article 39.2, the protection must apply to information that is secret, that has commercial value because it is secret and that has been subject to reasonable steps to keep it secret. The agreement does not require undisclosed information to be treated as a form of property, but it does require that a person lawfully in control of such information must have the possibility of preventing it from being disclosed to, acquired by, or used by others without his or her consent in a manner contrary to honest commercial practices. "Manner contrary to honest commercial practices" includes breach of contract, breach of confidence, and inducement to breach, as well as the acquisition of undisclosed information by third parties, who knew, or were grossly negligent in failing to know that such practices were involved in the acquisition.

The agreement also contains provisions on undisclosed test data and other data, whose submission is required by governments as a condition of approving the marketing of pharmaceutical or agricultural chemical products, which use new chemical entities. In such a situation, the member government concerned must protect the data against unfair commercial use. In addition, members must protect such data against disclosure, except where necessary to protect the public, or unless steps are taken to ensure that the data are protected against unfair commercial use.

Control of anti-competitive practices in contractual licences

Article 40 of the TRIPS agreement recognizes that some licensing practices or conditions pertaining to intellectual property rights, which restrain competition, may have adverse effects on trade and may impede the transfer and dissemination of technology (paragraph 1). Member countries may adopt, consistently with the other provisions of the agreement, appropriate measures to prevent or control practices in the licensing of intellectual property rights, which are abusive and anti-competitive (paragraph 2). The agreement provides for a mechanism whereby a country seeking to take action against such practices involving the companies of another member country can enter into consultations with other member and exchange publicly available non-confidential information of relevance to the matter in question and of other information available to that member, subject to domestic law and the conclusion of mutually satisfactory agreements concerning the safeguarding of its confidentiality by the requesting member (paragraph 3). Similarly, a country whose companies are subject to such action in another member can enter into consultations with that member (paragraph 4).

15.9. WORLD INTELLECTUAL PROPERTY ORGANIZATION (WIPO)

The World Intellectual Property Organization (WIPO) is an international organization dedicated to ensure that the rights of creators and owners of intellectual property are protected worldwide, and that inventors and authors are recognized and rewarded for their ingenuity.

The roots of the World Intellectual Property Organization go back to 1833. That year marked the birth of the Paris convention for the Protection of Industrial Property—the first major international treaty designed to help people of one country obtain protection in other countries for their intellectual creations in the form of industrial property rights.

The Paris convention entered into force in 1884 with 14 member states, which set up an international bureau to carry out administrative tasks.

In 1886, copyright entered the international arena, with the Berne convention for the Protection of Literary and Artistic Works. Like the Paris Convention, the Berne Convention set up an international bureau to carry out administrative tasks.

In 1893, these two small bureaus united to form an international organization called the United International Bureaux for the Protection of Intellectual Property (best known by its French acronym, BIRPI). Based in Berne, Switzerland, with a staff of seven, this small organization was the predecessor of the World Intellectual Property Organization of today—a dynamic entity with more than 170 member states and a staff that includes some 650 people from around the world.

As the importance of intellectual property grew, the structure and form of the organization changed as well. In 1960, BIRPI moved from Berne to Geneva to be closer to the United Nations and other international organizations in that city. A decade later, following the entry into force of the Convention Establishing the World Intellectual Property Organization, BIRPI became WIPO, undergoing structural and administrative reforms and acquiring a secretariat answerable to the member states.

In 1974, WIPO became a specialized agency of the United Nations system of organizations, with a mandate to administer intellectual property matters recognized by the member states of the UN. WIPO expanded its role and further demonstrated the importance of intellectual property rights in the management of globalized trade in 1996 by entering into a cooperative agreement with the World Trade Organization.

In 1898, BIRPI administered only four intranational treaties. A century later, WIPO administered 21 treaties (two of those jointly with other intranational organizations) and carried out a rich and varied program of work.

Through its member states and secretariat, WIPO seeks to:

- harmonize national intellectual property legislation and procedures. It provides services for international applications for industrial property rights. It exchanges intellectual property information. It provides legal and technical assistance to developing and other countries. It facilitates the resolution of private intellectual property disputes. Marshal information technology as a tool for storing, accessing and using valuable intellectual property information.

The World Intellectual Property Organization (WIPO) is an international organization dedicated to promoting the use and protection of works of the human spirit. These works—intellectual property—are expanding the bounds of science and technology and enriching the world of the arts. Through its work, WIPO plays an important role in enhancing the quality and enjoyment of life, as well as creating real wealth for nations.

With headquarters in Geneva, Switzerland, WIPO is one of the 16 specialized agencies of the United Nations system of organizations.

It administers 23 international treaties dealing with different aspects of intellectual property protection. The organization counts 182 nations as member states. The World Intellectual Property Organization (WIPO) was established by a convention of 14 July, 1967, which entered into force in 1970. It has been a specialized agency of the United Nations since 1974, and administers a number of international unions or treaties in the area of intellectual property, such as the Paris

and Berne Conventions. WIPO's objectives are to promote intellectual property protection throughout the world through cooperation among states, and, where appropriate, in collaboration with any other international organization. WIPO also aims to ensure administrative cooperation among the intellectual property unions created by the Paris and Berne Conventions, and subtreaties concluded by the members of the Paris Union.

The administration of the unions created under the various conventions is centralized through WIPO's secretariat, the "International Bureau". The International Bureau also maintains intranational registration services in the field of patents, trademarks, industrial designs, and appellations of origin. WIPO also undertakes development cooperation for developing countries through advice, training, and furnishing of documents. An agreement on cooperation between WIPO and the WTO came into force on 1 January, 1996. The agreement provides cooperation in three main areas:

- Notification of, access to, and translation of national laws and regulations
- Implementation of procedures for the protection of national emblems
- Technical cooperation

WIPO's Mandate to Assist Countries in TRIPS Implementation

In 1994, the WIPO General Assembly passed a resolution by which the International Bureau should be at the disposal of any state that expressly asks for the advice on issues of compatibility of its existing or planned national intellectual property legislation, not only with treaties administered by WIPO, but also with other international norms and trends, including the TRIPS agreement. In addition, during the WIPO General Assembly held in 1995, a new resolution on the subject matter was adopted. On that occasion, it was agreed *inter alia* that the International Bureau should assist developing country members of either WIPO or the WTO with translations of their intellectual property laws and regulations into one of the three WTO languages, and should also make the arrangements so as to be able to respond to requests from developing countries for WIPO's legal and technical assistance related to the TRIPS agreement. It was also agreed that WIPO should expand the technical assistance related to the TRIPS agreement in existing WIPO development cooperation program, as well as to allocate additional financing to meet the costs of additional activities to be taken in connection with the TRIPS agreement.

In addition, the agreement between WIPO and the WTO, concluded in December 1995, provides- *inter alia*-for the availability of legal-technical assistance by the two organizations to developing countries related to the TRIPS agreement.

WIPO provides legal advice in relation to the implementation of the TRIPS agreement—134 developing member states have received advice to date.

WIPO's advice takes into account all the flexibilities that are open to members under the TRIPS agreement, including those confirmed in the Doha Ministerial Declaration on the TRIPS agreement and Public Health ("the Doha Ministerial Declaration").

WIPO's advice takes into account the unique situation of each country, given that member states have different legal systems and different political and cultural structures.

WIPO follows up its written legal advice with an interactive process between the organization and major stake'iolders in the member state concerned. To strengthen the TRIPS implementation process, during the last four years, WIPO has promoted the interaction among stakeholders at the national level to include, for example, officials of Law Reform Commissions, Chambers of Commerce and Federation of Industries, Research and Development (R&D) institutions, Parliamentarians, high-level officials of Ministries of Trade, Agriculture, Health, Science and Technology, Culture, Justice, Environment, among others.

Sovereign member states make the final decisions concerning the content of their legislation, in accordance with their own political, legal and trade considerations at the national, regional, and international level.

15.10. PRACTICAL ASPECTS OF WIPO'S LEGAL AND TECHNICAL ASSISTANCE TO DEVELOPING COUNTRIES

In relation to the preparation of intellectual property legislation for the implementation of the TRIPS agreement, assistance is rendered by the Intranational Bureau, at the request of the countries concerned, in the form of the following:

* Preparation and submission of draft laws on all areas of intellectual property.
* Preparation and submission of draft provisions to amend and modernize existing laws in the field of intellectual property.
* Preparation and submission of comments and suggestions on draft laws prepared by (and received from) governments of developing countries and secretariats of regional organizations in developing countries.

Where the International Bureau receives reactions on the draft laws or comments sent, further assistance, in follow up actions, is also provided through the following:

1. Providing further written comments and suggestions on laws, which have been revised by the national authorities following their receipt of earlier comments [sometimes provided in discussions such as through (2) and (3)], below.
2. Holding discussions with government officials and other persons involved in the legislative reform process, in missions undertaken by WIPO staff members to the countries concerned, for that purpose.
3. Holding discussions with government officials

in Geneva (visits financed by the International Bureau).

Basis of WIPO's Legislative Advice

WIPO's legislative advice (the submission of draft laws or the submission of comments on draft laws prepared by governments, or the provision of written explanations or suggestions, or of clarifications in face to face discussions) is provided on the basis of WIPO's basic draft laws (which contain a commentary or explanatory notes), which are reviewed and updated by WIPO to take into account developments in the field of intellectual property. Such developments include, for example, TRIPS council review documentation and WTO Panel decisions. They also take into account developments in the field of intellectual property at the international level as reflected in meetings held in WIPO. In the interactive process, provisions of existing national laws of other countries are also utilized.

SUMMARY

The production of transgenic microorganisms, animals and transgenic plants seem to have much to offer to the productivity and quality of products for human consumption. However, small family farmers fear that allowing patents on transgenic animals and transgenic plants will push them out of the market. Because generally altered animals, plants and seeds may be more expensive, small farmers fear that small number of large corporations will corner the market on genetically engineered animals, etc. and thereby deprive the small family farms of their livelihood.

The National Wildlife Foundation fears that allowing patents will cause a greater number of transgenic animals to be created, thus increasing the risk to the environment. These issues are occuring not only at national levels but also at internatioinal levels. Hence, many numbers of issues raised in biotechnology, market concerned,

intellectual property rights, ethical aspects and socioeconomic are discussed in depth and solutions were found at the many international organizations such as World Trade Organization (WTO), General Agreement on Tariffs and Trade (GATT), Trade Related Aspects of Intellectual Property Rights (TRIPs) and World Intellectual Property Organization (WIPO).

Based in Geneva, the WTO oversees the rules of international trade. It is an arena for organizing multilateral trade negotiations and settling trade disputes between governments. The WTO was created in 1995, replacing the General Agreement on Tariffs and Trade (GATT). The GATT was organized after World War Two, at the instigation of the United States. It evolved from a set of guidelines for an institutional forum where trade negotiations took place and trade disputes were adjudicated. Since 1948, a series of eight multilateral trade rounds has led to wide-ranging trade concessions, largely tariff cuts on manufacturing goods but recently also nontariff barriers in farming, intellectual property rights, and government appropriations. The most recent round, in Uruguay, had mixed success, but the bigger picture is that GATT multilateral negotiations have dramatically lowered both tariff and non-tariff barriers around the world, creating freer trade and contributing to postwar world growth. One of the main differences between the WTO and the GATT is the WTO's tougher dispute settlement mechanism (individual countries can longer veto the dispute panel's ruling).

As did the GATT, the WTO works on the principle that free trade will allow all of its member countries to achieve higher levels of growth. The important norms within the WTO are non-discrimination and reciprocity. Nondiscrimination means that members cannot levy different duties on the same good coming from different countries. Reciprocity implies a give-and-take approach to creating mutually advantageous agreements. It makes trade agreements more politically palatable. Both reciprocity and nondiscrimination are useful tools for driving down trade barriers among a large and diverse group of countries. But the first step is getting the all members to agree to an agenda for a future trade round. Human barricades aside, this is turning out to be a difficult task.

The agreement on Trade-Related Aspects of Intellectual Property Rights (TRIPS) was concluded at the end of 1994 as part of the Uruguay Round multilateral trade agreement. It provides for establishment of standards for protection of a full range of intellectual property rights and the enforcement of those standards both internally and at the border through legal and administrative actions. The intellectual property rights covered by the TRIPS agreement are:

- Copyrights
- Patents
- Trademarks
- Industrial designs
- Trade secrets (undisclosed information)
- Integrated circuits (semi-conductors)
- Geographical indications

Following are highlights of the provisions:

The World Intellectual Property Organization (WIPO) is an international organization dedicated to ensure that the rights of creators and owners of intellectual property are protected worldwide and that inventors and authors are, thus, recognized and rewarded for their ingenuity.

It administers 23 international treaties dealing with different aspects of intellectual property protection. The organization counts 182 nations as member states. The World Intellectual Property Organization (WIPO) was established by a convention of 14 July 1967, which entered into force in 1970. It has been a specialized agency of the United Nations since 1974, and administers a number of international unions or treaties in the area of intellectual property, such as the Paris and Berne Conventions.

EXERCISE

1. Define genetically modified organisms and their applications.
2. What do you mean by genetic modifications? How does it proceed to the development of better food varieties?
3. Discuss the techniques used for genetic modifications of living organisms for the purpose of food exploitation.
4. Distinguish between genetically modified organisms and non-genetically modified organisms along with their uses.
5. Mention the risks and benefits associated with the genetically modified food.
6. Define social, ethical, and legal issues involved in the production of genetically modified organisms and their impact on future development.
7. Elucidate the negative effects of genetically modified organisms, and their use as food and resources.
8. Depict the controversies associated with the production of genetically modified organisms and foods.
9. Critically comment on the comprehensive and effective systems that are needed to find the solutions for the risks associated with the genetic modification technology.
10. What do you mean by 'genetically modified food'? Write a note on the importance of the safety assessment of GM food?
11. List out the various fields that are getting benefits from the genetic modification technology.
12. Mention the importance of education, certification and accreditation that are essential for the proper use of genetic modification technology and their products.
13. Differentiate between the allergens and intoxicants that are produced by the products of genetic modification technology.
14. Give an account of the environmental impact of genetic modification technology and their products used in the ecosystem.
15. Explain waiver of informed consent that are generally followed for the clinical trials and mention its merits in the research and testing.

16. What are the criteria necessary for conducting a genetic modification technology in plants at the present applied situation of the biotechnology.

17. Illustrate how genetic modification technologies in animals are used for the development and improvement of the animal derived foods.

18. What are the opinions of European Union on genetic modification technology and their products in the market?

19. Write about how safety is assessed for genetic modification technology and their end products for the human use.

20. Discuss the present status of the genetically modified tomato and its acceptance by the consumers in the world.

21. List out the ethical issues associated with the consumption of genetically modified food and their products in the different religious countries.

22. What role does the world health organization plays for the protection of the human health from genetically modified food?

23. In your opinion, what are the chances of the acceptance of genetically modified food and technology in the diversified world populations?

24. Is it possible to totally replace the conventional agriculture with the genetic modification derived organisms for the human use?

16

CHAPTER

Introduction to Patent and Process Involved in Patenting

16.1. INTRODUCTION

Biotechnology has revolutionized the fields of agriculture, medicine, industries and environment concerns. Large numbers of inventions have been made in the last three decades. This created a competition in the business market, and resulted in the copying, duplicating, and fake markets in a huge manner. Hence, the investments made by the original contributors were not fetching the returns. The patent system was designed to promote the progress of science by encouraging innovation, i.e., "inventiveness". Patent or an item of "intellectual property" (IP) is a contract between the inventor and the public (government); the latter being represented by the US Patent and Trademark Office (USPTO or PTO in USA), or the respective government of the different countries (for e.g., PTO in India). By granting a patent, the government gives rights to the inventor and excludes others from "practicing" the invention for a defined time. The meaning of the term "practicing" is defined in the statutes as making, using, selling, offering to sell, or importing.

The inventor, in return for this exclusive arrangement, must fully and clearly disclose the invention in a manner that allows others to practice it in the future. In doing so, the inventor is not given the right to practice the invention, but rather to prevent others from practicing (for example, a drug may be patented successfully, but the FDA may not approve its use in humans). Permission by the inventor for others to practice the invention is called a "license", which can be granted either exclusively or non-exclusively. Sometimes several patents overlap and the parties must "cross-license" all applicable IP. Licenses often have up-front fee and/or ongoing royalties that have to be paid pay to the inventor.

Patents are granted by the government for the commercial exploitation of an invention for a specific period of time in consideration of the disclosure of the invention, so that on expiry of the terms of the patent, the information can benefit the public at large.

In order to invent something new, and to develop it to a stage that is commercially viable, huge investments in terms of money and labor are necessary in addition to the intellectual activity. Nevertheless, when the invention is revealed, it comes to public domain. Hence, the inventor has to meet with competition from his rival manufacturers and therefore the return of his investment is not guarantied. This is the reason many people try to work secretly, but the risk of their invention being subjected to 'reverse engineering' persists. Therefore, the importance of patent system lies in the fact that it averts an impasse to technical advancement, which might occur otherwise.

In the present international business, the investor is always interested in the proper protection of its products. The patent system in any country also provides a social benefit, as it bestows monetary reward for revealing technological innovation along with accolades for the inventor (***Patent is an award for the inventor and a reward for the investor***). The grant of patent for an invention attracts investment because the commercial exploitation of the invention is possible during the term of patent. Another major advantage of the patent system is that it promotes *'invent around'* concept.

Definition

Patent can be defined as "the exclusive right to stop others (other than inventor) from making, using, or selling the invention". It does not necessarily mean that inventor has the right to make, use, or sell himself.

The patent is granted only when the invention and its operation, and the method by which it is to be performed are fully disclosed. When the patentee launches the product (in which the invention is incorporated) in the market, his competitors may lose the market if the product is technically advanced and cheap compared to the existing one.

The patentee can prevent others from manufacturing the same product without his authorization, and can resort to legal means to enforce his right. But the competitors have an option to 'invent around' the patented product by conducting further research to bring out a better invention, which may result in a cheaper and better product. It paves the way for healthy competition among manufacturers and results in day-to-day improvement of technology. Ultimately, it contributes to the economic growth of the country, thereby enhancing the living standards of the people.

By virtue of the grant, patentee gets the exclusive right to prevent the third parties (not having his consent) from the act of making, using, offering for sale, selling, or importing the patented product or process within the territory of grant.

History of Patent

In Europe and the United States, the system of intellectual property rights emanated as early as the 16th century. Britain's Statute of Queen Anne—the earliest copyright statute—was passed in 1709. The explosion in patents came with the development of the field of pharmacology in the early 1940s. Commercialization by the companies required IP protection to give some assurance that there was a way to provide a "return on investment" (ROI) to stockholders in publicly traded companies. Indeed, in the 1950s, the AMA changed its "canon of ethics" to allow patents on devices and drugs to comply with the shifting medical environment.

In the 1970s, Boyer and Cohen patented recombinant DNA technologies and revolutionized patents law. The subsequent US Supreme Court ruling (Diamond *vs.* Chakrabarty; 1980) allowing patents on live genetically engineered organisms (e.g., an oil-dissolving bacterium) was in the same year that the Bayh-Dole Act was passed to grant patent rights to universities and other grant recipients who used federal monies (i.e., provided for the retention of title to inventions, and allowed for exclusive licensing to commercialize the invention).

Prior regulations had only allowed nonexclusive license deals with the government and its scientists. Under certain circumstances, the government has "march-in" rights that may require the funding recipient to grant a license to a third party or take title itself in order to grant such a license. Post- 1980s patents have been issued for a broad spectrum of inventions: recombinant DNA and proteins, transgenomic animals, PCR and other enabling technologies, cell lines, enzymes, probes, antisense and gene sequences, as well as specific mutations and polymorphisms within genes.

16.2. HISTORY OF INDIAN PATENT SYSTEM AND LAW

India also has its own patent history. The first provision in the nature of patent rights in India, which was at that time under the British rule, was enacted in 1856. Under this Act, the monopolies were styled "exclusive privileges". This Act was repealed in toto in 1857, as it was introduced without the prior sanction of the Queen. Soon, in 1859, another Act free from the defects of the act of 1857 was passed. In 1872, the provisions of the Act of 1859 were further added by the provision of "The Patents and Designs Protection Act", which solely related to designs. The Act of 1872 was further supplemented in 1883. In 1888 all the Acts of 1859, 1872 and 1883 were superseded by Act V of 1888. This was further revised and replaced by the Indian Patents and Designs Act, 1911. This was amended from time to time in 1920, 1930, and 1945.

After independence, it was felt that the Indian Patents and Designs Act, 1911, was not fulfilling its objective. As such, a committee under the Chairmanship of Dr. Bakshi Tek Chand—a retired Judge of Lahore High Court—was appointed in 1949, with a view to ensure that the patent system was more conducive to national interest and to suggest modifications and alterations to the existing act, suitable for making the country self-reliant in technology. Based on the interim report, the act was modified, regarding working of the inventions in 1952, and in 1953, the controller was authorized to grant licenses on foods, medicines etc. A bill based on the recommendations of this

committee was introduced in the Parliament in 1953, but it lapsed due to the dissolution of Lok Sabha.

Later in 1957, Government of India further appointed Justice N. Rajagopala Ayyangar to examine and review the Patent law in India. He submitted his report in September 1959, recommending the retention of patent system despite its short-comings. The patent bill, 1965 based mainly on his recommendations and incorporating a few changes, in particular, related to patents for food, drug, medicines, was introduced in the lower house of Parliament on 21st September, 1965. The bill was passed by the Parliament, and the Patent Act 1970 came into force on 20th April, 1972, along with Patent Rules 1972. This law was suited to the changed political situation and economic needs for providing impetus to technological development by promoting inventive activities in the country.

After the number of changes, it is felt that patent system in India is not complete. The Patents Act, 1970 is a landmark in the industrial development of India. The basic philosophy of the act is that patents are granted to encourage inventions and to secure that these inventions are worked on a commercial scale without undue delay. Patents are not granted merely to enable patentee to enjoy a monopoly for the importation of the patented article into the country.

The said philosophy is being implemented through compulsory licensing; registration of only process patents for food, medicine, or pesticides; and substances produced by chemical processes, which, apart from chemical substances, also include items such as alloys, optical glass, semiconductors, inter metallic compounds etc. It may, however, be noted that products vital for our economy, such as agriculture and horticulture products, atomic energy inventions, and all living things are not patentable. Thus, the Patent Act 1970 was expected to provide a reasonable balance between adequate and effective protection of patents on the one hand, and the technology development, public interest and specific needs of the country on the other hand.

In accordance with the international development, such as Uruguay round of GATT, negotiations paved the way for WTO. Therefore, India was put under the contractual obligation to amend its patent act in compliance with the provisions of TRIPS. India had to meet the first set of requirements on 01.01.1995. This was to give a pipeline protection until the country starts giving product patent. It came to force on 26th March, 1999, retrospective from 01.01.1995. It laid down the provisions for the filing of application for product patent in the field of drugs or medicines with effect from 01.01.1995, and grant of exclusive marketing rights on those products.

India amended its Patent Act again in 2002 to meet with the second set of obligations (Term of Patent etc.), which had to be effected from 01.01.2000. This amendment, which provides for 20 years term for the patent, reversal of burden of proof etc. came into force on 20th May, 2003. The Third Amendment of the Patents Act 1970, by way of the Patents (Amendment) Ordinance 2004, came into force on 1st January, 2005. It incorporated the provisions for granting product patent in all fields of technology including chemicals, food, drugs and agrochemicals and this ordinance is replaced by the Patents (Amendment) Act 2005, which is in force now, having effect from 01.01.2005.

Under the Patents Act, the Government of India is empowered to make rules for implementing the Act and regulating the Patent Administration. Accordingly, the Government brought into force Patents Rules, 1972 w.e.f. 20.4.1972. These rules were amended on 2.6.1999 and replaced by the Patents Rules 2003 w.e.f. 20.5.2003. These were further amended by the Patent (Amendment) Rules 2005, which is in force now. This includes provisions related to time-lines with a view to introducing flexibility and reducing processing time gradually for patent applications,

and simplifying and rationalizing procedure for the grant of patent.

There are four schedules to the patents (Amendment) Rules 2005, which are as follows:

1. The first schedule prescribes the fee to be paid.
2. Second schedule specifies the list of forms, and the texts of various forms required in connection with various activities under the patents act are set out in this schedule. These forms are to be used wherever required, and if needed, they can be modified with the consent of the controller.
3. Third schedule prescribes patent form to be issued on the grant of the patent.
4. Fourth schedule prescribes costs to be awarded in various proceedings before the controller under the Act.

16.3. PATENTING AUTHORITIES

In addition to the USPTO, based in the Department of Commerce, there are patent offices throughout the world. Several treaties have been completed to facilitate the issuance of consistent patents between the countries. Uniform procedures in Europe were established in 1972 in order to obtain multinational protection of patent rights in member states. The Patent Cooperation Treaty (PCT) of 1978 is administered by the World Intellectual Property Organization (WIPO)—one of the sixteen agencies within the United Nations.

Thus, patent is a grant of the right by the US Patent and Trademark Office (or any other countries patent office) to *stop others* from *making, using or selling* an *invention* in the *United States* for a *limited period of time*. Patent

- is a *grant*, therefore, you do not have it until the government gives it to you (that is, you have no rights until the patent issues);
- is a right to *stop others,* and not necessarily a right to do anything yourself (someone else may have a patent, which would prevent you from using your invention, even though your invention is patentable);

- is a right to stop others from *making, using or selling* any one of these. Thus, even if an infringer were to make the invention in a foreign country, he could not sell it in the USA. Similarly, it is still an infringement if the invention is made in one country, but exported immediately, or if a person buys the invention overseas and uses it in the USA for his own user patent infringement;
- is only issued on *inventions*, as defined in the statute; and
- it is issued by the *US* Patent *and Trademark Office* (Indian patent office or any other countries patent office), therefore it can only be enforced against other people's actions within the USA. That is, a US Patent has no effect on people making, using and selling the invention entirely outside the USA (although it may serve as a basis for national patents in other countries, each of which would have effect in its respective areas).

It lasts only for a *limited period of time*, and once it has expired, the patented invention may be freely used by anyone (assuming, of course, no other patent would interfere).

16.4. ESSENTIAL REQUIREMENTS FOR PATENTING

There are three major requirements for patentability. The invention must be:
- novel;
- useful; and
- not obvious.

"Novel" means that the invention was never patented before you invented it, never described in a publication, never in public use or on sale, by others before your invention.

All of these also apply *to your own actions more than one year before you apply for a patent.* In other words, you have **one year** from the date you first described your invention in a publication, or first sold it, or publicly used it, within which you must apply for a patent if you ever want patent protection.

You should also be aware that most other countries do not give this one-year grace period. So, if you intend to apply for patents outside the USA, you should have your US application on file before your first publication, sale or public use.

"Useful" generally means that the invention does something, anything at all.

Very few applications are rejected on the grounds of "not useful"—mostly "perpetual motion" inventions and chemical compounds, and gene sequences with no known utility. Getting a patent does not mean that the USPTO has passed judgment on whether or not anyone really wants the product.

"Not obvious" means that the invention must not be an obvious development of what has gone before, in the judgment of an ordinary person skilled in the applicable field.

This is often the hardest to define. Every invention seems obvious to the inventor after it has been invented. Usually, "obviousness" is couched in terms of what a combination of references would have taught to the mythical *"Person Having Ordinary Skill In The Art" before* the invention was created. In other words, your invention is obvious if Mr. M.K. Girish, who knows everything there is to know, would have known to combine these previously-existing inventions to result in your invention, without having seen your patent application first.

16.5. TYPES OF PATENT

There are three main types of patents. **"Utility"** patents are directed to functional inventions, **"design"** patents are directed to ornamental manufactured objects, and **"plant"** patents are directed to asexually reproduced plants. Utility patents can have multiple claims, and each claim defines the limits of a single invention. The claims may appear in independent form or dependent form; the latter is a shorthand way to incorporate language from an earlier claim. Claims cover a "process, machine, manufacture of or composition of matter". Nucleic acids are considered a composition of matter, and can be covered if they have been isolated, purified or otherwise manipulated through the intervention of the "hand-of-man" (US Supreme Court, 1980; the original wording is "anything under the sun that is made by man"). A notice by the Commissioner of the US PTO in 1987 extended patents to "non-naturally occurring, non-human multicellular living organisms, including animals," which directly resulted in a patent granted to Harvard University in 1988 for the oncomouse".

In the US, there is a system that allows "provisional" applications to be submitted and held on file at the PTO. This is a useful way to extend the life of a patent, because they establish the "filing date" (called the "priority date"), and yet do not start the clock on the 20-year term because they are withheld from prosecution (examination) for upto one year. This is often used when there is a risk of imminent public disclosure, or when it is desirable to fix the priority date for a given amount of material. Filing serves as a constructive "reduction to practice", unless the court has ruled that there is no "conception without reduction to practice," as is the case with some genetic inventions. Several provisional applications may be combined into one application that goes forward into the prosecution stage. If the one-year time period is passed, provisional applications are considered automatically abandoned. They remain confidential at the PTO and are eventually destroyed. "Continuing" applications [e.g., "continuations," "divisionals" (i.e., the original application is divided into several applications)] and "continuations in part" (CIP) are applications that claim priority to a non-provisional application. Since the new information presented in a CIP will have its own priority date (*vs.* the date on the original application), any filing of a CIP must be carefully evaluated.

Numbers

In March 2003, over 50 million patents published were worldwide, with approximately 700,000 applications filed yearly (i.e., 20,000/week). Since 1980, when the US Supreme Court issued its ruling that living things (including genes) could be patented if "man-made" (see above), about 1800 patents have been issued for entire genes and several thousand applications are in active prosecution. The numbers are expected to go up exponentially with the completion of the human genome project. Private companies are aggressively seeking to protect their IP by filing multitudes of applications.

16.6. REQUIREMENTS OF PATENT

In order to have a patent issued, an invention must be a new and useful. It must meet the standards of utility, novelty, and non-obviousness. The invention must be something, which is not in public use, published in written form, or known to others in the field. To demonstrate these traits to the PTO, the body of the patent application must contain the following parts:

- An abstract (used for patent searches with key words).
- A specification, including a written description of the invention ("written description"); a description of the manner and process of practicing the invention without undue experimentation ("enablement"); and the "best mode" of the invention. Traditionally, there are several teaching "examples" used to explain the invention to the examiners:
- Drawings, if necessary
- At least one claim
- Gene or protein sequences, if necessary.

In the case of a new organism, plant, or animal, there may be a requirement for a specimen to be deposited or held at repository during the review of a patent (called the "prosecution" phase).

Claims

Claims are the most important part of any issued patent. They succinctly define the boundaries of the intellectual property. The PTO will sometimes work with (or negotiate with) the inventor and his, or her attorney to obtain allowable claims that do not cover claims in other patents already issued or currently in prosecution. Since patent applications are held confidential until issued, only the PTO knows the full extent of any overlap and thus has the upper hand WHILE revising claim language. They strive for narrow, limited claims rather than allowing broad, all encompassing claims. The applicant is entitled to claims as broad as the prior art will allow.

Inventorship

Inventors are designated when the individual has contributed to the conception of the invention and it is determined that the contribution to the invention was both substantial and well documented. Joint inventors do not have to be physically working together in the same laboratory at the same time, nor do they have to contribute equally. From the patent law perspective, a joint invention is a collaborative effort to solve a particular problem or identify a particular "composition of matter". Patents can be invalidated if there has been an incorrect designation of inventors when there was "deceptive intent," but corrections are allowed even after issue when inadvertent errors have been made.

Revised Interpretive Guidelines

In January 2001, new guidelines were issued that addressed the criteria for the patentability of gene sequences. Essentially, outlining what needs to be met to satisfy the "utility" requirement, the PTO stated that an application must provide only a single "specific, substantial, and credible" utility for each claimed invention. The PTO also requires that "a person of ordinary skill in the art would immediately appreciate why the invention is useful, based on the characteristics of the invention". Although a gene or an sequence may function in a certain way while in the human body, its utility, when purified away from the body (i.e., cloned), may be entirely different. An example would be a tumor antigen that is quantitatively up-regulated in a tumor cell, but is still a natural product of that type of cell, being used as a marker for the presence of malignancy (e.g., PSA) in an *in vitro* diagnostic test. The new interpretative guidelines also provide stricter rules for the body of the patent ("specification"), in presenting the written description of the invention and providing the "best mode" for the replication by others knowledgeable technologies.

Term

Under the current law, the term of a patent begins on the date on which the patent is issued, and ends 20 years from the date of filing of the patent application or from the date of application to which priority is claimed. The term may be calculated differently if certain circumstances apply. Patents that were issued prior to, or were pending on, June 8, 1995, are entitled to a term of at least 17 years from the date of issue. In addition, certain patents may receive extensions of term upon application to compensate for delays in obtaining regulatory approval to market a product covered by the patent. Filing a continuing application does not restart the clock on the term, rather the term is measured from the filing of the parent application.

Utility and plant patents are valid for a period beginning on the date the patent is issued by the Patent and Trademark Office, and ending, at the latest, 20 years after the filing date of the earliest non-provisional application upon which the patent is based. Maintenance fee has to be paid during the life of a utility patent, and if it is not paid on time, the patent will expire earlier. After the patent term is over, the invention is available to all.

Design patents are valid for a period of 14 years from the date of issue. No maintenance fee is due on design patents.

Infringement

Infringement is judged "claim by claim" during court review. It exists only if **all** the features stated within a claim are practiced by another. If all the features **plus** some new ones (i.e., not in the original claim) are practiced, generally, this will still constitute infringement, since the original claim is being practiced. On the other hand, dropping an element of a claim will not constitute infringement. The patent office does not hold any view regarding whether a license is needed; they only decide whether to issue patents. After issuance, one would look to what the courts have decided.

Special Cases

- Patents based on applications filed before December, 1980 (actual filing date, not priority date) were not subject to the payment of maintenance fee.
- Any utility patent based on an application, which was pending on June 8, 1995 (actual filing date, not priority date), and any utility patent which was issued and not expired as of June 8, 1995, is valid for the longer of 17 years from the date of issue or 20 years from the date of earliest non-provisional application in its chain of parentage (priority date; again, so long as maintenance fee are paid).
- Patents issued based on applications filed on or after May 29, 2000 (actual filing date, not priority date) will have their terms extended for patent office delays beyond certain limits. Applicant's delays, including requests for extensions of time to reply, will be deducted from the allowance for PTO delays. The term extension, if any, is printed on all new patents.

- Some patents have had their terms extended based on extreme delays in government approvals outside the patent office. This is very unusual, and applies almost always to pharmaceuticals (for example, Claritin® or Prozac®), food products (Aspartame), or medical devices or procedures, where FDA approval can sometimes eat up most of the patent term before the drug can be brought to market.
- Some patents have less than usual life span, because their terms are limited to the terms of earlier-issued patents. This is called a "terminal disclaimer" and is a result of filing two applications which claimed essentially the same invention. Terminal disclaimers will be marked on the later-issued patent.

The patent once expired cannot be renewed, nor can one pick up the rights to an expired patent. Once a patent expires, the invention is in the public domain.

16.7. THINGS THAT ARE PATENTABLE

Utility patents cover:
- machines;
- articles of manufacture;
- methods (processes);
- composition of matter (chemicals, cell lines); and
- improvements to any of the above.

Design patents cover the *appearance* of useful objects. They do *not* cover the function or construction of the object.

Plant patents cover certain plants.

16.8. THINGS THAT ARE NON-PATENTABLE

Many things are not allowed for patenting under the available patent laws, and they are as follows:
- Purely mental processes
- Mathematical algorithms or formulas (is, just a formula without a real-world effect—a

formula or algorithm may be claimed as part of a method, as long as there is an effect on the real world).

- Arrangements of printed matter (printed matter may be part of a patentable invention, but if the "invention" is just words, copyright protection is more appropriate).
- Naturally occurring things (unaltered—these are not "inventions").
- Scientific principles (a device or method which operates based on a new scientific principle can be patented, but the underlying principle cannot).
- Inventions solely useful in making atomic weapons.
- Human beings.

Since 1952, when the current Patent Act went into effect, "everyone knew" that methods of doing business were not patentable. However, in late 1998, the State Street Bank case was decided, in which the court stated that not only were methods of doing business patentable, but that they always had been. The effect of this decision is a flood of patent applications dealing with implementation of business methods, from methods of writing patents to methods of selling products and services. It seems safe to predict that we will soon see many cases testing the limits of the "methods of doing business" patent. The patent office department incharge of these patents [Technology Center 3600 (formerly part of TC2100)] has its own website.

16.9. THE PROCEDURE INVOLVED IN PATENT APPLICATION AND GRANTING OF A PATENT

The patent applications are generally drafted by a patent attorney (or patent lawyer or patent agent—a non-attorney with a science education, licensed to practice at the PTO), and filed by mail with the PTO in the country. Generally, the patent attorney will need the inventor(s) to sign an **inventor's declaration** (which states, among other things, that the inventor made the invention claimed in the

application) and an **assignment** (which evidences the inventor's duty to assign the patent to the inventor's employer (such as the University, Institutions). Patent applications are required to be held confidential by the PTO during the time of their processing and granting.

Usually, within a few months, the patent applicant will receive a written notice from the PTO, which states whether the application and its claims have been accepted in the form in which they were filed. More often than not, however, the PTO, represented by a patent examiner, rejects the application because either certain formalities need to be cleared up, or on the basis that the claims are not patentable over the "prior art" (anything that prior workers in the field have made or disclosed in the past. Prior art may include printed publications, conference handouts (sometimes even presentations within a university), books, newspaper articles—often regardless of where the material was published, and in what language it was published. Prior art may also include orally presented material, such as discussions at conferences, disclosures to competitors, certain disclosures to colleagues in a field, and other public statements). The letter sent by the patent examiner is referred to as an office action or official action **(Fig. 16.1)**.

If the application is rejected, the patent attorney will then need to file a written response (generally within three to six months) to any rejections. In such cases, generally the attorney will amend the claims and/or point out why the PTO's position is incorrect. This procedure is referred to as patent prosecution. Often it will take the action by two PTO officials, and two responses (and sometimes more) by the patent attorney before the application is resolved. The resolution is hopefully in the form of a PTO notice that the application is allowable (i.e., the PTO agrees to issue a patent). During this process, the patent attorney will often need the input of the inventor(s) in order to confirm the attorney's understanding of the technical aspects of the invention and/or the prior art cited against the application.

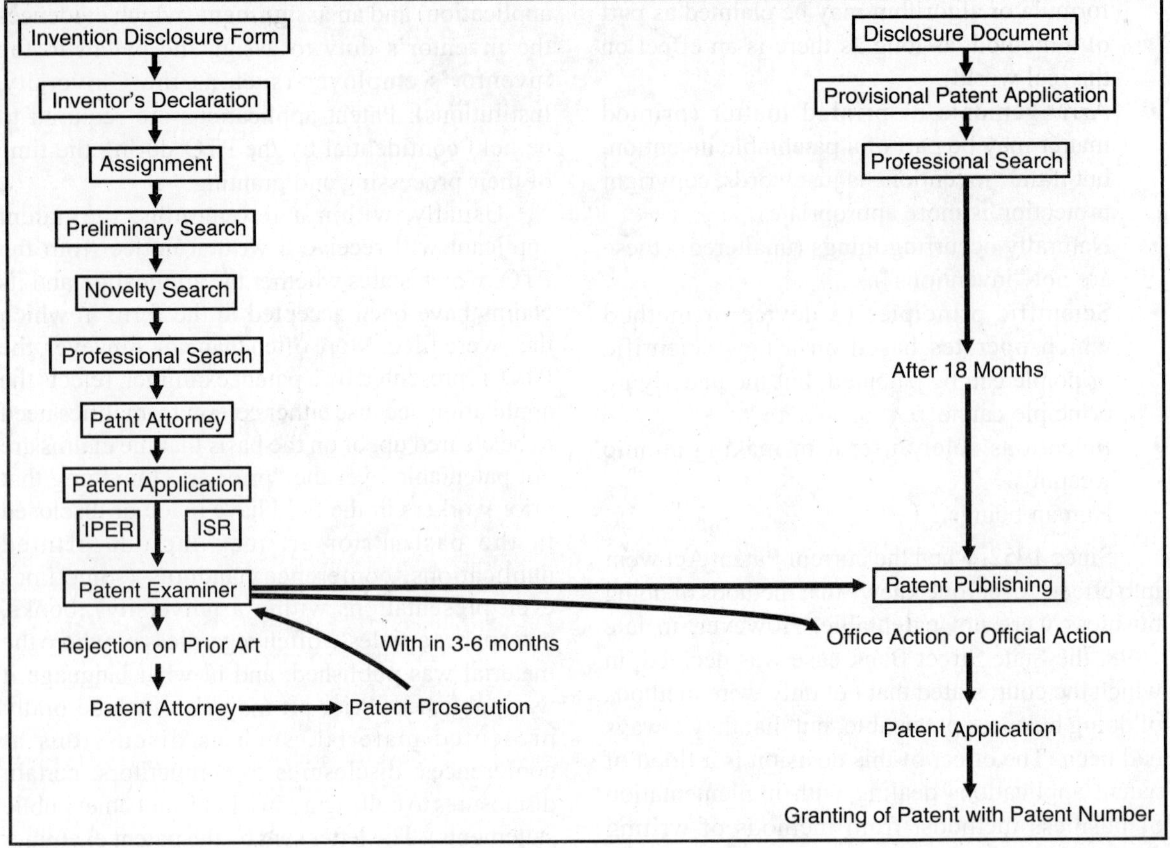

Fig. 16.1. Procedure involved in granting of patent. International preliminary examination report (IPER); International research report (ISR)

Currently, the average patent application is pending for just over two years in USA and over five years in India. Even though inventors in the fields of biotechnology and computer should plan on their applications taking longer. Once a patent is issued, it is enforceable until 20 years from the initial filing of the application that resulted in the patent, assuming that PTO-mandated maintenance fee is paid timely.

Pursuant to an international treaty—the Paris convention—foreign patent applications must be filed within a year of the filing of the US application in order to receive the filing date of the US application.

The procedure involved in patent application and granting of a patent can be summarized as follows:

- The inventor fills an Invention Disclosure Form, which provides the patent attorney with the description of the invention and other important information.
- The patent attorney conducts a novelty search at the Patent and Trademark Office in the respective country. This search will reveal US patents or the patents of other countries (if any) that cover inventions similar to the inventor's application. If the inventor wants to do his own preliminary search before ordering a professional search, he can do so.

- If, after reviewing the results of the novelty search, the inventor decides to proceed further, the patent attorney will prepare and file an application for him. The inventor may now mark the product "patent pending". The patent application will be held in confidence by the patent office, so only the inventor and the inventor's attorney will know what it contains for 18 months from the earliest filing date, at least.
- The patent examiner at the patent office will then review the application and do his or her own search.
- The examiner will issue an "office action," usually rejecting the application. The rejections are most often based on format or wording (sometimes called "112 rejections", after the section of the patent law which applies), novelty ("102 rejections"), or obviousness ("103 rejections").
- The inventor's attorney can reply to the action by modifying the application, arguing with the examiner, etc. The one thing that cannot be done later is adding any new matter to the application. Due to this reason, it should be made sure that the application is complete when it is filed. If, in any case, new materials have to be added, then it is necessary to file a new "Continuation In Part" patent application, and start the process all over again.
- If the examiner is satisfied by the reply given, then the patent will be issued to the inventor upon payment of an issue fee.

Provisional Application

Provisional application is a way to file a patent application without some of the formalities involved in the usual patent process. It is somewhat cheaper initially because of the lessened formality, but it does not substitute for a conventional utility patent application. A filing date is assigned to the application, but it will not be sent to an examiner, or reviewed in any way by the patent office. No patent will ever be issued on a provisional application. It is necessary to file the conventional application within a year of the provisional filing date, claiming the priority of the provisional application, or the application will lapse and will not receive the benefit of the earlier filing date.

Disclosure Document

The disclosure document program was started many years ago before the provisional application option became available in the US. It was a way of the patent office to take the disclosures people were sending them anyway, and charge for filing them. They are kept on file for two years and then discarded unless the people file a utility application within two years referring to the disclosure document. At most, a disclosure document will serve as evidence of your date of conception of the invention. Unlike a patent application, it does not give the filing date or set the date of invention.

While they are cheap to file, disclosure documents are of no positive value, and have, in some cases, been used to limit the scope of a later-issued patent. So, it is not recommended.

16.10. PATENTS IN INDIA

In March, 2005, parliament approved the Third Patent (Amendment) Bill 2005. The bill was introduced in parliament to replace the Patents (Amendment) Ordinance, 2004, promulgated on December 26, 2004, to meet the January 1, 2005 WTO deadline to move to a new patents regime.

The bill amends the Indian Patent Act, 1970 for the third time (the earlier two amendments were enacted in 1999 and 2002).

The salient features of the third amendment to the 1970 Act are as follows:

- Software patents excluded.
- Substituted 'new use' for 'mere new use', thus strengthening the provision that can be used to deny patents on the new use of a known substance.

- Refers the question of whether a pharmaceutical substance should be defined as 'a new entity involving one or more inventive steps' or 'a new chemical entity' to an expert committee.
- Prevents frivolous claims, and clarifies that the 'mere discovery of the new form of a known substance, which does not result in the enhancement of a known efficacy', is not patentable. But there is an ambiguity on what qualifies as 'enhancement of a known efficacy'.
- Salts, esters, ethers, polymorphs, metabolites, particle size, isomers, mixtures of isomers, complexes, combinations and other derivatives of known substances cannot be patented, unless they 'differ significantly in properties with regard to efficacy'. This has left the otherwise specific provision ambiguous and open to interpretation.
- The applicant has to comply with an 'inventive step', which means that the invention has to involve technical advances as compared to the existing knowledge or having economic significance or both. This requirement for technical advance has been diluted, as a patent can now be granted on economic significance alone.
- Patenting of microorganisms has been referred to an expert committee.

Exemptions for Patenting in India

Exemptions India, at present, does not allow the patenting of the following:

- Frivolous claims contrary to well established natural laws.
- Anything contrary to law or morality, or injurious to public health.
- Mere arrangement or rearrangement or duplication of known devices; each functioning independently of one another in a known way.
- A method or process of testing applicable during the process of the manufacture of rendering the machine, apparatus, or other equipment more efficient or for the improvement or restoration of the existing machine, apparatus, or other equipment, or for the improvement of control or manufacture.
- A method of agriculture or horticulture.
- Inventions related to atomic energy.
- Computer software.
- Aesthetic creations.
- Discoveries, scientific theories, mathematical methods.
- Schemes, rules or methods for performing mental acts, playing games, or doing business.
- Presentation of information.
- Methods of treating humans or animals through surgery or therapeutical diagnostics.
- Animals and plants, and biological methods of rearing/growing them (however, microorganisms are patentable in India).
- Products made by chemical synthesis—foods, medicines.

16.11. DRUG PATENTS IN INDIA

During 1947 to 1972, with a strong product patent regime under the Indian Patent Act of 1911, most of the effective drugs patented by foreign companies were not produced in India and the drug prices in the country were the highest in the world. Under the Indian Patent Act 1970, product patents were not allowed and only process patents were granted in respect of inventions related to drugs and medicines. This enabled the indigenous drug industry to manufacture the products patented in other countries by developing and using a different process.

In 1995, India agreed to adopt the product patents regime by 2005, as a part of its WTO commitments. This has encouraged our pharmaceutical companies to adopt a strategy of R&D based innovative growth. This can transform the Indian chemical and pharmaceutical industry, particularly the biotechnology sector, in which it has good prospects.

Parliament passed the patent (Amendments) Bill 2004 in March 2005 to further amend the Patent Act, 1970. The bill amends the Indian Patent Act, 1970 for the third time (the earlier two amendments were enacted in 1999 and 2002) to introduce product patents for drugs, food, and chemicals.

The salient features of the third amendment to the patent law are as under:

- Extension of product patent protection to all the fields of technology, by extending it to drugs, food and chemicals.
- Deletion of the provision related to exclusive marketing rights (EMRs), which would now become redundant, and introduction of a transition provision for safeguarding EMRs already granted.
- Introduction of a provision for enabling grant of compulsory licence for the export of medicines to the countries, which have insufficient, or no manufacturing capacity, to meet emergent public health situations (in accordance with the Doha declaration on TRIPs and public health).
- Modification in the provisions related to opposition procedures, with a view to streamlining the system by having both pre-grant and post-grant opposition in the patent office.
- Addition of a new provision to circumscribe rights in respect of mailbox applications, so that patent rights in respect of the mailbox shall be available only from the date of grant of patent, and not retrospectively from the date of publication.
- Strengthening the provisions related to national security, and to guard against patenting abroad of dual use technologies.
- Clarification of the provisions related to the patenting of software related inventions when they have technical application to industry or are in combination with hardware.
- Rationalization of provisions related to time-lines with a view to introduce flexibility and reduce the processing time for patent applications, and simplify and rationalize the procedures.

Following are the important public interest provisions in the patent law:

- Conditional grant of patent—the amendment empowers the government to import, make or use any patent for its own purpose. For drugs, it also empowers import for public health distribution.
- Revocation of patent in public interest—the amendment empowers the government to revoke a patent where it is found to be mischievous to the state or prejudicial to the public.
- Grant of compulsory license—the amendment deals with the general principles and circumstances for the grant of compulsory licences in order to protect public interest, particularly public health and nutrition. These provisions check the abuse of patent rights. They can be invoked if the reasonable requirements of the public with respect to patented inventions have not been satisfied, and the patented invention is not available for public at a reasonable price, and if the patented invention has not worked in the territory of India. Section 92 of this law provides for the action in case of national emergency, extreme urgency and public non-commercial use, and can be invoked without the grace period of three years from the grant of patent.
- Use of invention for the purpose of government.
- Acquisition of invention and patent for public purpose—the amendment empowers the government to acquire a patent to meet national requirements.
- Bolar provision—the amendment facilitates production and marketing of patented products immediately after the expiry of the term of patent protection by permitting preparatory action by non-patentees during the life of the patent.

- Parallel import—the amendment provides for import, so that patented product can become available at the lowest international price.

Apart from the manufacture of drugs, the product patent regime will help the pharmaceutical industry to tap outsourcing of clinical research. By participating in the international system of IPR protection, India, with its vast pool of scientific and technical personnel and well-established expertise in medical treatment and health care, has unlocked vast opportunities in both exports and outsourcing, and has the potential to become a global hub in' the area of R&D based clinical research. The patent legislation also provides adequate safeguards to protect the interest of the domestic industry and the citizens from any increase in the prices of drugs.

Apprehensions

There have been apprehensions from a few quarters that the patent amendment will drive up drug prices by ruling out access and availability of medicines at low cost. However, such apprehensions are unfounded. In the first place, the fact remains that 97% of all drugs manufactured in India are off-patents, and so will remain unaffected. These cover most of the life saving drugs, as well as medicines for common ailments. In patented drugs also, in most of the cases, there are always alternatives available. Further, the legislation has strong provisions for outright acquisition of patents to meet national requirements. Besides, there is also the Drug Price Control Order administered by the National Pharmaceuticals Price Authority. With such a framework in place, the concerns and fears related to rise in drug prices are misplaced, Besides, there are adequate safeguards to protect the interests of domestic industry and the common man from any increase in the prices of drugs.

16.12. VARIOUS TYPES OF PATENT APPLICATIONS IN INDIA

- Ordinary application
- Convention application
- PCT international application
- PCT national phase application
- Application for patent of addition
- Divisional application

Ordinary Application

An application for the patent made in the patent office without claiming any priority of application made in a convention country or without any reference to other application under process in the office is called an ordinary application.

Convention Application

When an applicant comes to the patent office with an application, claiming a priority date based on a similar application filed in one of the convention country, it is called a convention application (By virtue of Paris convention). To get a convention status, an applicant should file the application in Indian patent office within twelve months from the date of first filing of a similar application in the convention country. The priority document and its English translation (if required) should also be submitted by the applicant. A convention application should be accompanied by a complete specification. When two or more applications for the patents constituting one invention have been submitted in one or more convention countries, one application may be submitted within twelve months from the date on which the earlier or earliest of those applications was given. Multiple fee has to be remitted for multiple priorities, so that other applications filed earlier in the convention countries will be deemed to have been published in India. Applicant of convention application should furnish, when required by the Controller, copies of specification or documents (priority documents) certified by the official chief of the patent office of the convention country. If any such specification or document is in a foreign language, it should be translated into English.

PCT International Application

PCT is an international filing system for patents in which the applicant gains an international filing date in all the designated countries conferring the late entry (up to 31 months) to the national offices without affecting the priority date. This is a simple and economical procedure for the applicants seeking protection for the inventions in many countries. Indian patent office is a receiving office for international applications by nationals or residents of India. *An international application should be filed with the appropriate office in triplicate in respect of head office, and quadruplicate in respect of branch offices, either in English or in Hindi language (Rule 19(1)).*

PCT National Phase Application

An international application made according to patent co-operation treaty designating India can enter national phase within 31 months from the international filing date. This application, filed before the controller in the Indian patent office, claiming the priority and international filing date, is called PCT national phase application. Applicant can enter national phase with a request made on white paper. But Form 1A is preferred by the Indian patent office during National Phase Entry. The title, description, drawings, abstract and claims filed with the application should be taken as complete specification for the purpose of filing in India [S.10(4A)(i)]. The filing date of the application should be the international filing date accorded under the Patent Cooperation Treaty [S. 10(4A)]. It is not mandatory for the applicant to submit the documents while entering the national phase for filing the application in the designated or elected member countries, as it is obligatory on the part of WIPO to send those things to the designated offices. However, for convenience and faster processing, the applicant may submit the necessary documents. Office may ask for any other documents, which are necessary in addition to what was submitted along with the application.

Application for Patent of Addition

When an applicant feels that he has an invention, which is a slight modification of the invention for which he has already applied patent in India, the applicant can go for a patent of addition. The only benefit he gets is that there is no need to pay separate renewal fee for the patent of addition during the term of the main patent. Patent of addition expires along with the main patent unless it is made independent according to the provisions in the application. The complete specification of that application should include specific reference to the number of main patent or the application for the main patent as the case may be, and a definite statement that the invention comprises an improvement in, or a modification of the invention claimed in the specification of the main patent granted or applied for.

Divisional Application

When the application made by an applicant claims more than one invention, the applicant on his own request or to meet the official objection raised by the controller may divide the application and file two or more applications as applicable for each of the inventions. This type of application is called divisional application. The priority date for all the divisional applications will be same as that claimed by the parent application (ante dating). The complete specification of the divisional application should not include any matter disclosed in the complete specification of the first application. The reference of parent application should be made in the body of the specification.

16.13. PATENT SEARCH

A novelty search is required. There are two parts to such a search—a **literature search** (catalogs, websites, magazines, etc.,), which the inventor or patent attorney can perform; and a **search of issued patents and published applications**, which the applicant should do.

Patents may be searched at no cost on the website of the United States Patent and Trademark Office (http://www.uspto.gov/,), more particularly www.uspto.gov/patft/index.html

The details of the patent search can be obtained from http://www.uspto.gov/main/profiles/acadres.htm and http://www.uspto.gov/web/menu/helpfaq.htm#a20). The web page, searching "Issued Patents", will uncover patents issued (i.e., approved) in the United States since 1976. "Patent applications" are unissued (i.e., unapproved) patents, which are generally published 18 months after they are filed. Since applications have only been published since 2001, searching "Patent Applications" will uncover different numbers of (and limited number) of documents. Searching may also be performed at http://ipdl.wipo.int/

Patent Search on the Internet

A preliminary search of the invention on the US Patent and Trademark Office's web site, among others, which can supplement, if not replace, is a conventional search. Such a search is limited in several ways:

- A person can only do keyword searching, which means one has to try and guess what name the inventor or his attorney might have given to the features of the invention. One man's "screw" is another man's "advancing helix" or "threaded rod". A patent using the British term "adjustable spanner" may be the closest art to one's "monkey wrench".
- Keyword searching is an art in itself. Often a person will retrieve so many patents that the search becomes meaningless, and one needs to be conversant with boolean connectors to narrow things down. For example, a search in March, 2001, on "vacuum cleaner" resulted in 4,184 "hits"—"vacuum and cleaner" got 11,748—"vacuum or cleaner" got 346,223. A person needs to specify the search properly in the first place, and narrow down the search without throwing out relevant references.
- The database has a time limit. Although most patents issued by the USPTO since 1790 can be retrieved from the USPTO database (in image form), only the patents issued after 1976 are keyword searchable. Just because a patent has expired does not mean that it is no longer relevant for patentability purpose. Any patent, whenever or wherever issued, is considered "prior art".
- A person will not be able to easily look at the drawings. The USPTO database does permit viewing drawings, but one can only do it one page image at a time, and the process can be slow and cumbersome. Also, a special viewer must be downloaded and installed.
- If one is familiar with the terminology, patent language can be very confusing. It is easy to miss a close patent if you're not used to translating the language into the real world.

In the end, it is a good idea to check out your idea on the web and on the available free databases. If you find a "knock out", that's that. If you don't, however, these resources are no substitute for having an experienced professional doing a manual search.

More than 300,000 US patent applications are filed each year by people and companies from all over the world. The US government has issued more than 6,500,000 patents (The current patent numbering system began in 1836). Roughly 170,000 patents are issued each year in the United States.

16.14. PATENT COOPERATION TREATY (PCT)

The patent cooperation treaty is an agreement for international cooperation in the field of patents. It is the most significant advancement in this field, since the adoption of the Paris convention itself. It is, however, largely a treaty for rationalization and cooperation with regard to the filing, searching and examination of patent applications and the dissemination of the technical information contained therein. The PCT does not provide for the grant of "international patents". The task and responsibility for granting patents remain exclusively in the hands of the patent offices of, or acting for, the countries where the protection is sought (the "Regional Offices"). PCT is a special agreement under the Paris Convention open only to states, which are also party to the Paris convention. The PCT does not compete with, but, in fact, complements the Paris convention.

Introduction

On 7th September, 1998, India deposited its instrument of accession to the PCT, and on 7th December, 1998, thus became a member of the PCT, as the 98th contracting state of PCT. Furthermore, residents of India are entitled to file international applications for patents under PCT at the Patent Office, Kolkata, with effect from 9th November, 1999. The branches of the patent office at Mumbai, Chennai, and New Delhi are also receiving the PCT applications.

Principal Objectives of the PCT

The principal objective of the PCT is to simplify and render more effective and economical, in the interests of the users of the patent system and the offices that have responsibility for administering it, the previously established means of obtaining the protection for inventions in several countries. Before the introduction of the PCT system, the only means by which protection of an invention could be obtained in several countries was to file a separate application in each country. These applications, each being dealt within isolation, involved repetition of work of the filing and examination in each country.

To Achieve its Objective, the PCT

- establishes an international system which enables the filing, with a single patent office (the "Receiving Office"), of a single application (the "International Application") in one language having effect in each of the countries, which are party to the PCT that the applicant names ("designates") in his application;

- provides for the formal examination of the International Application by a single patent office—the receiving office;

- subjects each international application to an international search, which results in a report citing the relevant prior art (mainly published patent documents relating to previous inventions), which may have to be taken into account in deciding whether the invention is patentable. The report is first made available to the applicant and is published later;

- provides for centralized international publication of international applications, with the related international search reports, as well as their communication to the designated offices; and

- provides the option of an international preliminary examination of the international application, which gives to the offices, that has to decide whether or not to grant a patent, and to the applicant, a report containing an opinion as to whether the claimed invention meets certain international criteria for patentability. The procedure described in the preceding paragraph, is illustrated by two timelines such as chapter I and chapter II of

PCT. It is commonly called the "International Phase" to describe the first part of the patenting procedure, whereas one speaks of the "National Phase" to describe the last part of the patent granting procedure, which, as explained in the paragraph above, is the task of the designated offices, that is, the national offices of, or regional offices acting for the countries which have been designated in the international application. (In PCT terminology, a reference to "national" office, "national" phase and "national" fee, includes the reference to the procedure before a regional patent office). Under the PCT system, by the time the international application reaches the national office, it has already been searched by the International Searching Authority and possibly examined by an International Preliminary Examining Authority, thus providing the national patent offices with the important benefit of reducing their workloads, since they have the benefit of international phase procedures and thus need not duplicate those efforts. Further objectives of the PCT are to facilitate and accelerate access by industries and other interested sectors of technical information related to inventions, and to assist developing countries on gaining access to the technology.

Functions of the Receiving Office

- In the first step, the receiving office receives the international application from the applicant.
- In the second step, the receiving office checks the international application to determine whether it meets the prescribed requirements such as form and contents of the international application. This check is of a formal nature only and does not go into the substance of the invention. It, therefore, extends only to a certain number of rather elementary requirements specified in the treaty as forming part of that check.

— The receiving office shall accords as the international filing date, the date of receipt of the international application, provided the office has found that in accordance with article 11 at the time of receipt:

  (a) if the office finds that the international application did not, at the time of receipt, fulfill the requirements listed in paragraph (1), it shall, as provided in the regulations, invite the applicant to file the required correction.

  (b) if the applicant complies with the invitation, as provided in the regulations, the receiving office shall accord, as the international filing date, the date of receipt of the required correction.

- Receiving office checks certain formal and physical requirements (article 14). This may show that the international application does not meet certain requirements such as the form and content, and that the fees are, or not, fully paid. In that case, the receiving office communicates with the applicant in order to give him an opportunity to correct any defect. If the language of the filing of the international application is acceptable by the receiving office, but is not acceptable by the International Searching Authority, which is to carry out the international search, the applicant is required to furnish, within one month from the filing date of the application, a translation into a language which is as follows:

  — A language accepted by the International Searching Authority to carry out the international search.

  — A language of publication.

  — A language accepted by the receiving office (unless the international application is filed in a language of publication). In cases where the applicant fails to furnish, within the applicable time limit, a translation for the purpose of international search, the receiving office invites the applicant to furnish the missing translation, which in certain cases, is subject to the payment of a late furnishing fee. A separate invitation procedure is provided where the request does not comply with language requirements. When the applicant does not furnish the missing translation within the time limit fixed in the invitation, the international application, subject to certain safeguards for the applicant, will be considered withdrawn and the receiving office will so declare.

- Not all the requirements of the international application are required to be examined by the receiving office. For instance, the receiving office does not deal with substantive questions, such as whether the disclosure of the invention in the application is sufficient and whether the requirement of unity of invention is complied with. It also does not check many of the detailed physical requirements of the international application. These requirements are only checked to the extent that compliance with such requirements is necessary for the purpose of reasonably uniform international publication.

- Typical examples of defects, which may be corrected without affecting the international filing date, are:
  — Non-payment or partial payment of fee.
  — Lack of signature of the request.
  — Lack of a title of the invention.
  — Lack of an abstract.
  — Physical defects.

- As stated, in all such cases, lack of correction leads to the application being considered withdrawn, except where a physical defect would not prevent reasonably uniform international publication, and except for the payment of fee. With regard to the later, Rule 16b provides that the receiving office must invite the applicant to pay the missing fee together with a late payment fee. If the

applicant still does not pay the fee within the time limit fixed in the invitation, the receiving office will declare that the international application is being considered withdrawn. This solution protects the applicant against any loss of his application due to an erroneously delayed or incomplete payment of fee.

- In the third step, the receiving office transmits the *"record copy"* of the international application to the International Bureau, and the *"search copy"* to the International Searching Authority. The receiving office keeps a third copy, the *"home copy"*. The transmittals do not take place if and as long as national prescriptions concerning national security apply. The receiving office will then declare that national security provisions prevent the international application from being treated as such.

- The receiving office must mail the record copy promptly to the International Bureau, and in any case not later than five days prior to the expiration of the 13th month from the priority date. In many cases, the international application claims the priority of an earlier national application and is filed at the end of the 12-month priority period; the receiving office has only a few weeks for its processing tasks.

- The search copy must be transmitted by the receiving office to the International Searching Authority at the time of the transmittal of the record copy to the International Bureau, except where the search fee has not been paid on time, in which case the transmittal of the search copy takes place after the fee has been paid.

- If an applicant, who is a resident or national of a PCT contracting state, erroneously files his international application with a national office, which acts as a receiving office under the treaty, but is not competent under Rule 19.1 or 19.2, with regard to the applicant's residence and nationality, to receive that international application, or if an applicant files his international application with the competent receiving office in a language, which is not accepted by that office under Rule 12.1(a) but is in a language accepted under that Rule by the International Bureau as receiving office, the international application will be considered to have been received by the national office on behalf of the International Bureau as receiving office on the date on which it was received by the national office, and will be promptly transmitted to the International bureau as receiving office (unless such transmittal is prevented by national security prescriptions). The transmittal may be subjected by the national office to the payment of a fee equal to the transmittal fee. All other fee already paid to that office will be refunded by that office to the applicant, and the applicable fee will have to be paid to the International bureau as receiving office.

Conditions which Should be Fulfilled for According an International Filing Date

The applicant should be a resident or national of the contracting State for which the receiving office acts, and has the right to file with that receiving office (note, however, that the international application is to be transmitted to the International Bureau as receiving office under Rule 19.4(a):

(i) If that condition is not fulfilled.

(ii) The international application should be in the language, or one of the languages, accepted by the receiving office for the purpose of filing international applications (note, however, that the international application is to be transmitted to the International Bureau as receiving office under Rule 19.4(a)(ii) if that condition is not fulfilled).

(iii) The international application should contain at least the following elements: (a) an indication that it is intended to be an international application, (b) the designation of at least one contracting state (c) the name

of the applicant in a form allowing the applicant's identity to be established, (d) a part which on the face of it appears to be a claim or claims.

(iv) If one of these requirements is only complied with after correction, the international filing date will be the date on which the correction was received. In other words, in these cases a defect, which is corrected later, affects the international filing date. If all such defects are not properly corrected, the application will not be treated as an international application.

(v) For all the other cases, non-compliance with the formal requirements does not affect the international filing date. In other words, if the applicant corrects a defect in such cases, the international filing date remains unchanged. If the applicant does not correct, the defect properly, the international application will, however, be considered withdrawn by the receiving office. Extension of the time limit fixed by the receiving office for the correction of defects under article 14 may be requested.

Monitoring of Time Limits

Easy supervision and monitoring of only a few time limits and events is required by applicants, namely:

(i) Monitoring the receipt of the confirmation of receipt of the international application by the receiving office.

(ii) Monitoring the time limits for payment of fee.

(iii) Checking the notification from the International Bureau, confirming the receipt of the International Bureau confirming the receipt of the record copy (Form PCT/IB/301) for correctness of the designations indicated on the form, and if needed, confirm precautionary designations within 15 months from the priority date.

(iv) Deciding, after receipt of the international search report, whether or not to file a demand for international preliminary examination (which must be filed prior to the expiration of 19 months from the priority date).

(v) Deciding, after receipt of the international search report, whether or not to file amended claims under article 19, within the applicable time limit. This will usually be considered only if demand for international preliminary examination is not filed.

(vi) Monitoring the receipt, during the 19th month from the priority date, of the notice from the International Bureau (Form PCT/IB/308) that the publication of the international application and its communication to the designated offices (article 20) has been effected.

(vii) Entering the national phase before the expiration of 30/31 months from the priority date by paying the national fee and furnishing (if required) a translation of the international application.

16.15. PATENTS AND COMPULSORY LICENSING

Patents are granted for the purpose of exploitation, which will enhance industrial development and therefore should be worked in its fullest extent within the territory of India. The Patents Act provides that *"patents are granted to encourage inventions and to secure that the inventions are worked in India on a commercial scale and to the fullest extent reasonably practicable without undue delay and they are not granted merely to enable patentees to enjoy a monopoly for the importation of the patented article"*. The Patentee and every licensee should furnish the details of working of the invention at every six months and whenever required by the Controller.

Compulsory Licenses

The provisions for compulsory licenses are made to prevent the abuse of patent as a monopoly, and to make the way for commercial exploitation of the invention by an interested person. Any person, who is interested, can give an application for the grant of compulsory license for a patent after three years from the date of grant of that patent on any of the following grounds:

- The reasonable requirements of the public with respect to the patented invention have not been satisfied.
- The patented invention is not available to the public at an affordable price.
- The patented invention is not worked in the territory of India. The request for grant of a compulsory license can also be made by a licensee of the patent. Application for compulsory license should be made in Form 17 with the prescribed fee of Rs.1500 for natural person, and Rs. 6000 for other than natural person with a statement setting out the nature of the applicant's interest and the facts upon which the application is based.

Some Guidelines for Compulsory Licenses

- Failure to make patented inventions available to public at a reasonable price has to be considered keeping in mind that the reasonable price for a patented article will depend upon the circumstances of each case.
- Failure to work the patented invention within the territory of India will be considered with respect to the facility available in India for the working of the invention. Provision of importation of patented article is allowed. But the mere importation of patented articles, when there is a possibility of manufacturing within India, will be a factor that will receive consideration.
- Reasonable requirements of the public will be deemed as not satisfied, if due to the refusal of the patentee to grant a license on reasonable terms, an existing trade or industry or the development thereof, or the establishment of any new trade or industry in India or

the trade or industry of any person or class of persons in India is prejudiced. Other conditions are also mentioned in *S. 84(7)*.

- In determining whether to order the grant of a license, the controller also takes account the considerations of the following [S. 84(6)]:

  — The nature of the invention, the time since the grant of the patent, and the measures already taken by the patentee or any licensee to make full use of the invention in order to recognize whether the invention required longer time to establish adequate working. This is reflected by the power under section 86(1) to adjourn an application to allow sufficient time for working.

  — The ability of a potential licensee (the applicant) to work the invention to the public advantage, and his capacity to undertake the risk in providing the capital and working the invention.

  — Whether the applicant has made any genuine efforts to obtain voluntary license from the patentee. This will demonstrate the intention of the person-seeking license that he is really interested in manufacturing the product, which is protected under the patent. This provision will not apply in times of national emergency. If upon consideration of the evidence, the controller is satisfied that a prima facie case has not been made out for the making of an order under section 84, he shall notify the applicant accordingly, and unless the applicant requests to be heard in the matter, within one month of the date of such notification, the controller shall refuse the application. But if the applicant requests for a hearing within one month of the notification, the controller shall, after giving the applicant an opportunity of being heard, determine whether the application may be proceeded with or whether it shall be refused.

Revision of the Grant of Compulsory License

- An application under sub-section (4) of section 88 for the revision of the terms and conditions of a license, which has been settled by the controller, shall be in form 20 along with the prescribed fee of Rs. 1500/- for natural person, Rs. 6000/- for other than natural person in duplicate, and shall state the facts relied upon by the applicant and the relief he seeks, and shall be accompanied by evidence in support of the application. Such application can be made only after twelve months from the compulsory, licensee has started working the invention.

- If the controller is satisfied that a prima facie case has not been made out of the revision of the terms and conditions of the license, he may notify the applicant accordingly, and unless within a month the applicant requests to be heard in the matter, the controller may refuse the application.

- The controller, after giving the applicant an opportunity of being heard, shall determine whether the application shall be proceeded with or refused.

- If the ontroller allows the application to be proceeded with, he shall direct the applicant to serve copies of the application and the evidence in support thereof upon the patentee or any other person appearing in the register to be interested in the patent or upon any other person on whom in his opinion such copies should be so served.

- The applicant shall inform the controller the date on which the service of copies of application and of the evidence on the patentee and other persons referred to in the above para has been effected.

- The patentee or any other person to whom copies of the application and evidence have been served, may give to the controller notice of opposition in Form 14 in duplicate, within 2 months from the date of such service. Such notice shall contain the grounds relied upon by the opponent, and shall be accompanied by evidence in support of the opposition.

- The opponent shall serve copies of the notice of opposition and his evidence on the applicant, and inform the controller the date on which such service has been effected.
- No further evidence or statement shall be filed by either party without special leave of or on requisition by the controller.
- On completion of the above proceedings, or at such other time as he may deem fit, the controller shall appoint a date and a time for the hearing of the case, and shall give the parties not less that 10 days notice of such hearing.
- If the controller decides to revise the terms and conditions of license, he shall amend the license granted to the applicant in such a manner as he may deem necessary.

Compulsory Licensing of Patents Related to the Manufacture of Pharmaceutical Products for Export to Countries with Public Health Problems

Section 92 A of the Patent Act 1970, as amended by The Patents (Amendment) Act 2005, has been recently introduced, and states—92 A. Compulsory licence for export of patented pharmaceutical products in certain exceptional circumstances:

- Compulsory licence shall be available for manufacture and export of patented pharmaceutical products to any country having insufficient or no manufacturing capacity in the pharmaceutical sector for the concerned product to address public health problems, provided compulsory licence has been granted by such country, or such country has, by notification or otherwise, allowed importation of the patented pharmaceutical products from India.

- The controller shall, on receipt of an application in the prescribed manner, grant a compulsory licence solely for the manufacture and export of the concerned pharmaceutical product to such country under such terms and conditions as may be specified and published by him.

- The provisions of sub-sections (1) and (2) shall be without prejudice to the extent to which pharmaceutical products produced under a compulsory license can be exported under any other provision of this act. (Explanation—for the purpose of this section, "pharmaceutical product" means any patented product, or product manufactured through a patented process, of the pharmaceutical sector needed to address public health problems, and shall be inclusive of the ingredients necessary for their manufacture, and diagnostic kits required for their use). In order to get compulsory licence under 92 A, the applicant has to file application in Form 17 along with the prescribed fee of Rs. 1500/- for natural person, and Rs. 6000/-for other natural person.

This provision is to be construed in wider sense to allow export to any country having insufficient or no manufacturing capacity in the pharmaceutical sector, whether it is a member of WTO or not. As this section is intended to address the public health problems faced by a country having insufficient or no manufacturing capacity in the pharmaceutical sector, and to facilitate access to affordable medicines for the people in these countries, it should be used in good faith and not with the primary purpose of addressing other objectives, in particular, of a purely commercial nature. It may be noted that this section is an "enabling provision" for the export of pharmaceutical products to any country having insufficient or no manufacturing capacity in the pharmaceutical sector in certain exceptional circumstances to address public health problems. The term "provided compulsory licence has been granted by such country" has therefore been given a liberal meaning, and it should be considered to include 'a licence in any form' from such countries also, where there is no patent protection or where the drugs in question are not patented.

Government Use of the Inventions

- Anytime, after filing or grant of patent, government or any person authorized by it can use the patented invention for the purpose of government.

- If an invention is used before the priority date of the relevant claim of complete specification by the government or any person authorized by it for the purpose of government, no royalty or remuneration need to be paid to the patentee.

- If an invention is to be used at any time after the grantance of complete specification by government or any person authorized by it for the purpose of government, its use should be made only on terms agreed between the government or any authorized person and the patentee, or in default of agreement be decided by High Court of India under section 103.

- Government can authorize any person in respect of an invention, either before or after the grant, whether or not the patentee authorizes that person.

- Where government authorizes any person for using an invention for government purposes, then, unless it is contrary to the public interest, the Central Government shall inform the patentee from time to time the extent of the use of the invention for the purpose of government. In case of use by the undertaking, government may call for such information from the undertaking.

- The right to use the invention for the purpose of government includes the right to sell the goods, and the purchaser has the power to deal with the goods as if the government or the person authorized were the patentee of the invention.

- In case, of an exclusive licensee, as per section 101(3), or an assignor, Central Government should also inform the exclusive licensee or assignor as the case may be, regarding the extent of the use of invention for the purpose of government.

- In respect of an invention used by the government for the purpose of government, any agreement, license of assignment etc. between the patentee or applicant and any person other than the government shall have no effect, if the agreement restricts the use for the purpose of government, or instructs any payment in respect of any use for the purpose of government.

- In relation to any use of the invention made for the purpose of government by the patentee to the order of Central Government any sum payable by virtue of section 100(3) shall be divided between the patentee and the assignor in such proportion as may be agreed or in default be decided by High Court of India under section 103.

- In case there is an exclusive licensee authorized under his license to use the invention for the purpose of government, the patentee shall share any payment, and such licensee in such proportion as agreed upon or in default is decided by High Court under section 103.

- If necessary central government can acquire an invention from the applicant or patentee for a public purpose, by publishing a notification to that effect in the official gazette.

- Notice of acquisition shall be given to the applicant or patentee, as the case may be, and other persons appearing in the register as having interest in that patent.

- Compensation should be given by the Central Government to the concerned person, as agreed upon between them or in default be determined by High Court of India under section 103.

Any dispute arising out of the use of an invention by the Central Government for the purpose of Government may be referred to the High Court by either party to the dispute in such manner as may be prescribed by rules of High Court under section 103 government may ask for revocation of patent or raise an issue regarding the validity of the patent. In case the government thinks disclosure of any document regarding the invention be prejudicial to the public interest, then the government can disclose confidentially to the advocate of other party in any proceeding at any time at the High Court.

16.16. PROCEDURE FOR THE GRANT OF COMPULSORY LICENCE

The procedure for the grant of compulsory licence is not cumbersome. An application under rule 96 of the Patents Rule 2003, as amended by the Patents (Amendment) Rules, 2005, shall be filed on the prescribed form 17, with the prescribed fee of Rs. 1500 for natural person, and Rs. 6000 for others, setting out the nature of applicant's interest, and terms and conditions of the licence, that the applicant is willing to accept. The controller, upon consideration of the evidence, if satisfied that a *prima facie* case has been made out, as prescribed in section 87 of the Patents Act, shall direct the applicant to serve the copies of the application upon patentee, and will publish the application in official journal. The patentee, if so desires, may oppose the application. Thereafter, the controller shall order the grant of licence upon such terms as he may deem fit, keeping in view the provisions of section 90 of the patents act. The time for the grant of compulsory licence in normal cases may not exceed six months. Further in the case of availability of compulsory licence by way of notification by the Central Government under section 92 of the Act, an application also on form 17 will be filed. In the case of section 92(1), the provisions of section 87 will apply and the compulsory licence shall be granted on such terms and conditions as the controller think fit, securing that the articles manufactured under the patent shall

be available to the public at the lowest prices, and the licensee may also export the patented product, if need be in accordance with the provisions of sub-clause (iii) of clause (a) of section 84. In the case of semi-conductor technology, the licence granted is to work the invention for public non-commercial use. In case the licence is granted to remedy a practice determined after judicial or administrative process to be anti-competitive, the licensee shall be permitted to export the patented product, if need be. However, in the circumstances of National Emergency or Extreme urgency or public non-commercial use including public health crises, relating to Acquired Immuno Deficiency Syndrome, Human Immuno deficiency virus, tuberculosis, malaria or other epidemics, to avoid any delay the procedure under section 87 will not apply. The compulsory licence will be granted, immediately under section 92(3) also with the terms and conditions that the articles manufactured under the patent shall be available to the public at the lowest prices.

In case of compulsory licence for export of patented pharmaceutical products to any country having insufficient or no manufacturing capacity to address public health problem, the application will also have to be made on Form 17, and the compulsory licence will be granted, immediately under section 92A of the Act.

Notification by Central Government— Availability of Compulsory Licence

Section 92(1) opposition for the settlement of terms and conditions.

Immediate grant of compulsory licence for export of patented pharmaceutical products to the countries having insufficient or no manufacturing capacity to address public health problems. **Patent** section 92 A three years after grant section, 84 any time after grant, section 92 Any time after grant Immediate grant of compulsory licence under section 92(3) in the circumstances of national emergency or extreme urgency or public non-commercial use including public health crises, relating to AIDS, HIV, tuberculosis, malaria or other epidemics.

Biotechnological Inventions

In the field of biotechnology the invention may be related to living entity of natural origin, such as animal, plant, human beings including parts thereof, living entity of artificial origin, such as microorganism, vaccines, transgenic animals, and plants etc.; biological materials such as DNA, plasmids, genes, vector, tissues, cells, replicons etc.; process related to living entities; process related to biological material; methods of treatment of human or animal body, and biological processes. The living entities of natural origin such as animals, plants, in whole or any parts thereof, plant varieties, seeds, species, genes and microorganism are not patentable. Any process of manufacture or production related to such living entities is also not patentable. Any method of treatment such as medicinal, surgical, curative, prophylactic, diagnostic and therapeutic of human beings or animals or other treatments of similar nature are not patentable.

Any living entity of artificial origin such as transgenic animals and plants, any part thereof are not patentable. The living entity of artificial origin such as microorganism, vaccines are considered patentable. The biological materials such as organs, tissues, cells, viruses etc. and process of preparing thereof are not patentable under section 3(c). The biological material such as recombinant DNA, Plasmids and processes of manufacturing thereof are patentable provided they are produced by substantive human intervention.

Gene sequences, DNA sequences without having disclosed their functions are not patentable for lack of inventive step and industrial application. The processes related to microorganisms or producing chemical substances using such microorganisms are patentable. Essentially biological processes for the production of plants and animals such as method of crossing or breeding etc. are not patentable. Any biological material and method of making the same which is capable of causing serious prejudice to human, animal or plant lives or health or to the environment including the use of those would be contrary to public order and morality are not patentable such as terminator gene technology. The processes for cloning human beings or animals, processes for modifying the germ line, genetic identity of human beings or animals, uses of human or animal embryos for any purpose are not patentable as they are against public order and morality. In case of use of biological material in the invention disclosed in the patent application the source or geographical origin of such material is required to be mentioned in the specification. In case of use of new biological materials in the invention disclosed in the patent application, such materials are required to be deposited in any of the International Depositary Authorities (IDA) recognized under the BUDAPEST treaty on or before filing of the application in order to supplement the description for sufficiency of disclosure of the invention and reference of such deposit to be made in the patent specification. Any invention which in effect is traditional knowledge or which is an aggregation or duplication of known properties of traditionally known components is not patentable.

16.17. IMPACT OF INTELLECTUAL PROPERTY RIGHTS OR PATENTING ON BIODIVERSITY RICH DEVELOPING COUNTRIES

The revolutions in biotechnology and intellectual property protection began in the developed world. The USA led the global transformation of intellectual property protection, and has been the leader in commercialization of biotechnology in agriculture. Now all members of the World Trade Organization are committed to offer intellectual property protections for agriculture. Will the benefits of agricultural biotechnology proliferate globally as a result? Can we now rely on the dynamism and focus of the private sector to exploit the potential of biotechnology to meet the needs of developing nations? In this chapter, we look forward, drawing some inferences from the record thus far about the future relevance of agricultural biotechnology for developing countries.

To date, the large and overwhelmingly public agricultural research effort of developing countries has, with a few notable exceptions, made relatively little progress in developing and commercializing agricultural biotechnology innovations. Some applications, such as virus elimination using *in vitro* propagation, marker-assisted breeding, and use of genetic cultivar identification to improve the efficiency of germplasm conservation, have been used for years in many countries. But here we shall concentrate on the technology that has most captured the interest of farmers, breeders and private firms—the development of new cultivars with transgenic traits. As in North America, transgenic traits offering tolerance to broad-spectrum herbicides or resistance to insects involving *Bt* genes, have dominated agricultural biotechnology innovations adopted by farmers in other countries. Although contributions from China are becoming increasingly significant in terms of variety of traits and the range of species of transgenic plants and animals under development, achievements in development of new transgenic plant varieties elsewhere have been very modest (Pardey *et al.*, 2005).

Until recently, agricultural research in developing countries, which is predominantly public, has been less constrained by valid intellectual property rights than many international researchers and other observers appear to have believed. As intellectual property rights proliferate in the most important export and consumer markets, the problems of freedom to operate in developing economies will become similar to those encountered by the public and non-profit researchers who account for almost half of agricultural research effort in developed

economies. Experience thus far in countries that are leaders in applications of biotechnology offers important insights for the dozen or so developing countries with substantial near-term potential for the development and application of agricultural biotechnology. First, scientists in the public laboratories have been relatively unconstrained by most intellectual property claims not embodied in materials, because they ignore them. For them, freedom to operate tends to mean ability to get the tools needed to do the project. Conversely, lack of freedom to operate is perceived as lack of timely access to materials physically controlled by others. Material transfers are the intellectual property claims that dominate their attention.

For research policy makers, the situation is quite different (Graff *et al.*, 2004). The private firms who tend to acquire most claims on key technologies originating in the public and private sectors are focused on a handful of crops, but their technologies are much more broadly applicable. Although they cannot, and often have no incentive to, control infringers in the lab, they can prevent the commercialization of new cultivars and other innovations achieved using their proprietary technology. This is the major freedom to operate problem for the stronger public agricultural researcher programs (as distinct from individual scientists in these programs), in rich and poor countries alike. It can become evident only in that tiny but important subset of lines of research that are successful enough to have been commercialized (In the survey by Cohen 2005, only seven cases, of 201, had reached the commercialization stage).

It is likely that one or more patented technologies will become embodied in the product by the scientists oblivious to IPR claims. When this has happened in the course of public-sector transgenic innovations for use in a developing country, failure of the patentee(s) to consent to negotiate seems to have been a more prevalent roadblock than demands for excessive royalties. Concerns with control, reputation and the cost of negotiation seem to outweigh the modest licensing revenues obtainable from crops of no direct interest to the private sector. Even a single 'train wreck' in public sector development can mean frustration, withdrawal of funding sources, and indeed the failure of public research programs.

Nations with strong agricultural research capacity, like China, Brazil, India and Argentina have shown that they need little or no overseas investment to support rapid diffusion of the most profitable transgenic technologies originating in the USA and elsewhere. Strong domestic intellectual property protection is probably not in their interests. This might change, if they succeed in developing and patenting new transgenic rice technology, for example, that is attractive to other countries. It may be that they will demonstrate a comparative advantage in transgenic cultivar development for some crops in the medium term. In any event, these countries should give serious attention to the initiatives such as PIPRA and BIOS, which also serve the needs of public agricultural research in the rich countries. Beyond the effects on the choice of programs and their successful completion, the new IPR regime is also changing the perceived mission and funding structure of the stronger public research institutions in developing countries. For example, Koo *et al.*, (2006) report that EMBRAPA, the large Brazilian federal agricultural research organization, has negotiated contracts with private seed producers that restrict the latter from pursuing further research on the germplasm they receive, in sharp contrast with the traditional notion of the role of public research institutions in fostering private applied research based on their own innovations.

Finally, it is important to recognize that the dominant constraint on agricultural innovation for the majority of developing countries is not at present IPR. It is a lack of sufficient and sustained funding in the face of well-documented high social returns on research using non-transgenic technologies, a problem also observed in developed economies with strong IPR, for all but

a handful of crops. Especially in poor countries, another serious constraint at present is strict and costly biosafety regulation. For many, if not most, developing countries, these problems will not be solved in the near future. For such countries, applications of biotechnology are likely to be concentrated in non-food cash crops, and in particular in corn, soybean, and cotton, often using cultivars developed from transgenic events produced in the rich countries. Where substantial adaptive research or market development investment is required, strong intellectual property rights might be necessary to encourage rapid development and diffusion of imported technology by private seed producers. Where local public sector biotech research capacity continues to be weak, local development of biotech for staple food crops is unlikely to occur, regardless of intellectual property protection.

As noted above, a major issue of importance to the future of agricultural research is the conservation of genetic resources held in fields and in national and international collections, along with guaranteed access for researchers on terms that recognize the contribution made by farmers in the developing world in conserving, improving and making available these resources. The foundation for international action to ensure the conservation, use and availability of plant genetic resources was the FAO Undertaking on Plant Genetic Resources agreed in 1983.

Subsequently, the concept of Farmers' Rights arose in debates in the FAO, where it was recognized that there was an imbalance between the IP rights afforded to the breeders of modern plant varieties and the rights of farmers who were responsible for supplying the plant genetic resources from which such varieties were mainly derived. A second concern was the consistency between making plant genetic resources available as the common heritage of mankind, and the taking out of private IP rights on varieties derived from them.

In 1989, the FAO agreed to recognize these concerns by incorporating Farmers' Rights arising from the past, present and future contributions of farmers in conserving, improving, and making available plant genetic resources, particularly those in the centres of origin/diversity in the undertaking. Farmers' rights were to be implemented through an International Fund for Plant Genetic Resources, which would finance relevant activities, particularly in developing countries. Subsequently, the FAO agreed that Plant Breeders' Rights, as provided for under UPOV...are not incompatible with the International Undertaking, a choice of words that reflected the continuing ambivalence felt by some developing countries about the underlying consistency between the undertaking and UPOV.

The agreement of the CBD in 1992, was the basis on which the process of transforming the undertaking into the treaty (ITPGRFA), finally agreed in 2001, was undertaken. The ITPGRFA has the specific objective of facilitating access to plant genetic resources held by contracting parties, and those in international collections, for the common good, recognizing that these are an indispensable raw material for crop genetic improvement, and that many countries depend on genetic resources which have originated elsewhere. This represents an implementation of the CBD principles taking account of the specific characteristics of plant genetic resources. Most varieties now in existence, in particular those derived from public breeding programmes, contain genetic material from many sources, often derived from genetic material in gene banks, which themselves may have diverse origins.

The ITPGRFA also recognizes the contribution of farmers in conserving, improving and making available these resources, and this contribution is the basis of Farmers' Rights. It does not limit in any form whatsoever rights that farmers may enjoy under national law to save, use, exchange and sell farm-saved seed. It also sets out the right to participate in decision-making about, and to derive fair and equitable benefits from, the use of these resources.

16.18. FARMERS' RIGHTS

The ITPGRFA leaves it entirely up to national governments to implement farmers' rights. Thus, implementing specific farmers' rights is not an international obligation like that imposed under provisions in TRIPS.

The rationale for farmers' rights combines arguments about equity and economics. Plant breeders and the world are at large benefit from the conservation and development of plant genetic resources undertaken by farmers, but farmers are not recompensed for the economic value they have contributed. Farmers' rights may be seen as a means of providing incentives for farmers to continue to provide services of conservation and maintenance of biodiversity. As noted, the protection of plant varieties contains an inherent tendency to encourage uniformity and reduce biodiversity, to which the traditional practices of farmers are an essential counterweight. Farmers should be supported in recognition of the economic value they conserve, which is not recognized in the market system, and is to some extent threatened by technical change and the extension of plant breeders' protection. Moreover, the extension of intellectual property protection does carry the risk of restricting farmers' rights to reuse, exchange and sell seed, the very practices which form the basis of their traditional role in conservation and development.

Farmers' rights are not an intellectual property right, but they need to be viewed as an important counterbalance to the rights accorded to breeders in the formal sector under PVP or patents. However, defining how to implement these rights at national level is complex, as we discuss in the next chapter in the context of CBD. The Treaty provides for a financing mechanism to be set up, financed by contributions and the share of the proceeds of commercializations, which will enable the implementation of agreed plans and programmes for farmers, who conserve and sustainably utilize plant genetic resources for food and agriculture.

Farmers' Rights in ITPGRFA

- The contracting parties recognize the enormous contribution that the local and indigenous communities and farmers of all regions of the world, particularly those in the centres of origin and crop diversity, have made and will continue to make for the conservation and development of plant genetic resources, which constitute the basis of food and agriculture production throughout the world.
- The contracting parties agree that the responsibility for realizing Farmers' rights, as they relate to plant genetic resources for food and agriculture, rests with national governments. In accordance with their needs and priorities, each contracting party should, as appropriate, and subject to its national legislation, take measures to protect and promote farmers' rights, including:
  - protection of traditional knowledge relevant to plant genetic resources for food and agriculture;
  - the right to equitably participate in sharing benefits arising from the utilization of plant genetic resources for food and agriculture; and
  - the right to participate in making decisions, at the national level, on matters related to the conservation and sustainable use of plant genetic resources for food and agriculture.
- Nothing in this article shall be interpreted to limit any rights that farmers have to save, use, exchange and sell farm-saved seed/propagating material, subject to national law and as appropriate.

Multilateral System

Under the treaty, countries have agreed to provide facilitated access to plant genetic resources from an agreed list of crops listed in an annex, which are important for food security. By signing the treaty, governments agree to put such resources under their direct control into the "Multilateral System". They will also encourage institutions, not under their direct control, to do likewise. Of particular importance is the large collection of genetic material of interest to developing countries under the aegis of the CGIAR, but there are of course many national collections of worldwide importance in both developed and developing countries, as well as the store of genetic diversity in farmers' fields.

With regard to IPRs, the potentially contentious part of the treaty is that referring to the protection of resources accessed from the multilateral system. As finally agreed, the treaty states: "Recipients shall not claim any intellectual property or other rights that limit the facilitated access to the plant genetic resources for food and agriculture, or their genetic parts or components, in the form received from the Multilateral System".

This wording is inevitably a diplomatic compromise, reflecting a desire on the part of many developing countries to avoid a limitation on access being imposed by the grant of IP rights, and of some developed countries to allow patenting of genetic material according to existing criteria applied nationally. The crucial words "in the form received" mean that material received cannot be patented *as such*, but they do allow patents to be taken out on modifications (however defined) to that material.

The compromise wording clearly excludes the patenting of seeds as obtained from the seed bank. But the extent to which patents can be taken out on a gene isolated from that material is controversial. During the negotiation of the Treaty, some countries were of the opinion that this article should be read as precluding such patenting. Others thought that the isolated form of a gene (for which a function has also been determined) is different than the "form received", and hence should be patentable. Thus, the wording raises the important general issue of what are the appropriate rules for patenting genetic material, both for developed and developing countries. This revolves around the nature of the inventive step required for patenting, the nature of the claims for the invented use of that material, and the extent to which those claims might limit the use of the underlying genetic material.

The Treaty has also established an important principle that any user of material will sign a standard Material Transfer Agreement (MTA), to be devised by the governing body of the treaty, which will incorporate the conditions for access agreed in the Treaty, and provide for benefit sharing of the proceeds of any commercialization arising from the material through a fund established under the treaty. This significantly goes beyond the provisions of CBD in suggesting a concrete mechanism for benefit sharing, based on multilateral rather than bilateral arrangements.

Developed and developing countries should accelerate the process of ratification of the FAO International Treaty on Plant Genetic Resources for Food and Agriculture, and should, in particular, implement the Treaty's provisions relating to the following:

- Not granting IPR protection of any material transferred in the framework of the multilateral system, in the form received.
- Implementation of farmers' rights at the national level, including:
  — protection of traditional knowledge relevant to plant genetic resources for food and agriculture.
  — The right to equitably participate in sharing benefits arising from the utilization of plant genetic resources for food and agriculture.
  — The right to participate in making decisions, at the national level, on matters related to the conservation and sustainable use of plant genetic resources for food and agriculture.

The importance of the agricultural sector in developing countries as a source of food, incomes, employment and often foreign exchange cannot be overstated. As much as good health, a productive and sustainable agricultural sector is critical to achieving economic growth and poverty reduction. About three quarters of the world's poor people live and work in rural areas. Apart from its direct role in sustaining incomes and employment, the role of agriculture, and in particular technological change in agriculture, in stimulating overall economic growth has been much discussed by economists and policymakers. Raising productivity in agriculture can directly increase the incomes and employment levels of the majority of poor people dependent on agriculture. It can also help to reduce food prices (relatively or absolutely) for poor people in both rural and urban sectors.

Historically agriculture has been seen, sometimes controversially, as a source of food, labor and finance to supply a growing urban and industrial sector on which sustained growth in incomes will depend. Achieving this transition usually depends on achieving productivity increases if food prices are not to rise, and stifle both industrial growth and poverty reduction. In developed countries changes in technology and institutions in the agricultural sector are regarded as having been instrumental in the industrial revolution.

In developing countries, technical progress traditionally occurred through a process of on-farm experimentation, selection, and adaptation of traditional landraces of crops. Subsequently, this was supplemented by purposive breeding of new varieties of crops, mainly through crossing varieties with desirable characteristics. This process of research was largely conducted in the public sector by national research institutes, supported by a network of international research institutes for the last thirty years under the umbrella of the Consultative Group on International Agricultural Research (CGIAR). It was this network which led to the Green Revolution of the

1960s, based initially on high yielding semi-dwarf varieties of rice and wheat. In spite of criticisms of its environmental and distributional impact, this technology is widely credited with having had a favourable impact on nutrition, employment and incomes, albeit mainly in the areas of developing countries capable of reasonably assured irrigation. Subsequently, further breeding efforts have been tried, but with less success, to extend these technologies to new crops and to rainfed and dryland areas.

More recently, significant changes have occurred in both the technology and the structure of research in agriculture. First, the advent of biotechnology, and in particular genetic engineering in the last twenty years has vastly expanded the possibilities of what can be achieved in agricultural research (for example, introducing new genetic traits in plants). Secondly, while public investment in public research, at least through the CGIAR, has tended to stagnate in recent years, investment by the private sector has gone up rapidly. Market forces have increasingly guided the direction and purpose of additional research spending.

16.19. PLANT BREEDER'S RIGHTS

1. Subject to this Act, the holder of the plant breeder's rights respecting a plant variety has the exclusive right:

    (a) to sell, and produce in the country for the purpose of selling, propagating material, as such, of the plant variety.

    (b) to make repeated use of propagating material of the plant variety in order to produce commercially another plant variety if the repetition is necessary for that purpose.

    (c) where it is a plant variety to which ornamental plants or parts thereof normally marketed for purposes other than propagation belong, to use any such plants or parts commercially as propagating material in the production of ornamental plants or cut flowers.

    (d) To authorize, conditionally or unconditionally, the doing of an act described in (a) to (c).

2. **Exemption:** Paragraph (1)(a) does not apply to the sale of propagating material that is not in Canada when it is sold, but if any such propagating material, the sale of which to any person is exempted from that paragraph by this subsection, is used as propagating material, in Canada by that person, an infringement of the exclusive right conferred by virtue of that paragraph is constituted by the purchase and subsequent use of the propagating material by that person, who shall be liable to be proceeded against in respect of that infringement.

3. **Implications:** The sale of propagating material in the exercise of any exclusive right conferred by subsection (1) does not imply that the seller authorizes the purchaser to produce, for the purpose of selling, propagating material as such but, subject to any terms or conditions imposed by the seller. The sale implies that the seller authorizes the purchaser to sell anything sold, in that exercise of the exclusive right, to the purchaser.

4. **Royalty:** Without limiting the generality of paragraph (1)(d), and without prejudice to any rights or privileges of the Crown, where authority is conferred subject to conditions pursuant to that paragraph, whether or not the holder of the plant breeder's rights, the conditions may include a requirement to pay royalty to the holder.

Term of Plant Breeder's Rights

•   The term of the grant of plant breeder's rights shall, subject to earlier termination pursuant to this Act, be a period of eighteen years, commencing on the day the certificate of registration is issued under paragraph 27(3)(b).

Payment of Annual Fee

- The holder of plant breeder's rights shall, during the term of the grant of those rights, pay to the Commissioner the prescribed annual fee in respect of those rights.

16.20. APPLICATIONS FOR PLANT BREEDER'S RIGHTS

Section 7: Entitlement to apply for plant breeder's rights.

1. Subject to section 8, the breeder of a new variety or a legal representative of the breeder may make an application to the commissioner for the grant of plant breeder's rights respecting that variety if:

    (a) in the case of a new variety of a recently prescribed category, neither the breeder nor a legal representative of the breeder sold or concurred in the sale of that variety in Canada before the commencement of such period prior to the date of receipt, by the Commissioner, of the application as is prescribed for the purposes of this paragraph.

    (b) in any other case, neither the breeder nor a legal representative of the breeder sold or concurred in the sale of that variety in Canada before the effective date of the application.

    (c) subject to any prescribed exemptions, neither the breeder nor a legal representative of the breeder sold or concurred in the sale of that variety outside Canada before the commencement of such period prior to the date described in paragraph (*a*), as is prescribed for the purposes of this paragraph.

Application where breeder's participation unobtainable for joint application, where

    (a) a new variety is bred by two or more breeders otherwise independently of each other; and

    (b) either or any of the persons entitled to make an application for the grant of the plant breeder's rights respecting that variety refuses to do so, or information of the whereabouts of either or any of those persons cannot be obtained through diligent inquiry on the part of the remainder of them, the remainder of those persons may make an application for that grant.

Section 8: Required citizenship, residence, or location of registered office.

A person is only eligible to apply for the grant of plant breeder's rights if the person is a citizen of, or is resident or has a registered office in Canada or a country of the Union or an agreement country.

Section 9: How should application be made:

1. An application for the grant of any plant breeder's rights must:

    (*a*) be made in the prescribed manner.

    (*b*) be accompanied by the prescribed fee.

    (*c*) be supported by the documents and any other material prescribed.

    (*d*) include any request referred to in sub-paragraph 75(1)(*k*)(i) that the applicant makes.

Agent required for non-resident applicant:

2. An applicant who, in the case of an individual, is not resident in Canada or that, in the case of a corporation, does not have its registered office in Canada, shall submit the application through an agent resident in Canada.

Section 10: Priority and dating of application:

1. Subject to subsections (2) and 11(1), the effective date of an application is the date on which the application is received by the Commissioner, and in the case of receipt by the Commissioner of two or more applications respecting a new variety the breeders of which bred it independently of each other, priority shall be given to the application first received by the Commissioner.

Applications of same date:

2. Where the effective dates of applications described in subsection (1) are the same, priority shall be given to the application pertaining to the breeder, who was first in a position to apply for the plant breeder's rights, respecting the new variety, or who would have been first in the position to do so if the provision made by or under this Act for so doing had always been in force.

Section 11: Priority based on preceding application in country of Union or agreement country:

1. Where an application made under section 7 is preceded by an application made in the appropriate manner, in a country of the Union or an agreement country, for protection pursuant to the breeding of the same new variety by the same breeder as in the case of the application made under section 7, the date that is the effective date of the application made in that country shall be deemed to be the effective date of the application made under section 7, and the applicant is entitled to priority in Canada accordingly, notwithstanding any intervening use, publication or application respecting the new variety, if:

   (*a*) the application is made under section 7 in the prescribed form within twelve months after the date on which the application was made in that country; and

   (*b*) the application made under section 7 includes or is accompanied by a claim respecting the priority and is accompanied by the prescribed fee.

Confirmation of claim to priority:

2. A claim respecting priority based on a preceding application made in a country of the Union or an agreement country shall not be allowed unless, within three months after the date on which the claim is submitted to the Commissioner, it is confirmed by filing with the Commissioner a copy, certified as correct by the appropriate authority in that country and accompanied by an English or French translation of the certified copy, if made in any other language, of each document that constituted the preceding application.

Supporting application given priority:

3. An application given priority under subsection (1) shall be supported by the required material furnished pursuant to this Act and the regulations before the expiration of the prescribed period, not exceeding four years, after the last day of the twelve months within which the application is submitted in accordance with paragraph (1)(*a*).

Cases of two or more preceding applications:

4. Where an application made under section 7 is preceded by two or more applications made, in the appropriate manner, in different countries of the Union or agreement countries, for protection pursuant to the breeding of the same new variety by the same breeder as in the case of the application made under section 7, reference in subsection (1) to an application made in any such country shall be construed as reference to whichever of those applications was first made.

Section 12: Priority Conditional on Residence, etc.

- No claim referred to in paragraph 11(1)(*b*) shall be based on any preceding application, unless it was made by a person who, at the time of the application, was a citizen of, or a person resident or having a registered office in, Canada or a country of the Union or an agreement country.

When previous application is disregarded:

- for the purposes of subsection 11(1), no account shall be taken of an application that was made in a country outside Canada at a time when the plant variety, to which the application relates, did not belong to a prescribed category.

Priority Established over Previous Grant

Where priority for an application is established pursuant to this Act, the Commissioner should refuse any application against which the priority is established, or if the priority against it is established after granting on it any plant breeder's rights, the Commissioner should annul the grant and section 36 and paragraph 70(3)(*b*) apply, with such modifications as the circumstances require, in respect of the annulment.

Under TRIPS, countries may exclude from patentability plants and animals and essentially biological processes for producing them, but not microorganisms. And they are required to apply some form of protection, either by patents or a *sui generis* system to plant varieties.

There are many legal complexities about definitions arising from the wording of TRIPS, such as the exact meaning of a plant variety, a "microorganism", or an essentially biological process. But it is important to note here that TRIPS does not mention whether or not genes should be patentable, whether derived from plants, humans or animals. The issue raised by TRIPS is what constitutes an invention in relation to genetic material. For instance, should genetic material identified in nature be patentable on the grounds that isolating and purifying it differentiates it from an unpatentable discovery? This is a matter for national legislation. The only specific requirement, other than for microorganisms, is that plant varieties should be protected.

Some people object altogether to the patenting of life forms on ethical grounds, considering that the private ownership of substances created by nature is wrong, and inimical to cultural values in different parts of the world. The sequencing of the human genome also raises specific concerns. The ethical and legal issues in respect of patenting DNA are discussed in a recent report of the Nuffield Council on Bioethics. Our task here is to consider the practical and economic consequences of patenting in agriculture, and how this affects the livelihoods of poor people and the implications for policy.

Intellectual property protection can be conferred in relation to plant materials in a number of ways:

- The US model of plant patents, which are distinct from normal (utility) patents.
- Through allowing normal patents on plants or parts thereof, such as cells.
- Through patenting plant varieties as is the practice in the US and few other countries (for example, not in the EU).
- Through applying a *sui generis* form of plant variety protection (PVP), such as plant breeders' rights (as in the EU or the US) or other modalities.
- Through allowing patents on DNA sequences and gene constructs, including the gene, plants transformed with those constructs, the seed, and progeny of those plants.

In addition, patents are widely used to protect the technologies, which are employed in research on plant genomics. Apart from the use of patents and PVP, the intellectual property in plants can be appropriated by technological means. For instance, crops such as commercial hybrid maize cannot be reused if hybrid yield and vigour are to be maintained. This characteristic of some hybrids confers a natural form of protection by which seed companies can more readily capture a return on their investment through repeat seed sales. By contrast, other types of seed variety can be replanted each year without deterioration in yield, so that farmers may replant their own seed without repurchasing.

The Green Revolution varieties were of this nature, which is one reason why they were so successful. It is only more recently that hybrid varieties of rice and wheat have been developed. Genetic Use Restriction Technologies (known as GURTs) is a term used to describe different forms of controlling the action of genes in plants. The so-called "terminator" technology, which would render the seed sterile so that it is not physically possible to grow a second crop, is well-known, but other characteristics can also be controlled, either for agronomic or commercial reasons. The effect of technological protection is similar to that of IP protection, but possibly cheaper and certainly more effective in the sense that it is self-enforcing.

Research and Development

As compared to medical research, there is a great deal more agricultural R&D undertaken by, and of relevance to, developing countries. For instance, it is estimated that in 1995, total expenditure by the public sector on agricultural research in developing countries, although unevenly distributed, amounted to $11.5 billion (at 1993 international dollar values) compared to the $10.2 billion spent in developed countries. The majority of research is conducted in the more technologically advanced developing countries in Asia and Latin America. Moreover, research expenditures by these countries grew at 5-7% annually between 1976 and 1996, while they stagnated in Africa. By contrast, of worldwide private research expenditure totalling $11.5 billion, only $0.7 billion is attributable to developing countries.

This means that globally about one-third of all agricultural R&D is spent in developing countries in marked contrast to the maximum of 5% estimated for health research for developing countries. Three points should be noted here. First, global R&D on agriculture is only a little more than that estimated for health R&D. Secondly, there is almost twice as much agricultural R&D in the public sector as the private sector. In medicine, expenditure by the private sector is proportionately larger, as we have seen. Thirdly, and partly as a result, the developing countries are relatively better served in the case of agricultural research.

Nevertheless, current trends give cause for concern. Although the CGIAR spends only about $340 million per year, its role is strategically important. For instance, the CGIAR centres played a crucial role in the Green Revolution, and now act as the guardian of the world's largest collection of genetic resources of relevance to developing countries, which is the major source of crop improvements for the future. But funding for the CGIAR system, which is provided by the donor community, has fallen in real terms since 1990, and this threatens both its research effort and ability to maintain its gene banks, or assist developing countries in maintaining their own collections. The FAO and CGIAR have launched an endowment specifically to ensure that these genetic materials across the world can be properly maintained. While funding from the aid donor community is stagnating, the private sector is the dynamic element in agricultural R&D, but little of its effort is of direct relevance to poor farmers in developing countries.

16.21. IMPACT OF PLANT VARIETY PROTECTION

In this section we examine the evidence on the impact of plant variety protection (PVP) in developed and developing countries and what PVP systems might have to offer developing countries. Most of the evidence related to the impact of patent or plant variety protection on research is from developed countries, and even that is quite sparse. Before IP protection was introduced, private sector breeding initiatives focused on hybrid varieties, particularly of maize in the US, because inherent in these varieties is an element of "technological protection". In the US, a study from the 1980s suggested there was no evidence that total R&D activity had increased as a result of the introduction of PVP, though it appeared to have had some impact on soya beans, and perhaps wheat. The latter crops also accounted for the majority of PVP certificates issued. There was also an evidence that PVP was used as a marketing strategy for product differentiation, and that it had contributed to the large number of mergers that took place in the seed industry. But the evidence is inconclusive, in particular, because of the difficulty in isolating the effect of protection from other ongoing changes.

Even now research spending on hybrid crops as a share of sales continues to exceed than that on non-hybrid crops, which are the principal objects of PVP. A recent study found that PVP on wheat in the US had not contributed to increased investment in private sector wheat breeding, but may have done so in the public sector. Nor had it contributed to an increase in yields. But the share of wheat acreage sown to private varieties had increased markedly, reinforcing the suggestion that the main impact of PVP was as a marketing tool.

A major study conducted in middle income developing countries found little evidence of an increased range of plant material available to farmers or increased innovation as a result of PVP protection. Access to foreign genetic material had improved, but its use was sometimes subject to restrictions, for example on exports. Generally speaking, commercial farmers and the seed industry were perceived as the principal beneficiaries. Poor farmers had not benefited directly from protection, but could potentially be adversely affected by restrictions on seed saving and exchange in the future.

Under TRIPS, developing countries may choose an "effective *sui generis*" PVP system. A major decision is to identify a system that is suitable to their particular agricultural and socio-economic circumstances. The UPOV convention is one system that they may adopt, based on the legislation introduced in Europe and the US. A consideration is that it provides a ready-made legislative framework, but a disadvantage is that it was designed with the commercialized farming systems of the developed countries in mind. There are therefore concerns expressed about the application of the UPOV model in developing countries, some of which apply to any form of PVP.

The criteria for awarding a PVP certificate involve lower thresholds than the standards required for patents. There are requirements for novelty and distinctness, but there is no equivalent of non-obviousness (inventive step) or utility (industrial applicability). Thus, PVP law allows breeders to protect varieties with similar characteristics, which means that the system tends to be driven by commercial considerations of product differentiation and planned obsolescence, rather than genuine improvements in agronomic traits. Developing countries might consider raising the threshold, in particular so that protection is only given for significant or important innovations with particular characteristics that are deemed socially beneficial (for example, increase in yield, or traits of nutritional value). Thus, the criteria for distinctness may be strengthened, and also criteria formulated defining utility in terms of the objectives of agricultural policy. Alternatively, countries may decide to retain lower standards for certain categories of plant in order to facilitate access by nascent domestic breeding industries to PVP protection from which may flow commercial and export benefits.

Similarly, the requirement for uniformity (and stability) in UPOV type systems excludes local varieties developed by farmers that are more heterogeneous genetically and less stable. But these characteristics are those that make them more adaptable and suited to the agro-ecological environments in which the majority of poor farmers live. Again it would be open to developing countries to devise systems that would offer protection for varieties that meet criteria suited to the circumstances and crops on which poor farmers depend. But such criteria may be difficult to devise, and the system costly to operate. And governments may consider that extending such a system would not play a positive role in the development of their farming systems.

Another concern is about the criterion for uniformity. While proponents argue that PVP, by stimulating the production of new varieties, actually increases biodiversity, others claim that the requirement for uniformity, and the certification of essentially similar varieties of crops, will add to the uniformity of crops and loss of biodiversity. Of course, this concern goes wider than PVP. Seed legislation in many countries imposes strict uniformity requirements, sometimes stricter than PVP legislation. Moreover, similar concerns have arisen in respect of greater uniformity arising from the success of Green Revolution varieties, leading to greater susceptibility to disease and loss of on-field biodiversity. But, as plant breeding becomes an increasingly private sector activity, and new varieties displace traditional varieties on a large scale, there is the crucial issue of how genetic resources are to be conserved and maintained for possible future use, whether in fields or in "gene banks".

There may also be a need to differentiate standards of protection between different kinds of crop. For instance, countries with significant commercial and export sectors might adopt UPOV-type standards for the relevant crops in those sectors to encourage innovation and commercializations. However, they might adopt other standards for food crops grown by farmers to protect their practices of saving, trading and exchanging seeds, and informal systems of innovation. For instance, in Kenya. PVP rights seem to have been predominantly applied for by the foreign-owned commercial exporters of flowers and vegetables to underpin commercializations and exporting. This may be beneficial to the expansion of Kenya's export industries and commercial agriculture, and indirectly to poor people. PVP may facilitate the availability of new varieties in Kenya (which might have been withheld in the absence of protection) but appears to play little part in stimulating local research. The system has not appeared to be very relevant to the direct concerns of Kenya's poor farmers and the crops they grow.

16.22. INTERNATIONAL UNION FOR THE PROTECTION OF NEW VARIETIES OF PLANTS (UPOV)

The internationally recognized agreement on PVP protection is UPOV. The UPOV convention dates back to 1961, and has been revised thrice subsequently. Apart from South Africa, the first developing countries to join UPOV were Uruguay and Argentina in 1994, when there were 26 members in total. Since 1994, 24 further developing countries have joined UPOV. Although TRIPS only specifies that there should be a *sui generis* regime, UPOV has been an obvious choice, as it provides an off-the shelf solution to developing such legislation. In addition, pressure has been put on various countries to join UPOV in the context of bilateral trade agreements (for instance, the recently concluded Vietnam-US trade agreement obliges both parties to be members of UPOV, of which the US is already a member). The purpose of the UPOV Convention is to ensure that the member States of the Union acknowledge the achievements of the breeders of new plant varieties, by making available to them exclusive property right, on the basis of a set of uniform and clearly defined principles.

As UPOV has been revised successively (1978 and 1991), the scope and length of protection has been extended. The minimum period of protection has been increased to 20 years (25 years for vines and trees) in the 1991 version (from 15 and 20 previously). Unlike patents, the criteria for protection do not involve an inventive step as such. Rather, to be eligible, varieties must only be distinctive, uniform and stable (DUS in the jargon), and novel (in terms of prior commercialization). The 1978 Act allowed breeders to use protected varieties as a source for new varieties, which could then be protected and marketed themselves. The 1991 Act has preserved the breeders' exception, but the right of the breeder extends to the varieties, which are "essentially derived" from the protected variety that cannot be marketed without the permission of the holder of the original variety.

The 1978 Act provided the breeder with protection in respect of the production for the sale of seed, its offer for sale, and its commercialization [article 5(1)], and it therefore implicitly allowed farmers to replant and exchange the seed (although this right is not spelt out). The 1991 Act is more restrictive of the rights of farmers. The right of the breeder now extends to production or reproduction, in addition to the marketing of propagated or harvested material [article 14(1)]. This is mitigated by an optional farmers' exception, which allows "farmers to use for propagating purposes, on their own holdings, the product of the harvest which they have obtained by planting on their own holdings, the protected variety or [an essentially derived variety]" [article 15(2)].

Thus developing countries should consider basing their PVP legislation on a realistic appreciation of how it could benefit their agricultural development and food security, taking account also the role of agriculture in generating exports, foreign exchange and employment. In particular, they need to consider possible modifications to the UPOV model to adapt it to their circumstances. A number of countries have passed or are considering legislation, which incorporates elements described above.

An important aspect of *sui generis* systems is the scope of the farmers' exception. Unlike patents, PVP legislation generally allows an exception, as in UPOV 1978, which permits farmers to reuse on their own holding harvested seeds without the permission of the rightsholder. In the US, this exception was expanded to allow limited sale of harvested crops for seed purposes to other farmers. And, in the developing world, in the absence of legal rules, farmers exchange and sell their seeds informally. As we have noted, this is a practice which is still very widespread amongst poor farmers in developing countries, and even still common in developed countries. These systems of sale and exchange are an important mechanism by which farmers have traditionally selected and improved their own varieties, and the restriction of this right may impede this process of improvement.

Although UPOV (1991) permits nations to allow farmers to reuse their own crop for seed purposes on their own holdings, it does not allow for informal sale or exchange. In contrast, TRIPS only requires that there should be some form of IP protection for plant varieties, and does not define in any way the exceptions that may be provided to the rights of owners of protected varieties. Thus countries and organizations have experimented with a number of alternatives in this area. For instance, the OAU (now the African Union) has produced model legislation which it recommends African countries adapt in their own legislation. This provides for the right to save, use, multiply and process farm-saved seed, but not to sell it on a commercial scale. The Indian government, which has recently decided to seek admission to UPOV, has incorporated in its PVP legislation (2002) a clause [39(1)(iv)] that states: "a farmer shall be deemed to be entitled to save, use, sow, re-sow, exchange, share or sell his farm produce including seed of a variety protected under this Act in the same manner as he was entitled to before the coming into force of this act, provided that the farmer shall not be entitled to sell branded seed of a variety protected under this Act". The breeders' exception under PVP also differs from patent law in that breeders may, without authorization, use a protected variety as the basis for breeding another variety (which itself may then gain protection). Thus PVP provides less protection than patents, and as we have argued little incentive for research, but correspondingly is less restrictive of incremental follow-on innovation than patents. Again developing countries are free to choose exactly what exceptions they provide. At one extreme, PVP could be conferred as a superior kind of seed certificate or seal, giving the holder the sole rights to sell seed with this seal. But there would be no rights to protect subsequent use or sale of the seed, as long as it was not sold under the certificate. This right would be superior to a trademark or seed certificate, but would not restrict subsequent reuse of harvested material in any way. Such a system might be a way to tailor the PVP system to the needs of poor farmers, but it would offer less incentive for breeders.

Impact of Patents

Patents on plant varieties, as such, are only allowed in the US, Japan and Australia, and are most frequent in the US. The 1930 US Act introduced a special kind of Plant Patent for vegetatively propagated materials, but in the US, standard utility patents can now be granted on plant varieties. Patents are the strongest form of intellectual property protection in the sense that they normally allow the rights holder to exert the greatest control over the use of patented material by limiting the rights of farmers to sell, or reuse seed they have grown, or other breeders to use the seed (or patented intermediate technologies) for further research and breeding purposes. However, patent law can provide for exceptions similar to those in PVP systems. For example, the EU Biotechnology Directive, while not permitting the patenting of plant varieties, provides for a farmer's exception where a patent on genetic material would otherwise prevent reuse on the farm.

It also contains a provision for compulsory licensing, subject to certain conditions, where a breeder's use of material would otherwise infringe the patent right. In the US, the patenting of plant varieties is particularly important because, with appropriate claims in the patent, the holder of the patented variety can prevent others from using it for breeding purposes. This is a significant difference from PVP. Proving that a new variety meets the criteria for patentability is more difficult and more costly than obtaining plant variety protection, where the criteria for protection are lower. Patent protection is also frequently obtained through a broad patent, which claims the gene, the vector or carrier for effecting the transformation and so on, which may cover a number of potential varieties or crops incorporating the gene. For practical purposes, this may have the same effect as patenting the whole plant, because the patent normally extends to all material in which the product is incorporated.

Whatever the incentives provided by patenting, market forces will tend to direct research efforts by the private sector to where there is the most substantial potential return. However, in contrast to medicines, there is the potential for companies to become attracted to crops that are widely grown in developing countries. The investment costs are correspondingly lower than for medical research, and the potential markets correspondingly larger. For instance rice, where the value of production in India alone exceeds that of the US maize market, has hitherto been a crop where breeding has been the preserve of the national or international public sector (principally the CGIAR). Since then the private sector has become increasingly interested in rice research. Monsanto and Syngenta have worked on sequencing the rice genome of two major rice varieties. The number of patents relating to rice issued annually in the US has risen from less than 100 in 1995 to over 600 in 2000.

So far, about 80% of trials of transgenic crops have occurred in developed countries, where three-quarters of the world's GM crops are grown. The breeding strategies of the multinationals have been naturally oriented to the needs of developed world markets, and the commercial sectors of middle income developing countries (for example, Brazil, Argentina or China). The development of genetic traits, such as herbicide tolerance has been determined principally by the search for commercial advantage, rather than for characteristics useful to poor farmers in developing countries. But companies are introducing GM varieties which, although controversial both in developed and developing countries, are considered by some developing countries to be of potential benefit to them (for example, the Bt gene which confers insect resistance). Bt Cotton or Bt maize is now grown in at least five developing countries, and other countries may also be interested, if they can resolve environmental concerns. For instance, India has approved the planting of Bt Cotton. Companies have also donated technologies of relevance to developing countries (for example, through royalty free licences), including those related to vitamin A enriched rice (Golden Rice) and cassava. Some companies have published scientific articles based on their genomic research, but have aroused controversy by not depositing the raw data in public databanks.

Negotiations about the deposit in public databanks have been complicated by the companies' desire to limit access to components of data with the greatest potential commercial value. Thus there is the potential for agricultural technologies developed by the private sector to spill over to the benefit of the commercial sectors in developing countries. Nevertheless, if the Green Revolution, which was developed and applied with public sector funding, failed to reach the poor farmers living in agro-ecologically diverse rainfed environments, it is apparent that biotechnology-related research led by the private sector will be even less likely to do so. For that, more public sector research specifically oriented to such farmers will be required. In 1998, the CGIAR system spent $25 million on such research compared to the $1.26 billion invested by Monsanto.

Apart from the problem of incentives for research relevant to poor farmers, there is evidence that patents, and to some extent PVP, have played a part in the major consolidation of the global seed and agricultural input industries. The consolidation appears to be driven by technological change, with an objective of vertical and horizontal integration, so that the appropriability of investment in research can be maximized through better control of distribution channels, including those of complementary agricultural inputs (such as herbicides).

Companies acquire patent rights to protect their own investment in research, and to prevent the encroachment of others. But by the same token, other companies' patent rights can impede one's own research. For instance, there are several hundred overlapping patent rights for the Bt technology, and at least four companies obtained patents that cover Bt-transformed maize. Recently,

Syngenta filed two law suits in the US against a number of its competitors, alleging infringement of several of its patents relating to this technology, although the companies involved have been using these technologies and selling seeds incorporating them for several years. Cross licensing or strategic alliances can also be used as mechanisms to overcome problems of conflicting patents. However, merger or acquisition may be the most effective means of obtaining the freedom to use required technologies in a particular field of research. All of these approaches, not just the last, reduce competition. And the major multinational agrochemical companies, with their growing control over essential proprietary technologies, also represent a formidable barrier to the entry of innovative start-ups. In the 1980s, the university and public sector accounted for 50% of the total of granted US patents related to *Bacillus thuringensis* By 1994, independent biotechnology companies and individuals held 77% of the patents, but by 1999 the big six companies (which became five with the merger of the agricultural arms of Astra Zeneca and Novartis to form Syngenta) held 67% of them. Moreover, the growing control of these companies was demonstrated by the fact that 75% of their Bt patents in 1999 had been obtained by the acquisition of smaller biotechnology and seed companies.

In developing countries, there is an evidence of similar trends with an extremely rapid process of merger and acquisition by the multinational companies. For instance, in Brazil, following the introduction of plant variety protection in 1997 (but presumably also related to the expected permission to grow GM crops), Monsanto increased its share of the maize seed market from 0% to 60% between 1997 and 1999. It acquired three locally based firms (including Cargill as the result of an international deal), while Dow and Agrevo (now Aventis) also increased their market share by acquisition. Only one Brazilian-owned firm remained with a 5% market share. This trend appears widespread in developing countries.

Thus, the speed of concentration in the sector raises serious competition issues. There are considerable dangers to food security if the technologies are overpriced to the exclusion of small farmers, or there is no alternative source of new technologies, particularly from the public sector. Further, the increase in concentration, and the conflicting patent claims when both the public and private sectors have patented plant technologies, may have had an inhibiting effect on research. In the private sector the response has been alliances or acquisitions, but a problem for the public sector is how to access the technologies they need to undertake research without infringing IP rights, and if they develop new technologies, the terms on which they may be made available. A recent review published by the US Department of Agriculture concludes that whether the current intellectual property regime is stimulating or hampering research is unclear.

Historically, intellectual property laws are rooted in the principle of national sovereignty. IP was a domestic policy issue that was based on each country's level of development and technological needs. Until recently, the World Intellectual Property Organization (WIPO) was considered the most important intergovernmental forum on intellectual property rights. All of that changed with the creation of the World Trade Organization (WTO) in 1994. One reason is that WIPO's treaties on IP include few mechanisms for enforcement, dispute settlement, or compliance. By contrast, the WTO places far more pressure on countries to adopt minimum standards of IP; member countries must assume the obligations of WTO agreements (including intellectual property) in order to gain WTO membership. The WTO's dispute settlement procedure creates a strong mechanism for compliance, including the power to impose trade sanctions against member states that fail to abide by its binding agreements.

The World Trade Organization today administers the most comprehensive multilateral agreement on IP. WIPO continues to provide technical assistance on national IP laws and institutions, and administers 19 international treaties concerning intellectual property (such as the Paris Convention on Protection of Industrial Property and the Berne Convention on Protection of Literary and Artistic Works). But although it remains an important body for international standard setting on IP, it is overshadowed by the more powerful WTO.

1994 saw the conclusion of the Uruguay Round of the General Agreement on Tariffs and Trade (GATT) and the creation of the World Trade Organization (WTO), which came into being in January 1995. The WTO operates on the principle that a liberalized system of international trade based on non-discrimination and the elimination of trade barriers is essential for global economic well-being. The WTO is based in Geneva. By May 1999, 134 countries were its members. The WTO's primary functions are to:

- administer WTO trade agreements;
- act as a forum for multilateral trade negotiations;
- handle trade disputes;
- monitor national trade policies;
- provide technical assistance and training for developing countries; and
- cooperate with other international organizations.

TRIPs brings intellectual property to centre stage in multilateral trade negotiations. Prior to the Uruguay Round, the GATT dealt mainly with trade in merchandise, and was not concerned with services and intellectual property. The inclusion of IP reflects, in part, the explosive growth in information technology and biotechnology in international trade, and the strong desire of some industrialized countries, particularly the US, Europe and Japan, to protect products from intellectual 'piracy' in foreign markets. Between 1980 and 1994, the share of high technology products in international trade doubled, from 12% to 24%.

The TRIPs agreement ensures that all signatories provide minimum standards of protection in a number of different areas of intellectual property law (e.g., patent, copyright, geographical indications, and so on). In this way, TRIPs makes these minimum standards universal, at least as far as the cumulative body of signatory states is concerned. The TRIPs agreement requires member countries to make patents available for inventions, whether products or processes, in all fields of technology without discrimination, subject to the standard patent criteria of novelty, inventiveness and industrial applicability. TRIPs requires that patents be available without discrimination as to the place of invention and whether products are imported or locally produced (article 27.1). However, TRIPs does not aim at identical national IP laws, nor does it establish an international patent system, rather, it sets minimum standards for member countries to follow. Under the WTO TRIPs agreement, all members are obligated to adopt minimum standards for intellectual property rights and a mechanism for their enforcement.

Of particular relevance to biological diversity, TRIPs article 27 requires that all member states adopt national-level IP systems for all products and processes, including pharmaceuticals, modified microorganisms and microbiological processes. During the negotiations, however, members had a difficult time reaching consensus on the controversial area of biotechnological inventions. While some industrialized countries pushed for no exclusions to patentability, members of some developing country preferred to exclude all biological diversity-related inventions from IP laws. Still other members preferred something in between.

The text that prevailed is found in article 27.3(b). It states that plants and animals as well as essentially biological processes may be excluded from patentability. However, WTO members must offer protection for plant varieties either by patents and/or by an effective *sui generis* system. *Sui generis* means a system of rights that is unique, 'of its own kind', for a specific item or technology.

However, the agreement on TRIPs does not define *sui generis*. Although several attempts have been made, the term has not been defined in TRIPs negotiations.

Due to the difficulty in reaching consensus, it was agreed that the controversial sub-paragraph would be reviewed in 1999. Of course, 1999 has come and gone, and the review of 27.3(b) did not get started. It is hard to predict at this point when it will get addressed, and in what form. No doubt, a great deal will be said about the effect of article 27.3(b)'s current wording in the context of the general review of the implementation of TRIPs scheduled to take place in 2000 (as mandated by article 71.1). In the end, however, WTO member states may find themselves too tied up with other business to be able to directly address the possibility of amending 27.3(b) in the next few years.

16.23. TRIPS ARTICLE 27.3(B)

Members may also exclude from patentability plants and animals other than microorganisms, and essentially biological processes for the production of plants and animals others than non-biological and microbiological processes. However, members should provide for the protection of plant varieties either by patents or by an effective *sui generis* system, or by any combination thereof. The provisions of this sub-paragraph shall be reviewed four years after the entry into force of the WTO agreement.

Ordre public or *morality* article 27.2 allows the exclusion from patenting of inventions contrary to ordre public or morality. This explicitly includes inventions dangerous to human, animal or plant life or health, or seriously prejudicial to the environment.

Can TRIPs' public morality exclusion be used to reject patents on life forms of controversial new technologies such as genetic seed sterilization?

This exclusion is subject to the condition that the commercial exploitation of the invention must also be prevented, and that this prevention must be necessary for the protection of *ordre public* or morality.

There is a disagreement among crucible members as to the scope of this optional exclusion and its application to specific technological developments, such as genetic use restriction technology (GURT).

At a time, when the global economic system is facing severe crisis, the short-term cost of implementing TRIPs, including considerations such as structural capacities, financial resources, and technical expertise, is a significant factor for many LDCs. A 1996 study by UNCTAD estimates that in order to comply with the TRIPs agreement, Bangladesh will need to spend $250 000 in one-time costs for legislative drafting and over $1.1 million in annual costs for judicial work, equipment, and enforcement measures—not including the cost of training personnel. For the United Republic of Tanzania, the costs of implementing TRIPs are estimated at $1-1.5 million.

There is a concern that implementation of TRIPs could divert scarce resources away from basic social programs in the developing world. Article 67 of TRIPs calls for bilateral technical assistance to be given by industrialized countries to developing countries, which might help to reduce some of these costs. African countries, in particular, have expressed concern that not enough has been done under article 67 to provide technical assistance for the implementation of the TRIPs agreement.

Implementation of Article 27.3(b)

Options for Implementing Article 27.3(b)

The obligations imposed by the TRIPs agreement require many developing countries to enact IP legislation for plant varieties (and other biological materials) for the first time. With deadlines for implementation and enforcement of new legal regimes fast approaching, members have a finite time period to explore options and implement patent and/or *sui generis* systems for plant variety protection. The TRIPs agreement does not specifically define what an 'effective *sui generis* system' might look like, and thus allows for many different systems to emerge.

Discussion on how to implement article 27.3(b) has tended to focus on the following questions: Is it best for developing countries, most of which have not yet offered any IP protection for plant varieties, to follow the model of plant breeders' protection as foreseen in the UPOV Acts? Or should developing countries explore alternative protection systems for plant varieties rather than following the ready-made protection systems currently being used in many industrialized countries? Which of the different options will better serve a country's particular national interests? It is in this context that we examine UPOV and the UPOV conventions, and other alternative forms of plant variety protection.

Although some argue that the existing system of plant breeders' rights is by far the best way to implement the TRIPs obligation to provide for the effective protection of plant varieties, it is recognized that the TRIPs agreement does not oblige WTO member states to adopt a protection system along the lines of existing Plant Breeders' Rights. There is also an agreement that WTO member states are under no obligation to become members of UPOV.

Union for the Protection of New Varieties of Plants (UPOV)

Founded in 1961, UPOV is an intergovernmental body that establishes international rules under which countries grant intellectual property rights to the developers of new plant varieties (individuals or institutions). To qualify for protection, a new variety must be novel, distinct, uniform and stable. The original UPOV Convention was revised in 1972, 1978 and 1991. Today, the vast majority of UPOV members are party either to the 1978 or the 1991 Act.

Initially, UPOV membership consisted primarily of a handful of industrialized nations. But recently there has been some changes. With the recent accession of the People's Republic of China, Kenya, Bolivia, Brazil and Slovenia, the total number of UPOV members is 44.

Policy Primer

UPOV is gaining prominence as a legislative model for Plant Breeders' Rights because article 27.3(b) of TRIPs obligates WTO members to adopt patents and/or 'an effective *sui generis* system' for plant varieties. Although no such system has been defined, UPOV asserts that it is the only internationally recognized *sui generis* system for the protection of plant varieties. In addition, a number of influential bodies, including the WTO, are pushing for a narrowing of the *sui generis* option to the legislative model provided by UPOV.

At the June 1999 Congress of the International Association of Plant Breeders for the Protection of Plant Varieties (ASSINSEL), private and public plant breeders from 31 industrialized and developing countries representing over 1000 seed companies met to define their position on the protection of intellectual property. ASSINSEL concluded that the type of protection for plant varieties varies according to the technical, legal, and socioeconomic status of the country. Developing country members concluded that it was premature to develop protection of plant varieties through utility patents. The ASSINSEL Congress recommended that developing countries adopt a *sui generis* system based on the 1991 Act of the UPOV convention.

States that are Party to the International Convention for the Protection of New Varieties of Plants (Status as of 1 July 1999)

- Party to the 1961 Act, revised 1972: Belgium and Spain.

- Party to 1978 Act: Argentina, Australia, Austria, Brazil, Bolivia, Canada, Chile, China, Colombia, Czech Republic, Ecuador, Finland, France, Hungary, Ireland, Italy, Kenya, Mexico, New Zealand, Norway, Paraguay, Poland, Portugal, Slovakia, South Africa, Switzerland, Trinidad and Tobago, Ukraine, Uruguay.

- Party to 1991 Act: Bulgaria, Denmark, Germany, Israel, Japan, Netherlands, Republic of Moldova, Russian Federation, Slovenia, Sweden, United Kingdom, United States of America. Many developing countries currently do not confer protection on plant varieties in any manner. While there is an ample scope for national discretion in interpreting the *sui generis* option, deadlines for the implementation of TRIPs article 27.3(b) are fast approaching. As a result, many developing countries are under considerable pressure to consider UPOV's *sui generis* model for IP protection of plant varieties.

Members of the Crucible Group do not agree on whether or not and to what extent UPOV 1991 restricts the right of farmers to save protected seed for their own use.

Some CSOs and developing countries view the 1991 Act as a more stringent form of plant variety protection that dramatically strengthens the rights of commercial breeders while narrowing the rights of farmers. They believe that the 1991 Act is biased towards the interests of industrial breeders and does not adequately protect the rights of farmers and community innovators. UPOV 1991 significantly extends the rights of breeders and the scope of protected material (see discussion below). The cumulative effect, insist some CSOs, is that plant variety protection laws based on UPOV 1991 increasingly resemble the 'industrial strength' protection afforded by the patents.

UPOV asserts that strong intellectual property protection is necessary to ensure an acceptable return on research investment, and to encourage further plant breeding research that is essential to meet the challenges of increasing food production in the coming years. Proponents of UPOV insist that critics are misinterpreting the 1991 Act. While UPOV 1991 clearly strengthens the breeder's position in specific ways, it also preserves the ability of member states to allow farmers to save and replant protected seed (this is described by UPOV as the farmers' exemption or as the 'farmers' privilege').

Why the sharp difference of opinion surrounding the interpretation of UPOV 1991, article 5(1) of the 1978 Act is generally interpreted by governments and CSOs to allow farmers to save and replant protected seed for their own use (without payment of royalties). The critical clause does not mention farm-saved seed at all. It simply indicates that the authorization of the breeder is not required for the production and non-commercial marketing of protected material. UPOV proponents insist there is really no difference between the 1978 that Act and the 1991 Act of UPOV in this regard, as neither Act, they argue, mandates an exemption allowing farmers to fully use farm-saved seed.

The 1991 Act leaves each member state free to include the farmers' exemption in national legislation ('within reasonable limits and subject to the safeguarding of the legitimate interests of the breeder'). Under the 1991 Act, the farmers' exemption explicitly becomes an option for national laws. Some interpret this to mean that the farmers' exemption is no longer an automatic feature of the international rules governing plant protection under UPOV. Without positive action by member states on behalf of farmers, some fear that the farmers' right will be lost or significantly restricted. Who will determine the 'legitimate interests' of breeders? Under pressure from commercial breeding interests, some CSOs are concerned that the rights of commercial breeders will take precedence over the rights of farmers in national laws.

The 1991 Act of UPOV is interpreted as allowing a mandatory provision for subsistence farmers to use farm-saved seed. In those cases, where subsistence farmers are farming for a non-commercial purpose to produce food for their families, they would fall squarely within that exclusion. However, virtually all subsistence farmers market some portion of their harvest. If subsistence farmers are using protected varieties for commercial purposes, the exemption for farm-saved seed would not apply. Some CSOs believe that denying poor farmers the option of trading or selling planting material within their customary markets could restrict their role in maintaining genetic diversity and enhancing local plant breeding, and ultimately threaten food security. Proponents of UPOV point out that it is very unlikely that plant breeders would sue a subsistence farmer even if he markets a portion of his harvest.

The 1991 Act extends breeders' rights to the harvested product of the protected variety, if that variety has been used without authorization of the breeder (infringement of breeders' rights). However, if the variety has been used with the authorization of the breeder (that is, if a royalty has been paid on the seed), the breeder cannot claim rights on the harvested material.

Some CSOs are critical of provisions extending the scope of the breeders' rights beyond the reproductive material to the harvested material. Under article 14(1) of the 1991 Act, for example, explicit authorization of the breeder is required to sell, export or import propagating material of the protected variety, among other activities. If a country, which is not a member of UPOV, cultivates a UPOV-protected variety (without authorization of the breeder), then the breeder could prevent that country from exporting its harvest into UPOV territory. Without these provisions, assert UPOV proponents, commercial plant breeders operating in global commodity markets would be unable to protect their protected material from country to country. Some CSOs assert that the wider scope of these provisions will give seed companies too much control over the rights of farmers and the food system.

UPOV 1991 extends coverage to all plant genera and species. Under UPOV 1978, plant varieties could be protected under plant breeders' rights or by patent law, but countries could not allow both patents and plant variety rights for the same species of plant. The 1991 Act allows for double protection under both breeders' rights and patents, if countries so elect.

UPOV 1991 introduces the concept of 'essential derivation', which is designed to prevent the practice of cosmetic breeding. If a new plant variety differs from an older one by a minor modification—the insertion of a single gene, for example—the new variety is deemed 'essentially derived' from the older one. The concept of essential derivation aims to protect breeders from piracy. The functional and legal application of the term 'essentially derived' has yet to be implemented. It is not certain what minimum genetic distance will be required to distinguish between protected varieties, especially for minor crops.

Advocates of the concept of essential derivation say that it will foster the breeding and development of new and increasingly productive varieties. Cosmetic breeding should be eliminated, because it has the negative effects of reducing genetic diversity, increasing genetic vulnerability, and makes no positive contribution to improved productivity. Advocates believe that the concept of essential derivation will protect farmers, as well as breeders, from piracy.

Some CSOs are concerned that the criteria for determining essential derivation will work in favour of corporate breeders, and will put farmers and breeders in the South at a disadvantage. There is concern that the use of high-technology mechanisms (molecular marker profiles, pedigree distance data) to determine the level of genetic distance and innovative skill in plant breeding will discriminate against community plant breeders and give too much control to IP regimes.

Alternative Forms of Plant Variety Protection that may Implement 27.3(b)

Although the TRIPs agreement does not give any details on what elements an effective *sui generis* system would have to include, certain minimum requirements that such a system would have to fulfil may be drawn from the context of Article 27.3(b), the context of the agreement as an integral part of the WTO agreement, and finally from the objectives of the TRIPs agreement itself.

Since the TRIPs agreement elaborates no further upon the term 'plant variety', member states might have to provide for the protection of plant varieties of all species and botanical genera. The *sui generis* system has to be an intellectual property right. It needs to comply with the basic principle of national treatment. Thus, members have to accord to the nationals of other member states treatment no less favorable than that which they accord to their own nationals with regard to the protection of plant varieties. Furthermore, any advantage, favor, privilege or immunity granted by a member to the nationals of any other country has to be accorded immediately and unconditionally to the nationals of all the other member states (most-favored-nation treatment).

Finally, the *sui generis* system has to be effective. While some argue that the term 'effective' refers to a certain minimum level of protection, others argue that the *sui generis* system is effective if it provides for an enforcement procedure, so as to permit *effective legal* action against any act of infringement of the *sui generis* right.

Thus, although the *sui generis* system has to comply with certain basic requirements, it allows countries to develop national plant variety protection laws that are distinct from the UPOV model. Alternative proposals being discussed at the moment include the proposal to provide for the protection of farmers' varieties, to protect varieties only if the country of origin of the breeding material is disclosed, or to grant rights that are weaker than the rights granted to breeders under the 1978 or the 1991 UPOV Acts.

Many countries and civil society organizations are engaged in processes to elaborate national plant variety protection laws that also embrace Farmers' Rights, community rights and other systems for managing genetic resource use and access. Some of these aim to construct mechanisms for the exchange and transfer of biological resources, and to remunerate local communities for their contributions. In most countries, these laws have not yet been adopted. The elaboration of plant variety protection laws that seek to incorporate the rights of farmers and local communities is evolving rapidly and experience is very recent.

Genetic Resources Action International (GRAI), a CSO based in Barcelona, has recently compiled a summary of non-UPOV *sui generis* plant variety protection laws under debate in developing countries. The following are just a few examples:

- At the Organization of African Unity's (OAU) Summit in June 1998. The OAU has since developed the 'African Model Legislation for the Recognition and Protection of the Rights of Local Communities, Farmers and Breeders, and for the Regulation of Access of Genetic Resources. The model legislation will be available in a final draft form in early 2000.

- The Zambian government has drafted a plant variety protection law that seeks to protect the innovations of local communities and indigenous peoples, in keeping with its obligations under the CBD. The draft is now undergoing a process of wide public consultation.

- India's draft plant variety protection act attempts to balance recognition for Plant Breeders' and Farmers' Rights. It states that nothing contained in this Act shall affect the farmer's traditional right to save, use, exchange, share or sell his farm produce of a variety protected under this Act, except where a sale is for the purpose of reproduction under a commercial marketing arrangement. The draft PVP law also provides for communities to register collective rights.

- In Thailand, two bills have been drafted that aim to recognize traditional knowledge and the rights of local communities. The draft Plant Variety Protection Bill would combine recognition for the rights of plant breeders to their newly developed varieties, with the protection of native varieties that have been conserved and developed by the farmers and local communities. The draft Traditional Medicine Bill would recognize rights to traditional healers and medicinal genetic resources, based on the concept of collective rights. It includes the registration of traditional medicines and some form of benefit-sharing, in case medical or scientific researchers make use of the protected knowledge.

- The Plant Varieties Act of Bangladesh, drafted by the National Committee on Plant Genetic Resources, recognizes community rights and Farmers' Rights, and proposes the establishment of a fund to support communities in the conservation and development of plant varieties. The draft law is undergoing a process of public debate.

- Costa Rica introduced a plant breeders' rights bill at the end of 1999 for the purpose of complying with TRIPs. Because of concerns raised by civil society organizations, the bill is undergoing a process of public review and consultation. However, Costa Rica's 'Biodiversity Law' was signed into law in May 1998. The Biodiversity Law aims to comply with the mandate of the CBD. Under the term 'sui generis community intellectual rights', the law recognizes and expressly protects the practices and innovations of indigenous peoples and local communities related to the use of biodiversity components, and their associated knowledge. The National Seed Office and the Intellectual Property Office are obliged to consult with the National Commission for Biodiversity Management (CONAGBIO) on innovations, which involve biodiversity components prior to granting IP protection. The law obliges CONAGBIO's technical office to reject any request for the recognition of intellectual or industrial rights for biodiversity components or knowledge that is already recognized by community rights.

Some members of the Crucible Group ask whether or not it is possible for governments to pursue policies that genuinely stimulate innovative research and still to craft a responsible legislative answer to article 27.3(b) of WTO's TRIPs chapter. The divide between those who see intellectual property as a cost-effective 'win-win' for society and inventors, and those who see its exclusive monopoly provisions as antithetical to innovation and society's needs, is too great. In the light of this divide, the Crucible Group does not provide a definitive answer, but lists legislative options along with possible implications of the proposed choice. The Group stresses that the discussion should not lead readers to conclude that Crucible considers any legislative option to be sufficient or desirable at this time. Some Group members regard all of the options to be fundamentally flawed.

Could Plant Breeders' Rights (Plant Variety Protection) Include Farmers' Varieties?

This issue highlights the widely divergent views that exist on the desirability of intellectual property. It is controversial also because there is little agreement about what constitutes a variety; whether such a system can be enforced; and the balance between the production and distribution of new ideas. Many agree that the innovative activities of farmers and their communities must be recognized and encouraged to ensure food security and improve agricultural productivity, to conserve and create agricultural biodiversity. However, it is less certain whether a sui generis system that provides protection for farmers' varieties will help realize these objectives.

The discussion over the inclusion—or not—of informal IPRs for farmers' varieties raises

questions regarding the potential to amend existing conventions. However, many observers would agree that existing legislation and conventions could be amended to better accommodate the special concerns of farmers and indigenous peoples.

Farmers' Varieties and Intellectual Property Protection

There are fundamentally different views on the desired scope of the farmers' exemption (that is, the right of farmers to save and exchange proprietary planting material under IP laws). A *sui generis* system for protection of plant varieties may be different from the UPOV model with regard to the farmers' exemption. While Crucible Group members do not agree on the scope of the farmers' exemption, the Group believes that an element-by-element examination of the issues is helpful and can lead to progress.

One perspective argues that every farmer and farming community has the inalienable right to save and exchange any breeding material in any way they wish, along with commercial sale of IP-protected germplasm. Another view is that such an unrestricted use of IP material is a violation of their rights and a threat to the future of world food security. Still others are prepared to recognize the rights of certain groupings of farmers, possibly defined by economic status, land holdings, or their use of planting material to utilize seed in ways that would be prohibited to other groups of farmers.

Some members of the Group believe that plant breeders with proprietary varieties are, in general, prepared to accept that farmers, who customarily plant back harvested seed because they lack access to—or financial resources for—new seed every growing season, should be allowed to continue to do so. Further, such farmers should be permitted to exchange planting material with their neighbors, since this activity represents a traditional form of community plant breeding that could readily fall under the normal breeders' exemption. Further still, farmer/breeders could be granted the privilege of trading and even selling seed within their customary market area. Proprietary breeders stress that such activities would have to be consistent with traditional practices and not merely an evasion of the rights of proprietary plant breeders to control commercial exploitation of protected propagating material. It is recognized that among poor or primarily subsistence farmers, denial of the option to trade or sell planting material within their customary market could deny farmers an important mechanism through which they are able to maintain genetic diversity and enhance local plant breeding.

Proprietary breeders, in the main, also recognize that many poor farmers and poor countries work within geographic and economic environments within which it is difficult for the commercial sector to provide full and consistent services. In the same way, breeders accept that from time to time, national emergencies and other threats to food security may arise that require the suspension of their rights, so that urgent human needs can be addressed. Such situations are adequately covered under normal intellectual property conventions and other international undertakings.

Proprietary breeders point out, however, that any suspension of rights that they regard to be appropriate and consistent with IP conventions must be monitored carefully, and that they should have the opportunity to challenge and seek redress where they consider their rights to have been abused. Further, breeders feel that much of the international furore over this issue has incorrectly pitted farmers as 'David' versus multinational enterprises as 'Goliath'. Even subsistence farmers on marginal lands in poor economies could benefit from the development of local entrepreneurship in both plant breeding and seed multiplication and distribution through private or cooperative organizations that might improperly be constrained if farmer plant-back goes unmonitored and opportunities for commercialization are restricted unduly. Governments should seek to encourage appropriate local entrepreneurship at every opportunity.

Others in the Crucible Group regard the right of farmers and farming communities to save and exchange, including sell, planting material as vital to food security and essential to the conservation and enhancement of plant genetic resources. Any effort to constrain this right should be scrupulously studied and challenged. Limitations to this right based upon land holdings, economic status, or purpose are not valid within the understanding of customary use or in recognition of the importance of this practice for the conservation and enhancement of diversity for future generations.

1999 Review of Article 27.3(b)

A much-anticipated review of article 27.3(b), the provision calling for patents or *sui generis* rights on plant varieties, was to be placed in 1999. However, there is growing uncertainty about the nature of the review itself. When the TRIPs Council met in April 1999, there was disagreement about whether the exercise was merely a review of member states' efforts to implement article 27.3(b), or if it should also include a review and renegotiation of the text. Throughout 1999 the TRIPs council collected information from industrialized and developing country members to determine the status of implementation of article 27.3(b). Some industrialized countries, including the US and EU members, prefer a simple information-gathering exercise, rather than a renegotiation of the text. However, a number of developing countries are proposing to reopen the negotiations and amend the language of article 27.3(b).

Final decisions on the scope of the review, and whether or not article 27.3(b) is to be reopened and renegotiated, are still pending. Attention shifted away from the review of article 27.3(b) (or at least away from the idea that it might occur in isolation) as member states prepared to launch a new round of trade negotiations at the WTO

Ministerial Conference in Seattle, 29 November–4 December 1999. For a while, it looked as though the review might be dealt with simultaneously, along with a host of other issues. Given member states' failure to launch a new round of negotiations, they will once again be faced with deciding what to do about the review on its own.

Sharply contrasting proposals are emerging for the renegotiation of article 27.3(b). The US and other industrialized countries may push for deletion of the entire subparagraph, thus eliminating exceptions from patentability. WIPO, UPOV and some UPOV member countries suggest that UPOV 91 should be explicitly named as an (or the) 'effective' *sui generis* system. By contrast, some developing countries and CSOs are advocating for an expansion of what may be excluded from patentability under TRIPs to provide member states with the option of excluding all biological materials from patentability. While the US did not anticipate or favor new negotiations on TRIPs at the Seattle Ministerial meeting, some developing countries submitted proposals to the WTO TRIPs Council for new negotiations.

At the 20-22 October 1999 meeting of the WTO TRIPs Council, the US and India each submitted papers concerning article 27.3(b). The US favors the US-style patent-based model for protection of plant varieties, and warns that any *sui generis* model for plant variety protection that is not modeled on UPOV 1991 would need to be examined on a case-by-case basis. By contrast, India asserts that the TRIPs agreement conflicts with the CBD, and that the two must be reconciled before they can be implemented at the national level. India's paper focused on the limitations of IP regimes in adequately addressing aspects of indigenous knowledge. India advised developing countries to consider different models of protection before rushing to implement *sui generis* systems.

Ultimately, consensus will be required on any amendments to article 27.3(b). Given the fast-approaching deadlines and the contentious nature of plant and animal patenting, consensus may be difficult to reach.

SUMMARY

The WTO's TRIPs review has provided a sharp focus for the international debate surrounding intellectual property claims involving biological material. Among many farm and indigenous organizations, opposition to 'life patenting' is mounting. In between lies a group that believes it is too early to extend obligations to patent life-forms.

A rapidly changing IP environment and increasing privatization of agricultural research has forced the CGIAR to develop policies and procedures on IP over the past decade. The process has been complicated by the fact that the CGIAR System has no legal status, and its members often represent opposing sides of the highly politicized IP debate. In addition, there are at least 14 'policy-making' bodies within the CGIAR. After years of discussion and debate by numerous committees, the CGIAR System is still in the process of developing a coherent, comprehensive policy on IP. Given the rapidly changing international policy environment and ongoing debate in many international fora, the CGIAR decided in 1996 to endorse 'Guiding Principles for the CGIAR Centres on Intellectual Property and Genetic Resources' as an 'interim working document that will be continually reviewed and revised'. Among other principles: 'The Centres will not claim legal ownership nor apply intellectual property protection to the germplasm they hold in trust, and will require recipients of the germplasm to observe the same conditions, in accordance with the agreements signed with FAO'.

The use of proprietary materials and technologies is an increasingly complex policy challenge for the CGIAR. In 1998-99 the CGIAR Panel on Proprietary Science and Technology was charged with distilling the complex policy issues of intellectual property and its role in the future of CGIAR. The FAO/CGIAR Agreement obliges centres to exclude IP protection over 'in-trust' germplasm. That much is clear. But what about IP protection for technologies and materials developed by CG scientists? Should CGIAR centres seek proprietary rights, and, if so, under what circumstances? A survey conducted by the International Service for National Agricultural Research (ISNAR) in late 1997 found that IARC scientists are routinely using proprietary technologies (i.e. selectable markers, gene promoters, transformation systems, etc.) in biotechnology research. What are the legal implications of using someone else's proprietary inputs in CGIAR research? Are IARC scientists free to disseminate results derived from proprietary science? Does the use and development of proprietary science within the CGIAR distort or advance its mission to promote food security for the poor? These are among the complex questions facing the CGIAR, as it struggles to define its role as the world's premiere public sector agricultural research body in an era of rapidly evolving IP regimes, proprietary science and declining research budgets.

EXERCISE

1. Define patent.
2. Give an account of patent, and the need for patent in the business of biotechnological products in the competitive world market.
3. Write briefly about copyrights. What forms of intellectual property rights are covered under the copyright issues?
4. Explain the importance of patents in the present day biotechnological developments, and their protections for the proper returns to the inventor.
5. Describe the history of patent and its relevance to research communities.
6. What is the criteria used for obtaining the patent from the government?
7. Enumerate different patent laws that are used for the protection of the inventions and ideas in the Indian system.
8. Elucidate the applications of patent protections and its need for the research and development.
9. Write a note on different types of patents in the biotechnology, and add a note on its importance to the multinational companies.
10. Mention the recent patent law amendments made in India, and their importance in the biotechnology.
11. What is meant by essential requirements of patents? Mention how is it important for biotechnology, along with few examples.
12. How do utility patents and plant patents differ form each other from the biotechnological researcher's point of view?
13. Distinguish between the inventions and discovery in the biotechnological research and development for the natural product.
14. Explain in detail about periods of patent in different countries for biotechnological inventions by the researchers.
15. Give a note on the infringement of patents in the biotechnology research and developments, and the punishment given for that.
16. List out the essential requirements that are necessary for obtaining the patents in India for the new inventions.
17. Give an explanation of the materials that are not allowed in the patent category, and their impact on the third world countries.
18. Critically comment on the different types of patent applications in biotechnology research and developments.
19. Discuss drug patents in India, and their importance in the pharmaceutical companies along with processes involved.
20. Elucidate various types of patent applications available in India.
21. What do you mean by patent search? What is its importance in the patent office for the grant of patents to the inventors?
22. Explain the role played by the patent cooperation treaty in normalization of patent granting procedures throughout the world.
23. List out the conditions that should be fulfilled for according an international filing date in the patent cooperation treaty.
24. Define official action and its importance in the procedures involved in the granting of patent to the invention.
25. Give an account of the compulsory licensing and its importance in the economical conditions of the developing countries.
26. Under what situations does the government use the patents?
27. Discuss the procedures involved in the grant of compulsory licensing applications in biotechnology research and developments.
28. How do the patent laws function in the biotechnological inventions? Write about their disputes in the third world countries.

17

CHAPTER   **Patenting Living Organisms**

17.1. INTRODUCTION

Biotechnology is developing at a greater phase and is encompassing wider aspects of the biology. By the close of the 20th century, humankind had acquired the power to transform the processes of all living species, including its own. The potential benefits of these powers can be exciting and the perils ominous. Consider, for example, the recent announcement that scientists have successfully produced cultures of embryonic stem cells. This breakthrough offers the potential to grow any type of human tissue and may eventually be used to repair damaged hearts, blood vessels or brains. This creates greater business opportunities in the commercial products and their returns. Patents on living organisms are not unique to genetic modification (GM) and biotechnology. Microorganisms, in particular, have commonly been patented. As far back as the nineteenth century, Pasteur was granted a patent on a strain of yeast in both France and the US. Plant patents were occasionally granted prior to the first UPOV Convention in 1961 that specifically excluded the granting of both patents and plant variety rights for the same plant variety.

Patent can be defined as the exclusive right to stop others from making, using or selling the invention. As in the biological field and molecular biology, a large number of genetically modified organisms are produced that are not similar to the natural organisms. Hence, it has a scope for applying for the patent for their creations in the diverse field of biotechnology from environmental biology to the medicine.

17.2. BEGINNING OF PATENTING LIFE

Prior to the development of genetic modification, patents on plants were not widely granted in the US or Europe. In the US, the patent act defines the patent law. The broad classes of patentable matter, which were framed in the eighteenth century to encourage a liberal approach to patentability, remain in the Act. According to the Product of Nature Doctrine, any naturally occurring material or law of nature is excluded from patent protection. The patent applicant was only allowed to claim materials or processes, which were novel and inventive. In Europe, the European Patent Convention (EPC) explicitly forbids patents on plant or animal varieties.

The development of techniques for the genetic modification of microorganisms, animals, and plants has challenged the concepts of what is and what is not patentable. In general, the US has been more responsive to these new developments, and has led to the creation of legal precedents that have broadened the scope of patentable materials. But in June, 1980, the situation altered with the decision of the US Supreme Court in the case of *Diamond vs. Chakrabarty*. Chakrabarty, a scientist who worked for General Electric, submitted an application to the US Patent Office in 1972 for a new strain of the bacterium, *Pseudomonas* (Fig. 17.1). The novel bacteria were intended to clean up oil spills in water by degrading the crude oil, and ingesting the degraded material, with the bacteria themselves, in turn, forming part of the normal food chain. Dr. Ananda Chakrabarty (Fig. 17.2) did not use rDNA techniques in producing the new strain; he relied on other techniques. Plasmids from separate organisms, each able to degrade one of the important hydrocarbons, which constitute crude oil, were bred into a single bacterium, thus combining all their superior properties in a single strain of bacteria called super bug.

Fig. 17.2. Dr. Ananda Chakrabarty with his patent certificate for Pseudomonas species.

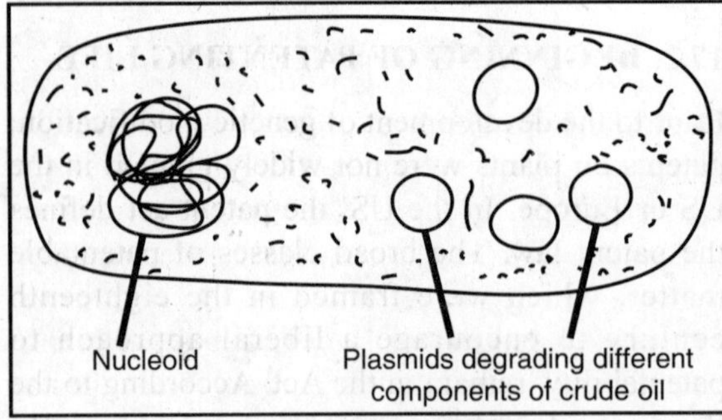

Fig. 17.1. A species of Pseudomonas that has the capacity to degrade oil spills

Two assumptions stood behind such recognition of patentability—artefacts can be biotic or abiotic, and the products of biogenetic technology qualify as biotic artefacts. Given these assumptions, it was expected that a favourable decision in the Chakrabarty case would clear the way for the numerous products of the biotechnology revolution, which were rapidly coming on stream but were held up until the Supreme Court had pronounced on the Chakrabarty case. After all, the rDNA organisms, animals, plants or microbes, compared with those produced by other forms of biogenetic technology, embody even a greater degree of artefacticity. It stood to reason that rDNA organisms, their products, and procedures are patentable as they are paradigmatically designed by humans and are not the products of nature.

In 1978, the *Diamond vs. Chakrabarty* case in the US allowed a patent on a novel genetically engineered microorganism. This was widely interpreted as a signal to the nascent biotechnology industry that it could expect strong and broad intellectual property protection for inventions involving GM organisms. As a result of the Chakrabarty decision, the US Patent and Trademark Office (USPTO) received over 7000 patent applications for inventions involving biotechnology. In 1987, the USPTO announced that 'the PTO now considers non-naturally occurring, non-human, multicellular living organisms, including animals, to be patentable subject matter within the scope of section 101. In *re Hibberd* resulted in a patent on maize plants, which had raised levels of the amino acid tryptophan.

Over 200 US patents in the 'plant biotechnology' category have now been granted. Many of these include claims for plants themselves, as well as for plant DNA, proteins, or other biochemical compounds. In 1988, a patent was granted for a mouse called oncomouse or popularly known as Harvard Mouse, which had been genetically engineered for a predisposition to develop cancer by group of scientists in Harvard University. The claims in the Harvard 'oncomouse' patent (USP 4736866) referred to 'non-human' mammals'.

In Europe, the development of patents for living organisms has been slower. The only means by which plant varieties can be protected in Europe has been under the UPOV convention (Union for the Protection of New Varieties of Plants). Under UPOV, which was founded to provide international protection to the plant breeding industry, the breeder is awarded an exclusive right to sell the reproductive material for 20–25 years. There are two important exemptions to the plant variety protection afforded by UPOV. First, other breeders may use the variety to develop new varieties under the research exemption provision. Secondly, farmers may save seed for crop production, though not for sale to other farmers, under the farmer's exemption provision.

Nevertheless, a number of plant patents have been allowed in Europe after protracted debate over whether the plants concerned were varieties or not. However, a decision in 1995 by the European Patent Office (EPO) somewhat reversed this emerging policy by refusing a patent on a GM crop, restricting instead the allowable claims to GM cells. No further EPO patents on plants have since been issued, though a test case under consideration should resolve the issue. The recent European Directive on the protection of biological inventions allows patents on plants and animals if the invention is applicable to more than one variety. Generic inventions such as wheat modified with the Bt gene are not plant varieties eligible for protection under UPOV, and are therefore patentable.

The 1956 US Patent Act provides that Whoever invents or discovers any new and useful process, machine, manufacture or composition of matter, or any new and useful improvement thereof, may obtain a patent therefore. Based on this the living organisms are considered for patenting. In **1980,** the United States Supreme Court first faced the issue of whether a live, human-made microorganism is patentable under 35 U.S.C. §101 as a new "manufacture" or composition of matter. The case dealt with microbiologist Ananda Chakrabarty's successful invention of a bacterium from the genus ***Pseudomonas,*** which contained at least two stable energy-generating plasmids; each plasmid providing a separate hydrocarbon degradative pathway. The bacterium was capable of breaking down multiple components of crude oil. Chakrabarty and his licensee **general electric** believed that the bacterium could "have significant value for the treatment of oil spills".

In a 5-4 decision, Chief Justice **Burger** determined that the patent statutes should be read broadly to allow the patenting of living microorganisms. The court held that Chakrabarty's bacterium was not a hitherto unknown natural phenomenon, but a nonnaturally occurring manufacture or composition of matter—a product of human ingenuity having a distinctive name, character, and use. The court added that his discovery was not nature's handiwork, but his own; accordingly, it was patentable subject matter under article 101. And so, he was given patent with patent number US 303, 206.

Justice Brennan, writing for the four dissenters, argued that the Plant Patent Acts of 1930 and 1970 evidenced Congress' understanding that 35 U.S.C. §101 does not include living organisms. If living organisms were included, Brennan contended, then the Plant Patent Acts would have been unnecessary and superfluous. He further emphasized that Congress had expressly excluded bacteria from the coverage of the 1970 Act. Justice Brennan concluded, "It is the role of Congress, not of this Court, to broaden or narrow the reach of the patent laws. This is especially true where, as here, the composition sought to be patented uniquely implicating matter of public concern".

The Chakrabarty decision has ignited criticism from diverse sources. Researcher, Dr. Leon Kass, for example, believes the Court's decision was legally and morally incorrect. What is most distressing, however, is that Congress has not spoken since the decision, but simply has allowed the Chakrabarty decision to become the law. In effect, Congress has taken the easy way out, and abdicated its constitutional authority to five unelected Supreme Court Justices in an area that uniquely implicates matters of public concern. Legislation by abdication does not serve the interests of Americans in such profound questions as whether a single individual or corporation can own new forms of life. As citizens, we must demand more from our elected officials.

Following the work done by the research teams of Stanley Cohen at Stanford University and Herbert Boyer of the University of California in San Francisco in 1973 and 1974, Stanford University, in 1974, filed an application to patent rDNA techniques for transforming cells with recombinant plasmids, using antibiotic-resistance genes on plasmids as genetic markers, *in vitro* genetic recombination techniques for producing recombinant plasmids as well as for the recombinant plasmids themselves. But in 1978, the submission was divided into two applications, one for a process patent and the other for a product patent. The process patent was granted in December 1980, following the Chakrabarty decision. The product patent was issued in 1984, covering as well products produced by bacterial plasmids in bacterial hosts.

The first Cohen-Boyer patent is registered as No. 4,237,244 and issued for the process for producing biologically functional molecular chimeras. The patent office spelt out the novelty of the procedure leading to the production of novel biotic artefacts very clearly indeed: 'The ability of genes derived from totally different biological classes to replicate and express in a particular microorganism permits the attainment of interspecies genetic recombination. Thus, it becomes practical to introduce into a particular organism... functions, which are indigenous to other classes of organism' (US Patent Office 1980, 1). In other words, the patenting of transgenic organisms recognizes that paradigmatically they are biotic artefacts. Whether one disapproves of it or has reservations on other grounds is another matter, but successful patenting is the logical conclusion from the fundamental premises that they are undoubtedly artefacts, and novel artefacts.

The success caused a good deal of ill will within the scientific community, as it credited Cohen and Boyer to be the inventors. Patent law after all assumes co-authors to be co-inventors, yet co-authors were left out. The two patents together were estimated to be worth more than a 1,000 million dollars, but as the earnings in the end, in the main, went to the two universities for research purposes rather than the two individuals named as the inventors, the acrimony eventually subsided.

Fig. 17.3. First mammal patented—onco mouse, transgenic mouse having breast cancer gene of human

In 1984, Harvard University applied to the US Patent and Trademark Office to patent its 'oncomouse'—a strain of laboratory mouse in which a gene, involved with the onset of breast cancer in humans, has been inserted **(Fig. 17.3)**. In 1988, the oncomouse was granted patent, the first bestowed on a transgenic vertebrate, whereas a year earlier, the US PTO had already granted patent on a transgenic oyster. In 1993, the European Patent Office granted patent on the oncomouse. However, 16 legal oppositions have been lodged with the EPO against it and hearings had been scheduled. European patent applications on animals have been filed, but only three have been granted. The objections, on the whole, have come from animal

welfare and animal rights groups. Their main argument appears to be based on the suffering caused to such transgenic animals.

So, from the point of view of patentability, one might be tempted to argue that the transgenic organism is not a suitable candidate for patentability, as its extent of artefacticity is really quite minor or limited. Before rushing to this plausible conclusion, perhaps one should ponder other aspects, which may be relevant to the debate. First, does the point above involve nothing more than a purely empirical issue? True, the examples of transgenic organisms usually cited seem to involve only a specific limited change, like the ability to produce a human protein in their milk or whatever. But in principle, are biotechnological methods and procedures thus restricted? As far as one can ascertain, the answer seems to be "no". Provided they can get away with it, genetic engineers in the agro-industries could well produce a non-sentient, wingless, featherless, beakless organism with avocado-colored flesh tasting like strawberries. Such a transgenic organism, given the extensive range of its unique characteristics, has a *sui generis* identity which, nevertheless, relies on the mechanisms possessed by the original bird to carry out its various biological functions, like digestion, respiration, defecation, etc.

The degree of artefacticity of such a product of genetic engineering would then be both deep and extensive. In other words, what should be a condition for patentability in this context? Should one rely:

- solely on extensiveness of change to the organism in question?
- solely on depth at which genetic material is manipulated?
- or on both extensiveness and depth?

In invoking extensiveness alone, then, on the whole, domesticated organisms produced by the less radical breeding technologies would be covered (This scenario, however, has been included in this discussion solely for completeness and for the purpose of clarification, as it can have no policy implication—modern patent law, by and large, have left domesticated organisms of such kinds outside the domain of patentability). But it would exclude transgenic organisms having one alien DNA sequence inserted into its genome and displaying only a specific and limited change in its phenotype—let us call this type A transgenic organism. *Vice versa,* in invoking depth alone, domesticated organisms bred in relatively traditional ways would then be excluded. But if both depth and extensiveness were invoked, then transgenic organisms in principle would certainly qualify.

If organisms could, indeed, be manipulated at the deeper and more radical level of their genetic material, crossing both species and kingdoms barriers, in such a way as to display a suite of phenotypical changes attendant upon genotypical ones—let us call this type B transgenic organism—then it is merely academic to confine discussion only to the majority of transgenic organisms. In any case, as a matter of fact, some transgenic organisms have already been produced, such as the 'liger' (or the 'tiglon'), whose genome shares the genetic components of both the lion and the tiger, and which correspondingly exhibit extensive phenotypical changes from its respective parents.

The same holds true for the 'geep' or 'shoat', which incorporates the genetic material from the sheep and the goat (The main technological procedure used in these examples of genetic manipulation is *in vitro* fertilization, rather than the insertion of specific alien DNA sequences into the genome of either the lion or the tiger).

To date, type B transgenic organisms are not so common simply because, for the moment, the climate and the market are not quite ready for them. But agro-industries would not be averse to opt for this kind of manipulation if the circumstances turn out to be propitious. Furthermore, what might be the response of such industries if type A transgenic organisms are denied patentability on the grounds that they are not sufficiently novel. This would immediately prompt these industries to change

ploy; their genetic engineers would be instructed to design and manufacture only type B transgenic organisms.

Just to take one hypothetical example: the cow with the alien DNA sequence to produce a human protein in her milk could then be the recipient of other transgenic DNA sequences which might alter her skin pigmentation to blue, or render the animal luminescent in the dark, etc. so long as these other sequences do not interfere with the capability of the transgenic cow to produce the human protein in question or to cause it to suffer in the way the Beltsville pig did. The obstacle to patentability encountered by type A transgenic organisms could in practice be overcome by simply pursuing the strategy of manufacturing only type B transgenic organisms. The ability of biotechnology to confine itself to effecting only one limited specific change in type A transgenic organisms is testimony to its powers of precise control and not, necessarily, to an inability on its part to bring about more extensive changes, should those wielding the technology so wish to do.

17.3. PATENT AND ITS IMPORTANCE IN BIOLOGY AND BIOTECHNOLOGY

Research scientists who work in public institutions often are troubled by the concept of intellectual property, because their norms tell them that science will advance more rapidly if researchers enjoy free access to knowledge. By contrast, the law of intellectual property rests on an assumption that without exclusive rights no one will be willing to invest in research and development (R&D).

Patenting provides a strategy for protecting inventions without secrecy. Patent grants the right to exclude others from making, using, and selling the invention for a limited term—20 years from the date of filing application in most of the world. To get a patent, an inventor must disclose the invention fully so as to enable others to make and use it. Within the realm of industrial research, the patent system promotes more disclosure than would occur if secrecy were the only means of excluding competitors. This is less clear in the case of public-sector research, which typically is published with or without patent protection.

The argument for patenting public-sector inventions is a variation on the standard justification for patents in commercial settings. The argument is that postinvention development costs typically far exceed preinvention research outlays, and firms are unwilling to make this substantial investment without protection from competition. Patents thus facilitate transfer of technology to the private sector by providing exclusive rights to preserve the profit incentives of innovating firms. Patents are generally considered to be very positive. In the case of genetic patenting, it is the scope and number of claims that has generated controversy.

Merits of Gene Patenting

- Researchers are rewarded for their discoveries, and can use money gained from patenting to further their research.
- The investment of resources is encouraged by providing a monopoly to the inventor and prohibiting competitors from making, using, or selling the invention without a license.
- Wasteful duplication of effort is prevented.
- Research is forced into new, unexplored areas.
- Secrecy is reduced, and all researchers are ensured access for to the new invention.

Demerits of Gene Patenting

- Patents of partial and uncharacterized cDNA sequences will reward those who make routine discoveries, but penalize those who determine biological function or application (in appropriate reward given to the easiest step in the process).
- Patents could impede the development of diagnostics and therapeutics by third parties because of the costs associated with using patented research data.

- Patent stacking (allowing a single genomic sequence to be patented in several ways such as an EST, a gene, and a SNP) may discourage product development because of high royalty costs owed to all patent owners of the sequence. These are costs that are likely to be passed onto the consumer.
- Since patent applications remain a secret until granted, companies may work on developing a product only to find that new patents have been granted along the way, with unexpected licensing costs and possible infringement penalties.
- Costs increase not only for paying for patent licensing but also for determining what patents apply and who has rights to downstream the products.
- Patent holders are being allowed to patent a part of nature—a basic constituent of life. This allows one organism to own all or part of another organism.
- Private companies, which own certain patents can monopolize certain gene test markets.
- Patent filings are replacing journal articles as places for public disclosure, thus reducing the body of knowledge in the literature.

Patenting Genes, Gene Fragments, SNPS, Gene Tests, Proteins, and Stem Cells

In terms of genetics, inventors must:
- identify novel genetic sequences.
- specify the sequence's product.
- specify how the product functions in nature— i.e., its use.
- enable one skilled in the field to use the sequence for its stated purpose.

Genes and Gene Fragments

The United States patent and trademark office (USPTO) has issued many patents for gene fragments of humans. William Haseltine of Human Genome Sciences claims that by 1995 his company had already isolated greater than 95% of all human genes. (Only about 3% of the DNA in the human genome codes for genes, and the rest is considered as non-coding DNA). Many commercial ventures are concentrating on the 'gene-rich' regions of the human genome, ignoring the non-coding DNA. On 17 September, 1998, HGS announced that patent applications published under the auspices of the Patent Cooperation Treaty include claims on 476 full-length human genes. According to HGS, each of the genes described in the patent applications represents a newly described human gene in the form of a corresponding complementary DNA (cDNA), the complete protein coding text of each gene and potential medical uses. Full sequence and function often are not known for gene fragments. On pending applications, their utility has been identified by such vague definitions as providing scientific probes to help find a gene or another EST, or to help map a chromosome. Questions have arisen over the issue of when, from discovery to development of useful products, exclusive right to genes could be claimed.

The 300 to 500 base gene fragments called expressed sequence tags (ESTs) represent only 10 to 30% of the average cDNA, and the genomic genes are often 10 to 20 times larger than the cDNA. cDNA molecule is a laboratory-made version of a gene that contains only its information-rich (exon) regions. These molecules provide a way for genome researchers to fast-forward through the genome to biologically important areas. The original chromosomal locations and biological functions of the full genes identified by ESTs are unknown in most cases. The furious pace of discovery in the field of genomics is reflected in the growing number of patent claims related to partial gene sequences or ESTs (expressed sequence tags). In 1991, the US Patent and Trademark Office had applications pending on 4000 EST sequences. In 1996, there was a total of approximately 350,000 EST sequences to be examined, and as of September, 1998, there were applications pending on over 500,000 EST sequences.

Patent applications for such gene fragments have sparked controversy among scientists, many of whom have urged the USPTO not to grant broad patents in this early stage of human genome research to applicants, who have neither characterized the genes nor determined their functions and uses.

In December, 1999, the USPTO issued stiffer interim guidelines (made final in January 2001) stating that more usefulness—specifically how the product functions in nature—must now be shown before gene fragments are considered patentable. The new rules call for specific and substantial utility that is credible, but some still feel the rules are too lax.

The patenting of gene fragments is contro-versial. Some say that patenting such discoveries is inappropriate because the effort to find any given EST is small compared with the work of isolating and characterizing a gene and gene product, finding out what it does, and developing a commercial product. They feel that allowing holders of such "gatekeeper" patents to exercise undue control over the commercial fruits of genome research would be unfair. Similarly, allowing multiple patents on different parts of the same genome sequence—say on a gene fragment, the gene, and the protein—adds undue costs to the researcher who wants to examine the sequence. Not only the researcher has to pay each patent holder via licensing for the opportunity to study the sequence, he also has to pay his own staff to research the different patents and determine which are applicable to the area of the genome he wants to study.

Single Nucleotide Polymorphisms (SNPs)

Single nucleotide polymorphisms (SNPs) are DNA sequence variations that occur when a single nucleotide (A, T, C, or G) in the genome sequence is altered. For example, a SNP might change the DNA sequence AAGGCTAA to ATGGCTAA. SNPs occur every 100 to 1000 bases along the 3-billion-base human genome. SNPs can occur in both coding (gene) and noncoding regions of the genome. Many SNPs have no effect on cell function, but scientists believe that others could predispose people to disease or influence their response to a drug.

Variations in DNA sequence can have a major impact on how humans respond to disease, toxins, chemicals, drugs, and other therapies. This makes SNPs of great value for biomedical research and developing pharmaceutical products or medical diagnostics. Scientists believe that SNP maps will help them identify the multiple genes associated with complex diseases such as cancer, diabetes, vascular disease, and some forms of mental illness. These associations are difficult to establish with conventional gene-hunting methods, because a single altered gene may make only a small contribution to the disease.

In April, 1999, ten large pharmaceutical companies and the UK Wellcome trust philanthropy announced the establishment of a non-profit foundation to find and map 300,000 common SNPs (they found 1.8 million). Their goal was to generate a widely accepted, high-quality, extensive, publicly available map using SNPs as markers evenly distributed throughout the human genome. The consortium planned to patent all the SNPs found, but to enforce the patents only to prevent others from patenting the same information. Information found by the consortium is freely available.

Gene Tests

When defective genes are found, complementary gene tests are developed to screen for the gene in humans who, they suspect, may be at risk for developing the disease. These tests are usually patented and licensed by the owners of the disease gene patent. Royalties are due to the patent holder each time the tests are administered, and only licensed entities can conduct the tests.

Proteins

Proteins are essential for the work of the cell. A complete set of genetic information is contained in each cell as genome. This information provides a specific set of instructions to the body. The body carries out these instructions via proteins. Genes encode proteins.

All living organisms are composed largely of proteins, which have three main cellular functions:
1. To provide cell structure and be involved in cell signaling.
2. Cell communication functions.
3. Enzymes are proteins.

Proteins are important to researchers because they are the links between genes and pharmaceutical development. They indicate which genes are expressed or are being used. Important for understanding gene function, proteins also have unique shapes or structures. Understanding these structures, and how potential pharmaceuticals will bind to them is a key element in drug design.

Stem Cells

Therapeutic cloning, also called "embryo cloning" or "cloning for biomedical research", is the production of human embryos for use in research. The goal of this process is not to create cloned human beings, but rather to harvest stem cells that can be used to study human development and treat diseases. Stem cells are important to biomedical researchers, because they can be used to generate virtually any type of specialized cell in the human body. Cell lines and genetically modified single-cell organisms are considered patentable material. One of the earliest cases involving the patentability of single-cell organisms was *Diamond vs. Chakrabarty* in 1980, in which the Supreme Court ruled that genetically modified bacteria were patentable.

Patents for stem cells from monkeys and other organisms already have been issued. Therefore, based on past court rulings, human embryonic stem cells are technically patentable. A lot of social and legal controversy has developed in response to the potential patentability of human stem cells. A major concern is that patents for human stem cells and human cloning techniques violate the principle against the ownership of human beings. In the US patent system, patents are granted based on existing technical patent criteria. Ethical concerns have not influenced this process in the past, but the stem cell debate may change this. It will be interesting to see how patent laws regarding stem cell research will play out.

17.4. SOCIAL CONSEQUENCES OF RECENT BIOTECH PATENTS

Since 1980, the evolution of intellectual property laws has allowed for the patenting of living organisms, enabling biotechnology companies to patent biological processes and products, thereby increasing incentives for the private sector to invest in this type of research. The WTO's TRIPs review has provided a sharp focus for the international debate surrounding intellectual property claims involving biological material. Among many farm and indigenous organizations, opposition to 'life patenting' is mounting. In between lies a group that believes it is too early to extend obligations to patent life-forms. The European Union, the European Parliament gave a final approval to the controversial biotechnology 'patent directive' in May 1998. The European Directive on the legal protection of biotechnological inventions aims to harmonize national legislation on the patenting of genetic material within the EU. The Directive creates, for the first time in European history, an explicit legal right to obtain patents for higher organisms, such as plants and animals. However, it does not create a European-wide patent, nor is it binding for the European Patent Office. The new law came into force on 30 July, 1998. The member states have two years to implement its provisions.

For one decade prior to its approval, the patent directive was a lightning rod for vigorous debate throughout Europe on the ethics and morality of biotechnology and patenting life. In 1995, European civil society organizations opposed to patenting life lobbied strenuously and effectively to defeat an earlier draft of the patent directive. Despite adoption of the EU patent directive, biotech patenting remains controversial in Europe, and the EU patent directive now faces legal challenges. The EU patent directive clarifies the scope of patentability for biotech-related inventions within 15 EU member states. It allows for the patenting of transgenic plants and animals, provided that they meet standard criteria for patentability. In a nod to ethical concerns, the directive outlaws transgenic animal patents on inventions likely to cause (animal) suffering without any substantial medical benefit to man or animal. The directive also says that plant and animal patent applications should specify the geographic origin of patented material. In the case of human materials, the person from whom the genetic material was taken must have had the opportunity of giving free and informed consent thereto 'in accordance with national law'. However, the directive provides no sanction for breach of these requirements.

Humans and human embryos are excluded from patentability, but Article 5 of the directive recognizes that human genes isolated from the human body or otherwise produced by means of a technical process can be patented even if the structure of that element is identical to that of a natural element. The directive also clarifies the EU position on the patenting of partial gene sequences, by requiring that the function or industrial application of a gene sequence be disclosed in all patent applications.

From the viewpoint of those opposed to the patenting of life forms, approval of the patent directive opened the floodgates to the 'industrial commodification' of all life forms, eliminating Europe's last symbolic barrier to legal resistance. For proponents in government and industry, the patent directive offers much-needed clarification in a controversial area of law and provides a 'sensible compromise' between the views of the biotech industry and the ethical concerns that are surrounding the direction of genetic research.

With globalization of trade and the disappearance of trade barriers, the market for biotechnology products has expanded, providing increased opportunities for large multinational corporations. The role of multinational corporations has both positive and negative implications for global science. From a positive point of view, these firms have strong research and development arms that are involved in both applied and basic research, frequently involving scientists from many countries and benefits that extend across borders. However, the products of private sector research are generally proprietary, and may not be accessible to the poor or those who need them.

Case for Broad Biotech Patents

As discussed earlier, the United States has recognized constitutionally, legislatively, and judicially that inventors should be rewarded for their efforts and ingenuity. The concept of "appropriability" increasingly is employed by economists seeking broader patent rights. Unless firms or individuals can appropriate or capture sufficient returns from their risky investments in developing new products or processes, they will lack the incentive to make such investments.

Patent rights proponents note that the average cost of developing a new pharmaceutical ranges from $500-$800 million. Only 26 new drugs were approved by the FDA for consumer use in 2000, and 13 have already been pulled off the market. If pharmaceutical companies were unable to recover their substantial sunk research and development costs through patent protections or successful drugs, research and development investment could drop radically endangering future discoveries. The biotechnology and pharmaceutical industries argue that the fast-pace of discovery and characterization of genes and their impact in the acceleration of

drug development, and the availability of new drugs and diagnostics is a testament to how well patents have worked in encouraging and buttressing successful research and development. Finally, patent rights proponents add that by requiring publication and public disclosure, patents catalyze new discoveries related to the original patented discovery.

Case Against Broad Biotech Patents

A strong moral case against patenting life in any form can be made. As Dr. Leon Kass asks, "Is it not clear, if life is a continuum, that there are no visible or clear limits once we admit living species under the principle of ownership?" Pragmatic arguments can be added such as who will be responsible for a new life form that has not co-evolved with the rest of the environment. What if a new bacterium or virus runs amok destroying other life forms and ecosystems? As Dr. Kass notes, no patent holder could ever guarantee complete safety.

The recent proliferation of blocking patents also raises grave concerns for future biomedical research and development. One must now wonder whether licensing, transaction and litigation costs will significantly restrict access to the research tools and materials necessary for technological advance. One legal commentator has argued that the current regime of biotechnology patent protections is likely to close off much of the genetic commons to small firms that have been the indispensable catalysts of most fundamental innovation in the biotechnology sector.

The biotechnology revolution also is sparking a technology transfer rage that could jeopardize the independence and pure research motives of universities and non-profits. As distinguished Harvard biologist Richard Lewontin notes: "No prominent molecular biologist of my acquaintance is without a financial stake in the biotechnology business. As a result, serious conflicts of interest have emerged in universities and government service. One wonders whether such a system can produce the independent thinkers, our society desperately needs, to guide the biotechnology revolution. As stated by Karen Charman: "Few academics are willing to openly criticize biotechnology, for the fear of retribution from biotech boosters."

Finally, the sheer cost of pharmaceuticals and emerging medical technologies must be considered. The recent "anthrax scare" highlighted the problems of gaining mass distribution for drugs like Bayer's Cipro, which is protected by a patent. Recently, however, on July 12, 2001, the United States House of Representatives rejected 269 to 157, after only one hour of debate, legislation that would have allowed American wholesalers and pharmacies to import FDA-approved American-made drugs which are sold overseas. Furthermore, the escalating battles involving generic and patented brand name pharmaceuticals show no signs of going away in the next decade.

Some of the leaders in IPR say that in order to minimize the costs of twenty-first century medical care and to protect the public against serious technological risks, it should first statutorily reverse the Supreme Court's Chakrabarty decision, and pass legislation stating that living organisms may not be patented. It should further legislate that genes and gene sequences are natural products of nature that are not patentable subject matter. And it should then dramatically increase the resources allocated to the PTO, so that its offices are not overwhelmed by impossible demands. To lower barriers to entry should require the mandatory licensing of key blocking biotechnical patents at reasonable royalty rates. As argued by Jonathon Barnett, "A patent regime that lacks the threat of compulsory licensing may allow large firms to integrate vertically upstream, acquire a patent portfolio that erects insurmountable entry barriers to small firms, and ultimately slow down the rate of fundamental innovation." Finally, such legislation should be supplemented by vigorous antitrust enforcement against mergers, acquisitions, patent pools, and other business activities that seek to consolidate market power over pharmaceutical patents, research techniques, and genomic data bases.

Patenting of living organisms poses special problems for the patent system. To enable the public to practice an invention embodied in a self-replicating organism, a deposit must be made in an acceptable depository. With regard to gene patents, the patent office requires the use of standard symbols and a format for sequence data in most sequence-type patent applications. This is a departure from general patent office practice that allows the inventor to be his own lexicographer. The patent office has also recently clarified the utility requirements for gene patents. These changes may have profound implications for some applicants. In clarifying the utility requirement, the patent office decided against developing a utility standard specifically for gene patents, and stated that the utility must be "specific and substantial". Statements of fact made by the applicant are treated as true unless one skilled in the art would doubt them. A lack of utility is also the basis for a rejection based upon a failure to disclose how to use the invention.

As with the utility requirement, the patent office decided to develop neutral standards for the written description requirement that apply across all arts. The written description must be sufficient so that one skilled in the art could practice the invention. In order to avoid confusion, the patent office has elected not to attempt to define the word "gene". Taken together these requirements will prevent applicants from obtaining patent protection on nucleotide sequences, with no known applications other than the subject of further research.

Plant Patents

A special type of patent is available for new varieties of plants found in cultivated areas. Section 161 provides "[w]hoever invents or discovers and asexually reproduces any distinct and new variety of plant, including cultivated sports, mutants, hybrids, and newly found seedlings, other than a tuber propagated plant or a plant found in an uncultivated state, may obtain a patent therefore, subject to the conditions and requirements of this title. The provisions of this title related to patents for inventions shall apply to patents for plants, except as otherwise provided." These are called plant patents and are available for asexually reproduced plants. Plants capable of reproducing by seed are also covered if they are capable of being reproduced asexually. Plant patents cannot be obtained on tuber crops such as Irish potatoes and Jerusalem artichokes. The new plant must be a distinct variety. No deposit is required for plants that are the subject of plant patents. Nonetheless, the applicant may be required to provide a specimen of the plant.

Certificates of Protection under the Plant Variety Protection Act

Certificates of protection are available through the Plant Variety Protection Office of the US Department of Agriculture. This patent-like form of protection is available where "the breeder of any sexually reproduced or tuber propagated plant variety (other than fungi or bacteria), who has reproduced the variety, shall be entitled to plant variety protection for the variety...". The term of the certificate of protection is 20 years for most crops, and 25 years for trees, shrubs, and vines.

Where the producers of developing country grow crops for export to the developed country markets, a US patent of dubious validity can be, and has been, used to disrupt exporters to the USA. For example, a Colorado firm patented a yellow bean, which it named the 'Enola bean', bred from beans purchased in Mexico. The firm had reproduced and selected the yellow beans over several self-pollinated generations. After receiving the patent, the firm proceeded to demand licenses from the importers of similar Mexican beans. This patent was challenged in the USA, and after several years was surrendered in 1995 for reissue/re-examination. The patent's long-run status is unclear, though plant variety protection still covers the Enola bean (Nottenburg, 2005).

In another case, an improperly obtained patent on a new and superior variety of pineapple was referenced in a letter from the Vice President of Del Monte Fresh Produce to a Central American researcher. The writer warned him against working on pineapple plant material developed by Del Monte Fresh Produce, which, he added, owned a US Plant Patent. Similar warnings, allegedly, were sent to potential US competitors, including Dole. Later it became clear that the variety in question was not patented, and indeed had been refused a US patent in 1992. The patented variety was another improved variety—a hybrid that was a sibling to the Del Monte Gold variety. The patent was later withdrawn by Del Monte Fresh Produce, which acknowledged pre-patent sales of the patented variety by a competitor.

Intellectual property protection can be conferred in relation to plant materials through:

- the US model of plant patents, which are distinct from normal (utility) patents;
- allowing normal patents on plants or parts thereof, such as cells;
- patenting plant varieties, as is the practice in the US and in few other countries (for example, not in the EU);
- applying a *sui generis* form of plant variety protection (PVP), such as plant breeders' rights (as in the EU or the US) or other modalities; and
- allowing patents on DNA sequences and gene constructs including the gene, plants transformed with those constructs, the seed, and progeny of those plants.

In addition, patents are widely used to protect the technologies, which are employed in research on plant genomics. Apart from the use of patents and PVP, the intellectual property in plants can be appropriated by technological means. For instance, crops such as commercial hybrid maize cannot be reused if hybrid yield and vigour are to be maintained. This characteristic of some hybrids confers a natural form of protection by which seed companies can more readily capture a return on their investment through repeat seed sales. In contrast, other types of seed variety can be replanted each year without deterioration in yield, so that farmers may replant their own seeds without repurchasing.

The Green Revolution varieties were of this nature, which is one reason why they were so successful. It is only recently that hybrid varieties of rice and wheat have been developed. Genetic Use Restriction Technologies (known as GURTs) is a term used to describe different forms of controlling the action of genes in plants. The so-called "terminator" technology, which would render the seed sterile so that it is not physically possible to grow a second crop, is well-known but other characteristics can also be controlled, either for agronomic or commercial reasons. The effect of technological protection is similar to that of IP protection, but possibly cheaper and certainly more effective in the sense that it is self-enforcing.

Controversy over the collection and patenting of human genetic material is not new. In 1993, the Human Genome Diversity Project, an informal consortium of universities and scientists in North America and Europe, proposed to collect human DNA samples from hundreds of so-called 'endangered' indigenous communities around the world. Many indigenous peoples' organizations protested vigorously, asking:

- Will profits be made from the genes of poor people whose physical survival is in question?
- Who will have access to stored DNA samples, and where will these collections be located?
- What benefits, if any, will accrue to the indigenous peoples from whom DNA samples will be taken?

In March, 1995, the US patent office issued a patent on a cell line containing unmodified DNA from a Hagahai tribesman in Papua, New Guinea. Indigenous peoples' organizations vocally denounced the patent as a threat to human dignity and a violation of human rights. The controversy generated by scores of indigenous peoples' organizations, together with civil society organizations and governments, eventually caused the US government to 'disclaim' the Hagahai patent in October 1996.

Commercial trade in human tissue is accelerating. Scientists, in both the public and private sector, are collecting human DNA samples from rural and urban communities across the globe. Of particular interest to genetic researchers are populations that are genetically homogeneous, or those that exhibit a genetic predisposition to an inherited disease. After pinpointing the location of so-called 'disease genes' genomic companies and their pharmaceutical partners hope to develop commercial products such as diagnostic tests and therapies that are based on proprietary human genes. That quest has taken gene prospectors to remote locations such as Tristan da Cunha in search of asthma genes; to Kosrae, Micronesia, in search of obesity genes, and to Tibet in pursuit of high-altitude genes, just to name a few.

In early 1998, the prospect of nationwide collection and commercialization of human DNA made headlines when Hoffman-La Roche (Switzerland) and DeCode Genetics Inc. (Iceland) signed a $200 million collaborative research contract to identify disease genes based on studies of Iceland's relatively isolated and strikingly homogeneous population. DeCode's goal is to amass the world's most comprehensive collection of genealogical family data for studying the genetic causes of common diseases. The company says that its studies could lead to new diagnostic tests and drugs for inherited diseases, which would be made available free to Icelanders if the research leads to a new therapy. The Icelandic situation has become an international test case for many of the ethical and intellectual property issues surrounding the collection and commercialization of human DNA. Despite opposition by growing numbers of

Iceland's scientific and medical community, a bill was passed by the Icelandic parliament on 17 December, 1998, that gives "DeCode Genetics" the right to collect current and retrospective medical information from Iceland's 270.000 inhabitants into a centralized, comprehensive database. The new law gives DeCode Genetics exclusive rights for the commercial exploitation of genetic information till 12 years.

A vocal minority of Iceland's scientific and medical community, including the Icelandic Medical Association, the Association of Icelanders for Ethical Science, and the Icelandic Mental Health Alliance oppose implementation of the law and are advising doctors and their patients to refuse participation in the collection of DNA samples. Opponents believe that the bill violates principles of privacy and informed consent, and they object to a single company gaining exclusive rights to a valuable scientific resource. For example, the law allows only for individuals to opt out of the database, but does not require any other form of consent. Although the database is supposed to be confidential and anonymous, critics charge that personal information can be deciphered and computer security measures proposed by the company are not adequate to ensure confidentiality.

By the close of the 20th century, humankind had acquired the power to transform the processes of all living species, including its own. The potential benefits of these powers can be exciting and the perils ominous. Consider, for example, the recent announcement that scientists have successfully produced cultures of embryonic stem cells. This breakthrough offers the potential to grow any type of human tissue, and may eventually be used to repair damaged hearts, blood vessels or brains. The very same week, the UK's Sunday Times reported that scientists can theoretically engineer deadly biological organisms to produce 'ethno-bombs' that are capable of targeting human victims by ethnic origin. Given the dizzying pace of technological advancements in genetics and biology, it is not surprising that society is grappling ever more urgently with the social, ethical and legal implications of humankind's ability to decipher and control the genetic blueprint of life. Opinions differ sharply on the implications of new biotechnologies, but everyone agrees that advances in technology are taking place at a rate far faster than social policies can be devised to guide them, or legal systems can evolve to address them. There is a growing recognition worldwide that the development of scientific knowledge must be accompanied by public debate on societal choices and the informed participation of citizens. 'Bioethics' attempts to identify the social and cultural implications of breakthroughs in life sciences, to anticipate its applications, and to ensure that progress in the life sciences benefits humanity as a whole. Bioethics acknowledges that there is a distinction between what is scientifically possible and ethically acceptable. What is good for society, what is equitable, and what is safe? Who will decide? These are among the questions that are fueling debate on the applications of biotechnology in health, agriculture and human development. At a meeting of the UK Food Group in April,... Indira Jaising of the Supreme Court of India noted a trend in TNCs (transnational companies) such as pharmaceuticals moving from the developing world to the US, with the contingent drain of capital, because the US allows patenting of life forms. Whether or not the rest of the world falls in line with the US in accepting life patents, researchers predict that with advances in biotechnology there will be a switch in centers of production away from the developing world, accompanied by the loss of export income.

17.5. CONTROVERSIES OVER THE PATENTING OF LIVING ORGANISMS

One issue relates to the nature of patents, especially patents awarded for elements of biodiversity or living organisms. Patents are awarded for novelty, inventiveness, and utility. Patents for biodiversity, whether at the gene or organismal levels, challenge these three main criteria, yet they have become increasingly frequent. For example, in 1980, 16 patents were awarded for gene sequences. In 1990, the number was more than 6,000, and in 2000 more than 355,000 (Dutfield, 2002). As stated by Demaine and Fellmeth (2003), "subtly and without fanfare, the prohibition on patenting products of nature has fallen into desuetude." Those in favor of patenting genes argue that locating, isolating, and describing genes require ingenuity.

The US Patent and Trademark Office (2001) agreed with this interpretation and extended to earlier findings of genes stating that purification of a compound (e.g., adrenaline, prostaglandins, etc.) outside of its natural state could make this compound eligible for the patent. In the case of genes, it is also argued that if the gene was previously unknown, isolation of this gene can now lead to an explanation of its function, mode of action, and possible industrial application. If the application relies not on the gene but on the corresponding cDNA, then a case could be made for an invention based on a synthetic chemical. Furthermore, supporters of gene patents argue that biotechnology research is risky and expensive, and the promise of a temporary monopoly provided by the patent is a necessary stimulus for this type of research.

By contrast, critics argue that DNA sequence isolation and characterization, including by reverse transcription, is now a routine operation even for those with "average skills in the art" and does, therefore, not qualify as an inventive step. After all, private companies sell kits that help achieve some of the steps of the procedure of the isolation of DNA sequence. DNA sequencing is such a routine operation that it generally done by public or private service labs, and whole genome sequencing has been achieved for an increasingly large number of species. To this objection, the US Patent Trademark Office (2001) replies that obviousness does not depend on the amount of work required to characterize the DNA molecule: "Patentability shall not be negatived by the manner in which the invention was made, and the existence of a general method of isolating cDNA or DNA molecules is essentially irrelevant to the question whether the specific molecules themselves would have been obvious."

In Europe, an isolated gene or gene sequence or other element isolated from an animal or plant is patentable, provided that its function is known and a suitable industrial application is derived from that product. However, no rights are given to that product when found in its natural environment. The European Commission Directive also allows for the patenting of plants and animals, provided that the application of the invention is not technically confined to a single plant or animal variety.

This means that the patent for a novel gene sequence that confers benefit on a plant will extend to any plant in which the gene has been artificially inserted (European Commission, 1999, 2002). It can also be argued, however, that a mere routine purification does not warrant novelty or nonobviousness. Instead, some have argued that the claimed product should represent a substantial modification of the natural substance to become eligible for the patent. Based on this test, called the Substantial Transformation Test, a product is substantially transformed only if it has a new and distinct character or use. There have also been questions about the proof of utility provided by patent applicants. The US Patent and Trademark Office (2001) states that "when a patent application claiming a nucleic acid asserts a specific, substantial, and credible utility, and bases the assertion upon homology to existing nucleic acids or proteins having an accepted utility, the asserted utility must be accepted by the examiner unless the office has sufficient evidence or sound scientific reasoning to rebut such an assertion. A 'rigorous correlation' need not be shown in order to establish practical utility; 'reasonable correlation' is sufficient". This statement raises the question as to how the USPTO would gain evidence to be able to rebut a utility claim.

Although the USPTO has access to published information (including the scientific literature and previous US and foreign patents), it does not have research capabilities of its own or may not be able to rely on outside expertise. It raises the question as to whether some utility claims are guided more by desirability than real-world utility. In effect, the burden of proof for utility is still fairly low even after the tightening of rules by the US Patent and Trademark Office (2001). The lightness of this original burden of proof contrasts with the burden associated with either patent infringement or challenges. Discussions on the patenting of biodiversity also involve patenting of whole organisms. Apart from Japan and Australia, the United States is the only country awarding utility patents for plant cultivars. In other countries, such as the European Union, plant cultivars can only receive IP protection under the PVP legislation.

The European Patent Convention explicitly prohibits patents on both plant and animal varieties. This was further clarified by the European Commission biotechnology directive (European Commission, 2002), which specifically states that the following are not patentable:

- Plant and animal varieties
- Essential biological process for the production of plants and animals
- The human body at various stages of formation and development.
- The simple discovery of one of its elements, including the sequence or partial sequence of a gene.
- Inventions, the commercial exploitation of which would be contrary to "ordre public" (roughly translated as "public good") or morality.

In the United States, animals are also patentable subject matter since the award of a patent to Harvard University for the "oncomouse"—genetically engineered to have an increased risk of malignancies. The oncomouse is a biological model for testing carcinogenicity of different compounds, and is a useful model for researchers trying to develop treatments and cures for cancer. It has been patented in the United States, Europe, and Japan, but not in Canada. Harvard University had asked not only to patent the process used to engineer the mouse, but also the mouse itself, as well as all other nonhuman mammals rendered susceptible to cancer. The Canadian Supreme Court refused to grant a patent on the grounds that a higher organism cannot be considered a "manufacture" or "composition of matter (Supreme Court of Canada, 2002)".

The Canadian oncomouse decision raises two issues. First, for certain patents such as the oncomouse patent, there are prominent ethical issues. In this case, one can raise the question whether living organisms should be a patent subject matter. Furthermore, the pain and suffering imposed on individual animals of the oncomouse strain, which were engineered to be susceptible to cancer, is also an ethical issue. A different issue is raised by the case of *Moore vs. Regents* of the University of California (Dorney, 1990), namely the issue of prior informed consent by a patient and fiduciary duty (breach of trust) on the part of the treating physician. In the US patent legislation, such considerations are not taken into account, and a review of the case law suggests that economic and policy concerns drive patenting first and foremost, for example, to promote certain industries such as the biotechnology industry, in line with the overall free-enterprise ideology of the country (Curci Staffler, 2002).

Presumably, there are other venues, such as the courts, where ethical concerns can be addressed in the United States. Yet, the European Patent Convention explicitly states that patents shall not be granted in respect of inventions, the publication or exploitation of which would be contrary to ordre public. The concept of ordre public is defined as covering the protection of public security and the physical integrity of individuals as part of society. It also encompasses the protection of the environment. Accordingly, inventions, the exploitation of which was likely to seriously prejudice the environment, are to be excluded from patentability as being contrary to ordre public (European Patent Office, 1999). It should be noted here that it is not patents per se that are construed as being against the ordre public, but it is their exploitation that can be objectionable (Curci Staffler, 2002).

The second issue associated with the Canadian oncomouse decision is that it constitutes an exception to the pattern of "imitation" (Curci Staffler, 2002). Since IPRs are generally believed to stimulate innovation, developed countries, including Japan and countries in Europe have followed the liberal attitude towards patenting of the United States by introducing stronger standards of IP protection and widening the patentable subject matter in order to maintain the competitiveness of their national industries. This type of arms race has an unexpected side effect in that it further widens the gap in technology and IPR between developed and developing countries.

SUMMARY

Genetic modification technology development influenced the living organisms patents. Patents on plants were not widely granted in the US or Europe. In the US, the Patents Act defines the patent law. The broad classes of patentable matter, which were framed in the eighteenth century to encourage a liberal approach to patentability, remain in the Act. According to the Product of Nature Doctrine, any naturally occurring material or law of nature is excluded from patent protection. The patent applicant was only allowed to claim materials or processes which were novel and inventive. In Europe, the European Patent Convention (EPC) explicitly forbids patents on plant or animal varieties.

The appearance and maturity of a new basic scientific discipline, that is, molecular biology, and in particular its sub-branch, molecular genetics, has engendered a technology, namely, biotechnology, which permits a quantum leap, so

to speak, in the kind and degree of control of biotic nature, over its predecessor technology based on the classical gene-chromosome theory.

For the first time ever, we are able to cross the species barrier and, in principle, to dispense with natural evolution in the production of novel organisms and new species. Biotechnology makes it possible for us to make over biotic nature to our will and design. Many of the things that were discussed as science fiction five years ago have already happened. This is not just a change of technique; it is a new way of seeing. Splicing organisms, combining functions, dovetailing abilities and linking together chains of properties, can transcend the limitations of species. The living world can now be viewed as a vast organic Lego kit inviting combination, hybridization, and continual rebuilding. Life is manipulability.

In the first half of the twentieth century, the technology of hybridization generated by the theoretical discoveries of Mendelian genetics produced with a greater degree of precision plants and animals possessing characteristics deemed to be desirable than the traditional methods of breeding. However, such Mendelian products, evertheless, may be said, in comparison, to embody a lower level of artefacticity than those produced by rDNA technology induced by the fundamental discoveries of molecular genetics in the second half of the last century. In the case of the latter, their greater degree of artefacticity is due to the fact that their mode of production involves the manipulation of, and indeed, the exchange of genetic material at the molecular level across species, and even, kingdoms.

The status of transgenic organisms as biotic artefacts may be further elucidated *via* the issue about their patentability. Patents are about inventions and the legal rights over their financial exploitation (for a limited period). To obtain a patent, the item for which application has been filed must, first and foremost, constitute an invention; furthermore, at least three other conditions must obtain: the invention must be novel, it must not be something obvious to an expert in the field, and it should have industrial application. The discussion to follow will concentrate on the first requirement, that is, that transgenic organisms are indeed inventions within the meaning of the modern patent law. But it will also consider their novelty. However, the condition of non-obviousness will not be explicitly touched upon, and the discussion will simply assume that transgenic organisms have, on the whole, been fabricated with industrial application in mind.

Up to 1980, no one could be sure in any country with a Western-type legal system whether patents could be granted to any living organism, which claimed to have been made by humans. Up to then, animal varieties and any biological processes which underpinned the production of animals and plants fell outside the ambit of patenting. However, there was legislation to protect plant varieties in several countries—for instance, in the USA, the 1930 Plant Patent Act (PPA) covers asexually reproducing plants, and the 1970 Plant Variety Protection Act (PVPA) covers sexually reproducing ones. The UK 1983 Plant Varieties Act comes under the aegis of the Ministry of Agriculture, Fisheries and Food, not the Patenting Office, and covers the reproductive materials of plants. There is also the 1968 International Union for the Protection of Plant Varieties.

But in June 1980, the situation altered with the decision of the US Supreme Court in the case of *Diamond vs. Chakrabarty.* Chakrabarty, a scientist, who worked for general electric, submitted an application to the US Patent Office in 1972 for a new strain of the bacterium, *Pseudomonas.* The novel bacteria were intended to clean up oil spills in water by degrading the crude oil, then ingesting the degraded material, with the bacteria themselves, in turn, forming part of the normal food chain. Chakrabarty did not use rDNA techniques in producing the new strain. He

relied on other techniques. Plasmids from separate organisms—each able to degrade one of the important hydrocarbons which constitute crude oil—were bred into a single bacterium, thus combining all their superior properties in a single strain of super bacteria.

EXERCISE

1. What do you mean by patenting living organisms? What is its importance in the biotechnology?
2. Discuss how living organisms patent came into existence, and explain its applications and the risk involved.
3. Define cloning of animals and its impact on the ethical, legal and social issues raised in the cloning of animals?
4. Enumerate the importance of Ananda Chakrabarty's case in the patent history, and its adaptations in the living organisms patent.
5. Write a note on the patent, and its importance in the field of biology as a whole, and biotechnology in particular.
6. Explain the different types of patents that are commonly, adopted for obtaining the patents on living organisms.
7. Describe various types of organisms that are patented and their applications.
8. Critically comment on the ethical, social and legal issues raised by the living organisms' patents.
9. Distinguish between the plant patents and gene patents, and discuss their ethical issues.
10. Briefly write about the advantages of the gene patenting and their applications in the genomic and proteomics.
11. What is meant by gene fragments patenting? Add a note on the ethical issues associated with the gene fragments patenting?
12. How single nucleotide polymorphisms are considered as patenting materials. Add a note on its importance in developing the knowledge of science.
13. Write about the social consequences of the recent biotechnology patents and their impact on the society and culture.

14. List out what religious groups are saying against the living organisms in the present circumstances of conserved society.

15. Explain different types of limitations that are associated with the technology of cloning technology.

16. What is meant by human cloning? Explain its importance in the present day patent rich gene technology world.

17. Give an account of the arguments for and against that are associated with living organisms patents and their applications.

18. Briefly explain how scientists and researchers are supporting the importance of living organisms' patents and technological need of the patent.

19. Elucidate the future of plant patents, and it impact on the developing countries that are in the economic poverty and resource crunch.

20. Give a note on the plant variety protections and its limitations and advantages. Add a note on the benefits of the plant variety protections for the developed countries.

21. What are the ethical issues associated with the different types of living organisms' patents? Add a note on its impact on transgenic organisms.

22. Describe how different countries following different criteria, methodologies and requirements are followed for living organisms patent.

18

CHAPTER

Traditional Knowledge, Commercial Exploitation and Protection

18.1. INTRODUCTION

In biotechnology, the information required by the researchers has increased as more and more secrets of the life processes are discovered. The scientific endeavors of the 21st century are fundamentally different from the single-laboratory work of the past. Some of the most exciting research is interdisciplinary, multinational, and data-intensive. From multibillion dollar nuclear colliders to intensive study of the genetic code, information transfer and collaboration has become a necessary part of the research landscape. In medicine, agriculture, education, and industry, advances in human knowledge drive improvements in the quality of human life. The twenty-first century offers the promise of even greater progress, as digital technologies drive down the costs of creation, reproduction, distribution, and consumption of knowledge. As the role of knowledge in promoting human progress gains importance, so does the question of who has access to that knowledge.

For decades now, the idea that stronger intellectual property protection is critical to economic development has been the driving motto of national and international law-makers. Recent studies have, however, begun to question the validity of this idea, given the absence of strong empirical validation. The dominant view postulates that stronger intellectual property (IP) protection allows a nation to promote and protect innovation and creativity, thereby driving economic growth and wealth creation. Property rights that enable the internalization of positive externalities are supposed to incentivize and reward investment of effort and resources in the innovation process.

This macroeconomic perspective forms the basis of the World Intellectual Property Organization's (WIPO) "international policy-making activities" that operate under the motto "IP as a tool for economic, social, and cultural development". Very often, however, this view tends to ignore the obvious reality that developing countries have socioeconomic structures that differ greatly from those in the developed countries. In the absence of basic human and technical capacity, strengthening IP protection can do very little for economic growth, and may, in fact, impede the development process. But does this imply that an intellectual property regime can never contribute towards economic development?

Recent studies have also begun to show that there may be alternative non-proprietary structures that can facilitate innovation without, in any way, impeding the overall development process. It remains to be seen if these structures can indeed be assimilated into the overall innovation policy of a country, and whether they can operate side-by-side with the beneficial elements of the extant IP regime. Conceptually, however, introducing such structures into a policy may face the additional challenge of an ideological lock-in at the institutional level in favor of the dominant view. A balanced economic model will, therefore, have to overcome challenges at various levels—the conceptual, the institutional, and the ideological.

Privacy and National Security are the most prominent notions of public concern that suggest circumstances where the regulation of information flow and the restriction on access to knowledge, may be better for the public interest. But there are also uneasy intersections of access to knowledge, (A2K) with the preservation of culture inspiring the recognition of traditional knowledge, the bioethical concerns about genetically modified organisms, and the social order promoted by fighting cyber crime.

18.2. DEFINITION OF TRADITIONAL KNOWLEDGE

Despite the global reach of the information economy, the dominant intellectual property paradigm is heavily biased against indigenous people throughout the developing world. Although Western intellectual property regimes offer strong protections for individual creators and inventors' "private knowledge", they offer no obvious ways to protect "traditional knowledge" or "indigenous knowledge". Traditional knowledge or indigenous knowledge means the cultural expressions, traditional practices, and genetic resources that

have been developed by indigenous communities through collective and incremental processes, which span generations. Indigenous knowledge is therefore easily exploited by those who would use dominant legal paradigms to appropriate and copyright unprotected indigenous knowledge as their own creations.

In other words, indigenous knowledge (IK) refers to the unique, traditional, local knowledge existing within and developed around the specific conditions of women and men indigenous to a particular geographic area. The development of IK systems, covering all aspects of life, including management of the natural environment, has been a matter of survival for the people who generated these systems. Such knowledge systems are cumulative, representing generations of experiences, careful observations, and trial-and-error experiments. *An old African proverb says that* **"When a knowledgeable old person dies, a whole library disappears"**.

Indigenous knowledge systems are also dynamic, new knowledge is continuously added. Such systems innovate from within, and internalize, use, and adapt external knowledge to suit the local situation.

Ruddle (1993) examined the transmission of traditional ecological knowledge for sites in Venezuela and Polynesia. Two to five-year-old children already knew the names and characteristics of the more common biota. By the age of 14, children were competent in household tasks, cultivation (plant identification, harvesting), seed selection, weeding, animal husbandry, fishing, and hunting. Overall, he found that the training was age specific, structured, and systematic. Specific times are allocated for training during the daily work routine.

All members of a community have **traditional ecological knowledge**—elders, women, men, and children. The quantity and quality of the IK that individuals possess vary depending upon:

- age;
- education;
- gender;
- social and economic status;
- daily experiences;
- outside influences;
- roles and responsibilities in the home and community;
- profession;
- available time;
- aptitude and intellectual capability;
- level of curiosity and observation skills;
- ability to travel and degree of autonomy; and
- control over natural resources are some of the influencing factors.

Indigenous knowledge is stored in peoples' memories, and is expressed in:
- stories;
- songs;
- folklore;
- proverbs;
- dances;
- myths;
- cultural values;
- beliefs;
- rituals;
- community laws;
- local language and taxonomy;
- agricultural practices, equipment;
- materials;
- plant species; and
- animal breeds.

Indigenous knowledge is shared and communicated orally by specific example and through culture. Indigenous forms of communication and organization are vital for local-level decision-making processes and in the preservation, development, and spread of IK.

Contents in IK Research

Although each IK system consists of an integrated body of knowledge, researchers interested in learning more about traditional knowledge systems tend to focus on discrete aspects. A diversity of topics are studied under the protocol of IK research. To convey an appreciation of the scope of the research area, examples are listed below.

Learning systems

- Indigenous methods of imparting knowledge
- Indigenous approaches to innovation and experimentation
- Indigenous games
- Indigenous specialists

Local organizations, controls, and enforcement

- Traditional institutions for environmental management
- Common property management practices
- Traditional decision-making processes
- Conflict-resolution practices
- Traditional laws, rights, taboos, and rituals
- Community controls on harvesting

Local classification and quantification

- A community's definitions and classification of phenomena and local flora and fauna
- Indigenous methods of counting and quantifying

Human health

- Nutrition
- Human-disease classification systems
- Traditional medicine and the use of herbal remedies in treatment of diseases
- The locations of medicinal plants, proper time for collection, most useful parts, and methods of preparing and storing medicines

Animals and animal diseases

- Animal breeding and production
- Traditional fodder and forage species and their specific uses

- Animal-disease classification
- Traditional ethnoveterinary medicine

Water

- Traditional water-management and water-conservation systems
- Traditional techniques for irrigation
- Use of specific species for water conservation
- Freshwater and saltwater fisheries and aquatic-resource management

Soil

- Soil conservation practices
- The use of specific species for soil conservation
- Soil-fertility enhancement practices

Agriculture

- Indigenous indicators for determining favourable times to prepare, plant, and harvest gardens
- Land-preparation practices
- Indigenous ways to propagate plants
- Seed storage and processing (drying, threshing, cleaning, and grading)
- Seed practices
- Indigenous methods of sowing (seed spacing and intercropping)
- Seedling preparation and care
- Farming and cropping systems (for example, complementary groupings)
- Crop harvesting and storage
- Food processing and marketing
- Pest-management systems and plant-protection methods

Agroforestry

- Indigenous techniques used for recognizing potential swidden farmland, and the criteria used for making choices regarding its use
- Criteria and techniques used for allowing a farm to go fallow
- Fallow management and uses

- Indigenous adaptations for intensification
- Changes adopted during the shift to sedentary agriculture
- The management and the productivity of forest plots
- The knowledge and use of forest plants (and animals)
- The interrelationship between tree species, improved crop yields, and soil fertility

Other topics

- Textiles and other local crafts
- Building materials
- Energy conversion
- Indigenous tools
- Changes to local systems over time

18.3. EROSION OF IK SYSTEMS

As outsiders have become increasingly aware of the value of IK, so has the awareness that IK systems, *biodiversity*, and cultural diversity (three interacting, interdependent systems) are threatened with extinction. Not withstanding the fact that some IK is lost naturally, as techniques and tools are modified or fall out of use, the recent and current rate of loss is accelerating because of rapid population growth, growth of international markets, educational systems, environmental degradation, and development processes—pressures related to rapid modernization and cultural homogenization. Below are given some examples, which illustrate these mechanisms:

- With rapid population growth—often due to immigration or government relocation schemes in the case of large development projects, such as dams—standards of living may be compromised. Due to poverty, opportunities for short-term gain are selected over environmentally sound local practices. With increasing levels of poverty, farmers, may also have less time and fewer resources to sustain the dynamic nature of IK systems through their local experiments and innovations.

- The introduction of market-oriented agricultural and forestry practices focused on monocropping is associated with losses in IK and IK practices, through losses in biodiversity and cultural diversity. For instance, policies promoting generic rice and wheat varieties devalue locally adapted species.
- With the ready availability of many commercial foods, some of the biodiversity seems to have become less relevant, such as seed and crop varieties selected over the years for their long-term storage attributes.
- In the short term, chemical inputs seem to reduce the need to tailor varieties to difficult growing conditions, contributing to the demise of local varieties. (However, the failure of green-revolution technology strongly suggests that uniformity is a poor long-term strategy).
- With deforestation, certain medicinal plants become more difficult to find (and the knowledge or culture associated with the plants also declines).
- More and more knowledge is being lost as a result of the disruption of traditional channels of oral communication. Neither children nor adults spend as much time in their communities anymore (for example, some people travel to the city on a daily basis to go to school, to look for work, or to sell farm produce. Many young people are no longer interested in, or do not have the opportunity for, learning traditional methods). It is harder for the older generation to transmit its knowledge to young people.
- Since IK is transmitted orally, it is vulnerable to rapid change, especially when people are displaced or when young people acquire values and lifestyles different from those of their ancestors.
- Farmers traditionally maintained their indigenous crop varieties by keeping household of seed stocks and obtaining seed through traditional family and community networks and exchanges with nearby communities. Some of these traditional networks have been disrupted or no longer exist.

In the past, outsiders (for example, social, physical, and agricultural scientists, biologists, colonial powers) ignored or maligned IK, depicting it as primitive, simple, static, "not knowledge," or folklore. This historic neglect (regardless of its cause—racism, ethnocentrism, or modernism, with its complete faith in the scientific method) has contributed to the decline of IK systems, through lack of use and application. This legacy is still in evidence. Many professionals are still sceptical. Also, in some countries, official propaganda depicts indigenous cultures and methodologies as backward or out of date, and simultaneously promotes one national culture and one language at the expense of minority cultures. Often, formal schooling reinforces this negative attitude. Local people's perceptions (or misperceptions) of local species and of their own traditional systems may need to be rebuilt. Some local people and communities have lost confidence in their ability to help themselves and have become dependent on external solutions to their local problems.

18.4. OUTBURST OF INTEREST IN IK

The interest (of outsiders) in this "old" knowledge is recent, and emerged in tandem with the politicization of indigenous groups and indigenous-rights movements. Many indigenous people are demanding the right to be heard in development decisions.

This often includes demanding that their rights to land and resources be recognized and officially acknowledged. Concurrently, the international political system and many national governments are showing some willingness to listen to indigenous people. In a sense, this improved political climate supports a dialogue on IK. Some governments (Australia, Canada, Greenland, the United States) have mechanisms such as settled land claims and co-management resource boards that support IK systems by supporting self-government and the joint management of natural resources.

As well, the "life industry" (those industries that profit from the use of living organisms— agrochemical, pharmaceutical, food and seed industries) and critics of the life industry have done much to reveal the past, current, and future value of IK, and the implications for indigenous peoples of its unregulated theft from the South (see Chapter on Intellectual Property Rights).

Of late, IK has been lauded as an "alternative collective wisdom relevant to a variety of matters at a time, when existing norms, values, and laws are increasingly called into question". The need for some alternative wisdom in development initiatives is supported by the following observations:

- Green-revolution technology is associated with ecological deterioration.
- Economic decline (at the local level).
- Poorer diets and nutritional losses resulting from the eradication of traditional foods or from their substitution by nontraditional foods.

F.J. Tester, School of Social Work, The University of British Columbia, Vancouver, BC, Canada, noted the following problems for future development:

- Development as planned and implemented for the last 30 years has placed unprecedented pressures on the planet's soils, watersheds, forests, and other natural resources.
- Some development solutions from outside are based on incorrect assumptions, are not economically feasible or culturally acceptable, and are often abandoned (for example, techniques are too complex or require too much maintenance).
- Some technical solutions are introduced to solve problems, which are not perceived at the local level and are abandoned.
- Development interventions tend to benefit small numbers of people from relatively privileged groups.
- Some critics observe that communities receiving the most externally driven development assistance become less capable of handling their own affairs.

- Top-down planning fails to promote effective natural-resource management at the local level.

In short, development planning has often failed to achieve the desired result: **sustainable development**. In some cases dependencies have been created by an outside world that orders and demands (through laws and natural resource regulations), but does not truly contribute to development. Communities are often left to find their own means.

Development efforts that ignore local circumstances, local technologies, and local systems of knowledge have wasted enormous amounts of time and resources. Compared with many modern technologies, traditional techniques have been tried and tested, are effective, inexpensive, locally available, and culturally appropriate, and in many cases are based on preserving and building on the patterns and processes of nature.

Some farmers in Zimbabwe prefer to use a local strategy to combat termites and ants, rather than the commercial remedy, which is expensive and not readily available. Termites are the major destroyers of gum and orchard trees, especially during the early stages of growth. Through their informal experiments, the farmers discovered that either ashes or a mix of a small smelly plant ground together with onions and paraffin or used oil repels termites and ants.

Western technoscientific approaches are (in themselves) an insufficient response to today's complex web of social, economic, political, and environmental challenges. The paradigm in support of "one technology or one knowledge system fits all" has been debunked. IK systems suggest a different approach to the solving problem. Whereas Western science attempts to isolate a problem to eliminate its interlinkage with various other factors and to reduce a problem to a small number of controllable parameters, traditional approaches usually examine problems in their entirety, together with their interlinkages and complexities (Shankar 1996). For example, people in the field of medicine are realizing the importance of including the physical, spiritual, sociocultural, and psychological well-being of a person when considering matters of health. Although this is a fairly new concept for modern medicine, this holistic approach is the basis of many traditional systems.

Increasingly, development practitioners argue that paying attention to local IK can:

- create mutual respect, encourage local participation, and build partnerships for joint problem resolution;
- facilitate the design and implementation of culturally appropriate development programs, avoiding costly mistakes;
- identify techniques that can be transferred to other regions;
- help identify practices suitable for investigation, adaptation, and improvement; and
- help build a more sustainable future.

IK for Sustainable Development

Sustainable development is "a development that meets the needs of the present without compromising the ability of future generations to meet their own needs" (WCED 1987). Sustainable agricultural and natural resource development means the utilization, management and conservation of the natural resource base and the orientation of technological change to ensure the attainment and continued satisfaction of human needs, such as food, water, shelter, clothing, and fuel, for the present and future generations.

According to the World Commission on Environment and Development, sustainable development has the following nine objectives (WCED 1987):

- Reviving growth
- Changing the quality of growth
- Meeting essential needs for jobs, food, energy, water, and sanitation
- Ensuring a sustainable level of population

- Conserving and enhancing the resource base
- Reorienting technology and managing risk
- Merging environmental considerations and economics in decision-making
- Reorienting international economic relations
- Making development more participatory

Sustainable development at the local and national levels is a function of five variables (Matowanyika 1991):

- Biophysical and socioeconomic resources
- External factors, such as available technologies and development ideologies
- Internal factors, including sociocultural belief systems and local production and technological bases
- Population factors
- Political and economic factors

A sustainable development strategy will take into account all of the above variables, and will involve working, learning, and experimenting together at the local, regional, national, and international levels. The predominant focus of this guidebook is on the local level, and what IK can contribute to a local sustainable-development strategy, taking into account local circumstances, potential, experiences, and wisdom.

Sustainable development at the local level is dependent on the implementation of enabling mechanisms at the local, national and international levels.

Some corporations, particularly in the pharmaceutical and life science industries, have begun systematic efforts to collect, commercialize, and assert ownership over indigenous knowledge. When these efforts succeed, indigenous people are not only unable to reap any rewards from the benefits brought to countless other people, but may also find themselves unable to exploit the very knowledge they originally produced. In addition, such knowledge may also have a moral significance to its originating community that cannot be captured by the economics of copyright compensation. Proposals for protecting indigenous knowledge against misappropriation and exploitation must therefore also consider ways to regulate the dignitary costs imposed by such uses.

In crafting ways to safeguard traditional knowledge, we must reckon with an inherent tension with the broad goals of the access to knowledge movement. The project of the indigenous knowledge movement, i.e., giving voice to those who have been silenced by traditional legal systems, seems to require stronger, or at least additional, forms of intellectual property. Such forms will make it easier for marginalized people to claim rights over the fruits of their knowledge. Yet, instituting such reforms might undermine the A2K goal of removing the roadblocks to knowledge. Resolving this tension between access and protection is one of the major projects of both movements. These proposed modalities for establishing new intellectual property norms and enforcement regimes include:

- binding (and non-binding) international agreements;
- transnational model codes and coordination of individual nations' legislative developments; and
- innovations in technology and code (e.g., Creative commons licenses, restructuring internet domain name assignment, inexpensively distributed and maintained "appropriate technologies") that may supplement or circumvent legal frameworks altogether.

Some of the questions raised at present for the protection of TK:

- Is it possible to alter the current western patent system to permit protection of indigenous knowledge?
- If not, how could TRIPS should be amended to extend protection in this area? What are the risks of such an extension?
- Developing nations are themselves often starkly divided into 'haves' and 'have-nots' with respect to money, industrial capital, and education.
- How can the access to knowledge movement ensure that knowledge flowing to these nations overcomes the cultural, technological, and political barriers impeding its dissemination to the have-nots?
- Which modalities will permit the strongest protection for various forms of indigenous knowledge, such as crop and animal breeds, agricultural methods, medicinal and other scientific expertise, cultural practices, or folklore and artistic works?
- In light of the tension between "free culture" and "native culture," is the strongest protection necessarily the best choice?

18.5. NEED FOR THE PROTECTION OF TRADITIONAL KNOWLEDGE

Traditional knowledge (TK) holders face various difficulties. In some cases, the very survival of the knowledge is at stake, as the cultural survival of communities is under threat. External social and environmental pressures, migration, the encroachment of modern lifestyles, and the disruption of traditional ways of life can all weaken the traditional means of maintaining or passing knowledge onto the future generations. There may be a risk of losing the very language, which gives the primary voice to a knowledge tradition and the spiritual world-view that sustains this tradition. Either through acculturation or diffusion, many traditional practices, associated beliefs, and knowledge have been irretrievably lost. Thus, a primary need is to preserve the knowledge that is held by elders and communities throughout the world.

This kind of challenge arises in a host of immediate, practical ways. Some examples are as follows:

- A recent agreement would give traditional healers in Samoa a share of the benefits from a new AIDS drug drawing on their knowledge of the *mamala* tree.
- The Kani tribe of South India is to share the benefits from a new sports drug that is based on their knowledge of the medicinal plant *arogyapaacha*.

- Representatives of TK holders have opposed patents drawing on their TK (e.g., concerning the use of extracts from the *neem* tree, and the use of turmeric as a wound-healing agent).
- Traditional ecological knowledge held by Aboriginal communities in Canada has proven to be valuable in environmental planning and resource management.
- For some communities, TK provides a pathway to social and economic development, and new more culturally appropriate forms of tourism. The Seri people of Mexico use the *Arte Seri* mark to distinguish their craftworks based on their TK and associated genetic resources, and to support a sustainable trade in these products.
- Portugal recently passed a law to protect the TK and plant varieties of Portuguese farmers, adding this to a growing collection of so-called *"sui generis"* laws on TK in a range of countries around the world.
- In 2001, China granted more than 3000 patents on innovative developments within the field of Traditional Chinese Medicine.
- *Oryza longistaminata* is a wild rice, which grows in Mali. This rice variety *grows in the marshes and river banks of Mali.*

Local farmers considered it a weed, but the migrant *Bela* community developed detailed knowledge of its agricultural value. The *Bela* community developed systematic understanding of the distinct properties of this and other kinds of rice, and recognized that *Oryza longistaminata* has stronger resistance to diseases, such as rice blight, than many other local varieties of rice. Guided by this traditional knowledge, researchers subsequently isolated and cloned a gene, Xa21, which conferred this resistance in rice plants.

The wider significance of TK means that it arises in international discussions on a host of issues:

- Food and agriculture
- Biological diversity, desertification and the environment
- Human rights, especially the rights of indigenous peoples
- Cultural diversity
- Trade and economic development

TK has also moved towards the center of policy debate about intellectual property (IP). This leads to some challenging questions. Is the IP system compatible with the values and interests of traditional communities, or does it privilege individual rights over the collective interests of the community? Can IP bolster the cultural identity of indigenous and local communities, and give them greater say in the management and use of their TK? Has the IP system been used to misappropriate TK, failing to protect the interests of indigenous and local communities? What can be done, legally, practically, to ensure that the IP system functions better to serve the interests of traditional communities? What forms of respect and recognition of TK would deal with concerns about TK, and give communities the tools they need to safeguard their interests?

With these questions in mind, WIPO started to work on TK in 1998. The first step was to listen directly to TK holders, learning of the needs and expectations of some 3,000 representatives of TK-holding communities in sixty locations around the world. Their insights and perspectives still guide WIPO's work. The WIPO Intergovernmental Committee on Intellectual Property and Genetic Resources, Traditional Knowledge and Folklore ("IGC") was established in 2001 as an international policy forum. WIPO's work, therefore, ranges from the international dimension of TK and cooperation with other international agencies, to the capacity building and pooling of practical experience in this complex area. This booklet gives an overview of this work, discusses some key concepts and describes various national approaches to protecting TK against misuse or misappropriation.

The protection of TK is important for communities in all countries, particularly in developing and least developed countries. First, TK plays an important role in the economic and social life of those countries. Placing value on such knowledge helps strengthen cultural identity, and the enhanced use of such knowledge helps to achieve social and development goals, such as sustainable agriculture, affordable and appropriate public health, and conservation of biodiversity. Second, developing and least developed countries are implementing international agreements that may affect how the knowledge associated with the use of genetic resources is protected and disseminated, and thus how their national interests are safeguarded. Patterns of the ownership of TK, cultural, scientific and commercial interest in TK, the possibilities for beneficial partnerships in research and development, and the risk of the misuse of TK are not neatly confined within the national boundaries, so that some degree of international coordination and cooperation is essential to achieve the goals of TK protection. A comprehensive strategy for protecting TK should, therefore, consider the community, national, regional, and international dimensions.

The stronger the integration and coordination between each level, the more likely the overall effectiveness. Many communities, countries, and regional organizations are working to address these levels respectively. National laws are currently the prime mechanism for achieving protection and practical benefits for TK holders. For instance, Brazil, Costa Rica, India, Peru, Panama, the Philippines, Portugal, Thailand, and the United States of America have all adopted *sui generis* laws that protect, at least, some aspect of TK *(sui generis* measures are specialized measures aimed exclusively at addressing the characteristics of specific subject matter, such as TK). A WIPO background paper entitled *"Consolidated analysis of the legal protection of traditional knowledge"* analyses these laws in more detail. In addition, a number of regional organizations, such as in the South Pacific and Africa, have been working on defining the specific rights in TK and how to administer them. Various TK holders and other

stakeholders in different countries have already found existing IP rights useful, and their TK protection strategies make some use of the IP system.

While there are diverse national and regional approaches to protection, which reflect the diversity of TK and its social context, some common elements arise in policy debate. For instance, it is stressed that protection should reflect the aspirations and expectations of TK holders, and should promote respect for indigenous and customary practices, protocols, and laws as far as possible. Several *sui generis* measures, as well as conventional IP law, have recognized elements of such customary law within a broader framework of protection. Economic aspects of development need to be addressed, and the effective participation by TK holders is also important, in line with the principle of prior informed consent. TK protection should also be affordable, understandable, and accessible to TK holders. The view is widely voiced that holders of TK should be entitled to fair and equitable sharing of benefits arising from the use of their knowledge.

The international legal framework, within and beyond the IP system, is also an important consideration. Where TK is associated with genetic resources, the distribution of benefits should be consistent with measures established in accordance with the Convention on Biological Diversity (CBD), providing for the sharing of benefits arising form the utilization of the genetic resources. Other important international instruments include the International Treaty on Plant Genetic Resources for Food and Agriculture of the Food and Agriculture Organization (FAO), the International Union for the Protection of New Varieties of Plants (UPOV), and the UN Convention to Combat Desertification (UNCCD). Other areas of international law, notably human rights and cultural policy, are also part of the context for protection of TK.

18.6. TYPES OF TRADITIONAL KNOWLEDGE PROTECTION

Two key demands on the IP system, in particular, have arisen in policy debate. First is the call for recognition of the rights of TK holders related to their TK, and second concerns about the unauthorized acquisition by third parties of IP rights over TK. Two forms of IP-related protection have, therefore, been developed and applied:

A. Positive protection: giving TK holders the right to take action or seek remedies against certain forms of misuse of TK; and

B. Defensive protection: safeguarding against illegitimate IP rights taken out by others over TK subject matter.

Stakeholders have stressed that these two approaches should be undertaken in a complementary way. A comprehensive approach to protection in the interests of TK holders is unlikely to rely totally on one form or the other.

SUMMARY

Indigenous and local communities justly cherish traditional knowledge (TK) or indigenous knowledge (IK) as a part of their very cultural identities. Maintaining the distinct knowledge systems that give rise to TK can be vital for their future well-being and sustainable development and for their intellectual and cultural vitality. For many communities, TK forms part of an holistic world-view, and is inseparable from their very ways of life and their cultural values, spiritual beliefs and customary legal systems. This means that it is vital to sustain not merely the knowledge but the social and physical environment of which it forms an integral part.

Traditional knowledge also has a strong practical component, since it is often developed in part as an intellectual response to the necessities of life: this means that it can be of direct and indirect benefit to society more broadly. There are many examples of important technologies being derived directly from TK. But when others seek to benefit from TK, especially for industrial or commercial advantage, this can lead to concerns that the knowledge has been misappropriated and that the role and contribution of TK holders has not been recognized and respected. One of the challenges posed by the modern age is to find ways of strengthening and nurturing the roots of TK, even in times of social dislocation and change, so that the fruits of TK can be enjoyed by future generations, and traditional communities can continue to thrive and develop in ways consistent with their own values and interests. At the same time, TK holders stress that their TK should not be used by others inappropriately without their consent and arrangements for fair sharing of the benefits; more generally, it leads to calls for greater respect and recognition for the values, contributions and concerns of TK holders.

TK holders face various difficulties. In some cases, the very survival of the knowledge is at stake, as the cultural survival of communities is under threat. External social and environmental pressures, migration, the encroachment of modern lifestyles and the disruption of traditional ways of life can all weaken the traditional means of maintaining or passing knowledge on to future generations. There may be a risk of losing the very language that gives the primary voice to a knowledge tradition and the spiritual world-view that sustains this tradition. Either through acculturation or diffusion, many traditional practices and associated beliefs and knowledge have been irretrievably lost. Thus, a primary need is to preserve the knowledge that is held by elders and communities throughout the world.

This kind of challenge arises in a host of immediate, practical ways.

EXERCISE

1. Define traditional knowledge and its importance in Indian agriculture and medicinal practice.
2. Distinguish between the traditional knowledge and patents in the biotechnological research and development for the inventions.

3. What is meant by indigenous knowledge? Add a note on ethical issues associated with the use of such indigenous knowledge for intellectual property rights by the researchers.

4. How is traditional knowledge at present helpful for various biotechnological processes? Add a note on its social impact on the traditional people.

5. Briefly explain how the erosion of indigenous knowledge is occurring in the present day global village of information technology.

6. Write notes on different aspects of traditional knowledge as studied by the researchers in the field of biotechnology.

7. Describe various types of traditional knowledge used by the present day scientists for obtaining the patents.

8. Name the important legal protections that are introduced by the different governments around the world for the protection of traditional knowledge.

9. Enumerate the different criteria that have failed in achieving the proper development in the sustainable development.

10. Elucidate the applications of traditional knowledge for local communities when compared to the developed methodologies.

11. Write a note on the ethical issues raised by exploiting the traditional knowledge, and their impact on ethnic groups.

12. Mention how traditional knowledge contributed to the development of today's biotechnological methods that are used in agriculture.

13. List out how traditional knowledge is involved in the bioresources preservation and management in biodiversity rich countries.

14. Explain the history of traditional knowledge, and its usage in the different parts of the world for sustainable development.

15. What do you mean by traditional knowledge? Why now-a-days the outburst of traditional knowledge is seen in the various parts of the world?

16. Explain how technological advancement is becoming a negative factor for traditional knowledge. Add a note on its affects on India.

17. Critically comment on the ethical, social and legal issues raised by the broader usage and exploitation of traditional knowledge.

18. Give an account of the issues raised for the protection of the traditional knowledge in the technologically advanced world.

19. Write about the importance of the traditional knowledge. How does it affect the development of biotechnology without affecting its own values?

19

CHAPTER

Bioethics in Biodiversity and Resource Management

19.1. INTRODUCTION

Biosphere that supports the life forms is invaded by the ever growing explosive human population growth. The one species that is still globally expanding in number in human. The world population has gone up from 2.5 billion in 1950 to 6 billion today. It is expected, by the UN, to reach 8 billion in 2015, and 9-10 billion in 2050. Over 95% of the expected population increase will be in the developing countries. In those countries, most of the population growth will occur in the cities. The additional population will require more space to live in, more water, more energy, more food, and more services. For the years 1995-2020, the largest relative population increase (80%) is expected in sub-Saharan Africa. In absolute numbers, it is expected to go from 500 to 900 million. It is not clear whether the HIV/AIDS epidemic, with an estimated 25 million infected in this region in 2000, will substantially affect population dynamics within the next 25 years, as most of the infected are of reproductive age, as pointed out by Cohen in 2000. It is important to realize that the FAO estimates that there are over 800 million people globally, who don't have enough to eat today, and it is imperative that more food be produced for them where it is needed, namely in the developing countries. As pointed out by Lipton in 1999, poverty reduction in rural, agricultural areas of developing countries is also likely to have an indirect effect in favor of the urban poor. Improved seeds will have a positive effect, but many other factors such as education, access to microcredit, and reduction in farm subsidies are very important for the development of the third world countries.

19.2. PRESSURE ON THE RESOURCES AVAILABLE IN THE NATURAL ENVIRONMENT

There is a constant pressure on the resources that are available in the natural environment. There are some specific technical challenges, which are difficult to handle through traditional crop-breeding and animal breeding. And there, biotechnology may offer great possibilities for the breeding of crops and animals, microbes uses and exploitation and forestry. These include drought and heat-tolerance; improved nutrient uptake and rooting; biological nitrogen fixation; responses to carbon dioxide; and tolerance to key abiotic stresses, such as salinity, cold and drought. As part of a strategy of diet diversification, genetic modification may help us address protein, vitamin, and iron deficiencies (the famous example of the golden rice). The intrinsic value of species and ecosystems, in addition to their value as starting material for finding new products, is the basis for these measures. The biggest threat to biodiversity is habitat destruction. The ever-increasing spread of cities and the accompanying expansion of agriculture must be largely held responsible. Humid tropical forests are particularly valuable reservoirs of biodiversity, and are currently being seriously threatened. As the human population expands, the need for food is expected to double in the next 30 years, with the ensuing threat of massive habitat destruction, particularly in the less developed countries. Increasing crop productivity on the land already under cultivation would prevent or at least reduce habitat destruction. One of the several measures aimed at increasing yields is the use of better seeds, including those enhanced by modern biotechnology. Many other measures from the technical, socio-economic and political fields need to be taken at the same time in order to balance intensification and sustainability of modern agriculture. Modern biotechnology offers new means of improving rather than threatening biodiversity. If properly tested for both risks and benefits to humans and the environment, transgenic crops are more likely to increase agricultural biodiversity and help maintain native biodiversity rather than endangering it. Such applications need to be judged by the criteria of improved sustainability and compared to current as well as alternative farming practices.

19.3. DEFINITION

Biodiversity is the multitude of different living beings in a particular ecosystem or on the whole earth or biosphere. Biodiversity can be seen and studied at different organizational levels—genetic, organismal, and ecological. This means the types of plants, animals, microorganisms, and other living entities present in the specified area such as forest pond, or agricultural field. It touches upon both native environments on land and sea, as well as agricultural and other man-made surroundings.

The convention on biological diversity defines 'biological diversity' as a variability among living organisms from all sources including, inter alia, terrestrial, marine and other aquatic ecosystems and the ecological complexes of which they are part, WHO includes diversity within species, between species, and of ecosystems.

To address these issues, humans are looking for the variety of gene or genes present in the different organisms such as plants, animals or microbes. Unfortunately, most of the traits needed are governed by multiple genes, and it will probably be sometime before farmers and consumers benefit from such research. The perceived profit potential of genetically modified organisms (GMOs) has already changed the direction of investment in research and development in both the public and private sectors away from systems-based approaches to pest management, and towards a greater reliance on monocultures with potentially adverse effects on landscape biodiversity. The possible long-term environmental costs of such strategies may be overlooked to meet the demand at the present condition. These changes are always unpredictable, unexplained, and uncontrolled in nature.

Hence, better methodologies have to be adopted for greater control over the technology. If research is to address the challenges of agriculture in the future, we need to put genetic modification in context, and realize that it is but one of the many elements of agricultural change. Scientists must not be blinded by the glamour of cutting-edge molecular science for its own sake. It is also worried about the disconnection between the laboratory and the field, and the decline of agronomy as an integrative science. Governments need to be vigilant and not let this glamour to be ongoing. And the perception by private industry of major profit opportunities draws investment away from the research to more traditional fields such as water and soil management or ecology, and from public sector research. At the same time, the best science is developed in a climate of intellectual freedom, without much direct government interference. It is a difficult balance to strike! The cost of molecular technologies in plant and animal breeding, and the size of markets that are required to recoup the profits, lead to the growing use of "hard" intellectual property rights over seeds and plant or animal materials, and the tools of genetic engineering. This becomes stricter as more and more countries adopt the WTO agreements.

This condition creates greater consequences and changes the relationship between the public and private sectors, to the detriment of the public sector. A policy question that governments must take up, in both the national and international contexts, is how to ensure that public research is not the poor relation. In developing countries, in particular, it is important for the public sector to retain enough capacity, resources, and freedom of action to provide the services on which their national private sectors can build. They will also need to build their policy and regulatory capacities with regard to transgenic crops that originate elsewhere from known or unknown sources, and in this FAO has an "honest broker" role to play. Thus, we need a science that, in a globalized world, addresses local agricultural practices and food security and helps to keep the rural areas liveable and attractive to young people in search of employment. The constraints to doing so are more economic than technical in a sustainable manner for long term ecology.

Biotechnology is becoming important both in the business and investment industries worldwide, and has a huge impact on the economy. Globalization is not only a question of size, but also of kind. It is inextricably linked to privatization, and major economic restructuring in both developed and developing countries have changed the balance of public and private sectors, and the privatization of knowledge through intellectual property rights. This results in concentration, and there is often little economic incentive for global companies to address local needs and environment change. It has been estimated that the world's top ten companies account for about 85% of the global market for the seed and agro-chemical industries. In 1998, just four companies controlled 69% of the seed market in the USA. Of course, globalization also promises to create opportunities for millions of farmers throughout the world, offers opportunities to the poorer countries, and drives rapid development and local capital accumulation. But it may also aggravate existing differences among countries.

Developing transgenic and other biotechnological techniques implies massive investments and the need for massive returns. The small number of GM technologies currently in use suggests that there is a real danger that the scale of the investment may lead to selective concentration on species and problems of global importance. In this context, it must also give a note of warning about the rising costs of regulation. In the health industry, where regulation is heaviest, bringing a pharmaceutical to market now costs about $500 million; bringing a pesticide to market can cost about $200 million; and it has recently been reported that a GM crop costs about $30 million to produce, to which regulatory costs can add up to a further $5-6 million.

Developing the biotechnology based techniques is suitable for all types of crops. Which crops can bear such costs in the commercial market? How are we to serve local needs, small farmers, and poor consumers, and promote genetic and dietary diversity? These questions are very difficult to answer and are cost-based technologies. Hence, we always have to look for the living systems or biological diversities that are having beneficial characters inherent naturally with them. Researchers throughout the world are isolating, identifying, and developing such natural biodiversities, especially in the developing countries. Nevertheless, biodiversity of living organisms is a reservoir of useful genes, and isolation and use of such genes will remain of utmost economic significance in the next millennium. There has been a pronounced response to agricultural research based on biotechnology.

19.4. ETHICAL ISSUES OF BIODIVERSITY

There is a necessary situation aroused worldwide to make the biodiversity available for the future generations. The globalization and trade liberalization are fueling economic growth, increased prosperity, and new opportunities. World exports of goods and services nearly tripled between the 1970s and 1997. Foreign direct investment exceeded (US) $400 billion in 1997; seven times the level in the 1970s. However, there is growing disparity between poverty and privilege, both within and between countries and regions. Thus far, the benefits of globalization are uneven. The United Nations Development Programme (UNDP) concludes that the top fifth of the world's people in the richest countries accounts for 82% of the expanding export trade, and 68% of foreign direct investment. The bottom fifth of humanity in the poorest countries account for only 1%. While millions of people are integrated and empowered by knowledge and communication technologies such as the Worldwide Web, others remain isolated and marginalized.

The biodiversity is having long impact on the development of natural processes that occur in the ecosystem. The issues involved in the interaction between biodiversity and biotechnology have far-reaching consequences, and need to be subject to an open and knowledge-based dialogue in society. The heated public debate, seen primarily in Europe in the last couple of years, and which, in some circles, led to near hysteria, is not sufficient to solve the underlying problems.

The dialogue needs to include many different stakeholders, including farmers of developing countries, diverse scientists, policy makers and communicators. Cultural values involved in farming and food production needs to be taken into consideration, just as much as the emotional side of eating and drinking. A significant aspect of ethical behavior is openness. Transparent information is required both from scientists in non-commercial settings as well as from industry. Although competition from fellow researchers and industries does not allow the immediate divulgence of all research results, both groups should make concessions to the interested public and policy makers. This should be done by scientists personally, and would help improve mutual confidence.

Scientific Progress and Bioethics

The wide type of basic and applied research in the field of biotechnology on the living organisms increased the understanding of the biodiversity and their applications. Several large projects in biotechnology like HUGO, the Human Genome Organization, have an "ELSI" component, dealing with ethical, legal and societal implications of biotechnology. This shows that some scientists realize that science cannot be seen as a human activity taking place in a void, without any connection to social and political realities.

Bioethicists stress that it is the responsibility of scientists to be actively concerned with these issues, and that they should take special responsibility in communicating with non-specialists to explain what they know as well as what they don't know. Matrices of criteria have been set up, for instance, by Mepham to make it possible to ask relevant questions, such as who and what may be affected by agricultural biotechnology, and what properties like well-being, autonomy or justice may or may not be infringed upon. The questions put forward by ethicists need to be answered on the basis of concrete scientific knowledge, using comparisons with other, established agricultural practices. Ethicists will have critical and thought provoking questions.

However, their answers and attitudes may vary considerably. If they are positive as Comstock, ethicists will suggest pursuing research and application with appropriate caution. There is no easy answer to what the word "appropriate" means, beyond answering questions on a case-by-case basis to balance the benefits and risks. Actively shaping the future and attempting to solve problems always involves risk. It will be important to see ethics not as a stable set of principles, but rather as a discursive process, which will be able to cope with the growing speed of new developments. The Nuffield Council of **Bioethics** recently came to the conclusion that the moral imperative for making GM crops readily and economically available…is compelling, if well-informed governments of developing countries want to introduce them. This imperative should be taken seriously and treated with urgency. Everyday, 20,000 children die of malnutrition, and 30,000 ha of humid tropical forest are destroyed.

As we enter the new millennium, globalization and privatization are among the most obvious and fundamental trends that affect and influence policy debates on ownership, conservation, and exchange of biological materials. The past two decades have witnessed increasing privatization of agricultural research and development, expansion of the scope of intellectual property rights to cover biological products and processes, and liberalization of global markets.

These trends have stimulated the commercial development of biotechnology products for agriculture and human health, and the concentration of economic power among a handful of giant life sciences corporations. The process of globalization influences not just the economy, but also the culture, technology, and governance related to biological materials.

New actors, new roles, and new rules are redefining global governance. The World Trade Organization (WTO), for example, and the multilateral agreement on intellectual property it administers are reducing the scope for national policy in that arena. Multilateral rule-making in a global marketplace is changing the role of state sovereignty. There is concern that countries and local communities will be increasingly restricted in their ability to determine domestic standards for regulatory and environmental protection.

Developing Countries and their Needs

The combined pressures of poverty, population growth, and environmental degradation pose daunting challenges for agriculture and human development, especially in the developing world. Over 800 million people in the world today are chronically undernourished. An estimated 1.3 billion people live on incomes of less than one dollar a day. Within ten years, more than half of the world's population will be living in cities. By the year 2020, there will be an additional 2 billion people to feed. Members of the scientific group may not agree on the underlying causes of poverty and hunger, but it is clear that different visions of agricultural development are emerging to meet the challenge of global food security and sustainability.

Given limited possibilities for expanding cultivated land area, the scientific group agrees that our future food security depends upon a combination of carefully crafted production and distribution policies combined with scientific strategies that ally farmer-researchers with formal sector plant breeders, and laboratory-researchers to maximize germplasm enhancement and farming systems. Beyond this, however, policy and program choices often differ. The conventional approach to close the food gap emphasizes the role of high-input, industrial-scale farming, perhaps complimented by commercial biotechnologies to raise yield ceilings.

This approach stresses the contribution of farmer-led initiatives, the use of crop varieties developed by farmer/breeders in partnership with formal sector plant breeders, and the use of technologies that will lessen farmers' dependence on purchased inputs. Not without its critics, this approach is sometimes attacked for its perceived Malthusian naivety or absurd political correctness. Still others would see policy as by far the most important engine of change, and argue that sound, 'pro-poor' choices in land tenure, credit, and price subsidization are the key to food security.

No matter what combination of germplasm, technologies, farming systems and policies are employed to achieve food security in the 21st century, commercial biotechnology and subsistence farmers alike will need to make better use of a broader range of the world's plant and animal genetic diversity. Farmers will require crop varieties and livestock breeds capable of producing under diverse and rapidly changing conditions. Under any scenario, genetic resources for food and agriculture, the role of farming communities who nurture and develop diversity, and the vital contributions of formal sector plant breeders would assume critical importance.

This chapter introduces some of the major social, economic and environmental trends that influence or inform the larger debate on ownership, conservation and exchange of biological materials. What are these trends that influence the way society thinks about, values and uses biological diversity at the beginning of a new millennium?

19.5. ACCELERATION IN THE LOSS OF BIOLOGICAL DIVERSITY

There is an increasing awareness worldwide of the value, importance and fragility of biological diversity. In the final decade of the 20th century, greater numbers of people became aware of species extinction, erosion of genetic resources, and the threat of ecosystem destruction. The concept of biodiversity has entered the mainstream of government thinking, and in some cases, traveled beyond national ministries of environment. Despite heightened appreciation and awareness of biodiversity and the crafting of international conventions designed to conserve it, the loss of biological diversity continues. Forests are falling, fisheries are collapsing, plant and animal genetic diversity is eroding all over the world.

Loss of diversity in plant genetic resources for food and agriculture has been substantial, and these resources are disappearing at unprecedented rates. No one knows how much diversity once existed in domesticated species, so it is impossible to say exactly how much has been lost historically.

Domestic animal breeds are disappearing at an annual rate of 5%, or six breeds per month. According to FAO, the status of almost one-third of all livestock breeds is endangered or critical.

Tropical forests are falling at a rate of just under 1% per annum, or 29 hectares per minute. From 1980 to 1990, the loss was equivalent to an area of the size of Ecuador and Peru combined together. All of the world's main fishing grounds are being fished at or beyond their limits. About 70% of the world's conventional marine species are fully exploited, overexploited, depleted, or in the process of recovering from overfishing. During the 20th century, about 980 fish species became threatened. Nearly 60% of the earth's coral reefs are threatened by human activity. A new study conservatively estimates that 34,000 species of plants—12.5% of the world's flora—are facing extinction. At least one of every eight known plant species on earth is threatened. For every plant that becomes extinct, 30 other species go with it—many of them are microorganisms. Some biologists warn that plant pathogens (including fungi, viruses, and bacteria) should be conserved with the same urgency as other species because they play a vital role in the functioning of ecosystems.

The loss of biodiversity threatens food security, especially for the poor, who rely on biological products for 85-90% of their livelihood needs (i.e., food, medicine, fuel, fibre, clothing, shelter, energy, transportation, etc.). New estimates from FAO indicate that there are 828 million chronically undernourished people in the world—a small increase since the early 1990s.

Erosion of Cultural Diversity

The 1993 scientific group concluded that we cannot conserve the world's biological diversity unless we also nurture the human diversity that protects and develops it. Today, there is growing recognition that loss of cultural diversity—of traditional farm communities, languages, and indigenous cultures—is intricately linked to the loss of biological diversity. Many members of the scientific group are alarmed by the loss of the culturally-based knowledge represented by thousands of diverse cultures that are themselves endangered or disappearing.

Globally, language is one of the strongest indicators of cultural diversity. The highest levels of plant and animal diversity, as well as the world's richest linguistic life, are found close to the equator. Ten out of 12 'megadiversity' countries identified by the International Union for Conservation of Nature and Natural Resources (IUCN) rank among the top 25 countries for 'endemic' languages (i.e., languages spoken exclusively within a country's borders, which usually means the majority of the smaller languages of the world). Some cultures have many distinct names describing a single plant, its parts, and its uses. The number of languages and dialects spoken by distinct communities that use the same biological resource multiplies the diversity of names associated with distinct properties of a species. With the loss of their language, the community loses its ability to describe and, therefore, use the plant. While loss of knowledge does not imply the loss of the plant itself, this is commonly the case, since with the decline in the knowledge of the uses of the plant, the community may lose interest in its conservation.

As with biological diversity, the magnitude and pace of the current 'extinction crisis' in linguistic diversity is unprecedented. Linguists, who monitor the status of surviving languages, conclude that approximately 6-11% of the 6703 languages spoken in the world today are 'nearly extinct', and they predict that 50-90% will disappear during the 21st century.

Local and indigenous people who speak ancestral languages are severely threatened by loss of sovereignty over land, resources and cultural traditions and the promotion of linguistic assimilation. As they become increasingly marginalized, local people lose local scientific knowledge, innovative capacity, and wisdom about species and ecosystem management. As one scholar concludes, 'Any reduction of language diversity diminishes the adaptational strength of our species, because it lowers the pool of knowledge from which we can draw'.

The loss of traditional farm communities, languages, and indigenous cultures represent the erosion of human intellectual capital on a massive scale. It is tantamount to losing a road map for survival, the key to food security, environmental stability and improving the human condition. Thus, it is increasingly difficult to talk about the conservation and sustainable use of genes, species, and ecosystems separate from human cultures.

On-Farm Conservation and use of Plant Genetic Resources

Although many in the scientific community are used to measuring progress through precisely defined achievements, one of the genuinely significant 'breakthroughs' of the past five years owes more to rediscovery (by some) than discovery. Since 1993, agricultural research, including plant breeding and germplasm conservation, has been revitalized by the creative association of farmer-innovators and their communities, with formal sector scientists and research institutions.

The management and enhancement of plant genetic resources has been in the hands of farmers since the beginning of agriculture. Fortunately, there is far greater recognition today than there was five years ago that the contributions of farmers and indigenous peoples are critical to the conservation, use and active enhancement of biological diversity, and that these groups and individuals should be recognized and rewarded for their contributions. This principle is a prominent feature of the Convention on Biological Diversity (CBD) and of Farmers' Rights as discussed and supported in the FAO Commission on Genetic Resources for Food and Agriculture (CGRFA). The innovative activity of farmers with respect to the development of crop varieties and farming methods is described as research, plant breeding, ethnoscience, informal innovation, and so on. Not everyone agrees on which is the best term.

As a collective, the scientific group wants to avoid getting bogged-down in terminological wrangling. In the end, what is important is that the group recognizes the value of farmers' innovations. In the field of agricultural genetic resources, there is greater appreciation for the fact that *in situ* (on farm) conservation is a crucial element in the conservation of agricultural biodiversity, and that it must be complementary to gene bank collections. When *ex situ* germplasm is removed from its cultural and environmental context, it loses the ability to adapt to constantly evolving pests, diseases and the ever-changing needs of local communities. By placing greater emphasis on *in situ* and farmer/community level management of genetic resources, both the CBD and the Leipzig Global Plan of Action for Plant Genetic Resources for Food and Agriculture (PGRFA) emphasize that the future of world food security depends not just on stored crop genes, but on the people who use and maintain diversity on a daily basis. Leipzig's Global Plan of Action provides the first intergovernmental recognition and support for on-farm management and improvement of plant genetic resources. It recommends new on-farm conservation and participatory breeding initiatives, including the need for stronger links between

conservation and utilization of plant genetic resources.

'Participatory plant breeding' (PPB) is still in its youth, but is spreading quickly. PPB is a new approach to germplasm development, and conservation involving scientists, farmers and other end-users (i.e., rural cooperatives, consumers, extension workers, etc.). It is termed 'participatory' because users have a research role in all major stages of the breeding and selection process. PPB is a crop improvement strategy that especially seeks to involve disadvantaged user groups (i.e., women and other resource-poor farmers). Over 70 cases of participatory plant breeding are documented, involving a range of crops and geographic regions. These include, for example, pearl millet in India, barley in Syria, common beans in Brazil, rice in Nepal, and cassava in Colombia.

Support and recognition for on-farm conservation and farmer-driven breeding is growing. A range of strategies is being developed to enhance genetic materials on-farm, for and by farmers. Examples include: CGIAR's Systemwide Program on Participatory Research and Gender Analysis (and its working group on participatory plant breeding), the Community Biodiversity and Development Conservation Program (CBDC), the Seeds of Survival Program in Africa, the Academy of Development Sciences in India, and Projecto en Tecnologia Alternativa (PTA) in Brazil.

Efforts to conserve and enhance germplasm systems are found for both minor and major crops, and in 'marginal' farming areas (i.e., those with poor soil, little rainfall, steep hillsides) as well as irrigated lands. The increase in participatory plant breeding and other collaborative programs involving farmers, their communities and formal sector scientists raises new questions and challenges for recognizing collaborative innovation in plant breeding. Some observers believe that neither Farmers' Rights nor Breeders' Rights adequately address these issues. The International Development Research Centre (Canada) has recently funded work to address the property rights, best practice, and ethical dimensions of community-based and formal collaborations.

Many members of the scientific group agree that there is a need to strengthen the role of indigenous and local communities in order to ensure their full participation in germplasm conservation and enhancement.

Global Climate Change and Biodiversity

Although there is no consensus among scientists, there is a growing international opinion that global climate change will have profound impacts on biodiversity and will compromise the sustainability of human development on the planet. The Inter-governmental Panel on Climate Change (IPCC) predicts that the build-up of greenhouse gases in the atmosphere will cause global temperatures to rise by 1-3.5 degrees Centigrade during the next century; and that melting glaciers and thermal expansion of the ocean will bring an associated rise in sea level between 15 and 95 centimeters. Simulation models predict that each one-degree rise in temperature will displace the adaptation of terrestrial species some 125 km towards the poles, or 150 metres in altitude. Approximately, 30% of the earth's vegetation could experience a shift as a result of climate change. But since climate will change faster than the migration rate of many species, models predict a 'drastic reduction' in global species diversity. New research on the impact of global warming on vegetation offers especially grim predictions for the tropics.

According to the Institute of Terrestrial Ecology (Edinburgh, UK), by 2050 a warming of up to 8° centigrade in parts of the tropics will lead to higher evaporation rates, lower rainfall, and eventually the collapse of tropical ecosystems. The World Bank estimates that a 2-3 degree rise in global mean temperature will reduce the mass of mountain glaciers by one-third to one-half, and endanger at least one-third of all species surviving in forests. Changes in glacier mass and forest area will have a profound impact on agricultural productivity. Millet crop yields in Africa are expected to drop by between 6% and 8%. A Senegalese study predicts that millet yields in Senegal will decrease by between 11% and 38%. In South Asia, yields for rice and wheat are expected to fluctuate wildly. The maize crop in South Asia and Latin America may shrink by between 10% and 65%.

Not all scientists (and not all members of the Scientific group) agree with grim predictions for global climate change, and some point out that substitute crops may be developed to offset shrinking yields and increase productivity in some regions. Some models have shown that global warming could be neutral or favorable in temperate latitudes and disadvantageous for the tropics and sub-tropics. The discrepancies and uncertainties in climate models do not permit more accurate predictions. Regional projections suggest that Africa might be the continent worst hit by climate change. Ultimately, biological diversity is the key to adapt to global climate change. If we are to adapt food production systems to radically changing conditions in the coming decades, plant and animal diversity will be the single most critical resource for doing so.

Agricultural Biodiversity

In addition to biodiversity in the wild, there is the biodiversity of organisms used for farming and other human activities. In agriculture, 7,000 species of plants are used by farmers somewhere in the world, but only 30 species provide 90% of our calorific intake. as observed by Haywood in 2000. Within these dominant crop species, there are many hundred thousands of varieties (landraces, cultivars) adapted to local climates, farming practices, and cultural predilections like taste, color, structure, ability to store the products etc. Much of this large crop diversity is important for providing the initial material for breeding.

However, it must be recalled that the genetic diversity found in crops is much less broad than the genetic diversity observed in plants or animals living in the wild, which points to the importance of wild species for agricultural breeding programs. The top three crops are wheat, rice, and maize (corn) with around 500 million tons annual production each. Traditional breeding led us in to the trap of narrowing down genomes, and perhaps wisely used biotechnology could bring back at least that part of genetic diversity which enhances pest resistance and yield. There are many indications that mixtures of varieties of a crop or of different crops may give higher yields and are more resistant to pests and diseases than monocultures, as reported recently by Zhu for rice in China. However, even in mixed cultures, high quality, well defined varieties, and pure seeds are required, and the sustainability of mixed cropping related to pest management has still to be proven. In addition, it is still not clear, whether or not, in natural, non-agricultural habitats the yield is dependent on biodiversity. Based on experimental results, some researchers claim that the loss of species leads to a reduction in biomass, while others disagree, as demonstrated by Hector in 1999, and Kaiser in 2000. There simply may not be valid correlations between biodiversity and biomass yield, in either agricultural or non-agricultural settings.

Changing Roles of Public and Private Sector Agricultural Research

Until recently, agricultural research has been largely publicly financed and its products made freely available. Today, agricultural research is increasingly privatized. A rapidly changing IP environment and declining research budgets have marginalized the role of public sector agricultural research in both industrialized and developing countries. We are living in an era when, increasingly, science is subject to property rights and knowledge is commodified.

Given the demonstrated success of agricultural research in bringing science and technology-based solutions to agricultural production constraints, it is paradoxical that public agricultural research is now facing a global crisis. Many national agricultural research institutions in developing countries suffer from lack of money, even to buy the basic essentials or to pay salaries. Institutions in industrialized countries and the international research centres of the Consultative Group on International Agricultural Research (CGIAR) face similar financial constraints.

The crisis is not only financial, it is also one of confidence. In spite of the demonstrated effectiveness of agricultural research in the past, policy-makers today are not convinced that further investments are necessary or that they can accomplish the seemingly overwhelming tasks of assuring food security to millions of the poorest of the poor. The end result is that financial allocations to agricultural research are reduced at a time when its importance to economic development remains critical. Reduced support leads to low productivity, low visibility and, ultimately, less political support for the agricultural research system—a vicious circle.

Public funds for agricultural research have stagnated or declined, while private investments have increased at an unprecedented rate:

In the OECD, for example, private agricultural R&D totaled $7 billion in 1993 compared with $4 billion in 1981, an annual rate of growth of 5.1%. By contrast, publicly performed agricultural R&D rose by 1.7% per annum, from $5.7 billion in 1981 to $6.9 billion in 1991.

The relative importance of private R&D in total agricultural R&D varies across the OECD countries, but exceeds 50% in seven countries, including the US and Japan, which together account for over one-half of all private agricultural research throughout the OECD. The private share in the remaining OECD countries is smaller (about one-third in 1993), but has been growing rapidly. Private sector R&D in developing countries typically accounts for 10-15% of total agricultural R&D.

Private and public agencies perform different types of R&D. About 12% of private agricultural research focuses on farm-level technologies, while 80% of public research has that orientation. Food processing and post-harvest research is the dominant focus of private agricultural research, accounting for 30-90% of all private agricultural R&D.

The increased role of the private sector is propelled, in part, by dramatic advances in biotechnology and the strengthening of intellectual property protection for biological materials in many countries. Until about a decade ago, market failure in agricultural R&D seems to have been widely taken for granted. The main reason was inappropriability of benefits. Constraints to private investment in agricultural research have changed dramatically: with advances in biotechnology, the process of technology generation has accelerated, making it possible to reap the benefits of investments relatively soon.

Since 1980, the evolution of intellectual property laws has allowed for the patenting of living organisms, enabling biotechnology companies to patent biological processes and products, thereby increasing incentives for the private sector to invest in this type of research.

With globalization of trade and the disappearance of trade barriers, the market for biotechnology products has expanded, providing increased opportunities for large multinational corporations. The role of multinational corporations has both positive and negative implications for global science. From a positive point of view, these firms have strong research and development arms that are involved in both applied and basic research, frequently involving scientists from many countries and benefits that extend across borders. However, the products of private sector research are generally proprietary and may not be accessible to the poor or those who need them. The decline in public sector agricultural

research budgets is prompting new partnerships between the public and private sectors, especially in the field of agricultural biotechnology.

For example, Swiss agrochemical giant, Novartis, and the University of California at Berkeley signed a $25 million, five-year agreement in November 1998. Although the agreement specifies that Novartis cannot dictate what research will be performed with its money, the company will have first rights to negotiate an exclusive license on 30-40% of any inventions made in the Department of Plant and Microbial Biology. Private–public interactions can create fertile ground for knowledge generation and transfer of technology to the marketplace. But private financing of the public sector is not without controversy. Critics charge that such alliances will give private companies the ability to influence the research agenda at publicly funded institutions, allowing public goods to be appropriated for private profit.

The CGIAR's agricultural research centres are also discussing new models of collaboration with the private sector (joint research projects, strategic alliances, etc.) and policies relating to the use of proprietary materials and technologies. For example, a Bt (*Bacillus thuringiensis*) gene for insect resistance provided by Plant Genetic Systems (Belgium) has been transferred to potato varieties, and the International Potato Centre has permission to distribute them in ten developing countries.

Monsanto has entered into agreements with Kenyan and Mexican agricultural research institutes to develop virus-resistant crops. While these collaborations are successful, they are few in number, highly bilateral, and often components of philanthropic programs. The potential neglect of the public good is one of the primary issues raised by the dramatically changing roles of the public and private sector in agricultural research. There is concern that the formerly open exchange of materials and technologies to help the poor is being constrained and complicated by intellectual property. Over the past five years, the New York-based Rockefeller Foundation has invested $50 million in plant biotechnology for the developing world. In the words of Dr. Gordon Conway, President of the Rockefeller Foundation:

"As plant research in the industrialized world has come to be dominated by private companies, which closely guard their proprietary technologies, the process of innovation in the developing countries has slowed. Public sector plant breeders don't know how to respond, and when they try, they are handicapped by the huge disparity in resources and negotiating power between themselves and the companies."

Consolidation in the Life Sciences Industry

Consolidation is taking place in all sectors of the global economy. In 1997, the value of all mergers and acquisitions hit a staggering $1.6 trillion in worldwide deals, up from $454 billion in global activity recorded in 1990. In 1998, the total volume of mergers and acquisitions worldwide was a record $2.4 trillion—a 50% increase over 1997. According to the UN Conference on Trade and Development, over four-fifths of all foreign direct investment worldwide is now in the form of global mergers and acquisitions. The past decade recorded dramatic consolidation in the 'life sciences', with market shares of bioindustrial products related to agriculture, food, and health tightly concentrated in the hands of giant transnational enterprises. For example, the world's top ten agrochemical corporations account for 91% of the $31 billion agrochemical market worldwide; the top ten global seed companies control an estimated one-quarter to one-third of the $30 billion commercial seed trade; the top ten pharmaceutical companies account for 36% of the $251 billion global pharmaceutical market; and the top ten firms hold 61% of the animal health industry market valued at $16 billion.

Under the life sciences banner, many firms are using complementary technologies to become significant actors in all of these categories.

Traditional boundaries between pharmaceutical, biotechnology, agribusiness, food, chemicals cosmetics, and energy sectors are blurring and eroding. Major transnational companies are restructuring to take advantage of the molecular revolution and the complementary use of technologies such as high-throughput screening, combinatorial chemistry, transgenics, bioinformatics, and genomics. Life sciences companies are securing and protecting information and technology via patents, and in some cases, the quest is driving a restructuring of the industry. In today's knowledge-based economy, intellectual property assets have surpassed physical assets such as land, machinery, or labor as the basis of corporate value. At the end of 1995, for example, the Hoechst group held 86,000 patents and patent applications. According to Dr. Richard Helmut Rupp, head of Hoechst R&D, The most important publications for our researchers are not chemistry journals, but patent office journals around the world. Demonstrating the value of intellectual property assets, the cover of Novartis' 1997 annual report announces that the company holds more than 40,000 patents. Worldwide demand for patents shows a strong upward trend. At the end of 1995, approximately 3.84 million patents were in force worldwide. The patent offices of the US and Japan together with the European Patent Convention accounted for 81% of the total patents.

Bioethics and Society Choices

By the close of the 20th century, humankind had acquired the power to transform the processes of all living species, including its own. The potential benefits of these powers can be exciting and the perils ominous. Consider, for example, the recent announcement that scientists have successfully produced cultures of embryonic stem cells. This breakthrough offers the potential to grow any type of human tissue and may eventually be used to repair damaged hearts, blood vessels or brains. The very same week, the UK's Sunday Times reported that scientists can theoretically engineer deadly biological organisms to produce 'ethno-bombs' that are capable of targeting human victims by ethnic origin.

Given the dizzying pace of technological advancements in genetics and biology, it is not surprising that society is grappling ever more urgently with the social, ethical and legal implications of humankind's ability to decipher and control the genetic blueprint of life. Opinions differ sharply on the implications of new biotechnologies, but nearly everyone agrees that advances in technology are taking place at a rate far faster than social policies can be devised to guide them, or legal systems can evolve to address them. There is growing recognition worldwide that the development of scientific knowledge must be accompanied by public debate on societal choices and the informed participation of citizens. 'Bioethics' attempts to identify the social and cultural implications of breakthroughs in life sciences, to anticipate its applications, and to ensure that progress in the life sciences benefits humanity as a whole. Bioethics acknowledges that there is a distinction between what is scientifically possible and ethically acceptable. What is good for society, what is equitable, and what is safe? Who will decide? These are among the questions that are fueling debate on the applications of biotechnology to health, agriculture and human development. At the intergovernmental level, the United Nations Educational, Scientific and Cultural Organization (UNESCO) created the International Bioethics Committee in 1993 as the world's only international body to study the implications of human genome research and genetic engineering. In November 1997, UNESCO adopted a non-binding Universal Declaration on the Human Genome and Human Rights, the first international text on the ethics of genetics research. Since 1993, a growing number of countries have established national bioethics advisory committees to examine bioethics and to provide guidance to national governments. In 1998, the CGIAR adopted a set of ethical principles to guide its work on genetic resources. The Scientific group

recognizes and appreciates the vital contribution of ethical debates. There is concern, however, that the appointment of expert panels and commissions devoted to bioethics should not become a substitute for broad public debate and participation in the review and assessment of new technologies.

19.6. CONVENTION ON BIOLOGICAL DIVERSITY (CBD)

In decreased aerable land and industrialized world, there is ever increasing threat to the biological diversity, both in the wild and recent bioreserves. By recognizing that the biodiversity of organisms in the wild should be maintained for their own intrinsic value, but also on practical grounds, the United Nations prepared the Convention on Biological Diversity (CBD) and succeeded in having it adopted in 1992. It entered into force in 1993. This is the first time that a large majority of States have agreed to a legally binding instrument for biodiversity conservation and the sustainable use of biological resources. A radical change brought about by the CBD is the recognition that states have a sovereign right over biodiversity within their own territory. Previously, organisms were considered to be the common heritage of mankind. Living organisms or their products may, under the terms of the CBD, only be removed from a country under mutually agreed conditions. This ensures the better maintenance of the biodiversity.

Large number of organizations, both at country and the international level, are established to meet the conservation of biodiversity. The CBD is a comprehensive approach to biodiversity conservation of both wild and domesticated species at international level. It aims conservation at the genetic, species, and ecosystem levels. As reviewed by Buhenne-Guilmin, action is delegated to the national level, obliging States to assess biodiversity, enact legislation for its conservation *in situ* and *ex situ*, and to enforce legislation within national boundaries. The field of biotechnology is particularly affected by articles 16 and 19 of the

CBD, since they require a fair and equitable sharing of benefits derived from the use of genetic resources. This includes providing facilities and financial means for technology transfer and open access to scientific and technical information. The sovereignty over biological resources means that no one can remove specimens of plants, animals or microorganisms from a country without the prior consent of that country. One example of a joint effort in "bioprospecting" is a search for specific active ingredients of plants by the "Merck" Company in the tropical forests of Costa Rica. This brought the country 2 million dollars over a five year period, in addition to the potential for royalties it profitable products emerge. A small number of similar agreements have been undertaken elsewhere in the world. In India, Ministry of environment and forest, ICAR, IVRI, NPGR, IMTECH and other seed banks are storing the genetic resources of plants, animals, and microbes.

Continued efforts to cooperate, manage, exchange such genetic resources are made between the national and international organizations. The regulations of the CBD have been in operation for only a few years. It is too early to assess their long-term effects. As far as biotechnology and "bioprospecting" is concerned, it will take more time to establish smooth administrative procedures to allow simple routine implementation of close collaborations. This system will only be perpetuated if the national authorities from countries rich in biodiversity, and pharmaceutical or other companies see the mutual advantages to be gained from such collaboration. One of the real obstacles to this is the exorbitantly high costs of drug testing that needs to be done to meet strict and justified regulation before marketing. The expectations of some LDCs to make rapid earnings may have been too optimistic. The search for natural, highly active pharmaceuticals in wild plants may often be more cumbersome than laboratory searches using genomics, rational drug design, and combinatorial chemistry.

SUMMARY

Biodiversity is one of the major natural resources present on the earth. It is the multitude of different living beings in a particular ecosystem or on the whole earth. Biodiversity can be seen and studied at different organizational levels: genetic, organismal and ecological. It touches upon both native environments on land and sea as well as agricultural and other man-made surroundings. The biodiversity we observe today is the result of 3.5 billion years of evolution. Through the processes of mutation and selection, all living organisms we know today, as well as those that ever lived before, developed from one single-cell microorganism. How this first living being arose 3.5 billion years ago is still a matter for speculation. This unitary origin explains why all organisms share the same basic chemistry. DNA is always the storage molecule of genetic information and the complex process of protein biosynthesis is virtually the same in all organisms. Metabolic pathways are also similar in all organisms. The vast majority, probably more than 99%, of species that arose on this globe, disappeared again. This shows that evolution is an ongoing process, with species coming and going. The species are highly dynamic in nature.

The wide varieties of the species diversity form the major force acting on the environmental processes. This dynamic situation is important when discussing the conservation of species living here today. In the long-term view, there has never been any stability in life on earth-only change. However, these changes were very slow compared to the length of a human life, or even compared to the time humans have existed. Clearly, today, with the massive amount of human interference on the globe, changes are much faster than at any other time in the last 65 million years, the point at which the trilobites and later the dinosaurs and many other creatures vanished from the surface of the globe in a relatively short time period. One estimate made in 1992 indicated that the number of species of plants, animals and eukaryotic microorganisms is probably around 10 million today, but only 1.4 million have been characterized and given a name by scientists. Thus, still lot more taxonomic work is need for their identifications.

Biodiversity is the major guiding force in the natural history. There is much variation in what is known about the different groups. Virtually all of the 40,000 vertebrate animals are known and most of the 300,000 vascular plant species as well. On the other hand, there are likely to be over a million species each of fungi and nematodes, of which only 70,000 and 13,000 have been named. There are thought to be far more than a million different insect species as well. With prokaryotes, the situation is even more extreme: about 5,000 bacteria and viruses have been named individually, yet the total number in both of these two groups may well, according to Bull, be in excess of one million. Since many microorganisms and viruses are associated with specific plants and animals, Staley considers that their own biodiversity will depend on the biodiversity of their hosts, as well as the microorganisms' own host range.

All the different species of plants and animals do not have an independent existence, but are associated in specific communities and ecosystems to form more or less stable associations. One such association is, for instance, the humid (and dry) tropical forest, which is generally thought to have the highest degree of biodiversity, with more tree species per km^2 than there are tree species in North America or in Europe, as discussed by Burslem. Another example comes from specific types of alpine meadows or specific sorts of rivers or ponds. The Convention on Biological Diversity requires that all Member States take measures to preserve both native and agricultural diversities.

EXERCISE

1. What do you mean by ethics in biotechnology?
2. Discuss ethics in biodiversity and its importance for the maintenance of biodiversity for the future generations.
3. Elaborate how the scientific progress, human requirements, biodiversity and bioethics should go together without compromising the development.
4. What do you mean by biodiversity?
5. Elucidate different biodiversity among the biological organisms and their distribution in the world ecosystem.
6. Define agricultural biodiversity and its uses in the development of agriculture, agricultural biotechnology and society.
7. What do you mean by 'microbial biodiversity'? What is its impact on the research and development?
8. Distinguish native biodiversity and conserved biodiversity.
9. List out the ethical issues addressed in the animal biodiversity and agricultural biotechnology.
10. What business and consumer issues should addressed in the biodiversity, and what is their impact on third world countries?
11. How does the biodiversity erosion in the biotechnology fields affect the poorer nationals in the protection of their identity?
12. Mention large number of biodiversities and their ethical issues raised by the improvements made in the biology and biotechnology in particular.
13. Critically comment on the economic issues raised in the biodiversity and their exploitation in developing countries and their importance.
14. Mention how the new bioactivity identified with the particular species is affecting the developing countries and under developed countries to suffer from the biological resources.
15. Differentiate between the conventional and biotechnological methods of xenotransplantation in the animal biotechnology.
16. Illustrate how Commission on Genetic Resources for Food and Agriculture (CGRFA) is responsible for the maintenance of biological resources.
17. Write a note on accelerating trend of biodiversity loss in the worldwide ecosystem or biosphere by the industrialization.
18. Name the contributing factors of the climatic change and loss of biodiversity in the world biosphere and their impact on the environment.
19. Define convention on biological diversity (CBD) and its importance in the world scenario.

20

CHAPTER

Ethical Issues in Genetically Modified Organisms: Foods and Crops

20.1. INTRODUCTION

Biotechnologists and biotech industries all over the world claim that biotechnology and their products will help to increase crop and animal yields on huge scale than the conventional system of production processes. Therefore, the availability of food to feed an ever growing population can be made possible. They point to projections of population growth, increases in food demand, and declining rates of growth in yields from Green Revolution technologies to support their claims. They claim that modification of crop plants and animals can contribute to sustainability by reducing the amount of fertilizers, herbicides, and pesticides used in farming and improving the efficiency of water and fertilizer use **(Table 20.1)**.

Scientists in both the private and public sectors clearly see genetic modification as a major new set of tools. They are also participants and spectators in a major shift of research from the public to the private sector, which will undoubtedly influence the future direction of research and research investment. As shareholders in the genetically modified organism (GMO) debate, scientists must recognize that there is also a substantial public distrust of science. Obviously, the industry looks at GMOs as opportunities for corporate profit, yet, at the same time, recognizes that public acceptance may be a stumbling block. In turn, governments often lack coherent policies in relation to GMOs, and have not yet developed and implemented adequate regulatory instruments and infrastructures. As a result, in most countries, there is no consensus on how biotechnology and GMOs, in particular, can focus on the key challenges of the food and agricultural sector. Governments need to be more proactive in addressing these questions, and in this, the role of scientists in public service will be crucial.

Biotechnologists immediately started addressing the various fields of living organisms and their products, where these aspects can be improved by transgenic methodologies. A genetically modified organism (GMO) is an organism whose genetic material has been altered using recombinant DNA technology. Recombinant DNA technology is the ability to combine DNA molecules from different sources into the one molecule in a test tube. Thus, the abilities or the phenotype of the organism, or the proteins it produces, can be modified through the

Table 20.1: *History of transgenic food development from farming practice*

8000 BC — Farming begins in Mesopotamia	**1900** — **1953.** Watson and Crick publish structure of DNA
Breadmaking wheat grown in Egypt, rice cultivated in China	**1957.** Kornberg isolates DNA polymerase
4000 BC — Potatoes grown in Peru	**1960s-1970s.** 'Green revolution' leads to greatly increased crop yields based on the incorporation of dwarfing genes discovered by Norman Borlaug and the widespread use of agrochemicals
3000 BC — All major food crops in Eurasia being cultivated	**1966.** Weiss and Richardson discover DNA ligase
1000 BC — All major food crops in Americas being cultivated	**1969.** First commercial triticale (wheat/rye hybrid) release
700 BC	**1970.** Hamilton Smith discovers restriction enzymes
0 — Babylonians use selective breeding techniques with date palm	**1972.** Berg produces first recombinant DNA molecule
100	**1973.** Boyer, Chang, Cohen and Helling produce a recombinant plasmid DNA molecule
1000	**1977.** Gilbert and Sanger independently develop techniques to determine the sequence of nucleotides in a DNA molecule Nester, Gordon and Chilton show that *Agrobacterium tumefaciens* genetically modifies host plant cells
1600 — Potato, maize and tomato introduced into Europe from the Americas	**1981.** Insulin produced in GM E. coli used in medicine
1753. Linnaeus publishes 'Species Plantarum', effectively beginning the science of plant taxonomy	**1983.** Groups in Ghent, Leiden, St. Louis and Cambridge use A. tumefaciens to introduce bacterial genes into plant cells
1600	**1983.** Hall produces GM sunflower plants containing a gene from French bean
1843. John Lawes patents superphosphate, the first artificial fertiliser	**1985.** First GM plants in field in the UK
1859. Darwin publishes 'On the Origin of Species by Means of Natural Selection'	**1988.** Grierson and Schuch report antisense inhibition of gene expression in a GM plant
1866. Mendel publishes 'Versuche über Pflanzen-Hybride'	**1994.** Slow-ripening GM tomatoes approved for food use
1989. Miescher discovers DNA	**1996.** First large-scale cultivation of GM soybean and maize
1990. Mendel's work rediscovered	**2000.** Nucleotide sequence of entire Arabidopsis genome
1902. Garrod links inherited trait with protein function	**2001.** Word-wide GM crop area exceeds 50 million hectares
1923. Russet Burbank hybrid potato launched	**2003.** GM crops grown on 65 million hectares in 18 countries
1933. First hybrid maize variety launched in USA	
1923. Russet Burbank hybrid potato launched	
1941. Beadle and Tatum publish one gene-one enzyme hypothesis. First modern herbicide (2,4-D) synthesized	
1944. Avery, Macleod and McCarthy show that DNA is the material of inheritance	
1950s. First release of commerical crops produced by chemical and radiation mutagenesis	

cont....

modification of its genes. The term generally does not cover organisms whose genetic make-up have been altered by conventional cross breeding or by "mutagenesis" breeding, as these methods predate the discovery of the recombinant DNA techniques. Examples of GMOs are diverse, and include transgenic experimental animals such as mice, several species of fish, transgenic plants, or various microscopic organisms altered for the purposes of genetic research or the production of pharmaceuticals.

As soon as the wide varieties of organisms are recognized for their applications in the human life, they are considered for the improvement by using biotechnological processes. The term "genetically modified organism" does not necessarily imply, but does include, transgenic substitution of genes from another species. For example, genes for fluorescent proteins can be co-expressed with complex proteins in cultured cells to facilitate study by biologists, and modified organisms are used in researching the mechanisms of cancer and other diseases.

Since their evolution, organisms modified themselves through the process of natural selection. Humans have been altering the genetic make-up of animals and plants for centuries. Selective breeding has been directed towards producing desirable characteristics for a variety of reasons, for example, to increase yields, modify the food quality and content (starch, protein or lipid), and to confer resistance to disease. For many years, the genetic make-up of seeds has been altered by exposing them to radiation. This causes mutations leading to new characteristics, which can then be selected and bred to produce desirable characteristics. For all this, human had to wait for long periods.

Hence, this made them to look for other alternative processes that could fetch them the faster and better qualities in his products. The difference between practices like conventional and modern approach of genetic modification lies in the way in which humans can now influence these processes of development, growth, and yield in the animal and plant kingdom. With the dramatic increase in our basic understanding of the genetic make-up and biochemistry of living organisms, we are now in the position to use this knowledge in a more considered manner for the improvement of plant and animal for agriculture and food purposes. The returns are always higher, and the need based techniques in the new methodologies.

Large number of research organizations started working on these technologies for the higher competitive edge. The laboratory processes that are used to manipulate this genetic code can be likened to the process of cutting and pasting. Strands of DNA, the basic chemical of life, which are found in the nucleus of all organisms, and which produce a particular effect in one living organism, can be 'cut' out and then 'pasted' into the nucleus of another living organism. The genetic strands are 'trimmed' so that only a precise, fully defined piece of DNA is pasted into the recipient organism.

Hence, the urgent need of the world is to produce ever increasing populations. Apart from the fact that the process uses advanced molecular techniques, there are three basic differences between modern genetic modification and traditional animal and plant breeding methods, which are as follows:

- Genetic modification enables single, well defined genes to be isolated and transferred, whereas with traditional methods many thousands of genes are 'crossed' at one time.
- Genetic modification allows the introduction of a desired gene from one plant species into another. In addition genes can also be introduced from other organisms such as microorganisms and animals or synthetic source.
- Genetic modification gives the possibilities for making multigene based modifications for more than one characters.

20.2. HISTORY OF GENETIC MODIFICATION

Genetic selection started as early as 8000 BC, when farming practice began in Mesopotamia. The technology of bread making started during 4000 BC in Egypt. Biologists and biotechnologists considered their interests in the production of genetically modified organisms in different species. The first GMO was created in 1973 by Stanley N. Cohen and Herbert Boyer, demonstrating the creation of a functional organism that combined and replicated genetic information from different species. In mid-1974, very soon after the first GMO was created, scientists called for and observed a voluntary moratorium on certain recombinant DNA experiments. One goal of the moratorium was to provide time for a conference that would evaluate the state of the new technology and the risks, if any, associated with it. The conference concluded that recombinant DNA research should proceed, but under strict guidelines. Such guidelines were subsequently promulgated by the National Institutes of Health in the United States and by comparable bodies in other countries. These guidelines form the basis upon which GMOs are regulated throughout the world including the third world countries. In 1977, Nester, Gordon, and Chilton showed that *Agrobacterium tumefaciens* can deliver bacterial genes into plant cells. Hall in 1983 produced GM sunflower plants containing a French bean gene. Later, first GM plants were released in the fields of United Kingdom. Grierson and Schuch, during 1988, reported the antisense inhibition of gene expression in a GM plant. In 1994, slow ripening GM tomatoes were approved for the food purpose. After two years, in 1996, large scale cultivation of GM soybean and maize started worldwide for their advantages over the conventional varieties.

Animal sciences are also attracted towards GM techology. Large number of animals are considered for their utility in the different fields of biotechnology. The first transgenic animals were mice, created by Rudolf Jaenisch in 1974. Jaenish successfully managed to insert foreign DNA into the early-stage mouse embryos. The resulting mice carried the modified gene in all their tissues. Subsequent experiments, in which leukemia genes were injected into early mouse embryos using a retrovirus vector, proved that the genes integrated not only in the mice, but also in their progeny.

20.3. TECHNIQUES OF GENETIC MODIFICATION

Genetic Modification of Bacteria

With the advent of molecular biology and genetic engineering, a large number of techniques were developed for the genetic modification of organisms. Genetic modification involves genetic engineering, also known as *gene splicing*—a technique to splice DNA fragments from more than one organism and thus preparing a "recombinant" DNA molecule in a test tube, producing a single piece of genetic material containing the original information from multiple fragments which can then be inserted into another organism. This is achieved by cutting up DNA molecules with restriction enzymes and splicing these fragments together using DNA ligase. A *transgenic* organism that contains such DNA sequences from a foreign organism integrated into its own genome; the term "transgenic" literally means *across gene*. An example of a transgenic organism is an animal that is not a jelly fish that expresses the green fluorescence protein, such as mice or fish. The gene coding for the protein originated from jelly fish.

Presently, transgenic organisms can be produced with only a small proportion of extraneous DNA. For example, the genome of most mammals contains three billion basepairs of DNA, while it becomes relatively difficult to insert more than 10,000 to 20,000 basepairs of foreign DNA. More sophisticated techniques using yeast artificial chromosomes and bacterial artificial chromosomes allow insertions of up to 320,000 base pairs—approximately 0.01% of the total genome. In concept, multiple rounds of transgenesis or interbreeding of transgenics could lead to organisms with a higher proportion of foreign DNA, but cost and time considerations prevent this.

For genetic modifications, the required gene has to be selected and introduced into the elite organism. In order to introduce new DNA into the receiving host, *vectors* are used. Vectors range from small circular pieces of DNA, such as plasmids, to various viruses that can carry and transmit genetic information. Three processes are known by which the genetic composition of bacteria can be altered. They are as follows:

- Transformation
- Conjugation
- Transduction

Transformation is a process by which some bacteria take up DNA naturally to acquire new genetic traits. This phenomenon was discovered by Frederick Griffith in 1928, though the fact that it was DNA molecule that carried the genetic information was not proven until 1944. Bacteria that are competent to undergo transformation are frequently used in molecular biology. The foreign DNA uptake is facilitated by the presence of certain cations, such as Ca^{2+}, or by the use of electric current (electroporation). Transformation does not normally integrate new DNA into the bacterial chromosome. Instead, it remains in a plasmid.

In conjugation, DNA is transferred from one bacterium to another via a temporary connecting protein tube called *pilus* (a process analogous to but biologically distinct from mating). *Plasmid* is transferred through the pilus. Conjugation is not widely used for the artificial genetic modification of bacteria, but happens often in nature.

Transduction refers to the introduction of new DNA into the bacterial cell by a bacteriophage—virus that infects bacteria.

Genetic Modification of Crop Plants

Plants are the major producers, and humans depend on them for varieties of food, medicine, clothes, firewoods, and other daily needs. The principal technique for the genetic modification of plants is based on a natural ability of the bacterium *Agrobacterium tumefaciens*. This bacterium infects plants and causes a tumor-like growth termed as crown gall. *A. tumefaciens* contains a plasmid (a circular piece of DNA) that transfers from the bacteria into the infected plant and integrates into the plant's genome. The transferred genes cause the plant to form the gall, which houses the bacteria and produces nutrients that support the growth of bacteria. A number of scientists contributed to this discovery throughout the late 1960s and the 1970s, with key discoveries by Jeff Schell, Marc Van Montagu, Georges Morel, Mary-Dell Chilton, and Jacques Tempé.

Later, molecular biology and genetic engineering has made greater understanding of life processes and their utility to the development of the living organisms. By 1983, biotechnology had reached the point where it was possible to insert additional genes of interest into *A. tumefaciens*, and than transfer those genes into plants. This process is commonly used to create transgenic crop plants for agricultural purposes. Another widely used process to create transgenic crops is the biolistic method (gene gun)—a method used for the creation of two most common transgenic crops—Roundup Ready soybean and Bt-corn. Biolistic techniques are generally more suited to monocots, whereas Agrobacteria are used primarily with dicots. However, newer techniques and strains of *A. tumefaciens* have also found utility in the transformation of monocots.

Genetic Modification of Animals

Humans also depend on the animals for their basic needs. Hence, researchers started concentrating on the animals for improvement and exploitation. Similarly to bacteria and plants, animals can also be genetically modified by viral infection. However, the genetic modification occurs only in those cells that become infected, and in most cases these cells are eventually eliminated by the immune system. In some cases, it is possible to use the gene-transferring ability of viruses for gene therapy, i.e., to correct diseases caused by a defective gene by supplying a normal copy of the gene. Permanent genetic modification of entire animals can be accomplished in mice.

The process begins by genetically modifying a mouse embryonic stem cell. This is normally done by physically introducing into the cell a plasmid that can integrate into the genome by a process known as transfection. During transfection, the DNA integrates into the animal genome via non-homologous recombination. This altered cell is implanted into a blastocyst (an early embryo), which is then implanted into the uterus of a female mouse. A pup born from this blastocyst will be a chimera containing some cells derived from the unmodified cells of the blastocyst and some derived from the modified stem cell. By selecting mice whose germ cells (sperm- or egg-producing cells) developed from the modified cell and interbreeding them, pups that contain the genetic modification in all of their cells will be born.

There is also an example of genetically manipulated bull Herman, which had 55 offspring. A human gene was built into his genetic code while in an early embryonic stage, in 1990. As a result, milk from his female descendants contained the human protein lactoferrine, which can be used as medicine, but it was present at such low levels that it was not profitable to extract it.

Insects can be genetically modified by injecting them with artificial transposons and a source of an enzyme "transposase". The transposon, which can include new genes, is then integrated into the genome. Such insertions are unstable and can 'jump-out' in the presence of transposase.

Transgenic fish are often created by microinjection. First generation is mosaic, but several lines have been produced with the transgene incorporated into the germ line, and transgenic fish can then be produced "the natural way" by crossing male and female gametes. Although many types of transgenic fish exists (e.g., for increased cold tolerance, antibiotic production, ornamental Glofish etc.,), the main focus has been on so-called growth hormone transgenic fish, mainly salmonids, tilapias and carps. These fish have an over-production of growth hormone, which results in increased growth rate from a few percent up to 30-40 times that of wild-types.

In some species, final size as well as growth rate is increased, providing an incentive for commercial breeders to farm such fish. However, ecological concerns over potential negative effects of transgenic fish in nature largely prevent the commencement of commercial production. A large and important portion of the research on transgenic fish today is, therefore, focusing on environmental risk-assessement of GH-transgenic fish.

Genetic modifications are not only used for their products, but also for understanding the basic functions of the life. In order to gain knowledge about a particular gene's function, researchers often use *knockout* organisms. These organisms have a specific gene that has been functionally destroyed or "knockedout". They are used extensively in disease research with model organisms. For example, while investigating the cause of cystic fibrosis, researchers identified the CFTR gene as a likely candidate for the disease, found the mouse equivalent, bred a mouse with this gene "knocked out", and noted that the knockout mouse also had cystic fibrosis.

20.4. USES OF GENETIC MODIFICATION

Genetic modification in the laboratory was first reported in the early 1970s. Since then, a wide range of applications in agriculture, medicine, environment, food production, manufacturing industry, and research have been developed. A number of examples are listed below:

In the Agricultural Area

* Herbicide tolerant crops
* Insect resistant crops
* Virus resistant crops
* Pest resistant crops
* New products and growing techniques
* Enhanced taste and quality
* Reduced maturation time
* Increased nutrients, yields and stress tolerance

In Animals

* Increased resistance, productivity, hardiness, and feed efficiency
* Better yields of meat, eggs, and milk
* Improved animal health and diagnostic methods

In the Field of Medicine

* Insulin to treat diabetes
* Production of human growth hormone
* Production of blood clotting factors VIII and IX
* The treatment of cystic fibrosis
* Research in human and animal diseases

In other Aeas

* Environmental clean-up of soil spills.
* Treatment of contaminated land and water.
* Manufacture of useful chemicals such as enzymes.
* Plants providing renewable sources of industrial chemicals.

Gene Modification and its Controversies

Since the development of the first transgenic microbes, plants, and animals, health and environment issues concerning the safety of genetically modified foods and feeds have been raised. The main con-cerns regarding genetically modified (GM) foods include toxin or allergen production, changes in nutrient levels, and development of antibiotic resis-tance. Gene flow and effect on non-target organism of GM crops are some of the environmental issues. Globally-debated issue of genetically modified organisms (GMOs) is considered in a stricter scene all over the world. Genetic modification (GM) is the subject of controversy in its own right. Some see the science itself as intolerable meddling with "natural" order, despite known examples of natural genetic crossings occurring throughout history. While some would like to see it banned, others push simply for required labeling of genetically modified food. Other controversies include the definition of patent and property pertaining to products of genetic engineering, and the possibility of unforeseen global side effects as a result of proliferating modified organisms. The basic ethical issues involved in genetic research are discussed in the article on genetic engineering.

Considering the potential hazard of the GMOs, some states in the USA have made stricter regulatory issues. In 2004, Mendocino County, California became the first county in the United States to ban the production of GMOs. The measure passed with a 57% majority. In 2005, a standing committee of the government of Prince Edward Island in Canada began work to assess a proposal to ban the production of GMOs in the province. This is a largely symbolic and empty gesture, as PEI has already banned GMO potatoes, which account for most of its crop. In California, the Trinity and Marin counties have also imposed bans on GM crops, while ordinances to do so were unsuccessful in Butte, San Luis Obispo, Humboldt and Sonoma counties. Supervisors in the ag-rich counties of Fresno, Kern, Kings, Solano, Sutter and Tulare have passed resolutions supporting the practice.

Roles and responsibilities of relevant government agencies in the safety management of GMOs, GMO import procedures, GMO production procedures, GMO labeling, building a nation-wide GMO safety management system, and establishment of biosafety information center are very important in this biotechnologically advanced world. Currently, there is little international consensus regarding the acceptability and effective role of modified "complete" organisms such as plants or animals. A great deal of modern research that is illuminating complex biochemical processes and disease mechanisms makes vast use of genetic engineering.

The urge of various countries for the use of GM technologies in the world is ever increasing. The practice of genetic modification as a scientific technique is not restricted in the United States. Individual genetically modified crops (such as soybeans) are subject to intense study before being brought to market, and are common in the United States, but estimates of their market saturation vary widely. Some countries in Europe have taken the opposite position, stating that genetic modification has not been proven safe, and therefore they will not accept genetically modified food from the United States or any other country.

This issue has been brought before the World Trade Organization, which determined that not allowing modified food into the country creates an unnecessary obstacle to international trade. Consequently, genetic modification within agriculture is an issue of a strong debate in the United States, the European Union, and some other countries.

Large number of organizations, both government and private, raised their vioces against the GM technology and their products. They also have a valid scientific support for such criticisms. Some critics have raised the concern that conventionally bred crop plants can be cross-pollinated (bred) from the pollen of modified plants. Pollen can be dispersed over large areas by wind, animals, and insects. Recent research with creeping bentgrass has lent support to the concern when modified genes were found in normal grass up to 21 km (13 miles) away from the source, and also within close relatives of the same genus Agrostis GM proponents point out that

outcrossing, as this process is known as, is not new. The same thing happens with any new open-pollinated crop variety. Newly introduced traits can potentially cross out into neighbouring crop plants of the same species, and in same cases to closely related wild relatives.

But it also has some added advantages for the immediate need. Defenders of GM technology point out that each GM crop is assessed on a case by case basis to determine if there is any risk associated with the outcrossing of the GM trait into wild plant populations. The fact that a GM plant may outcross with a related wild relative is not, in itself, a risk unless such an occurrence has consequences. If, for example, a herbicide resistance trait was to cross into a wild relative of a crop plant, it can be predicted that this would not have any concequences except in areas where herbicides are sprayed, e.g., a farm. In such a setting, the farmer can manage this risk by rotating the herbicides. If patented genes are outcrossed, even accidentally, to other commercial fields and a person deliberately selects the outcrossed plants for subsequent planting, then the patent holder has the right to control the use of those crops. This was supported in Canadian law in the case of *Monsanto vs. Percy Schmeiser.*

An often cited controversy is a hypothetical Technology Protection technology (dubbed "terminator" by NGOs). This yet to be commercialised technology would allow the production of first generation crops that would not generate seeds in the second generation because the plants yield sterile seeds. The patent for this so-called "terminator" gene technology is owned by Delta and Pine Land and the USDA, despite being mis-associated with Monsanto. In addition to the commercial protection of proprietary technology in self-pollinating crops such as soybean (a generally contentious issue), another purpose of the terminator gene is to prevent the escape of genetically modified traits from crosspollinating crops into wild-type species by sterilizing any resultant hybrids. The terminator gene technology created a backlash amongst those who felt that the technology would prevent the re-use of seed by farmers growing such terminator varieties in the developing world, and was ostensibly a means to exercise patent claims. Use of the terminator technology would also prevent "volunteers", or crops that grow from unharvested seed—a major concern that arose during the Starlink debacle.

20.5. GENETICALLY MODIFIED FOOD

Genetically modified food (GM food) is a food product derived in whole or part from a genetically modified organism (GMO) such as a crop plant, animal, or microbe such as yeast. Genetically modified foods have been available since 1990s. The principal ingredients of GM foods currently available are derived from genetically modified rice, corn, soybean, potato, tomato, banana, maize, and canola.

Genetically modified food is defined in the EC Novel Foods Regulation as "a food which is, or which is made from, a genetically modified organism and contains genetic material or protein resulting from the modification". A list of products considered by the ACNFP is included in a separate technical Note A, which is available on request.

Combining genes from different organisms is known as recombinant DNA technology, and the resulting organism is said to be "genetically modi-fied", "genetically engineered", or "transgenic." GM products (current or in the pipeline) include medicines and vaccines, foods and food ingredients, feeds, and fibers. Some GM foods that have been developed are whole plants or parts of organisms that are eaten raw, such as tomatoes and fresh chicory (used as salad leaves). GM yeasts containing their own enzymes for breaking down sugar to produce more alcohol have been produced for use in brewing beer. GM baking yeasts have also been developed to allow better digestion of sugars obtained from the starch in flour to give a better texture. These yeasts are not yet being used commercially.

Locating genes for important traits, such as those conferring insect resistance or desired nutrients, is one of the most limiting steps in the process. However, genome sequencing and discovery programs for hundreds of different organisms are generating detailed maps along with data-analyzing technologies to understand and use them. Many crop plants that are used to produce food ingredients, are now being genetically modified, for example soya and maize.

Soya beans can be processed to yield many different food ingredients, from soya protein and flour, to oil and lecithins used as emulsifiers. Maize can also be processed to yield a variety of ingredients from starch and sugars to oil and flour. Some ingredients derived from crop plants are highly refined, for example, sucrose and vegetable oils, and these refining processes destroy and remove any genetic material and protein that might be present in the food ingredient.

The end product that goes into food is there-fore not itself modified and cannot be distinguished from that produced by conventional means. GM crops are grown commercially or in field trials in over 40 countries and 6 continents. In 2000, about 109.2 million acres were planted with transgenic crops; the principal ones being herbicide and insecticide-resistant soybeans, corn, cotton, and canola. Other crops grown commercially or field-tested are a sweet potato resistant to a virus that could decimate most of the African harvest; rice with increased iron and vitamins that may alleviate chronic malnutrition in Asian countries; and a variety of plants able to survive weather extremes.

Animals that have been genetically modified to produce pharmaceutical products for use in human therapy do not enter the food chain. No GM animals have been approved for food use.

Some governments have a very strong mutual disagreement over the labelling and traceability requirements for GM food products. For example, the European Union and Japan require labelling and traceability, while regulatory agencies in the United States do not believe that these requirements are necessary.

Biotechnology, refers to the use of living organisms or their components, such as enzymes, to make products that include wine, cheese, beer, and yogurt. Genetic engineering may more correctly be termed "genetic re-contextualisation", where genes can be transferred to new contexts in order to generate new characteristics. GM products (current or in the pipeline) include medicines (e.g., insulin, vaccines), foods, food ingredients, feeds, and fibers.

Development and Application

The origin of genetic engineering represent a series of sequential scientific advances from the Nobel prize-winning discovery of DNA to the production of the first recombinant *E. coli* bacteria.

The first commercially grown genetically modified food crop was a tomato created by Calgene called the FlavrSavr. Calgene submitted it to the US Food and Drug Administration (FDA) for assessment in 1992. Following the FDA's determination that the FlavrSavr was, in fact, a tomato, did not constitute a health hazard, and did not need to be labeled to indicate that it was genetically modified, Calgene released it into the market in 1994, where it met with little public comment. Considered to have a poor flavor, it never sold well and was off the market by 1997. However, it had improved solids contents, which made it an attractive new variety for canned tomatoes.

Between 1996 and 2001, the total surface area of land cultivated with GMOs had increased by a factor of 30, from 17,000 km² (4.2 million acres) to 520,000 km² (128 million acres). The value in 2002 was 145 million acres (587,000 km²), and in 2003 was 167 million acres (676,000 km²). Soybean crop represented 63% of total surface in 2001, maize 19%, cotton 13%, and canola 5%. In 2004, the value was about 200 million acres (809,000 km²), of which 2/3 were in the United States.

Four countries represent 99% of total GM surface in 2001: United States (68%), Argentina (22%), Canada (6%) and China (3%). It is estimated that 70% of products on US grocery shelves include GM ingredients. In particular, Bt corn is widely grown, as are soybeans genetically designed to tolerate glyphosate herbicides.

The US agriculture department estimated that 38% of the 79 million acres (320,000 km²) of corn planted in 2003 will be genetically engineered varieties, as well as 80% of the 73.2 million acres (296,000 km²) soybeans. The grocery manufacturers of America estimate that 75% of all processed foods in the US contain a GM ingredient.

Future envisaged applications of GMOs include bananas that produce human vaccines against infectious diseases such as Hepatitis B; fish that mature more quickly; fruit and nut trees that yield years earlier; and plants that produce new plastics with unique properties. The next decade may see an exponential progress in GM product development, as researchers gain increasing access to genomic resources that are applicable to organisms beyond the scope of individual projects.

Benefits and Risks

The majority of commercially available crops have an agronomic advantage like herbicide tolerance or insect resistance. These traits offer major benefits to the farmer. However, there are indirect benefits to the consumer from these traits: GM crops have been shown to contribute to significantly reduced greenhouse gas emissions from agricultural practices. This reduction results from decreased fuel use—about 1.8 billion liters in the past nine years—and additional soil carbon sequestration because of reduced ploughing or improved conservation tillage associated with biotech crops. In 2004, this reduction was equivalent to eliminating more than 10 billion kg of carbon dioxide from the atmosphere.

Controversies surrounding GM foods and crops commonly focus on human and environmental safety, labeling and consumer choice, intellectual property rights, ethics, food security, poverty reduction, and environmental conservation.

Controversies over Risks

Although no major health hazards have come to light since GM food was introduced 10 years ago, some fear for the long term health risks, which GM could pose, or that the risks of GM have not yet been adequately investigated.

Dangerous Potatoes

In August 1998, widespread concern, especially in Europe, was sparked by the remarks of nutrition researcher, Dr. Arpad Pusztai, regarding some of his research into the safety of GM foods.

Pusztai claimed his experiments showed that rats fed on genetically modified potatoes had suffered serious damage to their immune systems, and shown stunted growth. He was criticized by leading British politicians, the majority of scientific peers with expertise in the area, and by the GM companies because he announced his preliminary results in a television interview before the publication of his results in a scientific journal. When his studies were finally published in "The Lancet" 1999 354:1353-1354, no evidence of stunted growth or damage to immune system was substantiated. The Royal Society's review of the Pusztai data led to the conclusion that the study "is flawed in many aspects of design, execution, and analysis and that no conclusion should be drawn from it".

Dangerous Corn

Another controversy recently arose around biotech company Monsanto's data on a 90-day rat feeding study on a strain of GM corn. In May 2005, critics of GM foods pointed to differences in kidney size and blood composition found in this study, suggesting that the observed differences called into question the regulatory doctrine of substantial equivalence—that GM food with similar proteins and toxins is deemed no different than conventional food—without further investigation of the effects of any other differences. Some argued that this study suggested human health might be affected by eating GM food.

However, the European Food safety authority has examined the Monsanto data and concluded that the observed small numerical decrease in rat kidney weights were not biologically meaningful, and the weights were well within the normal range of kidney weights for control animals. There were no corresponding microscopic findings in the relevant organ systems, and all blood chemistry and organ weight values fell within the "normal range of historical control values" for rats. Thus, the experts concluded that there were no effects on the functioning of kidneys in rats fed on a diet of GM corn.

New Allergens

GMOs that induce allergies have been produced in the laboratory. In 1993, Pioneer Hi-Bred International developed a soybean variety with an added gene from the Brazil nut. This gene increased the levels of methionine—a nutrient commonly added to poultry feed—in the GM soybean. However, a preliminary pioneer funded study by the University of Nebraska indicated that the added gene could cause allergic reactions in humans. The completed study, published in the New England Journal of Medicine, later confirmed the preliminary results. Pioneer discontinued the further development of the GM soybean and had all material related to the modified soybeans destroyed. While this study indicates the possible risks of GM foods, some point out that it establishes the commitment the developmental community has towards consumer safety as well as the competence of current safeguards. A similar result was published in November 2005, when a pest resistant field pea developed by the Australian CSIRO for use as a pasture crop was shown to cause an allergic reaction in mice. The immunologist who tested the pea noted that the episode illustrated the need for each new GM food to be very carefully evaluated for potential health effects.

Environmental and Ecological Impacts

As discussed above, there is some evidence for positive impacts of the planting of GM crops on reduced greenhouse gas emissions and pesticide loads in the environment. However, there has been controversy over the results of a farm-scale trial in the United Kingdom, comparing the impact of GM and conventional crops on farmland biodiversity. Some claimed that the results showed that GM crops had a significant negative impact on wildlife.

Others pointed out that the studies showed that using herbicide resistant GM crops allowed better weed control, and under such conditions there were fewer weeds and fewer weed seeds. This result was then extrapolated to suggest that GM crops would have significant impact on the wildlife that might rely on farm weeds. In July, 2005, the British scientists showed that transfer of a herbicide resistance gene from GM oilseed rape to a wild cousin, charlock, and wild turnips was possible.

rBGH

rBGH, a genetically engineered version of bovine somatotropin, is a hormone used to increase the production of milk in cows. Although it is used in the United States, it is banned in Canada, the EU, and Australia. rBGH is a frequent target of anti-GMO groups, who claim that there is evidence of an increased risk of disease in consumers, who drink milk from rbST-injected cows.

Public Perception

Research by the Pew Initiative on Food and Biotechnology has shown that in 2005, Americans' knowledge of genetically modified foods and animals continues to remain low, and their opinions reflect that they are particularly uncomfortable with animal cloning. The Pew survey also showed that despite continuing concerns about GM foods, American consumers do not support banning new uses of the technology, but rather seek an active role from regulators to ensure that new products are safe.

20.6. POLICY AROUND THE WORLD

Although the growth of transgenic crops is expected to plateau in industrialized countries, it is increasing in developing countries.

There are two policy areas surrounding GM food:
1. The standards and regulation of testing for food safety
2. The requirements for labelling and traceability of GM products in the food chain

Countries vary on their approach to both these points.

United States

Genetically modified food is widely available in the United States. Some environmentalist groups believe that US should regulate GM food more closely, and have called for mandatory labeling and testing requirements.

Testing

The Food and Drug Administration regulates the safety of all foods sold, including the safety of GM foods. Under these regulations, substances added to food that do not meet the statutory definition of generally recognized as safe, and are not pesticides, are classified as food or color additives and must be pre-approved before they may be marketed. Prior to marketing food additives, manufacturers are required to submit documentation to FDA, demonstrating their safety and await approval for their use.

Section 402(a)(1) of the federal food, drug, and cosmetic act covers unintentional introduction of unsafe toxicants into foods. Producers of the foods are legally responsible *to ensure that no substance occurs in the food that "may render" the food injurious to health* FDA Statement of Policy: Foods Derived From New Plant Varieties. In order to comply with this part of the act, GM companies subject new GM foods to saftey testing in animal models.

Some people raise concerns that the safety of these novel food forms relies on the effectiveness of testing by food manufacturers. Defenders of the regulatory model point out that such a model has served the public well for many years with respect to non-GM food, drugs, and food additives. They further point out that the manufacturers data is subject to close scrutiny, and the law protects the consumers from dangerous foods. It is thus in the manufacturers' best interests to ensure that the safety testing of GM foods is done adequately.

European Union

In Europe, a series of unrelated food crises during the 1990s [e.g., the BSE (or 'mad cow' disease) outbreaks and foot and mouth disease] created consumer apprehension about food safety in general, and eroded the public trust in government oversight of the food industry. This has further fueled widespread public concern about GMOs, in terms of environmental protection (in particular biodiversity), health and safety of consumers, and the right to make an informed choice. The apprehension might also be due to the novelty of GM foods, as well as cultural factors related to food. The mishandling of the BSE crisis has left some

consumers unwilling to consider "science" to be a guarantee of quality.

European consumers are demanding that their "right to know" the content and origin of the food they consume should be respected. In the context of local food surplus, where current GM food has little added nutritional value, many European consumers are wondering why any risk should be taken. However, as a result of the high quantity of GMO crops, the presence of GM in imported food products (shipments of grain for food, feed and processing for example) is now thought to be inevitable and largely unavoidable, and usually not mentioned.

EU Regulation

For these reasons, the marketing of GM food is regulated in a manner that helps to provide the necessary levels of safety, transparency and reassurance. At the beginning of the 2000's, European officials insisted that new regulations were needed to "restore consumer confidence" in the technology. These new regulations required strict labeling and traceability of all food and animal feed containing more than 0.5% GM ingredients. Directives, such as directive 2001/18/EC, were designed to require authorization for the placing on the market of GMO, in accordance with the precautionary principle.

One of the features of the European system is a comprehensive pre-market risk assessment—a system trying to provide means for products to be followed at each stage of their production and distribution, by both transmission of accurate information and labeling. This traceability is a means to implement post-market measures, such as monitoring and withdrawals (recalls). This system is not only limited to GMO products, but should encompass any food product ultimately.

The original EU rules for labeling of GM products were limited to products where transformed DNA and/or transformed protein are detectable, not to products that have been produced from GMOs but no longer appears to contain modified DNA and/or proteins. New rules for traceability and labeling, which came into force in 2004, also require labeling of highly refined products made from GM ingredients like oil and corn syrup, even though the presence of recombinant DNA or protein cannot be proven. The labeling rules do not apply to the products of microbial genetic engineering. So, the cheese made with the help of GM-chymosin doesn't have to be labeled. Officials stress that while traceability facilitates the implementation of safety measures, where appropriate, it cannot and should not be considered as a safety measure.

In April 1998, a five-year ban was pronounced on new genetically modified crops. At the end of 2002, European Union environment ministers agreed that new controls on GMOs could eventually lead the 25-member bloc to reopen its markets to GM foods. European Union ministers agreed to new labeling controls for genetically modified goods, which will have to carry a special harmless DNA sequence (a DNA code bar) identifying the origin of the crops, making it easier for regulators to spot contaminated crops, feed, or food, and enabling products to be withdrawn from the food chain if problems arise. A series of additional sequences of DNA with encrypted information about the company, or what was done to the product could also be added to provide more data.

The more information can be obtained from Trade war over genetically modified food on disputes and more recent developments between the United States and the EU arising from EU position on genetically modified organisms.

Japan

Japan, like Europe, maintains labelling standards for GM food products. Japanese demand and assistance has led to a small effort to set up separate processing facility for non-GM soybeans in the US.

Canada

Labelling is currently not required for GM food products sold in Canada. In 2005, a standing committee began work in the province of Prince Edward Island to assess a proposal to ban the production of GM foods within the province.

China and other Developing Countries

China is currently the producer of GM cotton. Research published in Science claims that Chinese farmers growing GM cotton use significantly less pesticides, which reduces the cost and improve the health of the farmer. The Chinese government has also released safety certificates following field and laboratory testing, allowing the cultivation of GM tomato, pimiento, and a species of morning glory. Development of new GM crops for food is an active field of research in Chinese institutions.

In March 2002, China introduced biosafety rules that demanded strict labelling, extensive documentation, and government approval for food shipments. Under these new rules, all soybean shipments from the United States were briefly interrupted until interim safety certificates could be acquired.

In 2004, the Chinese Ministry of Agriculture announced its intention to assess the safety of GM rice lines developed by Chinese institutions for insect, disease, and herbicide resistance. With government approval, the crops were planted in 2006.

Agriculture officials from developing and other economically disadvantaged nations are receiving training courses on GMO at the American Agriculture Department, with instruction in the WTO rules on GM products and benefits of biotechnology. United States industry groups are also providing "technical assistance" to fund initiatives that promote "science-based and transparent biotechnology regulations" in countries such as China.

20.7. HEALTH IMPLICATIONS OF GENETICALLY MODIFIED FOODS

We have examined the process involved in genetic modification of food, the areas of human health that could be affected, the safety and regulatory mechanisms which are in place, and the need for further research.

Findings

While assessing the possible hazards of GM foods to human health, the main issues which need to be considered are:

- Whether there are any inherent hazards in the genetic modification process itself.
- Whether the products (i.e., the food itself) might be harmful.
- Whether GM food given to animals, which are then eaten by people, could pose a hazard to human health.

There is also the question of whether GM technology could lead to environmental change, which had a secondary effect on human health. This is beyond the scope of this chapter, which considers the possibility of direct effects upon human health.

It is important to identify key aspects of health, which would need to be monitored if effects on human health were to arise.

The main measures to provide safeguards against any real or hypothetical risks are:

- Rigorous pre-market assessment of safety.
- Research to improve the understanding of the science of genetic modification of food.
- Health surveillance to provide reassurance against any unexpected adverse effects on health.

Many of the issues raised by foods resulting from genetic modification are equally applicable to foods produced by conventional means. For example, potential nutritional imbalances or allergic effects could occur from either type of food.

- There is no current evidence to suggest that the GM technologies used to produce food are inherently harmful.
- We are reassured by the precautionary nature and rigor of the current procedures used to assess the safety of individual GM foods. This process could be strengthened by the development of a health surveillance system.
- Nevertheless, nothing can be absolutely certain in the field of rapid scientific and technological development. Genetic modification is a young science, and there is a need to keep a close watch on developments and continue to fund research to improve scientific understanding in this area.
- The recent steps to improve the openness of the regulatory procedures to public scrutiny are welcomed. This would encourage further such moves to help to inform public debate on the issues related to the health implications of GM foods.

20.8. STEPS FOR GM FOOD TECHNOLOGY REGULATIONS

Tracking Research and Acting on New Evidence

Government advisory bodies should continue to closely monitor development in scientific knowledge and regulation on an international basis, and provide advice on any fresh action which they consider necessary.

Promoting High Standards of Regulation

The United Kingdom's current system of regulation of GM food technology and other novel foods is rigorous. The government should offer its expertise and use its influence to promote high standards of regulation internationally.

Need for a Continuing Research Strategy

Government should continue to fund research to improve scientific knowledge and to fill gaps in current knowledge. The government should invite the Medical Research Council and other major research bodies to participate in the further development of this research strategy. Before any new research is acted upon by the government, it must be seen through the standard peer review process to ensure that it has scientific credibility. Government's own response to new data should be made in line with the Guidelines on the Use of Scientific Advice in Policy Making to allow the full scientific merits of new research to be assessed.

Instituting Population Health Surveillance

The development of robust population health surveillance in relation to consumption of GM foods is essential to ensure that the government is able to respond rapidly if any unexpected effects occur. The advisory committee on novel foods and processes and the medical research council are already discussing how this might be done. As part of this, consideration also needs to be given to the establishment of a national surveillance unit to monitor population health aspects of genetically modified and other types of novel foods. Surveillance could be used to examine trends over time to detect any early changes in the incidence of adverse health outcomes, while recognizing the difficulties in establishing causal relationships.

Antibiotic Resistant Marker Genes

The use of alternatives to antibiotic resistant genes as a part of the GM process is already stated a good practice by the Advisory Committee on Novel Foods and Processes (ACNFP). Those who are developing foods using genetic modification should be encouraged to phase out the uses of antibiotic resistance marker genes as soon as is feasible.

The Safety of Foods Obtained using Genetic Modification Techniques.

Of course, different types of GM food may raise different theoretical concerns, and the safety of any particular GM food needs to be considered on a case-by-case basis. Working from first principles, an assessment of the theoretical risks to human health must take account of the nature of the new technology and how it could adversely affect the human health. It must also take into account, in general terms, the types of human disease, which can occur, and whether they are more likely as a result of these developments.

Genetic Modification Process and their Affect on Human Health

The theoretical health implications arising from the use of new technologies to manipulate genetic material are as follows:

* The inserted gene may itself have adverse effects.
* The inserted gene may code for a protein that is toxic to human beings or produces an allergic reaction.
* The inserted gene may alter the way existing genes in a plant or animal express themselves, which may in turn increase the production of existing toxins or switch on the production of previously silent genes.
* The inserted gene may alter the behavior of a microorganism, which is carrying it, to make it potentially harmful.
* The inserted gene may be transferred from one microorganism to other microorganism in the human gut or respiratory tract, or to animals or humans.
* The consumption of a GM microorganism may alter the balance of existing microorganisms in the human gut.

Chances of GM Affecting Human Health

People are constantly exposed to foreign DNA from the food they eat, the microorganisms in the environment, skin, and in their digestive and respiratory tracts. DNA itself is not a toxic chemical, and consumption of this chemical does not, therefore, have direct toxic effects. In addition, genetic modification results in the transfer of only single genes or small groups of genes, which are well characterized and whose function is understood. Plants typically contain 20-40,000 genes, and the function of the majority of these genes is not yet understood. Traditional plant breeding increases the human exposure to some of the products of these genes in a random way that does not first involve their isolation and definition. The random nature of conventional plant breeding has produced potentially harmful products on a number of occasions.

Experiments have shown how difficult it is to introduce genes into human cells. For example, attempts to introduce genes into human cells in the body to replace defective genes, such as those leading to cystic fibrosis, have met with very limited success, even when conditions for transfer are optimised. This would tend to support the view that DNA from GM foods is unlikely to enter the human cells.

The human intestinal tract is an efficient digestive system, and DNA is rapidly broken down under normal conditions into pieces too small to be functional. Thus, intact foreign DNA is not thought to be available for transfer into human cells, though there is a remote possibility that DNA fragments may be taken up by bacteria in the gut. There is some recent evidence from one group of researchers to show the possible DNA uptake. Under certain conditions, direct feeding of free DNA (obtained from bacteriophages) to experimental mice result in some pieces of this DNA being taken up by the cells in the mouse intestine and other tissues. This has so far not been reported for other DNA sources such as foods, and the significance is unclear.

There is an evidence to support the exchange of genes between bacteria in the environment either by direct transformation or via natural vectors such as plasmids. There is also an evidence to support the transfer of free DNA (for example, in soil as a result of the breakdown of plant material) into the bacteria in the environment, and some recent evidence has shown the transfer of a marker gene used in genetically modified sugar beet into other plants. However, for this transfer to occur, the free DNA has to be relatively stable and persistent. In addition, the recipient bacterium needs to be able to take up the DNA, which then has to be integrated into its own DNA and expressed. For this transfer to have any possible health implications, this integration would need to result in a new form of the bacteria that is stable and can survive in the environment. The success of DNA transfer and survival of the recipient bacteria is dependent on environmental conditions such as temperature, pH, and pressure.

There are many natural factors that reduce the chances of successful gene transfer, including the breakdown of free DNA, the fragmentation of any foreign DNA that enters cells by protective enzymes, and host immune defence systems that recognize and destroy the invading bacteria.

Non-GM microorganisms have been used in agriculture in pathogen control and as a means of increasing nitrogen fixation, without any apparent adverse human health consequences. Data from field releases of GM microorganisms have not shown any evidence of transfer of selective marker genes from the modified bacteria, in which the gene construct was inserted in a stable way, to the bacteria in the environment. However, there is an evidence to support the transfer of DNA from plasmids under laboratory conditions. Studies of the rate of transfer of inserted marker genes from GM plants into soil microorganisms have shown that this happens only with a frequency of less than 1×10^{-13} (one in ten million million) under optimised laboratory conditions. The frequencies are less than 1×10^{-16} (one in a ten thousand million million) in field conditions.

Comparable Concerns of Non-GM Foods

Some of the issues raised in connection with GM foods are equally applicable to the foods produced by conventional means, and there are a number of examples of health concerns arising from the traditional plant breeding. These include the Lenape potato (increased solanine levels), vegetable squashes (increased levels of cucurbatin), and celery (increased levels of psoralens). These were not detected until the product was close to release into the market, as none of these developments were required to be subjected to a safety assessment.

20.9. POSSIBLE HEALTH OUTCOMES

Communicable Disease

Any consideration of the impact on communicable diseases must take account of: other likelihood of organisms producing more serious effects than they would have previously, other emergence of organisms which are more resistant to antibiotics, and other creation of new infective agents. Existing plant pathogens, such as viruses, normally only infect a limited number of plants and the metabolic processes in plants are typically very different to those in animals. This reduces the likelihood that such pathogens could 'jump' from one species to another.

Bacterial pathogens have a specialized lifestyle and they need to possess many properties to allow them to invade a host and to reproduce there. The safety evaluation of any GM microorganism that would be consumed in a live form (for example, in a yoghurt culture) includes a detailed evaluation of its ability to cause human infection. If the GM microorganism was to be used to produce food ingredients, such as enzymes or defined chemicals, the safety evaluation would include an evidence to show that there was no DNA or novel protein that might have health implications in the final food product. There is a theoretical possibility that a novel strand of DNA could be generated in GM plants from the promoters derived from plant viruses, such as the cauliflower mosaic virus, and that such a sequence could then transform a pathogen to express a novel virulence factor. However, given the widespread natural occurrence of this virus in many vegetables, it is more likely that pathogens could evolve novel virulence factors more easily by an exposure to the native virus.

Antibiotic Resistance

The possibility that antibiotic resistant genes might be transferred from GM organisms needs particular consideration. These genes were often used in the early years of the development of the genetic modification technology as 'selective markers'. However, the use of alternative marker systems, or subsequent deletion of the antibiotic resistance gene, is now becoming more common.

Clearly, if a GM microorganism was to be eaten in a live form, it would be unacceptable for it to contain an antibiotic resistant gene. However, the GM microorganism could be used to produce a food ingredient (such as an enzyme). In such a case, the main consideration would be the level of DNA present in the final food ingredient. It would also be important to consider whether that DNA was still functional and was likely to transfer and become active in gut microorganisms.

Several GM plant varieties contain the marker that codes for resistance to the antibiotic kanamycin linked with a plant promoter. This marker system is used to select those plants that have been successfully modified. Such genes would not confer additional kanamycin resistance on bacteria, because they are not linked to a bacterial promoter. However, in some crops submitted for approval, an ampicillin resistant gene linked to a bacterial promoter was used. When the Advisory Committee on Novel Foods and Processes (ACNFP) assessed such crops, it did not recommend their approval because of the small risk of transfer of resistance to an important clinical antibiotic.

It is considered that the transfer of complete antibiotic resistant marker genes from plant material into the bacteria present in the human gut in a functional form is very unlikely, but it cannot be ruled out. This needs to be considered in the context of other causes of the development of antibiotic resistance, such as was recognized by the House of Lords Select Committee on Antimicrobial Agents.

Antibiotic marker genes serve no useful purpose in the final modified plant. Indeed, it is now possible to remove such a marker after the initial multiplication step in bacteria, but before the novel DNA is introduced into the host plant. Therefore, the ACNFP recommends the removal of such intermediary marker genes after the initial modification step as 'best practice'.

Non-Communicable Diseases, Including Chronic Disease and Foetal Abnormalities

Some diseases are not caused by infection and are often referred to as chronic diseases. Most forms of cancer fall into this category, as do diseases like diabetes mellitus, heart disease, and arthritis. A proper health assessment of GM foods must examine any likelihood of the incidence of these diseases being increased.

Many chronic diseases have a genetic component, i.e., they relate to the genetic make-up of an individual, though other factors are also involved, including diet, environmental exposure to chemicals, and/or radiation and lifestyle factors. It is unlikely that the genetic component could be influenced by the consumption of GM foods, as the evidence suggests that DNA from foods is not likely to be incorporated into the human cells.

Some conventional plants produce chemicals as a defence mechanism against attack by insects, or as protection against adverse conditions, such as the glycoalkaloids produced in potatoes. Some of these naturally occurring chemicals may be harmful to humans, and can cause cancer and foetal abnormalities in animals. It is for this reason that those involved in the preparation of food must take care to avoid green potatoes and to boil red kidney beans.

With the increased awareness of genetically modified food, people need to consider the genetics characters of the host plant. The history of any host plant that is being genetically modified needs to be evaluated. This would include considering whether the production of toxins known to be associated with the plant or its close relatives was increased. Traditional plant breeding could also result in the changes in toxin-producing potential, and this needs to be considered in both types of plant breeding. The incidence of some chronic diseases may be influenced by the nutritional factors, and they are dealt with in the following section.

Effects on Nutritional Imbalance

The incidence of many human diseases is dependent on a number of risk factors, including dietary variables such as fat intake and antioxidant levels in food. Genetic modification of an organism used for food may result in the composition of the final food product being different from that of the conventional food it would replace. This may be the intended consequence of the modification (for example, altered starch composition in potatoes, altered levels of beneficial nutrients such as antioxidants in fruit and vegetables, or altered levels of fatty acids in oils from oilseed crops), or it may be unintended.

The safety evaluation of all GM foods includes the consideration of any possible nutritional effects of the novel food. The cumulative effect of insignificant changes in the composition of the overall diet needs to be considered, especially for the groups of population such as infants, whose diets are derived from a limited number of food items.

Immune Response and its Modification

Some chemicals can alter the immune responsiveness by either increasing it, leading to allergy, or decreasing it. This possibility needs to be considered in the assessment of any new GM food. The insertion of genes that code for novel proteins not normally present in traditional food products may result in increased allergic reactions in some consumers. The allergenic potential of the modified food product is evaluated as a part of the overall food safety assessment, particularly if either the source of the inserted genes, or the host are known to cause allergy. This includes comparing the structure of the gene products with the known allergens.

It is also important to consider how functional the protein gene product would be in the conformed food (after processing and/or cooking), the level at which it might be present, and the likelihood that it would resist the digestive process in the gut. Assessment of potential allergy is complicated by the lack of suitable animal models that can be used on a routine basis in safety evaluations, although research is underway to find mechanisms to identify proteins that are likely to cause an allergic reaction. It has been suggested from the results of the work by Dr. Pusztai on GM potatoes that consumption of genetically modified food may result in depression of the immune system. The mechanisms by which this could occur are unclear. This work has been reviewed by the Royal Society, who concluded, on the basis of the information made available to them, that the work appeared to be flawed in many aspects of the design, execution and analysis, and no conclusions could be drawn from it.

They found no convincing evidence of the adverse effects of the GM potatoes studied. They also concluded that it would be unjustifiable to draw any general conclusion about the safety of GM foods, in general, from the results of studies, however well conducted, on one particular product modified by the insertion of one particular gene by one particular method (see also GM potato below).

Indirect Effects

It is possible that indirect effects on human health may arise due to the effects of GM organisms on the wider environment, and such an assessment is made a part of the evaluation of the release of GMOs to the environment under the Deliberate Release Directive.

In addition, the possible consequences for human health by the consumption of GM feed products by farm animals are subject to the same rigorous assessment procedures as GM foods. The genetic modification of plants to introduce herbicide resistance or insect tolerance traits is intended to reduce the overall amounts of herbicidal and insecticidal chemicals sprayed onto the plants. This may result in final food products with lower concentrations of chemical residues.

Particular Examples of Health Concerns

Some specific health worries have been raised in relation to the use of genetic technology in food production, and these are discussed below.

Tryptophan

It has been argued that genetic modification was responsible for several deaths in the USA. Contaminated tryptophan, a food supplement, was implicated in the human disease known as Eosinophilia Myalgia Syndrome (EMS). The particular tryptophan involved was produced by the fermentation involving a GM bacterium. The contamination was linked to 37 deaths in the late 1980s in the US. Following an in-depth investigation, the US Food and Drugs Administration could not find any evidence to suggest that the contaminant was produced as a direct consequence of the genetic modification process.

An investigation reported in the New England Journal of Medicine in 1990 identified an association between the EMS cases and a reduced level of carbon in a purification step in the production of the supplement involved, in addition to the genetic change in the bacterium strain. In a 1992 report, the US Department of Health and Human Services noted that 3-5% of the EMS cases had not been definitively linked to the supplement involved, and at least eight cases were linked with the tryptophan obtained from ordinary plant sources. The report also noted that the cases of EMS had been occurring prior to the 1989-1990 epidemic.

Recent United States Food and Drug Administration reports have found the impurity linked with the development of EMS in a number of both synthetic and natural versions of the tryptophan on sale as supplements for insomnia. It is, therefore, inappropriate to conclude that the cases of EMS were only linked to tryptophan produced by GM bacteria.

This case illustrates the importance of strict quality control monitoring for all food products. Such an information is an essential part of the safety assessment procedure for GM foods.

GM potatoes

A work was undertaken at the Rowett Institute by Dr. Pusztai, in which potatoes were genetically modified to express an insecticidal lectin protein. The results have been widely reported to suggest that the process of genetic modification itself may be harmful for health. The results of this work have been reviewed by the Royal Society, who concluded, on the basis of the information made available to them, that the work appeared to be flawed in many aspects of the design, execution, and analysis, and no conclusions could be drawn from it.

They found no convincing evidence of the adverse effects of the GM potatoes studied. They also concluded that it would be unjustifiable to draw any general conclusions about the safety of GM foods, in general, from the results of studies, however well conducted, on one particular product modified by the insertion of one particular gene by one particular method. It should be noted that these potatoes were not intended for marketing and had not been submitted for marketing approval. If they had been submitted, they would have been subject to a detailed safety assessment,

concentrating on the safety implications of the expression of the lectin gene, as some lectins are well known to exert toxic effects on animals and humans.

Ladybirds

Some work has been reported, which indicates the adverse effects in ladybirds eating aphids that have been colonising potatoes genetically modified to express an insecticidal lectin protein. The evidence has been considered by the Advisory Committee on Releases to the Environment (ACRE), which concluded that the findings are consistent with the known toxic properties of the lectin. This work underlines the need to conduct a thorough testing of GM crops for indirect effects on non-target organisms.

Milk from cows treated with BST

Bovine somatotrophin (BST) is a hormone that is naturally produced by all cows. It stimulates milk production and minute quantities are present in the milk of all cows, particularly in high yielding dairy cows. The production of BST using genetic modification is widely used in the USA to increase milk production. In the UK, BST falls within the definition of a medicinal product and it cannot be marketed without a marketing authorization. Following evaluation by an expert committee in Europe, the EC has accepted that milk from treated cows may contain increased amounts of a growth factor, which may be associated with some adverse effects on humans.

For this reason and because of the animal welfare issues, there is a moratorium on the marketing and use of the product in the Community that expires at the end of December 1999. This is to allow for practical tests to be carried out to obtain further scientific data needed to enable the Council of Ministers to make a final decision. The UK Veterinary Products Committee has set up a working group to advise on the safety of BST for the target animal, and for the humans consuming milk from treated animals.

How Safety is Assessed using a Comparative Approach?

The safety of GM foods is assessed in comparison with the foods that they will replace. This concept of substantial equivalence developed by the World Health Organization and the Organization for Economic Co-operation and Development is used extensively as a tool in the process of the assessment of the safety of GM foods by expert assessment bodies worldwide. The fact that a GM food may be substantially equivalent to a conventional one does not, however, mean that it is 'safe'. Nor does it remove the need for a thorough assessment to be carried out to ensure that this is so before it can be allowed in the market.

In this assessment method, the GM food is compared to its conventional counterpart, and consideration is given to both the intentional effects of the modification and also to any possible unintended secondary effects. This comparison involves the assessment of a wide range of information. This includes agronomic data derived over a number of generations (such as crop height, yield, flowering pattern, disease resistance and climatic tolerance), a detailed compositional information on nutrients (proteins, fats, carbohydrates, vitamins and minerals), and possible toxicants in both the plant and any derived food product. This comparison can have three possible conclusions:

- The GM food or food ingredient is substantially equivalent to the conventional counterpart in all agronomic, compositional, and toxicological respects.
- The GM food or food ingredient is substantially equivalent to the conventional counterpart except for a few clearly defined differences.
- The GM food or food ingredient is not substantially equivalent because the differences cannot be defined or because no counterpart exists.

In the first and second categories above, a safety assessment is carried out with particular attention being focused on any differences between the GM food or food ingredient and its conventional counterpart. Where a food is not substantially equivalent, it does not mean that the food is unsafe, but extensive data would need to be provided to demonstrate its safety. A separate technical note describing the history of how the current UK and European safety assessment procedures was developed and produced for the information.

Information Requirements for Safety Evaluations

As a starting point, the safety assessment of a GM food involves a careful assessment of the following information:

- The amounts of the GM food that people are likely to consume, including both average and extreme consumption.
- A detailed description of what the food is and how it is produced.
- A history of any possible adverse health effects linked to the organism being modified.
- A detailed description of the genetic modification process.
- An evaluation of any possible nutritional effects of the modified food.
- An evaluation of any toxicological effects of the modified food.
- An evaluation of any adverse microbiological effects of the modified food.
- An evaluation of any data on people eating the modified foods under controlled conditions.

International Perspective

The safety considerations of GM food have been considered in many other countries and by international organizations such as the World Health Organization. Many GM foods are now being marketed in other countries such as the USA and Canada, following approval by their regulatory authorities. However these products still need to be assessed for safety under the EU regulatory framework before they can be marketed in Europe.

After such approvals have been given, the World Trade Organization Sanitary and Phytosanitary Agreement rules prevent countries from taking action to restrict the import of such products into their markets, unless an evidence of harm subsequently comes to light. The US and Canadian systems place the responsibility for ensuring safety on the GM food producer, and such foods do not, therefore, require prior assessment by the regulatory authorities before being allowed onto the market, as is the case in the EC. In this regard, the EC system may be considered to take a more precautionary approach for the approval of these materials. In July, 1998 the US Department of Agriculture and the Canadian Food Inspection Agency reached a bilateral agreement on the data required for the molecular genetic characterization of GM plants. This agreement was seen as a first step towards the harmonization of such data requirements.

Population Health Surveillance

Although there is a rigorous pre-market safety assessment of GM foods, no systematic population surveillance system currently exists to detect any effects on health. This is a flaw in the present system. Surveillance of health at a population level is essential, and there are existing databases covering key health outcomes, such as cancer, foetal abnormality, and mortality that could be used in this respect.

An ACNFP Working Group has been considering the feasibility of setting up a population surveillance system to provide additional reassurance on the long term safety of those GM foods that have been approved for marketing. This could involve case-by-case studies of individual GM foods or more general epidemiological studies to evaluate the unanticipated consequences for health (adverse and beneficial) of GM foods more generally. The Medical Research Council is also

setting up an expert group to consider the feasibility of large scale epidemiological studies to evaluate the potential health effects (both beneficial and deleterious) of GM foods more widely. We await their conclusions, but see merit in considering establishing a new unit to act as a surveillance, at population level, of any health impact of genetically modified and other types of novel foods. The unit would act as an early warning of serious problems, and would also continuously monitor the world's health literature and advise on the significance of any new research when it becomes available.

20.10. WORLD HEALTH ORGANIZATION AND ITS FUTURE PLANS

The application of modern biotechnology in food production presents new opportunities and challenges for human health. The potential benefits to the public health sector include altering the nutrient content of foods, decreasing their allergenic potential, and improving the efficiency of food production systems. On the other hand, the potential effects on human health due the consumption of food produced through genetic modification must be carefully examined. Modern biotechnology must be thoroughly evaluated if it is to bring about a true improvement in our way of producing food. The tendency to explain all concerns in this area as a problem of perceptions originating in the consumers incapacity to understand is simplistic. Instead, the consumers right to be concerned as well as to be informed should be acknowledged.

The future developments in this area will focus WHO work in four major areas:

Establishing Scientific Safety Assessment Frameworks Based on Sound Science

There is a need to facilitate the establishment of sound safety and risk assessment frameworks for foods derived from modern biotechnology. In addition to providing input to the codex inter-governmental task force on foods derived from biotechnology, WHO will develop the principles and guidelines for addressing emerging issues based on sound science.

Standardising Methods for Nutritional Aspects in Safety Assessments of Food Derived from Modern Biotechnology

There is a need for an increased focus on nutritional assessments of new foods derived from modern biotechnology, and from novel foods in general with specific nutritional traits. These types of considerations are already a part of the FAO/WHO expert consultations on food derived from modern biotechnology, but a strengthened effort is needed. This effort should be coordinated with efforts in other intergovernmental fora, such as the OECD Task Force on Novel Foods.

Linking Risk Assessments to Risk Management and Communication

Effective mechanisms and approaches are urgently needed at both national and international level to bridge the outcomes of risk assessments into risk management and risk communication efforts. New thinking towards involving consumers and other interested parties already at the planning stages of risk communication efforts need to be developed. And the experience gained in this area could be used in other aspects of food safety.

Broader Perspective of Health and Development Policy

The development of new foods through modern biotechnology has not always been perceived to be guided by a perspective of "public good". New products with potential health or production benefits in developing countries could change this. Such products will need a more holistic assessment, looking into all aspects of such production changes (health benefit, nutrition, safety, development and socioeconomical issues, etc.). For the preparation of this work, there is a need for WHO to strengthen its collaboration with inter-sectoral partners. Such efforts would not only entail technical issues related to areas such as production efficiency and safety assessments, but also more general issues of how foods derived from modern biotechnology could be made useful in addressing the needs of developing countries.

Dr. Chris Williams attempted to build a comprehensive picture of the evidence concerning the health effects of industrial hazards, and in particular how these are damaging people's brains (Williams 1997). He found that science has a great difficulty in rising to the challenge due to several reasons:

- Science generally analyses the effects of single substances, while many of the most serious problems involve interactions between different industrial hazards, such as lead and fluoride (when added to water, fluoride increases the absorption of lead from pipes). Few comprehensive studies of multisubstance hazards have been undertaken due to the lack of funding.
- The complexity of the issues means that the scientists, who undertake such studies, risk arriving at inconclusive results.
- Clear results in a more narrow field of study bring more recognition.
- There are often long time lags and large distances between cause and effect.
- Scientific evidence of hazard or harm is frequently ignored, suppressed, or attacked.
- Scientists themselves are unwilling to under-take such studies due to the above factors, and the effect that these might have on their careers.

These problems also affect the risks associated with GM foods. The lack of research about diffuse effects makes our prevailing 'single-substance science' appear an inadequate survival skill. Given the central role of science in creating, regulating, tracking, and controlling technologies, science itself needs to look to its own practices and commitments if it has to play a full and socially responsible role (Williams 1997 and 1998).

20.11. ETHICAL ISSUES ASSOCIATED WITH THE CONSUMPTION OF GM FOOD

Since the release of first GMO, Flavr Savr, in USA in 1994, genetically modified food has started appearing in the market. Such GM foods are referred to as "Crop plants or animals created for human or animal consumption that have been modified in the laboratory to enhance the desired traits or improved nutritional content". Later, more than 80 crops in 40 countries and 6 continents were involved in the production of GM foods. The major corn, cotton, canola, banana, water melon, soybean, potato, tomato, papaya, rice, sorghum, tobacco for greater yields, longer shelf lives and stronger resistance to disease and insects. Around 25,000 field led trials were conducted all over the world. First GM food, Maize (Resistance to European corn borer) and Soybean (Resistance to Glyphosate), came into the market at full strength. The USA is a leader in the GM technology. In 2000 99% of global transgenic was grown in major players of GM technology such as USA (68%), Argentina (23%), Canada (7%) and China (1%).

The important ethical issues associated with the consumption of GM food are summarized as follows:

- Unnatural food
- Vegetarians (Vegans) and some religious groups against their practice
- No barrier between animal and plants
- Breaking natural barriers
- Spread of diseases across species barriers
- Creation of new allergens or toxins.
- Emanated publicity to GM food by hiding real gene function, position effect
- Real safety is ignored in front of economic gain, misinformation, lack of information and ethics, responsibilities of government, industries to protect public.
- Released without proper risk assessment. Transfer of new and unidentified proteins from one food to another have a potential to set off an allergenic reaction.
- Removal of important. Food elements which is considered all not important. But having important. Functions—unknown, crucial qualities: cancer inhibiting and immunologic abilities
- Antibiotic marker gene, absorbed from the digestive tract-blood and then into the bacteria in blood and instestine, thus affects symbiosis between the humans and bacteria.
- Stress on host organisms and its effects on environment (violation of intrinsic values)
- Changes in physiology, biochemistry, cytology by consumption of such GM foods
- Unintended transfer of transgenes through cross pollination, unknown effect on soil microorganisms, and loss of flora and fauna biodiversity
- Misleading, ignorance, and/or corruption at the top level in BT company or government
- Uncertainty in technology and its effects
- Genes are a part of an extremely complex and interconnected network, and are constantly reacting to their environment. Thus, the changes can lead to unexpected results
- Tampering with nature by mixing genes beyond the species boundary
- Labeling not mandatory
- Objections to consuming animal genes or its products in plants and vice versa

GM Foods and their Effect on Rural Sustainability

Research in Europe, Brazil and the Caribbean has led Terry Marsden and colleagues to the conclusion that GM technology will lead to a further round of intensification in the food industry (Marsden and Drummond 1999a and Marsden *et al.,* 1999b). If GM crops lead to the reductions in costs in food supply chains that are promised, this will lead to further economies of scale in food enterprises. In other words, the existing 'technological treadmill' intensifies yet further, leading to even greater concentration of production amongst smaller groups of large-scale growers and manufacturers.

GM technologies are thus likely, particularly in the context of the WTO millennium round, to further speed up the structural change in agriculture and food supply, making it more difficult for smaller producers to stay on the land. GM foods could, therefore, provide another cause of inequality, unemployment and depopulation in rural areas around the world. Many of these structural changes would be irreversible in generational terms.

In addition, if reductions in costs lead to lower costs to the consumer, it is also likely that alternative food supply chains, such as those for organic food, will be marginalised on the basis of price. The true costs of industrial foods incorporating all of the social and environmental effects-are not passed onto the consumer, this is also likely to be the case with GM products.

In July 1999, the government announced that it was setting up two new strategic committees on GM food to consider cross-cutting issues: the Human Genetics Commission and the Agriculture and Environment Biotechnology Commission. While these committees and their anticipated broader membership are a positive step, many of the assumptions and working practices at the heart of the regulation of GM food remain intact.

This means that many shortcomings remain, including:

- How individual aspects of risk are characterized and measured. For example, what assumptions are adopted in operating environments, working practices, and regulatory compliance? Assumptions about pollen distribution and the size of ' buffer zones' is one example of a current controversial issue.

- The implicit priorities adopted in comparing different aspects of risk—how are health aspects compared to environmental effects, involuntary versus voluntary risks, or risks to different groups of people? How are issues of geographical scope and time horizons dealt with?

- The treatment of deep uncertainties and 'ignorance' about the possibility of entirely unforeseen effects.

GM Food and its Need for the World

There are strongly varying opinions about the contribution that GM foods will make to addressing the food needs of the poor and the global demand for food more generally. As Barbara Adam writes: 'Like previous technological innovations, it holds out the promise of cornucopia: the end to food shortages and world hunger, poverty and disease, weather and season dependence'.

At present, there is a lack of conclusive data about GM crop yields. Several recent US studies have pointed to the variability of yields across different crops and regions. It is therefore difficult at this stage to form a solid conclusion about the contribution that GM crops might make to increase world food production. Terry Marsden suspects that while GM food technologies may increase the quantity of foods available, and in some cases might improve the quality in relative terms, they will not reduce food poverty and scarcity. Indeed, he expects the current uneven development of food production and consumption to intensify both between the North and South and within each region. The issue is less about food production and more about food distribution.

Tim Dyson, in a Fellowship that examined the trends in global food production and consumption, came to the conclusion that world population growth is not outpacing food production. However, it is not going to be easy to raise agricultural production to meet world population growth. So while the world could probably feed itself without GM foods, they will nevertheless help.

GM food is already helping to feed the poor. Sizeable areas of land in sub-Saharan Africa are planted with rice varieties that have been made more resistant to blight through the application of GM techniques. Similar approaches could be applied to the problems of drought resistance, salt tolerance, some crop diseases, and to addressing vitamin deficiencies.

However, clearly the interests of the big GM companies are in making profits from farmers who can pay. So the task of developing 'better' GM crops for poor farmers will fall mainly to scientists working in international food research stations and universities. Yet, international investment in developing country agricultural research has been declining. Aid could therefore usefully be channelled to supporting pro-poor GM technology development, while at the same time linking this to a wider vision for sustainable agriculture, poverty elimination and sustainable livelihoods.

GM Time Machine

A time-based perspective can give useful insights into the logic behind, but also the potential hazardous effects of genetic engineering. This is the conclusion of a Fellowship undertaken by Barbara Adam as a part of the Global Environmental Change Programme.

In industrial systems, time is money. Speed is tied to efficiency because of competition and the need for returns on investments. The control and compression of time is central to the creation of profit. By contrast, in nature everything has its own time, rhythm and season. This natural time is a barrier to productivity and profit (Adam 1998).

GM food technology represents a powerful new way of controlling time in agricultural production. It promises:

- instantaneous reproductive change where before this had to be achieved from one generation to the next;
- speeding up of ripening and maturing processes;

- control of growth and decay for just-in-time production, retailing and consumption; and
- increased profits for a few companies through control over the resources of production and reproduction through patenting.

Genetic engineering is possible only because of the shared genetic evolutionary history of all organisms on Earth. This is what makes the transfer of genes from one breed to another possible. And it means that genetic engineering operates in a timescale from the beginning to the end of time.

Time also sheds light on the politics of GM food. With a five year mandate, politicians have only partial legitimacy to make decisions about the issues such as the release of GMOs, since the effects are potentially irreversible and infinite in time. Decisions on GM releases therefore need to be made more legitimate by a strong ethical debate across society, which includes substantial attention to the needs of future generations (Adam 1999).

SUMMARY

Entirely new types of plants and animals can be created by altering their genetic structures through the new technology of genetic engineering. This advanced biotechnology enables us to create plants and animals with the characteristics we want by cutting, joining and transferring genes between unrelated species genes from insects, animals and humans are added to crop plants to create unique qualities that we have never seen in the original plants. Genetic engineers and their proponents call the biotechnology a "miracle of science," saying the genetically modified foods can contribute to the reduction of hunger in the world and to transforming agriculture into a value added industry. Genetic engineering is certainly a field that will continue to provide new challenges and opportunities to us in the next century. Hopefully, genetically engineered foods may contribute to reducing world hunger. But to make our dream come true, researchers should continue their efforts to prove that they can create truly safe and useful foods through this new technology. The risk assessor and manager are strongly affected by the relevant legislation and guidelines which have established as a result of the legislation. Risk assessment process can be compared to a filter system and risk management to devices to control flow rate. Each layer of filtration material increases product purity but reduces output. Too few layers allow unwanted contaminants to pass, resulting in adverse effect on the product. Too many layers result in lowering the rate of flow, perhaps to the point where the process is not economically feasible. Balance between purity and flow rate is very important. The role of risk assessment is to determine the balance point. Risk management attempts to insure safety by use of procedural strategies without preventing the development of useful products.

Once the safety of genetically modified foods is fully proven, a revolutionary change can be expected in the global agricultural production, which help eliminate the world's food shortage problem and environmental pollution problem as well. Therefore, the risk management should be done in a way that it satisfies both scientists and consumers so that it promotes biotechnology and should not be a tool for hindering biotechnological research and development. If the cartagena protocol on biosafety enters into force, our domestic regulation will also come into effect, thus requiring some revision of current regulatory system. If current voluntary guidelines for environmental safety approval needs to be replaced by mandatory one, it is necessary to have grace period so that applicants will have sufficient time to prepare for the compliance of revised rule. How to regulate domestic GMOs being released into the environment for conducting field experiments is one of the important issues that should be addressed when revision of environmental safety rule is considered.

Therefore, distinction between LMO and LMO-FFP as described in the biosafety protocol should be considered in setting up of new regulations or in revising the existing rules in terms of full implementation of biosafety protocol. Adoption of an approach that can be used to determine what, if any, additional data should be required to assess the environmental risk associated with a LMO-FFP in an importing country is very important. In many cases, the data and information developed in the country where the product is grown, in combination with experience with the non-modified crop, will be adequate to conduct an environmental risk assessment in the importing country. Therefore, differentiated application of environment risk assessment process based on the cartagena biosafety protocol for the approval of LMO-FFPs and LMOs to be intentionally released into the environment is one of the important issue for the full implementation of biosafety protocol and facilitating the trade between nations.

Technologies for genetically modifying (GM) foods offer dramatic promise for meeting some areas of greatest challenge for the 21st century. Like all new technologies, they also poses some risks, both known and unknown. Controversies surrounding GM foods and crops commonly focus on human and environmental safety, labeling and consumer choice, intellectual property rights, ethics, food security, poverty reduction, and environmental conservation. On the horizon are bananas that produce human vaccines against infectious diseases such as hepatitis B; fish that mature more quickly; fruit and nut trees that yield years earlier, and plants that produce new plastics with unique properties. In 2000, countries that grew 99% of the global transgenic crops were the United States (68%), Argentina (23%), Canada (7%), and China (1%). Although growth is expected to plateau in industrialized countries, it is increasing in developing countries. The next decade will see exponential progress in GM product development as researchers gain increasing and unprecedented access to genomic resources that are applicable to organisms beyond the scope of individual projects.

EXERCISE

1. Define genetically modified organisms and their applications.
2. Write about the advantages of the genetic modifications.
3. What do you mean by genetic modifications? How does genetic modification proceeds to the development of better food varieties?
4. Discuss the techniques used for genetic modifications of living organisms for the purpose of food exploitation.
5. Define social, ethical and legal issues involved in the production of genetically modified organisms and their impact on future development.
6. Elucidate the negative effects of genetically modified organisms and their use as food and resources.
7. Depict the controversies associated with the production of genetically modified organisms and foods.
8. Critically comment on the comprehensive and effective systems that are needed to find the solutions to the risks associated with the genetic modification technology.
9. What do you mean by 'genetically modified food'? Write a note on the importance of the safety assessment of GM food.
10. How are potential risks associated with the GM food identified? What are their solutions?
11. Differentiate between the allergens and intoxicants that are produced by the products of genetic modification technology.
12. Give an account of the enviromental impact of genetic modification technology and their products that used in the ecosystem.
13. What is the impact of the genetically modified food on the health and environment of human beings?
14. How is quality of products affected by the genetic modification technology? What is its relevance to the food habits?
15. Explain waiver of informed consent that are generally followed for the clinical trials, and mention its merits in the research and testing.
16. Describe the criteria necessary for conducting genetic modification technology in plants in the present applied situation of the biotechnology.
17. Illustrate how genetic modification technologies in animals are used for the development and improvement of the animal derived foods?
18. What are the opinion of European Union on genetic modification technology and their products in the market?
19. How is the safety assessed for genetic modification technology and their end products for the human use?
20. Discuss the present status of the genetically modification tomato and its acceptance by the consumers in the world.
21. List out the ethical issues associated with the consumption of genetically modified food and their products in the different religious countries.
22. What role does the world health organization play in the protection of the human health from genetically modified food?
23. Does genetic modification technology give the complete solution to the malnutrition in the world?

21
CHAPTER

Labeling of Genetically Modified Foods and Crops

21.1. INTRODUCTION

Genetically modified (GM) plants and animals are approaching commercialization and widespread deployment in the world. Risk assessments supporting release applications have largely been based on assumptions that genetic modifications of plants and animals will not alter their behaviour or that of other organisms in the natural environment. These assumptions are made from limited information on actual levels of gene flow occurring between crops and wild species, and small scale experiments with GM plants or animals and untransformed plants or animals respectively. Large scale releases of GM plants occurring in North America and other countries provide some additional information on risks, but are not always relevant to other parts of the world. There is thus concern that risk assessments are based on limited experimental data that do not fully take account of the novelty of the transgenes, or the scale and scope of their ultimate commercial deployment in nature.

There is also concern at the large number of different releases that are being proposed in Europe and other countries. Information on the transgene interactions within and between GM plants and animals is extremely limited, as is information on the environmental impacts of multiple transformations in single plants and animals, many of which could arise unintentionally. The role of the government is to develop policy and legislation around the purpose, use, regulation and monitoring of food biotechnology. A major role of public health at the provincial and municipal levels can be to influence policy development in response to local concerns and needs, both rural and urban. Hence such genetically modified materials require a proper identification system.

21.2. OUTBURST OF GENETICALLY MODIFIED PLANTS AND ANIMALS

An additional concern is that GM plants and animals may require different agronomic management, or may have agricultural consequences that effect the environment, e.g., changes in agrochemical usage, effects on predators, effect on different organisms etc. Agricultural impacts are not always considered in environmental risk assessments, and yet agriculture is a significant component of the total European or world environment. A workshop in Cambridge, UK, in October 1997, sponsored by the European Science Foundation, brought together European scientists involved in environmental impact research, plant and animals breeders, and representatives of organizations involved in the regulation of GM plant and animals releases. They discussed the range of transformations and plant and animals species most likely to have environmental impacts. They agreed on a number of research priorities, and also agreed that research in Europe required coordinating and enhancing, so that scientific information could be collated and conclusions made more widely available to support risk assessments in European countries and elsewhere. Food biotechnology is a new, complex, controversial and rapidly growing science and industry. The work-group believes those it is a public health issue because:

- it affects the food supply of the entire population;
- it has a strong impact on the agriculture and environment; and
- it is a new technology with unpredictable consequences, and therefore requires reliable policies and methods for assessment, monitoring, and regulation.

The assessment of the impacts of genetically modified plants (AIGM) programme has been established to coordinate the activities of the principal research programmes in Europe; to enhance them by recruiting younger research personnel to study in them; and to encourage these research programmes to respond to the new research priorities identified by the programme. It publicises the results of the research through conferences and workshops to a wide range of audiences in Europe, particularly to countries with little experience with GM plants or animals and risk assessments. Members of this programme are available to give expert views on risk assessments and assist with the development of regulations based on sound scientific principles.

The importance of the agricultural sector in developing countries as a source of food, income, employment and often foreign exchange cannot be overstated **(Fig. 21.1)**. As much as good health, a productive and sustainable agricultural sector is critical to achieving economic growth and poverty reduction. About three quarters of the world's poor people live and work in rural areas. Apart from its direct role in sustaining incomes, and employment, the role of agriculture, and in particular technological change in agriculture, in stimulating overall economic growth has been much discussed by economists and policymakers. Raising productivity in agriculture can directly increase the income and employment levels of the majority of poor people dependent on agriculture. It can also help to reduce food prices (relatively or absolutely) for poor people in both rural and urban sectors.

Fig. 21.1. Genetically modified food product label indicating source fruits used to make sauce

Historically, agriculture has been seen, sometimes controversially, as a source of food, labour, and finance to supply a growing urban and industrial sector on which sustained growth in incomes will depend. Achieving this transition usually depends on achieving productivity increases if food prices do not rise, and stifle both industrial growth and poverty reduction. In developed countries, changes in technology and institutions in the agricultural sector are regarded as having been instrumental in the industrial revolution.

In developing countries, technical progress traditionally occurred through a process of on-farm experimentation, selection and adaptation of traditional landraces of crops. Subsequently, this was supplemented by purposive breeding of new varieties of crops, mainly through crossing varieties with desirable characteristics. This process of research was largely conducted in the public sector by national research institutes, supported by a network of international research institutes, for the last thirty years under the umbrella of the consultative group on international agricultural research (CGIAR). It was this network, which led to the Green Revolution of the 1960s, based initially on high yielding semi-dwarf varieties of rice and wheat. In spite of criticisms of its environmental and distributional impact, this technology is widely credited with having had a favourable impact on nutrition, employment and incomes, albeit mainly in the areas of developing countries capable of reasonably assured irrigation. Further breeding efforts have been tried, but with less success, to extend these technologies to new crops, and to rained and dry land areas.

While the public is not widely knowledgeable about biotechnology and genetically modified food, consumer awareness appears to be growing. When the pew initiative on food and biotechnology compared the findings of two polls, respectively conducted in January 2001 and June 2001, it found the consumer awareness of genetically modified foods sold in grocery stores had increased by 11% between the polls. At the same time, use of the technology is also increasing in American fields, with farmers planting more genetically modified

soybeans and cotton this year, according to a recent US Department of Agriculture survey. The government's report also states that currently 68% of soybean acreage, 69% of all cotton acreage, and 26 of total corn acreage planted is genetically modified.

FAO deals with environmental and food risk-assessment questions under a more general concept of "biosecurity", namely the application of sanitary and phytosanitary measures (SPS), and a methodology of risk analysis for food and agriculture, including fisheries and forestry. They include:

- The protection of animals or plants from pests, diseases, or disease-causing organisms.
- The protection of human or animal life against risks arising from additives, contaminants, residues toxins or disease-causing organisms in foods, beverages or feedstuffs.
- The protection of human life or health against risks arising from diseases carried by animals, plants or plant products, or from pests.
- The prevention or limitation of other damage from pests.

Moreover, FAO has specific standard-setting responsibilities that relate directly to GMOs under the Technical Barriers to Trade (TBT) and the Sanitary and Phytosanitary (SPS) Agreements of the World Trade Organization. There are several important international mechanisms in the context of GMOs.

Firstly, the cartagena protocol to the convention on biological diversity, which was adopted in January 2000. It will come into force after fifty ratifications and is the main international agreement in relation to GMOs ("LMOs", or "Living Modified Organisms", in the language of the protocol). It covers the transboundary movement, transit, handling, and use of all LMOs (except pharmaceuticals) that may have adverse effects on the conservation and sustainable use of biological diversity, taking also into account risks to human health. It allows for standard-setting in relation to the handling, transport, packaging, and identification of LMOs.

Secondly, food safety aspects of genetically modified organisms are, at international level, dealt with by the FAO/WHO *Codex Alimentarius*, which covers all aspects of food safety. The *Codex* is currently working on standards for risk assessment for labeling, and for several other food safety aspects of GMOs. The *Codex* standards are recognized by the SPS and TLC agreements.

Thirdly, the international plant protection convention (IPPC) has the objective of preventing the spread and introduction of plant and plant product pests, including weeds and other species that have indirect effects on both wild and cultivated plants, and to promote appropriate control measures. This also applies to risks associated with LMOs. The IPPC sets International Standards for Phytosanitary Measures (ISPMs), which are also recognized in the SPS agreement. It is also in the process of establishing practical cooperation with the CBD and its biosafety protocol. An IPPC expert working group met in September 2001, in coordination with experts from the CBD, to develop a detailed standard specification for an ISPM that identifies the plant pest risks associated with LMOs, and ways of assessing these risks.

The early work of genetic engineering occurred during the 1970's, when it was discovered that different DNA fragments could be spliced together to form a new functional hybrid DNA molecule, or recombinant DNA (rDNA). This process is now referred to interchangeably as "genetic engineering" (GE), "genetic modification" (GM), or "biotechnology". Genetically modified organisms (GMOs) are microbes, plants, or animals, whose cells contain and express the new rDNA molecule(s), or transgenes.

At first, work centered on GM microbial strains, which were commercially valuable for enhanced fermentation and the production of antibiotics, vaccines, insulin, amino acids, enzymes, rennet, etc. It involved major financial investment into research, coupled with the potential of highly profitable financial return. Precedent was set with the 1980 US Supreme Court decision to grant the first patent protection for a life form—in this case a GM bacterial strain. This set the stage for "intellectual property rights" over the research data as well as patents on the living products (microbes, plants, animals, seeds) of genetic engineering. Patenting of life forms is an on going source of debate internationally. The next stage of research involved the transfer of genes into the genomes of plant or animal. DNA is inserted into the nucleus of a plant cell by means of a "gene gun" (DNA fragments on gold or tungsten pellets) or via a bacterial vector. Numerous experiments are needed before the desired traits are created. This process may include an additional "marker gene" (although it has been strongly recommended that antibiotic resistance marker genes be discontinued). Using such new techniques, the first transgenic plant, an herbicide-resistant tobacco plant, was licenced in 1983. The number rose to 48 GM plants approved worldwide by 1997. United States of America, Argentina, and Canada are now the three largest growers of GM crops in the world. By now, Health Canada has issued letters indicating "no objection" to the use of about 45 GM food crops, textile crops and animal feeds. These include soybeans, cotton, corn, flax, canola, tomatoes, squash, and potatoes. Since soy, canola, corn, and their products are used so widely in the food system, an estimated 60% to 70% of the retail processed food in USA and Canada now contains GM ingredients. The same situations are also noticed in the number of European countries. So far, these plants have been genetically changed mostly for characteristics like insect resistance and herbicide tolerance, with no more than two or three added genes.

In the near future, many more types of GMOs (the so-called "second generation") will reach the commercial release stage; some with multiple, "stacked" gene transfer; some will address

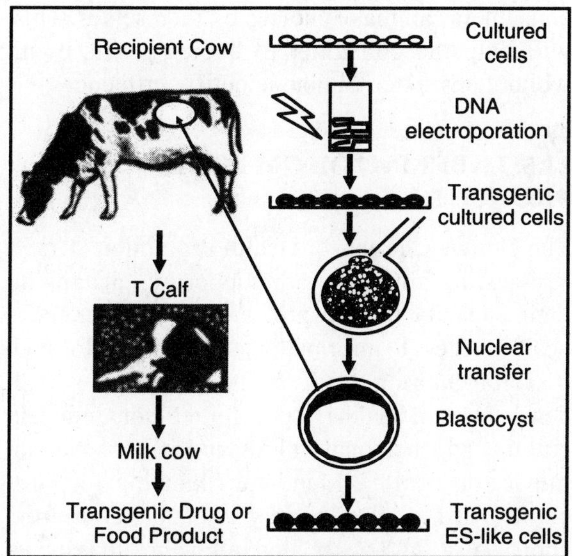

Fig. 21.2. Nuclear transfer: transgenic cell technology and end products in cow production for transgenic drugs or food

agricultural concerns (e.g., disease/pest/drought resistance; altered flowering or ripening time, fertility, susceptibility to spoilage, etc.). Others may alter micronutrient or phytochemical content of the plants. Transgenic (growth-enhanced) salmon and a few dozen other GM fish species have been developed. Research is well underway to improve methods of somatic cell nuclear transfer in livestock, so that more GM animals become a reality for both altered agricultural and nutritional traits **(Fig. 21.2)**.

To help clarify the issues in terms of public health, the following principles are useful. Originally, they were developed for a Health Canada policy review on the addition of vitamins and minerals to foods by a nationwide, multisectoral stakeholder consultation and an expert advisory panel. The food biotechnology workgroup has further modified them to take into account the unique characteristics of, and the issues surrounding GM foods. We believe that these principles should be the basis for policy development around the regulation of GM foods in Canada and other countries.

21.3. PUBLIC HEALTH PRINCIPLES REGARDING THE REGULATION OF GM FOODS

Principle 1

Policies and regulations must ensure that approval is granted only to those GM foods, which maintain and improve the quality of the food supply in a manner that is consistent with public health priorities and goals.

Principle 2

The policies related to GM foods must ensure that those foods do not result in health hazards.

Principle 3

The policies must ensure that decisions regarding approval of GM foods are based on on-going, rigorous, peer-reviewed evidence from multidisciplinary sources (including life sciences, environmental science, toxicology, agriculture, ethics, health law, etc.).

Principle 4

The "precautionary principle" must be applied in the regulation of GM foods.

Principle 5

The policies must help to prevent practices that may mislead, deceive or confuse the consumer, and promote practices, which facilitate consumer choice.

Principle 6

The policies must be responsive to the varied demographics, food habits, and expressed concerns of the Canadian population.

21.4. DEFINITION

Labeling of genetically modified foods and crops is one of the most controversial issues in agriculture biotechnology today. Currently, the United States does not require GM foods to be labeled, unless the new food product is substantially different from the original product (FDA).

Labeling means displaying information about the product on packaging materials in a distinct manner. It is necessary that the producer of the food product must reveal on a label all facts about material, with respect to consequences which may result from the use.

One of the primary duties of consumers is to know or acquire the knowledge about the source of any materials he or she is buying. Whatever the consumers are buying, they should have an idea about its impact on the health of their family including young childrens and aged parents. Even is there is a remote chance of health consequences of these modified materials, especially food, and health prouducts, consumer has the right to say "no" to such products. The consumers always prefer materials that do not pose any danger to any age groups, and the probability of the health impact is exceptionally nil or not having any type of variations in the quality of the materials at any conditions. This type of thinking and argument always made by the any type of the materials of consumer use and this choice based analysis is greater when it comes to the food categories.

In such a situation the consumer seeks the clear information by the producers of the product and instructions by the scientists and local government bodies or national authorities. Hence, all over the world, several attempts have been made to give such informations to the consumers at the primary level of buying the products. In the European Union, few genetically modified foods are allowed to sell in the market in the categories of value added foods. It has given clear instructions "not to be used by the children and aged peoples, as they are vulnerable for the contents of the food". Thus, the consumers can make choices by themselves. This will help the consumer to decide on its own, without any external non-scientific influences.

21.5. LABELING OF GM FOOD PRODUCTS

The Ottawa Charter for Health Promotion (1986) stresses the importance of supporting citizens in their capacity to take greater control over their health. Access to information is central to informed decision-making, and can be seen as a social determinant of health (respect for religious, ethical, cultural and personal values and beliefs has an impact on health). Mandatory labeling for GM content of food is necessary to enable consumer choice **(Fig. 21.3)**. A related public health issue is the capacity for epidemiological tracking in the case of food toxicity or increased incidence of allergy that may be associated with a GM product in the food system. Such an assessment is not possible without a system of food labeling. Moreover, it would be difficult to give consumers advice regarding food choices in such a situation.

The present system of food labeling laws in Canada is regulated primarily under the authority of the Food and Drugs Act. Labeling of GM ingredients is voluntary, and allows either a positive statement (the food/ingredients are genetically modified) or a negative one (the food

Fig. 21.3. Genetically modified food product label indicating source fruits used to make genetically modified soyabeans

is not genetically modified or is free of GM ingredients). Mandatory labeling is only required if there are safety concerns (e.g., the presence of a potential allergen), or is the food has undergone significant compositional change as a result of genetic engineering. To date, no GM food in Canada has required mandatory labeling. There is no way for consumers, epidemiologists, and others to know if food has originated from GMOs or not. Due to the sensitivity of DNA testing, some members of the food industry have suggested that the message "GM-free" cannot be proven true, as even stray GM pollens in a non-GM food would show up in tests. From this came the suggestion that the labeling message, "may contain..." would cover the liability issue. However, this message would be useless to help customers to make choices.

It has been suggested that mandatory labeling of GM food might not be necessary if rigorous regulatory procedures were in place. Others, including the Consumers Association of Canada, argue that there is no justification for a mandatory labeling requirement because:

- the knowledge of the origin of the food would not make any difference to people's health;
- most people lack sufficient knowledge of informed decision-making on food biotechnology, and mandatory labeling would only confuse, not inform them; and
- it is the safety of the food product itself that is relevant to health, not the method of production.

Industry members have also argued that mandatory GM food labeling would be prohibitively expensive due to the need for segregation of GM foods "from farm to fork" (i.e., separate harvesting equipment, storage, transportation, processing and packaging, etc., as well as the prevention of GM foods meant for animal feed to end up in human food). This cost would then be reflected in the price of food.

Cost estimates conducted for the industry by the international financial consulting firm KPMG are often quoted. For example, the Australia/New Zealand system of mandatory labeling was estimated at $394 million to implement and $55 million per year for compliance. However, this is an over-estimate, as it included labeling of GM ingredients plus an ingredient list, nutrition information and "best before" dates. A similar estimate by KPMG for a food industry group in Canada was $700-950 million per year. They warned that the cost of foods that contain major oilseed products could rise by as much as 40% of the current price. In British Columbia, a government discussion paper on GE labeling estimated that "the actual cost to BC business can be... substantially less, under $9 million per year" for GM food labeling only. Repeated surveys of Canadians have shown that the vast majority wants labeling of GM foods. And, in spite of all the obstacles, it is possible to have mandatory labeling laws. This is the case in the UK, for example, where the food standards agency policy states that it "supports consumer choice and recognizes that some people will wish to choose not to buy or eat GM foods, however carefully they have been assessed for safety. The current rules require that all foods that contain GM material should be clearly labelled".

The European Commission has developed a set of regulations for the mandatory labeling of novel foods and ingredients that were approved after 1997, as well as GM herbicide tolerant soy and GM insect-protected maize approved before that time. They set a threshold of maximum 1% GM contamination of non-GM material to waive the labeling requirement.

The issue of GM food labeling is vast, with global implications. In February, 2000, a joint Food and Agriculture Organization and World Health Organization (FAO/WHO) committee on food labeling—part of the codex alimentarius commission—released a draft document, and proposed draft recommendations for the labeling of foods obtained through biotechnology. Comments were invited with a deadline of April, 2000. Redrafting of the material was assigned to a working group (consisting of 23 member countries, the European Commission, and 9 international organizations) coordinated by the Canadian delegation. The next set of revisions is now being considered.

In April, 2001, an exposure bill called the Genetically Engineered Food Labeling Act (Bill 18) (for public input) was introduced in the BC legislature. Its purpose is to provide consumers with information for making choices respecting food composed of, containing, or derived from genetically engineered materials. It would enable the BC government to establish mandatory labeling for GE food sold in BC. The bill sets forth the requirements and standards, but does not set up the system for such labeling. It also establishes a GE food labeling advisory panel, consisting of experts in GE, consumers, and members from the food industry. Public opportunity to comment was provided until July 31, 2001.

In August, 2001, a draft standard (for public comment over 60 days) was released by a Canadian General Standards Board Committee on voluntary labeling of foods obtained or not obtained through genetic modification. If this standard were approved, it would set the stage to allow foods to be labelled "GMO free" if they contain upto 5% GM material. The draft standard was developed to ensure that any such claims on a label or in advertising made by manufacturers and/or distributors are consistent with an appropriate set of parameters.

In September, 2001, the Canadian House of Commons voted on the first stage of Bill C-287, which implemented the mandatory labeling of genetically engineered food in Canada. It comprised of a letter outlining reasons for mandatory labeling. It was prepared by the biotechnology caucus of the Canadian Environmental Network (CEN), and was introduced as a private member's bill.

The European Union, Australia, and New Zealand are among the areas that are now requiring mandatory labeling of genetically modified foods.

Australia and New Zealand created new policies that came into effect on December 7, 2001. They called for the mandatory labeling of GM foods. Foods that do not have to be labeled under the new rules are refined foods (such as oils), GM foods where no single ingredient is more than 1% genetically modified, and food prepared at the point of purchase (such as restaurants). The European Union also requires labeling on foods that are genetically modified (GM), and they have more stringent standards on the types of GM foods that can be allowed in the food supply. Certain types of GM foods, which critics feared may cause allergic reactions, were phased out.

The development of a new GM crop or food begins in the research laboratories, and is then passed onto the growth in the containment levels in glasshouse. It is then moved to the field trials with a certain degree of containment. Ultimately, it is released for commercial cultivation or production. In spite of the methods taken care to controlled containment the new variety escapes into the environment and enters the food chain. Hence GM crops and foods are regulated tightly by various specifications made by the governments around the world.

World health organization lists the essential guidelines followed for equivalence testing of GM crops and food when compared to its non-GM crops and food counterpart. Advisory Committee on Novel Foods and Processes (ACNFP) was established in 1988 by United Kingdom. It comprises of 14 academic members with expertise in allergenicity, genetics, immunology, microbiology, nutrition, and food toxins as well as consumer representative and an bioethicist. The growing GM food controversy and consumers' attempts to make better food purchasing decisions have pressed GM food labeling into an important public policy issue. Truthful labeling has been used to provide consumers with information on calories, nutrients, contents and food ingredients in the world market. In 1997, the European Commission adopted GMO food labeling that requires each member country to enact a law requiring labeling of all new products containing genetically modified organisms. In some EU countries, information technologies have made it economically feasible to encrypt large amounts of information on food package bar codes. Japan has also passed a law requiring GM labels for major foods. Labeling involves real costs—fixed costs of designing labels and testing, and variable costs of monitoring for truthfulness. One key issue is whether the social benefits from labeling will exceed the cost.

21.6. BENEFITS OF LABELING

The labeling of any products is having the following advantages to the consumers:
- Consumer's right to know the product
- Food and crop safety
- Consumer's right to say "NO"
- Education and raising awareness to products
- Storage and handling methods
- Environmental protection and precautionary approach

Under any countries act, a consumer must be informed by labeling if a food derived from a new plant variety differs from its traditional counterpart, with respect to saftey or usage issue. The FDA takes the view that the method of production of the new plant variety is not "material" to whether a food needs to be labled under the act. The FDA's policy is that foods be considered on case by case basis, and that labling would be required on individual GM food products if these foods were materially different.

In an increasingly technological society, the ability to manage the introduction of new technologies and practices is an essential skill for governments and firms. In our 'risk society', a range of potential risks and uncertainties are associated with new technologies that bring with them worries about human health, industrial competitiveness, ecological disruption, and the potential for trade disputes. Yet the evidence from research is that many of the public, far from requiring a better understanding of science, are well informed about scientific advance and new technologies, and are highly sophisticated in their thinking on the issues. Many 'ordinary' people demonstrate a thorough grasp of issues such as uncertainty, if anything, the public are ahead of many scientists and policy advisors in their instinctive feeling for a need to act in a precautionary way. What is more, our research calls into question the validity of the notion of 'sound science', on which politicians are inclined to rely. This is at a time when the implementation of the precautionary principle is a highly charged political, legal and scientific issue, not least with respect to trade issues. Science cannot always provide definitive answers in these cases, so the policy of relying on claims of 'sound science' may, ironically, itself be unsound. Ethical issues are central in such issues.

Greenpeace and Friends of the Earth advocate labels on GM foods to give consumers the right to choose whether or not to consume GM foods. Many environmental and consumer advocacy groups call for mandatory labeling, which they believe benefits consumers. The United States Department of Agriculture, Economic Research Service, has analyzed the potential benefits of labels on foods. One benefit is making it easy to find information, e.g., on nutritional content of foods. Thus, labeling of foods can lead to more informed choices on food and health by consumers. Also, some firms may want to avoid the prospect of placing a label that has negative connotations, and required labeling could lead them to improve their product.

Caswell and Padberg recommended a more comprehensive view of the benefits of labels on food products. These benefits can be above and beyond what are normally considered the typical benefits from labels. The benefits from food labels include increased consumer information, improved product design, and more consumer confidence in product quality. Also, labels can provide an option value, even for consumers who do currently read food labels. This option value exists because if a food is labeled, consumers always have the option to view the label, either now or in the future, and that option has some value.

Even though some companies started the labeling as a self practice, government also has the job of ensuring safety for consumers and the general public, and protection of the environment. This involves making informed decisions about potential effects on human health, ecology and the countryside more generally. The central question is how these decisions can be made when the technology, and therefore the information on which decisions must be based, are subject to such profound and open-ended uncertainties. The labeling system should pursue the policy of labeling for all GM crops and foods, at least until the consumers are familiar with the new technology.

Senior politicians frequently stress the need for decisions on GM food to be made in the light of 'sound science'. Their approach to public unease about the technology has often been to characterize the public as ignorant, irrational or even hysterical.

In 1995, ESRC researchers taking part in the global environmental change programme warned that the government would, sooner rather than later, run into serious public difficulties in its handling of genetically modified food. It took nearly four years for the prediction to come true, but when it did, the controversy rapidly became one of the government's biggest headaches. Ironically, this is a government that prides itself in its ability to stay in touch with public opinion. So, what went wrong? Why have successive governments seemed almost willfull to ignore the insights of research on these issues? Are there lessons to be learned from nuclear power, BSE and Brent Spar for the management of GM foods? In 1995, ESRC-funded researchers from the center for the study of environmental change (CSEC) at Lancaster University predicted that sooner or later

major problems with public acceptance of GM food would surface. To investigate the issues surrounding the regulation of GM food, CSEC and the consensus-building group "The Green Alliance" organized a series of three meetings with regulators, industry and environmental groups. These meetings identified serious problems with the regulator arrangements. None of the three main groups involved were confident in the regulation of GMOs, and major gaps in responsibility were identified.

Following these meetings, CSEC undertook a major study of public attitudes to GM food. The subsequent report has steadily made a significant impact on UK policy on GMOs. Entitled "Uncertain World: Genetically Modified Organisms, Food and Public Attitudes in Britain (1997)", it has been cited by the specialist journal ENDS as 'The study which has provided the deepest insights into public attitudes towards GM foods' (ENDS Report 283, August 1998). More recently, Sir Robert May, the Government's Chief Scientist said "I now have had a chance to read 'Uncertain World', which I wish I had indeed read earlier. It is in many ways a remarkably prescient document" (March 1999).

Much of the debate has been polarized between pro- and anti-GM voices, each side labeling the other as irresponsible in various ways. What has largely been lacking is a careful analysis of the way such issues can be handled better by government and firms, i.e., what this document sets out to do. It brings together the findings of research funded by the economic and social research.

Science never took place in vacuum, but now more than ever, it needs to stand up to public scrutiny, to engage with public opinion, and address technical and ethical issues in the application of science to the needs of the poor. The current debate over transgenics and the labeling of foods containing ingredients from GMOs has highlighted the need for transparency. There is a growing awareness of the public's ambivalence towards genetic engineering and its aims. The research sector should advertise the fact that it disposes of an important instrument for quality control. The scrupulou honesty of the peer review process, under the watchful eye of scientific institutions, including scientific academies, the great journals, is a guarantee for quality, if not always for relevance. The food and agriculture sector could well take the example of the declaration last week by twelve of the world's most prominent medical journals of a policy to prevent the corporate sponsors of research exercising control over the analysis, interpretation, and reporting of research results. However, the general public is largely unaware of peer reviews and standard setting (which takes place through similar processes), and public trust remains low.

21.7. GUIDELINES FOR LABELING OF GENETICALLY MODIFIED AGRICULTURAL PRODUCTS

Earlier to May, 1997, labeling was not a mandatory practice for GM products throughout the world. But some volunteers started on voluntary basis, e.g., the Co-op's vegetarian cheese prepared using GM Chymosin, to gain consumers confidence. The companies such as Sainsbury's (GM Chymosin) and Safeway's (GM tomato puree) are labeled voluntarily. Labeling guidelines were first started in UK in 1993 by Food Advisory Committee and the institute of Grocery Distribution in 1997. These guidelines made the labeling of novel GM product mandatory. Then other countries also made the labeling mandatory, especially European countries. Foods and crops having health implications and ethically sensitive genes should be labeled. Allergens, toxin content, anti-nutritional factors, nutritional components and genes from animals (pigs, cow, fish, insects, rat) and humans are necessarily included in such labeling system. On 15th May, 1997, European Union made regulations for novel GM foods or foods derived from GMO's under the labeling products. On 1st September, 1998, new council of

regulations included soybeans and maize under strict regulations. In 1999, GM soybean and maize were not only used for manufactured products, but also in restaurants, cafes, delicatessens, sandwich and other shops.

In such place, they should also inform through staff, indicated in menu or on a prominent notice Online or telephone orders should also informed, otherwise up to 5,000 Euro Fine was to be levied on the shops or stores.

Format for Labeling

There is a specific format, which has to be followed for the labeling purpose, and companies should adhere to the rules strictly. Some of the rules are as follows:

- Labels should be on front side of the package.
- Fonts should be easily readable.
- Proper letter size (>10 points).
- Clear color.
- Placed in an easily recognized way.
- Should not be removed easily or any other means.
- Manufacturers' responsibilities.

The labeling system introduced in the GM food and crops started the practice of positive labeling of GM/GMO's products, and negative labeling of GM/GMO's free materials. This gives the consumers a choice for the products that they dislike. This is very essential in the shops that sell both the types of products.

Positive Labeling of GM/GMO's Products

Positive labeling means that any product derived from genetic modification has to be clearly indicated on the package of the product.

Negative Labeling of GM/GMO's Free Products

Negative labeling means that any product which is not derived from genetic modification has to be clearly indicated on the package of the product.

Once the genetically modified foods and crops are produced, they should be labeled properly. The guidelines are established in many countries for such labeling **(Table 21.1)**. These guidelines aim at establishing genetically modified agricultural products subject to labeling standards and labeling methods, in accordance with the provisions agricultural and forestry and environmental ministry Product Quality Control Act. According to our mandatory labeling rules, GM commodities subject to labeling are soybean, corn, soybean sprout, and potato.

Table 21.1: *Genetically modified crops, organisms and GM foods that require labeling under the European Union regulations*

Genetically modified crops, organisms and GM foods	Labeling required	or not
GM plant or seed	Yes	—
Whole foods derived from a GM plant or seed	Yes	—
Refined products such as oil or sugar	Yes	—
Foods containing GMOs	Yes	—
Animal feed containing GMOs and their products	Yes	—
Fermentation products	Yes	—
Food made from GMOs enzyme	Yes	—
GM enzyme	Yes	—
Flavouring and other food and feed additives from GMOs	Yes	—
Foods catered from GMOs	Yes	—
Foods from animals fed with GM food	—	No
Food made from GM enzyme	—	No

GM Agricultural Products

GM agricultural products should be labeled as "Genetically modified (a name of agricultural product)". Bean sprouts grown with genetically modified soybean, however, should be labeled as "Genetically modified (bean sprouts grown with soybean)". Products containing GM agricultural products should be labeled as "containing genetically modified (a name of agricultural product)". If there is a possibility that GM agricultural products are contained, it shall be labeled as "It may contain genetically modified (name of the agricultural product)". For bean sprout grown with soybeans that may contain GM soybean, however, it shall be labeled as "It may contain genetically modified (bean sprouts grown with soybean)". Taking into consideration the possibility of GM agricultural products getting mixed together unintentionally, even when one segregates non-GM agricultural products in the production and marketing stage, if it contains less than 3% of GM agricultural products, labeling shall not be required. In this case, however, certificates proving that non-GM agricultural products are segregated and controlled should be furnished.

For the labeling methods, detailed labeling of GM agricultural products should be done on its package or at the point of selling. When it is sold after packaging, in principle, it should be directly labeled on package. It should be labeled with fonts that final purchasers can read easily; and in a place where it can be easily recognized. When it is sold without packaging, it should be marked at the sales point using a notice board, a post, etc. It should be marked with proper letter sizes so that purchasers can read easily. If is difficult to mark at the sales point when it is not sold to final purchasers, however, it can be marked on the invoice. For the labeling management system, the Ministry of Agriculture and Forestry should be responsible for the investigation of labeling of GM agricultural products. For the market investigation, the Ministry of Agriculture and Forestry or its affiliated body can collect necessary samples and conduct the verification of samples. The sample verification method should be the internationally recognized method, and the Head of such bodies should determine details on verification methods and sampling methods. Department of Biotechnology and its subsidiary bodies are responsible for the development of analytical method for GMO detection, to facilitate the implementation of labeling regulation.

Labeling Standard for Genetically Modified Foods

Mandatory labeling regulation on GM food went into effect from July, 2001 by Food and Drug Administration. This rule is intended to provide proper information to consumers, in accordance with the food sanitation act, by defining the matters concerning the labeling of foods or food additives manufactured or processed with ingredients obtained from agricultural, livestock, or fishery products grown or raised by recombinant DNA techniques, such as combining genes of an organisms with beneficial genes taken from other species. All foods or food additives, including foods or food additives subject to GM labeling requirements, come under the provisions of Article 16 Agricultural and Fisheries Products Quality Control Act if they contain any residual recombinant DNAs or foreign proteins in the final products after manufacture or processing. Soybean flour, corn flour, bean curd, soymilk and other 22 processed items are included in the mandatory labeling target.

Persons responsible for GM food labeling are those who operate foods manufacturing and processing business, instant food sales/manufacturing/processing business, food additives manufacturing business, food re-packing business, and food imports and sales business, in compliance with the provisions the Food Sanitation Act. The labeling of GM foods should be indicated using a 10-point or larger types of a character with a color clearly distinguishable from the background color of the container or package with indelible inks, stamp, brand and so forth, so that the consumers could easily recognize the label.

The labeling of GM foods should be indicated on the principle display panel in a way that the consumer may easily recognize the labels of "genetically modified food" or "food containing genetically modified XX"(XX refers to the name of the raw material). Or it should be indicated in parentheses besides the name of the genetically modified agricultural or fishery products used as the raw material of the food as "genetically modified" or " genetically modified XX".

Large number of countries including India make such laws for GM agricultural products and foods labeled in their container. This is essential because the products are identical, whether they come from a GM or non-GM source. Especially without the presence of DNA or protein, there is no way of confirming a GM origin. In such situations, the current market advantage in claiming that products come from a non-GM source, which will lead to the wide-spread fraudulent system. At present, labeling is exempted in refined vegetable oils, sugar and other products that do not contain DNA or protein and products containing less the 1% of GM materials as a result of the accidental mixing. Even though the stringent labeling laws undoubtedly make GM crops and foods less attractive for farmers and consumers, they may have the effect of forcing the food industry to face the reality. In future, large number of countries throughout the world will adopt the GM crops and food sources in a huge industrial scale. This can be very well observed in the recent figures of planting in 2004, indicating that over 85% of US soybean and 30% of US maize plantings were based on GM technology, whereas 100% GM soybeans were in Argentina.

SUMMARY

The importance of the agricultural sector in developing countries as a source of food, incomes, employment and often foreign exchange cannot be overstated. As much as good health, a productive and sustainable agricultural sector is critical to achieving economic growth and poverty reduction. About three quarters of the world's poor people live and work in rural areas. Apart from its direct role in sustaining incomes and employment, the role of agriculture, and in particular technological change in agriculture, in stimulating overall economic growth has been much discussed by economists and policymakers. While the public is not widely knowledgeable about biotechnology and genetically modified food, consumer awareness appears to be growing. When the pew initiative on food and biotechnology compared the findings of two polls conducted in January 2001 and June 2001, it found that consumer awareness of genetically modified foods sold in grocery stores had increased by 11% between the polls.

One of the primary duties of consumers is to know or acquire the knowledge about the source of any materials he or she is buying. Because what ever consumers ar buying, they should have an idea about their impact on the health of their family including young childrens and aged parents. Even if there is a remote chance of the health consequences of these modified materials especially food and health prouducts, consumer will always has a word to say no to such products. This is makes still higher degree of carefullness when it comes to the factor of gene based or genetically modified products. The consumers always prefer the materials that do not pole any danger to any age groups and the probability of the health impact is exceptioanlly nil or not having any type of variations in the quality of the materials at any conditions.

Labeling means displaying of information about the product on packaging materials in a distinct manner. It is a necessary requirement that a producer of a food product must reveal on a label "all facts that are material with respect to consequences which may result from use". The issue of GM food labeling is vast, with global implications. In February 2000, a joint Food and Agriculture Organization and World Health Organization (FAO/WHO) Committee on Food Labeling, part of the Codex Alimentarius Commission, released a draft document, proposed draft recommendations for the labeling of foods obtained through biotechnology. Comments were invited with a deadline of April, 2000. Redrafting of the material was assigned to a working group (consisting of 23 member countries, the European Commission and 9 international organizations) coordinated by the Canadian delegation. The next set of revisions is now being considered.

Benefits of labeling are immense and the labeling on any products is having the following advantages to the consumers. Consumer's have a right to know the product, food and crop safety, right to say "NO", education and raising awareness of products, storage and handling methods, environmental protection and precautionary approach. Hence, presently in biotechnology-based world, it always better to have labeling systems that are followed globally and have the customers choice.

EXERCISE

1. Define genetically modified organisms and their applications in the developing countries.
2. Describe the advantages of genetic modifications in food sector, and its importance to feed ever increasing world populations.
3. What do you mean by genetic modifications? How does the genetic modifications proceed to the development of better food varieties?
4. Distinguish between genetically modified organisms and non-genetically modified organisms for food productions.
5. Define social, ethical and legal issues involved in the production of genetically modified food and their impact on human health.

6. Elucidate the negative effects of genetically modified organisms, and their use as food. What are the methods involved in their identification? Add a note on requirement of labeling of genetically modified food.

7. Critically comment on the comprehensive and effective systems that are needed to find the genetically modified foods in the market.

8. What do you mean by 'labeling'? What is its impact on the food biotechnology in the present day market.

9. Briefly explain different foods produced by the genetic modification technology and their importance in the food sector.

10. How are the potential risks associated with the GM food identified in food biotechnology? What are the solutions to clean the mixing of non-GM food?

11. Give an account of the necessity of following labeling system for the genetically modified food.

12. Give an account of the principles that are essential for the regulations of the genetically modified food in food biotechnology. Add a note on benefits of labeling.

13. How can labeling help the consumer in choosing their foods in GM market?

14. How is the quality of products affected by the genetic modification technology?

15. Explain the guidelines followed for the labeling of genetically modified agricultural products in the market.

16. Write about the specific format that has to be followed for the labeling of the genetically modified food.

17. Give an account of genetically modified food products that require essential labeling.

22 CHAPTER

Stem Cell Research, Applications of Stems Cells and Ethical Issues Involved in Stem Cell Research and Use

22.1. INTRODUCTION

Stem cells are primal cells that have the ability to divide for indefinite periods in culture and can give rise to specialized cells. When stem cells divide, they perform two things—produce another cell similar to themselves, or create a number of cells with more specialized functions. For example, just one kind of stem cell in our blood can make new red blood cells, or white blood cells, or other kinds, depending on the needs of the body.

These cells are like the stem of a plant that spreads out in different directions as it grows. Because of their unique combined abilities of unlimited expansion and pluripotency, embryonic stem cells are a potential source for regenerative medicine and tissue replacement after injury or disease. Moreover, treatments that do not require destroying any human life are at least as promising—they are already healing some conditions, and are far closer to healing other conditions than any approach using embryonic stem cells. The choice is not between science and ethics, but between science that is ethically responsible and science that is not.

Stem cell research has a history of more than 26 years, and has made some significant and outstanding contributions to the understanding of haematopoiesis, mouse embryology and human cellular differentiations. Stem cells are truly remarkable. They bridge the gap between the fertilized egg that is of living animal or human origin, and the architecture that animal or human become. They supply the cells which construct the adult bodies, and as the animals or humans age, they replenish worn out, damaged and diseased tissues regularly. They renew themselves, resisting the powerful pull towards differentiation that overcomes more ordinary cells. Depending on the source, they have the potential to form one, many or all cell types of an organism. In the past few years, successes have been achieved in culturing human embryonic stem cells, and in manipulating their differentiation *in vitro*. In 1998, for the first time, investigators were able to isolate this class of pluripotent stem cell from early human embryos and grow them in culture. Evidence has emerged that these stem cells are, indeed, capable of becoming almost all of the specialized cells of the body, and thus, may have the potential to generate replacement cells for a broad array of tissues and organs, such as the heart, the pancreas, and the nervous system. Thus, this class of human stem cell holds the promise of being able to repair or replace cells or tissues that are damaged or destroyed by many of our most devastating diseases and disabilities.

The huge research on the cell culture oriented studies paved the way for the identification of new type of cells in animal biotechnology. A new era in stem cell biology began in 1998, with the derivation of cells from human blastocysts and fetal tissue having the unique ability of differentiating into cells of all tissues in the body, i.e., the cells are pluripotent. Since then, several research teams have characterized many of the molecular characteristics of these cells and improved the methods for culturing them. In addition, scientists are just beginning to direct the differentiation of the human pluripotent stem cells and to identify the functional capabilities of the resulting specialized cells. Although in its earliest phases, research with these cells is proving to be important for developing innovative cell replacement strategies to rebuild tissues and restore critical functions of the diseased or damaged human body.

22.2. DEFINITION OF STEM CELLS

Stem cells are also called magic seeds. A stem cell is a special kind of cell that has a unique capacity to renew itself and to give rise to specialized cell types. Although most cells of the body, such as heart cells or skin cells are committed to conduct a specific function, stem cells are uncommitted and remains uncommitted, until they receive a signal to develop into a specialized cell. Their proliferative capacity combined with the ability to become specialized makes stem cells unique. Since many years, researchers have looked for the ways to use stem cells to replace cells and tissues that are damaged or diseased.

Researchers and scientists interested in human development have been studying animal development for many years. This research yielded the first glimpse at a class of stem cells that can develop into any cell type in the body. This class of stem cells is called pluripotent, i.e., the cells have the potential to develop more than 220 different known cell types. Stem cells with this unique property come from embryos and fetal tissue.

In 1998 a breakthrough was seen in stem cell research. James Thomson at the University of Wisconsin-Madison isolated cells from the inner cell mass of the early embryo, called the blastocyst, and developed the first human embryonic stem cell lines. In the same year, John Gearhart at Johns

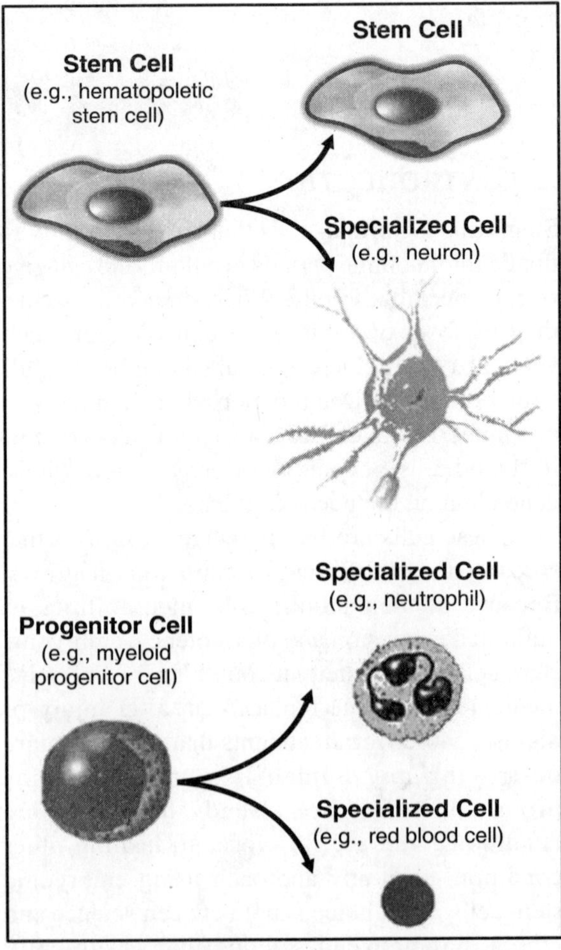

Fig. 22.1. Exclusive features of stem cells and progenitor/precursor cells

Hopkins University reported the first derivation of human embryonic germ cells from an isolated population of cells in fetal gonadal tissue, known as the primordial germ cells, which are destined to become the eggs and sperm. From both of these sources, the researchers developed pluripotent stem cell "lines", which are not only capable of renewing themselves for long periods, but can also give rise to many types of human cells or tissues **(Fig. 22.1)**. Human embryonic stem cells and embryonic germ cells differ in some characteristics, however, and do not appear to be equivalent.

22.3. STEM CELL LINE

It is an ongoing, living colony of stem cells in a laboratory, from which cells can be obtained for research or other uses. Sometimes these are called "immortal" cell lines, which is misleading, because eventually they do deteriorate. Embryonic stem cells are easier to grow in a stem cell line, but they also tend to develop serious genetic abnormalities associated with cancer.

Pluripotency of Human Embryonic Stem Cells and Embryonic Germ Cells

A truly pluripotent stem cell is capable of self-renewal and differentiating into almost all of the cells of the body, including cells of three germ layers. Human ES and EG cells *in vitro* are capable of long-term self-renewal, while retaining a normal karyotype. Human ES cells can proliferate for two years through 300 population doublings or even 450 population doublings. Cultures derived from embryoid bodies generated by human embryonic germ cells have less capacity for proliferation. Most will proliferate for 40 population doublings; the maximum reported is 70 to 80 population doublings. To date, several laboratories have demonstrated that human ES cells *in vitro* are pluripotent. They can produce cell types derived from all three embryonic germ layers. Currently, the only test of the *in vivo* pluripotency of human ES cells is to inject them into

immunodeficient mice, where they will generate differentiated cells derived from all three germ layers. These include gut epithelium (which, in the embryo, is derived from endoderm); smooth and striated muscle (derived from mesoderm); and neural epithelium and stratified squamous epithelium (derived from ectoderm).

Nevertheless, in stem cell biology, two aspects of *in vivo* pluripotency typically used in animals have not been met by human ES cells—evidence that cells have the capacity to be injected into a human embryo, and form an organism made up of cells from two genetic lineages; and evidence that they have the ability to generate germ cells—the precursors to eggs and sperm in a developing organism. These are theoretical considerations, however, because such tests using human ES cells have not been conducted. In any case, these two demonstrations of human ES cell pluripotency are not likely to be critical for potential therapeutic uses of the cells—in transplants or drug development, for example.

Comparisons Between Human Embryonic Stem Cells and Embryonic Germ Cells

The source of stem cells does not make much difference with respect to the characters of the cell. The ES cells derived from human blastocysts by Thomson and his colleagues, and from human EG cells derived by Gearhart and his collaborators, are similar in many respects. In both cases, the cells replicate for an extended period of time, show no chromosomal abnormalities, generate both XX (female) and XY (male) cultures, and express a set of markers regarded as characteristic of pluripotent cells. When the culture conditions are adjusted to permit differentiation (see below for details), both ES and EG cells spontaneously differentiate into derivatives of all three primary germ layers—endoderm, mesoderm, and ectoderm.

However, even with some similarities observed in the both types of stem cells, they still show dissimilarites. The ES cells derived from human blastocysts and EG cells differ not only in the tissue sources from which they are derived, but also with respect to their growth characteristics *in vitro*, and their behavior *in vivo*. In addition, human ES cells have been propagated for approximately two years *in vitro*, for several hundred population doublings, whereas human embryoid body-derived cells from cultures of embryonic germ cells have been maintained for only 70 to 80 population doublings. Also, human ES cells will generate teratomas containing differentiated cell types, if injected into immunocompromised mice colonies, while human EG cells will not. Several research groups are trying to grow human ES cells without feeder layers of mouse embryo fibroblasts (MEF), which are labor-intensive to generate. At a recent meeting, scientists from the Geron Corporation reported that they have grown human ES cell, without feeder layers, in a medium conditioned by MEFs and supplemented with basic FGF.

Human Embryonic Stem Cells, Embryonic Germ Cells and their Directed Differentiation *In Vitro*

Currently, a major goal for embryonic stem cell research is to control the differentiation of human ES and EG cell lines into specific kinds of cells—an objective that must be met if the cells are to be used as the basis for therapeutic transplantation, testing drugs, or screening potential toxins. The techniques now being tested to direct human ES cell differentiation are borrowed directly from the techniques used to direct the differentiation of mouse ES cells in vitro. (For more discussion on directed differentiation of human ES and EG cells).

22.4. ADULT STEM CELLS

Adult stem cells, like all stem cells, share at least two characteristics. First, they can make identical copies of themselves for long periods of time; this ability to proliferate is referred to as long-term self-renewal. Second, they can give rise to mature cell types that have characteristic morphologies (shapes) and functions. Typically, stem cells generate an intermediate cell type or types before they achieve their fully differentiated state. The intermediate cell is called a precursor or progenitor cell. These cells in fetal or adult tissues are partly differentiated cells that divide and give rise to differentiated cells. Such cells are usually regarded as "committed" to differentiating along a particular cellular development pathway, although this characteristic may not be as definitive as once thought. At about the same time, as scientists were beginning to explore human pluripotent stem cells from embryos and fetal tissue, a flurry of new information was emerging about a class of stem cells that have been in clinical use for years—so-called adult stem cells.

Most stem cell research uses cells obtained from adult tissue, umbilical cord blood, and other sources that pose no moral problem. Useful stem cells have been found in bone marrow, blood, muscle, fat, nerves, and even in the pulp of baby teeth. Some of these cells are already being used to treat people with a wide variety of diseases. For many years, researchers have been seeking to understand the body's ability to repair and replace the cells and tissues of some organs, but not others. After years of work, pursuing the how and why of seemingly indiscriminant cell repair mechanisms, scientists have now focused their attention on adult stem cells. It has long been known that stem cells are capable of renewing themselves, and that they can generate multiple cell types. Today, there is new evidence that stem cells are present in far more tissues and organs than once thought, and that these cells are capable of developing into more kinds of cells than previously imagined. Efforts are now underway to harness stem cells and to take advantage of this new found capability, with the goal of devising new and more effective treatments for a host of diseases and disabilities. What lies ahead for the use of adult stem cells is unknown, but it is certain that there are many research questions to be answered, and these answers hold great promise for the future.

The stem cell can yield the specialized cell types of the tissue from which it originated. During the past decade, scientists discovered adult stem cells in tissues that were previously not thought to contain them, such as the brain. More recently, they reported that adult stem cells from one tissue appear to be capable of developing into cell types that are characteristic of other tissues. For example, although adult hematopoietic stem cells from bone marrow have long been recognized as capable of developing into blood and immune cells, scientists reported that, under certain conditions, the same stem cells could also develop into cells that have many of the characteristics of neurons. So, a new concept and a term emerged—adult stem cell plasticity.

In the adult organisms, the adult stem cells are rare. Their primary functions are to maintain the steady state functioning of a cell called homeostasis, and, with limitations, to replace cells that die because of injury or disease. For example, only an estimated 1 in 10,000 to 15,000 cells in the bone marrow is a hematopoietic (blood forming) stem cell. Furthermore, adult stem cells are dispersed in tissues throughout the mature animal and behave very differently, depending on their local environment. For example, HSCs are constantly being generated in the bone marrow, where they differentiate into mature blood cells. The primary role of HSCs is to replace blood cells. In contrast, stem cells in the small intestine are stationary and physically separated from the mature cell types they generate. Gut epithelial stem cells (or precursors) occur at the bases of crypts—deep invaginations between the mature, differentiated epithelial cells that line the lumen of the intestine. These epithelial crypt cells divide often, but remain part of the stationary group of cells they generate.

Adult stem cells pose a number of difficulties for the identifications purposes. Unlike embryonic stem cells that are defined by their origin (the inner cell mass of the blastocyst), adult stem cells share no such definitive means of characterization. In fact, no one knows the origin of adult stem cells in any mature tissue. Some have proposed that stem cells are somehow set aside during fetal development and restrain from differentiating. Definitions of adult stem cells vary in the scientific literature, ranging from a simple description of the cells to a rigorous set of experimental criteria that must be met before characterizing a particular cell as an adult stem cell. Most of the information about adult stem cells comes from the studies on mice. The list of adult tissues reported to contain stem cells is growing, and includes bone marrow, peripheral blood, brain, spinal cord, dental pulp, blood vessels, skeletal muscle, epithelia of the skin and digestive system, cornea, retina, liver, and pancreas.

In order to be classified as an adult stem cell, the cell should be capable of self-renewal throughout the lifetime of the organism. This criterion, although fundamental to the nature of a stem cell, is difficult to prove *in vivo*. It is nearly impossible, in an organism as complex as a human, to design an experiment that will allow the fate of candidate adult stem cells to be identified *in vivo* and tracked over an individual's lifetime.

Ideally, adult stem cells should also be clonogenic. In other words, a single adult stem cell should be able to generate a line of genetically identical cells, which then give rise to all the appropriate, differentiated cell types of the tissue in which they reside. Again, this property is difficult to demonstrate *in vivo*. In practice, scientists show that either a stem cell is clonogenic *in vitro*, or that a purified population of candidate stem cells can repopulate the tissue.

An adult stem cell should also be able to give rise to fully differentiated cells that have mature phenotypes, are fully integrated into the tissue, and are capable of specialized functions that are appropriate for the tissue.

- The term phenotype refers to all the observable characteristics of a cell (or organism).
- Its shape (morphology).

- Interactions with other cells and the noncellular environment (also called the extracellular matrix).
- Proteins that appear on the cell surface markers.
- The cell's behavior (e.g., secretion, contraction, synaptic transmission).

The majority of researchers, who lay claim to having identified adult stem cells, rely on two of these characteristics—appropriate cell morphology, and the demonstration that the resulting differentiated cell types display surface markers that identify them as belonging to the tissue. Some studies demonstrate that the differentiated cells that are derived from adult stem cells are truly functional, and a few studies show that cells are integrated into the differentiated tissue *in vivo,* and that they interact appropriately with neighboring cells. At present, there is, however, a paucity of research, with a few notable exceptions, in which researchers are able to conduct studies on genetically identical (clonal) stem cells. In order to fully characterize the regenerating and self-renewal capabilities of the adult stem cell, and therefore to truly harness its potential, it is important to demonstrate that a single adult stem cell can, indeed, generate a line of genetically identical cells, which then give rise to all the appropriate, differentiated cell types of the tissue in which they reside.

The question of human adult and embryonic stem cells equivalent in their potential for generating replacement cells and tissues is not fully answered. Current science indicates that, although both of these cell types hold enormous promise, adult and embryonic stem cells differ in important ways. What is not known is the extent to which these different cell types will be useful for the development of cell-based therapies to treat disease. Some considerations are noteworthy regarding this report. First, in recent months, there have been many discussions in the lay press about the anticipated abilities of stem cells from various sources and projected benefits to be realized from

them in replacing cells and tissues in patients with various diseases. The terminology used to describe stem cells in the lay literature is often confusing or misapplied. Second, even among biomedical researchers, there is a lack of consistency in common terms to describe what stem cells are and how they behave in the research laboratory. Third, the field of stem cell biology is advancing at an incredible pace, with new discoveries being reported in the scientific literature on a weekly basis.

22.5. IDENTIFICATION OF STEM CELLS

In recent years, scientists have discovered a wide array of stem cells that have unique capabilities to self-renew, grow indefinitely, and differentiate or develop into multiple types of cells and tissues. Researchers now know that many different types of stem cells exist, but they all are found in very small populations in the human body; in some cases, 1 stem cell in 100,000 cells in circulating blood. And, when scientists examine these cells under a microscope, they look just like any other cell. So, how do scientists identify these rare type of cells found in many different cells and tissues—a process that is much akin to finding a needle in a haystack? The answer is rather simple, thanks to stem cell "markers". This feature describes stem cell marker technology, and how it is used in the research laboratory. **Table 22.1** lists some the commonly used stem cell markers.

Stem Cell Markers

From the point of view of culturing and maintaining stem cells, their proper identification is a must criteria in the stem cell biology. Some specialized proteins, called receptors, coat the surface of every cell in the body, are specialized, and have the capability of selectively binding or adhering to other "signaling" molecules. There are many different types of receptors that differ in their structure and affinity for the signaling molecules. Normally, cells

Table 22.1: *Comparison of mouse, monkey and human pluripotent stem cells*

Marker name	Mouse EC/ ES/EG cells	Monkey ES cells	Human ES cells	Human EG cells	Human EC cells
SSEA-1	+	—	—	+	—
SSEA-3	—	+	+	+	+
SEA-4	—	+	+	+	+
TRA-1–60	—	+	+	+	+
TRA-1–81	—	+	+	+	+
Alkaline phosphatase	+	+	+	+	+
Oct-4	+	+	+	Unknown	+
Telomerase activity	+ ES, EC	Unknown	+	Unknown	+
Feeder-cell dependent	ES, EG, some EC	Yes	Yes	Yes	Some; relatively low clonal efficiency
Factors which aid in stem cell self-renewal	LIF and other factors that act through gp130 receptor and can substitute for feeder layer	Co-culture with feeder cells; other promoting factors have not been identified	Feeder cells + serum; feeder layer + serum-free medium + bFGF	LIF, bFGF, forskolin	Unknown; low proliferative capacity
Growth characteristics *in vitro*	Form tight, rounded, multi-layer clumps; can form EBs	Form flat, loose aggregates; can form EBs	Form flat, loose aggregates; can form EBs	Form rounded, multi-layer clumps; can form EBs	Form flat, loose aggregates; can form EBs
Teratoma formation *in vivo*	+	+	+	−	+
Chimera formation	+	Unknown	+	−	+

Key: ES cell = Embryonic stem cell; EG cell = Embryonic germ cell; EC cell = Embryonal carcinoma cell
SSEA = Stage-specific embryonic antigen; TRA = Tumor rejection antigen-1; LIF = Leukemia inhibitory factor
bFGF = Basic fibroblast growth factor; EB = Embryoid bodies.

use these receptors and the molecules that bind to them as a way of communicating with other cells and to carry out their proper functions in the body. These cell surface receptors are the stem cell markers.

Each cell, for example a liver cell, has a certain combination of receptors on its surface, which makes it distinguishable from other kinds of cells. Scientists have taken advantage of the biological uniqueness of stem cell receptors and chemical properties of certain compounds to tag or "mark" cells **(Fig. 22.2)**. Researchers owe much of the past success in finding and characterizing stem cells to the use of markers.

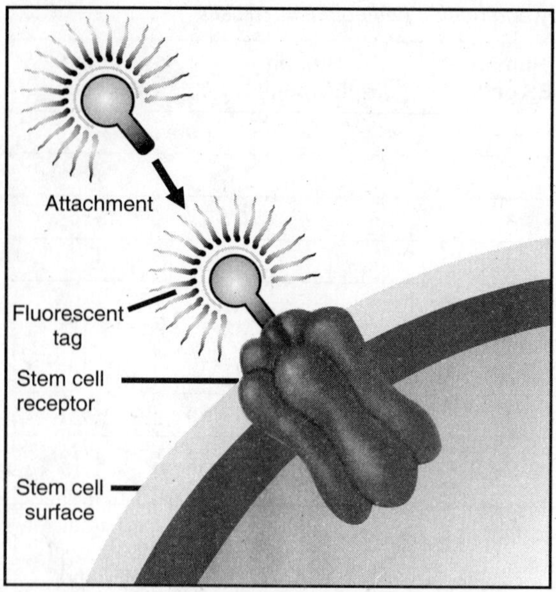

Fig. 22.2. Cell surface markers identification using fluorescent tags

Stem cell markers are given short-hand names based on the molecules that bind to the stem cell surface receptors. For example, a cell that has the receptor stem cell antigen-1 on its surface is identified as Sca-1. In many cases, a combination of multiple markers is used to identify a particular stem cell type. So now, researchers often identify stem cells in shorthand by a combination of marker names reflecting the presence (+) or absence (–) of them. For example, a special type of hematopoietic stem cell from blood and bone marrow called "side population" or "SP" is described as ($CD34^{-/low}$, $c\text{-}Kit^{+}$, $Sca\text{-}1^{+}$).

Cell culturists and researchers use the signaling molecules, which selectively adhere to the receptors on the surface of the cell, as a tool that allows them to identify stem cells. Many years ago, a technique was developed to attach another molecule, having the ability to fluoresce or emit light energy when activated by an energy source such as an ultraviolet light or laser beam, to the signaling molecule (or the tag). At the researchers' disposal are multiple fluorescent tags with emitted light that differ in color and intensity.

Similar to other cells, stems cells also have specific surface molecules on the plasma membrane. Two approaches of how researchers use the combination of the chemical properties of fluorescence and unique receptor patterns on cell surfaces to identify specific populations of stem cells are explained below.

The identified specific markers are exclusively present on the stem cells, but are absent on the other cells of the body. One approach for using markers as a research tool is to use a technique known as fluorescence-activated cell sorting (FACS). Researchers often use a FACS **(Fig. 22.3)** instrument to sort out the rare stem cells from the millions of other cells. With this technique, a suspension of tagged cells (i.e., bound to the cell surface markers are fluorescent tags) is sent under pressure through a very narrow nozzle—so narrow that cells can pass through one at a time. Upon exiting the nozzle, cells then pass, one-by-one, through a light source, usually a laser and then through an electric field. The fluorescent cells become negatively charged, while nonfluorescent cells become positively charged. The charge difference allows stem cells to separate from other cells. The researchers now have a population of cells that have same marker characteristics, and with these cells they can conduct their research with these cells. Their use in the culture and production of therapeutics is immense.

In the other method, stem cells are analysed by using intact tissues. This method uses stem cell markers and their fluorescent tags to visually assess cells as they exist in tissues. Often researchers want to assess how stem cells appear in tissues, and in doing so, they use a microscope to evaluate them rather than the FACS instrument. In this case, a thin slice of tissue is prepared, and the stem cell markers are tagged by the signaling molecule that has the fluorescent tag attached. The fluorescent tags are then activated either by special light energy or a chemical reaction. The stem cells will emit a fluorescent light that can be easily seen under the

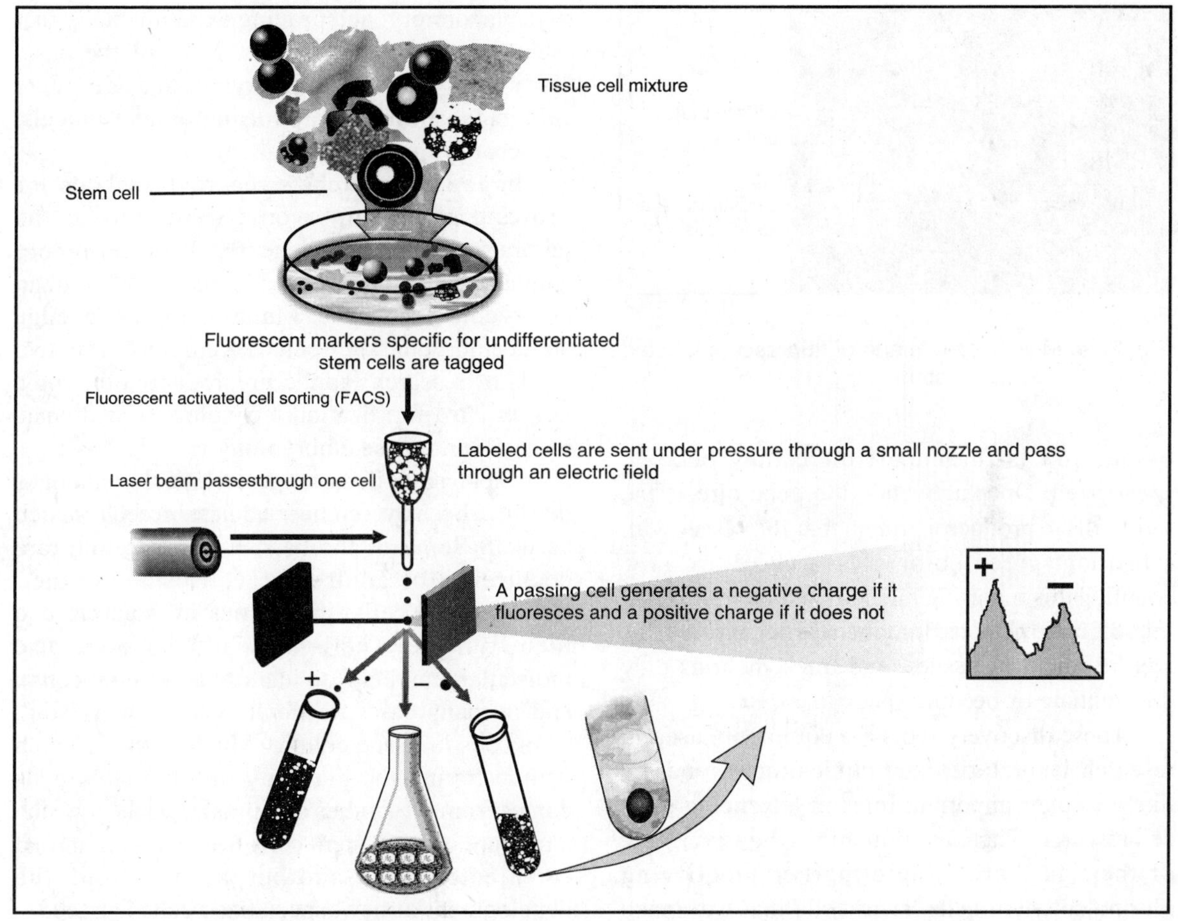

Fig. 22.3. Separation of stem cells from the mixture of source cells

microscope. This is useful for the identification of stem cells.

Extensive research on genetic and molecular biology techniques are used to study how cells become specialized during the organism's development. In such efforts, researchers have identified genes and transcription factors (proteins found within cells that regulate a gene's activity) that are unique in stem cells. Scientists use techniques such as polymerase chain reaction (PCR) to detect the presence of genes that are "active" and play a role guiding the specialization of a cell. This technique is helpful for the researchers to identify "genetic markers" that are characteristic of stem cells. For example, a gene marker called PDX-1 is specific for a transcription factor protein that initiates activation of the insulin gene. Researchers use this marker to identify cells that are able to develop islet cells in the pancreas.

However, recently, researchers have applied a genetic engineering approach that uses fluorescence, but isn't dependent on cell surface markers. The importance of this new technique is that it allows the tracking of stem cells as they differentiate or become specialized. Scientists have inserted into a stem cell a "reporter gene" called green fluorescent protein or GFP **(Fig. 22.4)**. The gene is only activated or "reports" when cells are undifferen-

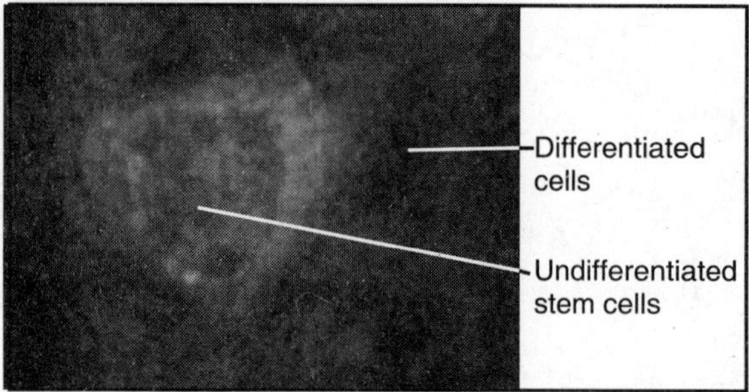

Fig. 22.4. Microscopic image of fluorescent-labeled stem cell

tiated, and is turned off once they become specialized. Once activated, the gene directs the stem cells to produce a protein that fluoresces with a brilliant green color. Researchers are now coupling this reporting method with the FACS and microscopic methods described earlier to sort cells, identify them in tissues, and track them as they differentiate or become specialized.

These discovery tools are commonly used in research laboratories and clinics today, and are likely to play important roles in advancing stem cell research. There are limitations, however. One of them is that a single marker identifying pluripotent stem cells, stem cells that can make any other cell, has yet to be found. As new types of stem cells are identified, and their research applications are become increasingly complex, more sophisticated tools will be developed to meet investigators' needs. For the foreseeable future, markers will continue to play a major role in the rapidly evolving world of stem cell biology.

22.6. STEM CELL RESEARCH

Large number of human disorders need the utlimate solutions for the life processes to continue. Stem cells can provide such solutions in the long run. It is important to understand some of the difficulties that researchers have had in isolating various types of stem cells; working with the cells

in the laboratory; and proving experimentally that the cells are true stem cells. Most of the basic research discoveries on embryonic and adult stem cells come from research using animal models, particularly mice.

In 1981, researchers reported methods for growing mouse embryonic stem cells in the laboratory, and it took nearly 20 years before similar achievements could be made with human embryonic stem cells. Much of the knowledge about embryonic stem cells has emerged from two fields of research: applied reproductive biology, i.e., *in vitro* fertilization technologies, and basic research on mouse embryology.

There have been many technical challenges that have been overcome in adult stem cell research as well. Some of the barriers include, the rare occurrence of adult stem cells among other, differentiated cells; difficulties in isolating and identifying the cells (researchers often use molecular "markers" to identify adult stem cells), and in many cases, difficulties in growing adult stem cells in tissue culture. Much of the research demonstrating the plasticity of adult stem cells comes from the studies of animal models in which a mixture of adult stem cells from a donor animal is injected into another animal, and the development of new, specialized cells is traced.

Stem cells, with their ability to divide for longer periods, can help in research work. The long-term proliferation ability and pluripotency of embryonic stem cells and embryonic germ cells are important in the stem cell biology. First, for basic research purposes, it is important to understand the genetic and molecular basis by which these cells continue to make many copies of themselves overlong periods of time. Second, if the cells are to be manipulated and used for transplantation, it is important to have sufficient quantities of cells that can be directed to differentiate into the desired cell type(s), and used to treat many patients suffering from a particular disease.

Presently, many investigators have been successful in using somewhat different approaches to derive human pluripotent stem cells. At least five laboratories have been successful in deriving pluripotent stem cells from human embryos, and one additional laboratory has created cell lines from fetal tissue. In each case, the methods for deriving pluripotent stem cells from human embryos and embryonic germ cells from fetal tissue are similar, yet they differ in the isolation and culture conditions as initially described by Thomson and Gearhart, respectively. It is not known to what extent US-based researchers are using these additional sources of embryonic stem and germ cells.

At present, there have been multiple human adult stem cell lines that have been created through a combination of public and private resources (e.g., hematopoietic stem cells). Substantial adult stem cell research has been underway for many years, and in recent years this has included basic studies on the "plasticity" of such cells.

Research Focus on Stem Cells

Stem cells are gaining much importance because of their wide applications in the biological research. Much has been written about the new discoveries of various stem cell types and their properties. Importantly, these cells are research tools, and they open many doors of opportunity for biomedical research. They find use in many fields, ranging from infertility to the gene therapy.

Transplantation research—restoring vital body functions

Transplantation is a highly discussed topic due to the absence or shortage of the organs required in this process. Stem cells may hold the key to replacing cells lost in many devastating diseases. Some of the human diseases, namely Parkinson's disease, diabetes, chronic heart disease, end-stage kidney disease, liver failure, and cancer are just a few for which stem cells have therapeutic potential. The populations affected by these diseases are increasing in the world. For many diseases that shorten lives, there are no effective treatments, but the goal is to find a way to replace what natural processes have taken away. For example, today science has brought us to a point where the immune response can be subdued, so that organs from one person can be used to replace the diseased organs and tissues of another. But, despite recent advances in transplantation sciences, there is a shortage of donor organs, which makes it unlikely that the growing demand for lifesaving organ replacements will be fully met through organ donation strategies.

Many of these diseases are life long diseases because of the absence of therapies. The use of stem cells to generate replacement tissues for treating neurological diseases is a major focus of research. Spinal cord injury, multiple sclerosis, Parkinson's disease, and Alzheimer's disease are among those diseases for which the concept of replacing destroyed or dysfunctional cells in the brain or spinal cord is a practical goal. This report features several recent advances that demonstrate the regenerative properties of adult and embryonic stem cells. Another major discovery frontier for research on adult and embryonic stem cells is the development of transplantable pancreatic tissues that can be used to treat diabetes. Scientists in academic and industrial research are vigorously pursuing all possible avenues of research, including ways to direct the specialization of adult and embryonic stem cells to become pancreatic islet-like cells that produce insulin and can be used to control blood glucose levels. Researchers have recently shown that human embryonic stem cells can be directly differentiated into cells that produce insulin.

There are common misconceptions about both adult and human embryonic stem cells. First, the lines of unaltered human embryonic stem cells that exist will not be suitable for direct use in patients. These cells will need to be differentiated or otherwise modified before they can be used clinically. Current challenges are to direct the differentiation of embryonic stem cells into specialized cell populations, and also to devise ways to control their development or proliferation once placed in the patients.

A second misconception is that adult stem cells are ready to use as therapies. With the exception of the clinical application of hematopoietic stem cells to restore the blood and immune system, this is not the case. The therapeutic use of this mixture of cells has proven safe, because the mixture is placed back into the environment from which it was taken, e.g., the bone marrow. In fact, many of the adult stem cell preparations currently being developed in the laboratory represent multiple cell types that are not fully characterized. In order to safely use stem cells or cells differentiated from them in tissues, other than the tissue from which they were isolated, researchers need purified populations (clonal lines) of adult stem cells.

In addition, the potential for the recipient of a stem cell transplant to reject these tissues as foreign is very high. Modifications in the cells, and immune system, or both will be a major requirement for their use. With the exception of the current practice of hematopoietic stem cell transplantation, much basic research lies ahead before the direct application of stem cell therapies is practiced.

Basic research applications

Basic understanding of developmental processes is a key to the success of the transgenic organisms or other life related activities. Embryonic stem cells are, undoubtedly, key research tools for understanding the fundamental events in embryonic development, which may one day, explain the causes of birth defects and approaches to correct or prevent them. Another important area of research that links developmental biology and stem cell biology is understanding the genes and molecules, such as growth factors and nutrients that function during the development of the embryo so that they can be used to grow stem cells in the laboratory and direct their development into specialized cell types.

Therapeutic delivery systems

In the therapy processes, the required compound, gene, or gene products has to be delivered in the right tissues at right time. Stem cells are already being explored as a vehicle for delivering genes to specific tissues in the body. Stem cell-based therapies are a major area of investigation in cancer research. For many years, restoration of blood and immune system function has been used as a component in the care of cancer patients who have been treated with chemotherapeutic agents. Now, researchers are trying to devise more ways to use specialized cells derived from stem cells to target specific cancerous cells and directly deliver treatments that will destroy or modify them.

Other applications of stem cells

Stem cells have applications in many fields, ranging from cell biology to the cancer biology. Future uses of human pluripotent cell lines might include the exploration of the effects of chromosomal abnormalities in early development. This might include the ability to monitor the development of early childhood tumors, many of which are embryonic in origin. Another future use of human stem cells and their derivatives include the testing of candidate therapeutic drugs. Although animal model testing is a mainstay of pharmaceutical research, it cannot always predict the effects that a developmental drug may have on human cells. Stem cells are likely to be used to develop specialized liver cells to evaluate drug detoxifying capabilities, and represent a new type of early warning system to prevent adverse reactions in patients. The coupling of stem cells with the information learned from the human genome project is likely to reap many benefits in future.

Comparison of Adult and Embryonic Stem Cells

Biomedical research on stem cells is at an early stage, but is advancing rapidly. After many years of isolating and characterizing these cells, researchers are now beginning to employ stem cells as discovery tools and a basis for potential therapies. This new era of research affords an opportunity to use what has already been learned to explore the similarities and differences of adult and embryonic stem cells. (In present text, embryonic stem cells derived from human embryos, and embryonic germ cells derived from fetal tissue, will be referred to equally as embryonic stem cells, unless otherwise distinguished).

Similarities between Adult and Embryonic Stem Cells

Stem cells have, in common, the ability to self-replicate and to give rise to specialized cells and tissues (such as cells of the heart, brain, bone, etc.) that have specific functions. In most cases, stem cells can be isolated and maintained in an unspecialized state. Scientists use similar techniques (i.e., cell-surface markers and monitoring the expression of certain genes) to identify or characterize unspecialized stem cells. They also use different genetic or molecular markers to determine that the cells have differentiated—a process that might be compared to distinguishing a particular cell type by reading its cellular barcode.

The stem cells are showing greater success in many fields of research. Stem cells from both adult and embryonic sources can proliferate and specialize when transplanted into an animal with a compromised immune system. (Immune-deficient animals are less likely to reject the transplanted tissue). Scientists also have evidence that differentiated cells generated from either stem cell type, when injected or transplanted into an animal model of disease or injury, undergo "homing"—a process whereby the transplanted cells are attracted by and travel to the injured site. Similarly, researchers are finding that the cellular and non-cellular "environment", into which stem cell-derived tissues are placed (e.g., whether they are grown in a culture dish or transplanted into an animal), prominently influences the differentiation of cells.

Immunological functions plays a critical role in either accepting or rejecting any cell in the body of an individual for the succesful usage. Another important area that requires substantially more research concerns the immunologic characteristics of human adult and embryonic stem cells. If any of these, stem cells are to be used as a basis for therapy. It is critical to understand how the body's immune system will respond to the transplantation of tissue derived from these cells. Presently, there is no clear advantage of one stem cell source over the other in this regard.

Differences between Adult and Embryonic Stem Cells

Even with a large number of similarities observed in both type of stems cells, differences still exist between the two. Perhaps, the most distinguishing feature of embryonic and adult stem cells is their source. Most scientists now agree that adult stem cells exist in many tissues of the human body (*in vivo*), though the cells are quite rare. In contrast, it is less certain that embryonic stem cells exist as such in the embryo. Instead, embryonic stem cells and embryonic germ cells develop in tissue culture after they are derived from the inner cell mass of the early embryo, or from the gonadal ridge tissue of the fetus, respectively.

Based on the culture conditions, embryonic stem cells may form clumps of cells that can differentiate spontaneously to generate many cell types. This property has not been observed in cultures of adult stem cells. Also, if undifferentiated embryonic stem cells are removed from the culture dish and injected into a mouse with a compromised immune system, a benign tumor called a teratoma can develop. A teratoma typically contains a mixture of partially differentiated cell types. For this reason, scientists do not anticipate that undifferentiated embryonic stem cells can be used for transplants or other therapeutic applications. It is not known whether similar results are observed with adult stem cells.

Stem cells in adult tissues do not appear to have the same capacity to differentiate as do embryonic stem cells or embryonic germ cells. Embryonic stem and germ cells are clearly pluripotent; they can differentiate into any tissues derived from all three germ layers of the embryo (ectoderm, mesoderm, and endoderm). But are adult stem cells also pluripotent? When they reside in their normal tissue compartments, the brain, the bone marrow, the epithelial lining of the gut, etc., they produce cells which are specific to that kind of tissue and have been found in tissues derived from all three embryonic layers. But can adult stem cells be taken out of their normal environment and manipulated, or otherwise induced to have the same differentiation potential as embryonic stem and germ cells? To date, there are no definitive answers to these questions, and the answers that do exist are sometimes conflicting.

In the culture conditions, both the cell types show variable capacity of growth. These sources of stem cells do not seem to have the same ability to proliferate in culture, and at the same time retain the capacity to differentiate into functionally useful cells. Human embryonic stem cells can be generated in abundant quantities in the laboratory, and can be grown (allowed to pro-liferate) in their undifferentiated (or unspecialized) state for many, many generations. From a practical perspective in basic research or eventual clinical application, it is significant that millions of cells can be generated from one embryonic stem cell in the laboratory.

In many cases, however, researchers have had difficulty finding laboratory conditions under which some adult stem cells can proliferate without becoming specialized. This problem is most pronounced with hematopoietic stem cells isolated from blood or bone marrow. These cells, when cultured in the laboratory, either fail to proliferate or do so to a limited extent, though they do proliferate if transplanted into an animal or human. This technical barrier has limited the ability of researchers to explore the capacity of certain types of adult stem cells to generate sufficient numbers of specialized cells for transplantation purposes.

The above mentioned differences in culturing conditions contribute to the contrasts in the experimental systems, which are used to evaluate the ability to become specialized under particular laboratory conditions. Much of the information on the directed differentiation of embryonic stem cells into cells with specialized function comes from studying mouse or human embryonic cell lines grown in laboratory culture dishes. In contrast, knowledge about the differentiation of adult stem cells comes from the observations of cells and tissues in animal models, in which mixtures of cells have been implanted.

Maturation of stem cells depends on many criteria in the culture conditions. Stem cells also differ in their capacity to specialize into various cell and tissue types. Current evidence indicates that the capability of adult stem cells to give rise to many different specialized cell types is more limited than that of embryonic stem cells. A single embryonic stem cell has been shown to give rise to specialized cells from all three embryonic layers. However, it has not yet been shown that a single adult stem cell can give rise to specialized cells derived from all three embryonic germ cell layers. Therefore, a single adult stem cell has not been shown to have the same degree of pluripotency as embryonic stem cells.

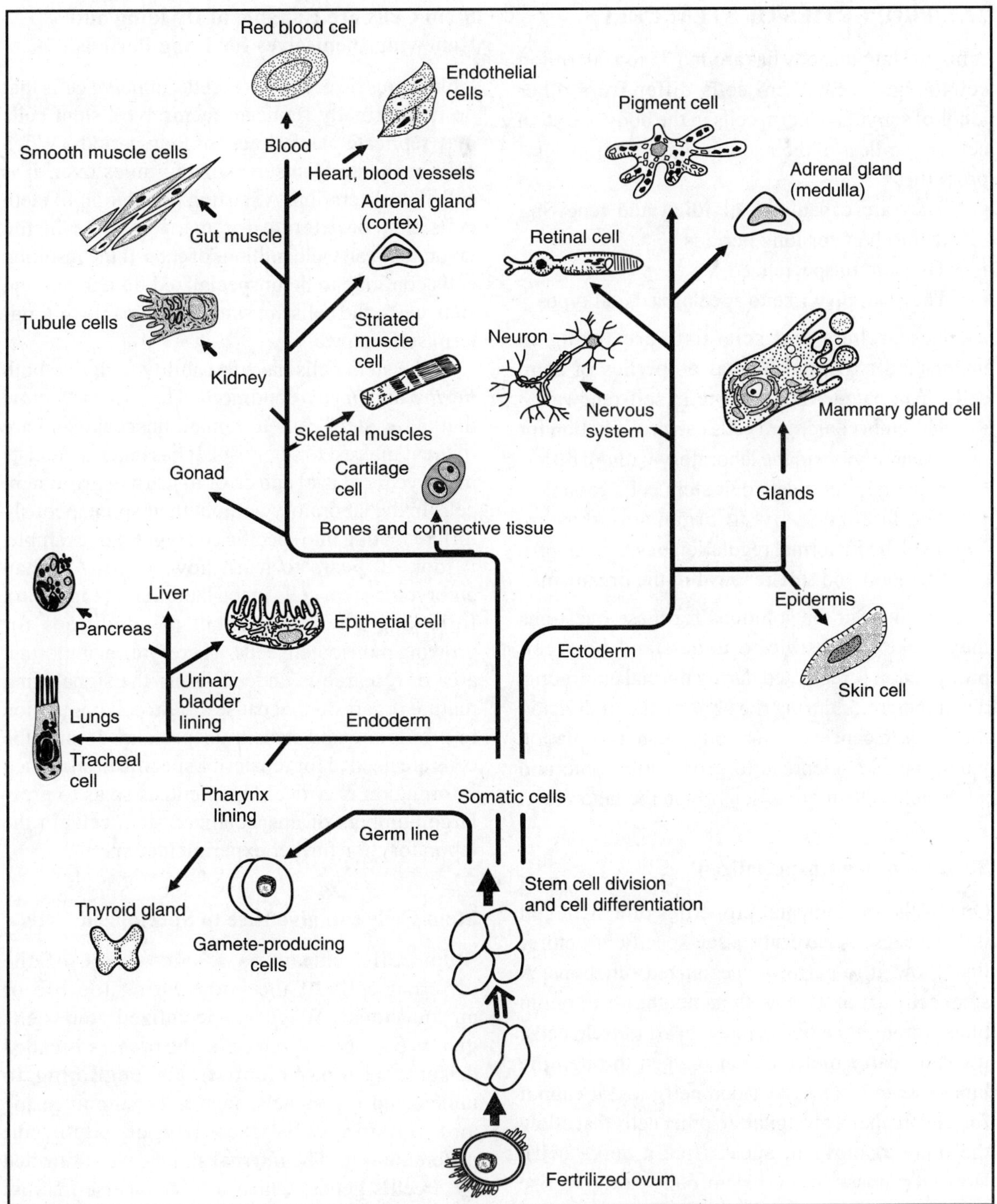

Fig. 22.5. Capacity of embryonic stem cells

22.7. PROPERTIES OF STEM CELLS

A mature human body has around 75 to 100 trillion cells **(Fig. 22.5)**. Stem cells differ from other kinds of somatic or germ cells in the body. All stem cells, regardless of their source, have three general properties:

- They are capable of dividing and renewing themselves for long periods.
- They are unspecialized.
- They can give rise to specialized cell types.

Researchers and scientists are trying to understand two fundamental properties of stem cells that relate to their long-term self-renewal:

- The embryonic stem cells can proliferation for a year or more in the laboratory without differentiating, but most adult stem cells cannot.
- The factors in living organisms that are involved in normal regulation of stem cell proliferation and self-renewal in the organisms.

Finding out the solutions for these questions may make it possible to understand how cell proliferation is regulated during normal embryonic development or during the abnormal cell division that leads to cancer. Importantly, such information would enable scientists to grow embryonic and adult stem cells more efficiently in the laboratory.

Stem Cells are Unspecialized

One of the fundamental properties of a stem cell is that it does not have any tissue-specific structures that allow, it to perform specialized functions. A stem cell cannot work with its neighbors to pump blood through the body (like a heart muscle cell); it cannot carry molecules of oxygen through the bloodstream (like a red blood cell); and it cannot fire electrochemical signals to other cells that allow the body to move or speak (like a nerve cell). However, unspecialized stem cells can give rise to specialized cells, including blood cells, heart muscle cells, islet cells, blood cells, or nerve cells.

Stem Cells are Capable of Dividing and Renewing themselves for Long Periods

Unlike muscle cells, blood cells, or nerve cells that do not normally replicate themselves, stem cells may replicate many times for long periods. When cells replicate themselves many times over, it is called proliferation. A starting population of stem cells that proliferate for many months in the laboratory can yield millions of cells. If the resulting cells continue to be unspecialized, like the parent stem cells, the cells are said to be capable of long-term self-renewal.

The stem cells have the ability to divide both *in vitro* and *in vivo* conditions. The specific factors that allow stem cells to remain unspecialized are of great interest to scientists. It has taken scientists many years of trial and error to learn to grow stem cells in the laboratory without their spontaneously differentiating into specific cell types. For example, it took 20 years to learn how to grow human embryonic stem cells in the laboratory **(Fig. 22.6)**, following the development of conditions for growing mouse stem cells. Therefore, an important area of research is understanding the signals in a mature organism that cause a stem cell population to proliferate and remain unspecialized, until the cells are needed for repair of a specific tissue. Such information is critical for scientists so as to grow large numbers of unspecialized stem cells in the laboratory for further experimentation.

Stem Cells can give Rise to Specialized Cells

Stem cells can act as a sole source of the different cells of the body during the life of an individual. When unspecialized stem cells give rise to specialized cells, the process is called differentiation. Scientists are beginning to understand the signals, factors or stimuli inside and outside cells that trigger stem cell differentiation. The internal signals are controlled by a cell's genes, which are interspersed across long strands of DNA, and carry coded instructions for all the structures and functions of a cell. The

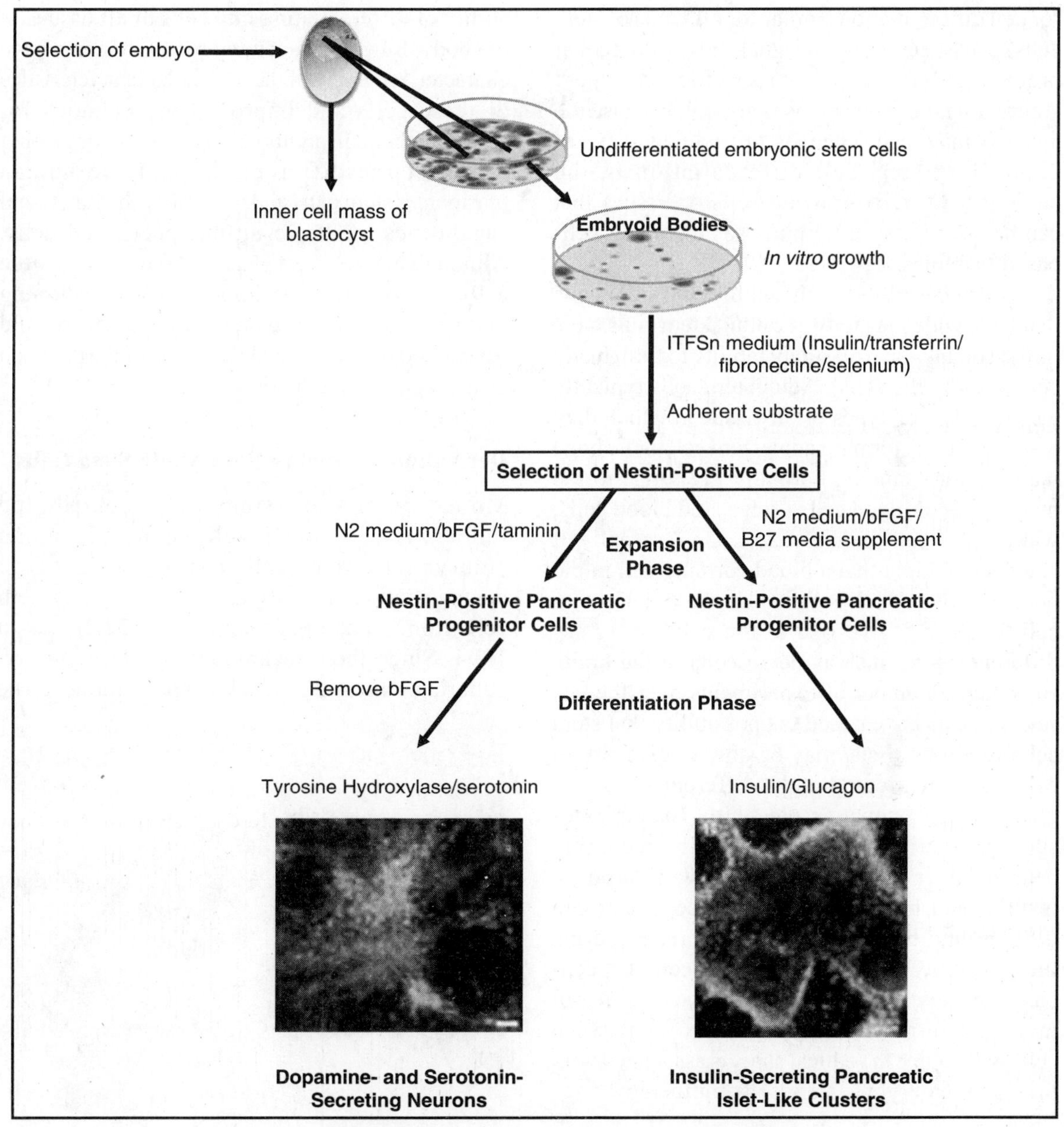

Fig. 22.6. Culture of stem cells in the laboratory

external signals for cell differentiation include chemicals secreted by other cells, and physical contact with neighboring cells and certain molecules in the microenvironment.

However, even with greater advancement made in the stem cell biology, many questions about stem cell differentiation remain unanswered. For example, are the internal and external signals for cell differentiation similar for all kinds of stem cells? Can specific sets of signals be identified that promote differentiation into specific cell types? Addressing these questions is critical, because the answers may lead scientists to find new ways of controlling stem cell differentiation in the laboratory, thereby growing cells or tissues that can be used for specific purposes including cell-based therapies.

When compared to the embryonic stem cells that have wide plasticity, the adult stem cells have limited or unknown plasticity (ability to form many cell types by the cells). Adult stem cells typically generate the cell types of the tissue in which they reside. A blood-forming adult stem cell in the bone marrow, for example, normally gives rise to the many types of blood cells such as red blood cells, white blood cells, and platelets. Until recently, it had been thought that a blood-forming cell in the bone marrow, which is called hematopoietic stem cell, could not give rise to the cells of a very different tissue, such as nerve cells in the brain. However, a number of experiments over the last several years have raised the possibility that stem cells from one tissue may be able to give rise to cell types of a completely different tissue—a phenomenon known as plasticity. Examples of such plasticity include blood cells becoming neurons; liver cells that can be made to produce insulin; and hematopoietic stem cells that can develop into heart muscle. Therefore, exploring the possibility of using adult stem cells for cell-based therapies has become a very active area of investigation for the researchers. In future, such cells will reduce the ethical aspects associated with the use of embryonic stem cells for therapy.

Human Embryonic Stem Cell and the Human Embryonic Germ Cell

Stem cell biology faced a lot of setback in the beginning. A new era in stem cell biology began in 1998, with the derivation of cells from human blastocysts and fetal tissue, having the unique ability of differentiating into cells of all tissues in the body. Since then, several research teams have characterized many of the molecular characteristics of these cells and improved the methods for culturing them. In addition, scientists are beginning to direct the differentiation of the human pluripotent stem cells and to identify the functional capabilities of the resulting specialized cells. Although in its earliest phases, research with these cells is proving to be important to developing innovative cell replacement strategies to rebuild tissues and restore critical functions of the diseased or damaged human body.

Derivation of Human Embryonic Stem Cells

Mouse embryonic stem cells contributed immensely to the development of human embryonic stem cell biology. The first documentation of the isolation of embryonic stem cells from human blastocysts **(Fig. 22.7)** was in 1994. Since then, techniques for deriving and culturing human ES cells have been refined. The

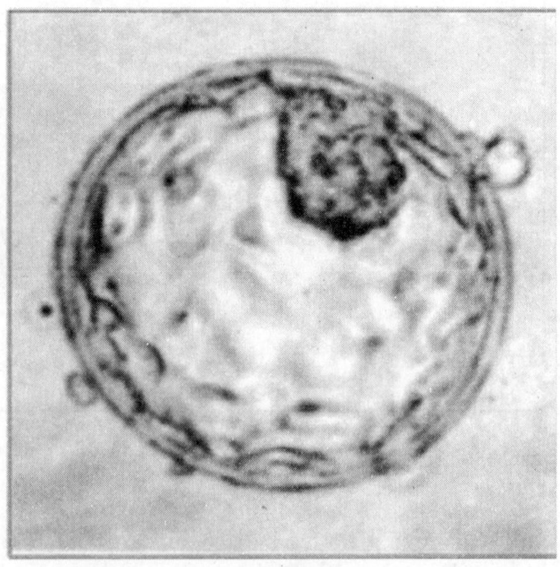

Fig. 22.7. Human blastocyst showing inner cell mass and trophectoderm

ability to isolate human ES cells from blastocysts and growing them in culture seems to depend, in large, on the integrity and condition of the blastocyst from which the cells are derived. In general, blastocysts with a large and distinct inner cell mass tend to yield ES cultures most efficiently.

Blastocyst development in vitro

An ova, or oocyte, or egg is the main cell for the development of the embryonic stem cells. About 10,000 to 50,000 motile sperms are used for fertilization and they are added to 100 µl—1000 µl of culture medium having the required oocyte or egg. The modified Ham's F10 medium, Earl's solution, modified Whitten's medium or Whittingham's T6 medium are used for the culturing and maintenance of oocyte and sperms. After a human oocyte is fertilized *in vitro* by a sperm cell, the following events occur according to a predictable time-line. The period of 18 to 24 hours, after *in vitro* fertilization of the oocyte, is considered day 1. By day two (24 to 25 hours), the zygote (fertilized egg) undergoes the first cleavage to produce a 2-cell embryo. By day 3 (72 hours), the embryo reaches the 8-cell stage called a morula. It is at this stage that the genome of the embryo begins to control its own development. This means that any maternal influences, due to the presence of mRNA and proteins in the oocyte cytoplasm, are significantly reduced. By day 4, the cells of the embryo adhere tightly to each other in a process known as compaction, and by day 5, the cavity of the blastocyst is completed. The inner cell mass begins to separate from the outer cells, which become the trophectoderm that surrounds the blastocyst. This represents the first observable sign of cell differentiation in the embryo. Many IVF clinics now transfer day-5 embryos **(Fig. 22.8)** to the uterus for optimal implantation—a stage of development that more closely parallels the stage at which a blastocyst would implant in the wall of the uterus *in vivo*. This represents a change, and a greatly improved implantation rate, from earlier IVF procedures in which a 2-cell embryo was used for implantation.

The fifth day blastocysts are used to derive ES cell cultures. A normal day-5 human embryo *in vitro* consists of 200 to 250 cells. Most of the cells comprise the trophectoderm. For deriving ES

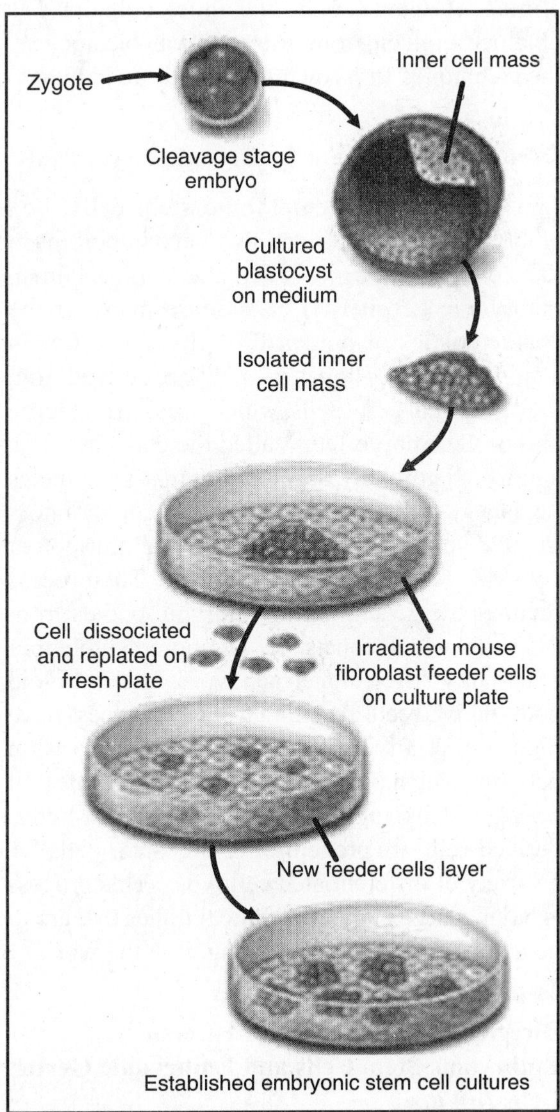

Fig. 22.8. Stages of development of embryonic stem cells

cell cultures, the trophectoderm is removed, either by microsurgery or immunosurgery (in which antibodies against the trophectoderm help break it down, thus freeing the inner cell mass). At this stage, the inner cell mass is composed of only 30 to 34 cells.

The *in vitro* conditions for growing a human embryo to the blastocyst stage vary among IVF clinics. However, once the inner cell mass is obtained from either mouse or human blastocysts, the techniques for growing ES cells are similar.

Derivation of Human Embryonic Germ Cells

The other types of embryonic stem cells, i.e., embryonic germ cells are also very important in the stem cell biology. As depicted earlier, human embryonic germ (EG) cells share many of the characteristics of human ES cells, but differ in significant ways. Human EG cells are derived from the primordial germ cells, which occur in a specific part of the embryo/fetus called the gonadal ridge, and which normally develop into mature gametes (eggs and sperm). Gearhart and his collaborators in 1998 devised methods for growing pluripotent cells derived from human EG cells. The process requires the generation of embryoid bodies from EG cells, which consists of an unpredictable mix of partially differentiated cell types. The embryoid body-derived cells resulting from this process have high proliferative capacity and gene expression patterns, which are representative of multiple cell lineages. This suggests that the embryoid body-derived cells are progenitor or precursor cells for a variety of differentiated cell types. This process is having large number of ethical issues that are to be addressed before proceeding with the work.

Directed Differentiation of Human Embryonic Stem Cells and Embryonic Germ Cells *In Vitro*

Stem cells are important to biomedical researchers because they can be used to generate virtually any type of specialized cell in the human body. Currently, a major goal for embryonic stem cell research is to control the differentiation of human ES and EG cell lines into specific kinds of cells—an objective that must be met if the cells are to be used as the basis for therapeutic transplantation, testing drugs, or screening potential toxins. The techniques now being tested to direct human ES cell differentiation are borrowed directly from the techniques used to direct the differentiation of mouse ES cells *in vitro*.

22.8. HEMATOPOIETIC STEM CELLS

In the category of stem cells, easily available cells are the hematopoietic stem cells derived form the blood tissues. Stem cells are undifferentiated cells that can make any cell type in the body. With more than 50 years of studying blood-forming stem cells called hematopoietic stem cells, scientists have developed sufficient understanding to actually use them as a therapy. Currently, no other type of stem cell—adult, fetal or embryonic—has attained such status. Hematopoietic stem cell transplants are now routinely used to treat patients with cancers and other disorders of the blood and immune systems. Recently, researchers have observed in animal studies that hematopoietic stem cells are able to form other kinds of cells, such as muscle, blood vessels, and bone. If this can be applied to human cells, it may eventually be possible to use hematopoietic stem cells to replace a wider array of cells and tissues than once thought.

Hematopoietic stem cells are studied from the point of identification of blood components and their divisions to yield many types of blood cells. Despite the vast experience with hematopoietic stem cells, scientists face major setbacks in expanding their use beyond the replacement of blood and immune cells. First, hematopoietic stem cells are unable to proliferate (replicate themselves) and differentiate (become specialized to other cell types) *in vitro* (in the test tube or culture dish). Second, scientists do not yet have an accurate method to distinguish stem cells from other cells recovered from the blood or bone marrow. Until scientists overcome these technical barriers, they believe it is unlikely that hematopoietic stem cells will be applied as cell replacement therapy in diseases such as diabetes, Parkinson's Disease, spinal cord injury, and many others.

For proper homeostatis, blood is a major contributing factor in the human circulatory system. Blood cells are responsible for constant maintenance and immune protection of every cell type of the body. This remarkable, relentless and brutal systematic work requires that blood cells, along with skin cells, have the greatest powers of self-renewal of any adult tissue.

Stem cells that form blood and immune cells are known as hematopoietic stem cells (HSCs). They are ultimately responsible for the constant renewal of blood—the production of billions of new blood cells each day. Physicians and basic researchers have known and capitalized on this fact for more than 50 years in treating many diseases. The first evidence and definition of blood-forming stem cells came from the studies of people exposed to lethal doses of radiation in 1945.

Basic research soon followed. After duplicating radiation sickness in mice, scientists found that they could rescue the mice from death with bone marrow transplants from healthy donor animals. In the early 1960s, Till and McCulloch began analyzing the bone marrow to find out which components were responsible for regenerating blood. They defined, what remain, the two hallmarks of a HSC—it can renew itself and produce cells that give rise to all the different types of blood cells.

Hematopoietic Stem Cell and its Sources

Hematopoietic stem cell can be isolated from the blood or bone marrow. It can renew itself, differentiate into a variety of specialized cells; mobilize out of the bone marrow into circulating blood; and undergo programmed cell death called apoptosis—a process by which cells that are detrimental undergo self-destruction.

A major thrust of basic HSC research since the 1960s has been identifying and characterizing these stem cells. Since HSCs look and behave in culture like ordinary white blood cells, this has been a difficult challenge and makes them difficult to identify by morphology (size and shape). Even today, scientists rely on cell surface proteins, which serve, only roughly, as markers of white blood cells.

Identifying and characterizing properties of HSCs began with studies in mice, which laid the groundwork for human studies. The challenge is formidable as about 1 in every 10,000 to 15,000 bone marrow cells is thought to be a stem cell. In the blood stream, the proportion falls to 1 in 100,000 blood cells. To this end, scientists began to develop tests for proving the self-renewal and plasticity of HSCs.

The "gold standard" for proving that a cell derived from mouse bone marrow is indeed an HSC is still based on the same proof described above and used in mice many years ago. That is, the cells are injected into a mouse that has received a dose of irradiation sufficient to kill its own blood-producing cells. If the mouse recovers and all types of blood cells reappear (bearing a genetic marker from the donor animal), the transplanted cells are deemed to have included stem cells.

These studies reveal that there are two kinds of HSCs. If bone marrow cells from the transplanted mouse can, in turn, be transplanted to another lethally irradiated mouse that can restore its hematopoietic system over some months, they are considered to be long-term stem cells that are capable of self-renewal. Other cells from bone marrow can immediately regenerate all the different types of blood cells, but under normal circumstances cannot renew themselves over the long term, and these are referred to as short-term progenitor or precursor cells. Progenitor or precursor cells are relatively immature cells that are precursors to a fully differentiated cell of the same tissue type. They are capable of proliferating, but have a limited capacity to differentiate into more than one cell type as HSCs do. For example, a blood progenitor cell may only be able to make a red blood cell.

The longevity of short-term stem cells for humans is not firmly established. A true stem cell, capable of self-renewal, must be able to renew itself for the entire lifespan of an organism. It is these long-term replicating HSCs that are most important for developing HSC-based cell therapies. Unfortunately, to date, researchers could not distinguish the long-term from the short-term cells when they are removed from the bloodstream or bone marrow.

The central problem of the assays used to identify long-term stem cells and short-term progenitor cells is that they are difficult, expensive, time-consuming, and cannot be done in humans. A few assays are now available to test the cells in culture for their ability to form primitive and long-lasting colonies of cells, but these tests are not accepted as proof that a cell is a long-term stem cell. Some genetically altered mice can receive transplanted human HSCs to test the cells' self-renewal and hematopoietic capabilities during the life of a mouse, but the relevance of this test to the cells in humans—who may live for decades—is open to question.

The difficulty of HSC assays has contributed to two mutually confounding research problems: definitively identifying the HSC, and getting it to proliferate or increase its numbers in a culture dish. More rapid research progress on characterizing and using HSCs would be possible if they could be readily grown in the laboratory. Conversely, there would be progress in identifying growth conditions suitable for HSCs and getting the cells to multiply if scientists could reliably and readily identify true HSCs.

Cell Markers and Identity of Hematopoietic Stem Cells

Cell surface markers play an important role in the identification of cells with different characters. HSCs have an identity problem. First, the ones with long-term replicating ability are rare. Second, there are multiple types of stem cells. And third, the stem cells look like many other blood or bone marrow cells. So how do researchers find the desired cell populations? The most common approach is through markers that appear on the surface of cells. These are useful, but not perfect tools for the research laboratory.

In 1988, in an effort to develop a reliable means of identifying these cells, Irving Weissman and his collaborators focused attention on a set of protein markers on the surface of mouse blood cells that were associated with increased likelihood that the cell was a long-term HSC. Four years later, the laboratory proposed a comparable set of markers for the human stem cell. Weissman proposed the markers shown in **Table 22.2** as the closest markers for mouse and human HSCs.

Table 22.2: Proposed cell-surface markers of undifferentiated hematopoietic stem cells

Mouse	Human
CD34$^{low/-}$	CD34$^+$
SCA-1$^+$	CD59^{+*}
Thy1$^{+/low}$	Thy1$^+$
CD38$^+$	CD38$^{low/-}$
C-kit$^+$	C-kit$^{-/low}$
lin^{-*}	lin^{-**}

*Only one of a family of CD59 markers has thus far been evaluated.
**Lin$^-$ cells lack 13 to 14 different mature blood-lineage markers.

Embryonic stem cells and embryonic germ cells

In 1985, it was shown that it is possible to obtain precursors to many different blood cells from mouse embryonic stem cells. Perkins was able to obtain all the major lineages of progenitor cells from mouse embryoid bodies, even without adding hematopoietic growth factors.

Mouse embryonic stem cells in culture, given the right growth factors, can generate most, if not all, the different blood cell types, but no one has yet achieved the "gold standard" of proof that they can produce long-term HSCs from these sources— namely by obtaining cells that can be transplanted into lethally irradiated mice to reconstitute long-term hematopoiesis.

The picture for human embryonic stem and germ cells is even less clear. Scientists from James Thomson's laboratory reported in 1999 that they were able to direct human embryonic stem cells— which can now be cultured in the lab—to produce blood progenitor cells. Israeli scientists reported that they had induced human ES cells to produce hematopoietic cells, as evidenced by their production of a blood protein—gamma-globin. Cell lines derived from human embryonic germ cells (cultured cells derived originally from cells in the embryo that would ultimately give rise to eggs or sperm) that are cultured under certain conditions will produce CD34$^+$ cells. The blood-producing cells derived from human ES and embryonic germ (EG) cells have not been rigorously tested for long-term self-renewal or the ability to give rise to all the different blood cells.

As sketchy as the data may be on the hematopoietic powers of human ES and EG cells, blood experts are intrigued by their clinical potential and their potential to answer basic questions on renewal and differentiation of HSCs. Connie Eaves, who has made comparisons of HSCs with fetal liver, cord blood, and adult bone marrow, expects cells derived from embryonic tissues to have some interesting traits. She says, actively dividing blood-producing cells from ES cell culture—if they are like other dividing cells— will not themselves engraft or rescue hematopoiesis in an animal whose bone marrow has been destroyed. However, they may play a critical role in developing an abundant supply of HSCs grown

in the lab. Indications are that the dividing cells will also more readily lend themselves to gene manipulations than do adult HSCs. Eaves anticipates that HSCs derived from early embryo sources will be developmentally more "plastic" than later HSCs, and more capable of self-renewal.

The cell surface markers found on mouse and human hematopoietic stem cells, as they exist in their undifferentiated state *in vivo* and *in vitro,* are indicated in **Table 22.2.** As these cells begin to develop as distinct cell lineages, the cell surface markers were no longer identified or lost during the cell development and differentiation.

Such cell markers can be tagged with monoclonal antibodies bearing a fluorescent label, and culled out of bone marrow with fluorescence-activated cell sorting (FACS).

The groups of cells thus sorted by surface markers are heterogeneous, and include some cells that are true, long-term self-renewing stem cells, some shorter-term progenitors, and some non-stem cells. Weissman's group showed that as few as five genetically tagged cells, injected along with larger doses of stem cells into lethally irradiated mice, could establish themselves and produce marked donor cells in all blood cell lineages for the lifetime of the mouse. A single tagged cell could produce all lineages for as many as seven weeks, and 30 purified cells were sufficient to rescue mice and fully repopulate the bone marrow without extra doses of backup cells to rescue the mice. Despite these efforts, researchers remain divided on the most consistently expressed set of HSC markers. Connie Eaves of the University of British Columbia says none of the markers tied to unique stem cell functions or truly defines the stem cell.

More recently, Diane Krause and her colleagues at Yale University, New York University, and Johns Hopkins University, used a new technique to home in on a single cell capable of reconstituting all blood cell lineages of an irradiated mouse. After marking bone marrow cells of a donor male mice with a nontoxic dye, they injected the cells into female recipient mice that had been given a lethal dose of radiation. Over the next two days, some of the injected cells migrated to the bone marrow of the recipients and did not divide. When transplanted into a second set of irradiated female mice, they eventually proved to be a concentrated pool of self-renewing stem cells. The cells also reconstituted blood production. The scientists estimate that their technique concentrated the long-term stem cells 500 to 1,000- fold compared with bone marrow.

22.9. SOURCES OF HEMATOPOIETIC STEM CELLS

Bone Marrow

Bone marrow is the site where different cells of the blood are generated. The classic source of hematopoietic stem cells (HSCs) is bone marrow. For more than 40 years, doctors performed bone marrow transplants by anesthetizing the stem cell donor, puncturing a bone—typically a hipbone—and drawing out the bone marrow cells with a syringe. About 1 in every 100,000 cells in the marrow is a long-term, blood-forming stem cell; other cells include stromal cells, stromal stem cells, blood progenitor cells, and mature and maturing white and red blood cells.

Peripheral Blood

Cells produced in the bone marrow come out of it and move into the circulation. As a source of HSCs for medical treatments, bone marrow retrieval directly from the bone is quickly fading into history. For clinical transplantation of human HSCs, doctors now prefer to harvest donor cells from peripheral, circulating blood. It has been known for decades that a small number of stem and progenitor cells circulate in the bloodstream, but in the past 10 years, researchers have found that they can coax the cells to migrate from marrow to the blood in greater numbers by injecting the donor with a cytokine, such as granulocyte colony

stimulating factor (GCSF). The donor is injected with GCSF a few days before the cell harvest. To collect the cells, doctors insert an intravenous tube into the donor's vein, and pass his blood through a filtering system that pulls out CD34$^+$ white blood cells and returns the red blood cells to the donor. Of the cells collected, just 5% to 20% will be true HSCs. Thus, when medical researchers commonly refer to peripherally harvested "stem cells", this is something of a misnomer. As is true for bone marrow, the CD34$^+$ cells are a mixture of stem cells, progenitors, and white blood cells of various degrees of maturity.

In the past three years, the majority of autologous (where the donor and recipient are the same person) and allogeneic (where the donor and recipient are different individuals) bone marrow transplants have actually been white blood cells drawn from peripheral circulation, not bone marrow. Richard Childs, an intramural investigator at the NIH, says that peripheral harvest of cells is easier on the donor—with minimal pain, no anesthesia, and no hospital stay—and also yields better cells for transplants. Childs points to evidence that patients receiving peripherally harvested cells have higher survival rates than bone marrow recipients do. The peripherally harvested cells contain twice as many HSCs as stem cells taken from the bone marrow, and engraft more quickly. This means the patients may recover white blood cells, platelets, and their immune and clotting protection several days faster than they would with a bone marrow graft. Scientists at Stanford report that highly purified, mobilized peripheral cells that have CD34$^+$ and Thy-1$^+$ surface markers engraft swiftly and without complication in breast cancer patients receiving an autologous transplant of the cells after intensive chemotherapy.

Umbilical Cord Blood

The waste produced during the birth of the child can now be used for the potential research. In the late 1980s and early 1990s, physicians began to recognize that blood from the human umbilical cord and placenta was a rich source of HSCs. This tissue supports the developing fetus during pregnancy, is delivered along with the baby, and, is usually discarded. Since the first successful umbilical cord blood transplants in children with Fanconi anemia, the collection and therapeutic use of these cells has grown quickly. The New York Blood Center's Placental Blood Program, supported by NIH, is the largest US public umbilical cord blood bank and has 13,000 donations available for transplantation into patients who need HSCs. Since it began collecting umbilical cord blood in 1992, the center has provided thousands of cord blood units to patients. Umbilical cord blood recipients, typically children, have now lived for more than eight years, relying on the HSCs from an umbilical cord blood transplant.

There is a substantial amount of research being conducted on umbilical cord blood to search for the ways to expand the number of HSCs, and compare and contrast the biological properties of cord blood with adult bone marrow stem cells. There have been suggestions that umbilical cord blood contains stem cells that have the capability of developing cells of multiple germ layers (multi-potent) or even all germ layers, e.g., endoderm, ectoderm, and mesoderm (pluripotent). To date, there is no published scientific evidence to support this claim. While umbilical cord blood represents a valuable resource for HSCs, research data have not conclusively shown qualitative differences in the differentiated cells produced between this source of HSCs and peripheral blood and bone marrow.

Fetal Hematopoietic System

Fetal blood is also one of the major source of HSCs. An important source of HSCs in research, but not in clinical use, is the developing blood-producing tissues of fetal animals. Hematopoietic cells appear early in the development of all vertebrates. Most extensively studied in the mouse, HSC production sweeps through the developing embryo and fetus in waves. Beginning at about day seven in the life of the mouse embryo, the earliest hematopoietic activity is indicated by the appearance of blood islands in the yolk sac. The point is disputed, but some scientists contend that yolk sac blood production is transient and will generate some blood cells for the embryo, but probably not the bulk of the HSCs for the adult animal. According to this proposed scenario, most stem cells that will be found in the adult bone marrow and circulation are derived from cells that appear slightly later and in a different location. This other wave of hematopoietic stem cell production occurs in the AGM—the region where the aorta, gonads, and fetal kidney (mesonephros) begin to develop. The cells that give rise to the HSCs in the AGM may also give rise to endothelial cells that line the blood vessels. These HSCs arise at around days 10 to 11 in the mouse embryo (weeks 4 to 6 in human gestation), divide, and within a couple of days, migrate to the liver. The HSCs in the liver continue to divide and migrate, spreading to the spleen, thymus, and near the time of birth, to the bone marrow.

Whereas an increasing body of fetal HSC research is emerging from mice and other animals, there is less information about human fetal and embryonic HSCs. Scientists in Europe, including Coulombel, Peault, and colleagues, first described hematopoietic precursors in human embryos only a few years ago. Most recently, Gallacher and others reported finding HSCs circulating in the blood of 12- to 18-week aborted human fetuses that was rich in HSCs. These circulating cells had different markers than cells from fetal liver, bone marrow, or umbilical cord blood.

Difference of HSCs from Varying Sources

As hematopoietic stem cells are derived from various sources, their clear cut identification is also an important factor before continuation of research work. Scientists in the laboratories and clinics are beginning to measure the differences among HSCs from different sources. In general, they find that HSCs taken from tissues at earlier developmental stages have a greater ability to self-replicate, show different homing and surface characteristics, and are less likely to be rejected by the immune system, making them potentially more useful for therapeutic transplantation.

Stem Cell Populations of the Bone Marrow

Development of different adult stem cells is a hot line of research work in the stem cell biology. When do HSCs move from the early locations in the developing fetus to their adult "home" in the bone marrow? European scientists have found that the relative number of $CD34^+$ cells in the collections of cord blood declined with gestational age, but expression of cell-adhesion molecules on these cells increased.

The researchers believe that these changes reflect the preparations for the cells to relocate—from homing in fetal liver to bone marrow.

The point is controversial, but a paper by Chen *et al.,* provides evidence that at least in some strains of mice, HSCs from old mice are less able to repopulate bone marrow after transplantation than are cells from young adult mice. Cells from fetal mice were 50% to 100% better at repopulating marrow than were cells from young adult mice. The specific potential for repopulating marrow appears to be strain-specific, but the scientists found that this potential declined with age for both strains. Other scientists find no decrease or sometimes increase in numbers of HSCs with age. Because of the difficulty in identifying a long-term stem cell, it remains difficult to quantify changes in numbers of HSCs as a person ages.

Effectiveness of Transplants of Adult Versus Umbilical Cord Blood Stem Cells

Stem cells from different sources have immense capacity depending upon the source and type of tissues used. A practical and important difference between HSCs collected from adult human donors and umbilical cord blood is simply quantitative. Doctors are rarely able to extract more than a few million HSCs from a placenta and umbilical cord—too few to use in a transplant for an adult, who would ideally get 7 to 10 million $CD34^+$ cells per kilogram body weight, but often adequate for the transplant of a child.

Leonard Zon says that HSCs from cord blood are less likely to cause a transplantation complication called graft-versus-host disease, in which white blood cells from a donor attack tissues of the recipient. In a recent review of umbilical cord blood transplantation, Laughlin cites evidence that cord blood causes less graft-versus-host disease. Laughlin writes that it is yet to be determined whether umbilical cord blood HSCs, in fact, live longer in a transplant recipient.

In lab and mouse-model tests comparing $CD34^+$ cells from human cord with $CD34^+$ cells derived from adult bone marrow, researchers found that cord blood had greater proliferation capacity. White blood cells from cord blood engrafted better in a mouse model, which was genetically altered to tolerate the human cells, than did their adult counterparts.

Effectiveness in Transplants of Peripheral Versus Bone Marrow Stem Cells

When compared to the other sources of stem cells, the peripheral stem cells have greater applications and access. In addition to being far easier to collect, peripherally harvested white blood cells have other advantages over bone marrow. Cutler and Antin's review says that peripherally harvested cells engraft more quickly, but are more likely to cause graft-versus-host disease. Prospecting for the most receptive HSCs for gene therapy, orlic and colleagues found that mouse HSCs mobilized with cytokines were more likely to take up genes from a viral vector than non-mobilized bone marrow HSCs.

As stated earlier, an HSC in the bone marrow has four actions in its repertoire:

1. It can renew itself.
2. It can differentiate.
3. It can mobilize out of the bone marrow into circulation (or the reverse).
4. It can undergo programmed cell death, or apoptosis. Understanding the how, when, where, which, and why of this simple repertoire will allow researchers to manipulate and use HSCs for tissue and organ repair.

Self-Renewal of Hematopoietic Stem Cells

Even though with the remarkable capacity of growth and division, stem cells are not easily evolved in the culture. Researchers and scientists have had a tough time trying to grow or even maintain efficient true stem cells in culture. This is an important goal because cultures of HSCs that could maintain their characteristic properties of self-renewal and lack of differentiation could provide an unlimited source of cells for therapeutic transplantation and study. When bone marrow or blood cells are observed in culture, one often observes large increases in the number of cells. This usually reflects an increase in differentiation of cells to progenitor cells that can give rise to different lineages of blood cells but cannot renew themselves. True stem cells divide and replace themselves slowly in adult bone marrow.

New tools for gene-expression analysis will now allow scientists to study developmental changes in telomerase activity and telomeres. Telomeres are regions of DNA found at the end of chromosomes that are extended by the enzyme telomerase. Telomerase activity is necessary for the cells to proliferate, and activity decreases with age, leading to shortened telomeres. Scientists hypothesize that decline in stem cell renewal will be associated with decline in telomere length and activity. Telomerase activity in hematopoietic cells is associated with self-renewal potential.

Cultural practice of long term culture of stem cells is very difficult to perform in the laboratory. Because self-renewal divisions are rare, hard to induce in culture, and difficult to prove, scientists do not have a definitive answer to the burning question: what puts or perhaps keeps the HSCs in a self-renewal division mode? HSCs injected into an anemic patient or mouse—or one whose HSCs have otherwise been suppressed or killed–will home to the bone marrow and undergo active division to replenish all the different types of blood cells and yield additional self-renewing HSCs. But exactly how this happens remains a mystery that scientists are struggling to solve by manipulating the cultures of HSCs in the laboratory.

Two recent examples of progress in the culturing studies of mouse HSCs are by Ema and coworkers and Audet and colleagues. Ema *et al.,* found that two cytokines—stem cell factor and thrombopoietin—efficiently induced an unequal first cell division in which one daughter cell gave rise to repopulating cells with self-renewal potential. Audet *et al.,* found that activation of the signaling molecule gp130 is critical to survival and proliferation of mouse HSCs in culture.

Different biosignaling molecules are found to be useful in the stem cell biology. Work with specific cytokines and signaling molecules builds on several earlier studies demonstrating modest increases in the numbers of stem cells that could be induced briefly in culture. For example, Van Zant and colleagues used continuous-perfusion culture and bioreactors in an attempt to boost human HSC numbers in single cord blood samples incubated for one to two weeks. They obtained a 20-fold increase in "long-term culture initiating cells".

More clues on how to increase numbers of stem cells may come from looking at other animals and various developmental stages. During early developmental stages—in the fetal liver, for example—HSCs may undergo more active cell division to increase their numbers, but later in life, they divide far less often. Culturing HSCs from

10- and 11-day-old mouse embryos, Elaine Dzierzak at Erasmus University in the Netherlands found that she can get a 15-fold increase in HSCs within the first 2 or 3 days after she remove the AGM from the embryos. Dzierzak recognized that this is dramatically different from anything seen with adult stem cells, and suggested that it is a difference with practical importance. She suspected that the increase is not so much a response to what is going on in the culture, rather, it represents the developmental momentum of this specific embryonic tissue. That is, it is the inevitable consequence of divisions cued by that specific embryonic micro-environment. After five days, the number of HSCs plateaus and can be maintained for upto a month. Dzierzak says that the key to understanding how adult-derived HSCs can be expanded and manipulated for clinical purposes may very well be found by defining the cellular composition and complex molecular signals in the AGM region during development.

In another approach, Lemischka and coworkers have been able to maintain mouse HSCs for four to seven weeks when they are grown on a clonal line of cells (AFT024) derived from the stroma— the other major cellular constituent of bone marrow. No one knows which specific factors secreted by the stromal cells maintain the stem cells. Ongoing gene cloning is rapidly zeroing in on novel molecules from the stromal cells that may "talk" to the stem cells and persuade them to remain stem cells, i.e., continue to divide and not differentiate.

If stromal factors provide the key to stem cell self-renewal, research on maintaining stromal cells may be an important prerequisite. In 1999, researchers at Osiris Therapeutics and Johns Hopkins University reported culturing and expanding the numbers of mesenchymal stem cells, which produce the stromal environment. Whereas, the cultured HSCs rush to differentiate and fail to retain primitive, self-renewing cells, the mesenchymal stem cells could be increased in numbers and still retained their powers to generate the full repertoire of descendant lineages.

Differentiation of HSCs into Components of the Blood and Immune System

Blood cells have to be supplied on a regular basis, whenever they are required by the system in the body. Producing differentiated white and red blood cells is the real work of HSCs and progenitor cells. M.C. MacKey calculates that in the course of producing a mature, circulating blood cell, the original hematopoietic stem cell undergoes between 17 and 19.5 divisions, giving a net amplification of between ~170,000 and ~720,000.

Through a series of careful studies on cultured cells—cells with mutations found in leukemia patients, or cells that have been genetically altered—investigators have discovered many key growth factors and cytokines that induce progenitor cells to make different types of blood cells. These factors interact with one another in complex ways to create a system of exquisite genetic control and coordination of blood cell production.

Migration of Hematopoietic Stem Cells into and out of the Marrow and Tissues

Many types of experiments have been performed to know the position of the various types of blood cells in the circulatory system. Scientists know that much of the time HSCs live in intimate contact with the stroma of bone marrow in adults. But HSCs may also be found in the spleen, peripheral blood circulation, and other tissues. Connection to the interstices of bone marrow is important both for the engraftment of transplanted cells and the maintenance of stem cells as a self-renewing population. Connection to stroma is also important for the orderly proliferation, differentiation, and maturation of blood cells.

Weissman said that HSCs appear to make brief forays out of the marrow into tissues, then duck back into the marrow. Scientists do not understand why or how HSCs leave bone marrow or return to it. Scientists find that HSCs, which have been mobilized into peripheral circulation, are mostly non-dividing cells. They report that adhesion molecules on the stroma play a role in mobilization, attachment to the stroma, and transmitting signals that regulate HSC self-renewal and progenitor differentiation.

Understanding the forces at play in HSC apoptosis is important for maintaining or increasing their numbers in culture. For example, without growth factors—supplied in the medium, or through serum or other feeder layers of cells—HSCs undergo apoptosis. Domen and Weissman found that stem cells need two growth factor signals to continue life and avoid apoptosis, one via a protein called BCL-2, the other from steel factor, which, by itself, induces HSCs to produce progenitor cells but not to self-renew.

22.10. POTENTIAL USES OF HUMAN EMBRYONIC STEM CELLS

Many uses have been proposed for human embryonic stem cells. The most-often discussed is their potential use in transplant therapy, i.e., to replace or restore tissue that has been damaged by disease or injury.

Using Human Embryonic Stem Cells for Therapeutic Transplants

Humans are suffering form large number of genetic diseases that are affecting populations. Diseases that might be treated by transplanting human ES-derived cells include Parkinson's disease, diabetes, traumatic spinal cord injury, Purkinje cell degeneration, Duchenne's muscular dystrophy, heart failure and osteogenesis imperfecta. However, treatments for any of these diseases require that human ES cells be directed to differentiate into specific cell types prior to transplant. The research is occurring in several laboratories, but is limited because only few laboratories have access to human ES cells. Thus, at this stage, any therapies based on the use of human ES cells are still hypothetical and highly experimental.

Stem cells have many greater applications in the medical field. One of the current advantages of using ES cells as compared to adult stem cells is that ES cells have an unlimited ability to proliferate *in vitro*, and are more likely to be able to generate a broad range of cell types through directed differentiation. Ultimately, it will also be necessary to identify the optimal stage(s) of differentiation for transplant, and demonstrate that the transplanted ES-derived cells can survive, integrate, and function in the recipient.

The potential disadvantages of the use of human ES cells for transplant therapy include the propensity of undifferentiated ES cells to induce the formation of tumors (teratomas), which are typically benign. Since these are undifferentiated cells—rather than their differentiated progeny, that have been shown to induce teratomas—tumor formation might be avoided by devising methods for removing any undifferentiated ES cells prior to transplant. Also, it should be possible to devise a fail-safe mechanism, i.e., to insert into transplanted ES-derived cells suicide genes that can trigger the death of the cells if they become tumorigenic.

Human ES derived cells would also be advantageous for transplantation purposes if they did not trigger immune rejection **(Figs. 22.9 to 22.11)**. The immunological status of human ES cells has not been studied in detail, and it is not known how immunogenic ES-derived cells might be. In general, the immunogenicity of a cell depends on its expression of class I major histocompatability antigens (MHC), which allow the body to distinguish its own cells from the foreign tissue, and on the presence of cells that can bind to foreign antigens and "present" them to the immune system.

The potential immunological rejection of human ES-derived cells might be avoided by genetically engineering the ES cells to express the MHC antigens of the transplant recipient, or by using nuclear transfer technology to generate ES cells that are genetically identical to the person

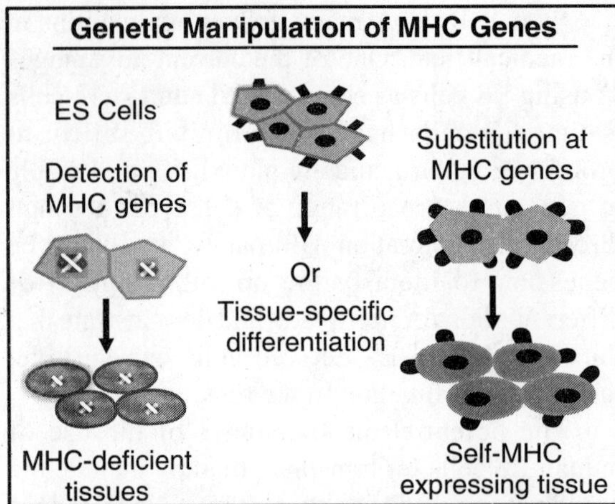

Fig. 22.9. Human embryonic stem cells and their genetic manipulation for major histocompatibility complex genes

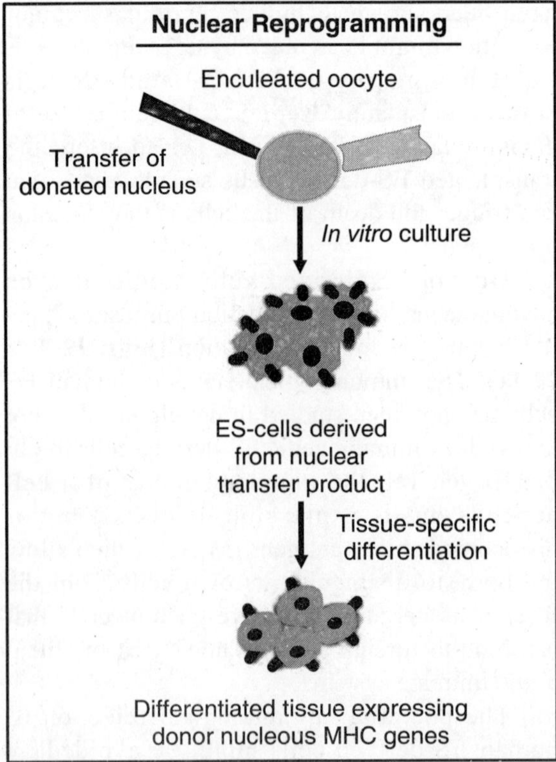

Fig. 21.10. Human embryonic stem cells and their genetic manipulation for major histocompatibility complex genes expression by nuclear reprogramming

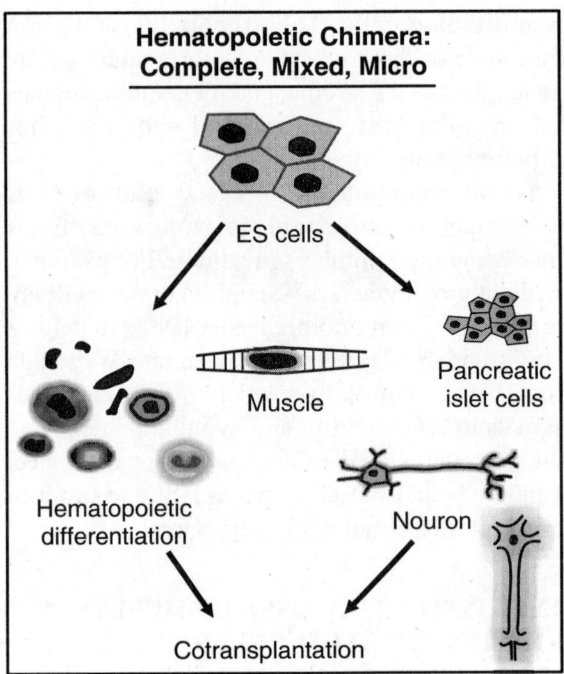

Fig. 22.11. Human embryonic stem cells and their multiple uses in the transplantation

who receives the transplant. It has been suggested that this could be accomplished by using somatic cell nuclear transfer technology (so-called therapeutic cloning), in which the nucleus is removed from one of the transplant patient's cells, such as skin cell, and injected into an oocyte. The oocyte, thus fertilized, could be cultured *in vitro* to the blastocyst stage. ES cells could subsequently be derived from its inner cell mass, and directed to differentiate into the desired cell type. The result would be differentiated (or partly differentiated) ES-derived cells that match exactly the immunological profile of the person who donated the somatic cell nucleus, and who is also the intended recipient of the transplant—a labor intensive, but truly customized therapy.

Other Applications

The advancement made in the stem cells showed that many potential uses of human ES cells have been proposed that do not involve transplantation. For example, human ES cells could be used to study early events in human development. Still, unexplained events in early human development can result in congenital birth defects and placental abnormalities that can lead to spontaneous abortion. By studying human ES cells *in vitro*, it may be possible to identify the genetic, molecular, and cellular events, which lead to these problems, and identify methods of preventing them.

Stem cells has pluripotency to form different cells of the individuals. These multifarious stem cells could also be used to explore the effects of chromosomal abnormalities in early development. This might include the ability to monitor the development of early childhood tumors, many of which are embryonic in origin.

Human ES cells could also be used to test candidate therapeutic drugs. Currently, before candidate drugs are tested in human volunteers, they are subjected to a barrage of preclinical tests. These include drug screening in animal models— *in vitro* tests using cells derived from mice or rats, or *in vivo* tests that involve giving the drug to an animal to assess its safety. Although animal model testing is a mainstay of pharmaceutical research, it cannot always predict the effects that a candidate drug may have on human cells. For this reason, cultures of human cells are often employed in preclinical tests. These human cell lines have usually been maintained *in vitro* for long periods, and as such often have different characteristics than *in vivo* cells. These differences can make it difficult to predict the action of a drug *in vivo* based on the response of human cell lines *in vitro*. Therefore, if human ES cells are directed to differentiate into specific cell types that are important for drug screening, the ES-derived cells are more likely to mimic the *in vivo* response of the cells/tissues to the drug(s) being tested, and offer safer and potentially cheaper models for drug screening.

Human ES cells could be employed to screen potential toxins. The reasons for using human ES cells to screen toxins closely resemble those for using human ES-derived cells to test drugs (above). Toxins often have different effects on different animal species. This makes it critical to have the best possible *in vitro* models for evaluating their effects on human cells.

Finally, human ES cells could be used to develop new methods for genetic engineering. Currently, the genetic complement of mouse ES cells *in vitro* can be modified easily by techniques such as homologous recombination. This is a method for replacing or adding genes, which requires that a DNA molecule be artificially introduced into the genome and then expressed. Using this method, genes to direct differentiation to a specific cell type or genes that express a desired protein product might be introduced into the ES cell line. Ultimately, if such techniques could be developed using human ES cells, it may be possible to devise better methods for gene therapy.

22.11. HEMATOPOIETIC STEM CELLS AND THEIR CLINICAL APPLICATIONS

Leukemia and Lymphoma

Leukemia and lymphoma are major causes of death in humans. Leukemia is a cancer originating in the blood cells that tend to remain as single cells in the lymph or blood liquid. Lymphoma is also a cancer of lymphoid cells and form a solid tumor. Among the first clinical uses of HSCs were the treatment of the cancers of the blood—leukemia and lymphoma, which result from the uncontrolled proliferation of white blood cells. In these applications, the patient's own cancerous hematopoietic cells were destroyed via radiation or chemotherapy, and then replaced with a bone marrow transplant, or, as is done now, with a transplant of HSCs collected from the peripheral circulation of a matched donor. A matched donor is typically a sister or brother of the patient who has inherited similar human leukocyte antigens (HLAs) on the surface of their cells. Other cancers of the blood include acute lymphoblastic leukemia, acute myeloblastic leukemia, chronic myelogenous leukemia (CML), Hodgkin's disease, multiple myeloma, and non-Hodgkin's lymphoma.

Thomas and Clift describe the history of treatment for chronic myeloid leukemia, as it moved from largely ineffective chemo therapy to modestly successful use of a cytokine interferon to bone marrow transplants—first in identical twins and then in HLA-matched siblings. Although there was a significant risk patient's death soon after the transplant, either from infection or from graft-versus-host disease, for the first time, many patients survived this immediate challenge and had survival times measured in years or even decades, rather than months. The authors write, "In the space of 20 years, marrow transplantation has contributed to the transformation of (chronic myelogenous leukemia) CML from a fatal disease to one that is frequently curable. At the same time, experience acquired in this setting has improved our understanding of many transplant-related problems. It is now clear that morbidity and mortality are not inevitable consequences of allogeneic transplantation, and that an allogeneic effect can add to the antileukemic power of conditioning regimens…".

In a recent development, CML researchers have taken their knowledge of hematopoietic regulation one step farther. On May 10, 2001, the Food and Drug Administration approved Gleevec™ (imatinib mesylate), a new, rationally designed oral drug for the treatment of CML. The new drug specifically targets a mutant protein, produced in CML cancer cells, that sabotages the cell signals, controlling orderly division of progenitor cells. By silencing this protein, the new drug turns off cancerous overproduction of white blood cells. So, doctors do not have to resort to bone marrow transplantation. Presently, it is unknown whether the new drug will provide sustained remission or a prolonged life for CML patients.

Inherited Blood Disorders

Another use of allogeneic bone marrow transplants is in the treatment of hereditary blood disorders, such as different types of inherited anemia (failure to produce blood cells), and inborn errors of metabolism (genetic disorders characterized by defects in key enzymes needed to produce essential body components, or degrade chemical by products). The blood disorders include aplastic anemia, betathalassemia, Blackfan-Diamond syndrome, globoid cell leukodystrophy, sickle-cell anemia, severe combined immunodeficiency, X-linked lymphoproliferative syndrome, and Wiskott-Aldrich syndrome. Inborn errors of metabolism that can be treated with bone marrow transplants include Hunter's syndrome, Hurler's syndrome, Lesch Nyhan syndrome, and osteopetrosis. Since bone marrow transplantation carries a significant risk of death, this is usually a treatment of last resort for otherwise fatal diseases.

Cancer Chemotherapy and Hematopoietic Stem Cell Rescue

Cancer has become a major cause for the death of humans. Cancer is technically called malignant tumor—tumor that continues to grow and becomes progressively invasive. Chemotherapy aimed at rapidly dividing cancer cells inevitably hits another target—rapidly dividing hematopoietic cells. Doctors give cancer patients an autologous stem cell transplant to replace the cells destroyed by chemotherapy. They do this by mobilizing HSCs and collecting them from peripheral blood. The cells are stored while the patient undergoes intensive chemotherapy or radiotherapy to destroy the cancer cells. Once the drugs have washed out of the patient's body, the patient receives a transfusion of his or her stored HSCs. Since patients get their own cells back, there is no chance of immune mismatch or graft-versus-host disease. One problem with the use of autologous HSC transplants in cancer therapy has been that cancer cells are sometimes inadvertently collected and reinfused back into the patient along with the stem cells. One team of investigators found that they can prevent reintroducing cancer cells by purifying the cells and preserving only the cells that are $CD34^+$, $Thy-1^+$.

Graft-Versus-Tumor Treatment of Cancer

In transplantations, the tissues can be exploited for the treatment purposes. One of the most exciting new uses of HSC transplantation puts the cells to work, attacking otherwise untreatable tumors. A group of researchers in NIH's intramural research program recently described this approach to treat metastatic kidney cancer. Just under half of the 38 patients treated so far have had their tumors reduced. The research protocol is now expanding to the treatment of other solid tumors that resist standard therapy, including cancer of the lung, prostate, ovary, colon, esophagus, liver, and pancreas.

This experimental treatment relies on an allogeneic stem cell transplant from an HLA-matched sibling, whose HSCs are collected peripherally. The patient's own immune system is suppressed, but not totally destroyed. The donor's cells are transfused into the patient, and for the next three months, doctors closely monitor the patient's immune cells using DNA fingerprinting to follow the engraftment of the donor's cells and regrowth of the patient's own blood cells. They must also judiciously suppress the patient's immune system, as needed, to deter his/her T-cells from attacking the graft and to reduce graft-versus-host disease.

A study by Joshi *et al.,* shows that umbilical cord blood and peripherally harvested human HSCs show antitumor activity in the test tube against leukemia and breast cancer cells. Grafted into a mouse model, which tolerates human cells, HSCs attack human leukemia and breast cancer cells. Although untreated cord blood lacks natural killer (NK) lymphocytes capable of killing tumor

cells, researchers have found that at least in the test tube and in mice, they can greatly enhance the activity and numbers of these cells with cytokines IL-15.

22.12. OTHER APPLICATIONS OF HEMATOPOIETIC STEM CELLS

With the increased research and studies in the field of stem cell biology, the new avenues will be opened. Substantial basic and limited clinical research exploring the experimental uses of HSCs for other diseases is underway. Among the primary applications are autoimmune diseases, such as diabetes, rheumatoid arthritis, and system lupus erythematosis. Here, the body's immune system destroy own body tissues. Experimental approaches similar to those applied above for cancer therapies are being conducted to see if the immune system can be reconstituted or reprogrammed. The use of HSCs as a means to deliver genes to repair damaged cells is another application being explored.

To date, only nonembryonic human stem cells have been used in cell-based gene therapy studies. The inherent limitations of these stem cells, as discussed below, have prompted scientists to ponder and explore whether human embryonic stem cells might overcome the current barriers for the clinical success of cell-based gene therapies.

22.13. STEM CELLS IN GENE THERAPY—PRINCIPLES AND PROMISE

Advancements made in the genetic engineering, molecular biology and medicine resulted in newer approaches for the treatment of various diseases of humans. Gene therapy is a relatively recent and highly experimental approach for treating human disease. While traditional drug therapies involve the administration of chemicals that have been manufactured outside the body, gene therapy takes a very different approach—directing a patient's own cells to produce and deliver a therapeutic agent. The instructions for this are contained in the therapeutic transgene (the new genetic material introduced into the patient). Gene therapy uses genetic engineering, the introduction or elimination of specific genes by using molecular biology techniques, to physically manipulate the genetic material—to alter or supplement the function of an abnormal gene by providing a copy of normal gene; to directly repair such a gene; or to provide a gene that adds new functions or regulates the activity of other genes. Clinical efforts to apply genetic engineering technology for the treatment of human diseases dates to 1989. Initially, gene therapy clinical trials focused on cancer, infectious diseases, or disorders in which only a single gene is abnormal, such as cystic fibrosis. Increasingly however, efforts are being directed towards complex, chronic diseases that involve more than one gene. Prominent examples include heart disease, inadequate blood flow to the limbs, arthritis, and Alzheimer's disease. Thus, newer promises can be seen with these technologies.

Even with the initial setback of gene therapy, the efforts to find better gene therapy protocols are continuing. The potential success of gene therapy depends not only on the delivery of the therapeutic transgene into the appropriate human target cells, but also on the ability of the gene to function properly in the cell. Both requirements pose considerable technical challenges.

In the medical biotechnology, researchers have employed two major strategies for delivering therapeutic transgenes into human recipients. The first is to "directly" infuse the gene into a person. Viruses that have been altered to prevent them from causing disease are often used as the vehicle for delivering the gene into certain human cell types, in much the same way as ordinary viruses infect cells. This delivery method is fairly imprecise and limited to the specific types of human cells that the viral vehicle can infect. For example, some viruses commonly used as gene-delivery vehicles can only infect cells that are actively dividing. This

limits their usefulness in treating diseases of the heart or brain, because these organs are largely composed of nondividing cells. Nonviral vehicles for directly delivering genes into cells are also being explored, including the use of plain DNA and DNA wrapped in a coat of fatty molecules known as liposomes **(Fig. 22.12)**.

In another gene therapy strategy, the researchers use the living cells to deliver therapeutic transgenes into the body of an individual. In this therapy, the vector or delivery cells, i.e., often a type of stem cell—a lymphocyte or a fibroblast—are removed from the body of patient, and the therapeutic transgene is introduced into them via the same vehicles used in the previously described direct-gene-transfer method. While still in the laboratory, the genetically modified cells are tested and then allowed to grow and multiply, and finally, are infused back into the patient.

Gene therapy, using genetically modified cells, offers several unique advantages over direct gene transfer into the body and cell therapy, which involves administration of cells that have not been genetically modified. First, the addition of the therapeutic transgene to the delivery cells takes place outside the patient, that allows an important measure of control for the researchers, because they can select and work with only those cells that contain the transgene and produce the therapeutic agent in sufficient quantity. Second, investigators can genetically engineer or program the cells' level and rate of production of the therapeutic agent. Cells can be programmed to steadily churn out a given amount of the therapeutic product. In some cases, it is desirable to program the cells to make large amounts of therapeutic agent, so that the chances that sufficient quantities are secreted and reach the diseased tissue in the patient are high. In other cases, it may be desirable to program the cells to produce the therapeutic agent in a regulated fashion. In this case, the therapeutic transgene would be active only in response to certain signals, such as drugs administered to the patient to turn the therapeutic transgene on and off.

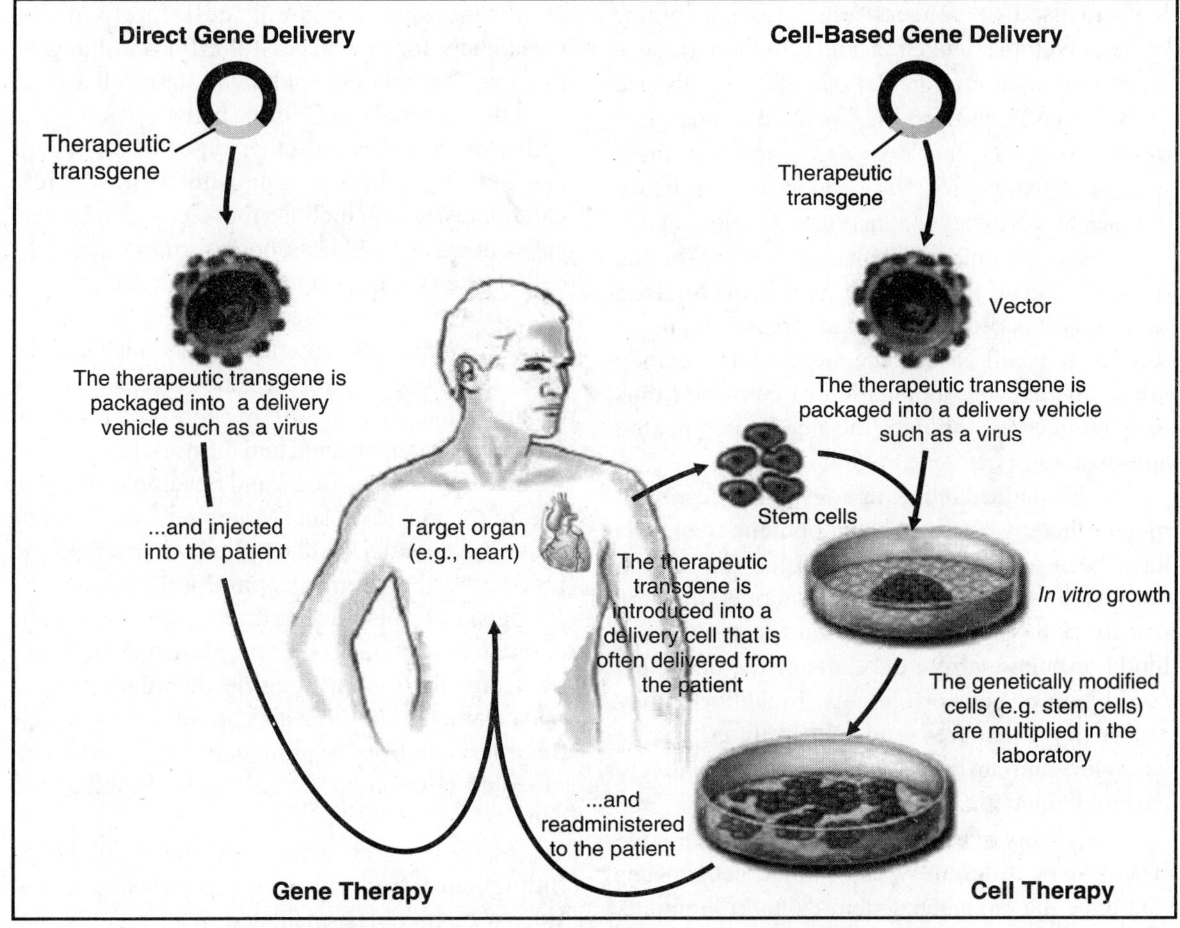

Fig. 22.12. Strategies for delivering therapeutic transgenes into patients

Necessity of Stem Cells in Some Cell-Based Gene Therapies

With the increased number of genetic diseases identified, and locating the required gene to their position on chromosome gave the way to make efforts for gene therapy. To date, about 40% of more than 450 gene therapy clinical trials conducted in the United States have been cell-based. Of these, approximately 30% have used human stem cells, more specifically, blood-forming, or hematopoietic, stem cells as the means for delivering transgenes into patients.

Researchers timely studied the understanding of various events in the role of stem cell in the organisms. Several of the early gene therapy studies using these stem cells were carried out not for therapeutic purposes per se, but to track the cells' fate after they were infused back into the patient. The studies aimed to determine where the stem cells ended up and whether they were indeed producing the desired gene product, and if so, in what quantities and for what length of time. Of the stem cell-based gene therapy trials that have had a therapeutic goal, approximately one-third have focused on cancers (e.g., ovarian, brain, breast, myeloma, leukemia, and lymphoma); one-third on human immunodeficiency virus disease (HIV-1); and one-third on so-called single-gene diseases (e.g., Gaucher's disease, severe combined immune deficiency (SCID), Fanconi anemia, Fabry disease, and leukocyte adherence deficiency).

But why should use stem cells be used for this method of gene therapy, and why hematopoietic stem cells in particular? The major reason for using stem cells in cell-based gene therapies is that they are a self-renewing population of cells, and thus may reduce or eliminate the need for repeated administrations of gene therapy.

In the medical biotechnology, with the advent of gene therapy research, hematopoietic stem cells have been a delivery cell of choice for several reasons. First, although small in number, they are readily removed from body via the circulating blood or bone marrow of adults or the umbilical cord blood of newborn infants. In addition, they are easily identified and manipulated in the laboratory and can be returned to patients relatively easily by injection.

The ability of hematopoietic stem cells to give rise to many different types of blood cells means that once the engineered stem cells differentiate, the therapeutic transgene will reside in cells such as T and B lymphocytes, natural killer cells, mono-cytes, macrophages, granulocytes, eosinophils, basophils and megakaryocytes. The clinical applications of hematopoietic stem cell-based gene therapies are thus also diverse, extending to organ transplantation, blood and bone marrow disorders, and immune system disorders.

In addition, hematopoietic stem cells "home", or migrate to a number of different spots in the body—primarily the bone marrow, but also the liver, spleen, and lymph nodes. These may be strategic locations for the localized delivery of therapeutic agents for disorders unrelated to the blood system, such as liver diseases and metabolic disorders such as Gaucher's disease.

Hematopoietic stem cells helped the researchers for the successful efforts in the gene therapy. The only type of human stem cell used in gene therapy trials so far is the hematopoietic stem cell. However, several other types of stem cells are being studied as gene-delivery-vehicle candidates. They include muscle-forming stem cells known as myoblasts, bone-forming stem cells called osteoblasts, and neural stem cells.

With the recent research on the cell types for the use in gene therapy, myoblasts appear to be good candidates for use in gene therapy, because of an unusual and advantageous biological property—when injected into the muscle, they fuse with nearby muscle fibers and become an integral part of the muscle tissue. Moreover, since muscle tissue is generally well supplied with nerves and blood, the therapeutic agents produced by the transgene are also accessible to nerves and the circulatory system. Thus, myoblasts may not only be useful for treating muscle disorders such as muscular dystrophy, but also possibly non-muscle disorders such as neurodegenerative diseases, inherited hormone deficiencies, hemophilia, and cancers.

Many research works have been conducted throughout the world on the promising animal studies of myoblast-mediated gene therapy. For instance, this approach was successful in correcting liver and spleen abnormalities associated with a lysosomal storage disease in mice. Investigators have also achieved stable production of the human clotting factor IX deficient in hemophilia at therapeutic concentrations in mice for at least eight months. Myoblasts engineered to secrete erythropoietin (a hormone that stimulates red blood cell production) were successful in reversing a type of anemia associated with end-stage renal disease in a mouse model of renal failure.

Another animal study of myoblast-mediated gene transfer involved a mouse model of familial amyotrophic lateral sclerosis (ALS, also known as Lou Gehrig's disease)—a fatal disorder characterized by progressive degeneration of the brain and spinal cord nerves that control the muscle activity. Investigators injected myoblasts containing the transgene for a human nerve growth factor into the muscles of the ALS mice before the onset of disease symptoms and motor neuron degeneration. The transgene remained active in the muscle for upto 12 weeks, and most importantly, the gene therapy successfully delayed the onset of disease symptoms slowed muscle atrophy and delayed the deterioration of motor skills.

A team of investigators with the series of experiments in rodents, has been testing neural stem cells as vehicles for cell-based gene therapy for brain tumors known as gliomas. Gliomas are virtually impossible to treat, because the tumor cells readily invade the surrounding tissue and migrate extensively into the normal brain. The researchers genetically modified human neural stem cells to produce a protein—cytosine deaminase—which converts a nontoxic precursor drug into an active form that kills cancer cells. The engineered neural stem cells were then injected into the brains of mice with human-derived gliomas. Within two weeks of the gene therapy and systemic treatment with the precursor drug, the tumors had shrunk by 80%. The animal studies also revealed that neural stem cells were able to quickly and accurately "find" glioma cells, regardless of whether the stem cells were implanted directly into the tumors, implanted far from the tumors (but still within the brain), or injected into circulating blood outside the brain.

Another medical biotechnology research for cell-based gene therapy system under investigation involves the use of osteoblasts, or boneforming stem cells. In a recent preliminary study examining gene therapy approach to bone repair and regeneration, researchers genetically engineered osteoblasts to produce a bone growth factor. The osteoblasts were added to a biodegradable matrix that could act as a "scaffold" for new bone formation. Within a month, after the cell-impregnated scaffold was implanted into mice, new bone formation was detectable. Although this work is in early stage, it offers the hope of an effective alternative to conventional bone-grafting techniques.

Role of Embryonic Stem Cells in Gene Therapy Research

A large number of disease oriented research laboratories are concentrating on gene therapy as a final option for treatment. With one notable exception, no therapeutic effects have been achieved in gene therapy trials to date. The first successful gene therapy occurred in a recent French study in which a therapeutic transgene for correcting X-linked severe combined immune deficiency was introduced into the bone marrow cells of children, resulting in improved function of their immune systems and correction of the disease.

This encouraging success aside, the generally disappointing results are due, in part, to the inherent limitations of adult and cord blood stem cells. In principle, at least, the use of human embryonic stem cells might overcome some of these limitations, but further research will be needed to determine whether embryonic stem cells are better suited to meet the needs of gene therapy applications than adult stem cells. In gene therapy, one important criteria of the optimal cell for delivering a therapeutic transgene would be its ability to retain the therapeutic transgene even as it proliferates or differentiates into specialized cells. Most of the cell-based gene therapies attempted so far have used viral vehicles to introduce the transgene into the hematopoietic stem cell. One way to accomplish this is to insert the therapeutic transgene into one of the chromosomes of the stem cell.

Retroviruses are able to do this, and for this reason, they are often used as the vehicle for infecting the stem cell and introducing the therapeutic transgene into the chromosomal DNA. However, mouse retroviruses are efficient only in infecting actively dividing cells. Unfortunately, hematopoietic stem cells are quiescent and seldom divide. The percentage of stem cells that actually receive the therapeutic transgene has usually been too low to attain a therapeutic effect. Because of this problem, investigators have been exploring the use of viral vehicles that can infect nondividing cells, such as lentiviruses (e.g., HIV) or adeno-associated viruses. This approach has not been entirely successful, however, because of problems relating to the fact that the cells themselves are not in an active state. One approach to improving the introduction of transgenes into hematopoietic stem cells has been to stimulate the cells to divide so that the viral vehicles can infect them and insert the therapeutic transgene.

Inder Verma of the Salk Institute has noted, however, that this manipulation can change other important properties of the hematopoietic stem cells, such as plasticity, self-renewal, and the ability to survive and grow when introduced into the patient. This possibility might be overcome with the use of embryonic stem cells if they require less manipulation. And in fact, some preliminary data suggests that retroviral vectors may work more efficiently with embryonic stem cells than with the more mature adult stem cells. For example, researchers have noted that retroviral vectors introduce transgenes into human fetal cord blood stem cells more efficiently than cord blood stem cells of newborns, and that the fetal cord blood stem cells also had a higher proliferative capacity (i.e., they underwent more subsequent cell divisions).

This suggests that fetal cord blood stem cells might be useful in cell-based in utero gene therapy to correct hematopoietic disorders before birth. In some cases, such as the treatment of a chronic disease, achieving continued production of the therapeutic transgene over the life of the patient is very important. Generally, however, gene therapies using hematopoietic stem cells have encountered a phenomenon known as "gene silencing," where, over time, the therapeutic transgene gets "turned off" due to cellular mechanisms that alter the structure of the area of the chromosome where the therapeutic gene has been inserted. Whether the use of embryonic stem cells in gene therapy could overcome this problem is unknown, though preliminary evidence suggests that this phenomenon may occur in these cells as well.

In the gene therapy, it is very essential that the introduced transgene is present for a long time. Persistence of the cell containing the therapeutic transgene is equally important for ensuring continued availability of the therapeutic agent inside the body. Verma noted that the optimal cells for cell-mediated gene transfer would be cells that will persist for the rest of the patient's life. These cells can proliferate and would make the missing protein constantly and forever. Persistence, or longevity of the cells can come about in two ways—a long life span for an individual cell, or a self-re-newal process whereby a short-lived cell undergoes successive cell divisions while maintaining the therapeutic transgene. Ideally, then, the genetically modified cell for use in cell-based gene therapy should be able to self-renew (in a controlled manner so tumors are not formed), so that the therapeutic agent is available on a long-term basis. This is one of the reasons why stem cells are used, but adult stem cells seem to be much more limited in the number of times they can divide compared to embryonic stem cells. The difference between the ability of adult and embryonic stem cells to self-renew has been documented in the mouse, where embryonic stems cells were shown to have a much higher proliferative capacity than adult hematopoietic stem cells.

Researchers are beginning to understand the biological basis of difference in the proliferative capacity between adult and embryonic stem cells. Persistence of cells and the ability to undergo successive cell divisions are in part, at least, a function of the length of structures at the tips of chromosomes called telomeres. Telomere length is, in turn, maintained by an enzyme known as telomerase. Low levels of telomerase activity result in short telomeres and, thus, fewer rounds of cell division—in other words, shorter longevity. Higher levels of telomerase activity result in longer telomeres, more possible cell divisions, and overall longer persistence. Mouse embryonic stem cells have been found to contain longer telomeres and higher levels of telomerase activity compared to adult stem cells and other more specialized cells of the body. As mouse embryonic stem cells give rise to hematopoietic stem cells, telomerase activity levels drop, suggesting a decrease in the self-renewing potential of the hematopoietic stem cells.

Human embryonic stem cells have also been shown to maintain pluripotency (the ability to give rise to other, more specialized cell types) and the ability to proliferate for long periods in cell culture in the laboratory. Adult stem cells appear capable of only a limited number of cell divisions, which would prevent long-term expression of the therapeutic gene needed to correct chronic diseases. "Embryonic stem cells can be maintained in culture, which is nearly impossible with cord blood stem cells," said Robert Hawley of the American Red Cross Jerome H. Holland Laboratory for Biomedical Sciences, who is developing gene therapy vectors for insertion into human hematopoietic cells. "So with embryonic stem cells, you have the possibility of long-term maintenance and expansion of cell lines, which has not been possible with hematopoietic stem cells."

The patient's immune system response can be another significant challenge in gene therapy. Most cells have specific proteins on their surface that allow the immune system to recognize them as either "self" or "nonself". These proteins are known as major histocompatibility proteins, or MHC proteins. If adult stem cells, for use in gene therapy, cannot be isolated from the patient, donor cells can be used.

Due to the differences in MHC proteins among individuals, the donor stem cells may be recognized as nonself by the patient's immune system, and rejected. Discoverer of embryonic germ cells, John Gearhart of Johns Hopkins University and Peter Rathjen of the University of Adelaide speculate that embryonic stem cells may be useful for avoiding such immune reactions. For instance, it may be possible to establish an extensive "bank" of embryonic stem cell lines, each with a different set of MHC genes. Then, an embryonic stem cell, which is immunologically compatible for a patient, could be selected, genetically modified, and triggered to develop into the appropriate type of adult stem cell that could be administered to the patient. By genetically modifying the MHC genes of an embryonic stem cell, it may also be possible to create a "universal" cell that would be compatible with all patients. Another approach might be to "customize" embryonic stem cells such that cells derived from them have a patient's specific MHC proteins on their surface, and then to genetically modify them for use in gene therapy. Such approaches are hypothetical at this point, however, and research is needed to assess their feasibility.

In medical biotechnolgy, ironically the very qualities that make embryonic stem cells a potential candidates for gene therapy (i.e., pluripotency and unlimited proliferative capacity) also raise safety concerns. In particular, undifferentiated embryonic stem cells can give rise to teratomas—tumors composed of a number of different tissue types. It may thus be preferable to use a differentiated derivative of genetically modified embryonic stem cells that can still give rise to a limited number of cell types (akin to an adult stem cell). Cautions Esmail Zanjani of the University of Nevada, "We could differentiate embryonic stem cells into, say, liver cells, and then use them, but I don't see how we can take embryonic stem cells per se and put genes into them to use therapeutically."

Presently, much more research is needed to determine whether the differentiated stem cells retain the advantages, such as longer life span, of the embryonic stem cells from which they were derived. Because of the difficulty in isolating and purifying many types of adult stem cells, embryonic stem cells may still be better targets for gene transfer. The versatile embryonic stem cell could be genetically modified, and then, in theory, it could be induced to give rise to all varieties of adult stem cells. Also, since the genetically modified stem cells can be easily expanded, large, pure populations of the differentiated cells could be produced and saved. Even if the differentiated cells were not as long-lived as the embryonic stem cells, there would still be sufficient genetically modified cells that could be given to the patient whenever the need arises.

Achieving greater clinical success with cell-based gene therapy requires new knowledge and advances in several key areas, including the design of viral and nonviral vehicles for introducing transgenes into cells; the ability to direct where, in a cell, the transgene is introduced; the ability to direct the genetically modified stem cells or the secreted therapeutic agent to diseased tissues; optimization and regulation of the production of the therapeutic agent within the stem cell; and management of immune reactions to the gene therapy process.

The ability of embryonic stem cells to generate a wide variety of specialized cell types, and to maintain them in the laboratory would make embryonic stem cells a promising model for exploring critical questions in many of these areas. "There are possibilities of long-term maintenance and expansion of embryonic stem cells and of differentiation along specific lineages that have not been possible with hematopoietic stem cells," Zanjani says. "And if they (embryonic stem cells) could be used (in the laboratory) as a model for differentiation, you could evaluate…vectors for gene delivery and get an idea of how genes are translated in patients." Cynthia Dunbar, a gene therapy researcher at the National Institutes of Health, similarly notes that embryonic stem cells could be useful not only in screening new viral and nonviral vectors designed to introduce therapeutic transgenes into cells, but also for testing levels of production of the therapeutic agent

after the embryonic stem cells differentiate in the culture. The major contribution of embryonic stem cells to gene therapy may be to advance the general scientific knowledge needed to overcome many of the current technical hurdles for a successful therapeutic gene transfer.

22.14. AUTOIMMUNE DISEASES AND STEM CELL-BASED THERAPIES

One of the many perplexing questions in biomedical research is, why does the body's protective shield against infections—the immune system—attacks its own vital cells, organs, and tissues? The answer to this question is central to understanding an array of autoimmune diseases, such as rheumatoid arthritis, type 1 diabetes, systemic lupus erythematosus, and Sjogren's syndrome. When some of the body's cellular proteins are recognized as "foreign" by immune cells called T lymphocytes, a destructive cascade of inflammation is set in place. Current therapies to combat these cases of cellular mistaken identity dampen the body's immune response and leave patients vulnerable to life-threatening infections. Research on stem cells is now providing new approaches to strategically remove the misguided immune cells and restore normal immune cells to the body. Presented here are some of the basic research investigations that are being guided by adult and embryonic stem cell discoveries. The body's main line of defense against invasion by infectious organisms is the immune system. To achieve this, the immune system must distinguish the cellular components of its own body (self) from the cells or components of invading organisms (nonself). "Nonself" should be attacked while "self" should not. Therefore, two general types of errors can be made by the immune system. If the immune system fails to quickly detect and destroy an invading organism, an infection will result. However, if the immune system fails to recognize self cells or components and mistakenly attacks them, the result is an autoimmune disease. Common autoimmune diseases include rheumatoid arthritis, systemic lupus erythematosis (lupus), type 1 diabetes, multiple sclerosis, Sjogren's syndrome, and inflammatory bowel disease. Although each of these diseases have different symptoms, they share the unfortunate reality that, for some reason, the body's immune system has turned against itself.

Immune System and Normal Health

The "soldiers" of the immune system are white blood cells, including T and B lymphocytes, which originate in the bone marrow from hematopoietic stem cells. Everyday the body comes into contact with many organisms such as bacteria, viruses, and parasites. These organisms have the potential to cause serious infections, such as pneumonia or AIDS. When a healthy individual is infected, the body responds by activating a variety of immune cells. Initially, invading bacteria or viruses are engulfed by an antigen presenting cell (APC), and their component proteins (antigens) are cut into pieces and displayed on the cell's surface. Pieces

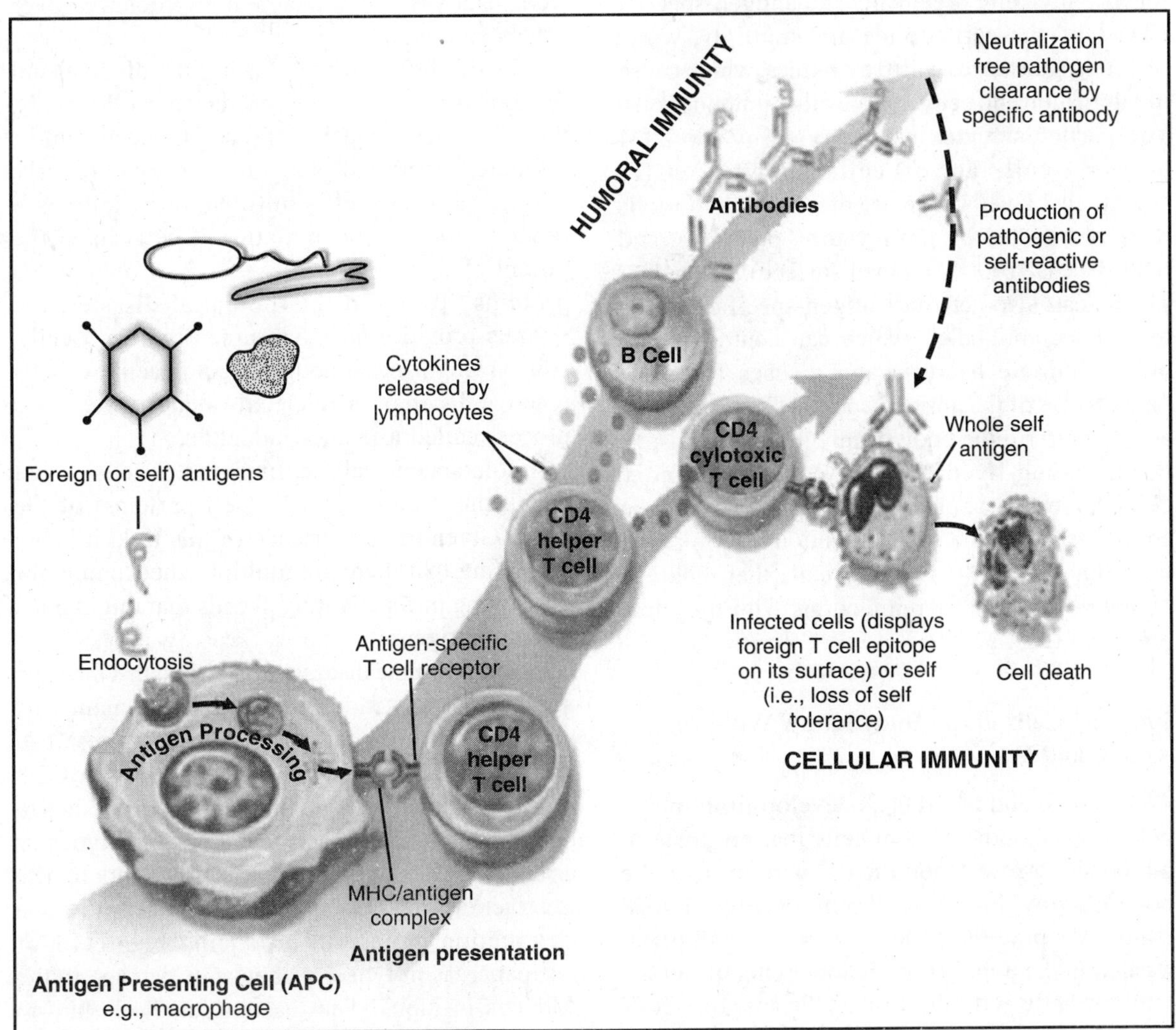

Fig. 22.13. Immune response to foreign or self antigens

of the foreign protein (antigen) bind to the major histocompatibility complex (MHC) proteins, also known as human leukocyte antigen (HLA) molecules, on the surface of the APCs. This complex, formed by a foreign protein and an MHC protein, then binds to a T-cell receptor on the surface of another type of immune cell—the CD4 helper T-cell. They are so named because they "help" the immune responses proceed, and have a protein called CD4 on their surface. This complex enables T-cells to focus the immune response to a specific invading organism. The antigen-specific CD4 helper T-cells divide and multiply, while secreting substances called cytokines, which cause inflammation and help activate other immune cells. The particular cytokines secreted by the CD4 helper T-cells act on cells known as CD8 "cytotoxic" T-cells (because they can kill the cells that are infected by the invading organism, and have the CD8 protein on their surface). The helper T-cells can also activate antigen-specific B-cells to produce antibodies, which can neutralize and help eliminate bacteria and viruses from the body. Some of the antigen-specific T and B-cells, activated to rid the body of infectious organisms, become long-lived "memory" cells. Memory cells have the capacity to act quickly when confronted with the same infectious organism at later times. It is the memory cells that make us "immune" from later reinfections with the same organism **(Fig. 22.13)**.

Immune Cells of the Body Know What to Attack and What Not to

All immune and blood cells develop from multi-potent hematopoietic stem cells that originate in the bone marrow. Upon their departure from the bone marrow, immature T-cells undergo a final maturation process in the thymus—a small organ located in the upper chest, before being dispersed into the body with the rest of the immune cells (e.g., B-cells). Within the thymus, T-cells undergo an important process that "educates" them to distinguish between self (the proteins of their own body) and nonself (the invading organism's) antigens. Here, the T-cells are selected for their ability to bind to the particular MHC proteins expressed by the individual. The particular array of MHCs varies slightly between individuals, and this variation is the basis of the immune response when a transplanted organ is rejected. MHCs and other less easily characterized molecules called minor histocompatibility antigens are genetically determined. This is the reason why donor organs from relatives of the recipient are preferred over unrelated donors.

In the bone marrow, a highly diverse and random array of T-cells is produced. Collectively, these T-cells are capable of recognizing an almost unlimited number of antigens. Since the process of generating T-cell's antigen specificity is a random one, many immature T-cells have the potential to react with the body's own (self) proteins. To avoid this potential disaster, the thymus provides an environment where T-cells, which recognize self-antigens (autoreactive or self-reactive T-cells), are deleted or inactivated in a process called tolerance induction.

Tolerance usually ensures that T-cells do not attack the "auto-antigens" (self-proteins) of the body. Given the importance of this task, it is not surprising that there are multiple checkpoints for destroying or inactivating T-cells that might react to auto-antigens.

Autoimmune diseases arise when this intricate system for the induction and maintenance of immune tolerance fails. These diseases result in cell and tissue destruction by antigen-specific CD8 cytotoxic T-cells or autoantibodies (anti-bodies to self-proteins), and the accompanying inflammatory process. These mechanisms can lead to the destruction of joints in rheumatoid arthritis; the destruction of insulin producing beta cells of the pancreas in type 1 diabetes; or damage to the kidneys in lupus. The reasons for the failure to induce or maintain tolerance are enigmatic. However, genetic factors, along with environmental and hormonal, influence, and certain infections may contribute to tolerance and development of autoimmune diseases.

22.15. STEM CELL BIOSAFETY

With the huge amount of research information and applications in the stem cell biotechnology, one should not ignore the essential aspect of biosafety measures. The isolation of human stem cells offers the promise of a remarkable array of novel therapeutics. Biologic therapies derived from such cells through tissue regeneration and repair, as well as through the targeted delivery of genetic material, are expected to be effective in the treatment of a wide range of medical conditions. Efforts to analyze and assess the safety of using human stem cells in the clinical setting are vitally important for this endeavor.

As the stem cells are derived from the living source and show all the characters of living cells *in vivo*, they have higher potential to be among other living cells. Transplanted human stem cells are dynamic biological entities that interact intimately with and are influenced by the physiology, immunology and biochemistry of the recipient individual. Before they are transplanted, cultured human stem cells are maintained under conditions that promote either the self-renewing expansion of undifferentiated progenitors or the acquisition of differentiated properties, indicative of the phenotype the cells will assume. After incompletely differentiated human stem cells are transplanted, additional fine-tuning occurs as a consequence of instructions received from the cells' physiologic microenvironments within the recipient. The capabilities to self-renew and differentiate, which are inherent to human stem cells, point simultaneously to their perceived therapeutic potential and the challenge of assessing their safety.

In such situations, assessing human stem cell safety requires the implementation of a comprehensive strategy. Each step in the human stem cell development process—beginning with identifying and evaluating suitable human stem cell sources—must be carefully scrutinized. Included in this global assessment are the derivation, expansion, manipulation, and characterization of human stem cell lines, as well as preclinical efficacy and toxicity testing in appropriate animal models. Being able to trace back from the cell population prepared for transplantation to the source of the founder, human stem cells also allows each safety checkpoint to be connected, one to the other.

Weaving a Stem Cell Safety Net

Similar to the other biotechnological success, the stem cell research also initiates a diversity of opinion among researchers about the feasibility of initiating pilot clinical studies using human stem cells. Some are of the view that it is reasonable to expect, within the next five years, that human stem cells will be used in transplantation settings to replace dead or dying cells within organs, such as the failing heart, or that genetically modified human stem cells will be created for the delivery of therapeutic genes. Others argue that a good deal more information about the basic biology of human stem cells needs to be accumulated before their therapeutic potential in humans can be assessed. Clinical studies involving the transplantation of blood-restoring, or hematopoietic stem cells have been underway for a number of years.

Reconstituting the blood and immune systems through stem cell transplantation is an established practice for treating hematological malignancies such as leukemia and lymphoma. Transplanting hematopoietic stem cells resident in the bone marrow or isolated from cord blood or circulating peripheral blood is used to counter the destruction of certain bone marrow cells caused by high-intensity chemotherapeutic regimens used to battle various solid tumors. Moreover, clinical trials are being conducted to assess the safety and efficacy of using hematopoietic stem cell transplantation to treat various autoimmune conditions including multiple sclerosis, lupus, and rheumatoid arthritis.

Although precedents exist for the clinical use of human stem cells, there is considerable reluctance to proceed with clinical trials involving human stem cells derived from embryonic and fetal sources. This hesitancy extends to adult human stem cells of non-hematopoietic origin, even though, by contrast, their plasticity is generally considered to be lower than their embryo- and fetus-derived counterparts. For human stem cells to advance to the stage of clinical investigation, a virtual safety net composed of a core set of safeguards is required.

Adequate Donor Screening and Safety Assurance

Immunerable number of organisms are associated with the persons body. Whether human stem cells are of embryonic, fetal, or adult origin, donor sources must be carefully screened. Routine testing should be done to guard against the inadvertent transmission of infectious diseases. Additionally, pedigree assessment and molecular genetic testing appear to be warranted. This is the case when human stem cells intended for transplantation are derived from an allogeneic donor, i.e., someone other that the recipient, and especially if the cells are obtained from a master cell bank that has been established using human embryonic stem or germ cells.

Source of stem cells are very important for use in the treatment processes. The purpose of pedigree evaluation and/or genetic testing is to establish whether the human stem cells in question are suitable for use in the context of a particular clinical situation. For example, embryos derived from a donor with a family history of cardiovascular diseases may not be the best suited for the derivation of cardiac muscle cells intended to repair damaged heart tissue. Similarly, the use of molecular genetic analysis could detect a mutation in the gene for alpha-synuclein. This gene is known to be responsible for the rare occurrence of early onset Parkinson's Disease. Detecting such genetic abnormality in neuronal progenitor cells derived from an established embryonic germ cell line could block the use of those cells as a treatment for a number of neurodegenerative conditions, including Parkinson's Disease.

With the genome sequencing efforts of human genome project throughout the world, huge number of genes and their functions are identified. The number of genes known to be directly responsible for causing disease or anomalous physiologic function is relatively small. Advances in techniques for identifying, isolating, and analyzing genes, coupled with the wealth of information destined to become available as one outcome of the human genome sequencing projects, will raise this number. Considerably, more will also be learned about how multiple gene products, each contributing an incremental quantity to the overall sum, predispose an individual to develop particular diseases. Clearly, it will eventually not be possible, or even necessary, to screen every source of human stem cells for the entire panoply of disease-associated genes. The screening of targeted genes will be conducted within the context of the relevant clinical population.

Enhancement of controlled, standardized practices and procedures for establishing cultured human stem cell lines

As large number of human stem cell lines are discovered in the stem cell biotechnology, it usage should be made in a stricter manner. To ensure the integrity, uniformity, and reliability of human stem cell preparations intended for clinical use, it is essential to demonstrate that rigorously controlled, standardized practices and procedures are being followed in establishing and maintaining human stem cell lines in culture. This helps in better and assured biosafety in the stem cell biotechnology.

In the human cell biotechnology, human stem cells from virtually every source other than blood-derived hematopoietic stem cells are maintained in tissue culture for some defined period of time. This is necessary to obtain a sufficient number of cells for use in clinical studies involving transplantation. Culturing human stem cells requires the use of formulated liquid media supplemented with growth factors and other chemical substances that promote cellular replication and govern the differentiation of the cultured human stem cells. Since human stem cells are a dynamic, biological entity, failure to standardize procedures for maintaining and expanding cells in culture could result in unintended alterations in the intrinsic properties of the cells. The initial seeding density of the cells, the frequency with which the culture medium is replenished, and the density cells are permitted to achieve before subdividing will all affect the characteristics of human stem cells maintained in culture. Altering the concentrations of supplemental growth factors and chemical substances, even switching from one supplier to another, may lead to changes in cell growth rate, expression of defining cell markers, and differentiation potential. Alterations in stem cell properties caused by the use of nonstandardized culture practices are likely to affect the behavior and effectiveness of the cells once transplanted.

The experimentation of human stem cells in the laboratories are also in high discussion. One particular concern is how safe it is to use serum derived from cows as a supplement to culture media. Due to the outbreak of bovine spongiform encephalopathy (BSE) in cattle herds, primarily those raised in the United Kingdom, only serum produced from cows reared in countries certified to be free of BSE should be used. Consumption of beef contaminated with the agent responsible for causing BSE has lead to the limited emergence of new variant Creutzfeldt-Jakob disease (nvCJD) in humans. This disease results in the relentless destruction of brain tissue and is invariably fatal.

Placing neural stem cells contaminated with the BSE infectious agent in a patient's nervous system to investigate cellular-replacement therapies for neurological disorders would be both irresponsible and devastating. Researchers are engaged in a vigorous effort to develop serum-free, chemically defined media that obviate risks associated with the use of bovine serum.

Substitution to culturing improve safety

It is well known that human stem cells requires an essential feeder layers for the successful culture initation. An issue unique to the culturing of human embryonic stem and embryonic germ cells involves the use of mouse embryonic fibroblast feeder cells to keep the embryonic cells in a proliferating, undifferentiated condition. Human embryonic stem and embryonic germ cells are seeded directly onto a bed of irradiated mouse feeder cells. Transplanting into humans stem cell preparations derived from founder cells that have been in direct, intimate contact with nonhuman animal cells constitutes xenotransplantation—the use of organs, tissues, and cells derived from animals to treat human disease. The principal concern of xenotransplantation is the unintended transfer of animal viruses into humans.

Human stem cell lines researchers are devoting considerable attention to developing culture conditions that do not use mouse feeder cells. In February, scientists from Geron Corporation a biotech company focusing on the development of embryonic stem cell technology for treating disease—presented findings at a scientific conference, demonstrating that human embryonic stem cells can be maintained without mouse feeder cells. Human embryonic stem cells seeded on a commercially available basement membrane matrix in media conditioned by feeder cells retain their proliferative potential and capacity to form all three embryonic germ layers (mesoderm, endoderm, and ectoderm). This suggests that human embryonic stem cells maintained in the absence of direct culture on a mouse feeder cell layer are comparable to human embryonic stem cells co-cultured with mouse feeder cells.

Extensive characterization and human stem cell populations

With the increased information and resources on various aspects of cultured cell lines, it makes easy to clear cut differentiation or separation of the stem cell lines. Detailed characterization of cell preparations intended for transplantation is critical to the development of human stem cells for clinical use. Identifying the cells that make-up an human stem cell population intended for clinical study requires identifying cells exhibiting the desired phenotype within the preparation, as well as those that do not. This poses considerable challenges, because human embryonic stem and embryonic germ cells have the capacity to give rise to all differentiated cell types, while adult human stem cells, though generally more restricted in their plasticity, are capable of generating all cell types that make up the tissue from which they were derived.

Characterization of cell lines are followed in almost all culturing work in the laboratory. On the basis of the complex biological properties of human stem cells, including their potential to differentiate along multiple lineages and give rise to a variety of cell types, it is expected that the characterization of stem cell preparations will require a panel of orthogonal assessments. Parameters that will prove useful in establishing identity include:

- cell morphology (visual microscopic inspection of cells to assess their appearance);
- expression of unique cell-surface antigens (as is the case for CD34+ hematopoietic stem cells);
- characterization of biochemical markers such as a tissue-specific enzymatic activity (e.g., enzymes that produce neurotransmitters for nerve cells); and
- expression of genes that are unique to a particular cell type.

Further, analysis of the nuclear chromosomal karyotype may be used to assess genetic stability of established human embryonic stem and embryonic germ cell lines maintained in culture for extended periods of time. Continued development and standardization of DNA microarray analysis (simultaneous screening for many genes) and proteomics (protein profiling) technologies will significantly enhance stem cell characterization.

Rigorous and quantitative identification of cell types within a heterogeneous population of differentiating human stem cells provides the means to gauge purity of a cellular preparation. In turn, this permits evaluation of the extent to which purity of a human stem cell preparation predicts efficacy after transplantation. It is not necessarily the case that homogenous populations composed of a single cell type will be more effective as a cell-replacement therapy than mixed populations of cells. It is conceivable that the reason differentiation of cultured stem cells obtained from the brain leads to formation of all the cell types found within the nervous system (namely, neurons, astrocytes, and oligodendrocytes) is that their coincidental presence is required to ensure maximum survival and functional capability. The interaction of various phenotypic cell types within a preparation of progenitor cells obtained after the controlled differentiation of cultured human embryonic stem cells is being actively investigated.

Once the stricter purity profile has been established for a population of human stem cells generated using standardized procedures, deviations that occur outside what is expected due to normal biologic variation serve as a harbinger that significant, and possibly deleterious, changes may have occurred. Such alterations could reflect the introduction of genetic mutations as a consequence of culture conditions used to promote expansion and to induce differentiation of the progenitor cell population.

Before clinical studies involving human stem cell transplantation can be done, it is essential to demonstrate that human stem cell preparations possess relevant biological activity. The bioassay provides a quantitative measure of the potency of a cell preparation and ensures that cells destined for transplantation are not inert. Assays may be based on a biologic activity such as insulin release from pancreatic islet-like cells, glycogen storage by cells intended for regeneration of liver tissue, or synchronous contraction in the case of stem cell-derived cardiomyocytes to be used for repairing damaged heart muscle. When cells that have not acquired fully differentiated functionality are to be transplanted, it may be appropriate to use surrogate markers that predict the acquisition of the intended biologic activity upon further differentiation. (For example, counting tyrosine hydroxylase-expressing neural progenitor cells in a mixed population of cells intended to provide dopaminergic neurons for treating Parkinson's Disease could predict the acquisition of relevant biologic activity after transplantation.)

Improved methods of assessment in human stem cell safety

A critical element of the safety net is the transplantation of human stem cells into animals to demonstrate that the therapy does what it is supposed to do ("proof of concept"), and to assess toxicity. Admittedly, animal models of human disease are imperfect because most human maladies do not spontaneously occur in animals. Chemical, surgical, and immunologic methods are used to damage neurons; induce diabetes; simulate heart attacks, stroke, and hypertension; or compromise organ function. In situations when focal genetic lesions are known to cause disease, the creation of transgenic mouse colonies in which the culpable gene is either eliminated or over-expressed results in disease models that are capable of faithfully reproducing human-disease-specific pathologies.

Human stem cells must be transplanted into animal models of human disease. Transplantation of neural stem cells should demonstrate measurable evidence of efficacy in models of neurodegenerative disease, such as Parkinson's Disease, Huntington's disease, and amyotrophic lateral sclerosis (ALS), Alzheimer's disease, as well as spinal cord injury and stroke. Improved liver function after transplantation of hepatocyte precursors should be observed in an animal model of hepatic failure. Normalization of blood insulin concentrations and amelioration of diabetic disease symptoms should result from the transplantation of pancreatic islet progenitors in a mouse model of diabetes. It is likely that in all cases, immuno-suppression will be required due to immunologic incompatibility between humans and the animal model species (usually mouse or rat).

In addition to efficacy, evidence for anatomic and functional integration of transplanted human stem cells should be assessed. Human stem cells destined for transplantation may be tagged with a marker, such as green fluorescent protein, that allows transplanted cells to be readily identified upon histological examination. A similar approach should be used to evaluate the migration of transplanted human stem cells from the site of injection into adjacent and more distant tissues. The migration of transplanted human stem cells to a nontarget site, and subsequent differentiation into a tissue type, which is inappropriate for that anatomic location, could be problematic.

Questions about the use of embryonic them cells compared with adult stem cells, with respect to robustness and durability should be addressed in animal transplantation models. Similarly, the issue of whether less differentiated cells will be more effective than more differentiated cells following transplantation should be investigated. Continued advancements in noninvasive imaging technologies, such as magnetic resonance imaging (MRI) and positron emission tomography (PET scanning), will allow these events to be observed in real time, with reasonable resolution and without having to use large numbers of animals.

From the perspective of toxicology, the proliferative potential of undifferentiated human embryonic and embryonic germ cells evokes the greatest level of concern. A characteristic of human embryonic stem cells is their capacity to generate teratomas when transplanted into immunologically incompetent strains of mice. Undifferentiated embryonic stem cells are not considered suitable for transplantation due to the risk of unregulated growth. The question that remains is, at what point during differentiation does this risk become insignificant, if ever? Identifying the stage at which the risk for tumor formation is minimized will depend on whether the process of stem cell differentiation occurs only in a forward direction or is reversible. Before the clinical trials being in humans, the issue of unregulated growth potential and its relationship to stem cell differentiation must be evaluated. It is essential that careful toxicology studies are performed that are of appropriate duration, and involve transplantation into immuno-compromised animals of undifferentiated or partially differentiated embryonic stem cells, as well as adult stem cells.

22.16. EMBRYONIC STEM CELL RESEARCH AND FUNDING

Embryonic stem cell research is allowed in the United States and many other countries. There is no government law (and almost no state law) against it. The government has only set some limits on the number of embryonic stem cell lines eligible for government funding. Supporters, disappointed at failures using these cells, sometimes blame this stem cell research "ban" (which is not really a ban at all). But, the much more serious obstacle lies in the nature of the cells, which are not working out as some predicted. The federal government gave $25 million to human embryonic stem cell research last year. But on August 9, 2001, President Bush said that federally funded research would use only embryonic stem cells already in existence (obtained by destroying embryos prior to that date).

In this way, he reasoned, federal funds could be used to explore this research, without encouraging researchers to destroy new embryos in order to obtain federal grants. Some of these existing stem cell samples have been used to create more than 20 cell lines for research, and others remain in storage for possible use in creating new cell lines in the future. There is no legal limit on the amount of funding that can be used for this avenue, if the total funding for it is relatively small, which is chiefly because researchers are not requesting the funds as they are finding other avenues more promising.

Many private foundations and for-profit biotechnology companies fund stem cell research, but the federal government (especially through the National Institutes of Health) remains the largest source of funds. The government's funding priorities have a large influence on the direction that medical research takes. Since available research funds are being diverted towards exploring embryonic stem cell research, some very promising adult stem cell avenues for treating juvenile diabetes, spinal cord injury, Parkinson's disease, etc., have been under-appreciated and underfunded. Many advances in these fields have emerged in other countries.

22.17. ETHICAL ISSUES OF STEM CELL RESEARCH AND USE

As seen in other area of biotechnology, stem cell research also poses many ethical issues with respect to use and source of stem cells. Stem cells show potential for many different areas of health and medical research, and studying them can help us understand how they transform into the dazzling array of specialized cells that make us what we are. Some of the most serious medical conditions, such as cancer and birth defects, are caused by problems that occur somewhere in this process. A better understanding of normal cell development will allow us to understand and perhaps correct the errors that cause these medical conditions.

Some have made this claim, but in fact this is largely speculation. Embryonic stem cells have never treated a human patient, and animal trials suggest that they are too genetically unstable and too likely to form lethal tumors to be used for treatment any time soon. Years ago it was said that stem cells from embryos would be the most useful because they are so fast-growing and versatile, able to make virtually any kind of cell. But those advantages become disadvantages when these cells make tumors, creating a condition worse than the disease. Yet many supporters remain wedded to this approach, having invested a great deal of money and effort and hoping they can still make it work. This kind of exaggerated "promise" has misled researchers and patient groups before— most obviously in the case of fetal tissue from abortions, which a decade ago was said to promise miracle cures and has produced nothing of the kind.

With huge predictions of the applications of stem cells in biotechnology has advanced the use of stem cells in research and development. Research on one kind of stem cell, i.e. human embryonic stem cells has generated much interest and public debate. Pluripotent stem cells (cells that can develop into many different cell types of the body) are isolated from human embryos that are a few days old. Pluripotent stem cell lines have also been developed from fetal tissue (older than 8 weeks of development).

As science and technology continue to advance, so do ethical viewpoints surrounding these developments. It is important to educate and explore the issues, scientifically and ethically.

The common people and religious groups are also stated speaking about the stem cells ethical issues. Stem cell research is controversial not because of its goals, but rather because of the means of obtaining some of the cells. Research involving most types of stem cells, such as those derived from adult tissues and umbilical cord blood, is uncontroversial, except when its effectiveness as an alternative to embryonic stem cells is debated. The crux of the debate centers around embryonic stem cells, which enable research that may facilitate the development of medical treatments and cures, but which require the destruction of an embryo to derive. In addition, because cloning is one method of producing embryos for research, the ethical issues surrounding cloning are also relevant.

As previously mentioned, the Bush administration, a group of Representatives, a group of Senators, and a group of Nobel Laureates have each presented their respective positions on embryonic stem cell research. In addition, various other organizations, individuals, and councils have issued opinions and reports on the topic. Some groups, such as the Christian Legal Society, Focus on the Family, and the Christian Coalition, support the 2001 Bush policy. Others, such as the National Academies, the Coalition for the Advancement of Medical Research (CAMR), former First Lady Nancy Reagan, and former Presidents Gerald Ford, Jimmy Carter, Bill Clinton favor more embryonic stem cell research than the Bush policy allows. Still others, such as the National Right to Life Committee and the United States Conference of Catholic Bishops, oppose all embryonic stem cell research.

Two presidential bioethics advisory panels have considered the issues involved in embryonic stem cell research. The President's Council on Bioethics (President's Council) published one report directly on the topic, *Monitoring Stem Cell Research*, in which it sought to characterize the issues. While the Council made no recommendations there, in two other reports it has recommended that "Congress should ... [p]rohibit the use of human embryos in research beyond a designated stage in their development (between 10 and 14 days after fertilization)," and unanimously recommended "a ban on cloning-to-produce-children," with a 10-member majority also favoring "a four-year moratorium on cloning-for-biomedical-research," and a seven-member minority favoring "regulation of the use of cloned embryos for biomedical research." A predecessor to the President's Council, the National Bioethics Advisory Committee (NBAC), recommended federal funding for stem cell research using "embryos remaining after infertility treatments," but not for the "derivation or use of embryos...made for research purposes."

Detailed review of the assorted reports and statements reveals that, while positions on embryonic stem cell research may be broadly categorized as *for* or *against*, there is an array of finer distinctions present. These finer distinctions in turn reveal the variation in ethical and moral as well as factual beliefs. The following discussion breaks down the arguments about embryonic stem cell research according to these finer distinctions, demonstrating both the complexity of the issues and the points of resonance among the groups.

Embryo Destruction and Relief of Human Suffering

Humans suffer from many diseases in their lifetime. Most positions on embryonic stem cell research rest, at least, on the relative moral weight accorded to embryos and that accorded to the prospect of saving, prolonging, or improving others' lives. For some, the inquiry begins and ends with this question. For instance, one opponent of the research, the American Life League, posits that human life begins at conception/fertilization, and there is never an acceptable reason for intentionally taking an innocent human life. Similarly, the United States Conference of Catholic Bishops states that the research is immoral because it relies on the destruction of some defenseless human beings for the possible benefit to others.

Years ago it was said that stem cells from embryos would be most useful because they are fast-growing, versatile, and able to make virtually any kind of cell. But these advantages become disadvantages when these cells produce tumors, creating a condition worse than the disease. Yet, many supporters remain wedded to this approach,

as they have invested a great deal of money and effort, and they hope that they can still make it work. This kind of exaggerated "promise" has misled researchers and patient groups—most obviously in the case of fetal tissue from abortions, which a decade ago was said to promise miracle cures and has produced nothing of the kind. Some groups explore the moral standing of human embryos, and also consider the duty to relieve the pain and suffering of others. Others take the position that embryos do not have the same moral status as people. They acknowledge that embryos are genetically human, but hold that they do not have the same moral relevance because they lack specific capacities, including consciousness, reasoning, and sentience. They conclude that performing research to benefit people justifies the destruction of embryos. Acceptance of the notion that the destruction of embryos can be justified in some circumstances forms the basis of pro-stem cell research opinions, and is usually modified with some combination of the distinctions and limitations that follow.

Viability of Embryos

Life of human being is precious and respected much. Some proponents of embryonic stem cell research draw distinctions based upon whether an embryo is viable. The idea behind this distinction is that it is morally preferable for embryos, which will not grow or develop beyond a certain stage, and/or those that would otherwise be discarded, to be used for the purpose of alleviating human suffering. This distinction has led some, though not all, pro-life advocates to support embryonic stem cell research that does not destroy embryos that are viable, i.e, preimplantation embryos; those created via cloning, which are incapable of full development; or those without a woman willing to carry them to term.

Most supporters of some type of embryonic stem cell research touch on the question of viability. The 2001 Bush policy requires, among other things, use of only excess (non-viable) embryos for federally-funded research. One report of the President's Council explores the moral significance of viability, which based upon human choices rather than an embryo's own intrinsic nature, but draws no conclusions. A second report broaches the subject of viability, recommending that Congress ban both the transfer of a human embryo to a woman's uterus for any purpose other than to produce a child, and also research conducted on embryos, 10 to 14 days after fertilization. The NBAC report touches on the moral status of embryos in utero and *in vitro*, though NBAC does not specify whether viability was a key rationale for its recommendations. A group of representatives, Senators, and CAMR implies, but does not state a distinction based on viability, by clearly calling for the use of "excess" embryos developed for IVF, and making no mention of those in utero. By contrast, the National Academies and the group of Nobel Laureates more broadly support research on embryos, making no mention of viability.

Purpose of Embryo Creation

Embryos are a must for the large scale use of stem cells in the research and development. A separate distinction that often leads to the same conclusions, as viability is the purpose for which embryos are created. This distinction draws an ethical line based upon the intent of the people creating embryos. In the view of some, it is permissible to create an embryo for reproductive purposes (such as IVF), but impermissible to create one with the intention of destroying it for research.

Most groups note the ethical significance of reproductive versus research motives for creating embryos. The 2001 Bush policy draws a distinction by including a requirement that federally funded research be conducted only on embryonic stem cell lines derived from embryos created solely for reproductive purposes. NBAC draws the same distinction by recommending that federal funding

can be used for embryos remaining after infertility treatment, but not for research involving the derivation or use of stem cells from embryos created for research purposes or from embryos created using cloning (SCNT). The President's Council recommends that Congress ban attempts at conception by any means other than the union of egg and sperm (essentially banning cloning via SCNT), but does not specify whether embryos might be created *in vitro* specifically for research purposes.

The National Academies support the creation of embryos for research purposes, via cloning (SCNT) to ensure that stem cell-based therapies can be broadly applied for many conditions and people by overcoming the problem of tissue rejection." Mrs. Nancy Reagan, her supporters, and the group of Nobel Laureates also take this position.

New and Existing Cell Lines

The researchers in the stem cell biology produced large numbers of stem cell lines. A distinction has been drawn based upon the timing of the creation of embryonic stem cell lines. Here, the premise is that it is unacceptable to induce the destruction of embryos for the creation of new lines. However, cases in which embryos have already been destroyed and the lines exist, it is morally preferable to use those lines for research to improve the human condition.

This was one central distinction drawn in the 2001 Bush policy, which limited the use of federal funding to research on lines derived on or before the date of the policy. Supporters of the Bush policy on both sides of the issue favor this distinction as a compromise. It allows research on some embryonic stem cell lines. It deters the future destruction of embryos for research. The President's Council writes that the Bush policy mixes prudence with principle, in the hope that the two might reinforce (rather than undermine) each other. The council notes that the policy is supported by what it titles a *moralist's* notion of when one may benefit from prior bad acts (referring to embryo destruction). It prevents the government from complying in the commission of or encouraging the act in the future, and reaffirms the principle that the act was wrong. The same report also contains analyses of the Bush policy that characterize distinction between new and existing cell lines as "arbitrary", unsustainable", and "inconsistent". The council itself takes no position on this or any other issue.

Opponents of the Bush policy, on both sides of the issue, view the distinction between new and existing stem cell lines with reproach. One side, which includes The National Right to Life Committee and the United States Conference of Catholic Bishops, objects because the distinction validates destruction of embryos, and in fact rewards those who did so first with a monopoly. The other side, which includes the National Academies, a group of representatives, Senators, Nancy Reagan, and her supporters, Gerald Ford, CAMR, and the group of Nobel Laureates, objects because the distinction limits the number of embryonic stem cell lines available for research, particularly since the number of authorized lines are dwindling and are "contaminated with mouse feeder cells". Like-wise, although NBAC recognized the distinction between destroying embryos and using ones previously destroyed (e.g., "derivation of [embryonic stem] cells involves destroying the embryos, whereas abortion precedes the donation of fetal tissue and death precedes the donation of whole organs for transplantation"), it still recommended future development of embryonic stem cell lines.

Consent of Donors

Source of the embryo and their derived cells are used in different types of research in stem cell biotechnology. There is a growing consensus throughout a wide array of viewpoints about embryonic stem cell research that embryos should only be obtained for research with the consent of their biological donors. This consent requirement necessitates that embryos be taken only with donors' knowledge, understanding, and uncoerced agreement. The donor consent requirement is consistent with the rules governing human beings' participation in research, and with individuals' general legal authority to make decisions regarding embryos they procreate. A drawback of the requirement is that it may restrict the number of embryos available for research purposes.

The 2001 Bush policy contains a donor consent requirement. It limits approved stem cell lines to those derived with the informed consent of the donors, and obtained without any financial inducements to them. The NBAC and the President's Council also favor donor consent requirements. The National Academy notes the importance of informed consent in its discussion of stem cell research oversight requirements.

Effectiveness of Alternatives

One factual distinction that has been used to support competing ethical viewpoints is the efficacy of alternatives to embryonic stem cell research. The promises of stem cell therapies derived from adult tissue and umbilical cord blood have buttressed opposition to embryonic stem cell research. These opponents argue that therapies and cures can be developed without the morally undesirable destruction of embryos. However, not all scientists agree that adult stem cells hold as much potential as embryonic stem cells. Most supporters of embryonic stem cell research believe that it is the quickest and, perhaps in some cases, the only path that will yield results. Supporters also stress that embryonic and other stem cell research should be conducted collaboratively, so that they can inform one another.

Findings regarding the effectiveness of alternatives to embryonic stem cell research are mixed. The President's Council notes that there is a debate about the relative merits of embryonic and adult stem cells. Focus on the Family cites promising non-embryonic stem cell research: "adult stem cells may be as "flexible" as embryonic ones, and equally capable of converting into various cell types for healing the body."

By contrast, the National Academies finds that the best available scientific and medical evidence indicates that research on both embryonic and adult human stem cells will be needed. NBAC finds in its deliberations that the claim that there are alternatives to using stem cells derived from embryos is not, at the present time, supported scientifically. CAMR supports both embryonic and adult stem cell research, and adds that many scientists believe and studies show that embryonic stem cells are likely to be more effective in curing diseases, because they can grow and differentiate into any of the body's cells and tissues and thus into different organs.

Use of Government Funding

Some division over the support for and opposition to embryonic stem cell research focuses on the question of whether the use of federal funding is appropriate. Those who oppose federal funding argue that the government should not be associated with embryo destruction. They point out that embryo destruction violates the deeply held moral beliefs of some citizens, and suggest that funding alternative research is morally preferable. Proponents of federal funding argue that it is immoral to discourage life-saving research by withholding federal funding. They point out that consensus support is not required for many federal spending policies, as it does not violate democratic principles or infringe on the rights of dissent of those in the minority. They argue that the efforts of both federally supported and privately supported researchers are necessary to keep the United States at the fore-front of what they believe is a very important, cutting edge area of science. Furthermore, supporters believe that the oversight, which comes with federal dollars, will result in a

better and more ethically controlled research in the field. Groups' positions on federal funding tend to mirror their positions on stem cell research generally.   The Bush policy authorizes federal funding for some of the embryonic stem cell research. The President's Council does not take a position on the issue, but notes the pros and cons, and stresses that there is a difference between *prohibiting* embryo research and *refraining from funding* it. Focus on the Family generally supports the President Bush and his policy, but is disappointed by its decision to allow federal funding of research on the existing stem cell lines. NBAC finds the arguments in favor of federal funding more persuasive than those against it. The National Academies, a group of representatives, senators, Mrs. Nancy Reagan and her supporters, CAMR, and the Nobel Laureates favor federal funding for embryonic stem cell research.

SUMMARY

A stem cell is a special kind of cell that has a unique capacity to renew itself and to give rise to specialized cell types. Although most cells of the body, such as heart cells or skin cells, are committed to conduct a specific function, a stem cell is uncommitted and remains uncommitted, until it receives a signal to develop into a specialized cell. Their proliferative capacity combined with the ability to become specialized makes stem cells unique. Researchers have for years looked for ways to use stem cells to replace cells and tissues that are damaged or diseased. Recently, stem cells have received much attention.

Two important points about embryonic and adult stem cells have emerged so far: the cells are different and present immense research opportunities for potential therapy. As research goes forward, scientists will undoubtedly find other similarities and differences between adult and embryonic stem cells. During the next several years, it will be important to compare embryonic stem cells and adult stem cells in terms of their ability to proliferate, differentiate, survive and function after transplant, and avoid immune rejection. Investigators have shown that differentiated cells generated from both adult and embryonic stem cells can repair or replace damaged cells and tissues in animal studies.

Scientists upon making new discoveries often verify reported results in different laboratories and under different conditions. Similarly, they will often conduct experiments with different animal models or, in this case, different cell lines. However, there have been very few studies that compare various stem cell lines with each other. It may be that one source proves better for certain applications, and a different cell source proves better for others.

For researchers and patients, there are many practical questions about stem cells that cannot yet be answered. How long will it take to develop therapies for Parkinson's Disease and diabetes with and without human pluripotent stem cells? Can the full range of new therapeutic approaches be developed using only adult stem cells? How many different sources of stem cells will be needed to generate the best treatments in the shortest period of time?

Predicting the future of stem cell applications is impossible, particularly given the very early stage of the science of stem cell biology. To date, it is impossible to predict which stem cells—those derived from the embryo, the fetus, or the adult— or which methods for manipulating the cells, will best meet the needs of basic research and clinical applications. The answers clearly lie in conducting more research.

There are many ways in which human stem cells can be used in basic research and in clinical research. However, there are many technical hurdles between the promise of stem cells and the realization of these uses, which will only be overcome by continued intensive stem cell research.

Studies of human embryonic stem cells may yield information about the complex events that occur during human development. A primary goal of this work is to identify how undifferentiated stem cells become differentiated. Scientists know that turning genes on and off is central to this process. Some of the most serious medical conditions, such as cancer and birth defects, are due to abnormal cell division and differentiation. A better understanding of the genetic and molecular controls of these processes may yield information about how such diseases arise and suggest new strategies for therapy. A significant hurdle to this use and most uses of stem cells is that scientists do not yet fully understand the signals that turn specific genes on and off to influence the differentiation of the stem cell.

Human stem cells could also be used to test new drugs. For example, new medications could be tested for safety on differentiated cells generated from human pluripotent cell lines. Other kinds of cell lines are already used in this way. Cancer cell lines, for example, are used to screen potential anti-tumor drugs. But, the availability of pluripotent stem cells would allow drug testing in a wider range of cell types. However, to screen drugs effectively, the conditions must be identical when comparing different drugs. Therefore, scientists will have to be able to precisely control the differentiation of stem cells into the specific cell type on which drugs will be tested. Current knowledge of the signals controlling differentiation fall well short of being able to mimic these conditions precisely to consistently have identical differentiated cells for each drug being tested.

Perhaps the most important potential application of human stem cells is the generation of cells and tissues that could be used for cell-based therapies. Today, donated organs and tissues are often used to replace ailing or destroyed tissue, but the need for transplantable tissues and organs far outweighs the available supply. Stem cells, directed to differentiate into specific cell types, offer the possibility of a renewable source of replacement cells and tissues to treat diseases including Parkinson's and Alzheimer's diseases, spinal cord injury, stroke, burns, heart disease, diabetes, osteoarthritis, and rheumatoid arthritis.

For example, it may become possible to generate healthy heart muscle cells in the laboratory and then transplant those cells into patients with chronic heart disease. Preliminary research in mice and other animals indicates that bone marrow stem cells, transplanted into a damaged heart, can generate heart muscle cells and successfully repopulate the heart tissue. Other recent studies in cell culture systems indicate that it may be possible to direct the differentiation of embryonic stem cells or adult bone marrow cells into heart muscle cells. In people who suffer from type I diabetes, the cells of the pancreas that normally produce insulin are destroyed by the patient's own immune system. New studies indicate that it may be possible to direct the differentiation of human embryonic stem cells in cell culture to form insulin-producing cells that eventually could be used in transplantation therapy for diabetics.

To realize the promise of novel cell-based therapies for such pervasive and debilitating diseases, scientists must be able to easily and reproducibly manipulate stem cells so that they possess the necessary characteristics for successful differentiation, transplantation and engraftment. The following is a list of steps in successful cell-based treatments that scientists will have to learn to precisely control to bring such treatments to the clinic. To be useful for transplant purposes, stem cells must be reproducibly made to: Proliferate extensively and generate sufficient quantities of tissue, Differentiate into the desired cell type(s), Survive in the recipient after transplant, Integrate into the surrounding tissue after transplant, Function appropriately for the duration of the recipient's life, Avoid harming the recipient in any way. Also, to avoid the problem of immune rejection, scientists are experimenting with different research strategies to generate tissues that will not be rejected. To summarize, the promise of

stem cell therapies is an exciting one, but significant technical hurdles remain that will only be over-come through years of intensive research.

Stem cell research and use of these cells in various types of applications creates large number of ethical issues that are unexplained, unjustified and unacceptable characters. These have to be further studied in detail and answer has to be found in an ethically sound and scientifically correct methods.

EXERCISE

1. What do you mean by stem cells? Give their importance in the fields of medicine.

2. Write a note on the therapeutic uses of stems cells in the medical biotechnology, and mention the demerits, at present, in understanding stem cell biology.

3. Explain the different types of stem cells used in the medical field and their applications in the treatment of genetic diseases.

4. Briefly write about the present situation of stem cell biology in the developed and developing countries, with an emphasis on India.

5. What is meant by embryonic germ cells? Write a note on the ethical issues associated with the use of embryonic stem cells in research and development.

6. Define embryonic stem cells. What is their impact on the ethical, legal and social issues raised in the cloning of animals and humans.

7. What is meant by ethical issues? Mention some of those which are associated with the different sources of stem cells, and add a note on its impact on future of stem cell biology.

8. How are the stem cells cultured *in vitro*? Add a note on their importance in the knowledge of science for the development of cloning methodologies.

9. Enumerate the importance of the usage of stem cells in the organ transplantations.

10. Distinguish between the embryonic stem cells and adult stem cells, and discuss their ethical issues associated with source of cells.

11. Discuss how stem cells are isolated in animals, and explain their applications and risk involved.

12. Explain the different types of limitations associated with the stem cell biotechnology and governments regulations in the different countries.

13. What is meant by cell based therapy? Explain its applications for therapeutic uses, and ethical issues associated with human stem cell biology.

14. Give an account of the arguments for and against the stem cell biology.

15. Critically comment on the ethical, social and legal issues raised by the human stem cell biology and the sources of stem cells.

16. Briefly explain how scientists and researchers are supporting the technological need of the stem cell biotechnology and its continued research.

17. Describe the saying of religious groups against the human and animal stem cell biology in present circumstances.

18. Elucidate the future of stem cell applications in the treatment of autoimmune disorders. Add a note on the treatment of autoimmune disorders adopted in the animals.

19. Give a note on the stem cells benefits and limitations and depict ethical issues raised by the stem cells technology.

20. Describe how human stem cell cloning is becoming more complicated because of the ethical and social issues raised by the different organizations.

21. What are the safety issues considered with the stem cell biology, and its regulations in the progress of biotechnolgy.

22. Name the research organizations that are involved in the stem cell research, and their progress in the stem cell biotechnology.

23. Depict your views about the importance of stem cell biotechnology in near future for better understanding of biological science.

23

CHAPTER

Use of Animals in Research and Testing, and Alternatives for Animals in Research

23.1. INTRODUCTION

Research using animals has made an important contribution to the advances in medicine and surgery, resulting in major improvements in the health of human beings and animals. Currently, for certain types of medical research, there are no viable alternatives to animal models. Even though supporting the ongoing development of reduction, refinement, or replacement (the 3Rs), it is opened that animal experimentation will continue to play a key role in research towards tackling major medical problems. International research supported by the animal use in test is expected to be carried out in the spirit of the Government medical legislation, as well as being compliant with all local legislation and ethical review procedures.

23.2. ANIMAL RESEARCH AND TESTING

The earliest references to animal testing are found in the writings of the Greeks in the third and fourth centuries BC, with Aristotle (384-322 BC) and Erasistratus (304-258 BC) among the first to perform experiments on living animals. Galen, a physician in second-century Rome, dissected pigs and goats, and is known as the "father of vivisection". Although the term "vivisection" means literally the "cutting up" of a living animal, and originally referred only to experiments that involved dissection of, or surgery on, live animals, it is now commonly used to refer to any experiment on a living animal, which is also known as *in vivo* testing (Croce 2000).

Animals are considered as major experimental organisms, and have played a role in numerous well-known experiments. In 1796, Edward Jenner extracted pus from pox-infected cows to inoculate James Phipps against smallpox. In the 1890s, Ivan Pavlov famously used dogs to describe classical conditioning. On November 3, 1957, a Russian dog named, "Laika" became the first animal to orbit the earth. In 1996, Dolly, the sheep, was born. It was the first mammal to be cloned from an adult cell. This paved the way for cloning of other mammals including humans.

Branches of Research using Animals

The various branches that use animals for research and testing are as follows:

- Anatomy
- Physiology
- Biochemistry
- Biotechnology
- Microbiology
- Pathology
- Biophysics
- Pharmacology
- Surgery
- Cardiology
- Neurology
- Cardio-thoracic surgery
- Lab medicine
- Neuro surgery
- Orthopedics
- Psychiatry
- Opthalmology
- Pediatric
- Paed. surgery
- Endocrinology
- Immunology
- Medicine
- Gastroentrology
- Gastrointestinal surgery
- Gene therapy
- Stem cell biology
- Toxicology

The most commonly used animals are:
- Mouse
- Rat
- Frog
- Hamster
- Guinea pig
- Rabbit
- Cat
- Monkey
- Sheep
- Dog

Animals and their Care

The various species of animals required for research and testing should be procured from recognized sources. Local procurement sources should be identified by the research and testing organizations for the supply of non-laboratory bred animals. These organizations, with breeding facilities, should procure the breeding stock from a reliable source for initiating a colony, ensuring that genetic makeup and health status of animals is known. Additionally, the following aspects have to be taken care of:

- Healthy animals should be obtained from a recognized source.
- Acceptable methods and norms of transportation should be followed, considering the distance, seasonal, and climatic conditions, and the species of animals.
- The animals should be given a reasonable period for physiological, psychological, and nutritional stabilisation before their use.

Food and water

Animals should be fed palatable, non-contaminated, and nutritionally adequate food, which should be procured from a reliable source. Good quality food and water should be provided ad libitum.

Areas in which food is processed or stored should be kept clean and enclosed to prevent entry of insects and wild rodents.

Watering devices, such as drinking tubes, should be examined routinely to ensure their proper operation.

Feeders should allow easy access to food and water while minimising contaminating by urine and faeces.

Sanitation and cleanliness

Animal rooms, corridors, storage spaces, and other areas should be cleaned with appropriate detergents and disinfectants.

Animals should be kept dry except for those species whose natural habitation need water. Soiled litter material should be removed routinely from the places where larger animals and non-human primates are housed.

Cages should be cleaned each time before animals are placed in them. Animal cages, racks and accessory equipment, such as feeders and watering devices, should be washed and cleaned frequently to keep them free from contamination.

Deodorisers or chemical agents other than germicides should not be used to mask animal odours.

Veterinary care

Wherever required, adequate veterinary care must be provided under the supervision and guidance of a registered veterinarian or a person trained and experienced in laboratory animal sciences.

Animals should be observed regularly, and problems of animal health and behaviour, recorded and addressed.

For the animals kept for experiments of longer duration, the following steps should be adopted:

- All animals should be observed for signs of illness, injury, or abnormal behaviour by the animal house staff, and reported to the attending veterinarian.
- Diseased animals should be isolated from healthy ones.

Numbers of Animals used for Research and Testing

Accurate global figures of animals used for testing and research are difficult to obtain. The British Union for the Abolition of Vivisection (BUAV) estimates that 100 million animals are experimented around the world every year; 10 to 11 million of them in the European Union and 1,101,958 in the United States in 2004. The Nuffield Council on Bioethics reports that estimates of the total number of animals used annually in research around the world are difficult to obtain, and range from between 50 and 100 million animals. Animals bred for research, which are then as surplus, or used for breeding purposes, are not included in the figures. According to the US Department of Agriculture, the total number of animals used in that country in 2002 was 1,137,718, not counting birds, mice, and rats, which make-up around 85% of research animals. The laboratory primate advocacy group has used these figures to estimate that 23 to 25 million animals are used in research each year in America. Figures released by the British Home Office show that in 2004, 2,854,944 procedures were carried out on 2,778,692 animals; an increase of 63,000 from 2003. The term procedure refers to an experiment, which might last several months or even years. The figures show that most animals are used in only one procedure: animals either die because of the experiment, or are killed and dissected afterwards.

Over half the experiments (e.g., behavioral tests, breeding stock, controlled dietary intake) in Britain in 2004—1,710,760—either did not require anesthetic or anesthesia was not used because this would interfere with the experimental results. 880,897 experiments were conducted in connection with pure research; 114,081 were toxicology tests, 982,640 were for breeding, and most of the rest were for applied studies in human medicine, veterinary medicine, or dentistry. 9,035 experiments involved the deliberate infliction of "psychological stress".

Advocates of Animal Testing

Arguments of the testing advocates

- It would be unethical to test substances or drugs with adverse side-effects on human beings.
- Controlled experiments involve introducing only one variable at a time. This is the reason why animals are experimented on while confined inside a laboratory. Human beings could not be confined in this way.
- There is no substitute for the living systems necessary to study interaction among cells, tissue and organs. Animals are good surrogates because of their similarities to humans.

- There is no substitute for psychiatric studies (e.g., antidepressant clinical trials) that require behavioral data.
- Animals have shorter life and reproductive spans, so several generations can be studied in a relatively short time.
- Animals can be bred especially for animal-testing purposes, meaning they arrive at the laboratory free from disease.
- Humans in some parts of the world are healthier due to advances in medical research derived from animal testing.
- Animals receive more sophisticated medical care because of animal tests that have led to advances in veterinary medicine.
- There have been several examples of substances causing death or injury to human beings because of inadequate animal testing.

Opponents of Animal Testing

Arguments of the opponents

- The animal-testing industry is a multi-million dollar concern. Advocates of testing may argue that their interests are scientific, but they are just as often commercial.
- Even with medical and non-commercial research, tests are often conducted to produce academic papers in order to acquire a Ph.D., academic tenure, or more funding, and not because the research is beneficial. The suffering of the animals is excessive in relation to whatever benefits may be reaped.
- Animal-testing facilities are not properly regulated or inspected, and several undercover investigations by activist groups have uncovered evidence of animal abuse.
- Animal testing is regarded by opponents as bad science because they believe:
  - Some animal models of disease are induced and should not be compared to the same disease in humans. Activists claim, that Parkinson's disease in humans cannot be reproduced by causing brain damage in an animal, though genetic and toxin-mediated animal models are now widely used.
  - Some drugs have dangerous side effects that were not predicted by animal models. Opponents often claim "Thalidomide" as an example of this. In spite of being tested on pregnant animals, birth defects are seen in mice, rats, hamsters, rabbits, macaques, marmosets, dogs, cats, fish, baboons, and rhesus monkeys.
  - Some drugs appear to have different effects on human and non-human animals. Aspirin, for example, is a teratogen when given to certain animals in high doses, but there is a conflicting evidence regarding its effect on human embryos.
  - The conditions in which the tests are carried out may undermine the results, because of the stress the environment produces in the animals. BUAV argues that the laboratory environment and the experiments themselves are capable of affecting every organ and biochemical function in the body. Noise, restraint, isolation, pain, psychological distress, overcrowding, regrouping, separation from mothers, sleeplessness, hypersexuality, surgery, and anaesthesia can increase mortality, contact sensitivity, tumour susceptibility and metastatic spread, as well as decrease the viral resistance and immune response.
- Some opponents, particularly the supporters of animal rights, argue further that even if animal testing did reap benefits to human beings, these could not outweigh the suffering of the animals, and that human beings have no moral right to use individual animals in ways that do not benefit that individual.

Diseases and Use of Animals

The development of disease frequently involves complex interactions between different systems or organs within the body. In these instances, studies using alternatives to animals (including cell lines, tissue cultures or whole organs, computer modeling and human studies) can provide only limited information, and a fuller understanding requires studies at the whole-body level. For some types of fundamental research, such as research on the control of physiological processes or where organ-to-organ communication plays an important role, there are no alternative approaches to using animals.

Use of alternatives to animals can often provide useful information in research or toxicity testing, particularly where a whole animal approach might not be essential. However, when researching complex diseases or testing new treatments in the form of drugs or therapies, it is obviously not possible to assess the effect on the whole animal by using only one cell or tissue type. Again, much of the fundamental research that underpins more applied research depends on observing what happens in whole animals.

The animals can be used in research and testing after the following requirements are satisfied:

- It is fully compliant with current government office legislation.
- It has been approved by a local ethics committee.
- It has been successfully peer reviewed.
- Due consideration has been given to the refinement, reduction or replacement of the animals in the experiment and no viable non-animal alternatives exist.

The use of animals in test considers, therefore, that for the kind of research that provides the most hope for understanding health and disease and the development of new treatments, it is currently difficult to foresee a time when it might be possible to conduct this research without using animals. Some people argue that animals are misleading as models of human disease; however, as knowledge about genetics has increased, it has now been shown that humans and animals have a similar genetic makeup. For example, mice share over 85% of the sequences of their genes with humans, and there are analogous genes in the mouse for nearly every human gene. This means that many biochemical pathways are identical, or similar, in animals and humans. These are summarized under the following four major headings:

- Anatomical and physiological similarities
- Similarities in diseases
- Easy in controlling the external conditions
- Use of animals in large numbers

Anatomical and Physiological Similarities

Many animals have similar anatomy and physiology as that of humans. The best similarities can be noticed in the animals belonging to the mammalian group. Hence, in the research and testing, animals are preferred in the preliminary studies before directly testing on the humans. This helps in avoiding the usage of large number of humans in the initial stages of research and testing, and reduces the cost.

Similarities in Diseases

Since animals have structure and functions of various tissue system similar to that of humans, certain categories of animals get diseases similar to the humans. So, these animals are used as a model system for the in depth study of the disease development, symptoms, progression and control/treatment methods.

Easy in Controlling the External Conditions

All the external and some times internal conditions are completely controlled according to the requirement of research and testing. But such situations are not completely possible in the humans because of their independent thinking capacity. In case of animals, the modifications can be made in their external environment such as temperature, humidity, air phase, and other physical factors.

Use of Animals in Large Numbers

For the purpose of proper scientific data generation, large number of animals can be used without much drawbacks. These animals can be easily bred and can be obtained at the required period without any delay or shortage. Such situations are very difficult, or some times it is totally impossible to obtain required number of individuals at right time.

23.3. ANIMAL MODELS

Model is a copy of the original object. For example, the model of an airplane looks like the real areoplane, but may be smaller, less detailed, made of different materials, and may or may not be able to do everything a real airplane can. However, by studying that model, a person can learn a lot about airplanes. In fact, airplane designers make models of the planes they are building to test them before they make the actual airplane. Similarly, it is practical to make a model of a disease to learn about it before attempting to diagnose and treat the patients.

There are many different ways to model a disease and they are as follows:

- Mathematical or computer models are used to predict the way a disease may behave.
- Some normal or diseased tissues can be grown in a Petri dish and studied outside the body.
- Patients suffering from a disease can be studied.
- An animal that has a similar disease can be studied.
- The disease can be induced in an animal.

The last two are considered "animal models" of disease. Animals can be used as models of human or animal diseases. Examples of animal models include:

- Induction of cancer in a mouse to simulate cancer in a human.
- Studying a genetic disease in a pure-bred dog that is similar to human disease.
- Transplanting an organ between two pigs.

Scientists can then use the animal model to study the way the disease progresses and what factors are important in it. The model can also be used to study disease treatment. This is done through the use of controlled experiments **(Fig. 23.1)**.

Of course, no animal can provide a perfect model for human disease. That is why all

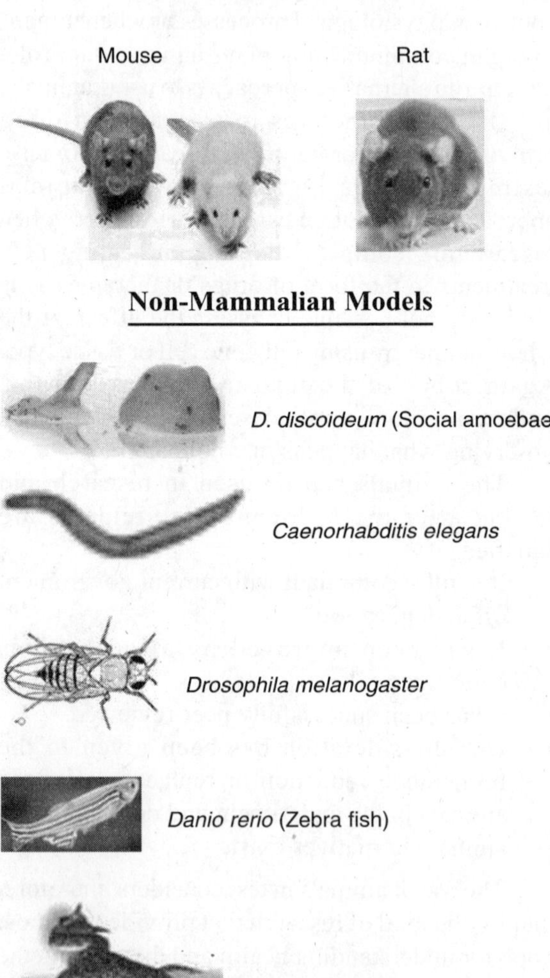

Fig. 23.1. Different animal models used for human disease research and testing

discoveries based on animal research, once they are in a well developed stage, are eventually tested on human volunteers before being used for treating patients. However, animal research allows us to understand the dynamics of health and disease in a whole organism, which increases the chance of developing suitable treatments for human patients.

23.4. GENETICALLY MODIFIED ANIMALS AND THEIR USE IN RESEARCH AND TESTING

Genetically modified (GM) animals provide valuable models of human disease, and they help the researchers to explain disease pathways and assess potential new therapies. The result of our ability to genetically manipulate the animals (mostly mice and rats) to mimic a variety of human diseases is likely to lead to the treatments of many of the diseases.

The need to deliberately create an animal with a condition that will inevitably cause discomfort to it should be justified in the same way as any other proposed research under the terms of the animals (Scientific Procedures) Act. The level of care of GM animals must always be appropriate to the needs of the animal.

The animal use in test does not consider that GM animals should be necessarily seen as "unnatural". Mutations arise spontaneously in animals in nature, and often, when humans consider the phenotypic result of these natural mutations advantageous or appealing, these mutated animals are selected and bred. Obvious examples are the various breeds of dogs that can be observed at Crufts, and the selection of traits in farm animals related to food production. All traits established by breeding have a genetic basis for that trait.

The same is true for inbred mouse strains used in research, where natural mutations or even induced mutations (for example by gene targeting or chemical mutagenesis) have been selected. For example, a non-obese diabetic (NOD) inbred mouse strain originated from breeding pairs of mice exhibiting a natural tendency to diabetes, can develop diabetes within four weeks of age. Similarly, a chemically induced mutated mouse strain is susceptible to deafness, and can lose the ability to hear within four to six weeks. The speed at which the disease develops in both these strains makes them extremely useful tools for studying the disease and developing potential treatments.

The main difference between the breeds of dog, inbred strains, and GM animals (the vast majority of which are mice) is that the mutations in the GM animals have been purposefully introduced by humans, as opposed to leaving it to nature, in order to develop a particular trait that makes them more or less susceptible to particular diseases, such as cancer, diabetes, rheumatoid arthritis etc. In fact, with GM animals, usually only one specific gene is targeted for modification, so the effects on physiology and development may be more accurately predicted than animals, which result from selective breeding or chemical mutagenesis.

23.5. ETHICAL ISSUES IN THE USE OF ANIMALS IN RESEARCH AND TESTING

At a practical level, all animal researchers have a duty to reduce suffering to a minimum. Good husbandry is as vital to the researcher as his/her experimental protocol. The law recognizes that animals be treated humanely and kept away from unnecessary suffering. It must also be noted that research using animals does not always take place solely for human benefit, but can also be of benefit to animals, for example in the development of veterinary medicines. The animal use in test considers that the existing cost/benefit assessment is reasonable, but the assessment of scientific validity of proposed experiments should remain primarily the responsibility of the expert scientific community.

The main strength of the cost/benefit assessment is that it ensures that the Home Office grants a project licence only if the potential benefits of the proposed research outweigh the costs. This assessment is necessarily subjective, but it does allow decisions to be made on a case-by-case basis. It is important to recognize that circumstances can change dramatically, particularly in a dynamic field such as biomedical science, for example, through the emergence of new diseases (e.g., BSE in cattle, and the emergence of new variant Creutzfeld-Jakob disease, nvCJD, in humans). This can lead to unforeseen justifications for the use of a particular species or procedure, which might not previously have been deemed to be justifiable under the existing cost/benefit assessment.

Since its introduction in April 1999, the animal use in test considers that the Ethical Review Process (ERP) has helped researchers to focus on the broader ethical issues relating to their work with animals. The process necessitates the researchers to spend more time justifying their research, including justifying all of the proposed procedures and why alternatives are not suitable. The presence of a lay member in the ethical review committee helps to demonstrate to the public that research using animals in the UK is governed by strict ethical guidelines.

The ERP was introduced with the intention that its implementation should not result in a duplication of effort or undue delay on behalf of both the local ethical review committee and the Home Office Inspectorate. However, the main weaknesses of the ERP seem to be the increased time spent by researchers on justifying their research, and the increased delays in the processing of new project licence applications and renewals of project licences.

Regulations on Research Involving Animals

All countries should have stringent regulations on the use of animals in research, particularly in the field of biomedical research. The animal use in test considers that the government legislation on animal procedures should not be made more restrictive or complex, as this may drive some of the countries' best scientists abroad, and deter overseas scientists from working in the country. This would have serious implications for research capacity in any country, particularly in the post-genomic era, and would have an adverse effect on the countries' economy.

The animal use in test is also concerned that the Home Office treats the production of all GM animals as independent procedures that require both authorization under a project licence and special welfare considerations. It should be noted that the inclusion of "normal" animals used as part of the breeding programme substantially increases the total numbers of animals used, as defined by the Home Office statistics. This might give misleading indications that the use of animals in research is increasing more substantially than is really so.

The animal use in test considers that even with the likely increased use of GM animals in biomedical research over the next decade or so, existing regulatory controls are satisfactory in terms of protecting the welfare of such animals. The animal use in test would be concerned if further regulations were imposed on the animal research generally, but particularly research that used transgenic technology.

The animal use in test is also aware that scientists are concerned about the stringent regulations associated with movement and importation of well-characterized transgenic animals. While they agree that, in some cases, there is a need for transgenic animals to have a 'superior' health status, or to be afforded a higher level of care, there is no reason why phenotypically 'normal' transgenic animals should be subject to the same stringent level of import/transport regulations. Reducing the barriers to the international movement of such animals could help to reduce both duplication of research programmes and the numbers of animals used in research.

The scientists consider that a continued dialogue on the use of animals in research is healthy and should be supported, and that the research community should explain the methods and aims of its research. The MORI public attitude studies in 1999, 2000, and 2002 indicate that animal use with regard to animal experiments increased substantially between 1999 and 2002, and that there is a link between levels of public animal use in test and perceptions of openness from the scientific community. The animal use in test therefore considers that those involved in biomedical research using animals need to be as open and transparent as possible, while at the same time should be aware of the need to protect researchers from intimidation from extremist groups.

23.6. ALTERNATIVES FOR ANIMAL MODELS IN RESEARCH AND TESTING

In 1959, "The Principles of Humane Experimental Technique" was published in London, defining the concept of animal testing alternatives as the "Three R's": Refinement, Reduction, and Replacement. The only viable choice for a true animal rights' supporter is the "Replacement" of animals used in tests; Refinement and Reduction are still viewed as morally wrong.

Mohandas K. Gandhi said it best in his autobiography "The Story of My Experiments": "To my mind, the life of the lamb is no less precious than that of a human being. I should be unwilling to take the life of the lamb for the sake of the human body. I hold that the more helpless a creature, the more entitled it is to protection by man from the cruelty of man."

Sophisticated alternatives to the use of animals in consumer product testing are readily available.

Most of the large producers of personal care and household products could adopt these methods, which are more cost effective, better predictors of human injury, produce far quicker results, and do not involve animal cruelty. Why don't all research and testing organizations become cruelty-free? The two main reasons are the fear of human safety, and the fear of product liability suits.

The development of alternatives to animals continues on many fronts and their use is increasing, but no method can satisfactorily replace animals in all cases. Animal use will continue, though the continued development of new techniques, such as cell lines and computer modelling, will reduce the need for animal research in some areas. The approval for the use of human embryos in research may allow the development of new human cell lines, which may reduce the need to use animals in some research areas.

However, the use of animals will be increasingly important in post-genome research, and it is expected that the number of animals used in research, particularly mice and rats, will increase substantially due to the increased use of GM animals.

Scientists have developed, and will continue to develop, non-animal techniques as a natural corollary to, or by-product of, their research. However, developing and promoting them as alternatives (to specific animal procedures) is not often the main objective of these scientists. Many of the non-animal methods in use in biomedical research were not developed specifically as "alternatives". For example, cell culture was developed originally as a method for studying cell biology and events at the cellular level, and not as an 'alternative to animals', but certain specific applications, such as monoclonal antibody production have become alternatives to animal methods.

While animal testing is still very much in existence at large corporations such as Procter and Gamble and Lever Brothers, there are now several hundred "cruelty-free" consumer product companies.

There are different types of alternative methods for the use of animals in research and testing that are in common use in the number of organizations. The most common type of alternative methods are:

- *In vitro* tests
- Computer software
- Databases of tests already done (to avoid duplication)
- Human "clinical trial" tests

Use of animal cells, organs, or tissue cultures is also deemed an alternative though, obviously, animal lives are sacrificed for the use of their parts.

Alternatives

It is unethical and illegal to experiment on humans. Human biological systems are complex, therefore it is necessary to research and test on animals that are similar to humans to obtain reliable and effective results. New drugs, devices or procedures must receive legal approval from the Federal Drug Administration (FDA) before being given to humans. In fact, almost every major medical advance in the last century is because of the research with animals. Research on animals provides necessary information on how a new drug or procedure will affect a human. Such medical advancements include, prevention of measles, diphtheria, mumps, whooping cough and polio. Medications to treat mental illness, chemotherapy to treat cancer, antibiotics to treat infectious diseases, heart and cardiovascular surgery, and organ transplants are some of the additional medical benefits from the animal research. Scientists must test medical treatments for effectiveness, and new drugs for safety before using them on humans. Small animals, especially bred for laboratory use, usually rats and mice, are used to identify any undesired side effects of new drugs, such as infertility, miscarriage, birth defects, liver damage and cancer-causing potential. Once new drugs are proven safe in animal studies, they may be used in clinical studies on human volunteers, who then have the assurance that they may fare better, and will not do worse, than if they were given standard treatment or no treatment at all. New surgical techniques must also be carefully devised and tested in living, breathing, whole organ systems with pulmonary and circulatory systems much like those in humans. The physicians and physicians-in-training who perform delicate cardiac, ear, eye, pulmonary and brain surgeries, must develop the necessary skills before patients' lives are entrusted to their care. Neither computer models, cell cultures, or artificial substances can yet simulate flesh, muscle, blood, bone and organs working together in the living system. If the use of live animals were abolished, it would be difficult to investigate the effects of how one system (for example, the nervous system) interacts with another (for example, the immune system or the endocrine system), while monitoring side effects (effects on respiration, kidney function, or heart rate). (For specific examples, read about "research on birth defects and implications for microsurgery"). Considering the negative impact of the usage of animals in the biotechnological fields, researchers started looking for the other means of studies in the research fields. The 'Three Rs' is the best substitute for the animals used in the research and testing. In 1959, British zoologist, William M.S. Russell, and microbiologist, Rex L. Burch published "The Principles of Humane Experimental Technique" in London. In this book they put forth the 'three Rs of animal research', and explained in detail how these three Rs can successfully substitute the use of animals in research and testing. The three RS are:

- **Refinement**—improve the experiment to reduce the suffering of the animal.
- **Reduction**—use statistics to reduce the number of animals that must be used for each experiment.
- **Replacement**—use alternative methods, e.g., testing on cell cultures (*in vitro*).

Russell and Burch defined the R's in the following way: reduction as a means of 'lowering the number of animals used to obtain information of a given amount, and precision'; 'refinement' as any development leading to a 'decrease in the incidence of severity of inhumane procedures applied to those animals, which have to be used'; and 'replacement' as any scientific method 'employing non-sentient material, which may replace methods using conscious living vertebrates'.

Refinement encompass all those methods described above (good housing, professionally qualified keepers and users), and includes the use of new techniques developed in human medicine aimed at decreased invasiveness of bodies, improved instrumentation for monitoring animals and analysis of body fluids, and better management of pain and distress in animals. A sophisticated instrument like magnetic resonance imaging can save the life of many animals, which are otherwise sacrificed at intervals to understand the time course of a drug action or a basic process. The use of such sophistication requires a single animal, which would provide all information along a given curve. With the advent of microelectronics, fibre optics, and laser technology, animals can be monitored externally and internally without restricting their movement in their primary enclosures. A variety of non-experimental variables associated with prolonged restraint, thus can be removed, contributory to defined and reliable data. Another improvement in instrumentation is the development of diagnostic clinical laboratory equipment, which requires only microlitter sample to perform a variety of diagnostic tests. The use of small sample size means the use of small animal species, and further it prevents the need to euthanise many of these species to obtain the required volume of blood. It is now possible to obtain serial blood samples from small laboratory rodents that reduces the number of animals necessary to obtain data over the length of the study.

The aim of any reduction strategy is to obtain comparable levels of information from minimum number of animals, or more information from a given number of animals, so that towards the end, a given project or test is completed with few animals than originally contemplated. Such reduction(s) can be attempted in several ways like animal sharing, changes in research strategy, use of proper statistical design, proper use of strains, mutants, transgenics etc. Animal sharing refers to an internal system adopted by any laboratory, wherein whenever the animal(s) is sacrificed at the end of the experiment, the organ, tissues, or body fluids not needed for that project can be shared amongst other researchers. Culling of morbid or very old animals (especially rodents) is a normal practice in many animal facilities. The organs and body fluids from such animals can be frozen and stored, and later made available to researchers whenever they require these materials for standardising certain biochemical tests/ procedures. Both these measures eliminate unwanted killings, and can bring about substantial reduction in the use of animals in absolute numbers. Research strategies adopted should be such that reduction in number occurs. For example, before starting a major experiment involving large number of animals, small pilot studies using few animals could be tried out to see the trend. In toxicity studies, non-mammals (insects, marine invertebrates, microorganisms, plants) or *in vitro* or test tube methods or even non-living systems (chemical or physical systems like computer simulation) can be tried as pre-screen strategies. There are enough studies to suggest that poor experimental design, together with inappropriate statistical analysis of experimental results, lead to an inefficient use of animals.

Proper statistical design, prior to undertaking any study, and appropriate analysis of the resultant data can give comparable or greater precision using fewer animals. Regarding the type of animals to be used, it should be remembered that a hetero-genous or out bred group of animals lead to large amount of variation, causing non-reproducible false data and the necessity to repeat the experiment in many cases. Inbred strains, specific mutants, transgenic, and knock out animals are made to order assembly products designed to give precise and uniform results in an experiment.

Replacement is the ultimate in animal welfare measures, where given purpose is achieved without conducting experiments or other scientific procedures on protected live animals. The replacement can be absolute or relative, direct or indirect, total or partial. When humane killing of vertebrate animal is done to provide cells, tissues and/or organs for the *in vitro* studies, it is a relative replacement for live animals. But if permanent human and invertebrate cell and tissue cultures are maintained for toxicity screening, that becomes absolute replacement. If a tissue (say skin) from human or dead animal is used *in vitro*, instead of the same tissue from live animals, it amounts to direct replacement. Replacement can be indirect as in cases where pyrogen test in rabbit is replaced by Limulus amoebocyte lysate (LAL) or a test based on whole human blood. Total avoidance of an animal procedure because of the lack of justification or reliability of the method; using a human volunteer instead of a laboratory animal; or testing a chemical *in vitro* instead of live animals—all tend to total replacement. But if on the contrary, non-animal methods are used as pre-screening in toxicity studies, so that most toxic compounds can be rejected before continuing with further tests in animals, one has achieved a level of partial replacement. Based on the various terms of replacements described above, different approaches can be practised by researchers.

Current Use of *In Vitro* Tests

Numerous *in vitro* tests have been developed. Many of these tests will be improved over time by the introduction of new scientific information and technological advances in *in vitro* toxicology and related fields, such as molecular biology and biotechnology. The Agency encourages the development and use of *in vitro* test systems for planning and interpreting the results from whole animal toxicity studies.

Significant advances have been made in the development of *in vitro* alternatives for ocular safety testing. Other *in vitro* systems have been proposed, which measure a broad range of end-points and are now in various stages of validation. The agency is currently part of an interagency regulatory groups evaluating these proposed alternative test methods.

In vitro approaches to toxicity testing can provide useful data when integrated with other information about the toxicity of food and color additives used in food. Results of *in vitro* tests can be used to optimize the design of conventional toxicity tests for a particular test substance by helping to determine appropriate dose levels and by deciding which species is the best model for man. Such improvements in the design of whole animal toxicity tests may reduce the number of test animals required to produce useful information about the safety of proposed food and color additives used in food.

In vitro tests can help elucidate the nature of the interaction between test substance and organism at the cellular, subcellular, and molecular levels. Thus, once the critical target organ or organ system has been identified in whole animal studies, *in vitro* tests can focus on the mechanism of action of the test substance at the target site. Information from these studies can assist the Agency in making decisions about the safety of proposed food and color additives used in food, by comparing responses observed in human and animal cells and by facilitating extrapolation from high-dose to low-dose responses.

At present, in evaluating a petition for the use of a food or color additive, the Agency considers *in vitro* tests to be useful in identifying the mechanism(s) of action of the test substance, and to provide information about subtle effects observed *in vitro* that may not be observed in *in vivo* studies

Although attempted on a primitive scale in 1885, tissue culture was not a practical methodology until the discovery of antibiotics, which inhibit the growth of bacteria. It is possible to culture both normal and abnormal tissue, and to harvest cells and tissues from humans and animals.

Cell culture is complicated by the tendency of isolated cells to "dedifferentiate" in culture, taking on the qualities of unspecialized cells instead of keeping the characteristics that define them as cells from a specific organ such as the liver. For that reason, the easiest type of cells to maintain in culture are the less differentiated cells such as fibroblasts. Cell lines can be developed from these cells and maintained continuously, eliminating the need for further animal or human donors. But due to their lack of specific functions, they are not really useful for many types of toxicity testing.

Primary cells, taken from a specific organ in an animal or human donor, are very useful, but maintain their differentiated functions only for a few days, or at most a few weeks. The challenge is to find a way to prevent the primary cells, with all their special structures and functions, from dedifferentiating. One possibility is by co-culturing them with different types of cells from the same organ. Studies have shown that in co-cultures, the cells survive longer and are better able to maintain their differentiated functions. Another potentially useful technique for long-term culture of differentiated cells is immortalization, in which the DNA encoding viral genes are transfected into primary cells. Ideally, the gene can be turned on and off, causing the cell to replicate indefinitely when the gene is turned on, and to redifferentiate when the gene is turned off.

Tissue slices are another *in vitro* option (one gaining in popularity), as the cellular architecture of the organ is preserved with all cell types present in the whole organism present in the tissue sample used for testing. However, given the status of current culture techniques, the slice is viable for a relatively brief period of time (hours to days). An additional technique involves the use of isolated organs. Presently, certain organs such as the liver can be maintained outside of the animal for several hours by perfusion—the pumping of blood or artificial media through the organ to nourish it. It is then possible to infuse chemicals into the organ and examine their effects.

Each of the above methodologies can be used to measure toxicity in a specific organ. However, to mimic better the effect of a toxic substance in a whole animal, researchers are beginning to co-culture cells from multiple organs. Many toxic substances are metabolized by the liver. Sometimes the metabolism of a nontoxic substance by liver enzymes will result in the formation of a toxic metabolite, which will then affect another organ in the body, for example the kidney. By co-culturing liver and kidney cells and then introducing a toxic chemical, the investigator can observe the process, wherein the liver cells metabolize the chemical and the kidney cells respond to the toxic effects of the metabolite.

While cell culture techniques have become increasingly sophisticated over the past decade, a great deal of work still needs to be done to define the optimum culture conditions for different types of cells, so that they may propagate indefinitely and still function as they do *in vivo*.

Physiologically-Based Toxicokinetic Modeling

Tissues and cell cultures alone cannot predict the effects of toxic substances in an animal or human. The cells and tissues must be placed in context within the organism. Physiologically-based toxicokinetic modeling attempts to do just that. Using computer simulation, researchers are attempting to relate the concentration of a chemical, which causes toxicity at the cellular level as measured with an *in vitro* test, to the corresponding *in vivo* exposure level that might have produced that effect. The simulations are performed using a computer to solve the differential equations that describe the model.

In this type of modeling, the toxicokinetics of various drugs and chemicals are determined by their tissue solubility characteristics, metabolic rates, and the physiology of the test species. The information required to construct the model can be gained through literature searches, *in vitro* studies, and *in vivo* experiments, which are limited to those designed to obtain very specific information relevant to the model. Due to the systemic nature of the model, much of the necessary information retrieved through literature search and databases is the result of past whole-animal studies.

The construction of a model for the first chemical in a class demands a great deal of time and effort. But once the prototype is created and validated, it can be extrapolated and applied to other related chemicals under various exposure conditions. Mathematical modeling offers a tool to predict the concentration of the active form of the toxicant at the site where it generates its effects. However, as researchers working in the field admit, physiologically-based toxicokinetic modeling is like much of *in vitro* toxicology, in its infancy with an enormous array of practical and theoretical challenges to be overcome before it is able to serve as a general basis for risk management.

Structure-Activity Relationships and Databases

The hypothesis on which the concept of structure-activity relationships is based states that the structure of a chemical inherently possesses all information necessary to predict its toxicity, including the manner in which the parent chemical and its metabolites interact with macromolecules in the cell. This principle has been successfully applied to certain classes of carcinogens; however, its broader applications to general toxicity have not yet been established.

While applying structure-activity relationships, biological effects are expressed in qualitative terms. A mathematical equation is prepared to correlate the toxicant's chemical properties with the biological effect. The relationship derived from the equation is used to make predictions about the toxicity of a chemical. Computers are used to establish these relationships. Advances in computer technology over the past twenty years have contributed greatly to the deve-lopment of both structure-activity relationships and physiologically-based toxicokinetic modeling.

The construction and use of comprehensive databases will also help reduce the number of animals used in testing and refine the testing process. John Frazier of the Johns Hopkins Center for Alternatives to Animal Testing, and Mildred Green of Technical Database Services are currently compiling a database, which will include descriptions of *in vitro* methodology and specific results obtained by testing chemicals with these methodologies. A collection of information about methods, together with a compendium of data produced by these methods, should prove useful for both validation and interpretation of *in vitro* testing data.

In Germany, ZEBET (The Center for Documentation and Evaluation of Alternative Methods of Animal Experiments) was established at the Institute of Veterinary Medicine of the Federal Health Institute in 1989. The ZEBET Data Bank offers a compilation of literature on alternative methods in relation to the three Rs (replacement, reduction, refinement); description of the animal experiment that can be replaced; reduced, or refined; names of scientists experienced in the area; and references of the specific alternative methods and animal experiment it is designed to replace, reduce, or refine. Most of the information in the ZEBET Data Bank is available only in German; however, both the summary and list of references are available in English. Fund for the Replacement of Animals in Medical Experiments (FRAME) in England, and Professor Nicola Loprieno and colleagues at the University of Pisa, Italy are also constructing databases that should prove useful to researchers implementing the three Rs.

In the United States, the Toxicology Data Bank developed in 1978 lists over 4,000 chemicals, and includes data on the production and use of each chemical—a description of its physical properties, and the results of *in vivo* pharmacological and biochemical experiments. As the Toxicology Data Bank and similar *in vitro* databases grow, researchers will be able to design and plan better experiments, based upon the knowledge gained from newly accessible data. Conversely, as an understanding of toxicological mechanisms increases, the data bank grows larger and more useful.

Validation and the Future of *in vitro* Toxicology

The future of alternatives research lies in validation. Validation is the door through which every alternative method must pass before entering the armamentarium of toxicity testing. In validation, a particular test is defined for a specific purpose. The test's relevance and reliability are established through intralaboratory and interlaboratory assessment, test database development, and evaluation.

In September 1992, the Validation and Technology Transfer Committee of the Johns Hopkins Center for Alternatives to Animal Testing completed a draft framework for validation and implementation of new *in vitro* toxicity tests. Noting that continuing advancements in both cellular and molecular biology, and bioanalytical and computer techniques had resulted in a proliferation of *in vitro* alternatives without a "formal administrative process to organize, coordinate or evaluate validation and implementation of these advancements, the committee proposed a validation plan aimed at increasing scientific and regulatory acceptance of alternative technologies.

Chronology of Alternatives for Research and Testing

1959: Book of Russell and Burch published; first enunciated the "Three Rs".

1962: Lawson Tait Trust (UK)—first research fund to support scientific development of alternatives.

1965: Littlewood Committee Report (UK)—reported that little would be gained by paying special attention to alternatives.

1967: USA—United Action for Animals, animal group that campaigned specifically for replacement alternatives.

1969: FRAME (UK)—new group to promote the idea of alternatives to the scientific community.

Lord Dowding Fund (UK)—new fund established to support alternative research.

1971: Council of Europe Resolution 621—suggested that an alternatives database be established. This was the first significant government initiative or recommendation on alternatives.

1975: USA—National Academy of Sciences Meeting—first major scientific meeting on the idea of alternatives in the US.

1977: Netherlands Animal Protection Law included a specific section on alternatives that has now grown into a program where the government provides hundreds of thousands of dollars to support alternative research.

1978: FRAME Meeting at the Royal Society on Alternatives in Drug Development and Testing, London—first big scientific meeting on alternatives in Europe.

Smyth book examining alternatives published in the research. (Smyth was president of the UK Research Defense Society, established to support animal research.)

1979: USA—HR 4805—Research Modernization Act—introduced by UAA (see above), directing that 30-50% of animal research funding be reallocated for alternatives. Gained wide public support and forced Congress to take an interest in the subject.

Sweden established $90,000 in Government funding for alternatives—first government funding for alternatives.

1980: USA—Henry Spira launches the Draize Campaign, this campaign against rabbit eye irritancy testing is credited with the establishment of a Rockefeller University alternatives research project (using $750,000 from Revlon) and the establishment of CAAT.

New England Antivivisection Society gives $100,000 for alternatives research on tissue culture, and a second animal-welfare consortium provides $176,000 for chorioallantoic membrane (CAM) test development.

1981: USA—Johns Hopkins University Center for Alternatives to Animal Testing (CAAT), established with $1 million fund from cosmetics industry (Avon and Bristol-Myers Squibb are leading donors—result of Draize campaign).

Swiss animal legislation—specifically requires consideration of alternatives.

1982: USA—Colgate Palmolive provides $300,000 to investigate chick chorio-allantoic membrane (CAM) system.

1983: Switzerland provides SFr 2 million over two years for alternatives research.

FDA formally announces that they no longer require classical LD50 data.

1984: FRAME (UK) receives £160,000 from Home Office—first UK government funding for alternatives research.

1985: Health Research Extension Act (USA) is passed, requiring NIH to develop a plan for alternatives.

Animal Welfare Act Amendments (USA) are passed that require greater attention to alternatives in research that cause pain and distress.

Index Medicus adds a subject heading: Alternatives to Animal Testing.

European Research Group on Alternatives to Toxicity Testing (ERGATT) is formed.

1986: The UK's Animals (Scientific Procedures) Act replaces the 1876 Act.

The US Congress's Office of Technology Assessment issues a landmark report "Alternatives to Animal Use in Research, Testing and Education".

The Council of Environmental Ministers of the European Community enacts EC Directive 86/609, requiring that member countries develop legislation promoting the Three Rs.

An FDA survey reports a 96% decrease in the use of classic LD50 tests in 1985 compared to 1975-1979.

Two new cell toxicology journals emphasizing non-animal methods, *Toxicology in vitro* and *Molecular Toxicology*, are established.

The Organization for Economic Cooperation and Development (OECD) announces changes in its guidelines for acute oral and dermal toxicity, and starts to discusses alternatives.

British Industrial Biological Research Association (BIBRA) increases funding of alternative research to £700,000 per annum.

The Industrial *in vitro* Toxicology Society (IVTS) is established in the United Kingdom.

1987: The HSUS publishes an analysis of the historical importance of alternative methods in biomedical-awarded Nobel Prizes.

The Dutch Alternatives to Animal Experiments Platform is established with participation from government, industry, and animal welfare organizations.

In vitro Toxicology: A Journal of Molecular and Cellular Toxicology is established.

The Swiss Foundation "Finanzpool 3 R" is established with one million Swiss francs to support alternative research.

1988: A government/industry workshop is held on alternatives to ocular irritancy testing, to review the Soap and Detergent Association's Alternative Program.

The Industrial *in vitro* Toxicology Group holds its first meeting.

The US Republican presidential platform encourages the implementation of alternatives to animal testing.

The J.F. Morgan Foundation for Alternative Research is established in Canada.

The Swiss government's Office for Animal Experiments and Alternatives is established.

1989: The Center for Documentation and Evaluation of Alternative Methods of Animal Experiments, known by its German acronym ZEBET, is established in Germany.

Procter and Gamble announces that it is contributing $450,000 per year for three years to its University Animal Alternative Research Program.

Avon announces that it will no longer use the Draize Test.

The Second International Conference on Practical *in vitro* Toxicology is held in the United Kingdom.

The Swedish Fund for Scientific Research without Animal Experiments invests 700,000 Swedish krona in alternative projects.

The American Anti-Vivisection Society establishes the Demeter Fund (later known as the Alternatives Research and Development Foundation) to support non-animal research, funding up to $50,000 annually for one or more projects.

1990: CAAT and ERGATT hold a workshop on validation of alternative methods.

The University of California Alternatives Center is established at UC-Davis.

The platform for alternatives to animal experiments in the Netherlands allocates the equivalent of $700,000 annually for the promotion and validation of research into the Three Rs, and the improvement of housing and care systems.

The Japanese Society for alternatives to animal experimentation publishes the journal Alternatives to Animal Testing and Experimentation (AATEX).

1991: The Interagency Regulatory Alternatives Group holds a workshop "Eye Irritation Testing Alternatives: Proposals for Regulatory Consensus" in Washington, DC.

The HSUS presents Alan Goldberg, director of CAAT, with the first Russell and Burch Award,

established to recognize scientists who have made outstanding contributions to alternative methods.

The Organization for Economic Cooperation and Development (OECD) accepts the Fixed Dose Procedure as an alternative to the LD50 Test.

Representatives of regulatory agencies in Japan, Europe, and the United States agree to drop the classic LD50 as a required measure of acute toxicity.

The UK Home Office announces a grant program for the funding of alternatives research.

The second report of the FRAME Toxicity Committee is published in ATLA.

The Swiss Institute for Alternatives to Animal Testing (SIAT) is established in Zurich.

1992: The European Centre for the Validation of Alternative Methods (ECVAM) is established.

The European Parliament amends the Cosmetics Directive 76/768 to ban the marketing of cosmetics tested on animals after January 1, 1998 (a decision on the ban is later postponed until June 30, 2000).

CAAT hosts a tenth anniversary conference in Baltimore, Maryland, giving Founders' Awards to Dr. D.A. Henderson, the CTFA, and Henry Spira.

1993: The NIH Revitalization Act directs the NIEHS to establish criteria for the validation and regulatory acceptance of alternative testing, and to outline a process for regulatory review of potential alternative methods. It also directs the NIH director to establish an alternative program and to report on its progress annually.

The first World Congress on Alternatives and Animal Use in the Life Sciences: Education, Research and Testing takes place in Baltimore.

Member states of the European Union agree that everything possible should be done to achieve a 50% reduction in the use of vertebrate animals for experimentation and other scientific procedures by the year 2000.

The Interagency Regulatory Alternatives Group holds its second meeting on alternatives in Washington, D.C.

1994: The U.S. government establishes the Interagency Coordinating Committee on the Validation of Alternative Methods (ICCVAM), co-chaired by William Stokes of NIEHS and Richard Hill of EPA, in response to the 1993 NIH Revitalization Act.

The Netherlands Centre for Alternatives to Animal Use (NCA) is established as a national information center on alternatives.

1995: The Gillette Company and the HSUS launch a program to fund research and development of alternative methods; two methods of $50,000 each are awarded annually.

1996: The second world Congress on Alternatives and Animal Use in the Life Sciences is held in Utrecht, The Netherlands.

The OECD holds a workshop to develop internationally harmonized criteria on validation and regulatory acceptance.

CAAT, The HSUS, Procter & Gamble, and other organizations establish Altweb, a web site devoted to information on alternative methods.

1997: ICCVAM issues guidelines on criteria for validation and regulatory acceptance of alternative methods.

The Institute for *in vitro* sciences is established in Gaithersburg, Maryland.

1998: ECVAM accepts the following alternative methods: 3T3 NRU PT test as an alternative for assessing phototoxicity, TER (transepithelial electrical resistance) test, and Episkin and similar methods for assessing skin corrosivity.

ECVAM endorses *in vitro* methods as alternatives to the ascites methods for the production of monoclonal antibodies.

The National Toxicology Program Interagency Center for the Evaluation of Alternative Toxicological Methods (NICEATM) is established to provide support to ICCVAM.

1999: The third world Congress on Alternatives and Animal Use in the Life Sciences is held in Bologna, Italy.

CAAT holds a TestSmart workshop to discuss alternatives to animal testing in the Environmental Protection Agency's High Production Volume (HPV) chemical testing program.

The EPA announces major changes in its HPV program, including funding for alternative methods, following the TestSmart workshop and negotiations with animal protection organizations.

ICCVAM endorses Corrositex for the assessment of skin corrosivity and the Murine Local Lymph Node Assay for the assessment of allergic contact dermatitis.

2000: ICCVAM and NICEATM organize the International Workshop on *in vitro* Methods for Assessing Acute Systemic Toxicity.

The OECD officially announces its plans to delete the LD50 Test (Test Guideline 401) from its testing guidelines, in favor of three alternative methods.

The ICCVAM Authorization Act is signed into law, changing ICCVAM's status from ad hoc committee to a permanent entity.

2001: CAAT receives a donation of $900,000 to establish a grants program devoted to refinement alternatives.

The NIEHS suggests that the use of human and/or nonhuman animal cell lines for the screening of chemicals can reduce the need for animals in the acute systemic toxicity tests and may eventually replace a large amount of animal testing altogether.

Congress approves an Environmental Protection Agency (EPA) funding bill that appropriates $4 million for the development of non-animal test methods.

Two expert scientific panels nominated by the OECD approve an *in vitro* (non-animal) cell and tissue test as replacements for two internationally accepted animal tests used in assessing the phytotoxicity and corrosivity of industrial chemicals.

Possible Applications of Alternative Tests

Isolated cells, tissues, and organs can be prepared and maintained in culture by methods that preserve properties characteristic of the same cells, tissues, and organs *in vivo*. Such *in vitro* systems will permit data to be generated under controlled experimental conditions, and in the absence of many complicating factors characteristic of experiments with whole animals. For example, the use of cell culture systems will enable the metabolism of a toxicant that occurs in one type of cell (i.e., hepatocyte cells) to be studied separately from a toxic endpoint that occurs in a different cell type.

Several toxic endpoints may lend themselves to quantification in an *in vitro* test system. Relevant endpoints could be identified by comparing the action of a toxicant at cellular, subcellular or molecular sites, with the toxic effects observed in the target organ or tissue *in vivo*. Analysis of a broad spectrum of *in vitro* cellular events may provide information about the *in vivo* progression of a toxic response as a function of toxicant concentration and time.

Since *in vitro* procedures have the potential to yield reproducible measurements, they theoretically lend themselves to standardization. However, interpreting data obtained from the standardized *in vitro* toxicity test with a reasonable degree of confidence can only occur after potential confounding factors, such as interactions between the test agent and non-cellular components of the test system have been identified or eliminated.

The process of validation appears to be key to the full acceptance of alternative tests, where the reliability and relevance of procedures are established for specific purposes. While there is much discussion about the framework for this process, several components appear essential for the overall coordination of the validation process, including scientific consensus on the definition of a validated test; reference chemicals with defined toxicity and general availability; a central repository for test performance data and protocols; an established network of laboratories having the capabilities of method validation; and scientific understanding of the mechanism of the toxicological process involved.

An impartial and competent group of scientists from regulatory agencies and the research community could facilitate the implementation of the validation process.

Limitations of Alternative Tests

Limitations of *in vitro* tests are well known.

In vitro test systems are not available for all tissues and organs. In addition, normal systemic mechanisms of absorption, penetration, distribution and excretion are absent from *in vitro* test systems.

In vitro systems lack the complex, interactive effects of the immune, blood, endocrine systems, nervous system, and other integrated elements of the whole animal. Thus, *in vitro* tests cannot be used to study the complex nature of systemic toxicity.

Validation of new methods is time-consuming and expensive. Acceptance of *in vitro* tests as alternatives to traditional toxicity testing in whole animals is expected to be slow. While many schemes have been proposed to expedite these processes, no alternative *in vitro* test, presently, can replace the *in vivo* toxicity studies.

Funding for the Research of Alternative Animals

Throughout the world, different organizations started funding for alternative methods of animal usage in the research and testing. Humanitarian organisations and governments have funded studies into alternative methods since the 1960s. For the past 15 years, Germany has given £4.2 million a year in research grants, while the Netherlands spends £1.4 million a year. It is estimated that the total spent by the UK government is in the region of £2 million a year. The European Centre for the Validation of Alternative Methods was set up in 1992 by the European Commission, and contributes £6.3 million annually. EU regulations state that researchers must assess the pain that an animal may feel during an experiment, and justify its suffering by what the research can achieve.

Revlon Cosmetics was one of the first large companies to fund research for alternatives, with a $750,000 contribution to the Rockefeller University in 1979. Several organizations such as the Johns Hopkins Center for the Alternatives to Animal Testing (CAAT); the International Foundation for Ethical Research; the Cosmetic, Toiletry, and Fragrance Association; and the Soap and Detergent Association followed suit and started their own programs to validate alternatives. However, it should be kept in mind that while companies search for alternatives, the use of animal actually increases, because the old test (using animals) must be done alongside the new test (without animals) to ensure consistent results.

Some of the methods to reduce the animal use in the research and testing are as follows:

Reducing death rates

In the past, the toxicity of a new substance was measured by an 'LD50' (lethal dose 50%) test. This test required up to 200 rats, dogs or other animals to be force-fed different amounts of the substance, to determine the dose that would kill exactly half that group of animals **(Fig. 23.2)**.

Fig. 23.2. Examination of rat for the lethal effect of drug

Recent changes in protocol have put a ban on the LD50 test, save in exceptional circumstances. In addition, the Organisation for Economic Co-operation and Development says that if a substance kills the first three animals it is tested on, further trials are unnecessary.

Statistics

Vaccine is only considered effective if at least 80% of the vaccinated animals survive after being exposed to a particular disease. However, the disease must also kill 80% of a control group not protected by the vaccine. Using statistical methods, Coenraad Hendriksen of the National Institute of Public Health and the Environment in the Netherlands has developed a method to test diphtheria and tetanus vaccines, which requires measuring the level of antibodies in an animal.

Apart from greatly reducing their suffering, it also uses half the number of animals. Other statistical techniques can use patient data to understand how a disease spreads, without testing it on animals.

Fewer mammals

Horst Spielmann of ZEBET has surveyed decades of industry data on pesticides. He concluded that if mice and rats prove sensitive to a chemical, it does not have to undergo further tests on dogs. Spielmann anticipates that 70% of dog tests can now be dispensed with.

There is a general effort by researchers to use lab animals that are less likely to suffer the sensations of pain or discomfort. In Canada, many studies have replaced mammals with fish, and now researchers are even trying to use bacteria in tests instead of rats.

Cell cultures

Today, *in vitro* (meaning, "in glass") as opposed to *in vivo* (meaning "whole animal") has flourished because of advances in tissue culture techniques and other analytical methods.

In the 1970s, the Netherlands used 5,000 monkeys a year to make polio vaccines. Now kidney cell cultures from just 10 monkeys provide enough vaccine for everyone in the country. Hormones or vaccines manufactured in cell cultures are also purer than those made within the animals themselves. This further reduces the need for animal tests to check the safety of the vaccines.

Synthetic membranes

The Department of Transportation became the first US agency to accept tests for skin corrosivity conducted on artificially-grown cells in 1993. The traditional test measured how far a corrosive substance ate into an anaesthetised rabbit's shaved back. Instead, the replacement uses reconstructed human skin or a synthetic material called 'Corrositex'. Similar solutions are being developed for many other types of experiments, which currently use animals.

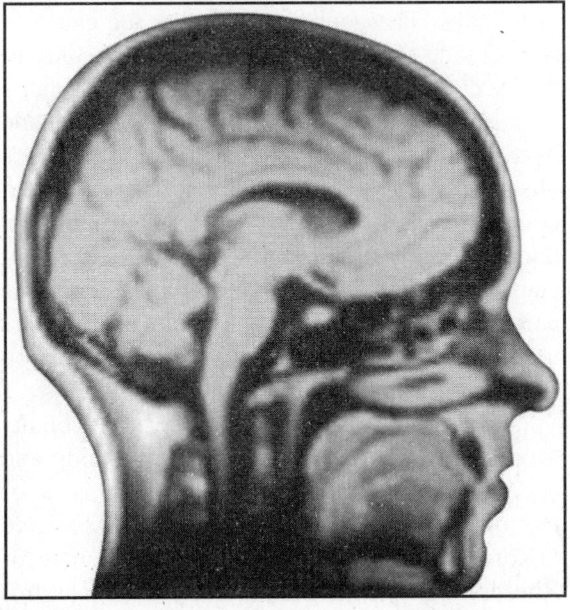

Fig. 23.3. Advanced scanning technology is useful in replacing animal tests

New technologies

New scanning technologies (such as Magnetic Resonance Imaging) can help doctors learn about diseases from human patients, without the need for invasive surgery or animal testing **(Fig. 24.3)**.

Similarly, autopsies and cell culture studies can reveal a great deal of information without having to simulate the disease in a lab animal. Computer models have also been developed that simulate an animal's response, removing the need for live animal tests.

In the Future

Animal researchers say that it will be impossible to eliminate all animal tests. But most scientists accept that it is extremely important to minimise the suffering of laboratory animals, and to use as few animals as possible. The specific tests are:

Eytex

Produced by the National Testing Corporation in Palm Springs, California, Eytex is an *in vitro* (test-tube) procedure that measures eye irritancy via a protein alteration system. A vegetable protein from the jackbean mimics the reaction of the cornea to an alien substance. This alternative is used by Avon instead of the cruel Draize eye irritancy test.

Skintex

An *in vitro* method to assess skin irritancy that uses pumpkin rind to mimic the reaction of a foreign substance on human skin (both Eytex and Skintex can measure 5,000 different materials).

EpiPack

Produced by Clonetics in San Diego, California, the EpiPack uses cloned human tissue to test potentially harmful substances.

Neutral red bioassay

Developed at Rockefeller University and promoted by Clonetics, the Neutral Red Bioassay is cultured human cells that are used to compute the absorption of a water-soluble dye to measure relative toxicity.

Testskin

Produced by Organogenesis in Cambridge, Massachusetts, Testskin uses human skin grown in a sterile plastic bag, and can be used for measuring irritancy, etc. (this method is used by Avon, Amway, and Estee Lauder).

TOPKAT

Produced by Health Design, Inc. in Rochester, New York, TOPKAT is a computer software program that measures toxicity, mutagenicity, carcinogenicity, and teratonogenicity (this method is used by the US Army, the Environmental Protection Agency, and the Food and Drug Administration).

Ames test

It is a test for carcinogenicity, which is done by mixing the test culture with *Salmonella typhimurium* and adding activating enzymes. It is able to detect 156 out of 174 (90%) animal carcinogens, and 90 out of 100 (88%) non-carcinogenes.

Agarose diffusion method

Test for the toxicity of plastic and synthetic things used in medical devices such as heart valves, artificial joints, and intravenous lines. Human cells and the test material are placed in a flask, and are separated by a thin-layer of agarose (a derivative of seaweed agar). If the material tested is an irritant, an area of killed cells appears around the substance.

The main disadvantages of animal tests, according to John Frazier and Alan Goldberg, of CAAT, are animal discomfort and death, species-extrapolation problems, and excessive time and expense. Animal protection advocates stress that the main disadvantage is the inhumane treatment of animals in tests due, in part, to the fact that anaesthesia for the alleviation of pain is often not administered. Scientists allege that using anaesthesia will interfere with test results.

Progress towards the widespread use of alternatives to animal testing will continue to gain strength, as awareness of, and support for, alternatives is made known. As consumers, we can make a difference in the lives of innocent animals by purchasing only the products deemed "cruelty-free", and writing to the companies that still perform animal testing and letting them know why you will not purchase their products.

A Novel Animal Model for Alzheimer Vaccine

A study describes a promising new primate model for testing a potential Alzheimer's disease vaccine. This may enable scientists to study the vaccine in an animal model of Alzheimer's that is very similar to humans. The goal is to discover the cause of serious side effects that halted an earlier study of the vaccine in people, according to the Alzheimer's Association. Tremendous progress has been made by the National Institute on Aging, the Alzheimer's Disease Centers, universities, pharmaceutical companies and the Alzheimer's Association in understanding Alzheimer's disease. The Association's goal of delaying the disabling symptoms and eventually preventing Alzheimer's appears to be a feasible objective, which the research community can achieve in the next decade.

Alzheimer Aß Vaccination of Rhesus Monkeys (*Macaca mulatta*) by Sam Gandy and colleagues, appeared in the March 2004 issue of the journal "Alzheimer's Disease and Associated Disorders". According to Gandy, because of Rhesus monkey's genetic similarity to humans, this study was initiated to determine whether the brain changes associated with vaccination of humans could be modeled in aged nonhuman primates. He believes he and his team were successful.

The monkeys responded to the vaccination by producing lots of antibodies to Alzheimer amyloid, and showed elevated plasma amyloid that is presumably on the way being cleared from the body. Brains were also examined after six months of vaccination, and there was no evidence of inflammation.

Alzheimer Vaccination

Mouse models of Alzheimer amyloid vaccination have shown excellent results. Vaccination of plaque-forming transgenic mice with aggregated Alzheimer amyloid beta peptide has shown that it can promote clearance of existing deposits of Alzheimer-type cerebral amyloid and reverse amyloid associated behavioral deficits in mice. The trial in humans was stopped in 2002, because a small percentage of participants developed symptoms of inflammation of the brain and spinal cord. However, still much has to be learned from the trial.

A preliminary report published on a small subgroup of the vaccine trial population indicated positive results by stopping or significantly reducing cognitive decline in those participants who developed antibodies to beta amyloid. And, in the first autopsy report of a participant in the vaccine trial, scientists found an evidence that significant areas of her brain were free from the amyloid plaques targeted by the vaccine—a phenomenon not seen in the brains of seven unvaccinated individuals with Alzheimer's who participated in the study. In a review article published in the same issue of the journal, Gandy and colleagues said that "These recent data clearly demonstrate that tackling beta amyloid plaques by vaccination is a feasible approach for preventing or even treating Alzheimer's disease-associated pathology. However, more fine-tuning in guiding the immune response is needed to circumvent detrimental side effects."

In this study, Gandy and colleagues vaccinated two rhesus monkeys with beta amyloid—a protein fragment considered a prime suspect in disrupting and destroying nerve cells in the Alzheimer's brain. The vaccinated monkeys developed significant levels of antibodies to beta amyloid as well as high levels of beta amyloid circulating in their blood; much of it attached to antibodies.

Two other monkeys were vaccinated with a "control" amyloid, in this case, the islet amyloid that builds up in diabetes. They did not develop antibodies against beta-amyloid and had much lower circulating beta-amyloid levels. None of the four monkeys showed any evidence of brain inflammation. The National Institute of Neurological Diseases and Stroke funded the study.

23.7. DIFFERENT ANIMAL MODELS OF DISEASE

These systems are becoming increasingly important secondary screens of *in vitro* hits. *In vitro* assays typically rely on simple interactions of chemicals with a drug target, such as receptor binding or enzyme activity inhibition. However, *in vitro* results often poorly correlate with *in vivo* results, because the complicated physiological environment is absent in the *in vitro* testing system. Although cell-based assays can provide some information, cultured cells still do not provide physiological conditions and complex interactions among different cell types and tissues. Moreover, cell lines are usually transformed, exhibiting different gene expression and cell cycle profiles than those of cells in the living organism.

There is a growing trend of using human tissues for drug discovery research. Tissues, however, only provide an isolated *ex vivo* condition, which is not completely representative of *in vivo* response, because drug action often involves metabolism and interplay among different tissues. For example, the effects of a drug on muscle may involve absorption by the intestine and metabolism by the liver. Therefore, results in animal studies are essential to validate HTS (high-throughput screening) hits and exclude compounds with unfavorable ADMET (absorption, distribution, metabolism, excretion, and toxicity) properties, which are responsible for more than half of compound attrition in costly clinical trials.

Currently, *in vivo* assays are not usually performed until or after the lead optimization stage. This is partly due to the low speed and high cost of conventional animal models (typically rodents) and the relatively high number of preliminary hits from HTS. With alternative small-animal models emerging, however, it is now possible to perform *in vivo* testing earlier in the process. Thus, researchers have developed model systems using both vertebrates (zebrafish) and invertebrates (the fruit fly, *Drosophila melanogaster*; and the nematode *Caenorhabditis elegans*) for drug screening. The small size, high fecundity, and experimental tractability of these animals enable cost-effective and rapid screening of numerous compounds.

Zebrafish (*Danio rerio*) and its Use

Even with technologies such as biophotonic imaging, using mammals for drug discovery is hampered by high cost and relatively low throughput. Recently, researchers have created assays based on the zebrafish (*Danio rerio*), a

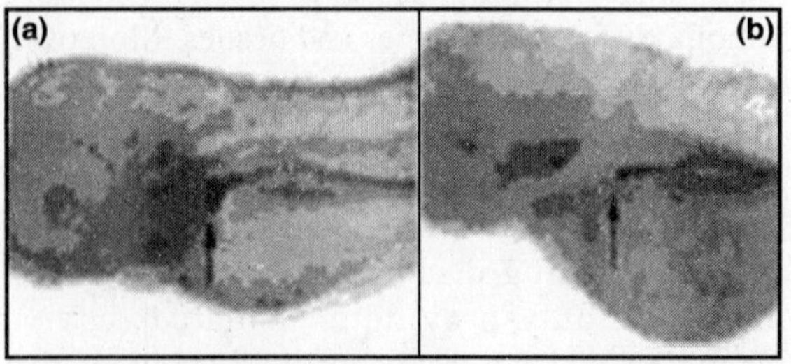

Fig. 23.4. Transparent crystal clear analysis in the zebrafish. The see-through nature of the zebrafish allows researchers to distinguish a normal specimen (a) stained with an antibody that highlights the kidney (arrow) from a drug-treated individual. (b) with smaller kidneys (arrow)

small freshwater teleost. Zebrafish embryos are transparent and develop externally from the mothers, permitting direct assessment of drug effects on internal organs and tissues *in vivo* (**Fig. 23.4**). The fish is easy to maintain and breed, and its fecundity is high. Each female can produce 100-200 eggs per mating, providing large numbers of animals for high-throughput assays.

Because of their size, zebrafish embryos and early larvae can be raised in only 100 mL of water in the wells of a 96-well plate for high-throughput whole-animal assays requiring only small amounts of compounds. Drug administration is also simple because researchers can dissolve small compounds in the water, where they diffuse into the embryos. Alternatively, researchers can micro-inject larger molecules, such as proteins, directly into the embryos. To knock down specific genes for target validation, morpholino antisense molecules can be injected into one- or two-cell-stage embryos, resulting in uniform distribution of the oligonucleotides across the embryos in several days.

Since the zebrafish has the same organs as found in mammals, it is a much more useful model than *Drosophila* and *C. elegans*. Most human genes have homologues in zebrafish, and the functional domains of proteins, such as ATP-binding domains of kinases, are almost 100% identical between homologous genes, although the similarity over the entire protein is only about 60%. Since protein function largely resides in functional domains where drugs often bind, the zebrafish is a highly valid model for studying gene function and drug effects in humans.

Indeed, many zebrafish versions of mammalian genes have been cloned and found to have similar functions, and numerous drugs tested in zebrafish caused effects similar to those observed in humans or other mammalian models. For example, research groups have tested anti-angiogenic compounds effective in mammals on zebrafish, and found them to have similar effects.

The zebrafish is the only vertebrate species for which large-scale genetic screens have been carried out, and many mutants obtained from these genetic screens display phenotypes that mimic human disorders, including cardiovascular disease, neurodegeneration, cancer, and blood diseases. These mutants not only identify genes that may be involved in diseases, but also can be used for drug screening.

Trolling Tumorigenesis

Zebrafish are responsive to carcinogenic chemicals and develop neoplasms that are histologically similar to human cancers. In addition, because of the rapid development of the zebrafish embryo, researchers can use it for testing drugs that affect cell proliferation. For example, Moon *et al.,* discovered novel microtubule inhibitors by zebrafish embryo screening. The zebrafish is also an excellent model for the screening of more specific cancer therapeutics, such as angiogenesis and apoptosis modulators. Angiogenesis pathways in zebrafish and mammals are highly conserved, and zebrafish homologues of several important mammalian angiogenesis regulatory genes are expressed in patterns similar to those in mammals. When these genes are mutated or knocked down, they also cause expected effects on angiogenesis.

Researchers can easily visualize blood vessels in zebrafish by their endogenous alkaline phosphatase activity, vessel-specific antibody staining, and, in transgenic zebrafish, by a vessel-specific promoter linked with a reporter, such as green fluorescent protein. After the lumen is formed, blood vessels can also be visualized by injecting fluorescent microbeads into the circulatory system as a functional assay to assess the integrity of the vascular system.

Researchers at Phylonix Pharmaceuticals have tested a number of compounds that showed anti-angiogenic effects in mammals, including SU5416 and flavopiridol. They found that a drug's effect on vessel inhibition in zebrafish correlated well with the effects seen in mammals, indicating that the zebrafish is a predictive model for testing angiogenesis inhibitors. Since angiogenesis is involved in other diseases, such as diabetic retinopathy and macular degeneration, the angiogenesis assays are also useful for discovering therapies for these diseases.

Angling at Apoptosis

Inducing apoptosis is also a promising approach for cancer treatment. The apoptotic processes in zebrafish and mammals are similar, and zebrafish homologues of most mammalian apoptosis-related genes have been identified. Screening for apoptosis inducers can be performed by looking for their effects in the zebrafish embryo. Researchers can easily detect apoptosis in live embryos by using acridine orange and fluorescence-conjugated caspase substrate (e.g., PhiPhiLuxG1D2). Researchers at Phylonix have developed quantitative assays that can be performed in 96-well microplates in conjunction with a fluorescent plate reader, enabling high-throughput screens.

Apoptosis is an important mechanism for morphogenesis and homeostasis, and abnormalities in apoptosis are involved in many other diseases in addition to cancers, such as neurodegenerative diseases. Using forward genetic screens, researchers have identified many zebrafish mutants that display abnormal apoptosis, and can serve as models for anti-apoptotic drug screening.

Researchers can also induce apoptosis in specific populations of cells to mimic certain disorders. For example, aminoglycoside antibiotics can cause degeneration of hair cells in zebrafish neuromasts, which is similar to the adverse effects of these antibiotics in human hair cells. The induced hair cell apoptosis can then serve as a model for screening agents to prevent or reverse the damage.

Tackling Toxins

Zebrafish and several other teleost species have been used to test environmental toxicants for a long time. Zebrafish-based toxicity assays for drug candidates have also been developed recently. These include assays for organ toxicity, developmental toxicity, and acute toxicity (LC_{50}). Changes in organ morphology and the occurrence of necrosis can be directly assessed under a dissecting microscope. Adverse drug effects on cardiac function can be detected by direct observation of heartbeat in the transparent zebrafish embryo, while other problems, such as neurotoxicity can be thoroughly examined by staining with cell-type-specific antibodies. Histology of the small zebrafish embryo is simple, because serial sections can be mounted on a single slide and quickly processed for staining. The zebrafish embryo has been used as a model for studying human fetal alcohol syndrome (FAS). The characteristic features exhibited in human FAS, such as brain defects, are also observed in zebrafish exposed to ethanol. Teratogenic effects on other organs and structures are also amenable for thorough assessment, making the zebrafish a useful preclinical model for predicting drug toxicity in humans.

With the dramatic rise in the number of potential but poorly validated targets and preliminary hit compounds, small-animal models are increasingly important for validating these targets and profiling the hits. Although several model systems exist, each with its own advantages, zebrafish can bridge the gap between invertebrate and mammalian models. Wider adoption of this small-vertebrate model organism in drug discovery research will help accelerate the drug development process.

Drosophila melanogaster and its Use

The most popular invertebrate model organisms, *Drosophila* and *C. elegans*, have been used extensively in many areas of biological research, especially genetics and development. The use of these models is supported by the existence of highly conserved molecular pathways between invertebrates and humans, such as the MAP kinase pathway. Combined with the powerful genetics, cellular and molecular biology tools available, these model systems are very suitable for drug discovery research.

Using the fly, researchers have developed models for many complicated pathologies. For example, Raymond Pagliarini and Tian Xu of the Yale University School of Medicine used genetics techniques to explore the complex, multistep processes of oncogenic transformation and metastasis in *Drosophila*. Similarly, researchers have generated transgenic *Drosophila* lines that overexpress mutated proteins, which cause flies to undergo neurodegeneration and allow them to serve as models for disorders such as Alzheimer's and Huntington's disease. Compared to transgenic mice, transgenic *Drosophila* are much easier to construct.

One *Drosophila* organ that researchers extensively analyze is the compound eye, which develops from a monolayer precursor tissue in the larva and consists of about 800 unit eyes (ommatidia) arranged in a highly accurate pattern **(Fig. 23.5)**. The large number and stereotypic

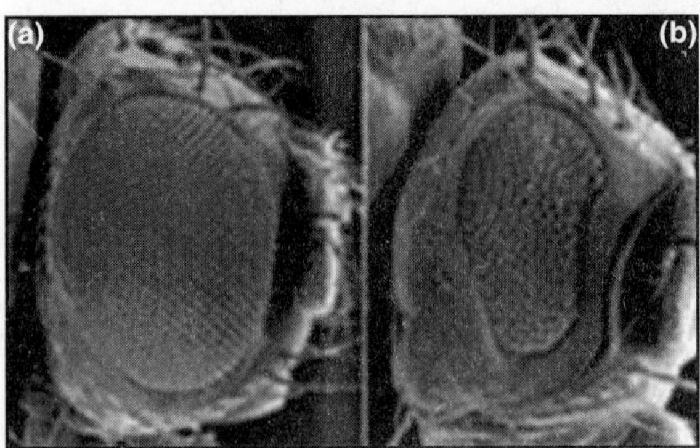

Fig. 23.5. Researchers use scanning electron microscope imaging to view differences between wild-type D. melanogaster (a), which has a normal compound eye with accurate arrangement of unit eyes (ommatidia), and a mutant fly (b), which has a small and rough compound eye with dramatically fewer ommatidia

pattern of the unit eyes make it easy for researchers to use the eye phenotype to identify the genes interacting with the defective gene or drugs affecting components of the pathway.

Several biotechnology companies have developed *Drosophila*-based drug discovery technology platforms. Exelixis has a gene-knockout array generated from a large collection of stocks that contain a single transposon insertion in the fly genome. This allows researchers to quickly characterize a gene's function using a reverse genetics approach. EnVivo Pharmaceuticals has developed models for neurodegenerative diseases based on the expression of human disease genes in the fly. These models mimic the characteristic neuropathology and specific symptoms of the diseases, and they can be used for drug screening. The Genetics Company has developed a high-throughput *in vivo* system that can be used to screen small-molecule compounds in flies to identify drug leads with better ADME properties. Study on the life cycle of *D. melanogaster* is also helping in understanding developmental biology and disease development **(Fig. 23.6)**.

Caenorhabditis elegans and its Use

The other well-characterized invertebrate model is *C. elegans*, which contains fewer than 1000 somatic cells **(Fig. 23.7)**. Its simple structure and transparency allows for the direct observation of cellular phenotypes. Researchers have extensively studied apoptosis in *C. elegans*, and found its regulation similar to that of mammals. Thus, the worm provides a convenient model for studying genes involved in apoptosis, and screening for compounds that modulate this process. This has important applications in developing treatments for cancers and neurodegenerative diseases. The nervous system of *C. elegans* is also very simple, making it a good model for studying neurons. NemaRx Pharmaceuticals is using the worm to test drugs for disorders affecting the nervous system, including pain and Alzheimer's disease.

Because of their small size and simple structure, *C. elegans* are very suitable for high-throughput *in vivo* screenings. Devgen company has developed and is using one such system to search for therapeutics for diseases such as diabetes. Union Biometrica has fashioned a

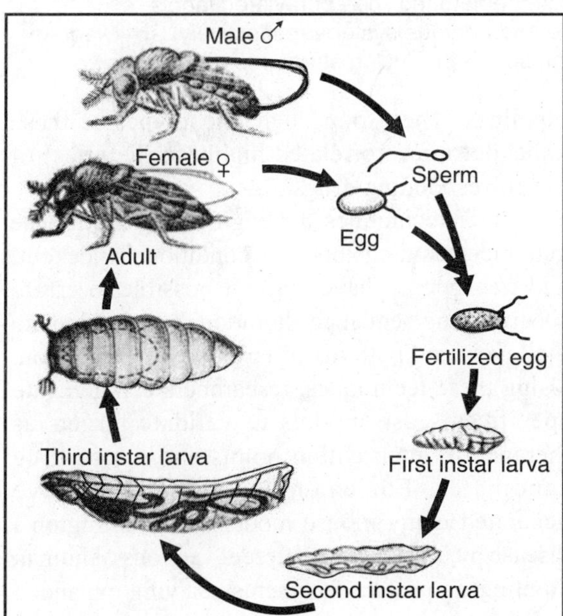

Fig. 23.6. Drosophila development.
A Drosophila passes through several stages as it develops from the egg to the sexually mature adult fly. About 12 days are required to complete the life cycle at 25°C. (The dotted lines within the pupa represent the animal undergoing metamorphosis)

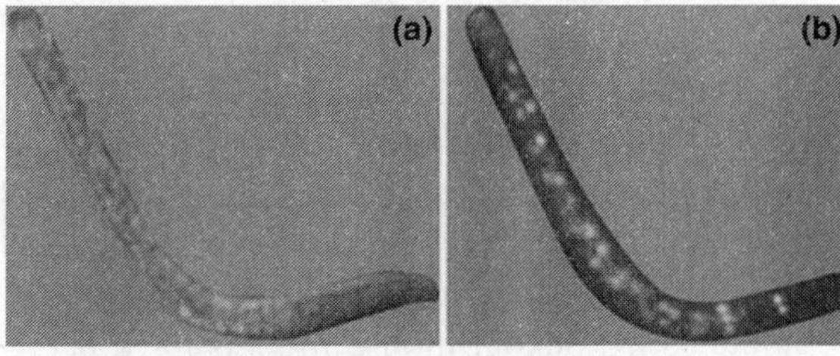

Fig. 23.7. A worm with a view.
Researchers use bright-field imaging (a) to see individual cells in C. elegans or fluorescent imaging (b) to see particular cells expressing green fluorescent protein

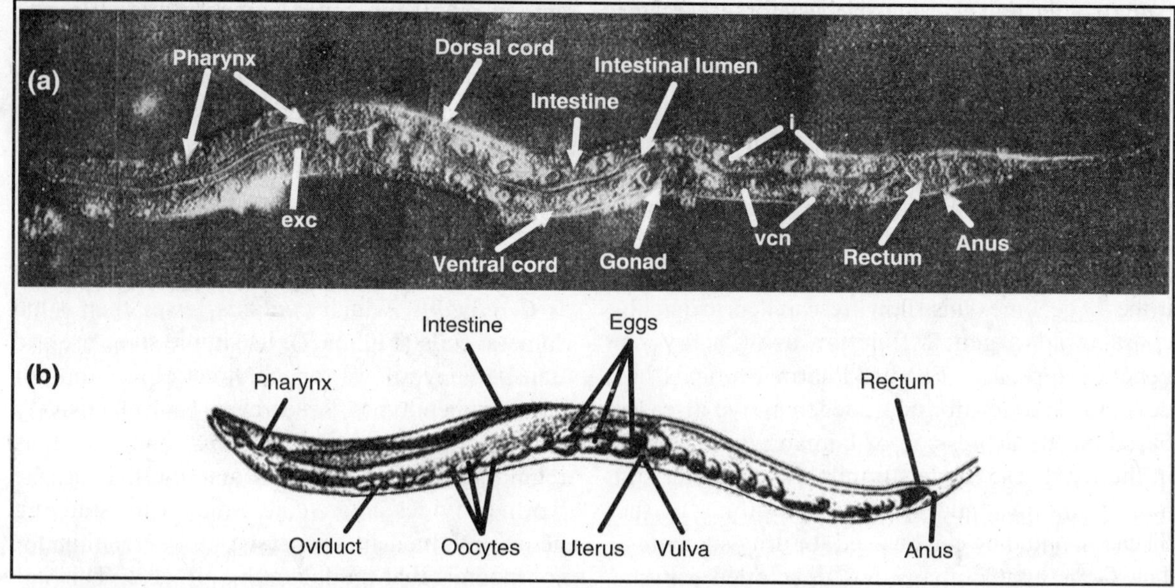

Fig. 23.8. Caenorhabditis elegans. This transparent organism has a fixed number of somatic cells. (a) Nomarski interference LM of the adult hermaphrodite nematode (The abbreviated labels "exc," "i," and "vcn" refer to certain cells of the excretory, digestive and nervous system, respectively). (b) Diagram illustrating structures in the adult hermaphrodite. The sperm-producing structures are not shown

particle dispenser useful for automatic manipulation and analysis of *C. elegans*. Researchers have also created imaging algorithms for automated real-time analysis of living worms, enabling high-throughput drug screening based on large-scale behavioral analysis **(Fig. 23.8)**.

Marking Mammals

Early-stage animal testing is typically conducted in rodents, followed by drug safety testing and certain efficacy evaluations in larger mammals, such as rabbits and dogs. Technologies for engineering the mouse genome have made it possible to create various disease models for use in the screening of corresponding therapeutic compounds. Knockout mouse models have been shown to be highly predictive of the effects of drugs that act on target genes. Lexicon Genetics researchers recently created knockout mice for the genes targeted by the top 100 drugs on the market, and 100 drugs in pharmaceutical companies'

pipelines. They found that phenotypes of these knockout mice correlated highly to the effect of the corresponding drugs.

Other techniques for engineering the mouse genome, including knock-in, conditional knockout, and transgenics, have made it possible to create specific gene-sequence alterations and manipulate the levels and patterns of target-gene expression. Using these techniques, researchers can generate specific disease models to validate targets as therapeutic intervention points and screen drug candidates. For example, researchers have generated many mouse models for Huntington's disease by introducing different versions of human Huntington protein fragments carrying expanded polyglutamine repeats. These transgenic or knock-in mice showed phenotypes characteristic of Huntington's disease patients, including the formation of neuronal inclusion bodies and apoptosis in certain regions of brain.

Rodent models are widely used by researchers in the pharmaceutical and biotechnology industries. To meet this demand, Lexicon Genetics has developed technologies for the high-throughput generation of knockout mice. The technology uses genetically engineered retroviruses that infect mouse embryonic stem cells *in vitro*, integrate into the chromosome of the cell, and disrupt the function of the gene into which it integrates. Although rodent models are powerful, traditional methods of analysis are slow, relying on *ex vivo* data collection from the tissues removed from sacrificed animals. Xenogen has developed *in vivo* biophotonic imaging technologies that enable real-time analysis of drug effects on biological processes in living animals. This allows the researchers to collect the data for the effects on internal organs without the need of surgery, and for time-course effects from the same animals. However, image resolution using this method is relatively low, limiting its application.

SUMMARY

Animals and their use have made an important contribution to advances in biotechnology, medicine and surgery, resulting in major improvements in the health of human beings and animals. Currently, for certain types of medical research there are no viable alternatives to using animal models. The animal use in test is of the opinion that animal experimentation will continue to play a key role in essential research towards tackling major biotechnological and medical problems. International research supported by the animal use in test is expected to be carried out in the spirit of the Government medical legislation, as well as being compliant with all local legislation and ethical review procedures.

The earliest references to animal testing are found in the writings of the Aristotle (384-322 BC) and Erasistratus (304-258 BC) among the first to perform experiments on living animals. Galen is known as the "father of vivisection". Although the term "vivisection" literally means the "cutting up" of a living animal, and originally referred only to experiments that involved dissection of, or surgery on, live animals, it is now commonly used to refer to any experiment on a living animal, which is also known as *in vivo* testing. Animals are considered as major experimental organisms and have played a role in numerous well-known experiments. In 1796, Edward Jenner extracted pus from pox-infected cows to inoculate James Phipps against smallpox, and in 1996, Dolly the sheep was born, the first mammal to be cloned from an adult cell. This paved the way for cloning of other mammals including humans. The use of animals in research and testing is contributing immmensely to the biotechnology.

A model is something that resembles some other things. A model of any thing looks like the real thing, but may be smaller, less detailed, made of different materials, and may or may not be able to do everything a real system can. However, by studying that model, a person can learn a lot about original system. It is practical to make a model of a disease to learn about it and test treatments before attempting to diagnose and treat patients with the disease. There are many different ways to model a disease and they are by using mathematical or computer models, which are used to predict the way a disease may behave, some normal or diseased tissues *in vitro*, patients with the disease can be studied, or an animal that has a similar disease can be studied.

In biotechnology, at a practical level, all animal researchers have a duty to reduce suffering to a minimum. Good husbandry is as vital to the researcher as his/her experimental protocol. The law recognizes that animals have interests in being treated humanely and in being kept from unnecessary suffering. It must also be noted that research using animals does not always take place solely for human benefit, but can also be of benefit to animals, for example in the development of veterinary medicines. The animal use in test is a major funder of research for the benefit of animal health. The animal use in test considers that the existing cost/benefit assessment is reasonable, but that the assessment of scientific validity of proposed experiments should remain primarily the responsibility of the expert scientific community.

To reduce the use of animals in research and testing in 1959, "The Principles of Humane Experimental Technique" was published in London defining the concept of animal testing alternatives as the "Three R's": Refinement, Reduction, and Replacement. The only viable choice for a true animal rights' supporter is the "Replacement" of animals used in tests; Refinement and Reduction are still viewed as morally wrong.

The development of alternatives to animals continues on many fronts and their use is increasing, but no methods can satisfactorily replace animals in all cases, as explained in response to the first question. Sophisticated alternatives to the use of animals in consumer product testing are readily available. Most of the large producers of personal care and household products could adopt these methods, which are more cost effective, better predictors of human injury, produce far quicker results, and do not involve animal cruelty. Large numbers of both mammalian and non-mammalian animals are used as models in varieties of studies in biotechnology. The major ones include *Drosophila melanogaster*, rat, mice, *Caenorhabditis elegans*, Zebra fish and others.

Animal use will need to continue, although the continued development of new techniques, such as cell lines and computer modelling, will reduce the need for animal research in some areas. The approval for the use of human embryos in research may allow the development of new human cell lines that may reduce the need to use animals in some research areas.

EXERCISE

1. Define animal models.
2. Name different types of animal models used in the biotechnology.
3. Explain the usage of animals in the science and biotechnology from the beginning of science.
4. Describe the importance of animals research and testing in biotechnology.
5. Give an account of how advocates of the use of animals in the research and testing proposed for its continued use.
6. What do you mean by regulatory systems of animals used in the research and testing in the biotechnology?
7. Name different types of animals used as models in the animal biotechnology, and present scenario in the research.
8. Discuss how genetically modified animals are helping in understanding the basic processes of biotechnology.
9. What do you mean by alternatives for animals use?
10. Elucidate different alternatives for the use of animals in the research and testing.
11. What is meant by 'replacement' in the use of animals in the research and testing?
12. Distinguish between reduction and replacement.
13. Briefly explain the characters to which the animals are considered as disease models in research and testing.
14. How are animal alternatives helpful in replacing the use of animals in the research and testing in the field of biotechnology?
15. Critically comment on the ethical issues in the use of animals in the research and testing.
16. Give examples for the substitute of animals in the research and testing.
17. Differentiate between the *in vitro* and *in vivo* studies followed for the research and testing in the animal biotechnology.
18. List out the different animal models diseases and their progress made in biomedical research and testing.
19. How are the studies on Alzheimer disease progressing by the use of animals? What are the alternative methods for better understanding of the disease?
20. Define apoptosis, and add a note on its studies on animals.
21. Explain the problems the animals face during the experimental studies.
22. Explain the applications of zebra fish in the research and testing.

24

CHAPTER

Animal Cloning, Human Cloning and their Ethical Aspects

24.1. INTRODUCTION

By the close of the 20th century, humankind had acquired the power to transform the processes of all living species, including its own. The potential benefits of these powers can be exciting and the perils ominous. Consider, for example, the 1998 announcement that scientists have successfully produced cultures of embryonic stem cells. This breakthrough offers the potential to grow any type of human tissue and may eventually be used to repair damaged hearts, pancreas, blood vessels or brains. The very same week, the UK's Sunday Times reported that scientists can theoretically engineer deadly biological organisms to produce 'ethno-bombs' that are capable of targeting human victims by ethnic origin. The techniques of biotechnology namely, tissue culture, recombinant DNA technology, and cloning have naturally attracted the attention of people worldwide.

Cloning is the production of one or more individual animals that are genetically identical to another animal or source animal. Cloning is an ambiguous term in biology. It basically implies the production of genetic copies (i.e., genetically identical copies). However these genetic copies can be strands of DNA, cells in culture or the entire organisms. But, the procedure is still referred to as cloning. In February, 1997, Dolly, the cloned sheep, became a global news sensation. Scientists at the Roslin Institute **(Fig. 24.1)** and PPL Therapeutics outside Edinburgh had rewritten the laws of biology in producing a live sheep by cloning from the cells of the udder of an adult ewe by nuclear transfer.

Fig. 24.1. Aerial photograph of Roslin Institute, Edinburgh

This unusual biotechnological technique attracted the scientists, researchers and general public throughout the world. The Director of the Church of Scotland's Society, Religion and Technology Project quickly became a focus for ethical comment, with numerous TV, radio, newspaper and magazine interviews and articles. The ethical issues involved in assessing the cloning, genetic engineering, stem cells research and other biotechnological processes are very important, because these techniques have direct impact on the humans and its environment. Dr. Bruce had written article on the ethics of cloning and mentioned important issues at the time of the first Roslin cloning discovery. The four main issues—the basic genetic engineering work; whether we should clone animals; whether we might one day clone humans; and how such research should be controlled and kept accountable to the public are to addressed immediately. The feat, cited by Science magazine as the breakthrough of 1997, also generated uncertainty over the meaning of "cloning"—an umbrella term traditionally used by scientists to describe different processes for duplicating biological material.

Cloning raised worldwide interest and concern because of its scientific and ethical implications. There are different types of cloning, and cloning technologies can be used for other purposes besides producing the genetic twin of another organism. A basic understanding of the different types of cloning is the key to take an informed stance on current public policy issues and make the best possible personal decisions. It is also helpful in understanding the cloning processes clearly, and making appropriate decisions on their usefulness and applications.

24.2. DEFINITION

Cloning is the process of producing genetically identical organisms through various techniques, including culture of specific cells, artificial division of a single embryo, or cell nuclear transfer, i.e., transferring the nucleus of a somatic cell into an oocyte (the mature female germ cell or egg) from which the nucleus has been removed—the technique used to create "Dolly " the sheep. Cloned animals, having improved characteristics, can be used in agriculture. They can also be used to model human diseases and to manufacture pharmaceuticals for use in health care. In medical research, cloning may also involve the artificial production of particular tissues or organs from embryonic or adult (e.g., bone marrow) cells for the repair of diseased or damaged tissue.

A clone is a genetically identical individual grown from a single donor cell. Mammals (mice and sheep) were successfully cloned from embryonic cells in the 1980s, and frogs decades earlier. The first report of a clone from an adult somatic (non-reproductive) cell was in 1997. Dolly—the sheep—was produced by nuclear transfer from an adult sheep cell at the Roslin institute in Edinburgh. Since then, pigs, cows, mice and cats have been cloned from either cultured foetal cells or adult somatic cells.

24.3. BIRTH OF THE CLONING TECHNOLOGY

Using nuclear transfer to produce a clone was first suggested in an experiment in 1938. The first animals to be produced in this manner were tadpoles, which were created by transfer of a nucleus from an embryonic cell into an enucleated egg fourteen years later. A succession of cloning experiments followed with various claims of success and breakthroughs, but only a few of these experiments were reproducible, and innumerable animals died in the course of this research. The year 1986 saw the first sheep cloned from undifferentiated embryonic cells using nuclear transfer. This was followed in 1995 by the production of the famous Megan and Morag **(Figs. 24.2 & 24.3)** at the Roslin Institute near Edinburgh—the first sheep to be cloned from

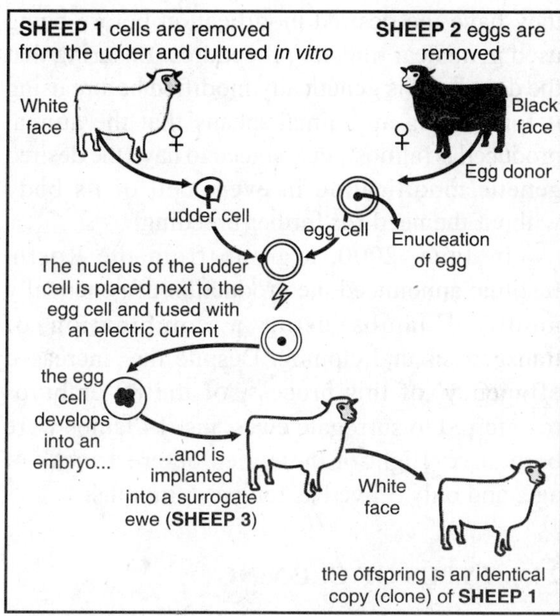

Fig. 24.2. Process involved in the cloning of Dolly

Fig. 24.3. Ray Ansell and Jim McWhir, who grew cells to produce Megan & Morag, cloned sheep in 1995

differentiated embryonic cells (i.e., cells that had already taken on the characteristics of a specific body tissue rather than the generalized cells of early embryos). In 1996, Dolly **(Fig. 24.4)** was also born at the Roslin Institute—the first mammal to be cloned from adult rather than embryonic tissue. The production of Dolly sparked a cloning frenzy. The year 1998 saw the arrival of cows cloned using the same technique that produced Megan and Morag. The Roslin institute cloned the first pig in 2000. In the US, in 2001, a rhesus macaque monkey called Tetra was cloned; in 2002, a cloned kitten was reported and in the same year as the first cloned rabbits in France. In the year 2003, the first cloned mule, horse and rat were reported in animal biotechnology.

Despite the problems encountered in all of these experiments, and the inefficiency of the cloning technique with a mere 0-3% of 'reconstructed embryos' resulting in live births, research in this area is expanding. The principles of cloning are being considered for racehorses, livestock, pets, rescue dogs, laboratory animals and rare or endangered species.

Fig. 24.4. Dolly with the Institute's researcher, produced by cloning from adult cell in 1996

Need of Animals Clones

Researchers are interested in cloning the animals for several reasons.

Firstly, cloning animals allows copies to be made of 'valuable' animals. These could be farm animals with unusually high production of products like meat, milk or wool. They could be naturally occurring mutants or genetically modified animals that are of interest to biomedical researchers. They could be animals genetically modified to produce valuable proteins in their milk or urine. They could also be pets, working animals (like guide dogs or rescue dogs), and rare or endangered wild or captive animals.

Secondly, the cloning process itself could make the production of transgenic animals more efficient. Current methods of incorporating foreign genes into animals, or disrupting specific genes in animals are incredibly inefficient, i.e., only a very small number of live animals are ever produced.

The embryonic stem cell technique of trans-genesis involves inserting a genetically modified cell into an early embryo and letting the embryo develop into an adult. Some of the adult's cell will be transgenic, because they derive from the genetically modified cell in the early embryo. The remainder of the adult's cells will not be transgenic, because they are derived from the 'normal' cells in the early embryo. An animal having some genetically modified cells and some normal cells is called a **chimera**. If the cells which make the chimera's sperm, are transgenic, his offspring will carry the modified gene in each cell of their body, i.e., they will be transgenic as all of their cells will be derived from their fathers sperm. If the chimera's sperm were not transgenic, his offspring would be of 'normal' genetic constitution and would be killed, as they would not be considered to be 'useful'. If a single cell that has been grown in culture can be used to make a whole animal (as is done in cloning), the transgenesis process is simplified considerably. The cells in culture can be genetically manipulated and checked to ensure

they have the desired modification before being used in nuclear transfer procedures. Ensuring that the donor cell is genetically modified before using it for cloning an animal means that the animal produced is (almost) guaranteed to have the desired genetic modification in every cell of its body without the need for further breeding.

In June, 2000, a group from the Roslin Institute announced the production of genetically modified lambs using a combination of transgenesis and cloning. Despite the 'increased efficiency' of this process, of the 80 embryos transferred to surrogate ewes, just 14 lambs were born alive. Eight of these died before 1 week of age, and only 3 lived as long as 6 months.

24.4. TYPES OF CLONING

The following three types of cloning technologies are normally used by the scientific communities:

- Recombinant DNA technology or DNA cloning
- Reproductive cloning
- Therapeutic cloning

Recombinant DNA Technology or DNA Cloning

This is also known as recombinant DNA technology, molecular cloning. or gene cloning. It has been around since 1970s, and is common in molecular biology laboratories. In this method a specific DNA segment get multiplied for many copies.

Reproductive Cloning

Reproductive cloning is also called adult DNA cloning. Its purpose is to produce a genetic duplicate of an existing or previously existing organism.

Therapeutic Cloning

This is also known as biomedical cloning, embryo cloning, or artificial embryo twinning. It essentially mimics the natural process of producing identical twins or triplets. Identical twins are created when the fertilized egg tries to divide into a two-cell stage, but the two cells separate instead. The two separated cells continue dividing on their own and eventually develop into two genetically identical individuals.

Based on the either cells or genetic materials used for cloning, it is classified into two types *viz.,* Cell cloning and DNA cloning.

Cell cloning

Cell cloning, as the name indicates, involves the cloning of the cells i.e., producing genetically identical copies of a particular cell. For example, this is used in producing clones of cancer cells. Individual cancer cells can be isolated and allowed to reproduce in a culture to form colonies. These colonies will contain many colonies of the original cancer cell. Cell cloning helps in the study of defective gene in cancer cell.

DNA cloning

This technique is used to reproduce a specific gene. First the DNA fragment containing the gene of interest is isolated and combined with a plasmid to form a recombinant DNA. It is then introduced into the host cell, and a particular gene is expressed. This is used in genetic engineering to produce transgenic plants and animals.

Cell cloning and DNA cloning have revolutionalized the fields of biology and medicine. They have helped people acquire better understanding of structure and function of cells and genes. Recently, scientists have gone one step forward and have succeeded in cloning animals. On 2001, August 14, Nobel Prize-winning scientist. Sir Joseph Rotblat said that our understanding of cloning is "too meager" at this time to succeed. He suggested that many of the scientists involved are motivated by money. He said that inevitably the problems will be overcomed, and it's then that the real ethical problems begin. He

referred to the *in vitro* fertilization and said that these tests were originally considered unethical, but are now widely accepted.

24.5. DNA CLONING

The different terms such as "recombinant DNA technology", "DNA cloning," "molecular cloning" or "gene cloning" refer to the same process—the transfer of a DNA fragment of interest (gene or length of DNA) from one organism to a self-replicating genetic element such as a vector, bacterial plasmid. The DNA of interest can then be propagated in a foreign host cell.

Scientists studying a particular gene often use bacterial plasmids to generate multiple copies of the same gene irrespective of their source. Plasmids and other types of cloning vectors are used by Human Genome Project researchers to copy genes and other pieces of chromosomes to generate enough identical material for further study and analysis.

DNA cloning is used to produce many copies of a particular segment of DNA containing one or more genes, to be studied in the laboratory. The DNA fragment of interest from an organism such as human is incorporated into the "plasmid DNA" of a bacterial cell. To clone a gene, the DNA fragment containing the gene of interest is isolated from the chromosomal DNA using restriction enzymes, and then united with a plasmid that has been cut with the same restriction enzymes. Plasmid is a circular self-replicating DNA molecule that is separate from the bacterial DNA **(Fig. 24.5)**. When the fragment of a chromosomal DNA is joined with its cloning vector in the lab, it is called a recombinant DNA molecule. The plasmid containing the genes or DNA of interest is now a piece of "recombinant DNA" made up of human and bacterial DNA. It is then put into a cell that will act as a host. Following introduction into suitable host cells, the recombinant DNA can be reproduced along with the host cell DNA. As the host cell is copied over and over again, the

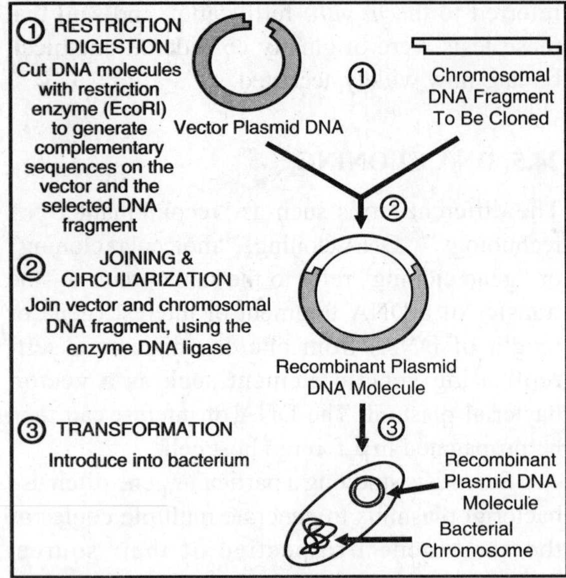

Fig. 24.5. DNA Cloning.
In this method, by fragmenting DNA of any origin (human, animal, or plant) and inserting it in the DNA of rapidly reproducing foreign cells, billions of copies of a single gene or DNA segment can be produced in a very short time. The DNA fragment of interest from an organism such as human is incorporated into the "plasmid DNA" of a bacterial cell. A plasmid is a circular selfreplicating DNA molecule that is separate from the bacterial DNA. When the recombinant plasmid is introduced into bacteria, the newly inserted segment will be replicated along with the rest of the plasmid

recombinant DNA is copied as well. Bacteria, yeast and mammalian cells can be used as often as the host cells. The end result is multiple identical copies of the same human DNA fragment or gene.

Plasmids can carry upto 20,000 bp of foreign DNA. Besides bacterial plasmids, some other cloning vectors include viruses, bacteria artificial chromosomes (BACs), and yeast artificial chromosomes (YACs). Cosmids are artificially constructed cloning vectors that carry upto 45 kb of foreign DNA, which can be packaged in

lambda phage particles for infection into *E. coli* cells. BACs utilize the naturally occurring F-factor plasmid found in *E. coli* to carry 100 to 300 kb DNA inserts. A YAC is a functional chromosome derived from yeast that can carry upto 1 MB of foreign DNA. Bacteria are most often used as the host cells for recombinant DNA molecules, but yeast and mammalian cells can also be used.

Benefits

Cloning is essential to enable enough copies of the gene to be analyzed. This may lead to human genetic testing to diagnose genetic conditions and predictive or presymptomatic genetic testing for genetic conditions in asymptomatic individuals.

24.6. REPRODUCTIVE CLONING

Reproductive cloning is also called adult DNA cloning. The purpose of this type of cloning is to produce a genetic duplicate of an existing or previously existing organism. Dolly, the cloned sheep produced in 1996, is an example of an animal produced by this kind of cloning (**Fig. 24.6**). Dolly was produced by a process called somatic cell nuclear transfer (SCNT). Somatic cell is any cell in the body other than the sperm or egg cells.

In the process called "somatic cell nuclear transfer" (SCNT), scientists transfer genetic material from the nucleus of a donor adult cell to an egg whose nucleus, and thus its genetic material, has been removed. The reconstructed egg containing the DNA from a donor cell must be treated with chemicals or electric current in order to stimulate cell division. Once the cloned embryo reaches a suitable stage, it is transferred to the uterus of a female host where it continues to develop until birth.

Dolly or any other animal created using nuclear transfer technology is not truly an identical

Fig. 24.6. Dolly produced in 1996 with her lamb, Bonnie produced in 1998

clone of the donor animal. Only the clone's chromosomal or nuclear DNA is the same as the donor. Some of the clone's genetic materials come from the mitochondria in the cytoplasm of the enucleated egg. Mitochondria, which are power sources of the cell, contain their own short segments of DNA. Acquired mutations in mitochondrial DNA are believed to play an important role in the aging process.

Dolly's success is truly remarkable, because it proved that the genetic material from a specialized adult cell, such as an udder cell programmed to express only those genes needed by udder cells, could be reprogrammed to generate an entire new organism. Before this demonstration, scientists believed that once a cell became specialized as a liver, heart, udder, bone, or any other type of cell, the change was permanent, and other unneeded genes in the cell would become

inactive. Some scientists believe that errors or incompleteness in the reprogramming process can cause high rates of death, deformity, and disability observed among animal clones.

This was an important step in the development of whole animals, as the genes that are needed for an embryo to grow and develop are inactivated soon after their job has been done. Adult cells, therefore, had to have these genes "turned on" again **(Fig. 24.7)**. The difference between an embryo created by SCNT versus an embryo created naturally lies in the chromosome. We all normally have 46 chromosomes in our body's somatic cells; 23 from our mother via the egg, and 23 from our father's sperm. In SCNT, the egg cell's single set of chromosomes is removed and replaced with the nucleus of a somatic cell, which already has 46 chromosomes. Thus, the embryo receives all the chromosomes from one somatic cell.

CLONING IN CASES OF MALE INFERTILITY

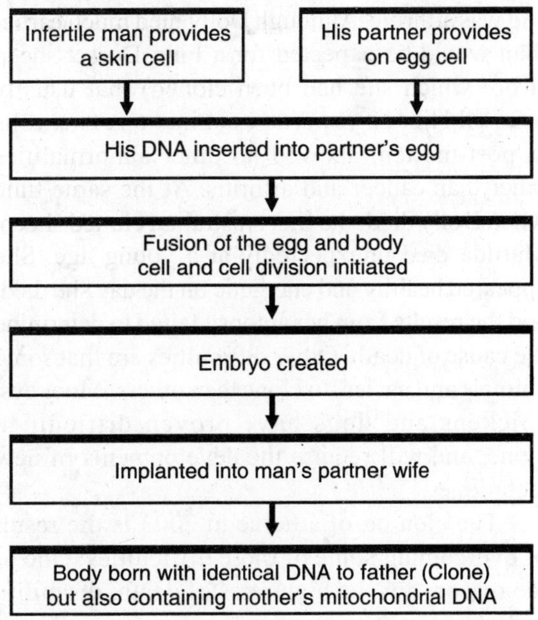

Fig. 24.7. The production of human clones in the laboratory

Advantages and Limitations

Reproductive cloning can have many uses. If the low success rates and issues of safety could be improved as discussed below, the technology can be used to mass-produce animals with special qualities, such as animals that are important agriculturally or are able to produce helpful drugs for human use.

SCNT has environmental uses too. It can be used to repopulate endangered species, as has been shown with the wild ox—the gaur. Some supporters of human reproductive cloning see it as a way of overcoming male infertility, where other methods of assisted reproduction have failed. The difficulties in producing cloned animals are reflected in the number of abnormalities seen in the experiments done to date. For example, Dolly was the only survivor from among 277 embryos cloned in the experiment conducted by Wilmut at the Roslin Institute in Scotland. Dolly died in 2003 at the age of six.

Earlier to her death, Dolly had lung cancer and was arthritic. Although Dolly died much earlier than would be expected for a Finn Dorset sheep (from which she had been cloned) that usually lives till 11 or 12 years, an examination of her cells at post-mortem showed no other abnormalities other than cancer and arthritis. At the same time when Dolly died, the first Australian cloned sheep, Matilda died unexpectedly at a young age. She appeared healthy and energetic on the day she died, and the results from her autopsy failed to determine the cause of death. Other difficulties are that some animals are "easier" to clone than others. Monkeys, chickens and dogs have proven difficult to clone, and will require the developments of new techniques.

The cloning of a horse in 2003 is the result of overcoming some of these difficulties. One of the reasons for the high rates of death, disability and abnormalities in cloned animals is perhaps in the reprogramming of the genetic make-up from an adult cell into an embryonic cell that must take place. Some genes require to be passed to an embryo through the egg or sperm for them to be active. Imprinted genes are involved in fetal growth, and may explain why some cloned fetuses have "overgrowth" syndromes. In November, 2001, scientists from Advanced Cell Technologies (ACT)—a USA biotechnology company—announced that they had cloned the first human embryos for the purpose of advancing therapeutic research. To do this, they collected eggs from women's ovaries and then removed the genetic material from these eggs with a very fine needle. A human adult skin cell was inserted inside the enucleated egg, which served as a new nucleus. The egg began to divide after it was stimulated with a chemical. This process was carried out with eight different eggs, of which only three began dividing, and only one was able to divide into six cells before the experiment was stopped.

In January, 2003, a biotechnology group called "Clonaid" associated with the Raeilian movement, a religious sect, announced the birth of a cloned baby named "Eve", but this has never been confirmed independently. There was universal condemnation of the possibility of a cloned human.

Ethical Considerations

Human reproductive cloning has raised many concerns, not the least of which are safety considerations. More than 90% of offspring from cloning are not viable: only 1 or 2 offspring are viable for every 100 cloning attempts. In addition to this low success rate, cloned animals have higher rates of cancer and infection, and a range of other disabilities. Also, while the animal may seem healthy at birth, problems often appear later. This is the reason why mature clones have often undergone sudden, unforeseen, and unexplainable deaths.

While these technical difficulties can be overcome in future and may enable cloning of animals with a greater success rate, the use of reproductive cloning of humans is viewed differently. Intellectual and emotional development is essential for human growth and health. It is also important to realise that a person is much more than a product of their genes. If it were possible to produce a cloned human, the clone would not be a duplicate of that person, other than the nuclear DNA. As explained in Genetic Fact Sheet 1, a person is a product of the environment as well as its genetic make-up. Even identical twins have subtle differences between them. The concepts of parenthood, "family" and views of social responsibilities are all challenged by human reproductive cloning.

24.7. THERAPEUTIC CLONING

This type of cloning is also known as biomedical cloning, embryo cloning or artificial embryo twinning. It essentially mimics the natural process of producing identical twins or triplets.

The goal of this process is not to create cloned human beings, but to harvest stem cells that can be used to study human development and to treat diseases **(Fig. 24.8)**. Stem cells are important for biomedical researchers because they can be used to generate virtually any type of specialized cell in the human body. Stem cells are extracted from the egg after it has divided for five days. The egg at this stage of development is called a blastocyst. The extraction process destroys the embryo, which raises a variety of ethical concerns.

In November 2001, scientists from Advanced Cell Technologies (ACT), a biotechnology company in Massachusetts, announced that they had cloned the first human embryos for the purpose of advancing therapeutic research. To do this, they collected eggs from women's ovaries, and then removed the genetic material from these eggs with a needle less than 2/10,000th of an inch wide. A skin cell was inserted inside the enucleated egg to serve as a new nucleus. The egg began to divide after it was stimulated with a chemical called ionomycin. Although this process was carried out with eight eggs, only three among them began dividing, and only one was able to divide into six cells before stopping.

The cells are stem cell that are undifferentiated cells capable of developing into most of the 220 types of cells found in the human body (e.g., blood cells, muscle cells, nerve cells etc.) that have different functions. The cells of an embryo are an excellent source of stem cells because they have not yet been differentiated into the cells of specific tissues or organs of the body. While there are around 30,000 genes in each body cell, only those genes that are needed for the function of the cells of the tissue or organ are turned on; the remainder are inactive. For example, brain cells have different genes sending instructions to the cells than liver cells. Embryonic stem cells have all the genes needed for the function of the boy's cells able to be turned on and so can be differentiated into any type of body cell. So, to generate cultures of specific types of differentiated cells (e.g., heart muscle cells, blood cells, or nerve cells), scientists try to control the differentiation of embryonic stem cells. They change the chemical composition of the culture medium, alter the surface of the culture dish, or modify the cells by inserting specific genes. Through years of experimentation, scientists have established some basic protocols for the directed differentiation of embryonic stem cells into some specific cell types. If scientists can reliably direct the differentiation of embryonic stem cells into specific cell types, they may be able to use the resulting, differentiated cells to treat certain diseases at some point in the future. Diseases that might be treated by transplanting the cells generated from human embryonic stem cells include Parkinson disease, diabetes, traumatic spinal cord injury, Purkinje cell degeneration, Duchenne muscular dystrophy, heart disease, and vision and hearing loss. **Fig. 24.8** is an example of the way in which

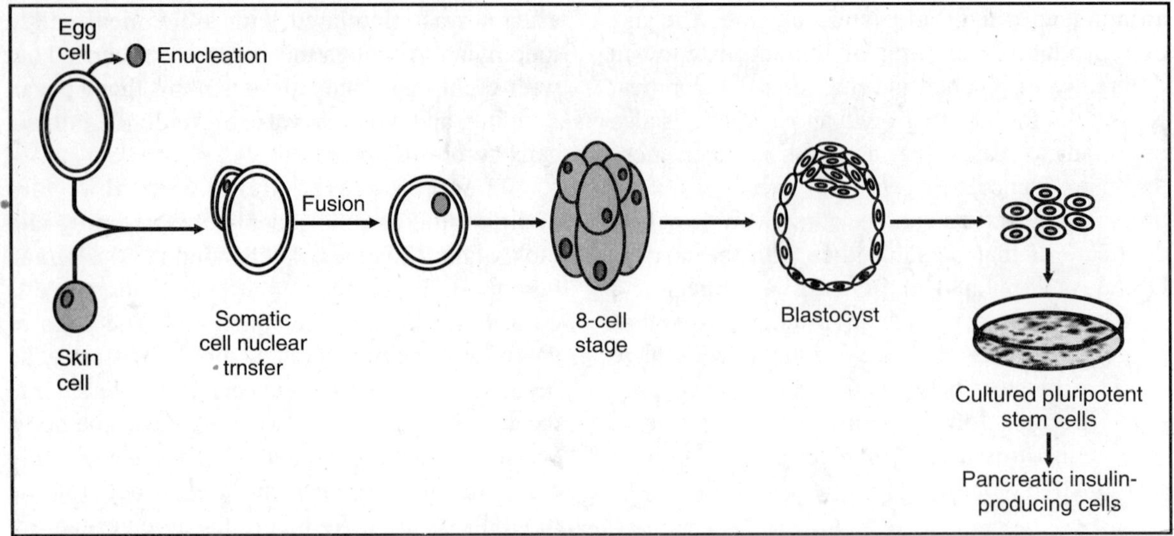

Fig. 24.8. Therapeutic cloning. The doctor takes a sample of skin cells from the patient and isolates their DNA. Next, a donor egg cell, emptied of its own genetic contents, is injected with the DNA from the patient. The embryo is nurtured to grow and divide into a blastocyst. Some blastocyst cells are harvested and coaxed with growth factors to mature into insulin-producing cells. Finally, millions of insulin-producing cells are injected back into the patient. In an ideal world, the patient's diabetes is temporarily 'reversed', with no side effects.

therapeutic cloning could be used in the treatment of diabetes type 1, where the damaged pancreatic cells that produce insulin are replaced, removing the requirement for daily administered insulin. An alternative to using embryonic stem cells for therapeutic cloning are adult stem cells that can be extracted from adult tissue such as the bone marrow, without harm to the person. An adult stem cell is an undifferentiated cell found among differentiated cells in a tissue or organ. It can renew itself, and can differentiate to yield all the major specialised cell types of the tissue or organ. The primary roles of adult stem cells in a living organism are to maintain and repair the tissue in which they are found. Some scientists now use the term "somatic stem cell" instead of adult stem cell. Unlike embryonic stem cells, which are defined by their origin (the inner cell mass of the blastocyst), the origin of adult stem cells in mature

tissues is unknown. Scientists have found adult stem cells in many more tissues and adult stem cells from bone marrow that form into blood cells have been used in transplants for 30 years. Certain kinds of adult stem cells seem to have the ability to differentiate into a number of different cell types, given the right conditions. If this differentiation of adult stem cells can be controlled in the laboratory, these cells may become the basis of therapies for many serious common diseases. There are often only a very small number of stem cells in each tissue. Stem cells are thought to reside in a specific area of each tissue, where they may remain quiescent (non-dividing) for many years until they are activated by disease or tissue injury. The adult tissues reported to contain stem cells include brain, bone marrow, peripheral blood, blood vessels, skeletal muscle, skin, and liver.

Advantages and Limitations

The purpose of this method of cloning is reflected in its name—to enable the rectification of health problems. The aim of this technique is to harvest stem cells from an embryo, so that the cells can be used to study human development and for the treatment of some diseases. Stem cells containing the person's own DNA could be grown in the laboratory and transplanted into them without the risk of tissue rejection. Before the technology can be used in this way, a number of barriers have to be overcome. First, the stem cells have to be isolated from a source and then grown in the laboratory. Then they have to be turned into the specific cell type needed for treatment. These two hurdles have been passed for most of the 220 cell types in the human body. Overcoming the next barriers of applying the technology clinically and ensuring that the new tissue or organ poses no risk to the patient is still a matter of much research. Skin has successfully been grown in the laboratory to create a self-compatible skin graft. Skin is relatively easy to grow because the mature, differentiated skin cells are still able to divide and produce more cells to repair the damage. The cells of other organs or tissues do not have this ability. An alternative is to create genetically modified pigs from which organs suitable for human transplantation could be harvested. Pig tissues and organs are somewhat similar to that of humans and the animal species. The transplant of organs and tissues from animals to humans is called xenotransplantation. However, the concern remains about the transmission of pig viruses to humans. Stem cells have the potential to be used to replace faulty cells in people affected with cancer, cardiac disease, diabetes type 1, and degenerative conditions such as Parkinson disease and Alzheimer disease. For example, recent work with stem cells inserted into damaged heart muscle has led to the regeneration of cells. Other work in mice has shown the potential to use stem cells to produce pancreatic cells for the production of insulin for the reversal of diabetes type 1. While adult stem cells may overcome some of the ethical considerations (see below), further studies are needed to determine their efficacy in differentiating into a variety of new types of cells. Initial work suggested that their potential may be limited, but recent work has suggested that this may be wider than first indicated.

Ethical Considerations

The therapeutic cloning technology used to harvest stem cells ultimately results in the destruction of an embryo. The creation of embryos specifically for the purpose of destroying them is of concern for some people. Currently, embryos that are used for the production of stem cells are those created by infertile couples during *in vitro* fertilization (IVF) programs. Usually more embryos, than are needed to have a baby, are produced. Some couples have agreed for their excess embryos to be used for stem cell research.

The new method would enable scientists to do a more precise genetic modification using less experimental animals, but the side effect is that the resulting sheep is a clone, genetically almost identical to its founder. PPL might clone 5-10 sheep from a single genetically modified animal. These sheep would then breed naturally to give flocks of varied sheep, all containing the desired genetic modification. On this limited scale, this would not seem ethically unacceptable. The possibility that farm animals might be cloned routinely for meat or milk production on a large scale is however a very different matter. In animal breeding, the need to maintain genetic diversity sets practical limits on how far cloning would make sense, but certain applications are already being considered. A breeder might wish to clone the best breeding stock to sell for feeding up for slaughter, or to found new nucleus herds. Would this be carrying our use of animals one stage too far?

This raises a last question on the control of such research. In many spheres of research, there is a deficit in public accountability in the existing procedures, whereby research priorities are set. There are no easy solutions to this problem, but, at the very least, it points to the need of a standing ethical commission on non-human biotechnology. The work of the commission is open to public comment and scrutiny, in which the areas of research, which are likely to have far reaching ethical implications, are first debated in public. According to the Christian, the world around us is God's creation. Variety is one of its characteristic features, and especially at the level of higher animals and humans. The overall picture in the Bible, in commandments, stories and poetry, is of a creation, whose sheer diversity is itself a cause of praise to its creator.

To reduce this diversity to a strict blueprint, and produce replica animals routinely on demand, would seem to go against something basic and God-given. The very fact that selective breeding has its limits reflects this fact. Some would argue that cloning is thus absolutely wrong, no matter what it was being used for. SRT argues that scale and intention play a part. PPL's (pharmaceutical proteins limited) limited context could be acceptable, since the main intention was not the clone as such but growing an animal of a known genetic composition, where natural methods would not work. What would be unacceptable would be in routine animal production, where natural methods exist, but would be side-stepped on the grounds of economics or convenience. This would represent one step too far beyond conventional selective breeding in the way we use animals as commodities. The approach that, whatever use we find for animals, we could clone them to do so more efficiently, brings the mass production principles of the factory too far into the animal kingdom. Just as in the Old Testament, an ox was not to be muzzled while treading out the grain, animals have certain freedoms which we should preserve. We may use animals to an extent, but we need to remind ourselves that they are firstly God's creatures, to whom we may not do everything we like.

24.8. ANIMAL CLONING

Animal cloning can be defined as the technique of production of genetically identical animals through sexual reproduction using somatic cells of an animal. It is the production of exact genetic replica of an individual, and the offspring derived or obtained from this technique constitutes clone. These clones contain the same genetic make-up as their parents from which they are derived.

Clones are formed naturally when identical twins or any other genetically identical multiple birth occur. Identical twins are produced when crossed zygotic embryo effects in to equal halves, and each half develops into a new individual, which is genetically identical to others and unique from parents.

The clones, which are naturally produced involves the parents, and the process involves the germ cell, which includes the fusion of male and female genetic to form zygote that later develops into an embryo. This is the normal process of sexual reproduction, which occurs in animals.

Scientists have invented a new technique, where they have proved that animals can be asexually reproduced using their body cells (i.e., somatic cells). In this technique, there is no fusion of male and female gamete, and the resulting offspring produced is genetically identical to the parent, from which somatic cells are taken. Much of the media attention focused on speculation about cloning humans, but missed the more immediate impact on animals, and the ethical questions on how far we should apply technology to them. The cloning arose from Roslin's search for more effective ways to do the existing work at PPL Therapeutics of genetically engineering sheep to produce therapeutic proteins in sheep's and cow's milk. The first product for emphysema and cystic fibrosis sufferers is undergoing clinical trials, and a range of other medical applications is in prospect.

The SRT Project working group had already found this work generally ethically acceptable. There were clear human benefits, with few animal welfare or other concerns once past the experimental stage.

24.9. HISTORY OF CLONING

Scientists have been cloning animals for many years. In 1952, the first animal, a tadpole, was cloned. Before the creation of Dolly, the first mammal cloned from the cell of an adult animal, was created from embryonic cells. After that researchers have cloned a number of large and small animals, including sheep, goats, cows, mice, pigs, cats, rabbits, and gaur. All these clones were created using nuclear transfer technology.

Hundreds of cloned animals exist today, but the number of different species is limited. Attempts at cloning certain species such as monkeys, chickens, horses, and dogs, have been unsuccessful. Some species may be more resistant to somatic cell nuclear transfer than others. The process of stripping the nucleus from an egg cell and replacing it with the nucleus of a donor cell is a traumatic one, and improvements in cloning technologies may be needed before many species can be cloned successfully.

Amphibian Cloning

The first animal in which cloning was performed where the amphibians. The technique of amphibian cloning began as early as 1950's, when the scientists began to clone amphibians such as frogs, salamanders etc. The technique, which the scientists used to clone amphibians was 'Somatic Cell Nuclear Transfer' method, which is abbreviated as 'SCNT'. This technique was invented by scientist called 'Spemann' in 1938. Later in 1952, scientists named Briggs and King worked on the American frog '*Rana pipiens*', and succeeded in cloning it. The experiment of Briggs and King was a success, and it proved that smaller animals such as frogs etc., and could be cloned through nuclear transfer. Later, this principle was applied to other species of frog, such as Xenopus, and other amphibians.

Mammalian Cloning

Mammals are higher animals, which give birth to their young ones (i.e., they do not lay eggs like amphibians). After the successful cloning of amphibians, scientists moved one step forward to clone mammals.

The first attempt to clone mammals started in 1975, but the success was achieved at a much later date. Mammalian cloning can be carried out by two methods:
- Embryo splitting
- Somatic cell nuclear transfer (SCNT)

SCNT is preferred more in mammalian cloning.

In 1995, two scientists, Ian Wilmut **(Fig. 24.9)** and Keith H.S. Cambel **(Fig. 24.10)** produced the twin-eyed lambs through 'embryo splitting'. This

Fig. 24.9. Ian Wilmut with his cloned and first transgenic sheep, Polly having human blood clotting factor IX expressed in the mammary gland

Fig. 24.10. Keith H.S. Campbell—co-worker of
Ian Wilmut

involves the artificial or natural way of splitting
embryo into two or more equal parts, such that
each part later develops into individual genetical
identical offsprings. These lambs were named as
Megan and Morag.

In 1997, the same team of scientists announced
that it had actually produced a 'Mammalian clone'
and named it "Dolly". It was produced by SCNT
at the Roslin institute near Edinburgh, Scotland.
Dolly was a great achievement of decades of
research into mammalian cloning.

History of Cloning Using Adult DNA

1997—Mammalian cloning was assumed to be
impossible, until July, 1996, when Dr. Ian Wilmut
of the Roslin Institute, Scotland, cloned a sheep.
News of the experiment was communicated to the
press on February, 23, 1997. "Dolly", a seven-
month-old sheep, was displayed to the media. She
was the first animal, which was cloned using DNA
from another adult by Roslin technique. Since
Dolly's conception, the institute has successfully
cloned seven sheep of three breeds. This technique
can probably be applied to other domesticated
mammals as well, including humans.

On March 4, 1997, apparently in reaction to
"Dolly," President Clinton ordered a widespread
ban on the federal funding of human cloning in
the US; research continues in other countries.

1998—On July 22, 1998, Dr. Ryuzo Yanagi-
machi of the University of Hawaii announced the
cloning of mice by Honolulu technique. The team
had produced 22 mice; seven of them are clones
of clones from the cells of a single mouse.

Japanese researchers from Kinki University
in Nara, Japan, cloned 8 calves from a single adult
cow's DNA. They used techniques similar to
that which produced "Dolly". Four died shortly
after birth due to, what the researchers called,
"environmental factors". Their study was
published in the issue of Science magazine
published on Dec. 9 1998.

On December 14, 1998, researchers at the
Infertility Clinic at Kyeonghee University, in
Korea, announced that they had successfully
cloned a human. Scientists, Kim Seung-bo and Lee
Bo-yeon took an ovum from a woman, removed
its DNA and inserted a somatic cell from the same
30 year old woman. Their report states: *"We were
able to confirm division upto the fourth cell stage—
the stage of embryo development, when a test tube
embryo is usually placed back in the uterus, where
it then further develops into a fetus."* The goal of
their research was not to clone humans, but to clone
specific, genetically identical organs for human
transplant. They did not implant the morula into
the human uterus because of ethical considerations;
they destroyed it. The *Korean Federation for the
Environmental Movement* (KFEM) immediately
issued a statement criticizing the study. Members
of the *Life Safety Ethics Association* held a protest
demonstration in front of the university.

Public opinion is having a chilling effect on cloning research in North America. Dr. Alan DeCherney of the University of California at Los Angeles said that if anyone tried to clone humans, they would *"become Dr. Kevorkian of reproduction"*.

2000—By the end of the year 2000, eight species of mammals had been cloned, including mice, cows, rhesus monkeys, sheep, goats, pigs, rabbits and rats. Between 3,000 and 5,000 cloned animals have been produced to date.

2003—On February 14, 2003 Dolly died after six years of life she gave birth to Bonnie in 1998.

2003—Home clone called prometea is produced. Rat clone called Ralph is produced.

2004—Fruit flies clones are produced.

2005—Korean scientists produced Afghan Hound clone called Snuppy.

Current speculation is that the cloning process seems to create random errors in the expression of individual genes. The egg must have its genes reprogrammed in minutes or hours during the cloning process. Ova normally take years to ripen naturally in the ovaries. It appears that the extremely fast rate of programming can produce random errors in the DNA of the clone.

24.10. CLONING AND ITS APPLICATIONS

Recombinant DNA technology is important for learning other related technologies, such as gene therapy, genetic engineering of organisms, and sequencing the genomes. Gene therapy can be used to treat certain genetic conditions by introducing virus vectors that carry corrected copies of faulty genes into the cells of a host organism. Genes from different organisms that improve taste and nutritional value or provide resistance to particular types of disease can be used to genetically engineer food crops. With genome sequencing, fragments of chromosomal DNA must be inserted into different cloning vectors to generate fragments of an appropriate size for sequencing. If the low success rates can be improved (Dolly was only one success out of 277 tries out of 277 multi

transformed, only 29 eggs developed into embryo and of which only 13 embryos are transplanted into surrogate mother. From 13 surrogate mother, only one gave a birth to offspring that is named as Dolly), reproductive cloning can be used to develop efficient ways to reliably reproduce animals with special qualities. For example, drug-producing animals, or animals that have been genetically altered to serve as models for studying human disease could be mass-produced.

Reproductive cloning could also be used to repopulate endangered animals or animals that are difficult to breed. In 2001, the first clone of an endangered wild animal was born—a wild ox called gaur. The young gaur died from an infection after 48 hours of birth. In 2001, scientists in Italy reported the successful cloning of a healthy baby "mouflon"—an endangered wild sheep. The cloned mouflon is living at a wildlife center in Sardinia. Other endangered species, which are potential candidates for cloning, include the African bongo antelope, the Sumatran tiger, and the giant panda. Cloning extinct animals presents a much greater challenge to scientists, because the egg and the surrogate needed to create the cloned embryo would be of a species different from the clone.

Therapeutic cloning technology may be used in humans to produce whole organs from single cells, or to produce healthy cells that can replace damaged cells in degenerative diseases such as Alzheimer's or Parkinson's. Much work still needs to be done before therapeutic cloning can become a realistic option for the treatment of disorders.

Cloning and Organ Transplantations

Scientists hope that therapeutic cloning can be used to generate tissues and organs for transplants. To do this, DNA would be extracted from the person in need of a transplant and inserted into an enucleated egg. After the egg containing the patient's DNA starts to divide, embryonic stem cells that can be transformed into any type of tissue would be harvested. The stem cells would be used to generate an organ or tissue, which is a genetic match to the recipient. In theory, the cloned organ would then be transplanted into the patient without the risk of tissue rejection. If organs could be generated from cloned human embryos, the need for organ donation would be significantly reduced.

Many challenges must be overcome before "cloned organ" transplants become reality. More effective technologies for creating human embryos, harvesting stem cells, and producing organs from stem cells would have to be developed. In 2001, scientists with Advanced Cell Technology (ACT) reported that they had cloned the first human embryos; however, the only embryo to survive the cloning process stopped developing after dividing into six cells. In February, 2002, scientists with the same company reported that they had successfully transplanted kidney-like organs into cows. The team of researchers created a cloned cow embryo by removing the DNA from an egg cell and then injecting the DNA into the skin cell of the donor cow's ear. Since little is known about manipulating embryonic stem cells from cows, the scientists let the cloned embryos develop into fetuses. They then harvested fetal tissue from the clones and transplanted it into the donor cow. In the three months of observation, following the transplant, no sign of immune rejection was observed in the transplant recipient.

One potential application of cloning in organ transplants is the creation of genetically modified pigs from which organs suitable for human transplants could be harvested.

Why pigs? Primates would be a closer match genetically to humans, but they are more difficult to clone and have a much lower rate of reproduction. Of the animal species that have been cloned successfully, pig tissues and organs are more similar to those of humans. To create a "knockout" pig, scientists must inactivate the genes that cause the human immune system to reject an implanted pig organ. The genes are knocked out in individual cells and are then used to create clones from which organs can be harvested. In 2002, a British biotechnology company reported that it was the first to produce "double knockout" pigs that have been genetically engineered to lack both copies of a gene involved in transplant rejection. More research is needed to study the transplantation of organs from "knock-out" pigs to other animals.

24.11. CLONING AND ITS RISKS

Reproductive cloning is expensive and highly inefficient. More than 90% of cloning attempts fail to produce viable offspring. At least 100 nuclear transfer procedures would be required to produce one viable clone. In addition to low success rates, cloned animals tend to have more compromised immune function and higher rates of infection, tumor growth, and other disorders. Japanese studies have shown that cloned mice have a poor health and die early. One-third of the cloned calves born alive have died young, and many of them were abnormally large. Many cloned animals have not lived long enough to generate good data about how clones age. Appearing healthy at a young age, unfortunately, is not a good indicator of long term survival. Clones have been known to die mysteriously. For example, Australia's first cloned sheep appeared healthy and energetic on the day she died, and the results from her autopsy failed to determine the cause of death.

In 2002, researchers at the Whitehead Institute for Biomedical Research in Cambridge, Massachusetts, reported that the genomes of cloned mice are compromised. By analyzing more than 10,000 liver and placenta cells of cloned mice, they discovered that about 4% of genes function abnormally. The abnormalities do not arise from mutations in the genes, but from the changes in the normal activation or expression of certain genes.

Problems also may result from programming errors in the genetic material from a donor cell.

When an embryo is created from the union of a sperm and an egg, the embryo receives copies of genes from both parents. A process called "imprinting" chemically marks the DNA from the mother and father, so that only one copy of a gene (either the maternal or paternal gene) is turned on. Defects in the genetic imprint of DNA from a single donor cell may lead to some of the developmental abnormalities of cloned embryos.

24.12. HUMAN CLONING

The birth of Dolly was a significant event for the science and society. It proved that an adult somatic cell contains, in usable form, all the genetic material needed to specify an entire animal. It also showed that scientists are now able to produce copies of a mammal having desirable traits. Finally, it awakened the world to the fact that the cloning of humans could become a reality, and gave debates about the ethics of human cloning a new urgency, as well as debates about its legal status and relevant public policy. The word "clone", although used in different contexts, refers to a precise genetic copy of a molecule, cell, plant, animal or human. Somatic cell nuclear transfer, or SNT is the process of taking genetic material from an existing organism and placing it inside an egg cell. The resulting embryo will have virtually the same genetic composition as the organism that donated the genetic material. Cloning, however, need not result in the gestation and birth of a clone. SNT technology has many non-reproductive uses in biological research and medicine. Cloning research has the potential to give us insight into the mechanisms of cellular and organismal development, and organ and tissue transplants.

The ethical issues surrounding human cloning extend far beyond the question of producing human beings. SNT's applications in pure scientific research and medical therapy are expected to advance science, and may provide treatment to those with previously untreatable or incurable diseases and disabilities. On the other hand, SNT research and therapeutic cloning requires scientists to work with human embryos in ways that some ethical and religious traditions find objectionable. There is a profound disagreement among the major religions on the morality of both reproductive and therapeutic human cloning. Several states have passed legislation that restricts human cloning and research on human cloning. Many more states and the federal government have considered, but not adopted, such legislation. A presidential directive currently bans Federal funding for human cloning in the United States. The implications and constitutionality of legislation prohibiting or limiting human cloning are not completely clear. In this context, it is essential to consider the biomedical, ethical, legal, and public policy implications of human cloning.

The creation of Dolly has proved that despite the biological differences between sheep and humans, it is technically possible to clone humans using the body cells. Due to the inefficiency of animal cloning (only about 1 or 2 viable offspring for every 100 experiments) and the lack of understanding about reproductive cloning, many scientists and physicians strongly believe that it would be unethical to clone humans. Not only do most attempts to clone mammals fail, about 30% of clones born alive are affected with "large offspring syndrome" and other debilitating conditions. Several cloned animals have died pre-maturely from infections and other complications. The same problems would be expected in human cloning. In addition, scientists do not know how cloning could impact mental development. While factors such as intellect and mood may not be as important for a cow or a mouse, they are crucial for the development of healthy humans. With so many unknowns concerning reproductive cloning, the attempt to clone humans at this time is considered potentially dangerous and ethically irresponsible. Physicians from the American Medical Association, and scientists with the American Association for the Advancement of Science have issued formal public statements advising against human reproductive cloning. Currently, the US Congress and other countries in the world are considering the passage of legislation that could ban human cloning.

People started recalling science fictions and takes in the 1970's book, and the movie "Boys from Brazil" where Nazis living in South America had cloned Hitler from preserved tissues. Such ideas lead to the belief that a nightmare was about to come true.

Regarding human cloning, the frequently raised question were:

- What if an individual started producing dozens of copies of himself or herself?
- What if parents desired to have a 'designer child', perhaps a super model or a superstar?
- What if parents stopped reproducing and instead reproduce themselves from ester cells?
- What if the clones produced are used for organ transplant?
- Would the clones be normal or defective?

24.13. ARGUMENTS AGAINST AND IN DEFENCE OF HUMAN CLONING

Road to Human Cloning

April 1998: Dolly had her first lamb, which was healthy and normal. ("Sheep clone has offspring." ABCNews.com, 23 April 1998).

April 1998: Researchers at the University of Colorado and the University of Massachusetts successfully treated Parkinson's disease in rats with the fetal brain cells of cloned transgenic cows. (Advanced Cellular Technology press release, 28 April 1998).

July 1998: Scientists at the University of Hawaii developed a new method for performing SNT with a higher success rate, and also demonstrated that clones of clones (mice, in this case) are born healthy and fertile. (Leutwyler, Kristin. "Send in the clones." Scientific American, 27 July 1998).

June 1999: The first human clone was created by a private biotechnology company in Massachusetts. The clone was created through SNT by inserting a human nucleus into a cow egg, and allowing the embryo to grow for 12 days. The company's goal was not reproductive cloning, but rather the cultivation of human stem cells. (Sung, Ellen. "First human embryo cloned." Policy.com, 21 June 1999).

March 2000: A litter of five piglets, all cloned from one adult, was born in Virginia ("This little piggy is a clone." AgBioTechNet.com, 15 March 2000).

Pigs have, to date been the most promising source of organs for transplant into humans. The ability to clone healthy pigs with predictable genes is one step towards successful transplantation. The other step is the ability to insert and knock out certain genes that would otherwise cause transplanted organs to be rejected by the recipient.

April 2000: Scientists in Europe have successfully performed nuclear transfers between human egg cells—a potential aid to women whose infertility is the result of a cytoplasm problem. The nucleus of a woman's egg can be placed inside a donor egg, and the resulting egg can then be fertilized in vitro and implanted into the womb of the nucleus donor. The resulting child would be almost 100% the genetic offspring of the mother and father. (Boseley, Sarah. "Fertility breakthrough raises human cloning fears." Sydney Morning Herald, 28 April 2000) While this procedure involves egg nuclei rather than somatic cell nuclei, it represents the kinds of innovations stemming from SNT research that could potentially be outlawed by overbroad legislation.

May 2000: Australian scientists produced their first cloned animal, a calf. The local development of this technology holds great promise for the Australian dairy and beef industries. The Australian technique used differentiated fetal cells as the nucleus source. Being able to develop their own procedure allowed Australians to avoid purchasing the technology from overseas. (Smith, Deborah. "Suzi leads herd as first in a cloning revolution." Sydney Morning Herald, 3 May 2000).

Table 24.1: Arguments against & in defence of human cloning

Against	*In defence of*
Cloning might lead to the creation of genetically engineered groups of people for specific purposes, such as warfare or slavery.	Cloning would enable fertile couples to have children of their own.
Cloning might lead to an attempt to improve the human race according to an arbitiary standard.	Cloning would give couples, who are at risk of producing a child with a genetic defect, the chance to produce a healthy child.
Cloning could result in the introduction of additional defects in the human gene pool.	Cloning could shed light on how genes work, and led to the discovery of new treatments for genetic diseases.
Cloning is unsafe. There are too may unknown factors that could adversely affect the offspring.	A ban on cloning would deprive people of the right to reproduce and restrict the freedom of scientists.
A clone might have a diminished sense of individuality.	A clone would not really be a duplicate, because environmental factors would mould him or her into a unique individual.
A clone might have fewer rights than other people.	A clone would have as much of a sense of individuality as do twins.
Doctors might use clones as sources of organs for organ transplant.	A clone would have the same rights as do all other people.
Cloning is against god's will.	Cloning is comparable in safety to a number of other medical procedures.
Some aspects of human life should be off limits to science.	Objections to cloning are similar to objections raised against previous scientific achievements, for example, heart transplants and test tube babies, that later came to be widely accepted.

June 2000: The first transgenic clones in a species other than mice were born in Scotland. The scientists successfully inserted new DNA into sheep nuclei and then placed those nuclei inside enucleated sheep eggs, resulting in two transgenic lambs. (Reaney, Patricia. "UK firm uses gene targeting to create cloned lambs." Reuters news release, 28 June 2000).

October 2000: SNT was used to help save an endangered species, the Asian gaur from extinction. A somatic cell nucleus from a living gaur was inserted into the egg cell of a cow and then implanted into that cow for gestation. The cow was near full-term of the time of the writing of the article.

Scientists have much hope for using SNT technology to preserve other endangered and recently extinct species. (Perlman, Heidi B. "Scientists close to cloning extinct animals." Chicago Tribune, 9 October 2000). The American Museum of Natural History's website suggests that as many as 100 species will become extinct every day as we head into the 21st century, and man's activities on the planet have accelerated the rate of extinction dramatically.

In 1990, UK pioneered legislation, making human cloning research illegal, but currently it has allowed the USA and several EU countries, and many other cultures with very different value systems. Some form of international treaty should be called for whereby no country would allow cloning research to be carried over from humans to animals. Realistically, there would be no way to stop a back street clinic or a dictatorship from ignoring such a treaty, but the lines need to be drawn. A second line of defence is also called for the notion of the ethical scientist, for whom it would be against all professional principles to pursue such research. Some have argued that research should be permitted into the possibility of cloning living transplant organs from body cells (**Table 24.1**). This would require more careful ethical consideration, but the danger of a "slippery slope" to full human cloning would loom large over such an enterprise.

This raises a last question of the control of such research. In many spheres of research there is a deficit in public accountability in the existing procedures whereby research priorities are set. There are no easy solutions to this problem, but, at the very least, it points to the need for a standing ethical commission on non-human biotechnology, whose work is open to public comment and scrutiny, in which those areas of research which are especially likely to have far reaching ethical implications are first debated in public.

In principle, cloning technology will allow derivation of stem cells from human embryos (**Fig. 24.11**), and implantation of these young stem cells into the brain or heart or other organs in which those stem cells will take on the properties of that tissue and presumably enhance its function.

However, the issues of using human embryos are fraught with far reaching ethical problems. Even using discarded embryos from *in vitro* fertilization procedures doesn't have consensus approval. The scientific and broader community will have to confront these issues as part of a larger public discussion. Legal and biomedical ethics experts at the Stanford Center for Biomedical Ethics are helping

Fig. 24.11. Human embryo

researchers address lab dilemmas in real time through a program called bench-side consultation. The program is a testing ground for integrating ethics into cutting-edge biomedical research as it unfolds, addressing topics from stem cells to genetic manipulation of disease causing microbes. The bench-side program was rolled out this fall by biomedical ethics center co-director, David Magnus, Ph.D, and associate director, Mildred Cho, Ph.D, to identify ethical or social impacts of biomedical lab research and suggest actions to minimize risks and maximize benefits to society.

Center members seek solutions acceptable to a range of stakeholders in science, policymaking and the public. They go on to contribute to state, national and international committees working on biomedical policy. The roots of the new program go back to the work Magnus and Cho did together before they joined Stanford. A researcher contacted them at the University of Pennsylvania, with concerns about the implications of creating a synthetic virus. Magnus said that they were able to lay out many issues in advance. The project prompted media discussion. He said that when the technology happened, it wasn't in a vacuum. This was very much in contrast to what happened with Dolly, where it came out of left field.

Stanford stem cell expert Irving Weissman, MD, came to the bioethics center for help with what became the bench-side program's first test case. Weissman, the Karel and Avice Beekhuis Professor of Cancer Biology and professor of pathology, developmental biology and, by courtesy, of biological sciences, was seeking animal models to study human brain diseases, such as Alzheimer's and Parkinson's, that do not normally occur in animals. To do that, he considered combining human and mouse stem cells to create a mouse with human brain cells. "But he recognized that the idea of having a mouse with human neurons is at the least unsettling and may have ethical implications," said Hank Greely, the C. Wendell So he came to them to ask for a review. Greely and Cho researched the case and advised Weissman that he could go forward, but only under certain conditions. They suggested that Weissman change the order of his experiments, conducting the least controversial work first and reevaluating future plans as results became available. Stanford bioethics work also helps to guide policy outside the university. Greely recently participated an a California advisory committee created to review a 1997 ban on human reproductive cloning. Based on the committee's January 2002 report, California enacted new legislation banning reproductive cloning indefinitely, and allowing, but regulating, non-reproductive cloning. "The recommendations were, by and large, enacted into law," Greely. He is now working with a new committee reviewing stem cell research. Bioethics creates bridges between science, policy and public discussion. It promotes a common understanding of terms like "clone", "embryo" and "stem cell" to help disparate groups discuss the implications of bench science and agree on policy and research conduct.

The Center for Biomedical Ethics provides education and consultation, and performs research on ethical problems in our health care system.

Community clinicians and researchers can request a panel to explore issues in depth or call the center with shorter-term problems, such as responding to grant agency concerns. Magnus said, "We have helped the researchers with response letters and helped to adapt consent forms to address issues". Magnus is also co-chair of the ethics committee for the Stanford Health Center, where he runs a bedside consultation program after which the bench-side program is modeled. "We're trying to give researchers the same kind of consultation service that we've been giving on the clinical side for a long time," he said.

Scientists wondered whether "Dolly" would be fertile. Some cloned frogs are infertile. Also, cells seem to have an internal clock that causes them to die off after a normal life. Since Dolly was conceived from a 6-year-old cell, her life expectancy may be reduced from about 11 to only 5 years. But this did not take place. Dolly, up to 5 years of age, continued to live a normal life, but died prior to the normal life span. Similar experiments to clone mice were initially unsuccessful. One speculation was that the DNA in sheep may not be used by the cells until after three or four cell divisions have completed. This would give the ovum sufficient time to reprogram the DNA from (its original) mammary cell functions to egg cell functions. Both human and mouse use the DNA after the second cell division. So, some researchers had predicted that humans as well as mice may not be "clonable". However, mice were successfully cloned later. Thus, cloning of humans might also be possible.

One of the many concerns with human cloning is that cloning of animals sometimes cause fetal overgrowth (aka large-offspring syndrome). The fetus grows unusually large and dies just before or after birth. They have under-developed lungs and reduced immunity to infection. Duke University researchers announced on August 15, 2001, that this particular problem would not exist in humans. The DNA of all primates, such as humans, monkeys and apes, have two copies of a gene that regulates fetal growth, whereas almost all other animals have only one. This spare copy should prevent fetal overgrowth in cloned human fetuses. Randy Jirtle, professor of radiation oncology at Duke University in Durham, NC, said, "It's going to be probably easier to clone us than it would be to clone these other animals because you don't have this problem—not easy, but easier." Kevin Eggan, of MIT's Whitehead Institute works with cloned mice. He called the Duke study interesting from the perspective of the evolution of imprinting genes". But he cautioned that there is no proof that abnormally large babies are born as a result of this one genetic difference. His lab has a *"four-times normal size"* mouse clone, which has normal IGFR2R genes. He suggests that there are other factors that can contribute to abnormal development in clones.

Richard Seed, a physicist from Illinois, attempted to establish a human cloning clinic. He claimed on January 7,1998, that he was 90% complete in hiring a team of experts to attempt the cloning of a human being **(Fig. 24.12)**, following the experiments on "Dolly". If successful, the resultant child would have identical DNA to one of its parents. Lord Robert Winston, a London based fertility expert who helped produce the first test-tube baby in 1978, said, "My first reaction is that here is somebody who is trying to make a quick buck off of self-advertising, because of course, there is no way you could clone a human being safely at this point. I think the man is clearly unhinged and I don't think he is to be taken seriously." Marian Damewood, a member of the board of the *American Society for Reproductive Medicine* said: "I have very serious reservations about cloning human beings." The Society has declared a 5 year voluntary ban on cloning humans. Mr. Seed responded, "I can't really answer the critics who think it's a bad idea. They'll never be persuaded. As far as I'm concerned, they have rather small minds and a rather small view of the world and a rather small view of God." Dr. Seed apparently did not succeed in his project.

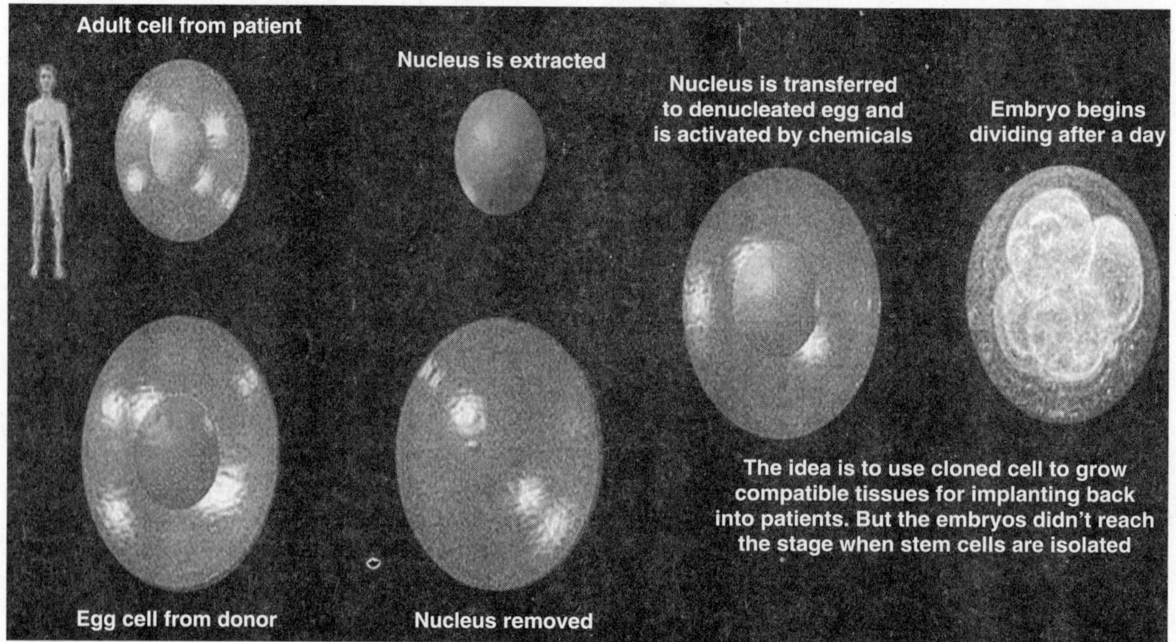

Fig. 24.12. Cloning of a human being

On August, 2001, there were two publicized projects underway to clone humans. There may be others which are secret.

Dr. Panayiotis Zavos of the *Andrology Institute* in Lexington KY, and Dr. Severino Antinori, a fertility doctor in Rome, announced in early 2001 that they want to proceed with the cloning of humans. Professor Antoniori announced in early August, 2001, that he intends to start cloning human embryos before the end of 2001.

".. a religious sect called the Raelians insists it will shortly undertake the same project. The Raelian sect believes, among other things, that human beings were created in laboratories by extra-terrestrials, and that the resurrection of Jesus was a cloning procedure! Donors and surrogate mothers have already lined up to pay $250,000 for the actual cloning experiment."

"In an effort to stall these attempts, US Congress is planning to pass a bill banning human cloning by any organization in the United States. However, it is unlikely that this will prevent these efforts from being made elsewhere." On August 14, 2001, Nobel Prize-winning scientist, Sir Joseph Rotblat said that our understanding of cloning is *"too meager"* at this time to succeed. He suggested that many of the scientists involved are motivated by money. He said, *"Inevitably, the problems will be overcome, and it's then that the real ethical problems begin."* He suggested that "Ethics are not absolute. Look at in-vitro fertilization. This was originally considered unethical but is now widely accepted..." I feel that this (cloning), too, will become acceptable." In April, 2002, Dr. Severino Antinori announced that a woman, who joined his program for infertile couples, is now eight weeks pregnant with a fetus derived by human reproductive cloning. Although such cloning is banned in Italy where he lives, he was allegedly able to go to another country to per-form the experimental technique. If the woman's pregnancy produces a live newborn, it will probably suffer from one or more serious genetic disabilities.

24.14. ETHICS OF ADULT HUMAN DNA CLONING

Few researchers argues for the continued research on the human cloning with their supporting in the following lines:

- Some talents seem to be genetically influenced.
- Musical ability seems to run in families.
- Cloning using the DNA from the cell of an adult with the desired traits or talents might produce an infant with similar potential.

A heterosexual couple in which the husband was completely sterile could use adult DNA cloning to produce a child. An ovum from the woman would be coupled with a cell from the man's body. Both would contribute to the child: the woman would provide the "factory" for creating cells; the man would provide the "genetic information." They might find this more satisfactory than using the sperm of another man.

Two lesbians could elect to have a child by adult DNA cloning rather than by artificial insemination by a man's sperm. Each would then contribute part of her body to the fertilized ovum: one woman would donate the ovum, which contains some genetic material in its mitochondria; the other woman the nuclear genetic material. Both would have parts of their bodies involved in the conception. They might find this more satisfactory than *in vitro* fertilization using a man's sperm.

There are many objections to the human cloning technology and the major ones are as follows:

There is no guarantee that the first cloned humans will be normal. The fetus might suffer from some disorder that is not detectable by ultrasound. They may be born disabled. Disorders may materialize later in life. Such problems have been seen in other cloned mammals. There is no reason to assume that they will not happen in humans.

Cells seem to have a defined life span built into them. "Dolly" was created from a cell that was about six years old; this is middle age for a ewe. There were some indications that Dolly's cells were also middle-aged. She was believed to be, in essence, about six years old when she was born. She was expected to live only for five years, which is shorter than the normal life span of 11 years. If this is also true of humans, then cloned people would have a reduced life expectancy. The cloning technique could take many years off their life. These fears proved to be unfounded. "Dolly" has grown into a comfortable middle age with signs of normal aging for her age.

Dolly was conceived using a ewe's egg and a cell from another ewe's body. It is noteworthy that no semen from a ram was involved. If the technique were perfected in humans, and came into general usage, then there would be no genetic need for men. All of the human males could be allowed to die off. The author of this essay is a male and does not think kindly of such a future. However, some readers might not object to this eventuality.

Large scale cloning could deplete genetic diversity. It is diversity that drives evolution and adaptation. It prevents an entire species from disappearing because of susceptibility to a disease. It is doubtful that cloning would ever be used at a level to make this a significant threat.

Some people have expressed concern about the effects that cloning would have on relationships. For example, a child born from an adult DNA cloning from his father would be, in effect, a delayed twin of one of his parents. That has never happened before and may lead to emotional difficulties.

Religious Objections to Cloning

There are religious objections to cloning. Most pro-life supporters believe that a fertilized ovum is a full human. When its nucleus is removed during cloning, that person is, in effect, murdered.

A secondary concern is the whole business of collecting surplus embryos and simply storing them in a deep-freeze as a commodity. Some claim that cloned humans may be born without souls. They speculate that the soul enters the body when a sperm fertilizes an ovum. Since there is no sperm involved

in cloning, perhaps the fetus would develop without a soul. There is no way to know whether a soul is present; it has no weight, it cannot be seen, touched, smelled, heard, or detected in any other way. In fact, many people believe that souls do not exist. Speculation on this topic can never be resolved.

At the current stage in cloning using adult DNA, the random appearance of genetic defects, noted above, appears to be an overwhelming problem. Such dangers would seem to put an indefinite halt on all ethical cloning of humans.

Dr. Keith Campbell and his colleagues at the Roslin Institute, Edinburgh, created dolly—the world's first clone of sheep. Cloning of humans for reproductive purposes, the form of cloning that most people think of first, is also the type of cloning, which the scientific community virtually unanimously agrees is abhorrent. "Human reproductive cloning," Campbell said, "is entirely out of the question." It is insupportable from both a scientific and an ethical viewpoint.

Technically, human cloning could be done, but from the scientific point of view it's too dangerous. Cloning is a fledgling science, and very unreliable—of the 277 original cloned cells, only one was fully viable, producing dolly. From the ethical stand-point, there's just no good reason to start cloning humans. Even if it were done, it would have horrible effects on all parties involved—the public would be furious, the clone's mother would face unbearable notoriety, and the clone would be under enormous pressure to behave like the person he or she was cloned from.

Despite the chorus of arguments against human cloning, it may have already happened. Rogue Italian physician, Dr. Severino Antinori has claimed to have impregnated a woman, somewhere in the world with the first human clone. His claims are vague, and the scientific community is skeptical. But if he has done what he claims, the implications would be enormous, breaching scientific, political and ethical standards. Dr. James Shapiro, the renowned researcher who co-developed the Edmonton Protocol—a treatment for Type-1 diabetes, and the person who brought Campbell in for the talk, doubts Antinori's claim. Shapiro says the science of cloning is so unstable that a human birth would be unlikely. "Scientists are all dead against the concept of making a person from a clone—unless you're this crazy guy in Italy, that is. It's not a reasonable thing to do." To clone a human would be to push cloning further beyond the ethical guidelines that are struggling to constrain it. According to Shapiro, cloning, or nuclear transfer as it is referred to by scientists, could be used to mass-produce insulin-secreting islet cells—the key to his diabetes treatment. Currently, islet cells are taken from deceased donors—a technique that isn't practical, as there simply aren't enough donors to meet the demands of the diabetes population.

Campbell, Shapiro, and one of Shapiro's colleagues, Dr. Jonathan Lakey, director of the USA Human Islet Isolation Laboratory, met in Singapore to discuss alternate islet cell sources that could be used in the future. Impressed by Campbell, Shapiro invited him to Canada to give the lecture. Since the meeting, the three have remained in frequent contact. "The amazing news now is the role of cloning in producing a stem cell," Campbell said, describing the realm of cloning that could be of benefit to the Edmonton Protocol team. Stem cells are undifferentiated cells that can make any cell type in the body. Shapiro and his colleagues hope that stem cells, which can produce islet cells, could be cloned and implanted into people with Type-1 diabetes, treating their disease and possibly even curing it.

But the ethical problems and public confusion surrounding the complexities of such therapeutic cloning may overshadow its huge potential benefits. Nowhere are these problems more evident than in the Canadian government's answer to the cloning debate. Bill C-56, entitled An Act Respecting Assisted Human Reproduction, bans human cloning for reproductive purposes, but its stance on therapeutic cloning is a little more difficult to interpret. While Shapiro is optimistic that the bill "leaves the window open so that scientists can continue their research within ethical guidelines".

Dr. Timothy Caulfield, research director of the USA-based Health Law Institute, and Dr. Glenn Griener of the John Dossetor Health Ethics Centre at the USA, both have serious misgivings about the bill. Caulfield thinks that under the new bill, human reproductive cloning and many forms of therapeutic cloning would be considered criminal offences—punishable with imprisonment. Although he recognizes that the bill is a step in the right direction from a government standing committee, which proposed last year to ban all human cloning whatsoever, Caulfield also says the proposed legislation makes a fundamental error. It brings cloning into the abortion debate. In an effort to be as broad as possible, the bill confuses a cloned human stem cell with an human embryo. Even if a cloned stem cell originally came from an adult, and if it started dividing, it would be considered under the bill to be an embryo. There have been several reports from around the world endorsing human cloning of this kind. This area is moving so quickly that it would be a mistake to criminalize it, and this certainly makes therapeutic cloning a criminal offence. Instead of such global legislation, Caulfield would like to see a regulatory body set up such as those that deal with the tax department and the security commission, a body that could deal with the nuances of cloning legislation. Griener echoes Caulfield's concern. "My own view is that this is a legislative mistake. I haven't seen any rationale from the government to back up their stance on therapeutic cloning." Griener supposes, though, that the government is attempting to reflect the views of a public that is wary of cloning. This, he says, is at least driven by the right ethical motivation. "Though, on the one hand, there's virtue in the government consulting scientists, I think the government also needs to worry about the public's perception (of cloning), as well. They need to take account of the variety of ethical opinions, and there are parts of the bill that clearly provide the necessary political compromise." Caulfield, though, is putting together a survey of public opinions about cloning from around the world, he doesn't find the public wariness one might expect. The support for embryonic stem cell research, he says, ranges from 60-86%. "The media," he said, "makes it sound like there's a divide amongst Canadians, and I don't see it at all." Instead, he says, the government is responding to a passionate and organized voice of a minority that is against such cloning. Whatever the outcome, the implications of the cloning debate, run deep. This whole debate raises very interesting questions about how we as a society are going to make decisions about complex science-based issues, and, as we move towards more science-based societies, it that raises some interesting questions for liberal democracies in the future.

The extraction of stem cells from a cloned human embryo represents a continuation of research rather than a radical discovery. The work announced in the journal Science is, undoubtedly, a very important step for medical researchers hoping to use stem cells as the basis of therapy for currently incurable diseases such as Alzheimer's and Parkinson's. But the road to these cures stretches far into the future, and today's news is just another milestone—one of several to have happened over the last few years. It was inevitable that the research was going to happen. Indeed, the scientists involved have been working on it for sometime. Scientists have cloned a human blastocyst (an early stage embryo containing just a few cells) before. They have also taken stem cells

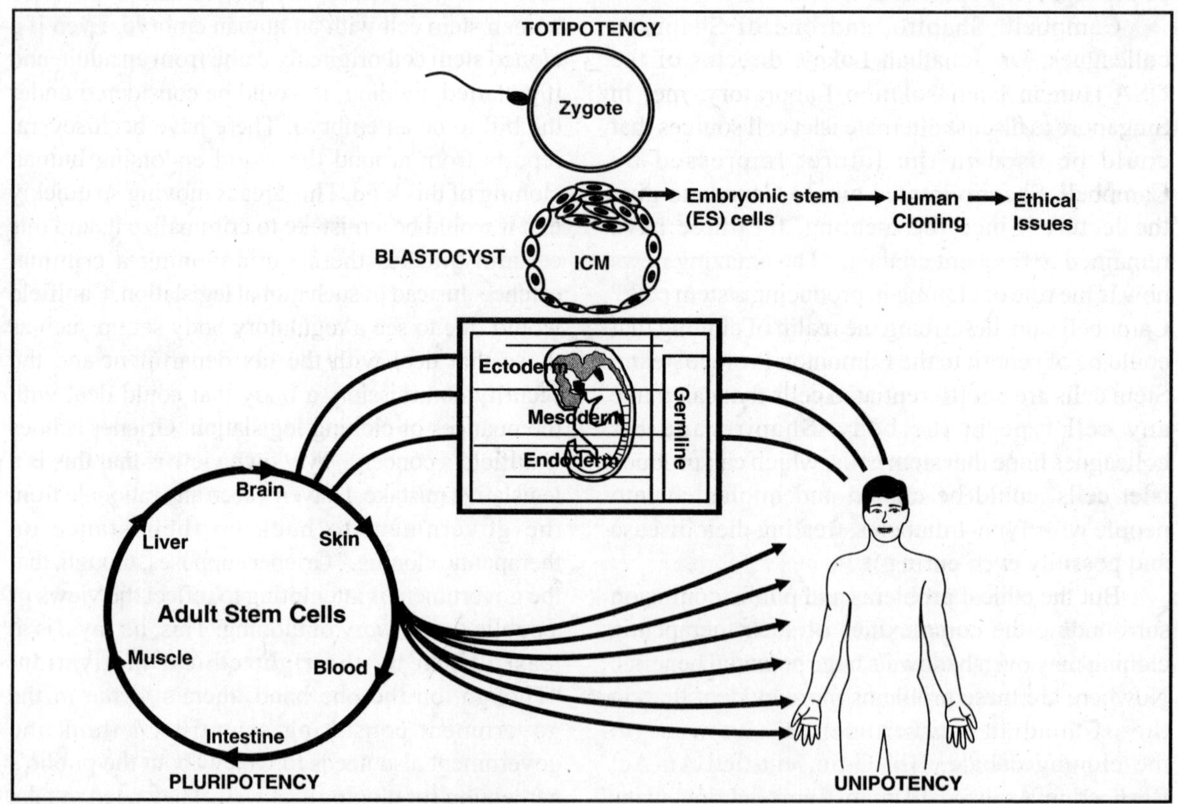

Fig. 24.13. Cell potencies in mammals.

Pure totipotency is a feature of the earliest embryonic stages only. Subsequently, only pluri- and unipotent cells are found in the embryo, fetus or adult

from embryos, left over after fertility treatment, for research purposes. The new work marries these two steps, and shows that embryonic stem cells can be taken from cloned embryos. It also demonstrates the potential for some of the more pitched ethical battles around stem cell research to be circumvented.

As well as cures for degenerative diseases, stem cell research raises the possibility that, sometime in the far future, doctors will be able to grow tissue that exactly matches a person's own, thereby reducing the risk of rejection in case they need a transplant. Of course, ethical problems abound. Many anti-abortion campaigners say that harvesting stem cells from embryos is wrong. They argue that adult stem cells, such as those present in bone marrow, should be used instead. But many scientists argue that embryonic stem cells **(Fig. 24.13)** are more versatile (or pluripotent), and can be grown into more different types of tissue than adult stem cells. These ideological arguments have taken place all over the world, resulting inseveral countries that are hostile to research on embryonic stem cells. In the US, for example, the George Bush has made his objections to the work clear. US embryonic stem cell research is therefore completely privately funded, and researchers are increasingly leaving the country to carry out their work overseas.

The British government has passed laws allowing the scientists to develop and research embryonic stem cells, and is trying to attract top foreign scientists in an attempt to bolster its research base. In general, research using embryonic stem cells is allowed in the UK, but researchers need to be licensed. The new research in South Korea may provide some hope for solving this dilemma. By using the nucleus of an adult cell and transferring it into an egg to create the blastocyte, the adult cell was effectively "reprogrammed" into becoming a pluripotent, just like an embryonic cell. If this can be repeated, and proves reliable, it could shortcut some of the ethical problems altogether. In any case, scientists in the UK are looking to restrict the impact of their

research on embryos. The Medical Research Council has set up a stem cell bank, where researchers can obtain samples of specific types of stem cells for their research. The MRC says that this will reduce the need for harvesting cells from the embryos in the future, because once a stem cell is deposited there, it can be cloned repeatedly to provide raw material for research.

24.15. CLONING TECHNOLOGY AND ITS PROBLEMS

The problems with cloning fall into three main categories—the suffering caused by the procedures involved in the cloning process, the suffering of the clones themselves due to abnormalities caused by the cloning process, and the ridiculous claims of those who stand to profit from cloning made in defence of their work.

Suffering Caused by the Cloning Procedures

The cloning technique involves a succession of procedures of varying severity. The eggs to be enucleated may come from the slaughterhouse. Alternatively, they could come from a live female animal, and this will involve a surgical procedure being carried out on her. Placing early embryos in the animal acting as a temporary recipient also involves a surgical procedure, and these animals would be killed to retrieve the embryos. The success achieved in cloning is not very high, and still has to be standardized **(Tables 24.2, 24.3 & 24.4)**.

Once retrieved, the embryos are placed in another surrogate female, which requires yet another surgical procedure. In addition to this, cloned animals tend to be abnormally large (a phenomenon known as "large offspring syndrome"). So, a caesarean is often required, as the clone may be too large to come through the birth canal. Surrogate mothers can suffer adverse effects, which are often severe. In one nuclear transfer study, pregnancy was confirmed in 18

Table 24.2: *Key developments in mouse cloning that helped in an improved cloning technology in animals and humans*

Year/Nucleus donora	Recipient cytoplast	Development	to researchers
1981 ICM	Zygote	—	Illmensee and Hoppe
TE	Zygote	—	Illmensee and Hoppe
1983 Zygote	Zygote	Term	McGrath and Solter
1984 Two cell to ICM	Zygote	< Blastocyst	McGrath and Solter
1986 Two cell	Two cell	Term	Robl *et al.*
Two cell	Two cell	Fetus	Robl *et al.*
1987 Two cell	Zygote	Term	Tsunoda *et al.*
Four to eight cell	Zygote	< Four cell	Tsunoda *et al.*
Eight cell	Two cell	Term	Tsunoda *et al.*
ICM	Two cell	< Four cell	Tsunoda *et al.*
1989 Male PGC	Oocyte	Implant	Tsunoda *et al.*
1990 EC	Oocyte	< Blastocyst	Modlinski *et al.*
1991 Late two cell	Oocyte	Term	Kono *et al.*
Eight cell to ICM	Oocyte	< Blastocyst	Kono *et al.*
1992 Four cell	Two-cell blastomere	Term	Kono *et al.*
1993 ES cell	Oocyte	Implant	Tsundo and Kato
Thymus	Oocyte	< Blastocyst	Kono *et al.*
Early four cell	Oocyte	Term	Cheong *et al.*
Early eight cell	Oocyte	Term	Cheong *et al.*
1995 Male PGC	Oocyte→two cell	Fetus	Kato and Tsunoda
1996 Late four cell	Oocyte→one cell	Term	Kwon and Kono
1997 Early morulae	Oocyte→two cell	Term	Tsunoda and Kato
1998 Early ICM	Oocyte→two cell	Term	Tsunoda and Kato
TE	Oocyte→two cell	Term	Tsunoda and Kato
Sertoli	Oocyte	Fetus	Wakayama *et al.*
Brain cell	Oocyte	Fetus	Wakayama *et al.*
Cumulus	Oocyte	Term	Wakayama *et al.*
1999 Tail cell	Oocyte	Term	Wakayama and Yanagimachi
ES cell	Oocyte	Term	Wakayama *et al.*
Oviduct cell	Oocyte→two cell	Late fetus	Kato *et al.*
2000 ES/Tail cells (X^{GFP})	Oocyte	Fetus	Eggan *et al.*
Sertoli cell	Oocyte	Term	Ogura *et al.*
Cloned cumulus cell	Oocyte	Term	Wakayama *et al.*
2001 ES cell (epigenetics)	Oocyte	Term	Humpherys *et al.*
Fetal fibroblast	Oocyte→two cell	Term	Ono *et al.*
Fetal gonadal cell	Oocyte	Term	Wakayama and Yanagimachi
ntES cell	Oocyte	Term	Wakayama *et al.*
2002 Mainly Sertoli cell (imprinted genes)	Oocyte	Term	Inoue *et al.*
Sertoli cell (early death)	Oocyte	Term	Ogonuki *et al.*

Abbreviations: ICM: inner cell mass; GFP: green fluorescent protein; EC: embryonal carcinoma; ES: embryonic stem; PGC: primoridal germ cell; TE: trophectoderm.

Table 24.3: *Donor cell developmental potency and NT embryo development in cloning of various animal species*

Species	Cell type	Morulae/blastocysts developing to term	Researchers
Mouse	Cumulus	2-3%	Wakayama *et al.*, 1998
	Sertoli	3.5%	Ogura *et al.*, 2000
	Fibroblast	1%	Wakayama and Yangimachi, 1999
	ES cells	5-21%	Wakayama *et al.*, 1999; Rideout *et al.*, 2000; Eggan *et al.*, 2001
	Blastomere	25%	Tsunoda and Kato, 1997
Sheep	Mammary	3.5%	Wilmut *et al.*, 1997
	Fetal fibroblast	7.5%	Wilmut *et al.*, 1997; McCreath *et al.*, 2000
	Embryo cells	4.5%	Wilmut *et al.*, 1997
	ICM ES-like	4%	Wells *et al.*, 1997
Cow	Cumulus	10%	Wells *et al.*, 1999
	Oviductal	15%	Kato *et al.*, 1998
	Fetal fibroblast	14%	Cibelli *et al.*, 1998
	ICM ES-like	12%	Sims and First, 1994
Pig	Granulosa (serial NT)	7%	Polejaeva *et al.*, 2000
	Fetal fibroblast (single NT)	1%	Onishi *et al.*, 2000; Betthauser *et al.*, 2000
Goat	Fetal fibroblast	3.5%	Baguisi *et al.*, 1999

Table 24.4: *Genetic influences on NT embryo development in animal cloning*

Cell type	Genetic background	Survival to term	Survival from term to about	Researchers
Cumulus	Various F1	1-2%	85%	Wakayama *et al.*, 1998; Wakayama and Yangimachi, 1999
	Inbred C57	0	N/A	Wakayama and Yanagimachi, 2001
	Inbred 129	1.5%	95%	Wakayama and Yanagimachi, 2001
ES cells	Various F1	8-17%	50-85%	Wakayama *et al.*, 1999; Rideout *et al.*, 2000; Eggan *et al.*, 2001
	Inbred C57	17%	0	Eggan *et al.*, 2001
	Inbred 129	1-18%	0	Eggan *et al.*, 2001
	Inbred 129	1.6%	20% (n = 1)	Wakayama *et al.*, 1999
ES cells into 4 N blasts	Various F1	18%	85%	Eggan *et al.*, 2001
	Inbred C57	4%	0	Eggan *et al.*, 2001
	Inbred 129	10%	0	Eggan *et al.*, 2001

surrogate cows. Six of these aborted and 4 more died as a result of the pregnancy. Of course in the smaller cheaper species the surrogate mother will have served her purpose by the time the clone is born, so she may be killed following the birth.

Suffering of the Cloned Animals

Since the birth of Dolly, scientists have been trying to apply cloning technology to other species also, but with limited success. As previously stated, less than 3% of 'reconstituted embryos' result in live births. These embryos fail in a number of ways—many fail to implant in the lining of the womb, but even when implantation is successful the pregnancies often end in miscarriage, apparently due to gross developmental abnormalities. For example, major abnormalities have been found in the heart and blood vessels, the placenta, and the lungs in dead foetuses.

Many more cloned offspring die in the perinatal period (i.e., shortly after birth). In a study of 80 genetically manipulated cloned lambs, only 14 survived, and all but three died before 12 weeks of age with abnormalities of the brain, kidneys, or liver. Similarly, in an experiment to clone genetically modified pigs suitable for xenotransplantation, only 7 of the approximately 100 implanted embryos were carried to term. They were all delivered by caesarean section. Two died shortly after birth due to severe breathing difficulties of a kind usually seen in premature babies; one of these piglets also had a cleft palate and both had joint deformities. Another piglet died 17 days after birth from a heart defect. Of the remaining 4 piglets, one had heart, lung and abdominal abnormalities, another had eye and ear abnormalities, and a third had joint defects.

As the first generation of clones are ageing, it is becoming apparent that they also suffer long term health effects that may only appear later in the animals' life. In January, 2002, just 5½ years after her arrival, it was announced by the Roslin Institute that Dolly was suffering from

premature arthritis. In February, 2003, Dolly was diagnosed with a progressive lung disease and was euthanased at just 6 years old. Genetically normal sheep can live upto 11-12 years. It is likely that Dolly's health problems were related to her status as a clone. Large scale studies on cloned animals have also revealed long-term health problems. A study published in 2002 revealed that cloned mice died much younger than non-cloned mice due to severe pneumonia and liver disease. The cause of these abnormalities is not fully understood, but much research is being carried out into why cloning methods produce such unhealthy animals. As the creator of Dolly has stated, "...*cloning by the present methods is a lottery... Several coins are thrown and all must come up as heads if normal life is to result.*"

24.16. USEFULNESS OF CLONING TECHNOLOGY

Given the astonishing loss of life and inevitable suffering of animals involved in the cloning process, scientists' relentless pursuit of cloning technology would imply the expectation of tremendous benefits if the technique is perfected **(Fig. 24.14)**. However, when claims for the potential benefits of cloning are examined, the outcomes driving this technology appear to be more to do with profit for the companies involved and prestige for the scientists. Other potential benefits often claimed in defence of cloning technology include the following:

Cloning of Pet Animals and 'Socially Useful' Animals such as Guide Dogs and Rescue Dogs

Although not yet established, this is an example of how humans value animals on the basis of how useful they are to humans, rather than valuing them as individuals in their own right. And cloning animals, even for what some argue "more socially responsible purposes", is totally unacceptable. If

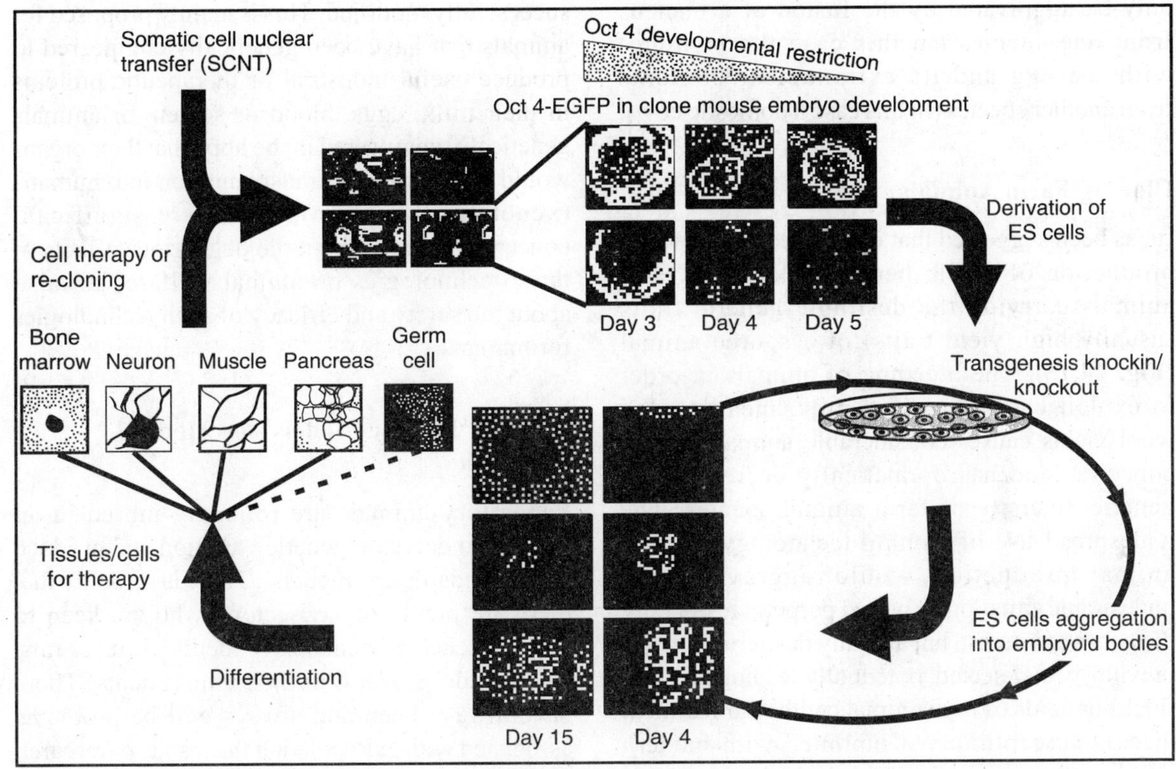

Fig. 24.14. Cloning can be used for different disease replacement therapy or research

society was aware of the catalogue of suffering, for example, dogs and puppies would endure, would this still seem so socially responsible? Furthermore, it is known that the ability of these animals is more dependent on appropriate training and environmental factors than genetic make-up.

Interestingly, the coat of cell cloning—the first cloned kitten is a different pattern to that of Rainbow (the animal she was cloned from) because the environment in the womb decides the pattern of coat pigmentation during development. This emphasises the importance of environment versus genetics in physical attributes let alone psychological ones.

Cloning of Endangered Species

Viable wild populations rely on a degree of genetic variation, which is impossible to achieve by cloning. Species most usually become extinct because of habitat loss. Preserving these habitats and all the plants and animals in them is a far more robust and responsible strategy for preserving key charismatic species than cloning animals, which are then destined to spend their lives in captivity. In addition, efforts to protect habitats are actually undermined by the proposition that cloning will 'save' endangered species, as people are given the false impression that extinction can be reversed. Recent attempts to clone an endangered ox, the Banteng, resulted in one of the two live clones being euthanased a week after its birth by caesarean section. The abnormally large calf was reluctant to feed and was described as *"grossly malformed"*. It would appear obvious that the difficulties encountered in the cloning process can only be aggravated by the fusion of a nucleus from one species (in this case the banteng) with the egg and its extranuclear contents from another species (in this case a domestic cow).

Cloning Farm Animals

It has been suggested that cloning could allow the production of whole herds or flocks of farm animals carrying the desirable genetic traits (usually high yield traits) of a source animal **(Fig. 24.15)**. The breeding of animals in order to exploit ever increasing milk, meat, egg or wool yields causes considerable animal welfare concerns, and has dramatically decreased the genetic diversity of farm animal species. The widespread use of cloning technology in farm animal production would aggravate this undesirable situation. It would perpetuate not only the genes of interest, but also any deleterious genes previously undetected, potentially leading to entire flocks or herds of farm animals with near identical disease susceptibility. Combined with modern intensive farming methods this is the perfect recipe for devastating disease epidemics.

Cloning Genetically Engineered Animals

Another suggested use of cloning technology is to produce copies of animals that have been

Fig. 24.15. Photograph showing many cloned calves from skin cells

successfully modified. This is mainly proposed for animals that have been genetically engineered to produce useful industrial or therapeutic proteins in their milk, eggs, blood or semen, or animals genetically engineered in the hope that their organs would be suitable for transplantation into humans (xenotransplantation). There are significant concerns not only about the deleterious effects of these technologies on animal welfare, but also about the safety and efficacy of such technologies for humans.

Cloning 'Disease Models' for Medical Research

Laboratory animals are routinely inbred in an attempt to decrease genetic variation and produce more standardized 'models'. There is no doubt that there are plenty of vivisectors, who are keen to order batches of genetically identical mice, rats, cats and dogs to use in their experiments. There have always been and always will be problems associated with extrapolating the results of research between physiologically distinct species, making animal experimentation at best a crude and blunt instrument. Although using 'identical' cloned animals may make research results more robust, it will not make them any more applicable to the human situation.

Looking behind the triumphant headlines at the facts and figures of cloning, coupled with the stories of individual clones, reveals a literature cataloguing a tremendous loss of life and unimaginable suffering for the victims of this unnecessary and unreliable technology.

SUMMARY

Cloning is the term mainly used to describe the production of one or more animals who are genetically identical to a source or 'donor' animal. The term is usually used to refer to taking a single cell from an adult animal and using it to build another complete and genetically identical animal.

This is possible because every individual cell in an animal contains all the information (in the form of DNA) required to make the entire animal.

The three types of cloning technologies will be normally followed in the scientific communities for their use and applications. They include recombinant DNA technology or DNA cloning, reproductive cloning, and therapeutic cloning. DNA Cloning is the production of many copies of required DNA materials. This type of cloning is also known as recombinant DNA technology, molecular cloning and gene cloning. It is a form of cloning that has been around since the 1970s and the technique is common practice in molecular biology laboratories. Reproductive cloning is also called adult DNA cloning. The purpose of this type of cloning is to produce a genetic duplicate of an existing or previously existing organism. Therapeutic cloning is also known as biomedical cloning, embryo cloning or artificial embryo twinning. This type of cloning essentially mimics the natural process of producing identical twins or triplets. Identical twins are created when the fertilized egg tries to divide into a two-cell stage but the two cells separate instead. The two separate cells continue dividing on their own and eventually develop into two genetically identical individuals.

Based on the either cells or genetic materials used for cloning, it is classified into two types viz., cell cloning and DNA cloning. The cell cloning as the name indicates this involves the cloning, i.e. producing genetically identical copies of a particular cell. For example, this is used in producing clones of cancer cells. Individual cancer cells can be isolated and allowed to reproduce in a culture, to form cell colonies. There, colonies will contain many colonies of the original cancer cell. Cell cloning helps in study of defective gene in cancer cell. The DNA cloning technique is used to reproduce a specific gene. Here, first the DNA fragment containing the gene of interest is isolated and combined with a plasmid to form a recombi-nant DNA and introduced in to the host cell and particular gene is expressed. This is used in genetic engineering purpose to produce transgenic plants and animals.

The first animals to be produced in this manner were tadpoles created by transfer of a nucleus from an embryonic cell into an enucleated egg in 1952. The year 1986 saw the first sheep cloned from undifferentiated embryonic cells using nuclear transfer. This was followed in 1995 by the production of the famous Megan and Morag at the Roslin Institute near Edinburgh, the first sheep to be cloned from differentiated embryonic cells (i.e. cells that had already taken on the characteristics of a specific body tissue rather than the generalized cells of very early embryos). In 1996 Dolly was born, also at the Roslin Institute, the first mammal to be cloned from adult rather than embryonic tissue. The production of Dolly has sparked a cloning frenzy. 1998 saw the arrival of cows cloned using the same technique that produced Megan and Morag. The Roslin institute cloned the first pig in 2000. In the US in 2001 a rhesus macaque monkey called Tetra was cloned. Also in 2002 in the US a cloned kitten was reported and in the same year as the first cloned rabbits in France. In the year 2003 the first cloned mule, horse and rat were reported in animal biotechnology.

Animal cloning also showed that it is possible are now to produce copies of a mammal that has desirable traits. Finally, it awakened the world to the fact that the cloning of humans could become a reality and gave debates about the ethics of human cloning a new urgency, as well as debates about its legal status and about relevant public policy. The word "clone," although often used in different contexts, refers to a precise genetic copy of a molecule, cell, plant, animal or human. Due to the inefficiency of animal cloning (only about 1 or 2 viable offspring for every 100 experiments) and the lack of understanding about reproductive cloning, many scientists and physicians strongly believe that it would be unethical to attempt to clone humans. Not only do most attempts to clone mammals fail, about 30% of clones born alive are affected with "large offspring syndrome" and other debilitating conditions. Several cloned animals have died prematurely from infections and other complications. The same problems would be expected in human cloning. In addition, scientists do not know how cloning could impact mental development. While factors such as intellect and mood may not be as important for a cow or a mouse, they are crucial for the development of healthy humans.

There is wide spread debut about the human cloning possibilities. Some of the proponents say that the some talents seem to be genetically influenced such as musical ability seems to run in families. Cloning using the DNA from the cell of an adult with the desired traits or talents might produce an infant with similar potential again and again. In such cases human cloning offers the method to maintain desired traits. A heterosexual couple in which the husband was completely sterile could use adult DNA cloning to produce a child. An ovum from the woman would be coupled with a cell from the man's body. Both would contribute to the child: the woman would provide the "factory" for creating cells; the man would provide the "genetic information." They might find this more satisfactory than using the sperm of another man.

Two lesbians could elect to have a child by adult DNA cloning rather than by artificial insemination by a man's sperm. Each would then contribute part of her body to the fertilized ovum: one woman would donate the ovum, which contains some genetic material in its mitochondria; the other woman the nuclear genetic material. Both would have parts of their bodies involved in the conception. They might find this more satisfactory than *in vitro* fertilization using a man's sperm.

But an opponent of the human cloning argues that there is no guarantee that the first cloned humans will be normal in all respect of human life. The fetus might suffer from some disorder that is not detectable by ultrasound. They may be born disabled. Disorders may materialize later in life. Such problems have been seen in other cloned mammals. There is no reason to assume that they will not happen in humans.

Cells seem to have a defined life span built into them. "Dolly" was created from a cell that was about six years old; this is middle age for a ewe. There were some indications that Dolly's cells were also middle-aged. She was believed to be, in essence, about six years old when she was born. She was expected to live only for five years, which is shorter than the normal life span of 11 years. If this is also true of humans, then cloned people would have a reduced life expectancy. The cloning technique could take many years off their life. These fears proved to be unfounded. "Dolly" has grown into a comfortable middle age with signs of normal aging for her age. Thus, now it is not right to attempt for the human cloning to made practice. Still large number of technological advancement has to be made to understand all aspects of the human life and their development.

EXERCISE

1. What do you mean by cloning of animals? What is its importance to the society?
2. Write a note on the therapeutic cloning for the production of similar clones in biotechnology, and mention its demerits.
3. Explain the different types of cell cloning used in the medical field and its applications in genetic diseases.
4. Describe various types of cloning and their applications for understanding the fundamental biological processes.
5. Briefly write about the present situation of prenatal diagnosis in the developing countries, with emphasis on India.
6. What is meant by reproductive cloning? Add a note on ethical issues associated with the reproductive cloning.
7. Define cloning of animals and its impact on the ethical, legal and social issues.
8. Describe the ethical issues associated with the different types of cloning. Add a note on its impact on transgenic organisms.
9. How was dolly produced? Add a note on its importance in the knowledge of science for the development of cloning methodologies.

10. Write about the history of cloning, and mention how human cloning has become possible in the near future.

11. Enumerate the importance of cloning, and its usage in the organ transplantations.

12. Distinguish between the reproductive cloning and therapeutic cloning, and discuss their ethical issues.

13. Discuss how cloning is done in animals. Explain the applications and risk involved in the cloning technology.

14. Explain different types of limitations associated with the cloning technology.

15. What is meant by human cloning? Explain the therapeutic uses and ethical issues associated with human cloning.

16. Give an account of the arguments associated with human cloning and their applications.

17. Critically comment on the ethical, social and legal issues raised by the somatic cell nuclear transfer and cloning.

18. Briefly explain how scientists and researchers are supporting the technological need of the human cloning.

19. Describe the sayings of religious groups against the human and animal cloning in the present circumstances.

20. Elucidate the future of somatic cell nuclear transfer (SCNT) adopted in the animals.

21. Give a note on the therapeutic cloning benefits and limitations, and depict ethical issues raised by this technology.

22. Describe how human cloning is becoming more complicated because of the ethical and social issues raised by the different organizations.

25

Testing of Drugs on Human Volunteers

25.1. INTRODUCTION

In the recent past, scientific investigation has extended and enhanced the quality of life and increased our understanding of human life processes, our relationships with others, and the natural world. It is one of the foundations of our society's material, intellectual, and social progress. For many citizens, scientific discoveries have alleviated the suffering caused by disease or disability. Earlier, animals were considered for the testing of various therapeutic compounds before practically using the drugs. But now, to be at most safer side, humans are considered for this purpose. Nonetheless, the prospect of gaining such valuable scientific knowledge need not and should not be pursued at the expense of human rights or dignity. In the words of philosopher Hans Jonas, "progress is an optional goal, not an unconditional commitment, and...its tempo... compulsive as it may become, has nothing sacred about it."

Unfortunately, history has also demonstrated that researchers sometimes treat participants not as persons, but as mere objects of study. As Jonas observed, "Experimentation was originally sanctioned by natural science. There it is performed on inanimate objects, and this raises no moral questions. But as soon as animate, feeling beings become the subject of experiment....this innocence of the search for knowledge is lost and questions of conscience arise."

In the medical field, the allied research involving human participants has become a vast academic and commercial activity, but the protection of human participants has not kept pace with that growth. On one hand, the system is too narrow to protect all participants, while on the other hand, it is often so unnecessarily bureaucratic that it stifles responsible research. Although some reforms by particular government agencies and professional societies are under way, it will take the efforts of both the executive and legislative branches of government to put in place a streamlined, effective, responsive, and comprehensive system that achieves the protection of all human participants and encourages ethically responsible research.

Since the formation of the "National Commission for the Protection of Human Subjects of Biomedical and Behavioral Research" 1974, and the activities of the US President's Commission for the Study of Ethical Problems in Medicine and

Biomedical and Behavioral Research, in 1980's American leaders have consistently tried to enhance the protections for human research participants. The research community has, in large part, supported two essential protections for human participants—independent review of research to assess risks and potential benefits, and an opportunity for people to voluntarily and knowledgeably decide whether to participate in a particular research protocol.

25.2. ORIGIN OF CLINICAL TRIALS

Ideas for clinical trials usually come from researchers. After the researcher tests new therapies or procedures in the laboratory and in animal studies **(Fig. 25.1)**, the experimental treatments with the most promising laboratory results are moved into clinical trials. During a trial, more and more information is gained about a experimental treatment, its risks, and how well it may or may not work.

Clinical trial (also clinical research) is a research study in human volunteers to answer specific health questions. Carefully conducted clinical trials are the fastest and safest way to find treatments and ways to improve health. Interventional trials determine whether experimental treatments or new ways of using known therapies are safe and effective under controlled environments. Observational trials address health issues in large groups of people or populations in natural settings.

Participants in clinical trials can play a more active role in their own health care, gain access to new research treatments before they are widely available, and help others by contributing to medical research.

All clinical trials have guidelines about who can participate. Using inclusion/exclusion criteria is an important principle of medical research that helps to produce reliable results. The factors, which allow someone to participate in a clinical trial, are

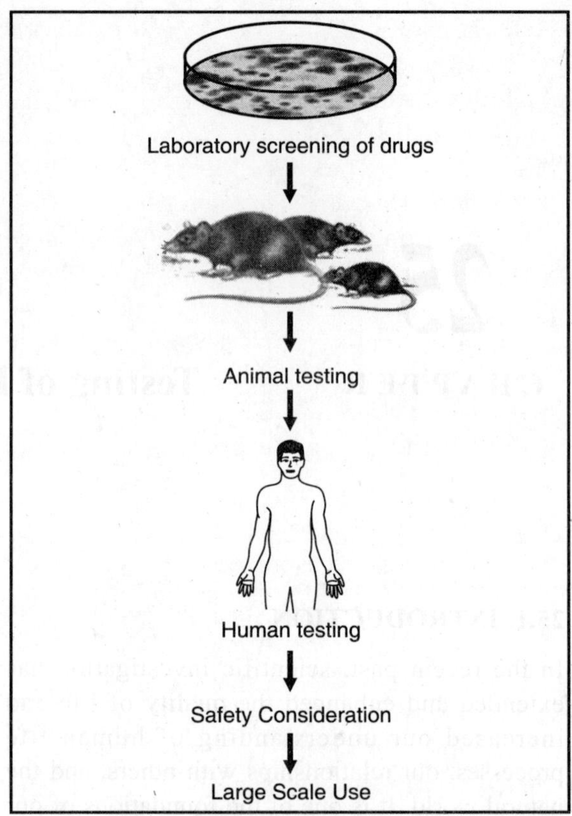

Fig. 25.1. Screening of drugs for use in treatment of diseases

called "inclusion criteria", and those that disallow someone from participating are called "exclusion criteria". These criteria are based on factors such as age, gender, the type and stage of a disease, previous treatment history, and other medical conditions. Before joining a clinical trial, a participant must qualify for the study. Some research studies seek participants with illnesses or conditions to be studied in the clinical trial, while others need healthy participants. It is important to note that inclusion and exclusion criteria are not used to reject people personally. Instead, the criteria are used to identify appropriate participants and keep them safe. The criteria help ensure that researchers will be able to answer the questions they plan to study.

The clinical trial process depends on the kind of trial being conducted. The clinical trial team includes doctors and nurses, as well as social workers and other health care professionals. They check the health of the participant at the beginning of the trial, give specific instructions for participating in the trial, monitor the participant carefully during the trial, and stay in touch after the trial is completed.

For all types of trials, the participant works with a research team. Clinical trial participation is most successful when the protocol is carefully followed, and there is frequent contact with the research staff.

25.3. INFORMED CONSENT

Informed consent is the process of learning the key facts about a clinical trial before deciding whether or not to participate. It is also a continuing process throughout the study to provide information for participants. To help someone decide whether or not to participate, the doctors and nurses involved in the trial explain the details of the study. If the participant's native language is not English, translation assistance can be provided. Then the research team provides an informed consent document that includes details about the study, such as its purpose, duration, required procedures, and key contacts. Risks and potential benefits are explained in the informed consent document. The participant then decides whether or not to sign the document. Informed consent is not a contract, and the participant may withdraw from the trial at any time.

25.4. BENEFITS AND RISKS OF PARTICIPATING IN A CLINICAL TRIAL

All clinical trials, based on the type of products (drugs) used, have many benefits and risks associated with them. Some of these are listed below:

Benefits

- Clinical trials that are well-designed and well-executed are the best approach for eligible participants to play an active role in their own health care.
- Participants can gain access to new research treatments before they are widely available.
- They can obtain expert medical care at leading health care facilities during the trial.
- They can help others by contributing to medical research.

Risks

- There may be unpleasant, serious, or even life-threatening side effects of an experimental treatment.
- The experimental treatment may not be effective for the participant.
- The protocol may require more of their time and attention than would a non-protocol treatment, including trips to the study site, more treatments, hospital stays, or complex dosage requirements.

25.5. SIDE EFFECTS

Side effects are any undesired actions or effects of the experimental drug or treatment. Negative or adverse effects may include headache, nausea, hair loss, skin irritation, or other physical problems. Experimental treatments must be evaluated for both immediate and long-term side effects.

The ethical and legal codes that govern medical practice also apply to clinical trials. In addition, most clinical research is regulated with built in safeguards to protect the participants. The trial follows a carefully controlled protocol—a study plan which details what will the researchers do in the study. As the clinical trial progresses, researchers report the results of the trial at scientific meetings, to medical journals, and to various government agencies. Individual participants' names will remain secret, and will not be mentioned in these reports.

People should know as much as possible about the clinical trial, and feel comfortable asking the members of the health care team questions about it, the care expected while in a trial, and the cost of the trial.

25.6. CONTENTS OF INFORMED CONSENT

It would be helpful for the participants to discuss the following questions with the health care team, answers to some of which are also given in the informed consent document:

- What is the purpose of the study? Who is going to be in the study?
- Why do researchers believe the experimental treatment being tested may be effective? Has it been tested before?
- What kinds of tests and experimental treatments are involved?
- How do the possible risks, side effects, and benefits in the study compare with my current treatment?
- How might this trial affect my daily life?
- How long will the trial last?
- Will hospitalization be required?
- Who will pay for the experimental treatment?
- Will I be reimbursed for other expenses?
- What type of long-term follow up care is part of this study?
- How will I know that the experimental treatment is working? Will results of the trials be provided to me?
- Who will be in charge of my care?
- What kind of preparation should a potential participant make for the meeting with the research coordinator or doctor?

Plan ahead and write down possible questions to ask; ask a friend or relative to come along for support and to hear the responses to the questions; bring a tape recorder to record the discussion to replay later.

Every clinical trial in the country should be approved and monitored by an Institutional Review Board (IRB) to make sure the risks are as low as possible and are worth any potential benefits. An IRB is an independent committee of physicians, statisticians, community advocates, and others, which ensures that a clinical trial is ethical, and the rights of study participants are protected. All institutions that conduct or support biomedical research involving people must, by federal regulation, have an IRB that initially approves and periodically reviews the research.

Participant has to continue work with a primary health care provider while in a trial. Most clinical trials provide short-term treatments related to a designated illness or condition, but do not provide extended or complete primary health care. In addition, by having the health care provider work with the research team, the participant can ensure that other medications or treatments will not conflict with the protocol. A participant can leave the clinical trial at any time. While withdrawing from the trial, the participant should let the research team know about it, and the reasons for leaving the study.

One should be familiar with the following terminolgy such as protocol, placebo, and control group in the clinical trials or testing of a drug in human beings.

Protocol

Protocol is a study plan on which all clinical trials are based. The plan is carefully designed to safeguard the health of the participants as well as answer specific research questions. Protocol describes what types of people may participate in the trial; the schedule of tests, procedures, medications, and dosages; and the length of the study. While in a clinical trial, participants following a protocol are seen regularly by the research staff to monitor their health and to determine the safety and effectiveness of their treatment.

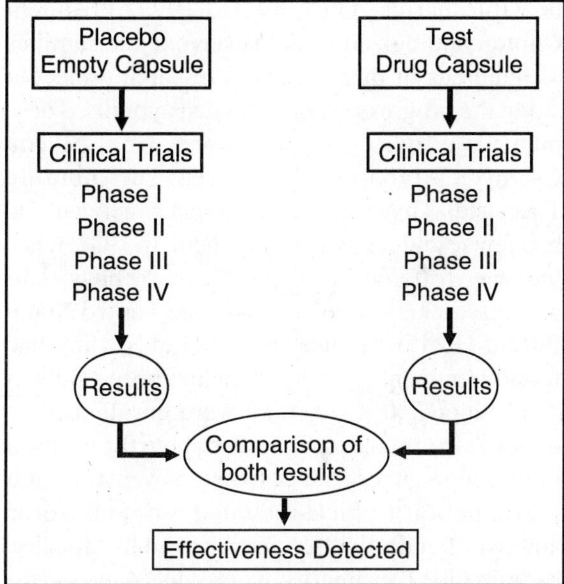

Fig. 25.2. Use of clinical data by using human voluneers for deciding effectiveness of drug

Placebo

Placebo is an inactive pill, liquid, or powder that has no treatment value. In clinical trials, experimental treatments are often compared with placebos to assess the effectiveness of experimental treatment. In some studies, the participants in the control group will receive a placebo instead of an active drug or experimental treatment **(Fig. 25.2)**.

Control or Control Group

Control is the standard by which experimental observations are evaluated. In many clinical trials, one group of patients is given an experimental drug or treatment, while the control group is given either a standard treatment for the illness or a placebo.

25.7. TYPES OF CLINICAL TRIALS

Treatment trials test experimental treatments, new combinations of drugs, or new approaches to surgery or radiation therapy.

Prevention trials look for better ways to prevent disease in people, who have never had the disease, or to prevent a disease from returning. These approaches may include medicines, vitamins, vaccines, minerals, or life-style changes.

Diagnostic trials are conducted to find better tests or procedures for diagnosing a particular disease or condition.

Screening trials test the best way to detect certain diseases or health conditions.

Quality of life trials (or supportive care trials) explore ways to improve comfort and the quality of life for individuals with a chronic illness.

25.8. PHASES OF CLINICAL TRIALS

Clinical trials are conducted in different phases. The trials at each phase have a different purpose and help scientists to answer different questions:

In **Phase I** trials, researchers test a experimental drug or treatment in a small group of people (20-80) for the first time to evaluate its safety, a safe dosage range, and side effects.

In **Phase II** trials, the experimental study drug or treatment is given to a larger group of people (100-300) to see if it is effective, and to further evaluate its safety.

In **Phase III** trials, the experimental study drug or treatment is given to large groups of people (1,000-3,000) to confirm its effectiveness, monitor side effects, compare it to commonly used treatments, and collect information that will allow the experimental drug or treatment to be used safely.

In **Phase IV** trials, post marketing studies delineate additional information including the drug's risks, benefits, and optimal use.

Expanded Access Protocol

Most human use of investigational new drugs takes place in controlled clinical trials conducted to assess safety and efficacy of new drugs. Data from the trials can serve as the basis for the drug marketing application. Sometimes, patients do not qualify for these carefully-controlled trials because of health problems, age, or other factors. For patients, who may benefit from the drug use but do not qualify for the trials, FDA regulations enable manufacturers of investigational new drugs to provide for "expanded access" use of the drug. For example, a treatment investigational new drug (IND) application or treatment protocol is a relatively unrestricted study. The primary intent of a treatment IND/protocol is to provide access to the new drug for people with a life-threatening or serious disease, for which there is no good alternative treatment. A secondary purpose for a treatment IND/protocol is to generate additional information about the drug, especially its safety. Expanded access protocols can be undertaken only if clinical investigators are actively studying the experimental treatment in well-controlled studies, or all studies have been completed. There must be evidence that the drug may be an effective treatment in patients like those to be treated under the protocol. The drug cannot expose patients to unreasonable risks, given the severity of the disease to be treated.

Some investigational drugs are available from pharmaceutical manufacturers through expanded access programs listed in ClinicalTrials.gov. Expanded access protocols are generally managed by the manufacturer, with the investigational treatment administered by researchers or doctors in office-based practice. If you or a loved one are interested in treatment with an investigational drug under an expanded access protocol listed in ClinicalTrials.gov, review the protocol eligibility criteria and location information and inquire at the contact information number.

25.9. ETHICAL ISSUES IN RESEARCH INVOLVING HUMAN PARTICIPANTS

Presently, it is emphasized that research must respect the autonomy of participants, must be fair in both conception and implementation, and must maximize potential benefits while minimizing possible harms. On May 16, 1997, President Clinton apologized to the survivors and families of hundreds of men used in a research project to study the progression of untreated syphilis. These men were mainly sharecroppers from Macon County, Alabama, (the area surrounding Tuskegee). They were poor, African American, and had few resources available to them. In 1932, when they were offered free medical care by physicians and researchers involved with the United States Public Health Service, they believed they had found treatment for what they had been told was "bad blood". Instead, they were enrolled in an observational research study, without their knowledge or consent. In exchange for their participation (which included agreeing to an autopsy after death), the men received free medical examinations (primarily to provide data for the study), transportation, and hot meals on the days of their examinations, and $50 for burial expenses. Even though some rudimentary remedies for syphilis were available in the early years of the study, they were not offered to these men, due to the fear that the study of the natural history of the untreated disease may get jeopardized.

The study (the Tuskegee Study of Untreated Syphilis in the Male Negro) did not end until 1972; forty years after it had begun, and twenty years after penicillin had been identified as an effective treatment for syphilis. The *New York Times* headline exposing the experiment read, "Syphilis victims in US study went untreated for 40 years." The article revealed the details of the study and called it "the longest nontherapeutic experiment on human beings in medical history." According to the report of the Tuskegee Syphilis Study Legacy Committee, "In almost 25 years, since its disclosure, the Study has moved from a singular historical event to a powerful metaphor. It has come to symbolize racism in medicine, ethical misconduct in human research, paternalism by physicians, and government abuse of vulnerable people."

According to President Clinton, "The legacy of the study at Tuskegee has reached far and deep, in ways that hurt our progress and divide our nation. We cannot be one America when a whole segment of our nation has no trust in America. An apology is the first step, and we take it with a commitment to rebuild that broken trust. We can begin by making sure there is never again another episode like this one. We need to do more to ensure that medical research practices are sound and ethical, and that researchers work more closely with communities."

In his remarks, the President promised that several steps would be taken to assure that such research tragedies would not happen again. The plan, laid out by the President, contained five elements and was designed to begin the process of restoring the public trust which had been severely damaged, particularly in the African American community. The first step was to award a planning grant to Tuskegee University, so that the University could establish a Center for Bioethics in Research and Health Care. This center would serve as a museum of the Tuskegee Syphilis Study, and support the university's efforts to strengthen bioethics training. The second element was to increase community involvement in research, so that the lost trust of some segments of society could begin to be restored. A third element of the plan was to strengthen researchers' training in bioethics to help increase their understanding of the core ethical principles of respect for individuals, beneficence, and justice, as set forth in the Belmont Report. Next, a commitment was made to provide postgraduate fellowships to train more bioethicists, especially among African Americans and other minority groups. Finally, by executive order, the President extended the charter of the National Bioethics Advisory Commission till October, 1999.

Following the apology, the Department of Health and Human Services charged several of its agencies with implementing the President's plan. In support of the first two elements, funds have been awarded to Tuskegee University for the establishment of a Center for Bioethics, and several meetings have been held to explore how communities, especially minority communities, can be involved in the planning and implementation of research. In response to the third and fourth elements, interagency program announcements soliciting research and training grant applications have been developed, issued, and are ongoing. They are designed to:

- support research to examine ethical issues surrounding research involving human participants;
- provide training opportunities in the form of mentored scientist awards in research ethics; and
- develop short term courses in research ethics.

Although regulatory safeguards for protecting research participants have improved significantly in recent years, historical cases such as the Tuskegee Syphilis Study stand as examples of the potential for ethical problems in research involving human participants, and of the need for vigilance in protecting the rights and welfare of these individuals. Contemporary safeguards such as requirements that federally funded research protocols be submitted to an institutional review board (IRB) are important, but by themselves are insufficient. Educating researchers and public about research ethics is critical for the protection of research participants.

To facilitate the identification and dissemination of materials to support the education of researchers on the ethics of research involving human participants, an interagency working group (the Bioethics Education Materials and Resources Subcommittee) was established. It included representatives from five department of Health and Human Services (DHHS) agencies. The agencies co-leading this effort—the National Institutes of Health (NIH) and the Centers for Disease Control and Prevention (CDC)—were joined by the Food and Drug Administration (FDA), Health Resources Services Administration (HRSA), and Substance Abuse and Mental Health Services Administration

(SAMHSA). This working group was charged with identifying existing bioethics educational materials and resources, and making these materials more accessible to the entire research and research training community. In addition, the working group was asked to identify gaps in these resources and make recommendations about how to fill them.

In response to this charge, the members of the Bioethics Education Materials and Resources Subcommittee have identified and drawn together an extensive bibliography of books, audiovisuals, and journal article citations. The focus of these materials is the inclusion and protection of human participants from diverse populations in biomedical research, and not on other areas of research ethics and scientific integrity. It provides information about topics critical to understanding ethical issues surrounding research involving human participants, and provides resources to support the further integration of these topics into research and research training settings. These materials have been compiled and are now available in a two-part publication, which will be disseminated to all DHHS-funded research and training sites.

The first part of this publication is a citation bibliography, consisting of more than 4000 journal and book entries on topics relevant to the involvement and protection of human participants in research. The first section provides a comprehensive overview of this topic. Next is a section of historical perspectives on research ethics, which includes references to some of the most striking lapses in research ethics (e.g., the Tuskegee Syphilis Study, the Willowbrook experiment, the human radiation experiments, etc.). This section also includes a number of citations regarding the egregious Nazi "experiments," the subsequent Nuremberg Trial, the resulting Nuremberg Code, and their role in the development of protections for research participants. This is followed by sections on the fundamental research ethics, topics of informed consent, including community consent, privacy and confidentiality, and clinical trials. Next, the bibliography focuses on selected populations that may be recruited for research, including—women; minorities; children; elderly; cognitively impaired (including those who have mental disorders and whose capacity to consent may be diminished or questionable); prisoners; military personnel; newly deceased individuals; and researchers involved in self-experimentation. The final section focuses on special research topics, including: genetics research; research on gametes, embryos, and fetuses; cloning; research involving human biological materials (including fetal tissue research); xenotransplantation; AIDS/HIV research; cancer research; emergency, acute and critical care research; drug and device development; and finally, institutional review boards and ethics committees.

The second part provides bibliographic and other information and resources that can be used to enhance bioethics education in research and other settings. It includes over 500 journal and book citations on bioethics education, case study citations that may be useful for discussion in educational settings, and a list of audiovisual materials. It also includes a list of bioethics journals and bioethics databases, as well as relevant government and non-government documents on laws, regulations and guidelines for research.

The journal article citations selected for this bibliography were published between 1989 and 1998 in English-language journals from around the world, in the fields of biomedicine, ethics and history. Also included, especially in the History section, are some earlier references considered to be seminal articles in the field. Books published and audiovisuals produced in English in the last 10 years were selected from the National Library of Medicine catalog (Locator Plus). Some earlier, particularly significant, books are cited as well. Journal articles and books listed in the Source Material sections may also be listed under another subject. Materials citing both Women and Minorities as keywords are listed under both subject headings. All other citations are listed under only one topic. Thus, if the subject of interest is, for example, HIV research involving women, it would be useful to look both in the AIDS/HIV Research section and the section on Women. Items included in the bibliography were selected solely on the basis of relevance to the subject.

The documents cited in this bibliography provide a rich source of history as well as practical guidance for those who are involved in teaching and mentoring research trainees. They may also be of use to those who are planning or conducting research.

Respecting Research Participants

People should participate in such research only when the study addresses important questions, its risks are justifiable, and an individual's participation is voluntary and informed. Whether testing a new medical treatment, interviewing people about their personal habits, studying how people think and feel, or observing how they live within groups, research seeks to learn something new about the human condition.

Such type of research must respect the autonomy of participants, must be fair in both conception and implementation, and must maximize potential benefits while minimizing possible harms.

Ensuring Independent Review of Risks and Potential Benefits

A central protection for research participants is the guarantee that someone other than the investigator will assess the risks of the proposed research. No one should participate in research unless independent review concludes that the risks are reasonable in relation to the potential benefits. In such condition, the **Institutional Review Board, or IRB**, has been the principal structure responsible for conducting such reviews.

Independent review of research is essential because it improves the likelihood that decisions are made free from inappropriate influences that could distort the central task of evaluating risks and potential benefits. Certainly, reviewers should not have a financial interest in the work, but social factors may be just as crucial. Reviewers may feel constrained because they are examining the work of their colleagues or their supervisors, and they should not participate in protocol review unless they are able to separate these concerns from their task. All reviewers, who themselves are members of the research community, should recognize that their familiarity with research, and (perhaps) their predilection to support research are factors that could distort their judgment. Truly independent and sensitive review requires more involvement of individuals drawn from the ranks of potential research participants or those who can adequately represent the interests of potential research participants.

A critical purpose of independent review is to ensure that risks are reasonable in relation to potential personal and societal benefits. This is a precondition, offering people the opportunity to volunteer, since informed consent alone cannot justify enrollment. When reviewed for risks and potential benefits, research studies must be evaluated in their entirety. Studies often include different components, however, and the risks and potential benefits of each should also be examined separately, lest the possibility of great benefit or monetary enticement in one component cause potential participants or IRBs to minimize or overlook risk in another. No matter what potential benefit is offered to individual participants or society at large, the possibility of benefit from one element of a study should not be used to justify otherwise unacceptable elements.

Obtaining Voluntary Informed Consent

Even when risks are reasonable, no one should participate in research without giving voluntary informed consent (except in the case of an appropriate authorized representative or a waiver). Investigators must make appropriate disclosures and ensure that participants have a good understanding of the information and their choices, not only at the time of enrollment, but throughout the research.

Engaging in this process is one of the best ways researchers can demonstrate their concern and respect for those they aim to enroll in a study. It also serves as the best means for those who do not wish to participate to protect themselves. Recommendations from our previous reports are reinforced in this report, which emphasizes the process of providing information and ensuring comprehension rather than the form of documentation of the decision to give consent. Both the information and the way it is conveyed—while meeting full disclosure requirements—must be tailored to meet the needs of the participants in the particular research context. In addition, documentation requirements must be adapted for varying research settings, and the criteria for deciding when informed consent is not necessary must be clarified so that participants' rights and welfare are not endangered.

The decision to participate in research must not only be informed, it must be voluntary. Even when risks are reasonable and informed consent is obtained, it may nonetheless be wrong to solicit certain people as participants. Those who are not capable of resisting the request to become participants—such as prisoners and other institutionalized or otherwise vulnerable persons—should not be enrolled in studies merely because they are easily accessible or convenient. This historic emphasis on protecting people from being exploited as research participants, however, has failed to anticipate a time when, at least for some areas of medical research, people would be demanding to be included in certain studies because they might provide the only opportunity for receiving medical care for life-threatening diseases.

Making Research Inclusive While Protecting Individuals Categorized as Vulnerable

Vulnerable individuals need additional protection in research. Although certain individuals and populations are more vulnerable as human participants than others, people whose circumstances render them vulnerable should not be arbitrarily excluded from research for this reason alone. This includes those viewed as more open to harm (e.g., children), more subject to coercion (e.g., institutionalized persons), more "complicated" (e.g., women, who are considered more biologically complicated than men), or more inconvenient (e.g., women with small children, who are viewed as less reliable research participants due to conflicting demands on time).

Calling competent people intrinsically "vulnerable" can be both insulting and misleading. It is not their gender or other group designation that exposes them to injury or coercion, but rather their situation that can be exploited by ethically unacceptable research. That is, it is their circumstances that create the vulnerability. At other times it is the intrinsic characteristics of the person, for example, children or those with certain mental or developmental disorders, which make them generally vulnerable in the research setting.

The response, whenever possible, should not be to exclude people from research, but instead to change the research design, so that it does not create situations in which people are unnecessarily harmed. To do otherwise is to risk developing knowledge that helps only a subset of the population. To the extent that the results are not generalizable, the potential societal benefits that justify doing the research are attenuated. Research participants must be treated equally and with respect. Whenever possible, research should be designed to encourage the participation of all groups while protecting their rights and welfare.

Compensating for Harms

Despite all these precautions, however, some research participants might be harmed. Participants who are harmed as a direct result of research should be cared for and compensated. This is justice. The fact that they offered to participate in no way alters the view that mere decency calls for us to take care of these volunteers. Unfortunately, this is a greater challenge than it might appear. For those who endure harm while participating in research, it is often very difficult to separate injuries traceable to the research from those that stem from the underlying disease or social condition being studied. For others, appropriate care and compensation would be far beyond the means of the researchers, their sponsors, and their institutions.

Establishing a Comprehensive, Effective, and Streamlined System

A comprehensive and effective oversight system is essential to protect the rights and welfare of participants, while permitting ethically and scientifically responsible research to proceed without undue delay. A fundamental flaw in the current oversight system is the ethically indefensible difference in the protection afforded to the participants in federally sponsored research, and those in privately sponsored research that falls outside the jurisdiction of the Food and Drug Administration (FDA). As a result, people have been subjected to experimentation without their knowledge or informed consent in fields as diverse as plastic surgery, psychology, and infertility treatment.

Today's research protection system cannot react quickly to new developments. Efforts to develop rules for special situations, such as research on those who can no longer make decisions for themselves, have languished for decades in the face of bureaucratic hurdles, and there is no reason to believe that efforts to oversee other emerging research areas will be any more efficient. In addition, the current system leaves people vulnerable to new, virtually uncontrolled experimentation in emerging fields, such as some aspects of reproductive medicine and genetic research.

Indeed, some areas of research are not only uncontrolled, they are almost invisible. In an information age, poor management of research using medical records, human tissue, or personal interview data could lead to employment and insurance discrimination, social stigmatization, or even criminal prosecution. The privacy and confidentiality concerns raised by this research are real, but the federal response has often been illusory. There is almost no guidance and certainly no coordination on these topics. The time has come to have a single source of guidance for these emerging areas, one that would be better positioned to effect change across all divisions of the government and private sector, as well as to facilitate development of specialized review bodies, as needed.

Education as the Key to Promote Local Responsibility

Currently, the protections depend on a decentralized oversight system involving IRBs, institutions, investigators, sponsors, and participants. We endorse the spirit and intent of this approach, specifically its contention that the ethical obligation to protect participants lies first with researchers, their sponsors, and the IRBs that review their research. Protecting research participants is a duty that researchers, research institutions, and sponsors cannot delegate completely to others or to the government. In addition, merely adhering to a set of rules and regulations does not fulfill this duty. Rather, it is accomplished by acting within a culture of concern and respect for research participants.

It is unrealistic to think that ethical obligations can be fully met without guidance and resources. To help researchers and IRBs fulfill their responsibilities, the federal government should promote the development of education, certification, and accreditation systems that apply to all researchers, all IRB members and staff, and all institutions. These tools should help researchers craft and IRBs review studies that pose few problems and to know when their work requires special oversight. Today, investigators and IRBs are rightly confused over issues as basic as which areas of inquiry should be reviewed, and who constitutes a human participant.

Education is the foundation of the oversight system, and is essential for protecting research participants. In all our reports, we have highlighted the need to educate all those involved in research with human participants, including the public, investigators, IRB members, institutions, and federal agencies.

Clarifying the Scope of Oversight

Many areas of scientific inquiry are "research", and many of these involve human participants, but only some need federal oversight, while others might be better regulated through professional ethics, social custom, or other state and federal law. For example, certain types of surveys and interviews are considered research, but they can be well managed to avoid harms without federal oversight, as the risks are few and participants are well situated to decide for themselves whether to participate. On the other hand, certain studies of medical records, databases, and discarded surgical tissue are often perceived as something other than human research, even when the information retrieved is traceable to an identifiable person. Such research does need oversight to avoid putting people at risk of identity disclosure or discrimination without their knowledge.

Ensuring that the Level of Review Corresponds to the Level of Risk

Even within areas of research that need oversight, many individual studies will involve little or no risk to participants. Although current federal policies allow for some distinction between research involving minimal risk and research involving more than minimal risk, the distinction operates mostly in terms of how the research will be reviewed, i.e., how procedures are to be followed. But the distinction should be based on how the research is pursued, how the participants are treated, and how the work is monitored over time. Overall, the emphasis should be on knowing how to protect participants rather than on knowing how to navigate research regulations. Instead of focusing so much on the period during which a research design is reviewed, oversight should also include an ongoing system of education and certification that helps researchers to anticipate and minimize research risks. Oversight should also make it easier for researchers to collaborate with their colleagues here and abroad without the burden of redundant reviews.

Research review and monitoring should be intensified as the risk and complexity of the research increase, and at all times should emphasize protecting participants rather than following rigid rules. In addition, the review process should facilitate rather than hinder collaborative research among institutions and across national boundaries, provided that participants are protected.

Providing Resources for the Oversight System

Creating a system that protects the rights and welfare of participants and facilitates responsible research demands political and financial support from the federal government as well as the presence of a central coordinating body to provide guidance and oversee education and accreditation efforts. The oversight system should be adequately funded at all levels to ensure that research continues in a manner that demonstrates respect and concern for the interests of research participants.

Level of Review

Although the definition of research involving human participants should be applied to all disciplines, the risks differ both qualitatively and quantitatively across the spectrum of research. Therefore, the oversight system should ensure that all covered research is subject to basic protections, such as a process of informed consent, with the exceptions of the specified conditions for which these protections can be waived, including protection of privacy and confidentiality and minimization of risks.

Since the proposed oversight system may include more research activities, it is more critical than ever that review mechanisms and criteria for various types of research are suited to the nature of the research and the risks likely involved. More specific guidance is needed for the review of different types of research, including appropriate review criteria and IRB composition. For example, procedures other than full board review could be used for minimal risk research, and national level reviews could supplement local IRB review of research involving novel or controversial ethical issues.

Education, Certification, and Accreditation

Protecting the rights and welfare of research participants is the major ethical obligation of all parties involved in the oversight system, and to provide these protections, all parties must be able to demonstrate competence in research ethics— that is, conducting, reviewing, or overseeing research involving human participants in an ethically sound manner. Such competence entails not only being knowledgeable about relevant research ethics issues and federal policies, but also being able to identify, disclose, and manage conflicting interests for institutions, investigators, or IRBs. Finally, the oversight system must include a sufficiently robust monitoring process to provide remedies for lapses by institutions, IRBs, and investigators. Educating all parties in research ethics and human participant protections is effective only when it results in the necessary competence for designing and conducting ethically sound research, including analyzing, interpreting, and disseminating results in an ethically sound manner. Such competence, however, cannot be assumed to follow from the exposure to an educational course or program. As the complexion of research continues to change and as technology advances, new and challenging ethical dilemmas will emerge. And, as more people become involved in research as investigators or in roles that are specifically related to oversight, it becomes increasingly important for all parties to be able to demonstrate competence in the ethics of research involving human participants.

Although accreditation and certification do not always guarantee the desired outcomes, these programs, which generally involve experts and peers developing a set of standards that represents a consensus of best practices, can be helpful in improving performance. Therefore, the choice of standards for these programs and the criteria for evaluating whether an institution has met them are critically important. Accreditation and certification programs should emphasize providing education and assuring that appropriate protections are in place, while avoiding excessively bureaucratic procedures.

Assessing and Monitoring Compliance

Assessing institutional, IRB, and investigator compliance can help to ensure that standards are being followed consistently. Current mechanisms for assessment include assurances of compliance issued by DHHS and several other federal departments, site inspections of IRBs conducted by FDA, other types of site inspections for participant protection, and institutional audits. In addition, some institutions have established ongoing mechanisms for assessing investigator compliance with regulations. However, institutions vary considerably in their efforts and abilities to monitor investigator compliance, from those that have no monitoring programs to those that conduct random audits. Assessing the behavior of investigators is an important part of protecting research participants, and should be taken seriously as a responsibility of each institution. Investigators, IRBs, and institutions should discuss the practical issues involved in monitoring investigators, as they conduct their research studies and provide input into the regulatory process.

Managing Conflicts of Interest

A research setting that involves human participants necessarily creates a conflict of interest for investigators, who seek to develop or revise knowledge by enrolling individuals in research protocols to obtain that knowledge. Overzealous pursuit of scientific results could lead to harm if, for example, investigators design research studies that pose unacceptable risks to participants, enroll participants who should not be enrolled, or continue studies even when results suggest that they should have been modified or halted. Conflicts of interest can also exist for IRB members or the institutions in which the research will be conducted. Thus, it is important to address, prospectively, the potentially harmful effects of the conflicts of interest on participants.

Organizations, particularly academic institutions, should become more actively involved in managing investigators' and IRB members' conflicts of interest, and increase their efforts for self-regulation in this arena. IRB review of research studies is one method for identifying and dealing with conflicts of interest that might face the investigators. By having IRBs review research studies prospectively and following an IRB-approved protocol, investigators and IRBs together can manage conflict between the investigators' desire to advance scientific knowledge and to protect the rights and welfare of research participants. Financial and other obvious conflicts

for IRB members, such as collaboration in a research study, are often less difficult to identify and manage than some of the more subtle and pervasive conflicts. Guidance should be developed to assist IRBs in identifying various types of conflict.

IRB Membership

Appropriate composition of IRB membership ensures that research studies are reviewed with the utmost regard for protecting the rights and welfare of research participants. Current federal regulations require that each IRB has "at least one member who is not otherwise affiliated with the institution, and who is not part of the immediate family of a person who is affiliated with the institution".

The regulations also require that each IRB include "at least one member whose primary concerns are in scientific areas, and at least one member whose primary concerns are in non-scientific areas." Some have raised the concern of whether only 1 unaffiliated member on an IRB is sufficient to avoid institutional influence, especially when IRBs have 15 to 21 members on average. In addition, unaffiliated members do not have to be present for an IRB to conduct review and approve research studies.

Thus, IRBs can approve research with only institutional representation present as long as a nonscientist and a quorum are also present. IRBs should strive to complement their membership by clearly recognizable members who are unaffiliated with the institutions; members who are nonscientists; and members who represent the perspectives of participants. However, it is difficult to require that IRBs increase the presence and participation of more unaffiliated members to reduce the influence of institutional interests on IRB decision making, because finding them can be difficult.

Currently, there are no rules or guidance that describe criteria for meeting the definition of an unaffiliated member that specify how long such members should serve, or that provide guidance regarding under what circumstances they may be removed or what payment should be provided. Institutions should be careful to select unaffiliated members who are truly separated from the institution, except for their role in the IRB. Procedures for the selection and removal of unaffiliated members should be established in a way that empowers the independent voices of those members. In addition, providing reasonable payment to IRB members, who are otherwise unaffiliated with the institution, can be a valuable way to strengthen the role of the members.

Guidance for Assessing Risks and Potential Benefits

In addition to protecting the rights and welfare of research participants, it is equally important to protect them from avoidable harm. Thus, an IRB's assessment of the risks and potential benefits of research is central to determining whether a research study is ethically acceptable. Yet, this assessment can be a difficult one to make, as there are no clear criteria for IRBs to use in judging whether the risks of research are reasonable in terms of what might be gained by the individual or society.

IRBs should be able to identify whether a clear and direct benefit to society or the research participants might result from participating in the study. However, IRBs should be cautious in classifying procedures as offering the prospect of direct benefit. In fact, if it is not clear that a procedure also offers the prospect of direct benefit, IRBs should treat the procedure as one solely designed to answer the research question(s). A major advantage of this approach is that it avoids justifying the risks of procedures that are designed solely to answer the research question(s) based on the likelihood that another procedure in the protocol would provide a benefit.

Minimal Risk

Determining whether a study poses more than minimal risk is a central ethical and procedural function of the IRB. The definition of minimal risk in the regulations provides an ambiguous standard by which risks involved in a research study are compared to those encountered in daily life. However, it is unclear whether this applies to those risks found in the daily lives of healthy individuals, or those individuals who belong to the group targeted by the research. If it refers to the individuals involved in the research, then the same intervention could be classified as minimal risk or greater than minimal risk, depending on the health status of the participants and their particular experiences. According to this understanding, the standard for minimal risk is a relative one.

This report recommends that IRBs use a standard related to the risks of daily life that are familiar to the general population, for determining whether the level of risk is minimal or more than minimal, rather than using a standard that refers to the risks encountered by particular persons or groups. These common risks would include, for example, driving to work, crossing the street, getting a blood test done, or answering questions over the telephone. Thus, research would involve no more than minimal risk, when it is judged that the level of risk is no greater than that encountered in the daily lives of the general population.

Evaluating Vulnerability

All segments of society should have the opportunity to participate in research, if they wish to do so and if they are considered to be appropriate participants for a given protocol. However, some individuals may need additional protections before they can fully participate in the research study, otherwise they might be more susceptible to coercion or exploitation. Individuals might be considered vulnerable within the research context because of intrinsic characteristics (e.g., they are children or have mental illness or retardation), or because of the situation in which they find themselves (e.g., they are impoverished, unemployed, or incarcerated). Recognizing various types of vulnerability and providing adequate safeguards can prove challenging for IRBs.

Appropriate and specific safeguards should be established to protect persons who are categorized as vulnerable. Once safeguards are established, investigators should not exclude persons, categorized as vulnerable, from research involving greater than minimal risk, because this would deprive them of whatever potential direct benefits they might receive from the research, and deprive their communities and society from the benefit of the knowledge such research might generate.

Emphasizing the Informed Consent Process

Rather than focusing on the ethical standard of informed consent and what is entailed in the process of obtaining informed consent, IRBs and investigators have followed the lead of the federal regulations and have tended to focus on the disclosures found in the consent form. However, from an ethics perspective, the informed consent process, not the form of its documentation, is the critical communication link between the prospective participant and the investigator throughout a study, beginning when the investigator initially approaches the participant. Informed consent should be an active process through which both parties share information, and during which the participant at any time can freely decide whether to withdraw from or continue to participate in the research. Time has come to place the emphasis on the process of informed consent to ensure that information is fully disclosed; the competent participants fully understand the research in order to make informed choices; and that decisions to participate or not are always made voluntarily.

Waiver of Informed Consent

Obtaining voluntary informed consent should not be a requirement for every research study. In fact, waiving the informed consent process is justifiable in research studies that include no interaction between investigators and participants, such as in studies using existing identifiable data (e.g., studies of records) and in studies in which risks generally are not physical. In these kinds of research, risks are likely to arise from the acquisition, use, or dissemination of information resulting from the study and are likely to involve threats to privacy and breaches in confidentiality. The criteria for waiving informed consent in such instances should be revised, so that if such studies have protections in place for both privacy and confidentiality, IRBs may waive the requirement for informed consent.

Documentation of Informed Consent

Although the federal regulations may have been intended to reflect a legal standard for documentation of informed consent, NBAC is aware of no law in which a signed, written consent form is required. To fulfill the substantive ethical standard of informed consent, depending on the type of research proposed, it may be more appropriate to use other forms of documentation, such as audiotape, videotape, witnesses, or telephone calls to participants verifying informed consent and participation in the research study.

Protecting Privacy and Confidentiality

Privacy and confidentiality are complex and poorly understood concepts in the context of some research. Privacy refers to the ways and circumstances under which investigators access information from participants. Since privacy concerns vary by type and context of research and the culture and individual circumstances of the participants, investigators should be well informed about the cultural norms of the participants. In addition, investigators should be aware of the various research procedures and methods that can be used to respect privacy. A clear, comprehensive regulatory definition of privacy is needed along with the guidance for protecting privacy in various types of research.

Like privacy concerns, concerns about confidentiality vary by the type and context of the research. No one set of procedures can be developed to protect confidentiality in all research contexts. Thus, IRBs and investigators must tailor confidentiality protections to the specific circumstances and methods used in each specific research study. Further, IRBs and investigators are encouraged to consider the use of strong confidentiality protections, which can also reduce some of the violations associated with privacy. A clear, comprehensive definition of confidentiality is needed, along with the guidance for protecting confidentiality in various types of research.

Monitoring Ongoing Research

Continual review and monitoring of research that is in progress is a critical element of the oversight system. Such review is necessary to ensure that emerging data or evidence have not altered the risks/potential benefits assessment so that risks are no longer reasonable. In addition, mechanisms are needed to monitor adverse events, unanticipated problems, and changes to the protocol. IRBs can do a better job in this area with the appropriate guidance and some restructuring of the review and monitoring process.

Currently, the requirement of continuing review is overly broad. The frequency and need for continuing review vary depending on the nature of research, with some protocols not requiring continuing review. In research involving high or unknown risks, the first few trials of a new intervention may substantially affect what is known about the risks and potential benefits of that intervention. Even if the knowledge does not warrant changes in study design, it may warrant changes in the information presented to prospective and enrolled participants.

On the other hand, the ethics issues and participant protections necessary in minimal risk research are unlikely to be affected by developments from within or outside the research—for example, research involving the use of existing data or research that will no longer involve contact with participants because it is in the data analysis phase. Continuing review of such research should not be required because it is unlikely to provide any additional protection to research participants and merely increases the burden of IRBs. However, because minimal risk research does involve some risk, IRBs may choose to require continuing review. In these cases, other types of monitoring may be more appropriate, such as assessing investigator compliance with the approved protocol or reporting of protocol changes and unanticipated problems. Clarifying the nature of the continuing review requirements would allow IRBs to better focus their efforts on reviewing riskier

Adverse Event Reporting

Assessing adverse events reports can be a major burden for IRBs and investigators because of the high volume and ambiguous nature of such events and the complexity of the pertinent regulatory requirements. Investigators have reported frustration in attempting to understand what constitutes an adverse event, the required reporting times, and to whom adverse events should be reported. The regulations need to be simplified, and one set of regulations should be available for safety monitoring. Regulations and guidance should be written so that investigators and sponsors understand what constitutes an adverse event, what type of event must be reported within what time period, and to whom it should be reported. In addition, regulations and guidance should be clear regarding whose responsibility it is to analyze and evaluate adverse event reports and should describe the required communication and coordination channels for these reports among IRBs and safety monitoring entities, such as Data Safety Monitoring Boards, investigators, sponsors, and federal agencies.

Review of Cooperative or Multi-Site Research Studies

One of the greatest burdens on IRBs and investigators is the review of multi-site studies. Requiring multiple institutions to review the same protocol is unnecessarily taxing and provides no additional protection to participants. In addition, such review poses problems in the initial stages of review as well as in the continual review and monitoring stages, and is especially problematic in the evaluation of adverse events in clinical research.

Innovative and creative alternative mechanisms and processes for reviewing protocols in multi-site research are needed. To allow for such projects and to support a change in the current system toward a more flexible review system, federal policy should be clear about the functions that must be performed, but be less restrictive about who performs each function.

Compensation for Research-Related Injuries

Participants who volunteer to be in a research study and are harmed as a direct result of that study should be cared for and compensated. However, no adequate database exists that describes the number of injuries or illnesses that are suffered by research participants, the proportion of these illnesses or injuries that are caused by the research, and the medical treatment and rehabilitation expenses that are subsequently borne by the participants. It may be argued that regardless of the magnitude of the problem, the costs of research injuries should never be borne by participants. If individuals are injured by research participation, those who benefit from the research (e.g., institutions and sponsors) bear some obligation to compensate those who risked and suffered injury on their behalf. At this time, injured research participants alone bear both the cost of lost health and the expense of medical care, unless they have adequate health insurance or successfully pursue legal action to gain compensation from the specific individuals or organizations that were involved in conducting the research.

A comprehensive system of oversight of human research should include a mechanism to compensate participants for medical and rehabilitative costs resulting from research-related injuries.

Need for Resources

Adopting the recommendations made in this report will generate additional costs for institutions, sponsors, and the federal government (through the establishment of a new federal oversight office). Sponsors of research, whether public or private, should work together with institutions carrying out the research to make the necessary funds available.

Human drug testing raises many questions about ethical issues that cannot be answered because of insufficient or nonexistent empirical evidence. Current thinking about ethical issues in research—such as analysis of risks and potential benefits, informed consent, privacy and confidentiality, and vulnerability—would greatly benefit from additional research. Deserving of more study, for example, are questions regarding the development of effective approaches for assessing cognitive capacity, for evaluating what participants want to know about research, and for determining how to ascertain best practices for seeking informed consent. Clearer and more effective guidance could be developed from a stronger knowledge base. In general, understanding the ethical conduct of research would be advanced by increased interdisciplinary discussion that would include biomedical and social scientists, lawyers, and historians.

SUMMARY

Scientific investigation has extended and enhanced the quality of life and increased our understanding of human life processes, our relationships with others, and the natural world. It is one of the foundations of our society's material, intellectual, and social progress. In medical filed and allied research involving human participants it has become a academic and commercial activity, but protection of human participants has not kept pace with that growth. On one hand, the system is too narrow in scope to protect all participants, while on the other hand, it is often so unnecessarily bureaucratic that it stifles responsible research. Although some reforms by particular government agencies and professional societies are under way, it will take the efforts of both the executive and legislative branches of government to put in place a streamlined, effective, responsive, and comprehensive system that achieves the protection of all human participants and encourages ethically responsible research.

A clinical trial (also clinical research) is a research study in human volunteers to answer specific health questions. Carefully conducted clinical trials are the fastest and safest way to find treatments that work in people and ways to improve health. Interventional trials determine whether experimental treatments or new ways of using known therapies are safe and effective under controlled environments. Observational trials address health issues in large groups of people or populations in natural settings. The clinical trials are having both merits and demerits and informed consent poses major problems. This has to be worked in detail for proper adoptions of clinical trials.

Most human use of investigational new drugs takes place in controlled clinical trials conducted to assess safety and efficacy of new drugs. Data from the trials can serve as the basis for the drug marketing application. Sometimes, patients do not qualify for these carefully-controlled trials because of other health problems, age, or other factors. For patients who may benefit from the drug use but don't qualify for the trials, FDA regulations enable manufacturers of investigational new drugs to provide for "expanded access" use of the drug. For example, a treatment IND (Investigational New Drug application) or treatment protocol is a relatively unrestricted study. The primary intent of a treatment IND/protocol is to provide for access to the new drug for people with a life-threatening or serious disease for which there is no good alternative treatment. A secondary purpose for a treatment IND/protocol is to generate additional information about the drug, especially its safety. Expanded access protocols can be undertaken only if clinical investigators are actively studying the experimental treatment in well-controlled studies, or all studies have been completed. There must be evidence that the drug may be an effective treatment in patients like those to be treated under the protocol. The drug cannot expose patients to unreasonable risks given the severity of the disease to be treated.

The ethical issues raised by the clinical trials are to discussed at international meetings and solutions should be found by medical persons and bioethicists.

EXERCISE

1. Name different types of clinical trials and their prerequisites.
2. What do you mean by clinical trials in biotechnology? Add a note on its importance.
3. Discuss the contents of the informed consent needed for the inclusion in the consent list that is given to participants.
4. Distinguish placebo and test materials.
5. What do you mean by social, ethical and legal issues involved in the clinical trials of human volunteers in the third world countries?
6. Elucidate different social issues that are affected by the clinical trials made in the biotechnological fields in the poorer nations.

7. Mention the risks and benefits associated with the clinical trials of human volunteers.

8. What do you mean by 'placebo'? What is its impact on the clinical trials of human volunteers?

9. Briefly explain different phases of clinical trials of human volunteers in biotechnology and their importance.

10. How are potential risks associated with the clinical trials of human volunteers in biomedical fields identified? What is their solution?

11. List out the ethical issues that are to be addressed in clinical trials of human volunteers in biotechnology and their impact on the poorer people.

12. Critically comment whether the comprehensive and effective oversight system is essential to uniformly protect the rights and welfare of participants.

13. Mention the importance of education, certification, and accreditation for the clinical trials of human volunteers in biomedical fields.

14. Differentiate between the diagnostic trials and screening trials that are used in clinical trials of human volunteers in biotechnology.

15. Illustrate how Institutional Review Board, or IRB is must for the clinical trials of human volunteers in biotechnology.

16. Give an account of the prevention trials and diagnostic trials, with the importance of each one and their applications.

17. What do you mean by screening trials that are used in clinical trials, and its use in clinical trials of the different genetic diseases of human beings?

18. How quality of life trials in clinical trials performed in human beings? Add a note on its use for curing the chronic human diseases.

19. Explain waiver of informed consent that are generally followed for the clinical trials, and mention its merits in the research and testing.

20. What is the criteria necessary for conducting clinical trials in humans at the present applied situation of biotechnology?

26

CHAPTER — Organ Transplantation and Ethical Issues

26.1. INTRODUCTION

Transplantation (meaning trans-transfer; plantation-keep in a particular place) is not a new procedure. Transplantation is the procedure by which cells, tissues, or organs are transferred from a donor to a recipient. It is an alternative treatment for a variety of otherwise fatal conditions. The replacement of diseased organs by the transplant of healthy tissue has been an objective in medicine, but has been frustrated to no mean degree by the uncooperative attempts by the body to reject grafts from other individuals.

Immunological rejection reaction, which occurs after transplantation of an incompatible graft, represents at present the major barrier for further development of clinical transplantations. Therefore, a study of the cellular and molecular mechanisms of transplantation reaction and search for new approaches to the induction of transplantation tolerance are the main tasks of experimental and clinical transplantation immunology. The research of our group is focused on determination of the regulatory role of individual cell subpopulations, interleukins and other cytokines in modulation of transplantation immunity. For these purposes, a large scale of *in vitro* techniques of cellular and molecular immunology have been established.

26.2. DIFFERENT TYPES OF TRANSPLANTATION

Several different forms of transplantation are currently being performed and researched. These include autografts, isografts, allografts and xenografts **(Fig. 26.1)**.

- **Autograft**: tissue grafted back onto the original donor.
- **Isograft**: graft between syngeneic individuals (i.e. of identical genetic constitution) such as identical twins or mice of the same pure line strain.
- **Allograft** (old term, holograft): graft between allogeneic individuals (i.e., members of the same species but different genetic constitution), e.g., man to man, and one mouse strain to another.
- **Xenograft (heterograft)** graft between xenogenetic individuals (i.e., of different species), e.g., pig to man.

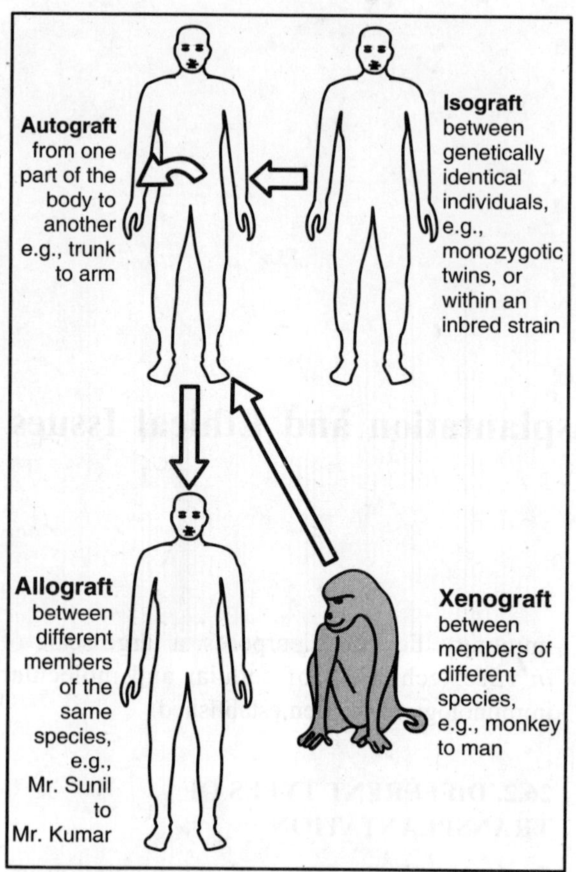

Fig. 26.1. Types of transplantations. The genetic relationship between the donor and recipient determines whether or not the rejection will occur. Autografts or isografts are usually accepted, while allografts and xenografts are not

Researchers have been working in the fields of tissue engineering, cloning, and stem cell research in attempt to maximize both the efficacy of the grafts and the benefits to the recipients of these transplants. Transplantation takes place when an organ from one person is surgically removed, and placed into the body of another person. It is used as a treatment in situations where a person's organ has failed because of illness or injury. Replacing the organ may be the only life-prolonging treatment choice for the patient or the best among several options. During the past few decades, the field of transplantation has undergone a dramatic expansion. Sophisticated surgical procedures have enabled the transplantation of almost any type of vital organ or tissue as an effective therapeutic modality. Since most transplants are done from genetically different donors, the success rate is also very low.

Demand for organ transplants has increased dramatically since the establishment of transplantation therapy. In fact, there is currently a substantial mismatch between allograft demand and sourcing. This has spurred the creation of an organ waiting list. In 2001, more than 23,000 patients received an organ transplant, 17,000 of which came from deceased donors and 6,500 of which came from living donors. Even though these numbers appear high, it is important to note that only less than 25% of individuals are able to receive vital organs from the waiting list. Despite the number of transplants performed in 2001, 6000 patients, who were on the waiting list, were reported to have died, still in need of a transplant.

26.3. HISTORY OF ORGAN TRANSPLANTATION IN HUMANS

Medical developments in immunosuppression, tissue and organ preservation, tissue typing and matching, and surgical technique have contributed to success in organ transplantation. Improvements in areas of general patient care, such as antibiotics, anaesthesia, blood banks, artificial kidney and heart-lung machines have also been instrumental in these advances. It is currently possible to graft all of the vital vascular organs and bone marrow cells in humans.

Medical technology has expanded the possibilities for organ transplantation. However, the supply of organs continues to exceed the demand. 76,124 individuals are currently on the "national patient waiting list for organ transplant". Every 16 minutes, a new name is added to the waiting list, and 15 people die each day, waiting for an organ.

The basic techniques necessary for the transplantation were understood and developed since the turn of the earlier century. Organ transplantation experiments using animals and humans have been conducted since the 18th century. In the early 1900s, a Viennese surgeon attempted the first organ removal and reinsertion in a procedure involving an autologous animal kidney graft. His success led him to investigate further the possibility of transplantation. He failed miserably because he crossed the boundary of the species unknowingly causing immediate and fatal graft rejection. This initial failure had the effect of stunting transplantation progress during the 1920s and 1930s. However, corneas were transplanted in 1905, though the procedure did not become routine until the 1940s. Skin was transplanted first in the 1920s. In the 1950s, a kidney was transplanted from one twin to another. In the 1960s, liver and heart were transplanted. Since then, lungs, hands, testicles, ovaries, and other body components have been transfered from one person to another. The age of modern transplantation was born in 1943, when Thomas Gibson and Sir Peter Brian Medawar **(Fig. 26.2)** demonstrated that allograft rejection is an immunological response. But in the 1940s, P.B. Medawar recognized that autologous skin grafts were accepted, whereas those given from one relative to another were rejected. This discovery led him to suggest that there was an immunological response. This barrier needed to be overcome for a successful allogeneic graft transplantation. He hypothesized that the human immune system, which is designed to attack foreign bodies, is able to determine that the grafts were non-self, and therefore enact effector mechanisms to ultimately reject the graft. In 1945, R.D. Owen discovered that fraternal twins could accept one another's skin and organs, if fusion of their placentas occurred in utero. The first successful solid organ transplant was accomplished in 1954. A kidney was transplanted successfully by Dr. Joseph E. Murray, working at Brigham and Women's Hospital, Boston,

Fig. 26.2. Professor P.B. Medawar

Massachusetts. These discoveries induced a new interest in this field.

The field of transplantation got a major boost in 1973, when researchers at a drug company in Switzerland extracted the first immunosuppressive drug, cyclosporin, from the fungus *Tolypocladium inflatum*, which they had isolated from soil on a hillside in Norway. In 1984, Japanese researchers extracted an even better drug, FK506, from another soil organism, *Streptomyces tsukubaensis*. With both cyclosporin and FK506, five-year gaps separated the discoveries of the drugs from their first uses in patients. These drugs revolutionized the field of transplantation. They greatly reduced the likelihood that transplanted organs would be rejected by the recipients, and also made it possible to transplant organs from donors to recipients whose tissue types were not perfectly matched.

Economics of Transplantation

The cost of the transplantation procedure and postsurgical care is often a source of great anxiety for the recipient of the graft. Several components relate to the cost of the transplant, including hospital costs, physicians' charges, medical supplies, diagnostic tests, postsurgical costs for follow-ups, and medications. The immuno-suppressive medications and follow-up care are vital aspects of the transplantation procedure, but the daily drug therapies can add up to thousands of dollars per year. The recipients of transplant and their families often struggle to find the needed coverage. So, several supporting transplant organizations have been developed to ensure that recipients continue to receive these vital medications.

Significance of Organ Transplantation

Human-to-human transplantation has become a successful way of treating various human diseases, such as heart disease or kidney failure. However, human tissue and organ transplantation usually depends on donations from people who have died; most often as the result of brain damage through stroke or accident. Over the past 20 years, transplants have become more frequent and successful. The number and scope of transplant procedures have also increased to include a broader range of tissue, cellular and organ transplants, such as transplants of insulin-producing pancreatic islet cells for the treatment of diabetes.

Transplantation has become the treatment of choice for many patients suffering from organ failure or complications arising from diseases of specific organs. This treatment option has been making rapid progress. Several factors have contributed to these improvements. These include increased efficacy of drug therapies used to treat and prevent organ rejection, technical advances in surgery, diagnostic test methods for monitoring the patients, improved histocompatibility testing, and enhancement in organ procurement procedures. Earlier and more accurate detection of rejection, as well as a more comprehensive understanding of the immune system have also played roles in improving patient and graft survival.

A good indication of the transplantation success over the past several decades is the improvement, over time, in one-year patient survival rates **(Table 26.1)**. The direct result of improved techniques and higher success rates is an increase in the number of transplants being performed in the United States each year. A major problem, as stated before, is the limited availability of organs.

Table 26.1: *Percentage of patients alive one year after transplant*

Percentage	Patients alive
93.8%	Kidney transplants (the most common transplant procedure)
90.5%	Pancreas transplant
82.4%	Heart transplant
76.7%	Liver transplant
68.4%	Lung transplant
57%	Heart-lung transplant

Table 26.2: *Growth in the number of waiting listed patients in 2000-2001*

Organ	Year		% increase
	2000	2001	
Total	72,393	78,265	8.1%
Kidney	44,966	48,405	7.6%
PTA	318	395	24.2%
PAK	457	675	47.7%
Kidney-pancreas	2,380	2.399	0.8%
Liver	16,253	18,173	11.8%
Intestine	150	177	18.0%
Heart	4088	4,076	−0.3%
Lung	3,580	3,758	4.9%
Heart-lung	201	209	4.0%

Statistics on Organ Transplantation

The department of health and human services, in a report entitled "Transplant center-specific graft and patient survival rates," found these statistics **(Table 26.2)**.

26.4. ETHICAL ISSUES

The ethical issues associated with transplantation are as follows:
- Is the human body a commodity?
- How should the decisions be made about who would receive scarce organs?
- Who should pay for transplants?
- Should one person receive several organs or should several people receive one?
- Should one person have a second transplant when the first one fails, or should a different person be given a first chance at a new organ?
- Should organs be given to people who have abused their bodies (smoking and drinking, etc.,), or only to people whose organs are damaged by disease?
- Should hands or other appendages, which are not essential for to life, be transplanted?
- Who can "donate" the organs of people who cannot give informed consent to the process?
- Should money now spent on transplantation be put to other uses?
- Is it possible to prevent the coercion of some donors?
- When should courts be involved in these questions?
- Should state and federal lawmakers be involved in transplantation?

The questions go on and on; the answers are never simple. The questions raised by the cases discussed here and in other transplant stories have medical, ethical, international, social, legal, cultural and technological overtones. The supply of organs is well below the demand. In a society where resources are not limitless, questions of how these resources should be allocated are fundamental.

Language use in the field of transplantation is somewhat unusual: organs are *harvested* and then *transplanted*; organs are *donated*; *donors* give a *gift* (often the *gift of life*); the *recipient* receives a *gift*; and so on. The gardening and gift-giving imagery add a strong nonmedical mystique to the whole transaction.

26.5. TRANSPLANTATION OF ORGANS

Solid organs that are transplanted include heart, lung, liver, kidney, pancreas, and intestine. In some cases, two organs are transplanted at once. Examples of this are heart-lung and kidney-pancreas. Tissues that can be transplanted include:
- cornea;
- bone;
- cartilage;
- skin;
- heart valves; and
- saphenous vein.

Time is very important in organ transplantation. Most organs need to be transplanted within a few hours after their removal from the donor. Some organs must be transplanted more quickly than others. **Table 26.3** shows the allowable time between retrieving the organ from the donor and transplanting the organ into a recipient. **Table 26.3** compares this allowable time (cold ischemia time) for the majority of transplanted organs.

Table 26.3: *Cold Ischemia time for solid organ transplantation*

Organ	Allowable time from donor to recipient (hours)
Heart	4 to 5
Kidney	15 to 18
Liver	12 to 18
Lung	5 to 6
Pancreas	12 to 15

26.6. SOURCE OF ORGANS

Cadaveric Versus Living Tissue Grafts

There are two current standard sources for organ transplants—cadaveric and living tissue graft donations. Most transplanted organs are taken from the bodies of people who have died (cadaveric or deceased donor transplantation). However, organs such as kidney and segments of liver can be taken from living donors, who may be family members or friends of the person who needs the transplant. From the point of view of the recipient, there are not many significant differences between living and cadaveric transplantations. The individual must still be evaluated in the same manner, and cross-matching with the donor is performed to ensure the immunological feasibility of the procedure. Most intraoperative practices are identical, except for the need to obtain greater exposure of the recipient vessels for patients receiving cadaveric grafts. Postoperative care will include immunosuppressant therapy and general monitoring of patient health (such as evaluation of fluid and electrolyte balance).

Organs are removed by surgeons in a sterile operating room just like any other operation. The identification of deceased donors and coordination of organ retrieval by the surgical teams is the responsibility of 59 organ procurement organizations (OPOs) across the US Each OPO has a geographic region within which it serves hospitals where potential donors have been identified.

Everybody who is a candidate for cadaveric organ transplant in the US is placed on a national waiting list. There are specific government regulations that guide the allocation process. Under a government contract, the nation's organ procurement and transplantation network formulates allocation policy based on input from transplant professionals, patients, donor groups, and the public.

Ultimately, these policies determine, for any given donor, which waiting patient has the highest priority. The rules are fairly complicated and vary from organ to organ.

Two major issues are considered: allocation rules and geographic distribution. Generally, when an organ becomes available, it is allocated to the person on the local OPO's waitlist who has the most urgent need (liver and heart), combined with other medical considerations and ranking of patients.

Organs are usually offered first to patients waiting at transplant centers within the OPO's service area, then to patients in that OPO's region (there are 11 defined geographic regions in the US), and then to patients anywhere else in the US.

Canada's donor rate of 14.4 per million populations is among the lowest of all developed countries. By of comparison, Spain is reported to have the highest human donation rate worldwide, with 33.6 per million in 1999. Canada's low organ donation rate is partly due to its low death rate from car accidents and gun shots—a major source of donated organs in other countries. However, Canada's organ donation rate could be increased—some say by as much as 80%—by ongoing efforts to promote donation through public education campaigns, and by more efficiently requesting, collecting and distributing the organs.

Another way to increase the number of donations is by adopting a policy of "presumed consent". This means that it is assumed that people agree to donate their organs upon death, unless they have indicated otherwise. Several countries have adopted a "presumed consent" policy. Governments adopting presumed consent would have to launch massive campaigns to make people aware of the policy. However, some people who do not want to be a donor would not get around to opting out, which could raise ethical concerns as their organs could be removed after death.

The ever-increasing demand of the organs, and the shortage of organs made the situations very critical in the medical field for patients. Other approaches to overcome this short-fall have been considered. The current demand for healthy cells, tissue and organs for medical transplants exceeds the available supply. Novartis, a biotechnology company, estimates that 180,000 people around the world are waiting for organ transplants, but only one-third will receive one. Besides patients suffering from organ failure, xenotransplantation could potentially treat those with hemophilia, AIDS, diabetes, Alzheimer's disease, Parkinson's disease, and Huntington's disease.

Disease Prevention

One approach is to prevent disease in the first place through exercise, healthy eating and lower alcohol and tobacco consumption. However, preventive therapies are unlikely to have an impact on organ failure rates for at least a decade.

Many experts believe that the demand for organs will never be met by prevention efforts combined with human donations.

Living Donors

The donation of kidneys and livers by living donors is now accepted as generally safe and effective. Almost one-third of the kidneys transplanted in the United States last year came from live donors. The liver, unlike other organs, can regenerate. Therefore, someone can donate a portion of his liver and his liver can still function. Living donations of lungs, pancreas and intestinal segments are generally considered risky for the donors.

Organ Markets

A controversial method for increasing donor rates involves permitting a legitimate market for human organs. A worldwide market for human organs currently exists, and an estimated 200 to 300 Americans a year buy organs from poor people in a year in developing countries. However, many people believe that an organ market is unethical, and would mark the end of the current organ donation system in which organs are donated as an act of generosity for strangers in need.

Artificial Organs

There have been significant improvements in mechanical and artificial hearts, but most are designed only to be temporary transplants, keeping patients alive until a human organ becomes available or until their own organs recover. More work is still needed to develop artificial devices that can last as long as donated human organs.

Other Novel Approaches

Human stem cell therapy is a promising treatment for Parkinson's and Huntington's disease. Human embryonic stem cells can develop into any type of tissue, including internal organs. For example, some researchers believe that stem cells could be cultured to grow unlimited supplies of human tissue for transplant. However, these approaches are in their initial stages, and are still theoretical.

26.7. KIDNEY TRANSPLANTATION

Kidney **(Fig. 26.3)** transplantation is the transfer of a healthy kidney from one person (the donor) into the body of another person, who has little or no kidney activity (the recipient). A person needs only one kidney to survive (some people are born with only one working kidney). So, the donor can be living, or recently deceased.

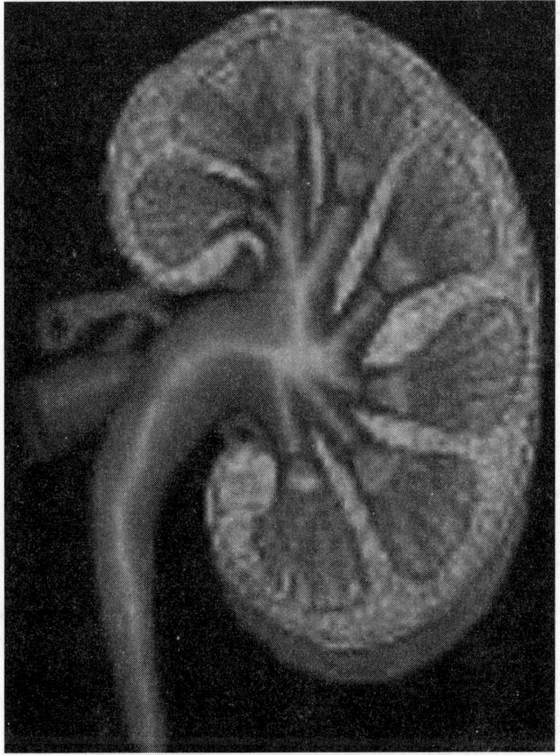

Fig. 26.3. The section of kidney showing anatomy

There are two main problems with kidney transplantation. First, there is a shortage of organs from the people who have died, so that not everyone with kidney failure receives a transplant, even if they are medically suitable. This shortage can be partially made up by using living donors, usually blood relatives or partners of the person with kidney failure. The second problem with kidney transplantation is that the recipient's body recognizes the transplant as if it were an invader, and tries to destroy it. This is called rejection. The recipient of the kidney transplant has to take a powerful cocktail of drugs to prevent rejection and treat the complications of the anti-rejection drugs.

There is some evidence that people with successful kidney transplants live longer than those on dialysis treatment. The main advantage of kidney transplant is that the quality of life is much better than on dialysis.

Dialysis is uncomfortable, has some risks, and disrupts the life style of the patient. Although a major operation has its own risks, and once success is achieved, the person is much freer to get on with a normal life than if staying on dialysis.

Causes of kidney failure requiring a kidney transplant include:

- Late effects of the kidney inflammation glomerulonephritis (see article on Glomerulonephritis).
- The kidney infection "pyelonephritis" (see article on Pyelonephritis).
- Polycystic kidney disease: inherited condition in which both kidneys are several times the normal size because of the masses of cysts, which gradually grow.
- Nephrosclerosis: hardening of the kidneys due to disease of the arteries.
- The failure of normal kidney development in an unborn baby, while developing in the womb (congenital).
- Kidney damage caused by diabetes.
- Systemic lupus erythematosus: disease of the immune system, where the body attacks the kidney as though it were a foreign tissue (see article on Lupus erythematosus).

In someone fit enough to have a major surgery, kidney transplant can be done when a suitable donor kidney becomes available. If someone has a living donor, it may be possible to perform the transplant just before the kidneys fail completely, so that dialysis is not necessary.

Necessity

Transplantation allows people to live without the need for dialysis, even after their own kidneys have stopped working, and with fewer restrictions on what they eat and drink. A transplant also works more effectively than dialysis in replacing the natural activity of a kidney.

Procedure of Kidney Transplantation

The transplanted kidney (known as kidney 'graft') requires a blood supply, and a way of draining urine into the bladder. It is usually placed in the lower abdomen via a cut, which looks similar to the scar for an appendix operation, except that it is longer. The artery and vein on the transplanted kidney are joined onto the main artery and vein on the leg. The ureter (which carries urine from the kidney to the bladder) is connected to the person's own bladder.

Usually, the person's own kidneys are left in place. When the donor kidney is from a living person, most hospitals have two surgical teams working simultaneously to minimize the time to take the graft out of the body.

One year after kidney transplantation, over 9 out of 10 transplants are still working, and over 9 out of 10 patients are still alive. Five years after a transplant, about 6 out of 10 transplants are working and 8 out of 10 patients are alive. Many kidney transplants, if there are no rejection episodes and the recipients takes their anti-rejection drugs regularly, work for 20 years or more. Recipients of kidney transplants can develop other problems. Diseases of the circulation are common, because of the effects of kidney failure and the anti-rejection drugs.

It is important that transplant patients stop smoking, keep to a healthy diet and have their blood pressure and cholesterol levels monitored. The anti-rejection drugs increase the risk of getting certain types of cancer, especially in the skin (avoid direct sunlight) and neck of the womb (cervix), so women should have regular cervical smears.

The results of kidney transplantation have improved immensely over the last 20 years, and will probably continue to improve with better understanding of anti-rejection drugs, and the introduction of newer anti-rejection drugs.

Research is very active into new ways of performing transplantation, by using animal organs, or growing new kidneys from 'stem cells'. It does not seem likely that either of these techniques will be widely used before 2008 at the earliest, and there is no guarantee that they will ever be successful.

Recovery

Kidney transplantation is a major operation, but the recipient is out of bed in the first couple of days, and usually home by 10 days. The first three months require clinic visits several times a week, and often readmission to hospital for investigations or to treat complications.

By three months, most recipients get back to work and are able to lead fairly normal lives. However, regular blood tests are needed to check that the kidney is working well, and that the anti-rejection therapy is not affecting the blood cells and immune system too much.

If someone has donated a kidney, he is usually out of hospital in a week, and back to full activities in six weeks, though it is often best to plan for three months off work.

26.8. HEART TRANSPLANTATION

Cardiac transplantation is a widely accepted therapy for the treatment of end-stage congestive heart failure. Most candidates for cardiac

transplantation have not been helped by conventional medical therapy, and are excluded from other surgical options because of the poor condition of the heart. About 45% of the candidates have ischemic cardiomyopathy; however, this percentage is rising because of the increase in coronary artery disease in younger age groups. Of the candidates, 54% have some form of dilated cardiomyopathy, which often has an unclear origin. The remaining 1% of candidates fall into the category of other diseases, including congenital heart disease, what is not amenable to surgical correction.

Candidacy determination and evaluation are key components of the process, as is post-operative follow-up care and immunosuppression management. Proper execution of these steps can culminate in an extremely satisfying outcome for both the physician and the patient **(Fig. 26.4)**.

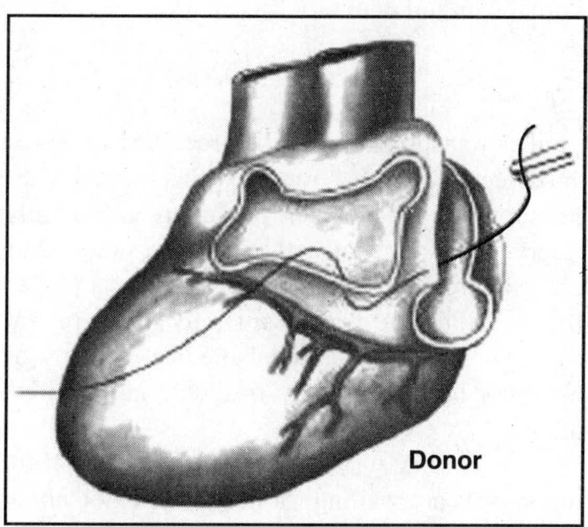

Fig. 26.4. Heart transplantation.
Suturing of the donor heart. Note that the left atrial anastomosis is performed first

History of the Procedure

In 1967, Christian Barnard performed the first successful heart transplant in a human in South Africa. The origins of the procedure date to 1905, when Alexis Carrel transplanted a puppy's heart into the neck of a dog. Due to the lack of immunosuppression, the experiment was unsuccessful; however, the work spurred numerous investigations that culminated in the success of the procedure. Early investigators included Frank C. Mann of the Mayo Clinic, V.P. Demikov of the Soviet Union, and Marcus Wong. The early efforts in transplantation were thwarted by the infancy of cardiopulmonary bypass and a lack of understanding of the immune system. As knowledge in these areas advanced, so did the field of cardiac transplantation. The clinical use of cyclosporine as an immunosuppressant revolutionized the field of cardiac transplantation in 1983. Recipient survival rates improved, thus producing an explosive increase in the number of transplant centers offering cardiac transplantation. The remaining limiting factor was the number of available organ donors.

Problem

Cardiac transplantation is the procedure in which the defective heart is replaced with a healthy heart from a suitable donor. The procedure is generally reserved for patients with end-stage congestive heart failure with a prognosis of less than a year to live without the transplant and who are not candidates for conventional medical therapy or have not been helped by conventional medical therapy.

As of this writing, approximately 4000 individuals are waiting for hearts. In 1999, about 2000 heart transplants were performed nationwide. Clearly, one issue surrounding the transplantation process is the availability of organs for this life-saving procedure.

Frequency

The annual frequency of the procedure is about 1% of the general population with heart failure, both candidates and noncandidates.

Etiology

The disease processes that require cardiac transplantation can be divided into the following categories:
- Idiopathic cardiomyopathy—54%
- Ischemic cardiomyopathy—45%
- Congenital heart disease and other diseases—1%

Pathophysiology

The pathophysiology of cardiomyopathy that may require cardiac replacement is dependent upon the primary disease process. Chronic ischemic conditions precipitate myocardial cell damage, with progressive enlargement of the myocyte followed by cell death and scarring. The condition can be treated with angioplasty or bypass procedures; however, the small-vessel disease is progressive in nature, resulting in progressive loss of myocardial tissue. This eventually results in significant functional loss and progressive cardiac dilatation. The pathologic process involved in the functional deterioration of the dilated cardiomyopathy is still unclear. Mechanical dilatation and disruption of energy stores appear to play a role. The pathophysiology of the transplanted heart is unique. The denervation of the organ makes it dependent on its intrinsic rate. As a result of the lack of neuronal input, left ventricular hypertrophy results. The right-sided function is directly dependent upon the ischemic time incurred prior to reimplantation and the adequacy of preservation. The right ventricle is easily damaged and may initially function as a passive conduit until recovery occurs.

The rejection process that can occur in the allograft is primarily of 2 forms—cellular and humoral. Cellular rejection is the classic form of rejection and is characterized by perivascular infiltration of lymphocytes with subsequent myocyte damage and necrosis if left untreated. Humoral rejection is much more difficult to characterize and diagnose. Humoral rejection is thought to be a generalized antibody response initiated by several unknown factors. The antibody deposition into the myocardium results in global cardiac dysfunction. This diagnosis is generally made on the basis of clinical suspicion and exclusion, because endomyocardial biopsy is of little value.

Coronary artery disease is a late pathologic process common to all cardiac allografts. The pathology is characterized by myointimal hyperplasia of the small, and medium-sized vessels. The lesions are diffuse in nature. They may appear any time from 3 months to several years after implantation. The inciting causes are still unclear, though cytomegalovirus (CMV) infection and chronic rejection have been implicated. The mechanism of the process is thought to be dependent upon growth-factor production in the allograft initiated by circulating lymphocytes. Currently, the process has no treatment other than retransplantation.

26.9. XENOTRANSPLANTATION AND ITS ETHICAL AND SOCIAL ISSUES

The term 'xenotransplantation' is derived from the Greek word xenos, meaning foreign. All the issues surrounding this procedure spring from the notion of associating something 'foreign' with the human body. Thus, xenotransplantation can be defined as grafting tissue or organs from one animal species to another; e.g., grafting pig organs to a human being **(Fig. 26.5)**.

In recent years the transplantation of human cells, tissues, and organs (allotransplantation) has dramatically improved the survival and health prospects of the people suffering from life-threatening illnesses. It is gaining importance because worldwide the demand for human cell, tissue, and organ donation far exceeds the supply. This would be true even if everyone agreed to donate their cells, tissues or organs when they die. Furthermore, the population is ageing, which will place increasing demand on donation as more people suffer from chronic degenerative diseases.

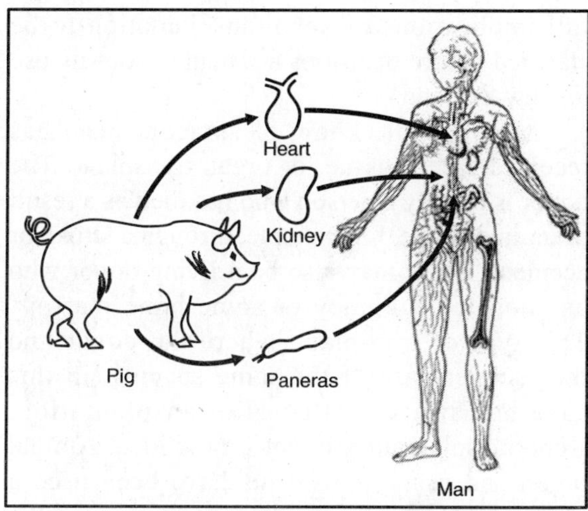

Fig. 26.5. Xenotransplantation of organs from pig to man

Currently, there are not enough human donors to go round, and as a result, people on waiting lists are dying.

The body's immune response is serious enough with donated human organs and tissues, but it is so severe with animal organs and tissues that xenotransplantation has not been seen as a viable option. However, interest in xenotransplantation has been rekindled as a result of recent advances in the genetic modification of animals to make their tissues less likely to be rejected. The development of increasingly sophisticated anti-immune drugs and the accelerating need for cells, tissues, and organs have also contributed to an intensification of research into the possibilities of xenotransplantation.

Along with the promise of potentially unlimited supply of organs and tissues, there are risks in xenotransplantation, especially from the transfer of diseases across species, as is thought to have happened with HIV. There are also questions as to whether we should be performing these procedures on animals and humans, with respect to the effects on animals' lives and the 'mixing' of different species. The Government wants these issues to be publicly debated and the full implications of xenotransplantation to be clarified before decisions are made about its use in New Zealand.

Many people know of someone who has received a cell, tissue, or organ transplant. The donor is usually a person who has died as a result of brain damage, for example, through a stroke or accident. But it may also be a living donor who has donated a kidney or some bone marrow. This type of transplant, where the donor and the recipient are of the same species (in this case human), is called allotransplantation. Xenotransplantation is not a new idea. Animal organ and tissue transplants have been tried a number of times over the centuries, but with little success. Now, however, researchers are working on this technology again. The technical definition of xenotransplantation is as follows: "The implantation or infusion into a human recipient of either live cells, tissues or organs from a nonhuman animal source, or human body fluids, cells, tissues or organs that have had *ex-vivo* (external) contact with live nonhuman animal cells, tissues or organs".

The use of non-living, processed biological products or materials, valves and insulins from animals such as pig, which has been in practice for decades, is not classified as xenotransplantation.

The current definition of xenotransplantation includes the grafting of cells, tissues or organs from non-human animal species into humans (although technically it can be the other way round or between any two species). It is obviously a subject that has fascinated people for a long time, because we find examples of this kind of organ grafting in the mythologies of many religions. Perhaps, the one best known is the grafting of the head of an elephant onto the body of the boy who went on to become the very popular Hindu god, *Ganesha*. The fascination now seems to have reached fever pitch, with high stakes for those concerned, particularly patients, scientists, the biotechnology industry, and infectious disease specialists.

The current demand for healthy cells, tissue and organs for medical transplants exceeds the available supply. Novartis estimates that 180,000 people around the world are waiting for organ transplants, but only one-third will receive one. Aside from patients suffering from organ failure, xenotransplantation could potentially treat those with hemophilia, AIDS, diabetes, Alzheimer's disease, Parkinson's disease, and Huntington's disease.

In Canada, upto 30% of people waiting for organ transplants die while still on the waiting list. In 1999, there were 3,544 people on waiting lists for donated organs (or 115.9 per million population). Most patients needed kidneys (78%); the rest needed livers (10%), hearts and lungs (4%), lungs (4%), and kidney-pancreas or pancreas (3%). The need for organs in Canada is expected to increase by almost 200% by 2020.

Canada's donor rate of 14.4 per million population is among the lowest of all developed countries. By way of comparison, Spain is reported to have the highest human donation rate worldwide, with 33.6 per million in 1999. Canada's low organ donation rate is partly due to its low death rate from car accidents and gun shots—a major source of donated organs in other countries. However, there is little doubt that Canada's organ donation rate could be increased—some say by as much as 80% by ongoing efforts to promote donation through public education campaigns and by more efficiently requesting, collecting and distributing organs.

26.10. HISTORY OF XENOTRANSPLANTATION

In the recent past, there have been efforts at xenotransplantation dating back to the early part of this century, before we knew anything about the immunological principles underlying transplantation in general. Most of these efforts have failed, though in one of Keith Reemtsma's patients in 1960s, a chimpanzee kidney did survive and work for about nine months. The reason for

Table 26.4: *Animal organs transplanted into humans, 1906-1995*

Donors organ	Transplants survival time	Author	Year
Pig kidney	3 days	Jaboulay	1906
Goat kidney	3 days	Jaboulay	1906
Macaque kidney	32 hours	Unger	1910
Sheep kidney	9 days	Neuhof	1923
Baboon kidney	4 days	Hitchcock	1963
Macaque kidney	12 days	Reemtsma	1963
Chimpanzee kidney	9 months	Reemtsma	1963
Baboon kidney	60 days	Starzl	1963
Chimpanzee kidney	1	Hardy	1964
Chimpanzee kidney	1 day	Hume	1964
Chimpanzee kidney	6 one 9 months	Reemtsma	1964
Baboon kidney	6 max 60 days	Starzl	1964
Chimpanzee kidney	49 days	Traeger	1964
Chimpanzee kidney	4 months	Goldsmith	1965
Chimpanzee kidney	31 days	Cortesini	1966
Pig heart	0 days		1968
Baboon heart	1	Barnard	1977
Baboon heart	20 days	Bailey	1985
Pig heart	< 1 day		1992
Baboon liver	70 days	Starzl	1993
Baboon liver	26 days	Starzl	1993
Baboon bone marrow	1	Gorman	1995

Other sources indicate that in fact there have been eight xenogeneic heart transplants and 11 **xenogeneic liver** transplants, of which one was with a pig liver (C.G. Groth, August 1998).

this success is not known, as there were no powerful immunosuppressive agents in the 1960s, and no sophisticated immunological or genetic manipulations of donor or recipient.

Table 26.4 gives the summary of these transplants. After 1980, there were a number of well publicized whole-organ (vascularized) xenotransplant attempts, as well as several less-publicized cellular and tissue transplants. These provide an opportunity to examine and define the issues of current concern in xenotransplantation.

In 1982, Loma Linda, California, a team of surgeons led by Dr. Leonard Bailey transplanted the heart of a baboon into Baby Fae—an infant born with hypoplastic left heart syndrome. Dr. Bailey's team proceeded because it had some laboratory evidence that xenotransplants would work, and because the new powerful immunosuppressive drug, ciclosporin A, became widely available to transplantologists in the United States at about this time. The operation was technically successful, and the child lived for about three weeks before the heart was rejected. In the early days after the transplant, the media were full of praise for the operation and its success, but this soon turned sour when the child died. Questions were asked about the adequacy of the information given to the parents.

The surgeons were faulted for not looking hard enough for a human heart to transplant, and for being too optimistic. The scientific evidence, in retrospect, was inadequate and many have come to view the Baby Fae episode as having had an overall negative effect in the field of xenotransplantation. In 1992, the team that was most advanced in the quest for success in xenotransplantation was the one led by Dr. Thomas Starzl in Pittsburgh.

In the 1960s, Starzl had performed about half a dozen baboon-to-human kidney transplants; all of which subsequently failed. This time his team had permission to perform four baboon-to-human liver transplants. The experimental nature of these attempts naturally leads to the selection of very sick persons. Thus, the first recipient was a patient with advanced AIDS and near-terminal hepatitis. This time, too, there was an extremely powerful new immunosuppressive drug called FK506, which is 100 times more effective.

Process of Developing Xenotransplantation

Researchers around the world are working out the science of xenotransplantation step by step. They start with laboratory studies on cells and tissues to work out the underlying science. Then they conduct studies on small animals (such as mice, rats or rabbits) to test possible procedures. The same approach is used in other medical research, such as cancer research or the development of new drugs. If these early studies are successful, further thorough research is needed to develop procedures that can be used to treat humans. This research covers two main areas.

- Animal-to-animal studies, in which the source and recipient animals are as similar as possible to the proposed human treatment (for example, from pig to baboon). These preclinical studies are needed to make sure that a procedure can be conducted safely and effectively on animals before it is tried on humans.

- Animal-to-human trials, in which animal cells, tissues or organs are used for human treatments in closely monitored clinical trials. These trials, which are most likely to involve pig-to-human transplants, would be attempted only if animal-to-animal studies showed a high likelihood of benefit to humans.

Clinical trials can be either therapeutic (the people participating are expected to benefit from their involvement in the study) or non-therapeutic (the study is designed to obtain further knowledge, but may not directly benefit the participants).

Because of the potential risks involved, only research with some prospect of helping people is considered to be acceptable for clinical trials of animal-to-human transplantation. As with other medical technologies, the process of testing new therapeutic procedures through clinical trials can take many years and involve several phases.

Scientists are working to overcome several major obstacles to xenotransplantation, including rapid immune rejection and structural or functional incompatibility of the animal transplant (it simply doesn't work properly in the human recipient). With the rapid development of genetic technology over the past decade, some scientists believe that it might be possible to overcome these problems by genetically modifying the source animals to make their tissues and organs more compatible with humans. Hence, the shortage of human donors, combined with developments in genetic technology, has further stimulated interest in xenotransplantation.

26.11. SOURCE ANIMALS FOR XENOTRANSPLANTATION

It would seem obvious, when searching for ways to reduce the rejection reaction to xenografts, to use animals physiologically close to humans, such as old world monkeys and the apes (for example, chimpanzees). However, this very closeness creates complex ethical problems as well as an increased likelihood of cross-species infection. There are strict regulations in New Zealand about the use of non-human primates in research, so it is unlikely that these species would be used for xenotransplantation research in this country.

At present, pig is the favoured animal for research in xenotransplantation, because it grows quickly to about the right size, produces large litters, and can be reared in specific pathogen-free conditions (where some but not all microorganisms are excluded). In terms of ethical concerns, the fact that pigs have long been used as a source of meat reduces, but certainly does not eliminate, the concerns of many people, especially when weighed against the possible benefits. However, recently it has become clear that there is the possibility of cross-species infection from pigs also. This takes the issue of whether to perform such a procedure out of the realm of an individual decision to take a personal risk, usually for the sake of a therapeutic effect, into the realm of the safety of the xenotransplantation recipient's close contacts and the community at large. The spectre of HIV and AIDS hangs over discussions on the possibility of cross-species infection, so we will now turn to look at the risks involved with xenotransplantation.

The PHS agencies and the Health Resource and Services Administration (HRSA) collaborated to develop the *Draft Public Health Service Guideline on Infectious Disease Issues in Xenotransplantation,* which was offered for public comment in September 1996. The guideline has now been revised to reflect comments received on the 1996 draft, advances in science, and international policy discussions.

The guideline goes on to describe general principles to prevent and control infectious diseases associated with xenotransplantation, placing emphasis on the following areas:

The xenotransplantation team should include:

- clinical transplant experts;
- infectious disease physician(s);
- veterinarian(s) with expertise in infectious diseases relevant to the source animal;
- specialist(s) in hospital infection control; and
- experts in research and diagnostic micro-biology laboratory methods.

The guideline describes the kind of clinical research plan, and type of clinical center appropriate for xenotransplantation trials.

- The informed consent process should include education of the recipient about infectious disease risks, the potential of transmitting diseases to others, and the need for life-long surveillance. This education and counseling should include recipients' family, close contacts and exposed health care workers.
- Only certain kinds of animals should serve as sources for xenotransplantation products, and the guideline explains how those animals should be raised to protect them from infections that may cause problems in xenotransplantation.
- Non-human primates should not be used as source animals, until scientists have sufficient information to address the risks of infectious diseases.
- Certain infection control practices should be implemented to reduce the risk of in-hospital infections.
- The guideline recommends the manner in which patients, some of their close contacts (such as health care workers), and the herd providing the source animal should continue to be monitored following the xenotransplantation procedure, so that possible emerging infections might be recognized and measures taken to contain them.
- The recipient and certain contacts of the recipient should not donate body fluids and other body parts.

- Specific records should be maintained on the xenotransplantation recipient—the source animal; the herd from which that animal came; the standard operating procedures followed by the animal facilities and the clinical centers involved in the xenotransplantation trial; and the occupational health service programs of relevant health care workers. All these records should be maintained for at least 50 years.

- Biological specimens (such as blood) should be collected from the recipient and the source animal, and stored for 50 years. This precaution should be taken in the event an infection emerges that requires the examination of such specimens.

- HHS is piloting a national database that would capture information from every clinical xenotransplantation. Data would be collected from all clinical centers that conduct xenotransplantation trials and all animal facilities providing source animals. This database would be an essential tool in performing post-transplantation surveillance of patients, and rapidly identifying and notifying all pertinent parties in case of adverse events.

- HHS is also considering options for creating a central biological specimen archive that would collect and store patient, source animal and xenotransplantation product specimens for use in public health investigations and related research.

- The Secretary's Advisory Committee on Xenotransplantation (SACX) has now been enhanced to serve as a forum to discuss scientific, medical, public health, ethical, social and legal issues surrounding xeno-transplantation, and to make recommendations to the Secretary of HHS.

Increased Awareness of Risk of Infection

It is impossible to fully understand or predict all the risks associated with xenotransplantation, because scientific knowledge is lacking. However, the most serious risk appears to be transferring undetected animal diseases to patients that could, in turn, be transmitted to the general public. The worst-case scenario would be a major new epidemic.

Early clinical attempts at xenotransplantation were conducted with limited consideration of animal-to-human (or "zoonotic") infection. Although there is no evidence that these attempts resulted in zoonoses, most patients died soon after treatment and did not have time to develop any symptoms. Much of the discussion around zoonoses began in 1996, when 40 scientists wrote a letter concerning xenotransplantation guidelines proposed by the American Public Health Services. In this letter, they pointed out the dangers of infection from non-human primates, noting that HIV almost certainly originated as an ape or monkey virus that infected humans.

Scientists now have a better understanding of viruses and infections. They know that "retro-viruses" present in animal cells can infect human cells and can exist for many years before any symptoms are displayed. With xenotransplantation, these animal infections could quickly become deadly because the human recipient's immune system would be suppressed.

Some animal viruses have jumped the species barrier and infected humans. The AIDS-causing HIV-1 and HIV-2 viruses are now known to have originated from non-human primate viruses, and jumped from monkeys to people after infectious events 60-some years ago. Two human leukemia viruses have also been shown to originate with non-human primates. All these examples involve transmissions from non-human primates to humans.

In 1997, the British government reviewed the current scientific knowledge and decided to ban experiments with humans and non-human primates. The United States decided not to ban experiments with non-human primates, but recommended that proposals for clinical trials using non-human primates should not be submitted to the Federal Drug Administration (FDA) until sufficient scientific information exists, addressing the risk posed by non-human primate transplants.

Because of the high risk of infection from non-human primates, pigs are the current animal of choice for xenotransplantation. But non-primates can also cause zoonotic infections. For example, tissues and cells of animals, including pigs, can act as a reservoir for various infectious agents that may then infect humans. In an effort to reduce the risk, pigs for xenotransplantation are being bred in germ-free environments. However, scientists are not able to produce pigs without any viruses.

Pig endogenous retroviruses (or "PERVs") can be found in all tissues of most pigs. In 1998, scientists discovered a PERV virus that can replicate in human cells. This discovery raised questions about other as yet unidentified infectious agents in pigs and other animals that could cause disease in human cells. It appears to be impossible to breed PERV-free pigs because it is in their genetic material. However, in 1999, a scientific paper considered 160 patients who had been treated with living pig xenotransplants of one kind or another over the previous 12 years, and the authors concluded that there was no evidence of a long term PERV infection in any of these patients.

It is impossible to fully understand or predict all the risks associated with xenotransplantation because scientific knowledge is incomplete. Certainly, no technology or medical procedure is risk free, and some say that in order to assess the risks, researchers should be allowed to conduct carefully controlled clinical trials. However, there is no consensus about the level of safety that would be necessary, and the degree of risk that would be acceptable for clinical trials to occur.

The "precautionary principle" says that when an activity threatens to harm the environment or human health, precautionary measures should be taken even if some information is not established with certainty from scientific studies. However,

the precautionary principle has been hotly debated, and if taken to the extreme, it could be argued that little or no scientific research would take place. It is therefore difficult to weigh the risks and benefits of xenotransplantation in the face of uncertain information, and it is not known that what level of risk the public would be prepared to accept.

The emphasis in discussion of xenotransplantation has changed from concern about the rights and welfare of potential non-human source animals to concern about the risk of xenozoonoses. One of several things that brought this change was a letter published by Dr. Jonathan Allan and about 40 other scientists, including a number of virologists, who pointed out the real dangers of infection from nonhuman primates. The authors of the letter and others again reminded us that, almost certainly, HIV came from simian sources, and that the incubation period of retroviruses can be many years before any clinical symptoms manifest.

It is also reminded that if these infections were transmitted to the recipient, they could be rapidly lethal because of the heavy immunosuppression, and could also be transmitted, theoretically, to immediate contacts and even to the public. This widely publicized letter was written in response to the draft guidelines for reduction of the risk of xenozoonoses published by the Public Health Services—a federal umbrella body, which encompasses both the FDA and the Centers for Disease Control and Prevention (CDC) in the United States. The main point of the letter was that these guidelines were not strict enough; the risk of infection from non-human primates was real enough; and we should therefore, boycott the use of primates completely. The public health service draft guidelines were also criticized by other organizations, including the American Society of Transplant Physicians. The criticisms included the vagueness of the document regarding the funding of the regulatory instruments and, again, its failure to exclude the use of primates.

26.12. XENOTRANSPLANTATION AND ITS GUIDELINES

In 1994 and 1995, something stirred those concerned with ethics and public policy in both Europe and the United States. In January, 1995, in the United Kingdom, the Nuffield Council on Bioethics set up a Working Party to look at many xenotransplant issues, and it produced a report, which came out in 1996. Subsequently the United Kingdom Government's Advisory Group on the ethics of xenotransplantation examined a similar range of issues, and came up in 1997 with the definitive Kennedy Report, named after its admirably capable chairman, Professor Ian Kennedy. The essence of the Kennedy Report was that it was worth pursuing xenotransplantation research, and that using pigs, but not non-human primates, as source animals would be ethically acceptable. Furthermore, it was ethically acceptable to alter the pig's genome to the extent foreseen, so as to facilitate the transplant, provided the pig remained recognizably a pig. Primates could ethically be used only as recipients, in small numbers, in animal-to-animal experiments.

Non-human primates were not ethically acceptable as source animals, partly because of the emotional attachment that human beings have to them, and partly because their being phylogenetically closer to man increases the risk of infection. In comparison to the pig, there were other disadvantages such as their depletability, slow breeding, small litters, and the fact that there are no specified-pathogen-free (SPF) colonies of primates, while there are such SPF colonies of pigs. These factors also mean that the cost of breeding non-human primates for xenotransplantation would be prohibitive. Furthermore, the organs of primates are often too small for humans.

The pig, on the other hand, has lived close to man for many generations without spreading any serious infections (leaving aside the swine-flu epidemic); its physiology is quite similar to man's; its organs are roughly the same size; it breeds very quickly; has large litters, and in any case is eaten by millions of people throughout the world. The matter of physiology, however, is not yet really resolved, particularly for major synthesizing organs such as the liver, which would be pouring out pig molecules into the human recipient's circulation.

Also, not much is known about the response of these organs to normal homoeostatic signals. This is a subject that has not received the attention that it really deserves. The most significant conclusion of the Kennedy Report was that the base of scientific knowledge was inadequate in 1997 to proceed to clinical trials, and that there should be a boycott until a formal regulatory authority was established, with an opportunity to review the evidence at the time of submission of any applications to it.

The government of the United Kingdom rapidly responded by setting up the Xenotransplant Interim Regulatory Authority under the Chairmanship of Lord Habgood. In the United States, the consultation process was initiated by the Institute of Medicine, which held a workshop in June 1995, and its report was published in 1996. Its conclusions differed from the Kennedy Report's in that non-human primates were not excluded as source animals, and the scientific base was considered to be adequate to "proceed with caution" to clinical trials. The American approach was emphatically to have an *advisory,* not a regulatory, body to deal with xenotransplantation.

International Scenario in Xenotransplantation

By 1997, it was obvious that there was risk to the public's health; the magnitude of this risk was not really known, and was perceived differently by different countries. Besides there were a number of ethical, social and cultural ramifications to xenotransplantation. There was a distinct need for an international and interdisciplinary dialogue. The World Health Organization responded to this global challenge by convening a consultation in Geneva in October 1997. Participants included experts from several countries in Europe, as well as from Canada, Cameroon, Japan, Oman, the Philippines, Sri Lanka and the United States. The main tasks given to the consultation were to work out technical and ethical guidelines to minimize the risk of infection, safeguard human dignity and human rights, and ensure animal welfare.

Despite the apparent differences of approach and perceptions amongst the participants, the consultation was very successful, and formulated a unified set of recommendations. One of the points that became clear was that if xenotransplantation succeeds, developing countries might be among the main beneficiaries. This is because in most of the developing world, organ replacement therapies such as haemodialysis are beyond the means of governments or individuals, and organ transplantation, especially through cadaveric donation of organs, has not really taken off.

For many of these countries, at the current level of expenditure, allotransplantation necessarily has lower priority than basic public health needs. Here, xenotransplantation holds out the hope that one day, when the ethical, technical, infectious disease and cost problems have been resolved, developing countries will be able to embark on this type of transplantation without having to divert too much of their scarce resources to it. The fundamental need for international cooperation in research, communication and standardization was recognized. The report of the WHO consultation was issued in February 1998.

Cellular and Tissue Xenotransplants

In 1993, a Swedish team transplanted 10 diabetic patients with pig islets of Langerhans. None of them succeeded in producing pig insulin in the long term, but the experiment is nevertheless important because the future of xenotransplantation may well lie in such cellular transplants for very common conditions such as diabetes mellitus. The other significant outcome of this experiment was the important data on the presence or absence of the risk of viral infection. All ten of the patients had developed antibodies to pig viruses; some to the influenza virus and some to picornavirus. According to the Swedish team, none of the patients has actually become sick with any pig viruses, and they believe that there is a possibility that the antibodies may simply be cross-reactive (Anne Tibell, 1998). Neural tissue transplants have been performed from pig to man to alleviate Parkinson disease, and more recently for intractable epilepsy. One of the patients treated for Parkinson disease died of other causes. The post mortem examination showed that connections had developed between the pig neural tissue and his brain, answering a critically important biological question, but at the same time raising philosophical and theological questions about the brain, identity and responsibility.

Current Status

The field is moving very rapidly. There are a number of important stakeholders, and the stakes are high for some of them. Major research is now being funded and carried out by venture-capitalized biotechnology companies, and one of the considerations in the current discussions is the part played by these companies in determining the timing, technology, and the development of xenotransplantation as a whole. There is a distinct difference in the approaches to xenotransplantation in Europe and the United States. In Europe, the feeling is that the scientific base is inadequate to proceed to clinical trials, and so there is at present a boycott. In the United States, the consensus seems to be that further laboratory and animal-to-animal experiments will not answer the key questions, and that the only way to advance the scientific base is to "proceed with caution" to clinical trials. In the United States, therefore, while the Public Health Services draft guidelines are becoming more strict in response to comments, criticisms and consultations, the Food and Drug Administration has been receiving applications. It has already authorized a number of clinical trials, most of which at present involve cells and tissues rather than whole vascularized organs, though a number of researchers are preparing for the latter.

Furthermore, rather than completely excluding the use of non-human primates, the approach in the United States is to set the requirements for their use at levels that are virtually impossible to achieve. This effectively boycotts their use without actually using the dreaded term. The question of consent to the clinical trials, especially in the early patients, is likely to be a difficult one. It is argued that some of the fundamental traditional principles of consent would need to be violated, and that because of the continuing need to monitor the patient and perform invasive investigations even if the graft failed, the agreement may need to be more of a contract, with specifically binding requirements, than the traditional consent whereby the patient has the right to withdraw at any time from the experiment.

Since the community is in a sense being put at risk, there is a real argument for considering some form of community consent as well. But, at present, with our inability to quantify the xenozoonotic risk (to calculate the risk–benefit ratio), it is not clear that how far we should push this point. In any case, there is little experience in obtaining such community consent.

On the purely scientific side, the evidence has accumulated rapidly on both sides of the divide. On one hand, our understanding of the hyperacute rejection phenomenon is increasing, and scientific enquiry is already being directed to the subsequent "delayed xenotransplant reaction". More and more animal models are being developed, and immunological manipulations are becoming more sophisticated. Animal models have been developed, which no longer express on their endothelium the alphagal molecule that is the main target of the antibodies mediating the hyperacute rejection response. Also, transgenic animals have been developed that express on their endothelium human molecules such as the decay accelerating factor, which helps to inactivate complement components in a species specific manner.

At the same time, however, we have growing evidence of viruses in pigs and in primates, which can theoretically cause xenozoonotic infection. New pig endogenous retroviruses, which would be very difficult to eradicate and which have been shown to infect human cell lines *in vitro,* have been described. While it is fair to say that there is little evidence that such viruses will be pathogenic in man, it cannot be assumed that they will not become so, especially if introduced into an immunocompromised host. Evidence has also accumulated of trans species transmission of viruses from pig to man, causing infection. For example, with paramyxovirus in Australia, and in the case of primates—the simian foamy virus (and other viruses) can be transmitted *via* scratches and bites to animal handlers.

The effect of xenotransplantation on the donation of cadaveric and living-donor human organs needs to be taken into account. Xenotransplantation itself can be seen as serving one of the several purposes: to be a complete substitute for human organs; to supplement human organs, thus alleviating shortage; or to be "bridging" rather than "destination" organs. Whatever the purpose, it would be a setback if the effect was to reduce the supply of human organs, because the public now perceives a lack of need since pigs are plentiful, or for whatever other speculative reasons. We think that xenotransplants will be very expensive in the first decade of their application, and so, for at least this reason, they will not be an adequate substitute for allotransplantation for at least a decade after xenotransplanatation is truly established.

One major issue for developing countries is the phenomenon of "expatriate" experiments. It is possible that, because of restrictions imposed in industrialized countries, researchers may perform these experiments in developing countries, with potentially disastrous results in terms of both safety and human rights. This is only one of the reasons for which developing countries need to be represented in formulating guidelines for this exciting, challenging and potentially useful new technology. Attempts are now being made to predict factors that would affect the response of the public to xenotransplantation, but surveys of public attitudes are beset by deficiencies and are cumulatively contradictory at present.

26.13. XENOTRANSPLANTATION AND ITS PROMISES

The considerable interest in xenotransplantation at present is linked to the persisting imbalance between the need for organ substitutes and their availability, which has no chance of being righted in view of the growing scarcity of so-called " brain deaths ", which represent less than 1% of hospital deaths. Because of progress achieved by modern medicine and biology, techniques for transplanting human organs have become routine, and have now amply demonstrated their life-saving and quality-of-life improvement value. The limit which constricts the use of these techniques in all the countries in which they are practised is the shortage of organs for transplantation. Turning to xenotransplantation would alleviate that scarcity and make healthy organs available at the exact time when they are wanted. It would be possible to eliminate the months of waiting, during which health deteriorates, and obviate the need for emergency surgery on an ill-prepared patient because an organ suddenly becomes available.

Furthermore, if it could confidently be said that xenotransplantation was devoid of risk, a large number of ethical issues raised by the use of human organs could be solved. The thought that one's life is saved by the death of a fellow human being, or the implicit gratitude, which a recipient owes to a living donor, are in some cases a major psychological handicap abundantly demonstrated

by various studies. In most cases, this should not arise with xenotransplantation. The use of xenografts should eliminate the ever-present risk of illicit trading of organs in spite of laws of prohibition in most countries.

Finally, the possibility of using embryonic animal tissues instead of human ones would solve a great many ethical problems about the appropriateness of using human foetuses. If the main technical difficulties of xenografting were solved, there would still remain a whole set of ethical matters to discuss before proceeding to the first clinical tests:

- How do you select the first patients for xenotransplantation?
- How do you inform these patients, and go about understanding and overcoming legitimate reluctance on their part?
- How can a public debate on a large scale be initiated to find out whether a society is ready to accept a technique, which may represent a potential danger of infection precisely for the members of society, who are not concerned by the graft?
- How can market forces be prevented from disturbing the present system in which human organs are voluntarily donated?

The idea of using xenotransplantation to alleviate the shortage of human organs is not new. Pig heart valves have been used in humans for nearly thirty years. They are processed so that they react like an inert material instead of a living **tissue**. More than eighty types of connective tissue (skin, bone), of porcine or bovine origin, are commonly used in human medicine. However, they only play a transitory role, as is the case for skin of porcine origin that is rejected very quickly. It should not be forgotten that an animal product, such as bovine or porcine insulin that was used for almost a century, saved the lives of millions of diabetics.

Various uses have been suggested for islets of Langerhans, neural foetal tissue, or bone marrow.

Using small masses of tissue devoid of blood vessels is less risky for a patient than whole organs,

Table 26.5: *Companies involved in xenotransplantation research*

Company	Location	Organs	Comments
Advanced cell technology	Worcester, MA	Kidney, heart	Using cloning and GM techniques.
Alexion pharma-ceuticals	New Haven, CT	Nerve cell based therapies	Focusing on Parkinson's disease and spinal cord damage using GM pigs.
Algenix	Shoreview, MN	Liver	Developing bio-artificial livers using pig cells for external use.
Circe biomedical	Waltham, MA	Liver, pancreas	Developing bio-artificial livers using pig cells.
Diacrin	Charlestown, MA	Nerve, liver and retina cell based therapies	In partnership with genzyme corp using tissue from GM pigs for treatment of neurological disorders.
Immerge bio-therapeutics	Charlestown, MA	Kidney, heart	Joint venture between Novartis and bio-transplant Inc. Agreement with infigen (an animal cloning company) to colla-borate on the production of GM miniature pigs for xenotransplantation.
Nextran/Baxter	Princeton NJ/Deerfield, IL	Liver	Uses GM pigs and has tested pig liver as an external support for liver failure.
PPL therapeutics	Edinburgh Scotland	Kidney, heart	Combining cloning and genetic modification technologies on pigs.
ReNeuron	England	Nerve cell thera-pies for stroke victims	Developing mouse stem cell lines.
Ximerex	Omaha, NE	Liver	Formed by a scientist from the University of Nebraska Medical Center. Uses GM pigs to produce human/pig hybrid liver by introducing human cells into foetal pigs.

and the psychological acceptance of tissue xenotransplantation is much better than for an organ to which the patient attaches symbolic value, such as the heart. However, it must be recognized that there is less shortage of tissue than whole organs, and so more ethical concern about the latter. The technical problem is that the living tissue of animal origin is rejected very quickly, even before the surgical procedure has been completed. It was only as late as 90s, and mainly in the last five years, that the causes of such rejection began to be better understood with the hope of overcoming the difficulties in the near future **(Table 26.5).** This explains the recent renewal of interest in the techniques of xenotransplantation.

26.14. SCIENTIFIC ASPECTS BEHIND XENOTRANSPLANTATION

Rejection Phenomena

When the human being receives a graft originating from another species, for instance pigs, three processes have to be overcome—hyperacute rejection, which occurs within minutes or hours;

acute rejection, which takes place seven to ten days later; and finally chronic rejection with long-term therapeutical consequences, (which are largely uncharted because of the brevity of survival of the transplants observed in animals). Acute rejection is of the same kind as occurs in the case of allotransplantation, and its effects can be limited with drugs that depress the immune system.

However, recent research suggests that the immune T-cell connected response against the xenograft differs slightly from what is observed with allografts, so that new immunosuppressants may need to be developed. But this is also an incentives for new research on this problem. Presently, most of the work is focused on hyperacute rejection. This occurs because of the presence of natural antibodies, which are constantly present in the blood stream. Throughout our lifetime, we produce and store antibodies directed against epitopes carried by the great number of molecules that we encounter, breathe, or eat. Furthermore, certain antibodies, described as preformed, are coded by our genes and protect us against foreign agents in our environment.

Thus, without the immune system needing any induction, we already possess antibodies directed against antigenic structures of other species. As soon as a graft, of porcine origin for instance, is irrigated by the blood stream, antibodies recognize antigen receptors located on the endothelial cells that line the blood vessels of the graft, and the endothelial cells are activated. A complex set of proteins called the "complement" circulate constantly in the blood stream, and one of its functions is to destroy the cells, which the immune system recognises as foreign. Systems that protected the graft's epithelial cells from the attack of complement's cease to operate, and the complement destroys the epithelial cells.

Simultaneously, systems, which prevented intravascular clotting, cease to function, and micro-thrombosis sets in. To control hyperacute rejection, researchers adopt several approaches. They can modify the natural antibodies of the recipient, the graft antigens, the complement, or the cascade of clotting reactions in the recipient. It would seem that the immuno-absorption of all natural antibodies will never suffice. However, some proteins, which differ from one animal species to another, are now known to be present on the surface of many cells, including endothelial cells, and they are able to inhibit the complement. The better known of these are called CD35, MCP (Membrane Cofactor or CD46), DAF (Decay Accelerating Factor or CD55), protectin (CD59), Homologue Restriction Factor (HRF). *In vitro*, it was found that these molecules protect the cells of another animal species from lysis by the human complement.

In practice, immunoabsorption of natural antibodies, or massive administration of drugs to prevent the complement's action, would lead to therapeutic protocols, which would be too burdensome to be usable; probably not totally effective, and possibly dangerous since they would expose patients to infectious diseases once the essential components of natural immunity start weakening. Physiological compatibility knowledge about physiological compatibility between human and animal organs (differences in enzymatic specificity, incompatibility between ligands and receptors, sensitivity to neurogenic stimulation, chronobiological cycles, etc....) is still very tentative.

Selection of Animal Donors and Immunology

The two animal species, which have mainly been used so far, are swine and non-human primates. The main argument in favour of primates is immunological. As noted above, hyperacute rejection begins by the fixing of natural antibodies on the corresponding antigenic epitopes. Epitope Gal-α-1-3-Gal, major epitope involved in the pig, is expressed over the whole surface of vascular endothelium. This epitope does not exist in primates. Through the process of evolution, the α-1-3-galactosyl transferase gene, which is the transport enzyme of the epitope concerned, has been deactivated by two deletions of the coding fraction in humans, chimpanzees, and baboons.

However, it has been found that the simple absence of epitope Gal-α-1-3-Gal is not sufficient in itself to preclude any risk of rejection. The first neonate to receive a baboon heart in 1985 survived only for twenty days. Primates also present other drawbacks. They are slow to develop and have a low reproductive rate, and they are difficult to rear in captivity. Chimpanzes are protected by law, as they are an endangered species. It would be difficult to cover all requirements with baboons only. Above all, risks of infection by viral disease transmitted by graft to recipients are very high (see below). Finally, the closeness in evolution between humans and non-human primates, the chimpanzee in particular, raises particular ethical objections.

The scientific community is therefore directing its thinking to use transgenic pigs. The aim is to obtain pigs whose endothelium is no longer activated by human blood. Several strategies are possible—non-expression of antigens recognized by natural human antibodies, or express on the endothelium of one or several molecules that inhibit either complement or clotting. These various approaches have been explored. As regards complement inhibitors, transgenic pigs have been produced, which express human DAF (CD55) or protectin (CD59), or both together. Transgenic pigs expressing DAF have raised great hopes. Since human DAF also inactivates the complement of other primates, transgenic pig hearts expressing this molecule have been used to transplant cynomolgus monkeys. The graft was attached in an abnormal site (the neck) to monkeys retaining their own hearts. Non-transgenic transplants survived on average 1.6 days, whereas average survival was 5.1 days for the transgenic transplant in recipients not treated with immunosuppressants, and 40 days on monkeys treated with a combination of cyclosporin, cyclophosphamide and corticosteroids, which are active in suppressing acute rejection.

These results were found encouraging initially to warrant moving on to human tests, but the authors seem to have given up the idea in the meantime. Protocols including both DAF and protectin genes do not seem to have produced better results so far than when only the DAF gene is used. A second strategy, aiming to suppress the Gal-α-1-3-Gal epitope is now being explored. For example, it should be possible to invalidate the α-galactosyl transferase gene, suppress its expression using an antisense messenger RNA, or use other strategies, but none have as yet proved effective. Therefore, even at the level of animal experimentation, research to produce transgenic donors for transplants that would be acceptable to the host, is still in its infancy.

With pigs, gene transfer is achieved with microinjections directly into the embryo pronucleus. The operation is not highly productive: 100 injected embryos are required to produce one transgenic pig. However, it is possible to produce lineages of transgenic pigs, which develop more or less as do ordinary pigs. Cost in 1999 remains high, about 150,000 FF per pig, not taking into account depreciation of the special pig-rearing facilities required.

Animal Selection and Risk of Infection

The characteristic aspect of xenotransplantation is putting into close and prolonged contact an animal organ and an entire recipient human who, at the time of transplant, is treated with powerful immunosuppressor drugs. These are ideal conditions for a microorganism in the graft to multiply in the host. Clearly, with allotransplantation, contamination from donor to recipient is also a risk, but within the same species so that there is a chance of being able to control infectious pathologies. The situation in xenotransplantation is more complex.

It is well-known that many animals may harbour microorganisms in their organs, cells, and genome for which they are healthy carriers, because, in the course of evolution, they have developed protective mechanisms, which render them resistant. Some of these microorganisms are capable of crossing the species, barrier and so express themselves in the immunodepressed host.

The appearance of "new" diseases after crossing the species barrier is not a myth. The HIV virus, for instance, is probably of simian origin, and is the cause of a pandemic in which the animal has ceased to play any part, but which we are still incapable of controlling as long as there is no vaccine. The most recent example, and one of the most alarming at this time, is the probable passage of prions, causing Bovine Spongi-form Encephalopathy, from cattle to humans.

This danger is, therefore, sufficiently serious to induce physicians and biologists to publicly raise the question of whether it is ethical to allow mankind to run the risk of devastating and uncontrollable pandemics, whereas xenotransplant techniques will never concern more than a limited group of patients. The real issue is whether the risk can be reduced to an acceptable threshold. In pigs, animal husbandry techniques have for some considerable time been able to eliminate bacteria, parasites, and viruses propagated by the environment. Animals are obtained in axenic state (totally devoid of microorganisms) and their digestive tract is inoculated with known flora. They are kept in sterile rooms, which are ventilated by overpressure, given sterile feed, and have minimum contact with the keepers. Multiple tests are used to check their specific pathogen free status.

There remains the problem of viruses and viral genomes, which are transmitted vertically to offspring. Most investigations have been focused on retroviruses. Many retroviral type sequences have been found in the genome of most animals. They were incorporated in the course of evolution following viral infection or transposition mechanisms not concerned by infectious processes. Various mechanisms prohibit permanently any expression of such retroviral

sequences. However, reactivation by some stimuli is possible. Some animal genomes harbour a great variety of retroviral sequences: in case of mice and primates. Recent information has come to light concerning the presence of proviral sequences in the pig genome, which can be transmitted to cultured human cells. However, on the other hand, there is no evidence of any identified swine pathology triggered by retroviral infection. Finally, since pigs and humans are two relatively distant species, transmission of a porcine retrovirus to human beings seems improbable. This is demonstrated by the history of two species in parallel, wounds inflicted on humans by pigs, widespread consumption of pork, and medication of porcine origin given to human beings (insulin, for example). None of these practices have ever led, apparently, to the transmission of infectious diseases to humans. There is already some experience of pig skin used as a graft for human beings, porcine pancreatic islets, or blood infused in porcine kidneys. Antibodies against porcine viruses have been found in the blood of some patients, but no new disease has been observed. However, not nearly enough time has elapsed for firm conclusions to be drawn about these ongoing observations.

The considerable capacity for the mutation of these proviral sequences should not be underestimated. This could change their tropism or mode of action. Furthermore, their capacity to integrate oncogenes or to settle close to an oncogene could trigger cancer in recipients. For all of these reasons, pigs devoid of such retroviral sequences would be ideal. This is a difficult achievement, but not impossible since there has been success with chickens for a certain type of retrovirus. Identification of the retroviral sequences in swine has begun and must continue to eliminate carriers. However, the number of copies of these retroviruses and time between generations are such that this is a long, costly, and uncertain approach. There is also one last problem with transgenic pigs, which do not express the Gal-α-1-3-Gal porcine epitope. In this case, viral particles from the cells of these animals will no longer express the antigen, and will therefore no longer be recognized as targets by the human complement.

In the same way, particles from pigs expressing the human DAF may be resistant to the human complement. This is why some authors have stated that transgenic pigs could be more dangerous than ordinary pigs in disease transmission. Altogether, it can be said that the known risk of transmission and the number of proviral sequences in their genomes preclude the use of primates as a source of xenografts for the time being. The second point also argues against the use of murine cells. Transgenic pig seems to be the least risky animal, as long as it is reared in strict specific pathogen free conditions. However, research must continue to gain a better understanding of proviral sequences existing in the animal, and to eliminate them to the greatest extent. It is, therefore, possible to subscribe to the opinion expressed by INSERM's Intercommission II, which considers that bio-risks connected to the xenotransplant of porcine organs can be limited by recent research results, and that they should not, *a priori*, prohibit all therapeutic testing in human beings.

Xenogenic cell therapy for human beings, using cells or cell masses of animal origin devoid of blood vessels as grafts for human beings, is only partly affected by the drawbacks described above for solid organs. Cells that have already been used in this way are:

- Pancreatic islets of Langerhans
- Hepatocytes, cell from the nervous system and the skin
- Cells from muscles
- Cartilage
- Endocrine glands
- Heart
- Kidneys
- Blood vessels are also being considered

In some cases, these cells are placed outside the body on inert supporting structures, in semi-permeable pockets, and put into contact with the blood in a cardiopulmonary bypass system. These are, in fact, bioartificial organs for which there is no immune rejection problem, but such systems can only be used for a brief period. When these cell xenografts are put into the body, they do not give rise to hyperacute rejection, since they are not vascularised, but are subject to the usual cellular rejection that is controlled more or less competently by immunosuppressants. In both situations, however, the risk of infectious contamination is identical to the one described for organs. This is the reason why even in the case of cell xenotransplantation, the pig is considered to be the best donor animal, and better than primates.

26.15. PROBLEMS ARISING OUT OF CLINICAL EXPERIMENTATION

Xenotransplantation raises a number of ethical issues. For example, how far is it acceptable to interfere with nature in order to prolong the human life? Would crossing the species barrier have an impact on what it means to be human? Similar questions were raised 40 years ago, when human-to-human transplants were being considered. At that time, questions were about crossing the frontier between self and non-self. Critics of xenotransplantation say that the procedure goes beyond medical innovation, crossing the very boundaries that define us as human. While few of the world's major religions have made formal pronouncements on xenotransplantation, some religious believers are willing to accept it in principle, provided unnecessary animal suffering is avoided. Others find it unacceptable, perhaps because of their relationship with the natural world.

Some people are opposed to altering the genetic make-up of animals—a procedure that would probably be necessary to prevent the human immune system from rejecting the xenotransplant. They argue that the boundaries that exist between species are for their protection, and tampering with the integrity of a species will have unpredictable repercussions. Others say that the boundaries between the species are not fixed but change with evolution. They argue that genetic modification is an extension of selective breeding that has gone on for centuries and, in any event, it would happen on a small and limited scale to benefit humans. Opponents argue that evolution has no direction, and for humans to arbitrarily use artificial means to control life is like 'playing god.' Attitudes to xenotransplantation could well change over time and it is difficult to say how different cultures and societies will respond.

The earliest xenotransplants to be performed on human beings are a topical controversy. On one hand, there was the English team from the Imutran pharmaceutical company (David White), who had announced their intention of performing transgenic pig organ and tissue xenotransplants in 1996, and on other hand, Thomas Starzl's team in the USA, who were the first to dare implant a baboon liver to a human being in 1992, and who, when it failed, requested a moratorium to provide enough time to gather new scientific data. After the wave of enthusiasm raised in 1995 by the publication of D. White's initial results (grafting a transgenic pig heart into the neck of a monkey) in early 1998, there was an energetic call for caution orchestrated by the magazine "Nature", which was echoed by mass media, in France particularly.

Several scientists, following in the steps of the American researcher Fritz Bach, now consider that to proceed to clinical trials should be regarded more as an ethical issue than a technical problem. It is indisputable that for a particular critically ill patient, the benefits to be expected from xenotransplantation, when rejection can be controlled, outweigh any risk of infection. However, for the population, as a whole, it is not possible to exclude completely the pandemic risk.

Therefore, Bach is in favour of a moratorium for any kind of clinical xenotransplantation, until such time as public debate on a large scale gives society a chance to say whether the non-zero risk of a new viral epidemic is acceptable. It is, therefore, clear that the present status of scientific competence, and also good clinical practices prohibit any intention of applying these techniques directly to human beings at this time. In these circumstances, three kinds of issues arise:

1. At which point in scientific progress will it be ethical to propose xenotransplation to the patient?
2. Who will be the first xenotransplant patients?
3. What precautions should be taken and what information should be given to these early patients?

Using xenotransplantation for human patients will require progress in at least the three areas listed above. The first requirement is overcoming acute rejection. Since we have a usable animal model, grafting transgenic pig organs to monkeys, it is essential that convincing results with graft survival times of more than just a few hours, or for that matter a few weeks, should be available and submitted for peer review in publications. Quite obviously, this will not constitute absolute proof that a kidney or liver graft will subsequently function exactly as would a human organ. But at least, there will be potent arguments in favour of patient survival extension. At this point there will arise an unsolved problem of controlling chronic rejection in the long term. The second field of necessary preliminary endeavour is patient infection by graft microorganisms.

As previously mentioned, even defective retroviruses integrated in pig genomes can be detected. As detailed as possible mapping of integrated viruses of porcine genomes is needed before animals are used, but this has not yet been done. Animals selected after testing must be reared in specific pathogen free conditions under supervision by independent health authorities that would have exclusive rights to deliver identification permitting their use for xenotransplantation.

Arguments in favour of choosing one or other type of graft as the first to be used in human experimentation are still inconclusive. Some opinions would prefer to begin with so called "mechanical" organs like the cardiac pump, rather than "personal" ones like neural tissue. In fact, using neural foetal pig cells to treat for Parkinson's disease is akin to introducing into the cranium a pump for the production of neurotransmitters such as dopamine.

Xenotransplants of nerve cells potentially capable of modifying the behaviour and therefore the personality of a functionally impaired patient (Parkinsonian syndrome, for example), if the procedure did become a working possibility, would only be ethically acceptable if the result was to reinstate a previously existing function. If, for whatever reason, the aim were to modify a personality, it would be totally unacceptable. Liver xenotransplants should probably not be the first to be attempted. The extraordinarily complex functions of this organ differ slightly from one animal species to another. It is difficult to predict, with any accuracy, the functions pig liver will carry out in a human body. Furthermore, we know absolutely nothing about human receptors to animal proteins. In fact, a totally transgenic liver still seems rather a utopian project. The technical difficulties inherent to initial clinical trials are such that of necessity they will be performed in one of a very restricted group of centres approved by national health authorities.

Once the necessary scientific progress has been made, it would seem logical that the first patients would be those for whom no other form of therapy is currently available—patients in a critical state and unable to wait for a human organ to be available, or other patients who cannot be put on a waiting list for whatever reason. It must therefore be clear that the technique's effectiveness will have to be judged at first on the results of particularly difficult cases.

A major ethical problem will frequently arise: the choice will not be so much "xenograft or death", as "xenograft now, or wait for a human organ for an indefinite length of time". As it seems likely that at first the reliability of a human organ will be considerably greater than that of a xenotransplant, this will be a difficult decision to take. The choice must therefore be perfectly unencumbered. If a patient turns down a xenograft, the possibility of obtaining a human graft must be retained with identical chances. In the same way, it is likely that in the initial phases, xenotransplants will be offered to patients in the acute phase of their sickness, as a possibility of gaining time before a human organ becomes available.

There again, it is important that the fact of having had the benefit of a xenotransplant does not in any way spoil one's chances of human organ transplantation at a later date (1). Patients' consent to xenografting, even initially, does not appear to be very different from what is generally practised now for human organs. Patients must be completely informed about the experimental nature of the procedure, the successive phases, the risks and the alternatives. Qualified personnel capable of coping with psychological, scientific, and ethical issues must definitely give such information.

Consent should be requested in the same manner as for allografts. During the experimental clinical phase, it would certainly be preferable to steer clear of all those who cannot give free and informed consent, such as children and unconscious patients, although the very young (less than one year old) have lower natural antibody counts than adults so that hyperacute rejection is likely to be less severe.

One special problem in connection with consent may well arise during the therapeutic trial phase. It will be absolutely essential during this phase to engage in detailed and prolonged epidemiological monitoring of these patients. It is also very likely that special precautions will be demanded of such graft patients to prevent any dissemination of some new pathogenic organism.

This means of course that patients will have to consent and even commit themselves in writing for this monitoring procedure to take place for some considerable period of time during which they would be obliged to accept the possible constraints of quarantine measures of the kind which are imposed on carriers of a risk of dissemination of an epidemic disease. Finally, it should be noted that even for those facing death, xenotransplantation should not be presented in over enthusiastic terms. In the event of defective xenotransplant function, burdensome therapy will be needed to prolong life to some extent, but quality of life will be very poor.

26.16. SOCIAL ACCEPTANCE OF XENOTRANSPLANTATION

Overall Attitudes to Xenotransplantation

Development of sophisticated therapy techniques such as xenotransplantation depends above all on the social context in which it takes place. It is very difficult to predict the attitudes of individuals when faced with routine use of a technique, which for the time being is still confined to the laboratory. This is, in fact, the heart of the problem. Xenotransplantation is not just a transition solution to the shortage of human organs, in the long term it would become a routine procedure, in approximately the same way as human kidney transplantation. Part of the reactions of the public to xenotransplantation will, in fact, be no more than an extension of its reactions to human organ grafts. Replacing a failing organ by another is part of the ongoing quest for prolonging life.

The importance of the quality of life versus a simple extension of life is certainly considerable, but the problem is in no way specific to xenotransplantation. Is it better to owe a "new" organ to a dead human or to an animal bred and sacrificed specifically for that purpose? The question is bound to arise, as this double possibility of transgression comes up against notions of sacred frontiers between the living and the dead on one hand, and humans and animals on the other.

At present, very little data is available on the impact of xenotransplantation concept on the public's imagination, and on the reactions of various segments of the population to such a subject. A survey was recently conducted in the United Kingdom to find out more about young peoples' attitudes to xenografting. Results were as follows: 55% of an 11-18 year old age group was considered; some of them enthusiastic, that research on the subject should be pursued, but 45% held an opposing view. Questions put to various sections of population revealed that only about 40% of them would accept a xenotransplant. However, the percentage is considerably increased if the individual is personally concerned, and in the case of risk to life, it goes up to 78%.

Nevertheless, about 75% of those questioned see xenotransplantation as a possibility for the future. In fact, answers depend very much on how the question was put. So the situation is full of contrasts and it is clear that total transparency on the progress of research will be useful for the idea of xenotransplantation to be acceptable to a majority of people.

Aversion to Xenotransplantation on the Part of Recipients

This rejection of the technique can be explained in various ways. Several religions consider pig as unfit for human consumption, but when religious authorities were asked, they were rather positive using pigs as graft donors. Some individuals hold philosophical beliefs to the effect that the life of an animal is as valuable as the life of a human being, and that we cannot claim for ourselves the right to sacrifice one to save the other.

However, such views are too uncommon for them to be the reason why in previous enquiries more than 50% of those polled were reticent. The main problem is probably connected to the individual's notion of identity as related to the perception of that individual's body. When an individual identifies with all the organs of his body, it is difficult enough to accept the thought of even a human graft. There will also be a tendency to establish a hierarchy in the importance of the organ—kidney is better than heart to which still symbolic emotional importance is associated. An animal organ will be thought of as even more destructive of identity. In fact, a graft of any kind breaks through the usually intact frontier between self and nonself, and the psychological repercussions of the "violation" has been studied with attention in the case of allografts. With xenografts, there is a further violation of the frontier between humans and animals, and this is of particular significance. An individual who manages to transcend the purely organic level of his being and who considers that the essence of humanity is thought, which precisely permits transcendence, will have little or no aversion to animal grafts.

On the contrary, those who will not, or cannot differentiate between their humanity and material being, will reject xenotransplantation. They will feel that the graft degrades them to the level of a human-animal chimera in which their humanity is dangerously diluted. The opposite line of reasoning is easily accessible. A human being deprived of kidney, colon, or even heart, while being sustained by extra-corporeal circulation, is no less human. Is a cancerous liver, which progressively destroys a man, likely to make him more human than the porcine liver that keeps him alive and thereby enables him to retain his human definition?

One could go so far as to say that awareness of the organic animality of humans will enable them to empower their neuronal, cortical, linguistic and relational capacities with more transcendental qualities than their livers, hearts, or other organs, and will teach them to reject any identity between humanity and bodily organs. The concept of human dignity signifies that the respect, which must be given to the integrity of the organs of the human body, does not *ipso facto* signify that the humanity of a human being is contained in those organs. There again, much more research is required to grasp the motivations of those, who would be among the first to receive xenotransplants, so that the experience gained might serve to better understand, inform, and counsel subsequent patients.

26.17. LEGISLATION AND XENOTRANSPLANTATION

Once the technical progress makes transplantation routine, the market for it will be gigantic, though it is difficult to estimate its size at this time. Global evaluations have been made ranging from $1.4 billion to $6 billion. Some authors have estimated that in the next few years, 50,000 pig hearts and 40,000 pig kidneys could be implanted in human beings every year. As a consequence, it appears that powerful biotechnological companies in the United Kingdom and the USA support most laboratories working in the field of xenotransplantation. These same corporations market immunosuppressants, which would be used to a much greater extent if xenografts were successful. It is clear that if the technique is a success, the present centralized system for the collection, allocation, and distribution of human organs that functions in most developed countries would be replaced by a commercial system, which would definitely need to be circumscribed by the legislation.

Although in other concerned countries there is no legislation to control access to transplants, France was the first country to have introduced it in its law dated July 1, 1998, which bears on the reinforcement of health supervision and sanitary checks on products for the use of human beings, and the use for therapeutic purposes of animal organs, tissues, and cells. This scientific anticipation provided by the law does not create *de facto* a judiciary anticipation, and although CCNE welcomes the *avant-garde* nature of the law, it finds that it is simply a guideline with specific recommendations which do not, obviously, imply the principle of authorisation to practise xenografts.

Rules of good practice for the use of animal cells, organs, and tissues are prepared by the French Agency for Sanitary Safety (*Agence Française de Sécurité Sanitaire*), after consulting the French Transplantation Establishment (*Etablissement Français des Greffes*), and approved by the Minister in charge of Health. The Minister's decisions set out rules of good practice regarding selection; production; and breeding of animals, sanitary conditions, which animals must meet; and rules of identification of animals so as to ensure final products can be traced. This is not a system of clinical trials that must comply with the *Huriet-Sérusclat* Law. Instead, the system is based on ministerial authorisation. In the United Kingdom, a working group was tasked with drawing up conclusions, which were adopted in 1995 by the Nuffield Council on Bioethics. In the United States, experts have drafted a code of good practices regarding xenotransplants that was published by the US Public Health Service, and approved by the FDA, the National Institutes for Health; and the Center for Disease Control and Prevention. As for the Parliamentary Assembly of the Council of Europe, they have proposed a moratorium, put forward by Mr. Plattner, for decision by the Council of Ministers. As a result, a working group was created.

Xenotransplantation raises the question of animal status. Animals have no judicial *persona*. Although they are considered as objects in law, they, nevertheless, are granted legal protection as living creatures (laws on cruelty to animals and vivisection). In societies, where sacrificing animals for various uses has been accepted since time immemorial, developing a new use, i.e., xenografts, is not likely to be a major problem. However, conforming with decent conditions for breeding and killing the animals implies that these conditions shall be considered as inherent to the practice of xenotransplantation. Respect for humanity implies that humans are duty-bound to treat animals with respect, though not necessarily the existence of animal rights as such.

This entails for donor pig rearing facilities, the obvious necessity of complying with European legislation as regards laboratory animals, and in particular with the rules governing the use of Genetically Modified Organisms (GMOs), in confined conditions during the research phase and in dissemination conditions later. Although transgenic animals express human proteins, which do not make them more human and those who breed and use them need do no more, but no less, than comply with these rules.

It is clear that the production of grafts, their use, and clinical trials will have to be strictly controlled by legislation and monitored by health authorities, in compliance with the existing law.

- The law on bioethics and sanitary safety will certainly need to deal with avoiding risks inherent to transplantation.
- A health authority such as the French Agency for Sanitary Safety would need to define rules to be followed by donor animal breeding facilities. He would have to make sure that good practices are followed while harvesting and transporting grafts, and give official approval to breeding centers.
- The national sanitary and health authorities concerned, in particular the Ministry of Health and the French Transplantation Establishment, will be required to approve centres best able to perform xenotransplants, and at least initially, to validate launching the first clinical trials and selection of early patients.
- Finally, the same bodies will need to initiate and monitor closely the essential pidemiological study of transplanted patients.

26.18. ETHICAL ISSUES

The main ethical issues raised by this technique are as follows:

1. Is the principle of using animal organs to improve the survival and well-being of humans acceptable?

Reply to this question is certainly affirmative, though the convictions of those who consider that all life has equal value and that therefore it is unacceptable to sacrifice an animal to ensure human survival, must be respected. CCNE is well aware that this issue is under debate. It considers that it is of sufficient importance to warrant further thought. Were xenotransplantation to become a general practice, it would also be possible to sidestep intricate ethical problems connected to harvesting them from living human donors, or cadavers, in particular as regards sampling embryonic tissue.

2. Is the risk of human contamination by hitherto unknown infectious organisms originating from the graft sufficient to refrain, albeit momentarily, from launching the clinical phase?

The question has not yet been answered. Some opinions are that the risk of triggering an irrepressible pandemic calls for a moratorium, at least temporarily. Others consider that if effectiveness and tolerance were no longer a problem, using the technique for humans would be acceptable immediately if the following drastic precautions are observed:

- Exclude murine tissue and particularly non-human primate organs, because their cells carry genomes that abound in retroviral sequences, which could be activated in humans.
- Prefer the use of transgenic pigs, since the species is further removed so that retroviral sequences are less likely to contaminate humans.
- Use swine reared in specific pathogen free environments, so as to be able to declare an absence of any transmissible infectious agent other than viral sequences integrated in the genome.

Intensive research must be encouraged to identify retroviral sequences in porcine genomes, so as to try and eliminate carriers, but much time and effort will be required to achieve this, because it will always be difficult to be certain of the presence or absence of late-appearing non-conventional agents. It could simply be said that, taking into account proximity between humans and pigs since the latter became domesticated and potential dangers involved, although the probability is very small, it is particularly disquieting.

When necessary trials are conducted involving inter-animal species xenografts (pigs/nonhuman primates), attempts should be made to identify events, which trigger the mobilization of porcine viral sequences and their recombination with primate sequences. Any new technique is risk-bearing. Clinical experiments can only start once the risks have been evaluated and benefits, expected with the consent of fully informed patients. However, xenotransplantation is different from allotransplantation, as the risk of infection is not confined to a single patient and extends to the population as a whole. This is no longer the classical "patient/physician" situation. There is a third partner and a major one—society as a whole—with an evaluation to be made of the individual benefit to collective risk balance. This goes to show that the ethics of xenotransplantation must be discussed in the most extensive public forum. Moreover, since epidemics do not respect national borders, debate must be international.

3. Is it now possible to control the immune rejection reaction sufficiently for early clinical trials to be allowed?

Transplant rejection occurring in the first few days may be reversed with the help of immunosuppressants, in approximately the same way as is the case for human allografts. However, hyperacute rejection, which appears as soon as the host blood irrigates the graft, is still incompletely controlled. Transgenic pigs that no longer express receptors causing rejection, or able to inhibit *in situ* the harmful activity of the complement, would seem to be the solution. Such avenues of research must continue to be explored actively. Furthermore, an intensification of experimental organ transplants from transgenic pigs to non-human primates is certainly desirable, since they are an excellent, albeit very costly, model for human xenografts. If acute or hyperacute rejections could be suppressed, there would still remain the problem of chronic rejection, which could allow infectious agents to make use of therapy to produce new diseases.

4. Does xenotransplantation raise specific social or individual acceptance problems?

There has been too little research so far to enable us to judge objectively the degree of acceptability of xenotransplantation, but it does seem that there is some widespread aversion. It is not that there is a fear of infection, but there is mental inability to transgress the man-to-animal barrier. A considerable amount of understanding and counselling will need to be deployed for prospective xenotransplant patients. Once that has taken place, obtaining free and informed consent must proceed on the same basis as for allografts. Let us also note that for quite a long time, xenotransplantation will remain nothing more than an alternative to the more reliable technique of human organ transplantation.

If xenotransplantation meets with growing success, we shall need to take care that its very success does discourage and demobilize organ donor volunteers. There is a real risk of moving from a situation where solidarity and a sense of responsibility prevail under the constraint of organ scarcity, with life saving objectives, into a situation of purely economic resource where convenience plays a growing role to the detriment of preserving survival.

5. Are there any new legal repercussions as a result of xenotransplantation?

It is only in France, so far, that the required legal framework has been drawn up to apply the technique to humans. Large biotechnological corporations are expecting a vast potential market (evaluated at 6 billion dollars for 2010) to which should be added the immunosuppressant market. A commercial system, based on market forces using resources strong enough to thwart state regulations, may well replace in the long run the benevolent gift of human organs. Therefore, it will need to be carefully ruled by legislative and health authorities, not just in our own country, but also by international standards because of the effects of globalisation.

The use of xenotransplantation could be included in the review of the law on bioethics and sanitary safety, aiming at, prior consent by national health authorities before implementing the first clinical trials; institutions authorised to conduct these early therapies approved by national health authorities; official approval and sanitary monitoring of donor animal rearing facilities, sampling procedures, and transport of grafts; monitoring of adherence to European legislation as regards the use of laboratory animals throughout all the phases of the procedure, taking into account legislation on genetically modified organisms when using transgenic pigs. It must be remembered that the existence of a law is not in itself a condition creating licit procedures. Ethical reflection fills the gap between law and the very principle of clinical trials involving human beings.

6. Is xenotransplantation a vital necessity in medical terms?

For certain patients awaiting an allograft, supposing its effectiveness has become patent, the answer is yes. However, there will never be many such cases. Were xenotransplantation to become routine, there would be an increase in indications so that the economic impact would be significant.

7. The issue of information.

The principle of applying xenotransplant techniques to humans supposes the utmost transparency as regards previous and necessary animal experimentation, and particular vigilance during follow-up. Once a xenograft has been performed, society would find it difficult to accept careless monitoring. Follow-up and pertinent information must be made known to the public without excessive media commotion, but so that it is freely available to anyone expressing an interest.

8. Decision to apply the technique.

A ranking of reasons for circumspection must be outlined. "Conventional" rules of caution: Lack of knowledge on existing chances of success. Lack of control over hyperacute and chronic rejection. Encouraging animal experimentation, but nowhere near sufficiently effective to apply to humans. "New" rules of caution, consideration of the ideology of an absolute boundary between humans and animals, even though such consideration is rejected by certain creeds, populations, or social groups. The potential risk of infection, which has by no means been validated, but the possible dangers of recombination are out of all proportion to the small number of patients who could benefit from xenotransplants.

CCNE is not requesting a moratorium on pre-clinical xenotransplantation research. However, it is felt that prior and obligatory success with animal models, follow-up of effectiveness, maximum evaluation of the possibilities for protection from the risk of infection, and psychosociological research, must be demanded before moving on to the clinical phase, which is unlikely to emerge in the near future. It advocates a continuation of scientific animal experimentation, providing such research is properly conducted, i.e. through transparency, high standards, and founded on the belief that saving human lives cannot be at the expense of presently accepted good practices respectful of the relationship between humans and animals.

Legal Issues that should be Considered

Xenotransplantation raises several legal issues; one of these issues is informed consent. It is generally agreed that those who participate in clinical trials would have to agree to a number of unusual restrictions on their freedom.

Several countries are developing guidelines for research and clinical trials, and patients may be required to consent to some or all of the following:

- Restricted travelling activities
- Never having children
- Identifying all their sexual partners to authorities
- Never donating blood
- Submitting to life-long surveillance
- Agreeing to a post-mortem after death

Since it could take years for an infectious disease to develop, it is unlikely that patients would be permitted to exercise the traditional right to withdraw their participation. New public safety laws might have to be created in order to enforce the strict monitoring requirements. Because of the potential risk of viruses to society at large, close contacts of xenotransplant patients might have to agree to life-long monitoring and would also have to give informed consent. Researchers, both in academic institutions and biotechnology companies, are the driving force behind xenotransplantation. Issues around business ethics, such as transparency, accountability and conflict of interest need to be considered. There are other issues such as liability and fair access to medical resources that would need to be explored.

Sources of Animal and their Issues

Non-human primates (e.g., apes, monkeys, chimpanzees, and baboons), pigs, dogs, rabbits and rodents (e.g., mice, rats) are being used in large numbers for xenotransplantation research all over the world. Currently, pigs are the source animals of choice. The preference for pigs stems largely from the more serious risks of infection from non-human primates. Non-human primates are also expensive to raise and breed slowly. Pigs, on the other hand, are not seen as being as developed as non-human primates, are easy to breed, and their organs are about the right size for humans (although it is not known how their organs will function in humans). Opinions about the rights of animals vary considerably. In general, there is a growing concern about the use of animals in experiments, particularly for testing cosmetics, pharmaceuticals, pesticides, chemicals and household products.

People for the Ethical Treatment of Animals (PETA)—one of the largest animal rights organizations in the world—operates under the principle that animals are not ours to eat, wear, experiment on, or use for entertainment. On the other side are those who say that humans have an absolute right to life that overrides any rights of other creatures, because humans rank highest in the natural order.

According to most opinion surveys, the majority of respondents agree that animals should be used in a controlled and regulated way, with due regard to their welfare, if there is a direct benefit to humans. The need to produce disease-free animals may have an impact on their welfare. In an effort to reduce the risk of infection in humans, pigs for xenotransplantation research are being bred and raised in sterile laboratory environments. The Canadian Council on Animal Care oversees the use of animals in publicly-funded biomedical research.

Cost Effectiveness

It is difficult to estimate how much xenotransplantation would cost if it becomes an accepted medical treatment. The costs associated with xenotransplantation at the clinical trial, post-clinical trial and implementation stages, would be borne by industry and federal and provincial governments. It is likely that a xenotransplant would be as expensive as other transplants.

The cost of a kidney transplant in British Columbia is approximately $20,000, with an additional yearly cost of about $6,000 for anti-rejection medication. The average cost of a heart transplant is reported to be $80,000. In Ontario, organ transplants cost about $47 million a year and that figure is predicted to increase to more than $121 million by 2005.

Additional costs for xenotransplants would include the actual cost of the animal organ (breeding, feeding, housing, slaughter, disposal of the rest of the body), long term surveillance of the patient and close contacts, mandatory autopsy on death and possible quarantine costs. Costs associated with xenotransplants may be balanced by the cost of treatments for people waiting for organ transplants. For example, about 700,000 patients around the world suffer from kidney disease and must undergo regular dialysis treatment at an annual cost of about $19 billion. It costs about 60% less to transplant a kidney than to keep the patient on life-long dialysis. Since any disease treated by human-to-human transplants could potentially be treated by xenotransplants, the demand for xenotransplants could be high if it becomes an accepted procedure.

Some fear that this increased demand could drive health costs up or divert monies from other health initiatives. Others say that in the long term, because of the ability to plan and coordinate transplantations, xenotransplantation may become an economical alternative to human transplantation.

Some Necessary Issues to be Addressed for Xenotransplantation

Ethics review

There is no federal legal obligation for ethics review and approval for research involving humans. However, in reviewing clinical trial submissions, it is at present necessary that the sponsors certify that a research ethics board approval of the protocol be approved before initiating the trial.

A condition for receiving funding from one of the three federal research councils is that any type of research involving human subjects be reviewed and approved by the institution's Research Ethics Board. It is further required that all researchers working within an institution, which receives public research funds, submit their protocols for review and approval by the Research Ethics Board.

Importation and exportation of organs, tissues or cells specifically designated for xenotransplantation

There are no explicit regulations covering the importation/exportation of organs, tissues, or cells specifically designated for xenotransplantation. In the absence of explicit regulations, xenotransplants are considered to be therapeutic products (drugs or medical devices) and are subject to the requirements of the *Food and Drugs Act and Regulations* and *Medical Device Regulations*.

Basically, the *Food and Drugs Act and Regulations* apply to the sale of all drugs or medical devices in Canada, whether manufactured in Canada or imported. There are no explicit requirements for importation.

Importation and exportation of genetically modified animals, organs, tissues or cells

The importation of animals into Canada is governed primarily by the *Health of Animals Act and Regulations*. The aim of this legislation is to prevent the introduction of animal diseases into Canadian livestock. Controls over the importation of genetically modified animals are only imposed at this level.

Infection risk control

There is currently no explicit federal legislation in Canada addressing xenotransplantation infection risk controls.

Animal protection

There is no explicit legal framework that applies to the protection of animals used for xenotransplantation. There is no clear cut regulating body for regulating the use of animals for experimental or other scientific purposes. If xenotransplantation becomes a recognized medical practice, there currently is no legal protection of animals that would be used, other than the general offences involving cruelty to animals and the general conditions governing the treatment of animals while they are transported.

Pre-clinical trials

Currently, studies are being carried out using laboratory animals in many countries. These pre-clinical or experimental trials do not involve human patients and are not regulated by the regulatory bodies of any country.

Clinical trials

Xenotransplants are considered to be therapeutic products (a drug or medical device) and are subject to the requirements of the *Food and Drugs Act and Regulations,* and *Medical Device Regulations*. In clinical trials, human patients are required to provide consent prior to taking part in authorized research under very controlled conditions.

A physician or sponsoring company can submit a request to health department to conduct a clinical trial. The decision on whether to approve a clinical trial is based on safety, quality, effectiveness and a benefit/risk evaluation. There are no explicit regulations covering clinical trials involving xenotransplants. The Minister of Health has the authority, under *Food and Drug Regulations* and in the *Medical Device Regulations*, to prohibit the use of xenotransplants in clinical trials if it is in the interest of public health to do so.

26.19. XENOTRANSPLANTATION AND ITS VIABILITY

The concept of xenotransplantation, in particular whole organ transplants, is not new and the procedure has largely been unsuccessful. The main scientific challenges to xenotransplantation are immune rejection and infection. While drugs have been available to sufficiently suppress the immune system for human-to-human transplants, they have not been successful in animal-to-human transplants.

To date, survival rates for the recipients of xenotransplants have been poor, though some cellular xenotransplants seem better able to resist immune rejection. In the 1990s, researchers began to focus on cell transplants, which in some cases are easier for a patient to tolerate. Several hundred cell transplants took place during the 1990s in Russia, the United States, Poland, Czech Republic, Switzerland, Sweden and New Zealand. Most were to treat patients with diabetes, Parkinson's, or Huntington's disease, though one case involved a baboon bone marrow transplant to a patient with HIV in the United States, and six patients with ALS (Lou Gehrig's disease) received hamster cells in Switzerland. Scientists are trying to overcome immune rejection by adding human genes to animal cells to make them more acceptable to a patient's immune system.

Transgenic pigs have been developed in various countries and shipped to researchers in Canada. The *National Post* recently reported on pre-clinical (animal-to-animal) studies by a researcher at the Robarts Research Institute in London, Ontario, who transplanted the kidneys of more than 20 genetically modified pigs into baboons.

These pre-clinical studies indicate the extent of progress being made in clinical studies involving humans. In December, 2000, the International Society for Heart and Lung Transplantation said that current experimental results indicate that a clinical trial (animal-to-human) at present would be premature, and these trials should be undertaken only when the risks have been shown to be minimal. The society said that clinical trials would be justified and should be encouraged when researchers achieve acceptable results in animal studies, but only if there is little possibility for the spread of animal viruses to humans.

The society says that pig-to-human heart transplants could be feasible within a few years. Some experts believe that moving to clinical trials is the only way the uncertainties surrounding xenotransplantation can be answered. Others say it is too early to consider clinical trials. And there are those who believe that even if animal-to-human transplants become a safe procedure, it does not necessarily mean we ought to do it.

Central Issue is the Ethics of the Decision to Launch Clinical Implementation

By assessing the balance between risk and benefit, the principle of caution applies and must first take into account effectiveness. But this must be a principle of caution, which rests more on the notion of responsible practitioners and researchers, than on a rejection of progress. A chimpanzee with a porcine liver graft enjoying a normal life would demonstrate that xenografting is technically possible. Once scientific, infectious, immunological and psychological problems have been solved, opening the way for applying xenotransplantation to humans, consideration would need to be given to the matter of human mobility in Europe and globally. OECD has raised the issue of importing genetically engineered animals and has requested a global communications network on risks involved. It is hard to imagine that one European country could authorise the principle of xenografting without consulting neighbouring countries. CCNE's present thinking is expressed in a time of anticipation. When moving on to clinical practice becomes based on acceptable scientific criteria, CCNE will wish to restate its position.

SUMMARY

Transplantation is not a new procedure in the medical history. Corneas were transplanted in 1905, although the procedure did not become routine until the 1940s. Skin was transplanted first in the 1920s. In the 1950s, a kidney was transplanted from one twin to another. In the 1960s, the first livers and hearts were transplanted.

Human-to-human transplantation has become a successful way of treating various human diseases and conditions, such as heart disease or kidney failure. However, human tissue and organ transplantation usually depend on donations from people who have died, most often as the result of brain damage through stroke or accident. Over the past 20 years transplants have become more frequent and successful. The number and scope of transplant procedures have also increased to include a broader range of tissue, cellular and organ transplants, such as transplants of insulin-producing pancreatic islet cells for treatment of diabetes. Transplantation has become the treatment of choice for many patients suffering from organ failure or complications arising from diseases of specific organs. This treatment option has been making rapid progress. Several factors have contributed to these improvements. These include increased efficacy of drug therapies used to treat and prevent organ rejection, technical advances in surgery, diagnostic test methods for monitoring patients, improved histocompatibility testing, and enhancement in organ procurement procedures. Earlier and more accurate detection of rejection as well as a more comprehensive understanding of the immune system have also played roles in improving patient and graft survival.

The term 'xenotransplantation' derives from the Greek word xenos, meaning foreign, and all the issues surrounding this procedure spring from the notion of associating something 'foreign' with a human body. In recent years the transplantation of human cells, tissues and organs (allotransplantation) has dramatically improved the survival and health prospects for people suffering from life-threatening illnesses. Xenotransplantation involves transplanting living cells, tissues or organs between species, and in particular, from animals to humans. It is being considered because worldwide the demand for human cell, tissue and organ donation far exceeds the supply. This would be true even if everyone agreed to donate their cells, tissues or organs when they die. Furthermore, the world's population is ageing, which will place increasing demand on donation as more people suffer from chronic degenerative diseases. Currently there just are not enough human donors to go round, and as a result people on waiting lists are dying.

It would seem obvious, when searching for ways to reduce the rejection reaction to xenografts, to use animals as physiologically close to humans as possible, such as old world monkeys and the apes (for example, chimpanzees). However, this very closeness creates complex ethical problems as well as an increased likelihood of cross-species infection. There are strict regulations in number of countries about the use of non-human primates in research, so it is unlikely that these species would be used for xenotransplantation research in these countries. At present the pig is the favoured animal for research into xenotransplantation, because they grow quickly to about the right size, produce large litters and can be reared in specific pathogen-free conditions (where some but not all microorganisms are excluded). In terms of ethical concerns, the fact that pigs have long been used as a source of meat reduces—but certainly does not eliminate—the concerns of many people, especially when weighed against the possible benefits. However, recently it has become clear that there is also the possibility of cross-species infection from pigs. This takes the issue of whether to perform such a procedure out of the realm of an individual decision to take a personal risk, usually for the sake of a therapeutic effect, into the realm of the safety of the xenotransplantation recipient's close contacts and the community at large.

Xenotransplantation raises a number of ethical issues. For example, how far is it acceptable to interfere with nature in order to try to prolong human life? Would crossing the species barrier have an impact on what it means to be human? Similar questions were raised 40 years ago when human-to-human transplants were being considered. Then, questions were about crossing the frontier between self and non-self. Critics of xenotransplantation say the procedure goes beyond medical innovation, crossing the very boundaries that define us as human. While few of the world's major religions have made formal pronouncements on xenotransplantation, some religious believers are willing to accept it in principle provided unnecessary animal suffering is avoided. Others find it unacceptable, perhaps because of their relationship with the natural world.

EXERCISE

1. What do you mean by transplantation? What is its importance in the society for the future generation?
2. Briefly write about the present situation of organ transplantation in the developed and developing countries with emphasis on India.
3. Define organ transplantation and its impact on the human generations suffering from various disorders. Add a note on its scenario in India.
4. What is meant by xenotransplantation? Add a note on ethical issues associated with the xenotransplantation performed in humans.
5. Write a note on autograft, allograft, isograft that are performed in the humans for the varieties of treatment procedures in the medical biotechnology.
6. Describe various types of organ transplantations used in the medical field and mention their significance in the present situations.
7. Explain the different sources of organ transplantations and write the demerits of each source.
8. Enumerate the different methods of organ transplantations commonly performed in humans and its essential requirements for well being of humans.
9. Give an account of genetic diseases associated with the humans that are identified by prenatal diagnosis.
10. What is meant by organ transplantations? Briefly write about its historical development.
11. How does the organ transplantation help the individual to recover from the surgery. Add a note on its social impact.
12. Write about the historical development of xenotrans-plantation.
13. Briefly explain how ethical issues are important before carrying out the xenotransplantation in case of humans.
14. Elucidate the guidelines that are necessarily adopted for xenotransplantation in the human beings.
15. Distinguish between isotransplantation (isograft) and xenotransplantation, and discuss their ethical issues in detail.
16. Discuss the risk factors considered for xenotransplantation in the human beings.
17. Explain the applications of xenotransplantation in the biotechnology business, and name the companies that are involved in research in xenotransplantation?
18. Explain how xenotransplantation is gaining much importance in the international scenario. Add a note on the developing countries set-back for regular use of xenotransplantation.
19. Critically comment on the ethical, social and legal issues raised by the newer technology of xenotransplantation in the present understanding of immunology in greater depth.
20. Give a note on the promises of the xenotransplantation in treating the affected individuals from the life threatening disorders that are not having any alternate therapies.
21. Describe how social acceptance is important for the success of xenotransplantation, and mention its present conditions in the world.
22. Illustrate some of the issues that are to be addressed for xenotransplantation for the powerful adaptations in the medical biotechnology.
23. Write notes on legal issues that are necessary for the xenotransplantation for the unregimented adaptations in medical surgery.
24. What are the immunological problems associated with the xenotransplantation? Write about the molecular solutions that are available to solve such problems.

27

CHAPTER

Public and Non-Governmental Organizations (NGOs) Participation in Biosafety and Protection of Biodiversity

27.1. INTRODUCTION

The overall aim of the protocol on biosafety to the convention on biological diversity is to ensure that countries importing, exporting and using LMOs have the opportunity and capacity to assess possible risks to the environment (taking into account risks to human health) posed by the products of modern biotechnology. Apart from a regulatory framework, an administrative structure and risk assessment methodologies, mechanisms of access to information and public participation form an integral part of most biosafety frameworks. There is an increasing precedent for all multilateral environmental agreements to contain provisions regarding public participation, and the responsibility of governments to engage in awareness-raising activities. The Cartagena Protocol on Biosafety is no exception. The emphasis on participation, and consultation is premised on the idea that the involvement of all stakeholders is critical to the effectiveness of any regulatory framework. It is also acknowledged that without higher levels of public consent or consensus than exist at present, decisions to allow the commercial growing of GM crops might provide a precarious basis for proceeding with GM crop development. At a more fundamental level, it is also possible to argue that people have a right to be informed about and consulted about decisions that have a direct impact upon their lives—in this case through the food they eat.

It is difficult to pick up a newspaper or listen to a news broadcast that does not include some mention of genetically modified organisms (GMOs). The application of genetic engineering to agriculture has raised considerable hopes for increased production and efficiency, but more notably it has also raised wide opposition. The controversy involves the future of biotechnology in both the North and the South. The lines of the controversy regarding the potential for the South are quite clearly drawn. Those who are optimistic about biotechnology argue for the need to increase food production and point to the possibility of addressing the problems of marginalised farmers. Opponents question the safety, relevance and equity of the new technology. This chapter examines the arguments on both sides, but pays particular attention to the case made by a number of NGOs who have been effective in bringing the issues to public attention.

In recent years, international agencies and governments have gained some experience in fostering public participation in policy processes in a range of sectoral and some cross-sectoral policy frameworks. These include the development of National Strategies for Sustainable Development (NSSDs) and national Poverty Reduction Strategies, both of which place strong emphasis on public participation in their design and implementation.

The establishment of National Strategies for Sustainable Development (NSSDs) by 2005 is one of the International Development Goals. Country ownership and broad consultation to open up the debate expose issues to be addressed and build consensus' are highlighted as key for strategy formulation. The principles to which the strategies should aspire are also relevant to the design national biosafety frameworks (NBFs).

- Being people-centred with effective participation.
- Having objectives tied to clear budgets.
- Being based on comprehensive and reliable analysis.
- Incorporating monitoring, learning and improvement.
- Being country-led and nationally owned.
- Having high-level government commitment to the process.
- Building on existing strategies, processes and capacity.
- Linking national and local levels.

A recent OECD guide on NSSDs notes that while sustainable development is a universal challenge, rather like developing biosafety regimes, the practical response can only be defined nationally and locally according to different values and interests. It notes 'A standardized blueprint approach is to be avoided, being at best irrelevant and at worst counter-productive' (OECD 2001). Instead, working with existing approaches and institutional arrangements

according to individual countries' needs, priorities and available resources is preferable. Efforts have been made to develop indicators for socially sustainable development, including indicators for levels and degrees of participation. For example, within the UK sustainable development strategy, the percentage of survey respondents reporting active participation in community events and the proportion of people satisfied with opportunities to participate in decision-making, are taken as indicators of participation (DFID 2002). Although most SSD indicators are national, participation is best measured at the local level where local indicators can be developed with local people. Indicators can be used to monitor strategy development and implementation. This is important for tracking progress, changing direction when necessary and promoting accountability for decisions taken (OECD 2001).

- Incorporating monitoring, learning and improvement.
- Being country-led and nationally owned.
- Having high-level government commitment to the process.
- Building on existing strategies, processes and capacity.
- Linking national and local levels.

A recent OECD guide on NSSDs notes that while sustainable development is a universal challenge, rather like developing biosafety regimes, the practical response can only be defined nationally and locally according to different values and interests. It notes 'A standardised blue-print approach is to be avoided, being at best irrelevant and at worst counter-productive' (OECD 2001). Instead, working with existing approaches and institutional arrangements according to individual countries' needs, priorities and available resources is preferable. Efforts have been made to develop indicators for socially sustainable development, including indicators for levels and degrees of participation. For example, within the UK sustainable development strategy, the percentage of survey respondents reporting active participation in community events and the proportion of people satisfied with opportunities to participate in decision-making, are taken as indicators of participation (DFID 2002). Although most SSD indicators are national, participation is best measured at the local level where local indicators can be developed with local people. Indicators can be used to monitor strategy development and implementation. This is important for tracking progress, changing direction when necessary and promoting accountability for decisions taken (OECD 2001).

Consument en Biotechnologie (Consumer and Biotechnology) [Netherlands] is a foundation set up in 1991 to enable consumer organizations to inform their grassroots support at an early stage about developments in the field of biotechnology (before they reach the market) and to mobilize them to influence policy developments when possible and appropriate. The organization functions as a centre of expertise on biotechnology and food production, tracking developments in the area of biotechnology and judging the risks and benefits associated with them according to criteria derived from the need to protect the environment and defend consumer interests. It also takes part in, and initiates discussion and consultation among groups in society including government and industry. Consument en Biotechnologie also undertakes information outreach through e-mail newsletters, publications and interviews in the media. They are supported in this task by working closely with international consumer groups on biotechnology issues.

Criteria of Organization for Economic Cooperation and Development (OECD) for Public and NGOs Participation

The Organization for Economic Cooperation and Development (OECD) guide highlights a number of lessons regarding effective participation. These are as follows:

- **Appropriate participatory** methods for appraising concerns, suggestions and ranking solutions.
- **A proper understanding** of all those with a legitimate interest in the framework and a concrete approach to include more disenfranchised groups.
- **Catalysts for participation**, non-governmental organizations (NGOs) and others to link national processes with the local level.
- **Specific activities and events** around which to focus participation.
- **A phased approach**, start modestly building on existing systems of participation and then seek to deepen participation, but do not think of design, implementation and monitoring as a linear process.
- **Adequate resources, skills and time**, effective processes often start slowly to build trust and require early investment of skills and resources. Costs can reduce over time, but a realistic budget and secured financial resources are key.

There is no single homogenous public, but multiple publics holding divergent views that need to be encompassed in a participatory exercise. The range of interests and views incorporated into the process needs to be appropriate to the issue in question, with all relevant stakeholders having the opportunity to take part on equal terms with other participants. The identification of stakeholders and analysis of their stakes is an important aspect of any participatory process, but it is more complex at the level of policy than in the context of projects, for which most stakeholder analysis techniques have been developed. Stakeholders tend to be more numerous and far-flung, more diverse, and their stakes harder to identify or predict. It is desirable for groups and individuals to determine for themselves what their stake is rather than have it proscribed from above.

Clearly, however, those initiating a process have to decide at the outset who should be involved. There is no simple, proven method for stakeholder analysis in the formulation and implementation of policy, but it is vital that actors convening participatory processes should consider carefully how stakeholders are identified and by whom; the heterogeneity of status, interests and knowledge that they bring to the process; the fact that different stakeholders will want or need to be involved at different stages in the process; and that it is probably neither desirable nor feasible for all stakeholders to be involved at all stages.

The framing of the problem to be discussed carries with it strong implications for the solution to be defined. Participatory processes have the potential to allow multiple forms of knowledge to enter the discussion, but there is a risk that the conventional ways in which dominant or influential players tend to think about and discuss the issue in question will constrain the deliberations within certain boundaries, rendering certain elements of a decision or policy non-negotiable or a foregone conclusion. Often, the way that multiple views are condensed and presented in a report at the end of the exercise still reflects the dominant ways of framing the issue. In other words, the diversity of opinions gathered through the consultation exercise may end up appearing as supplementary themes or divergent opinions, but remain within a conventional framework.

A report of the UK government suggests that clear objectives need to be defined for any consultation process, so as to establish why people are being asked to participate. In order to ensure a constructive outcome, these objectives need to be agreed by all participants, which means that participants must be involved in framing the questions to be considered.

Involving stakeholders in every aspect of the design process is crucial in order to achieve the best design and the highest level of commitment to the process.

27.2. TOWARDS THE LADDER OF PARTICIPATION IN POLICY

Representation and Intermediation

Since resources for information-sharing, consultation and participation are always finite and often scarce, governments tend to rely on a cascade mechanism in which they reach out only to those who claim to be intermediaries or representatives of broader groups (NGOs; trades unions; local political figures). These representatives are often then expected to do the rest. In some cases this is unproblematic; in many, however, it raises serious questions about who represents whom, how, and by what means they were selected or identified.

Again, resource constraints usually preclude governments investigating too closely the validity of 'representatives', but the general point remains that unless targeted efforts are made to extend information, consultation and participation to specific stakeholder groups, coverage will likely be patchy and will reproduce the spatial, educational, social, economic and political inequalities in a given society.

Providing a clear public record of the dialogue that takes place is important to the openness and transparency that participatory processes enable (Holmes and Scoones 2000). Openness and responsiveness between government and the organizers of, or participants in, participatory exercises are important in order to avoid communication problems or breakdowns in trust. Each side needs to have faith in the representativeness, honesty and openness of the processes and procedures set up by the other. For example, secrecy on the part of government, or sensationalism and distortion by any one group of participants (often through resort to the media) may inhibit such an outcome.

Transparency will also tend to enhance the perceived legitimacy and acceptability of the consultation process.

In this respect, it is particularly important to be honest about issues such as the source of funding or other support behind the participatory exercise. The UK Parliamentary Office of Science and Technology has suggested that participatory processes should include an evaluation element as a means of quality assessment. The evaluation needs to be tailored to the objectives sought and the methods used. As a prerequisite for any evaluation, a participatory process needs to have an agreed set of plausible indicators that help to reveal the successes and shortcomings of the process. For example, when people have been led to believe that their views will feed into government decision-making, there needs to be a clear way of demonstrating how this has happened.

Points of Procedure

Procedures need to be agreed by the participants, and should be designed to ensure transparency, accountability, and inclusiveness in order to encourage commitment to the process. Among the abundance of examples of public consultation and information-sharing in the PRS context, common procedural problems diminished the scope for good-quality interaction in the great majority of cases. What tends to go wrong with information-sharing, consultative and participatory procedures is summarized below.

Expectations

- Insufficient transparency on part of convening institution(s) as to their expectations and the parameters of the process.
- Insufficient attention to investigating interested parties' expectations and reconciling these with expectations of convening institution(s).

- Lack of clarity over who is accountable for the process and its outputs.

Timing

- Insufficient notice of forthcoming events or processes given to interested parties.
- Insufficient time allowed for genuine consultation or participatory processes to occur.

Information

- Not disseminated widely enough or in appropriate languages, styles or formats.
- Not disseminated in good time for interested parties to prepare their inputs in timely fashion, including consulting with constituencies if they are present as representatives.
- Not enough access to alternative, impartial analysis, produced by actors other than the principal institution(s) involved.
- Inadequate attention by convening institutions to provision of feedback to those consulted/participating on what happened to their inputs—on what basis these were/were not included.

Representation

- Consultation and participation are usually by invitation, using criteria which are not transparent nor devised on the basis of close knowledge of the full range of interested parties.
- Those parts of the population, which are hardest to reach—the poorest, furthest from capital city etc—are rarely represented.

Follow-up

- Insufficient provision made for conducting follow-up with all parties involved.

- Failure to take into account likelihood of changes in government etc., which could threaten the sustainability of the process.

Tools and Mechanisms of Biosafety Information Dissemination Strategy

A proactive information dissemination strategy requires a very different approach from the passive mode. A variety of tools and mechanisms, possibly in combination with each other, may be appro-priate. Examples might include the following:

Public databases or inventories of information on GM products, government notifications and permits, and current research and development projects that are open to public scrutiny. For countries with good ICT infrastructure, it may be practicable to make these available over the internet. Known examples include:

- A 'Gene Technology Book', which documents the state of science and technology (Austria).
- A publicly accessible Gene Technology Register, with information on products approved for market in the EU (Austria).
- The annual release of key information from government bodies to the public. In Germany, for example, yearly reports from the Central Commission for biological safety are published, and statements of the Bureau for Technology Assessment are made accessible to the public.

There have also been initiatives by various international bodies and private organizations to create publicly accessible databases and information gateways using the internet and other media. These are described below in more detail.

- **Workshops and seminars** targeted at particular stakeholders. For example, national and sub-national awareness-raising work-shops involving groups such as local councils, residents, consumers, farmers, industry representatives, journalists, teachers and so on. Training workshops on biosafety and biosafety regulation, to inform regulators, inspectors, laboratory workers and company officials about risk assessment, risk management and their legal responsibilities.
- **Printed information:**
  — Including technical fact-sheets on biotechnology or GMOs.
  — Leaflets targeted at the general public, consumers etc.
  — 'Use awareness' materials on biosafety precautions, and risk management for biotechnology practitioners, farmers etc.
  — Newsletters and magazines targeted in an accessible style and format.

- **Modern information and communication technologies** (ICTs) such as internet discussion forums and email news-groups. Where internet access is not a problem, such platforms can provide a channel for two-way communication.
- **National and local media** including newspapers, radio, and television, which can be used to inform people about biotechnology and biosafety issues, as well as to publicise new developments, meetings and events. Strategies for using the media to disseminate biosafety information are as follows:
  — Improving the quality and accessibility of the information released to the media is important. Workshops with journalists, to identify problems and potential solutions, may help in this regard.
  — Relationships with journalists may be improved if officials are helpful and cooperative with the media rather than secretive and defensive.
  — Information overload does not help to get messages across. Technical information presented in tables, charts, and figures may have less impact than a clear example or anecdote.

— It is more helpful to engage with public concerns and fears rather than dismiss them as ignorant or irrational.

- **Theatre** or other creative and performance methods may help to raise awareness and convey information in an accessible and engaging way.

- **Informal interest groups or 'learning communities'** may help to spread understanding of biotechnology and biosafety issues.

- **Supporting NGOs or civil society groups** to promote public awareness or mobilise public engagement and participation can be an effective way for governments to reach out to groups and stakeholders, which they may not easily reach by themselves. NGOs often have substantial networks, including contacts at grassroots level, and in some circumstances are more trusted than government organizations.

- **Public open days and demonstration projects:** Allowing public access to research stations and field trials may help to familiarize people with the science behind biotechnology, and enable them to see for themselves what risk assessment has been conducted and what precautions are in place. With a more promotional aim in view, some companies in the private sector have organized tours for journalists and other interested parties, 'farmer fielddays', as well as made use of video and internet channels to make people aware of their products.

- **Independent information bureaus** on LMOs may serve as a contact point for questions from members of the public. One such Bureau has been set up in Poland, and aims to respond to information requests received by telephone or e-mail by providing accurate scientific information on biotechnology for the public and media. In this case, the organization was set up by European.

Federation of Biotechnology's Task Group on Public Perceptions of Biotechnology is run by biotechnology students on a voluntary basis. The office also publishes a bulletin containing information on national as well as international developments in the area of modern biotechnology and food. At the international level there have been several initiatives to make biosafety information available.

27.3. EXAMPLES OF CONSULTATIVE PROCESSES ON BIOTECHNOLOGY AND BIOSAFETY

Brazil: Citizens' juries (facilitated by NGOs, trade unions and academics) in Fortaleza, Ceará, and Belem do Para, Para in 2001 and Rio de Janeiro in 2002.

Canada: Calgary, Alberta Citizens Panel, 1998. The federal government was also planning a multi-stakeholder consultation on the Biosafety Protocol in September 2002.

Denmark: Consensus conference, March 1999.

India: Citizens jury in Karnataka (facilitated by Action Aid India), and a scenario workshop in Andhra Pradesh (facilitated by academics and NGOs).

Kenya: National 'workshop' to discuss draft biosafety guidelines.

Namibia: Participatory stakeholders' workshops to draft the national policy on biotechnology and biosafety, 1998-99 (convened and coordinated by the Namibian Biotechnology Alliance).

Norway: Consensus conference on GM food, 1996.

Zimbabwe: Participatory 'brainstorming' workshops to discuss potential impacts of transgenic crops.

A theme we have sought to emphasise throughout this report is the importance of context to understanding, 'what works, when and why?' when it comes to promoting public participation, consultation, education, and awareness-raising around issues of biosafety. What works in some places is not going to work everywhere. Drawing out lessons of general relevance is therefore difficult. If applied in different countries, the same approach may yield significantly different outcomes. Nevertheless, reflecting on the range of different approaches and experiences reviewed in the case studies, it make the following six observations:

1. Public participation in the development of an NBF goes beyond the creation of an NBF document. It inevitably encompasses wider issues about the role of biotechnology, and requires ongoing participation in biosafety processes after regulations have been developed.

2. Despite the fact that parties face common challenges, there can be no universal prescription or standard formula for public participation and awareness-raising. What works in some places or in some circumstances will not work everywhere.

3. Governments have two roles. The first is to initiate participatory and awareness-raising activities. The second is to create an enabling environment for others, such as civil society groups and businesses, to take the initiative.

4. National biosafety processes involve development of a framework, implementation and monitoring. Currently, participatory efforts are not balanced across these stages.

5. To date, much more education and awareness-raising work is being undertaken than public participation and consultation.

6. There are plenty of participatory tools and approaches that have been effectively used in other policy domains and are currently underused in biosafety processes.

27.4. PUBLIC EDUCATION

A clear theme that comes through all the case studies is that information is a pre-requisite to effective participation. Therefore, the vital first step to promoting public participation must be to assess public awareness and information needs. Understanding what people know and what they want to know, in order to form opinions and make an informed judgement on biosafety issues, is key.

Furthermore, information needs will differ across a society. Therefore, if widespread participations from a spectrum of social groups is the goal, this requires techniques such as surveys and workshops to understand how to target different public with information through a range of media and in appropriate formats. In Estonia, for example, the Environment Ministry commissioned a survey, which revealed that there was a substantial level of ignorance about LMOs and a desire for more information, as well as a particular interest in receiving information in the Russian language for the Russian-speaking minority.

Basic information about the processes and mechanisms that exist for soliciting public views on biosafety issues is vitally important for promoting and facilitating public participation. People need to be aware what types of information are available and how they can be accessed. This type of information would include publicising official information sources, such as government information offices, official registers, and websites, and explaining the relevant procedures for accessing them, including contact information for responsible officials or departments. It would also need to include effective publicity for public meetings, the opening of public consultation periods, and invitations to contribute to public consultations.

Once the information needs have been assessed, it is possible to proceed to the phase of targeting, providing, and disseminating information. For many countries, lack of accurate and balanced, recorded information about biosafety capacity, and the status of particular approvals for release remains a key problem. Some countries, as in the Namibian case, may find that they have mechanisms for collecting and storing data, but lack an effective system for analysing and making it available. In other countries, there is a plurality of information sources, but their neutrality and independence is often questioned if the information comes from organizations that are seen to have a clear interest in presenting issues of biosafety in a particular way. Information from biotechnology companies and campaigning NGOs plays an important role in introducing different perspectives into the debate, but issues of credibility mean that governments have to take primary responsibility for establishing mechanisms for ensuring a steady flow of information that is, and is seen to be, accurate, complete, and balanced.

One possible method of striking an appropriate balance may be to work with journalists and media organizations, which, in principle, are in a position to provide balanced coverage of biotechnology and biosafety debates. For example, in the UK, the organisers of the 'public debate' on GM food and crops hope to achieve a broad reach by entering into a partnership with the media. In other countries, such as Norway and Estonia, the government has established or commissioned independent bodies to disseminate biosafety information and promote public awareness.

Although the provision of information is a vital precondition for informed participation, information alone is not enough. More information, even better quality information, does not guarantee a more engaged public audience, nor does it necessarily create the right conditions for active citizen engagement. The information needs to be appropriately targeted for particular purposes and audiences. For example, the various international web-based biosafety databases and clearing houses discussed in this report are better

placed to provide importers and exporters with useful information about biosafety regulations in order to facilitate the trade in LMOs, than they are to promote awareness among the general public. Information that circulates in technical databases on the internet and on CD-ROMs often serves this function. While such information sources play a useful and important part in ensuring the effectiveness of the protocol, their design, format and the medium through which they are communicated are generally not conducive to encourage public engagement.

Partly, the issue is one of access and capacity. Most people in the world do not yet have access to the internet, and cannot be expected to glean information from electronic sources. Many are also illiterate and living in areas remote from good communications networks. This situation applies to many of the cases we have looked at in developing countries in particular. The point is that it is precisely those people living in rural areas of developing countries, where testing may be taking place and where biosafety regulations are still in the process of being designed or implemented, that most urgently need relevant information and education on these issues.

27.5. PUBLIC PARTICIPATION

Public participation and transparency in decision-making processes are increasingly recognized as essential elements of good governance and sustainable development. When the views of all relevant stakeholders are taken into account through informed and meaningful participation, public concerns can contribute to the formation of more appropriate and acceptable policies and decisions.

Governance should involve all stakeholders, even though authority may remain with state bodies.

The 'Access Principles' in Principle 10 of the Rio Declaration of the 1992 Earth Summit articulated the following responsibilities of states:

- To provide appropriate access to publicly-held information.
- To give the opportunity to participate in decision-making processes.
- To facilitate and encourage public awareness and participation by making information widely available.
- To ensure justice through liability and accountability measures.

These access principles represent fundamental norms of transparent, participatory, and accountable governance that are essential in realizing sustainable development objectives.

It is not only governments, which play an important role in securing international agreements, but environmental non-governmental organizations (ENGOs) as well. This is reflected in the ever-growing body of literature concerning this topic. This article deals with the political influence of ENGOs on the biosafety protocol—a protocol linked to the UNEP convention on biological diversity. It concludes that these organizations did have an impact on the final agreement that was made, with one example being the inclusion of the precautionary principle in the final protocol (although not exclusively determined by ENGOs).

They were particularly able to influence policy outcomes by lobbying government delegates, cooperating with developing countries, and mobilizing public pressure. With regard to the timeframe of the biosafety negotiations, ENGOs succeeded in exercising most of their influence in the so-called '*pre*-negotiation' phase. The cartagena protocol on biosafety includes the right to public participation in article 23. Parties to the protocol are urged to promote and facilitate public awareness, education, and participation in dealing with GMOs. Governments are also mandated to consult the public in the decision-making process regarding GMOs, and to make the results of such decisions available to the public. Civil society has an important role to play in implementing article 23. For example, they could develop

indicators to assess the government's performance with respect to this provision.

Public participation in decision-making is not a single-method, simple activity with discrete steps leading to perfect decisions. There is no "universal prescription or standard formula. What works in some places or in some circumstances will not work everywhere. Instead, public participation is an adaptive, iterative process that will be unique to each setting. Governments and other bodies charged with or interested in implementing public participation in biosafety decision-making are faced with a number of questions, such as:

- What is necessary for effective public participation?
- When can and should public participation occur in the decision-making process?
- What forms can public participation take?

27.6. PUBLIC PARTICIPATION IN THE CARTAGENA PROTOCOL ON BIOSAFETY ARTICLE 23—PUBLIC AWARENESS AND PARTICIPATION

- The parties shall:
  - promote and facilitate public awareness, education and participation concerning the safe transfer, handling and use of living modified organisms in relation to the conservation and sustainable use of biological diversity, taking also into account risks to human health. In doing so, the parties shall cooperate, as appropriate, with other states and international bodies.
  - endeavor to ensure that public awareness and education encompass access to information on living modified organisms identified in accordance with this protocol that may be imported.
- The parties shall, in accordance with their respective laws and regulations, consult the public in the decision-making process regarding living modified organisms, and shall

make the results of such decisions available to the public, while respecting confidential information in accordance with article 21.

- Each party shall endeavor to inform its public about the means of public access to the biosafety clearing-house.

27.7. NGOs IN BIOSAFETY AND BIOTECHNOLOGY

Although the term NGO was not in use in the 19th century, it did come into use as early as 1920. For example, in that year, Sophy Sanger used the term "non-government organization" in a discussion of how such organizations had not been able to be present in 1906, during the first multilateral negotiations on labor treaties. Sanger contrasted this pre-war practice to the advent of the International Labour Organization (ILO) in 1919, which provided clear opportunities for the participation of non-Government Delegates and advisers chosen in agreement with the industrial organizations, if such organizations exist, which are the most representative of employers or work-people, as the case may be, in their respective countries.

Nevertheless, when it implemented this provision in 1950, the UN Economic and Social Council established a set of principles among which was that the consulted organization "shall be of recognized standing and shall represent a substantial portion of the organized persons within the particular field in which it operates."

This requirement, to a large extent, has been carried forward into the current ECOSOC credentialing rules, adopted in 1996. These rules state that the NGO "shall be of recognized standing within the particular field of its competence or of a representative character." These rules also state that "the organization shall have a representative structure and possess appropriate mechanisms of accountability to its members, who shall exercise effective control over its policies and actions through the exercise of voting rights or other appropriate democratic and transparent

decision-making processes." Thus, the claim that an ideal NGOs is representative was contributed to the United Nations by governments not by overreaching NGOs.

A final important historical development that should be noted was the open attitude by the League of Nations towards NGOs. Recognizing that the League would be dealing with both semi-public bureaux (i.e., groups containing government members) and private associations on a daily basis, and seizing on the spirit of article 24 of the League of Nations Covenant which, according to the Secretariat, required "that the League of Nations should follow closely and should encourage every international movement...," the Secretariat began publishing a handbook of International Organizations in 1921. The handbook was periodically updated and covered what we would today call intergovernmental organizations, transgovernmental networks, and NGOs. The handbook was organized with an excellent classification by function (e.g., "Politics and International Relations: Pacifism"), and provided valuable information on each organization's address, year of organization, object, membership, governing body, finances, activities, and history. The initiative taken by the League in 1921 to publish this handbook was one of the most important intergovernmental acts in the 20th century to recognize the legitimacy of NGOs and to improve their transparency.

NGO concerns about biotechnology cover a wide range of issues. Although there are differences between the stances of various NGOs, the following points represent some of the most widely-held views:

- Biotechnology holds little promise for improving poor people's access to food; the major constraint is not food production but rather distribution of resources.
- Transgenic crops present significant environmental dangers, and may also impose unacceptable health risks on consumers.

- The spread of transgenic crops will further increase the use of external chemical inputs, which are environmentally dangerous and inappropriate for resource-poor farmers.
- Transgenic crops will lead to dependence on seed companies, who will take advantage of farmers; emerging intellectual property regimes will limit farmers' ability to save their own seed.
- Transgenic crops represent a further step away from the traditional agricultural techniques and biodiversity that have served farmers well in the past.

There is little reason for most NGOs to become directly involved with biotechnology. The technology itself is not within most NGOs' purview; their comparative advantage lies more with building local organizations and capacities. So there is no reason to expect any significant NGO commitment to biotechnology, even if transgenic varieties were shown to be safe and beneficial for small-scale farmers. In present circumstances, of course, the incentives are strongly in the opposite direction. By demonstrating opposition to a technology that makes many northern consumers uneasy, NGOs can garner additional support for their own agricultural programmes.

Regardless of the strengths of various arguments or the ultimate fate of biotechnology, the current campaign raises a question about NGO strategy, which is related to larger concerns about globalisation. It is perfectly appropriate for NGOs to take advantage of their wide links to stimulate debate about a technology, which is global in character. But we need to know where NGOs stand in response to globalisation, and in particular, whether they choose to 'engage or de-link'.

In the case of agriculture, de-linking involves concentration on self-sufficient communities relying on 'traditional' production methods, isolated from larger markets or political processes that are judged to be inimical to the interests of marginalised farmers. This would not appear to be a realistic alternative, but if it is the path that is chosen, the northern public and donors that support this action deserve detailed reports on progress achieved. The opposite course, engagement, would seem to offer many more opportunities.

It makes no assumptions about the type of technology that is utilised. It may involve 'traditional' or 'modern' technology. It may develop methods that outperform anything that biotechnology has to offer, or it may complement or accelerate the use of biotechnology. The quality that distinguishes engagement is a clear commitment to ensure that marginalised farmers have a voice in agricultural markets, and that they are better represented in public activities such as agricultural research, extension, and regulation.

This type of engagement sees civil society not as a 'countervailing force' to expanding markets and disintegrating, but rather as the vital element that makes markets and governments perform for their citizens. NGOs need to contribute to strengthening states and markets in the South, so that people can debate and come to decisions about biotechnology as they do in the North.

In the summer of 1994, US environmental advocacy groups were getting ready to celebrate. The United States was about to join almost 90 other nations in ratifying the convention on biodiversity, which enjoyed broad support from US environmentalists, agro-business groups, and the biotechnology sector. After hearings characterized in the press as a "love fest", members of the Senate Foreign Relations Committee were almost unanimously prepared to back the treaty. Then a group of agricultural and trade nongovernmental organizations (NGOs) previously uninvolved in the debate weighed in, warning that ratification could, in effect, destroy US agriculture. As the Chicago Tribune reported in September 1994, evidence later surfaced that some of this opposition was based on a virulent misinformation campaign claiming, among other things, that treaty advocates were all foes of farming, logging, and fishing. But by then, the biodiversity treaty had been relegated to the back of a long line of treaties competing for congressional attention.

At a time, when NGOs are celebrating their remarkable success in achieving a ban on land-mines and creating an International Criminal Court (ICC), it may seem churlish to recall a four-year-old episode that many would likely regard as a defeat. But amid the breathless accounts about the growing power of NGOs, the failure of the biodiversity treaty is a useful reminder of the complexity of the role that these groups now play in international affairs. Embracing a bewildering array of beliefs, interests, and agendas, they have the potential to do as much harm as good. Hailed as the exemplars of grassroots democracy in action, many NGOs are, in fact, decidedly undemocratic and unaccountable to the people they claim to represent. Dedicated to promoting more openness and participation in decision-making, they can instead lapse into old-fashioned interest group politics that produces gridlock on a global scale.

The question facing national governments, multilateral institutions, and national and multinational corporations is not whether to include NGOs in their deliberations and activities. Although many traditional centers of power are fighting a rear-guard action against these new players, there is no real way to keep them out. Instead, the real challenge is figuring out how to incorporate NGOs into the international system in a way that takes account of their diversity and scope, their various strengths and weaknesses, and their capacity to disrupt as well as to create.

Importance of NGOs

Defining NGOs is not an exercise for the intellectually squeamish. A 1994 United Nations document, for example, describes an NGO as a non-profit entity, whose members are citizens or associations of citizens of one or more countries, and whose activities are determined by the collective will of its members in response to the needs of the members of one or more communities with which the NGO cooperates. This formulation embraces just about every kind of group except for private businesses, revolutionary or terrorist groups, and political parties. Other popular substitutes for the term NGO (private voluntary organizations, civil society organizations, and the independent sector) are likewise almost terminally vague. A better approach to understanding NGOs and what they are would focus on their respective goals, membership, funding sources, and other such factors.

Although there may be no universal agreement on what NGOs are exactly, there is widespread agreement that their numbers, influence, and reach are at unprecedented levels. In 1948, for example, the UN listed 41 consultative groups that were formally accredited to cooperate and consult with the UN Economic and Social Council (ecosoc); in 1998, there were more than 1,500 with varying degrees of participation and access. Until recently, NGOs clustered in developed and democratic nations; now groups sprout up from Lima to Beijing. They are changing societal norms, challenging national governments, and linking up with counterparts in powerful trans-national alliances. And they are muscling their way into areas of high politics, such as arms control, banking, and trade, that were previously dominated by the state. In general terms, NGOs affect national governments, multilateral institutions, and national and multinational corporations in four ways—setting agendas, negotiating outcomes, conferring legitimacy, and implementing solutions.

Biotechnology is a product of globalisation, and its critics also utilise global links in promoting a synergy between NGOs interested in development with those whose focus is environmental issues in the North. NGOs have long played a key role in forcing leaders and policymakers to pay attention. In the early 1800s, US and European bodies such as the British and Foreign Anti-Slavery Society were driving forces behind government action on the slave trade. By the turn of the century, groups such as the Anglo-Oriental Society for the Suppression of the Opium Trade were leading an influential antidrug movement that culminated in the 1912 Hague Opium Convention. In 1945, NGOs were largely responsible for inserting human-rights language in the UN Charter, and have since put almost every major human-rights issue on the international agenda. Likewise, NGO activism since the 1960s and 1970s successfully raised the profile of global environmental and population issues.

Instead of holding marches or hanging banners off buildings, NGO members now use computers and cell phones to launch global public-relations blitzes that can force issues to the top of policymakers' "to do" lists. Consider the 1997 Nobel prize–winning campaign by NGOs to conclude a treaty banning landmines over the objections of the United States. The self-described "full working partnership" between the Canadian government and a loose coalition of more than 350 humanitarian and arms-control NGOs from 23 countries was key to the negotiations' success. But what seized the attention of the public and policymakers was the coalition's innovative media campaign using the World Wide Web, faxes, e-mail, news-letters, and even Superman and Batman comic books. Treaty supporters won the signatures of 122 nations in 14 months. When several coalition members announced plans for a follow-on campaign against small arms, the US government sprang into action, meeting with 20 other countries in July 1998 to launch official talks on a possible treaty.

Negotiating outcomes

NGOs can be essential in designing multilateral treaties that work. Chemical manufacturing associations from around the world helped set up an effective verification regime for the 1997 Chemical Weapons Convention that could be supported by industries and militaries. Throughout the various sessions of negotiations on climate change, groups such as the Environmental Defense Fund and the World Business Council for Sustainable Development have helped craft compromise proposals that attempt to reconcile environmental and commercial interests; meanwhile, NGOs have been instrumental in helping government negotiators understand the science behind the issues that they seek to address.

NGOs can also build trust and break deadlocks when negotiations have reached an impasse. In 1990, a sole Italian NGO, the Comunità di Sant' Egidio, started the informal meetings between the warring parties in Mozambique that eventually led to a peace settlement. During talks in 1995 to extend the Treaty on the Non-Proliferation of Nuclear Weapons, NGOs from several countries working with the South African government delegation helped forge a compromise that led to the treaty's permanent extension.

Conferring legitimacy

NGO judgments can be decisive in promoting or withholding public and political support. The World Bank learned this lesson in the early 1990s, albeit the hard way. After decades of watching the bank do business with only a handful of NGOs and brush off demands for change, more than 150 public-interest NGOs took part in a sustained campaign to spur greater openness and accountability and to encourage debt reduction and development strategies that were more equitable and less destructive to the environment. Today, partly as a result of this high-profile pressure, about half of the bank's lending projects have provisions for NGO involvement—up from an average of only 6% between 1973 and 1988. The bank has even included NGOs such as Oxfam International in once sacrosanct multilateral debt relief discussions—against the wishes of many World Bank and International Monetary Fund (IMF) officials. Even the IMF is beginning to change its tune. In June 1998, the IMF Board of Directors met with several NGO leaders to discuss their proposals to increase the fund's transparency.

Making solutions work

NGOs on the ground often make the impossible possible by doing what governments cannot or will not. Some humanitarian and development NGOs have a natural advantage because of their perceived neutrality and experience. The International Committee of the Red Cross, for example, is able to deliver health care to political prisoners in exchange for silence about any human-rights violations that its members witness. Other groups such as Oxfam International provide rapid relief during and after complex humanitarian disasters—with and without UN partners. Moreover, as governments downsize and new challenges crowd the international agenda, NGOs increasingly fill the breach. Willy nilly, the UN and nation-states are depending more on NGOs to get things done. Total assistance by and through international NGOs to the developing world amounted to about $8 billion in 1992—accounting for 13% of all development assistance and more than the entire amount transferred by the UN system.

International NGOs also play critical roles in translating international agreements and norms into domestic realities. Where governments have turned a blind eye, groups such as Amnesty International and the Committee to Protect Journalists call attention to the violations of the UN Declaration on Human Rights. Environmental NGOs police agreements such as the Convention on International Trade Endangered Species, uncovering more accurate data on compliance than that provided by the member nations. Perhaps one of the most vital but overlooked roles of NGO is to promote the societal changes needed to make international agreements work. Signatories of the Organization for Economic Cooperation and Development's 1997 Bribery Convention, for example, are counting on the more than 80 chapters of Transparency International to help change the way their societies view bribery and corruption.

Increasingly, however, NGOs operate outside existing formal frameworks, moving independently to meet their goals and establishing new standards that governments, institutions, and corporations are themselves compelled to follow through force of public opinion. The UN moratorium on driftnet fishing in 1992, and the US International Dolphin Conservation Act of 1994, for example, largely codified changes in fishing practices that NGOs had already succeeded in promoting and then winning from commercial fisheries. More recently, even as governments and multilateral institutions slowly begin to consider measures to promote the sustainable use of forests, the environmental NGO "Greenpeace" led a European consumer boycott that persuaded a leading Canadian logging company to announce that it would change the way it harvests trees.

Rise of the "Global Idiots"

Despite the demonstrated capacity of NGOs to do good, their growing power on the ground has exposed them to heightened criticism; some of it justified. As salutary as international attention has been to the recent turmoil in Chiapas, Mexico, for example, it is hard to believe that the arrival there of 4,500 foreigners from 276 different organizations necessarily represents an unalloyed good. One recent study on NGOs and peacebuilding in Bosnia criticized the use of advertising (from signboards to t-shirts) by NGOs to promote their reconstruction programs to potential donors. Such advertising, the study noted, had the effect of denigrating local rebuilding efforts and raising questions about where NGOs were actually putting their money. In the Sudan and Somalia, NGOs have subsidized warring factions by making direct and indirect payments to gain access to the areas needing assistance. In other conflict settings such as Ethiopia and Rwanda, NGO constructed roads and camps for civilian assistance have instead been used by combatants.

Other longer-term concerns loom for these service-delivery NGOs. The UN High Commissioner for Refugees warned in 1996 that if national governments continue to favor NGOs over multilateral agencies in donor assistance, they may undermine important systems of coordination and cooperation in large-scale emergencies. Intense competition among NGOs in the relief sector has also pushed the sector towards a form of oligopoly that threatens to crowd out smaller players, especially local NGOs in developing countries. Eight major groups now control about 50% of the relief market.

But on balance, the record for such NGOs is surely no worse than that of governments. NGOs are increasingly aware of these weaknesses and are moving to address them by adopting codes of conduct and pledging to "do no harm". Moreover, given their origins as grassroots groups, NGOs tend to be wary of organizations that become too big. This innate suspicion can serve as a mechanism for self-regulation.

Yet, the greatest challenges created by the growing influence of NGOs are not in the field but in the arena of public opinion and the corridors of power. Today, in a phenomenon that one environmental activist bemoaned as the "rise of the global idiots," any group with a fax machine and a modem has the potential to distort public debate—witness the demise of not just the biodiversity treaty but the Multilateral Agreement on Investment (MAI), which for all its apparent shortcomings still deserved more reasoned consideration than it received. Even legitimate, well-established groups sometimes seize on issues that seem designed more to promote their own image and fundraising efforts than to advance the public interest. In 1995, for example, Greenpeace continued to attack the Royal Dutch/Shell Group for its plans to sink an oil rig (the Brent Spar) in the North Sea, even after independent scientific analyses showed that the environmental effects of doing so would be inconsequential. Steeped in a culture that encourages adversarial attitudes to the powers that be, many NGOs seem best suited to confrontation—a characteristic that some US policymakers seized on in noting that the NGO coalition against landmines might have won US

support (and hence a stronger treaty) if it had been more patient and willing to compromise.

Limits of Democracy

Governments, multilateral institutions, and corporations face inherent dilemmas in trying to work with NGOs. At their most fundamental level, these dilemmas hinge on two key questions: Who should participate, and how? On the one hand, opening up the floodgates to allow equal access to every group would frustrate decision-making. More than 1,500 NGOs were accredited at the 1992 United Nations Conference on the Environment and Development in Rio de Janeiro. Trying to include them all was impossible, so in the final days of the conference, government delegates increasingly retreated behind closed doors. On the other hand, narrowing the field fairly is extraordinarily difficult because no one algorithm or set of criteria can objectively rank the worth of an NGO to a participatory process. Should the World Trade Organization (WTO) consult and share information with groups that have large memberships but are sworn enemies of the WTO's existence? When seeking to devise a treaty on persistent organic pollutants, should negotiators insist that groups be representative of, and accountable to, large constituencies and thereby exclude reputable, experienced groups that represent future generations or nonhumans (the International Fund for Animal Welfare, for example)? Some traditional centers for power have done a better job than others at tackling these kinds of questions about participation. Among nation-states, Canada stands out for its role in forging an alliance with NGOs on the landmine ban. As Foreign Minister Lloyd Axworthy has said, "Clearly, one can no longer relegate NGOs to simple advisory or advocacy roles. They are now part of the way decisions have to be made." Nations such as Norway helped fund the NGO coalition for the ICC (On the other side of the funding fence, nations such as Sierra Leone and Bosnia received help at the ICC negotiations from legal advisers provided by the NGO No Peace Without Justice). The US record in working with NGOs is mixed, generally good in areas such as the environment and relatively poor in areas such as arms control and regional issues. China represents another extreme, having chosen to banish NGO delegates at the Fourth International Women's Conference in Beijing in 1995 to a site one hour's drive away from the main negotiations.

Most multinational corporations are still struggling to figure out how to handle NGO participation. One 1996 poll of 51 major European corporations found that although 90% of them believed that the impact of "pressure groups" would stay the same or increase over the next five years, only 20% had formal procedures in place for dealing with such groups, and only 12% for evaluating them. Still, there are grounds for optimism. A 1998 survey of 133 NGOs found that while many rated their current relationships with corporations as "antagonistic" or "nonexistent", most predicted the development of cooperative relationships in the future. The changing attitude towards NGOs of the Royal Dutch/Shell Group may be a case in point. Stung by fierce NGO campaigns on the Brent Spar episode and its operations in Nigeria, where its ties to the dictatorship of General Sani Abacha made it a target for human-rights groups, Shell has adopted a new Statement of General Business Principles that includes commitments on human rights and the environment. In regions such as Latin America, it now consults with NGOs to ensure that its oil operations take environmental and social factors into account.

The impact of NGOs on multilateral institutions (and national governments) may well be cyclical. As Yale professor Steve Charnovitz has observed, NGO involvement seems to depend on two factors: the needs of government and the capabilities of NGOs. During the nineteenth century, for example, governments had little experience with nonpolitical treaties and therefore needed NGOs. Although the young League of Nations later saw NGOs as allies in forming new international organizations, NGO participation diminished in the 1930s as the league's influence declined and governments garnered more practical experience in NGO–oriented issues. The same pattern holds true for the post-World War II years. An early burst of enthusiasm, followed by a gradual decline during the 1950s that reversed itself as issues such as the environment, development, population, and food aid became part of the international agenda. Today, in the wake of a powerful post–Cold War boost in NGO activity, the receptivity of multilateral institutions to NGO participation varies greatly, from relatively open ECOSOC bodies such as the Commission on Sustainable Development to the more closed UN Security Council and General Assembly, and from an increasingly receptive World Bank to the still recalcitrant IMF and WTO.

At bottom, however, the search for a form of perfect democracy that encompasses all NGOs makes about as much sense in the international system as it does in, say, the United States. Instead, institutions and NGOs should strive to create formal but flexible systems during negotiations to foster dynamism and self-adjustment. ECOSOC is a case in point. It ranks NGOs according to three tiers of status—category 1, category 2, and roster. The small percentage of those with Category 1 status, or top access, have more opportunities to attend meetings, submit written advice, and occasionally speak at conferences. The selection process deliberately favors NGOs operating in more than one country, with out barring national NGOs that cannot afford to travel. It also favors groups that are of a representative character, but is ambiguous enough to allow in groups that represent perennial values and scientific truths, not members. In 1998, more than 100 NGOs had top status, with nearly 1,500 enjoying some form of consultative status at ECOSOC, versus 978 in 1995, and 41 in 1948. NGOs want more, governments want less, but the system generally works.

Some advocates want to push more UN decision-making forums in the direction of the International Labor Organization (ILO), which gives formal voting rights to business and labor delegates as well as governments. A handful of variations on the ILO model exist that work well, particularly on technical issues and the setting of standards. (The International Organization for Standardization, or ISO, for example, is a hybrid NGO that brings together representatives of industry, government, consumer, and other bodies together on equal footing to resolve global standardization issues.) But trying to win voting rights for NGO representatives in more UN forums is by no means a sure way to improve the representative character of institutions. The unique ILO system works fairly well because national employer and labor umbrella groups already exist, making the identification of representative employer and labor delegates simpler than in other fields. But it is doubtful in most cases that such systems would actually improve the representative character of institutions. Even if NGOs elected their own representatives from among themselves, there would be insufficient guarantees that the pool of electors was truly representative of society or the public interest.

Letting NGOs be NGOs

Many governments and institutions, including the WTO, IMF, and several UN bodies, will continue to resist more public participation, arguing that their issues require great secrecy. But history offers a powerful argument that the holdouts suffer from a failure of imagination. Half a century ago, the architects of the postwar international trading system did not contest NGO involvement. The proposed charter for the International Trade Organization (ITO) included the very same language as the WTO charter, providing for "consultation and cooperation" with NGOs. The ITO framers had a different interpretation than their latter-day successors; however, they envisaged that commercial and public-interest NGOs would maintain regular contact with the ITO Secretariat, receive unrestricted documents, propose agenda items, and participate as observers and occasional speakers at conferences.

The spirit of these proposals faded away when the ITO failed in favor of an interim solution—the General Agreement on Tariffs and Trade, or GATT. Ironically, most multilateral institutions today face requests from NGOs that could be met by adopting the standards for participation that the ITO's framers proposed. The challenge facing NGOs is more subtle but no less important. As these groups acquire the access and influence that they have long sought, they must not lose the qualities that have made them a source of innovation and progress. Some analysts already fear that formerly independent NGOs may become more beholden to national governments, as they come to rely more on public-sector funding, which now accounts for around 40% of NGO budgets versus only 1.5% in 1970. And many of the schema for increasing NGO involvement may simply foster predictable and bureaucratic behavior among civil society representatives, potentially dulling the passion and richness of views that can emanate from narrowly focused groups. They may also cut off NGOs from the informal channels through which they have traditionally been most influential.

Instead, NGOs, governments, and multilateral institutions need to devise systems of public participation that draw on the expertise and resources of NGOs, their grassroots connections, sense of purpose and commitment, and freedom from bureaucratic constraints. Those NGOs that have seen the most rapid growth in their power will have to contend with inevitable limits on their influence and access. Those governments and institutions that have resisted the advance of these new players will have to permit an unprecedented level of public scrutiny and participation. Over time, this messy process of give-and-take promises to transform the way that international affairs are conducted. Yet as it plays out, both sides may realize that the new system that they have sought to create or resist is in many respects no different from the clash of competing interests that has characterized democracies since their inception.

27.8. ROLES OF NGOs

Among the wide variety of roles that NGOs play, the following six can be identified as important, at the risk of generalization:

Development and Operation of Infrastructure

Community-based organizations and cooperatives can acquire, subdivide and develop land, construct housing, provide infrastructure and operate and maintain infrastructure such as wells or public toilets and solid waste collection services. They can also develop building material supply centres and other community-based economic enterprises. In many cases, they will need technical assistance or advice from governmental agencies or higher-level NGOs.

Supporting Innovation, Demonstration and Pilot Projects

NGO have the advantage of selecting particular places for innovative projects and specify in advance the length of time which they will be supporting the project-overcoming some of the shortcomings that governments face in this respect. NGOs can also be pilots for larger government projects by virtue of their ability to act more quickly than the government bureaucracy.

Facilitating Communication

NGOs use interpersonal methods of communication, and study the right entry points

whereby they gain the trust of the community they seek to benefit. They would also have a good idea of the feasibility of the projects they take up. The significance of this role to the government is that NGOs can communicate to the policy-making levels of government, information bout the lives, capabilities, attitudes and cultural characteristics of people at the local level.

NGOs can facilitate communication upward from people to the government and downward from the government to the people. Communication upward involves informing government about what local people are thinking, doing and feeling, while communication downward involves informing local people about what the government is planning and doing. NGOs are also in a unique position to share information horizontally, net-working between other organizations doing similar work.

Technical Assistance and Training

Training institutions and NGOs can develop a technical assistance and training capacity, and use this to assist both CBOs and governments.

Research, monitoring and evaluation

Innovative activities need to be carefully documented, and shared-effective participatory monitoring would permit the sharing of results with the people themselves as well as with the project staff.

Advocacy for and with the poor

Non-governmental organizations (NGOs) are playing an increasingly important role in the process Foucault called "governmentality". Drawing on the study of biodiversity conservation as well as indigenous people's ancestral domain to show how two quite different NGO-led conservation agendas nonetheless share a common underlying purpose: persuading indigenous people to internalize state control through self-regulation. Ironically, it is this sort of NGO contribution to the elaboration of government that may turn out be the most significant and lasting contribution that NGOs make to social change.

In some cases, NGOs become spokespersons or ombudsmen for the poor, and attempt to influence government policies and programmes on their behalf. This may be done through a variety of means ranging from demonstration and pilot projects to participation in public forums and the formulation of government policy and plans, to publicizing research results and case studies of the poor. Thus, NGOs play roles from advocates for the poor to implementers of government programmes; from agitators and critics to partners and advisors; from sponsors of pilot projects to mediators.

Global Policy Forum is one such example for NGO at the international level. GPF is a non-profit, tax-exempt organization, with consultative status at the UN. Founded in 1993 by an international group of concerned citizens, GPF works with partners around the world to strengthen international law and create a more equitable and sustainable global society. GPF uses a holistic approach, linking peace and security with economic justice and human development, and place a heavy emphasis on networking to build broad coalitions for research, action and advocacy. Our energy will be put into well-focused and unique programs in which GPF has a special analytical and organizational edge. GPF's main office is strategically located across the street from UN headquarters in New York. GPF also has a European office located in Bonn, Germany. Global Policy Forum's mission is to monitor policy making at the United Nations, promote accountability of global decisions, educate and mobilize for global citizen participation, and advocate on vital issues of international peace and justice. GPF responds to a globalizing world, where officials, diplomats and corporate leaders take important policy

decisions affecting all humanity, with little democratic oversight and accountability. GPF addresses this democratic deficit by monitoring the policy process, informing the public, analyzing the issues, and urging citizen action. GPF focuses on the United Nations—the most inclusive international institution, offering the best hope for a humane and sustainable future. GPF analyzes the role of NGOs in the international arena and advocates for a broader NGO role in policy making at the United Nations. GPF's award-winning site contains over 25,000 text files, and posts 50 or more new documents each week. The site attracted almost six million visitors and more than fifty-one million "hits" during 2005.

27.9. ENVIRONMENTAL AND FOOD SAFETY

The major opposition to biotechnology in the North centres on environmental and food safety. The critics are concerned by the potential environmental damage that might be caused by transgenic crops crossing with related species, or by their effects on other parts of the ecosystem. Some also argue that transgenic crops have not been sufficiently screened for possible toxins or allergens that could be introduced to the diet. Supporters of biotechnology argue that the testing and regulatory processes applied to transgenic crops are capable of addressing these concerns. The battle over regulation is understandable, not only because regulatory failure has been at the heart of several recent food scares (e.g., BSE), but also because regulation is a valuable political territory. The two sides line up behind what sounds like comprehensive codes—"science-based regulation' and 'the precautionary principle". But it can be argued that these are simply formal expressions of subjective attitudes towards the regulatory process.

Despite the prominence of regulation in the debate, it is difficult to see contrasting perceptions of regulatory protocol as the primary force behind the controversy over biotechnology. For example, biotechnology's critics are selective in their regulatory concerns. In the past several decades, mutation breeding—exposing seeds to radiation or mutagenic chemicals—has produced many commercial crop varieties, including several popular malting barleys grown in the UK, but no alarm about food safety has been raised. Similarly, many of the properties of transgenic crops, such as disease resistance or even herbicide tolerance, have also been achieved by conventional plant breeding, but little attention is given to their possible environmental consequences. Such selectivity is typical of most risk perception: any society or group chooses what it wishes to identify as risks. The subjectivity is conditioned by various cultural, political and economic factors. Part may be explained by direct self-interest (e.g., a company denying that its products are potentially harmful), but risk perceptions are also related to broader world-views. This is certainly the case with agricultural technology, and it is thus worthwhile turning to some of the other arguments surrounding biotechnology, particularly for the South.

27.10. FOOD SUPPLY AND POPULATION

Perhaps, the most common defence of transgenic crops is the argument that population will soon outstrip world food supply, and that new technology is required to meet the challenge. The trends are indeed worrying, but biotechnology's defenders tend to use them as a threat, implying that no other strategies can possibly avert a crisis. The argument about food supply and population has been a familiar one since the advent of the Green Revolution. The counter-argument is that there is more than enough food in the world, and the challenge is distribution. This response draws attention to the role of inequality, rather than absolute production deficiencies as a primary cause of hunger.

Despite the undeniable substance of this counter-argument, it embodies several inconsistencies that significantly weaken its force.

In the first place, even if the agricultural production of some of the poorest countries were perfectly distributed, it would not provide adequate nutrition. In addition, the argument implies either massive population movements—far beyond what could be envisioned by any realistic opportunities for land reform—or expanding purchasing power for the poor—a more reasonable option, but one that includes the global food trade to which many NGOs object. Another problem with denying the need for production increases is that many of the NGOs make this case against biotechnology turn, often in the next breath, to promote their own agricultural production programmes, usually featuring low-input or traditional technology. Clarification requires attention to the types of farmers and the types of technology that appear in the debate.

Widespread resistance to GM food has resulted in a global showdown. US exports of genetically modified corn and soy are down, and hungry African nations won't even accept the crops as food aid. Monsanto is faltering financially, and is desperate to open new markets. The US government is convinced that EU resistance is the primary obstacle, and is determined to change that. On May 13, 2003, the US filed a lawsuit with the World Trade Organization (WTO), charging that the European Union's restrictive policy on GM food violates international agreements. When eminent scientist, Arpad Pusztai went public about his accidental discovery that genetically modified (GM) potatoes severely damage the immune system and organs of rats, he was suspended from the prestigious Scottish research institute, where he had worked for thirty-five years. He was silenced with threats of a lawsuit, while the Institute denied or distorted his findings.

In the ensuing war over public opinion, biotech advocates tried to spin the science in favor of GM foods, but were thwarted at each attempt by leaked documents and compelling evidence. Pusztai, who describes this chapter as "the most thorough and accurate report on the topic", was ultimately vindicated when his potato study was published in the *Lancet*. His remains the only independent safety assessment in a peer-reviewed journal. It contrasts sharply with the handful of published industry studies, an analysis of which reveals how they were designed to avoid finding problems.

The story of Arpad Pusztai made headlines throughout Europe for months, alerting readers to some of the serious health risks of genetically modified (GM) foods. It was barely mentioned, however, in the US press. The media watchdog group "Project Censored" described it as one of the ten most underreported events of the year. In fact, major US media avoided almost any discussion of the controversy over genetically modified organisms (GMOs) until May 1999. But that was all about saving the monarch butterfly from GM corn pollen, not about human food safety. Gerald Guest, the director of FDA's Center for Veterinary Medicine (CVM) sent a letter to the FDA's Biotech Coordinator, James Maryanski, saying that he and the other CVM scientists concluded that there is "ample scientific justification" to require testing and review of each GM food before it is eaten by the public. He stated, "CVM believes that animal feeds derived from genetically modified plants present unique animal and food safety concerns." He pointed out that, "Residues of plant constituents, or toxicants in meat and milk products may pose human food safety concerns."

Many scientists who understood the dangers, however, were not convinced by the FDA's assurances. Geneticist David Suzuki, for example, said, "Any politician or scientist who tells you that these products are safe is either very stupid or lying. The experiments have simply not been done." A January 2001 report from an expert panels of the Royal Society of Canada likewise supported the conclusions of the FDA scientists. The report said it was "scientifically unjustifiable" to presume that GM foods are safe. The report explains that the default prediction for any GM foods is that

expression of a new gene (and its products) will be accompanied by a range of collateral changes in expression of other genes, changes in the pattern of proteins produced, and/or changes in metabolic activities. This could result in novel toxins or other harmful substances. The report emphasized the need for safety testing, looking for short and long-term human toxicity, allergenicity, and other health effects. The panel began their comprehensive 245-page report by quoting the editors of the UK's *Nature Biotechnology*. "The risks in biotechnology are undeniable, and they stem from the unknowable in science and commerce. It is prudent to recognize and address those risks, not compound them by overly optimistic or foolhardy behavior.

27.11. ROLE OF PUBLIC AND NGOs IN PROTECTION OF BIODIVERSITY

At present, many environmental and ecological NGOs are registered in large number of countries. These NGOs are making the public understand the importance of biodiversity, and are involved in conservation of biodiversity in the respective countries. In general, these organizations focus on environmental education and training, review of new initiatives and their environmental impacts, and participate in discussion and development of key national environmental projects. However, the effectiveness of the NGO movement in many countries is currently constrained by a number of factors, including lack of co-ordination and co-operation with the Ministry of Nature Protection, lack of resources (inappropriate membership fees), and reliance on external sources of income (private sponsorship and grants from overseas organizations).

During 6 to 17 November, 1995, the 134 Governments that are Parties to the convention on biological diversity met in Jakarta, Indonesia, at the Second Session of the Conference of the Parties (COP), along with a number of observer States and numerous NGOs. For the uninitiated, such meetings often seem like little more than stylized posturing with government representatives, presenting their views in a dull monotone, translated into five other languages. But the participants were working hard in smaller contact groups and reached some important decisions, which will have a significant influence on biodiversity in the coming years.

The main achievements of the Second COP were as follows:

- A permanent Secretariat was agreed. It will be housed in Montreal, Canada. Its staff will be tripled and its budget doubled.
- A clearing-house mechanism will be established. It will use print and electronic media to enhance networking between international centres and other organizations. The Secretariat will develop a network of partners, and facilitate the transfer of technology relevant to the conservation and sustainable use of biodiversity.
- Steps to negotiate a protocol on biosafety were agreed. This will focus on the transboundary movement of living modified organisms resulting from modern biotechnology that may harm biodiversity. Many NGOs and some governments wanted invasive species to be included as part of biosafety, but the COP decided to cover invasive species in other ways and to maintain the biotech focus of this protocol.
- The COP extended the Global Environment Facility (GEF) as its interim financial mechanism, following heated discussion about how responsive the GEF has been to the COP. The COP recommended that GEF explore more effective ways of collaborating with the various interest groups involved. The Secretariat was also asked to explore possibilities for additional funds.
- On conservation issues, the COP agreed measures to promote *in situ* (on-site) conservation, and to conserve components of biodiversity under threat, including identification of the underlying causes. They requested the Inter-Governmental Panel on Forests to consider the role of forests in maintaining biodiversity and related issues, such as in situ conservation, sustainable forest management, access to the genetic resources in forests, and the role of indigenous and local communities. They established a panel on Conservation and Sustainable Use of Marine and Coastal Biological Diversity. Rosters of experts, including from NGOs, are to be established to contribute to these activities.
- On the question of access to genetic resources and intellectual property rights, only limited progress was made. The COP called for continued information-gathering and networking with institutions such as the World Trade Organization. The COP also recognized the special nature of agricultural biodiversity and sought to provide its views to the 4th FAO Conference on Plant Genetic Resources, was held in June 1996 in Leipzig, Germany.

In addition to its regular issues, the COP agreed to a further ambitious work programme. Its 1996 session, was held in Argentina, and addressed agricultural biodiversity, terrestrial biodiversity, indigenous and local knowledge, access to genetic resources, and biosafety. The 1997 work programme included linkages between *in situ* and *ex situ* conservation, sharing of benefits, and matters arising.

Throughout the conference, NGOs contributed substantively to many issues, though they were excluded from the biosafety drafting group. The decision on forests and biodiversity specifically called for the involvement of broad stakeholder interests, and many delegates welcomed the constructive role of NGOs.

Some may be concerned at the glacial pace of some of the negotiations and the tediousness of the plenary sessions, but the Second COP clearly demonstrated that the convention on biological diversity is entering a new and mature stage, where governments are squarely addressing issues of genuine interest and concern. Back home, different governments will use these proceedings in different ways, but it is reasonable to expect that they will be taking significant actions to implement the measures advocated in the convention in ways that are appropriate to their domestic needs.

The National Environmental Action Plan constituted by many countries reviews current issues for biodiversity conservation and identified priority areas for action, including improved legislation, inventory and monitoring, improved management *in-situ* (including protected areas), improved *ex-situ* conservation, sustainable use of biological resources, and development of educational and scientific programmes on biodiversity. These priorities have been developed into a detailed action plan.

Low-Resource Farmers

Some of biotechnology's strongest defenders claim that it offers the opportunity for public agricultural research to bring productive technology to the farmers in marginal environments that were excluded from the Green Revolution. Opponents say that the uneven distribution of benefits of the Green Revolution simply indicates what can be expected from biotechnology, and that emphasis should be given instead to other strategies that are more appropriate for marginalised farmers. Both sides tend to overlook some of the realities of the environments in which these farmers live. Differences in biological resources account for much of the differential impact of agricultural technology. Not surprisingly, good soils and adequate rainfall or irrigation are related to higher yields, and to the greater potential impact of agricultural technology of any kind. Part of the mandate of agricultural development programmes is to help level the playing field for marginalised farmers by overcoming these constraints, but there are limits. Whether transgenic varieties addressed at factors such as soil acidity or drought tolerance can address the inequality of agricultural assets remains to be seen, and many farmers in better environments could take advantage of such improvements as well. Alternative "low-input" technology to overcome these constraints faces its own challenges, particularly if it demands increased labour. For instance, many labour-intensive soil conservation technologies are rejected, because their returns are below what rural people can earn from other components of their livelihood strategies.

Another constraint to technology uptake in marginal environments is access to markets and information. Areas that benefited from the Green Revolution not only have good agronomic resources, but they also have relatively good markets, roads, and services. Extending this infrastructure to more remote areas could help marginalised farmers take advantage of whatever agricultural technology is on offer. Indeed, it may be argued that until adequate markets and better information provision are in place, there is little likelihood that transgenic varieties will reach farmers in remote or marginal areas.

Both sides of the controversy need to make some difficult choices about agricultural development resources. The high prevalence of poverty in marginal rural areas demands attention from governments, donors and NGOs. But the degree to which this kind of poverty is most effectively and realistically addressed by agricultural technology of any kind must be decided on a case-by-case basis. Investment in appropriate agricultural technology can make a difference to farmers in difficult environments, but promotion of unrealistic strategies—either high or low-tech—simply wastes people's time and diverts attention from strengthening skills and resources for non-agricultural opportunities.

External Inputs

There is considerable concern about the marriage between the chemical and biotechnology industries. Many of biotechnology's critics urge the reduction or elimination of agricultural chemicals. On the other hand, biotechnology proponents emphasise the potential for pest or disease resistant varieties to lower dependence on chemical inputs. The relation between agricultural chemicals and biotechnology deserves a brief review.

Insecticides have attracted particular attention in critiques of modern agriculture. Until recently, insecticide use was expanding rapidly, and dangerous, broad-spectrum insecticides have been responsible for many cases of illness or death. In addition, their indiscriminate application has often altered ecologies and made pest control more difficult. In some cases, such as in Green Revolution rice areas, the growth in pesticide use was caused at least as much by policies that subsidised these products as by biological need. The imperative to reduce and rationalise insecticide use has led to various types of integrated pest management (IPM) that include biological control methods and farmer training. Conventional plant breeding invests significant resources in pest and disease resistance. One of the reasons for the recent decline in pesticide use on rice is the improved resistance of many of the new varieties. Discussions of the possible contributions of biotechnology to insect control centre on somewhat controversial use of genes from the soil bacterium *Bacillus thuringiensis* (Bt), but there is a wide range of other genes controlling the production of toxins, enzymes and enzyme inhibitors that are also the subject of research. The challenge is to understand the degree to which such innovations can contribute to sustainable pest control strategies.

Approximately 80% of the transgenic crops currently grown incorporate herbicide tolerance, reflecting the importance of chemical weed control in the North's commercial agriculture. Few people foresee a prominent role for this trait in the South in the near future, though it must be acknowledged that herbicide use plays a larger role in small-farm agriculture than is often imagined. Discussion of potential herbicide use should consider several factors. First, herbicide is a labour-saving technology. In certain areas, its adoption could have adverse equity effects by reducing demand for rural labour, while in labour-scarce areas it can make significant and equitable, improvements in productivity. Second, in some hillside areas, particularly in Latin America, herbicide use is an integral part of soil conservation strategies for many smallholders. Finally, herbicides present many of the problems of other pesticides; some of them are quite toxic, and over-dependence on herbicides can lead to the emergence of resistant weeds. The degree to which herbicides are an appropriate choice for smallholders in the South, and whether relevant transgenic varieties might be useful is a question for location-specific research.

The most important external input is chemical fertiliser. The essence of the Green Revolution was the development of short-statured crop varieties that could respond to higher levels of fertiliser. The expanded utilisation of chemical fertiliser has been one of the major factors in the growth of world food production. This has not been achieved without a cost. Inappropriate fertiliser use can cause groundwater pollution and deterioration in soil quality. Critics of modern technology are uncomfortable with farmers' growing reliance on purchased fertiliser, and strive to develop low-input alternatives. Unfortunately, the realities of nitrogen metabolism bring sobering news for both sides in the biotechnology debate. Although it is possible to imagine productive agriculture without the use of chemical pesticides, synthetic nitrogen is an absolute necessity for maintaining world food supply. By the 1980s, China's national annual per hectare protein output was more than double that was achieved in areas of traditional intensive organic farming in the 1950s. This is due to the use of synthetic nitrogen almost four times than the US (Smil, 1991). On the other side, any hopes that biotechnology will find an efficient shortcut to biological nitrogen fixation remain a distant dream. No matter what type of technology is developed for resource-poor farmers in the foreseeable future, most of them will need access to chemical fertiliser.

This examination of external input use concludes that farming in the South will depend on the use of a wide range of biological and chemical inputs, combined with an increasing capacity in farm management. Labeling such inputs as "natural" (e.g., a virus introduced as part of an IPM programme) or "external" (e.g., synthetic nitrogen to supplement that available in the soil) does little help to make the pragmatic choices necessary for increasing production, lowering costs, and safeguarding health and the environment.

Markets

An important element in the opposition to external input use in small-farm agriculture is a concern about farmers' dependence on input markets. These fears are compounded by the advent of transgenic varieties. Even when they embody public technology, private seed companies will most likely deliver them. In the North, the commercial seed industry is well established, and although there are legitimate concerns about its recent concentration, spurred by investments in biotechnology, there is little discussion of the alternatives to commercial seed supply—even, for instance, in the conduct of organic agriculture.

For the South, however, a predominant assumption is that the commercial seed industry will take advantage of farmers. This assumption has complex origins. One element is quite reasonably rooted in the weakness of markets in remote areas. Another element is the image of peasants´ saving seed and the fear that commercial seed will erode their self-sufficiency. World over, including the North, the farmers will save seed if they are able to do so. Commercial seed has become predominant where it offers advantages in quality and convenience. Is it reasonable to deny this option to farmers in the South?

A further part of the argument concerns intellectual property rights (IPRs), particularly related to the establishment of the World Trade Organization (WTO). Although systems of plant variety protection have been available for many years, the advent of biotechnology certainly provides added incentives for companies to push for stronger IPRs. There is a considerable controversy over the breadth and types of IPRs that should be allowed. The notion that IPR regimes will deny farmers the right to save seed, even of their local varieties is alarming. It is important that a balance be established, so that IPRs provide sufficient incentives for research but that mechanisms are in place to allow small farmers to save seed of protected varieties. The situation is further complicated by instances of exceptionally broad patenting, particularly for some biotechnology—private annexation of public germplasm, and biopiracy. The challenge is to identify and control the excesses, while at the same time supporting the emergence of a commercial seed industry that can serve smallholders.

A final element in the argument is the assumption that although commercial seed companies in the North are acceptable, those in the South will behave irresponsibly. The history of the seed industry in the North has been characterised by many small firms and considerable competition. Despite the concentration of ownership of biotechnology in a few large firms in the US, for instance, hundreds of different transgenic varieties are available, offered by dozens of seed companies, many of them serving only a local market. The licensing of the technology allows it to be delivered by seed companies that have developed locally preferred varieties and have loyal customer bases. It will admittedly be a long time before biotechnology can be delivered to farmers in many developing countries in this fashion, but efforts towards diverse and responsive national commercial seed sectors certainly deserve support.

Traditional Agriculture

Past opposition to the Green Revolution and current concerns about biotechnology are partly related to the external input and commercialization issues discussed above, but they also draw on sentiments about the value of traditional agriculture. A Christian Aid document portrays the tragic suicides of indebted cotton farmers in India as a harbinger of biotechnology's potential impact, and laments that these farmers had abandoned their traditional crops for cotton. The image is a powerful one, but is quite misleading.

The recourse to traditional agriculture may be an expression of vague romanticism, or it may be part of a carefully crafted presentation. An outstanding example of the latter is Vandana Shiva, who has developed a considerable following through cleverly designed contrasts between an idealised traditional agriculture—diverse, feminine, self-sufficient, and the deficiencies of new technology. A comparison of this ideal with India's pre-Independence history, in which the majority of the peasantry was subject to a succession of empires, landlords and colonial rule, raises grave doubts about the supposition of an ecologically balanced, egalitarian rural society rent asunder by the imposition of foreign agricultural technology.

Supporters of traditional agriculture are concerned about the loss of biodiversity. The issue is of undeniable importance, but it is too often used to cover a vague defence of 'traditional' varieties rather than an examination of the dynamics and the trade-offs in diversity's contribution to agricultural production (Wood and Lenné, 1999). In addition, the emphasis on traditional agriculture promotes a paternalistic view of unchanging local customs, and overlooks the dynamic and adaptive qualities of virtually all agricultural systems. The origins of Green Revolution technology, fertiliser-responsive varieties, are to be found in 19th century Japan, where they were a response to land shortage, and not in the 20th century US, which is still behind countries like India or Nepal in its adoption of semi-dwarf wheat varieties.

The opposite tendency, of course, is to paint an overly optimistic picture of technological change in agriculture. The Green Revolution was responsible for a remarkable increase in food production that significantly lowered food prices. But its poverty impact is less straightforward. A comprehensive review of the literature concludes that the technology was responsible for 'massive rises in the yields of staple food crops eaten, grown, and worked mainly by poor people. There have been positive effects on employment, and on the availability, cheapness, and security of food. Yet, there have been only delayed, scanty and sometimes faltering and imperceptible improvements in the lot of the poor' (Lipton *et al.*, 1989). The new technology helped stem the growth of poverty, but it was not sufficient to reverse the trend. Instances of Green Revolution technology improving the production and incomes of the poorest farmers tend to be found, where there is greater equity of access to resources. The results confirm propositions held by both sides in the biotechnology debate. New technology can help alleviate poverty, but it is not a sufficient response.

The nature of agriculture makes it difficult to take decisions about technology that are based solely on efficiency criteria. More than almost any other enterprise, agriculture is both a productive activity and a way of life. Policies that affect agriculture in the South have an impact on the livelihoods of the majority of the rural population. Agricultural technology must be able to increase food production, but it must also provide employment opportunities and food security until economies grow and diversify. The equity considerations are less starkly drawn in the North, where agriculture provides employment for a tiny fraction of the population, but the choices are no less complex. For instance concerns about preserving family farms and 'traditional ways of life are important determinants of agricultural policy. We now turn to examine how policies related to agricultural technology are formed.

27.12. CONSERVATION AND SUSTAINABLE USE OF BIODIVERSITY

Sustainable Use

Proposed actions

- We applaud the proposed review of integration of biodiversity/environment issues into the different DG development sector policies, with a view to establishing and remedying deficiencies: this is absolutely critical.
- There is a need to refer to the (still unfinished?) environment manual with regards to the streamlining of EIA procedures. It has become clear that EIAs are still not being carried out properly. Furthermore, EIAs should be made available for public scrutiny before a decision for funding has been made, particularly to allow local participation in the assessment of EC funded projects.

Agriculture

- Issues relating to food security, TRIPs and bioprospecting/biopiracy are missing, or only mentioned in passing. It is important that the EU makes a strong stand on these issues. NGOs are happy to provide information.
- Higher yields are not the result of the GMOs presently used, quite the opposite in some cases (i.e., the RR soybeans).

proposed actions

- The area of natural habitat cleared for large scale industrial agriculture should be reduced. The focus on intensification of production is ambiguous, and may be contradictory with statements on protection of the biodiversity of agriculture. The role of

extensive agriculture and biodiversity should be addressed.
- EC development cooperation should support partner countries to build the necessary competence to prevent detrimental effects of GMOs on their environment and public health, and to implement and make full use of the biosafety protocol.
- Instead of supporting the development or even release of GMOs into the environment, EC development cooperation should aim at providing the potential to develop environmentally and socially adapted and safe methods of using biotechnology, including the use of latest technologies such as genomics and marker aided breeding technologies. It should not support the release of GMOs into the environment in partner countries.

Wildlife

- The negative impact of poaching is missing from the discussion. The traffic should be monitored, and actions should be developed to reduce poaching with the participation of local communities from inception to implementation.
- There is no reason to assume that wildlife equates to (large) mammals. Birds can generate considerable income from tourism, and butterfly farming has also been used for income generation.

proposed actions

- Participatory procedures should be developed and implemented. However, care needs to be taken not to assume that 'local' off-take will always be sustainable.
- There is a need to fund initiatives to support better benefits from non-consumptive uses of wildlife, such as training of local people to act as tourism guides. This links with work to reduce financial leakage of tourism revenues.

EC activities should ensure that the resource on which the activity (tourism or hunting) depends is being used sustainably.

Forestry

- Title is misleading and should refer to forests rather than forestry.

proposed actions

- An action plan must be developed to address the widespread problem of illegal logging—a regrettable omission in this document. Illegal logging results not only in forest destruction and loss of local livelihood, but also robs national governments as well as local communities of considerable revenues and benefits, distorts timber markets, and is generally an impediment to sustainable forest management. The EU and its member states must take innovative steps to supervise and control the overseas activities of EU companies and their affiliates; as well as to ensure the legality of timber imported into the EU market.
- Increased production of wood from industrial plantations to reduce pressure on natural forests is a theory not proven. There is no evidence that plantations reduce the pressure on natural forests. On the contrary, there is evidence that natural forests are cleared to make room for plantations, and furthermore that plantations provide a source of raw material to feed pulp mills, which then need more timber etc. (see developments in Sumatra, Chile and Brazil where plantations have led to the construction of pulp mills, which consequently have eaten up the rest of the forests).
- Like the World Bank's Forest Policy, the bio-diversity action plan should ensure that there will be no EC funding for logging, or indeed any other large scale industrial activities, road building, mining etc., in primary forests.

Fisheries

- The reform of the CFP in 2002 should be mentioned as a key opportunity to improve mainstreaming of biodiversity and securing policy coherence. In particular, there is the need for coherence between internal and external EC fisheries policies, and identifying clear objectives in EC Fisheries policy.

proposed actions

- The first action should address the need for EIA, not just fish stock assessments, before fisheries agreements are drafted. While the protection of the biodiversity of fisheries stocks is an issue, the impacts on wider biodiversity, for example non-target species and the ecosystem, are also vital.
- The proposed second action should stress the need to develop locally appropriate technical measures.
- EC funding should support innovative fishing practices, for example, provide incentive schemes for small scale but high value initiatives. Innovative processing and marketing techniques should also be supported.
- Support should also be aimed at local delivery of initiatives, even if this is done under the umbrella of a national scheme.
- Water abstraction and pollution also affect other aquatic biodiversity. This also needs to be addressed.
- One of the most important criteria for a proper cost-benefit analysis is that it chooses a sufficiently broad and equitable set of indicators to establish the impact on third country waters and economies, so that we can establish to whom these agreements are of most benefit, and at what cost.
- There is a need to integrate the management of fisheries resources into other national and regional plans.
- The EC should involve southern civil society groups in policy advisory groups.

Conservation

We applaud the statement that local communities need to be fully involved in nature conservation projects. There have been many cases in which local people have been evicted or dispossessed. Protected areas should only be established with the full and informed consent of local communities. Full participation in the decision-making process at all stages (e.g., from identification to implementation) is essential. There is a need for a policy framework on TRIPs. The removal of genetic resources from the lands of indigenous peoples should only take place with their full and informed consent. The EC should support land demarcation, and promote secure land tenure and access to resources by local communities and indigenous peoples, as local communities often hold the key to biodiversity conservation. All over the tropics, local people and their environment are threatened by government funded development schemes, logging and mining concessions, plantations etc.

Research and Education

We often know why biodiversity loss is occurring. Future research should, therefore, be focused on actions to halt or reverse this process. Much of the research mentioned does not have that clear intention. The proposed actions in this section should not be exclusively focused on internal audiences or decision makers. EC funding should support work on education for sustainable development. This could be through support for curriculum development, teacher training, and preparation of materials. Public awareness and understanding is vital. Biodiversity (and environment) needs to be mainstreamed into the EC's work on education.

Many issues are further considered for clarification—why is there no mention of actions to strengthen the capacity of relevant agencies as well as local people who hold, in many countries, the key to the protection of biodiversity. Despite the statements made in the biodiversity action plan that biodiversity issues should be well integrated in other EC policies, as well as the recognition that many of these policies (trade, agriculture) have a big impact on biodiversity loss in developing countries, there is no mention of concrete actions, targets, and deadlines to mainstream biodiversity conservation into development projects across different sectors of the economy in recipient countries. The EC's biodiversity strategies also requires actions in accession countries. Is DG enlargement developing a plan for this aspect?

How the Debate is Managed?

North

In a country like the UK, the debate over biotechnology has achieved a very high profile. The opposition focuses on environmental and food safety issues, and pushes for strict regulation or moratoria. Its principal emphasis is the changes to the countryside that could be caused by the introduction of exotic genes. Biotechnology's defence tends to be more vague; it can only express faith in the potential of new technology and confidence in the regulatory system. Economic arguments are difficult to establish. Farmers may be curious about transgenic varieties, but without the opportunity to test them, they can hardly be expected to lobby forcefully in their support. Even the most optimistic scenario for the utilisation of cost-saving biotechnology would be unlikely to lower food prices enough in the short-run to attract the attention of consumers.

The debate takes place in the media and in local political forums. The democratic process is necessarily imperfect, and decision-makers are subjected to pressures from corporate lobbyists and ecowarriors, but there is a good chance that a majority viewpoint will be heard. Because of a lack of immediate economic impact, popular opinion is largely influenced by competing images of the countryside and technology. If the opponents′ case continues to have appeal, it will result in strict labeling and segregation of transgenic produce, or even its outright prohibition. To illustrate how economic factors might come into play in the future, it is possible to imagine a scenario in which other countries benefit from the efficiency advantages of biotechnology, and evidence of environmental dangers fails to materialise, and the UK begins to find that its own agriculture is less competitive. These market effects could lead to pressure from UK farmers as well as consumers to rescind many of the restrictions that had been imposed on transgenic crops.

Another possible scenario, related more to the image of the countryside than to economics, involves the current tendency to see the debate as one between biotechnology and organic farming. The demand for organic produce is growing sharply despite its higher cost. In the UK, this demand is currently met in large part by imports. It receives increasing attention from large super-market chains, and is becoming part of the marketing strategy of industrialised agriculture. Organic farming is not necessarily small-scale; the rotation legume for organic rice production in California is sown by aeroplane. Future concerns about the fate of the countryside could turn towards the small family farm that wants to preserve its independence from the large marketing chains. Consumer support could develop for purchasing the—not necessarily organic, and perhaps even transgenic—produce of the small independent farm rather than the commercial products whose organic label is decreasingly redolent of traditional farming.

The point of these hypothetical scenarios is not to predict the fate of transgenic crops in the UK, but to show how public opinion, open political debate and responsive markets will determine policies on biotechnology in the North. We now turn to ask how these decisions are managed in the South.

South

How adequate are markets and political systems in the South for making democratic decisions about the future of transgenic crops? In the first place, attention must be given to regulatory mechanisms, in particular biosafety procedures. Even if the vast majority of transgenic varieties turn out to be environmentally benign, countries need access to adequate and impartial regulatory protocols to screen the range of biotechnology products that may be on offer. These mechanisms are in place in only a few countries in the South, and their establishment is a priority. But there are several other elements that affect the biotechnology debate in the South.

Various sectors have an interest in biotechnology. A number of countries in the South, especially those with well developed public research systems, are investing in biotechnology research. There is considerable sentiment among national agricultural researchers that transgenic varieties offer prospects for improving local agricultural production. In addition, local and multinational seed companies are eager to introduce transgenic varieties, because they can see opportunities for increased sales. On the other side, a small but growing number of middle-class consumers in the South have begun to hear about the food safety debate, and some of them have adopted the scepticism of their Northern counterparts.

Although farmers constitute a significant proportion of the population, their political voice is often not heard. Only a relatively few farmers, who have good connections to public research, may have an idea of the potential of biotechnology. Biotechnology may also become a highly visible rallying call for farmer organizations. In recent years, a number of farmer movements in India have been successful in calling attention to the declining profitability of agriculture and the political domination of the urban middle class. Some of these movements have found it effective to link their protests to the debate over India's liberalised trade policy. Most, but not all, of these groups are opposed to liberalisation; one well-known incident involved the occupation of a multinational seed company facility. More recently, several farmers' groups have protested against trials of transgenic cotton; one of the justifications was the erroneous rumour that the varieties incorporated the "terminator gene". Most of the Indian farmers involved in these movements regularly buy inputs (including seed). One can only speculate how these farmers might respond if a productive transgenic variety became available, offered by a local private or public seed company.

In some cases, local NGOs are also active in the debate, most often in opposition to biotechnology. Their concerns and alliances are similar to those of international NGOs, emphasising environmental protection, self-sufficiency, and traditional agriculture. In countries, such as the Philippines, NGOs' scepticism about transgenic varieties is consistent with their long-standing opposition to Green Revolution technology, which was promoted by the Marcos regime without concomitant political or land reform.

Our main interest, however, is the role of international NGOs. We have seen that many of them have launched a highly visible campaign against the possible use of biotechnology in the South. We need to examine the contribution of NGOs to ensure that the circle of the debate is closed, so that interests and consequences are apparent in political forums and market performance, as they are in the North. To what extent do the arguments, they present, allow the concerns of citizens in the South—which may differ from those of the Northern debate—to be the ones that matter? Do the activities of NGOs in the South strengthen people's capacities to make their voices heard? A recent initiative in "citizens" juries' in India attempts to move in that direction, by allowing small groups of farmers to participate in a debate about biotechnology with leading proponents and critics. But this is a long way from developing farmers' independent capacities to interact with public agricultural services and to have more control over markets.

Many NGO strategies are distinguished by their antipathy to markets. Several NGOs use the example of the cotton farmer suicides in India. The tenor of the argument is that it would be preferable for farmers to return to their traditional crops and practices, so that they could be protected from the capitalist system that lured them into cotton farming. There is little attempt to examine the complex conditions that led to the tragedy, the power of local moneylenders, the lack of adequate extension, the actions of unscrupulous merchants, and the failings of regulatory agencies. And there is little evidence in this case of any interest in helping make local markets work in favour of marginalised farmers, establishing credit facilities, strengthening farmers' technical capacities, or providing consumer education and organization.

It is easier for NGOs to make an example of such tragedies, and then wrap themselves in a cloak of innocence, promoting mixed, low chemical-use farming, which favours naturally improved and locally adapted plants (Christian Aid, 1999). It is difficult to object to such attractive proposals, as long as they provide an acceptable living for farmers. The problem comes when we begin to search for evidence regarding the outcomes of these alternatives.

This is not the place to examine the low-input agriculture movement, but many of its basic premises are correct. Excessive input use is responsible for environmental damage and inefficient farming. There is a wide range of possibilities for improving the productivity and safety of farming and these deserve careful testing and evaluation. There is a growing literature on alternative farming methods, and the management-intensive, small-scale nature of the technology makes it appropriate for NGO participation and support. But anyone looking for data on the impact of this type of NGO activity will be disappointed.

NGOs are particularly derelict in following up and evaluating the results of their agricultural projects. The deficiency is easy to understand. The vast majority of the public that contributes funds to these efforts is satisfied with an occasional brochure with pictures of smiling farmers, and the donors who provide funding are irresponsibly lax in organizing assessments that go beyond their own bureaucratic reporting requirements.

Many of low-input agriculture's most articulate supporters admit that there is relatively little solid evidence on the adoption and impact of the alternative production strategies that attract NGOs. This does not imply that the alternatives are inadequate, but rather that much more effort is required for evaluation and adjustment. In the cases where follow-up is done on NGO activities, visiting communities several years after project investment has been completed, the results may be disappointing (Cramb *et al.*, 2000). This is to be expected, as surely the majority of all agricultural projects fail to achieve their objectives. But it is ironic that the ability of NGOs and others to discuss and dissect the impact of the Green Revolution is due to the immense amount of data that is available about the adoption and effects of those technologies. Similar follow-ups of NGO work are in order.

A related issue is NGOs' role in agricultural technology generation. Despite the frequent tirades against the Green Revolution in NGO literature, it is not uncommon to find that their work in the South includes the introduction of new varieties, which are the products of the same public agricultural research system responsible for the Green Revolution. One may ask if NGOs see themselves as "reluctant partners" of public research, or rather as competitors for a decreasing volume of donor funding dedicated to agriculture. Do NGOs propose to contribute to building public research and extension systems that are more responsive to farmers' needs? The image of agricultural technology presented by NGOs has a significant influence on the nature and direction of donor support to agricultural development. In addition, if NGO positions are perceived as extreme or unrealistic, they thwart potential alliances with supporters of public agricultural research who are also alarmed by the growing power of the 'Life Science' industry.

SUMMARY

Integrating socioeconomic considerations into biotechnology and biosafety decisions is a difficult and complex challenge. But with objective, thorough and independent socioeconomic research and inclusive, transparent, and participatory regulatory processes, decision makers will be positioned to meet this challenge. Practical steps can be taken to integrate these considerations into decision-making.

This is clear that the complexity of the choices governments and societies must make about modern biotechnology and its products. Some people wish to move ahead with GM crops, believing that the world's poor cannot afford to wait for a final resolution of all environmental, health, or socioeconomic issues. Others, however, have called for moratoriums on the widespread adoption of agricultural biotechnology or the commercial release of genetically modified seeds while these issues are debated. In some cases, the debate over GMOs is based on a fundamental difference in people's visions of the roles of agriculture and technology in society, as biotechnology is often seen as an unwelcome extension of the industrial agricultural system that many feel is environmentally, economically, and socially unsustainable. While it may not be possible to reconcile such fundamental differences of opinion within regulatory systems, a balanced approach that acknowledges both the potential benefits and costs of modern biotechnology and calls for it to be developed and used with adequate safety measures should result in the best decisions not just for the environment and human health, but for society as a whole.

The problem of modern biotechnology is not necessarily the science itself, which could well be an important source of solutions to deal with poverty and hunger, but in the governance of its applications. The integration of socioeconomic considerations into biotechnology and biosafety decisions through analytically excellent and participatory research, and through regulatory processes that engage the public meaningfully, is an important and essential step toward the good governance of modern biotechnology.

There is little reason for most NGOs to become directly involved with biotechnology. The technology itself is not within most NGOs' purview; their comparative advantage lies more with building local organizations and capacities. So there is no reason to expect any significant NGO commitment to biotechnology, even if transgenic varieties were shown to be safe and beneficial for small-scale farmers. In present circumstances, of course, the incentives are strongly in the opposite direction. By demonstrating opposition to a technology that makes many Northern consumers uneasy, NGOs can garner additional support for their own agricultural programmes.

Regardless of the strengths of various arguments or the ultimate fate of biotechnology, the current campaign raises a question about NGO strategy, which is related to larger concerns about globalisation. It is perfectly appropriate for NGOs to take advantage of their wide links to stimulate debate about a technology which is global in character. But we need to know where NGOs stand in response to globalisation, and in particular whether they choose to 'engage or de-link'.

In the case of agriculture, de-linking involves concentration on self-sufficient communities relying on 'traditional' production methods, isolated from larger markets or political processes that are judged to be inimical to the interests of marginalised farmers. This would not appear to be a realistic alternative, but if it is the path that is chosen, the Northern public and donors that support this action deserve detailed reports on progress achieved. The opposite course, engagement, would seem to offer many more opportunities. It makes no assumptions about the type of technology that is utilised. It may involve 'traditional' or 'modern' technology. It may develop methods that outperform anything that biotechnology has to offer, or it may complement or accelerate the use of biotechnology. The quality that distinguishes engagement is a clear commitment to ensuring that marginalised farmers have a voice in agricultural markets and that they are better represented in public activities such as agricultural research, extension and regulation.

It is not that 'empowerment' as it is commonly understood fails to take place. It often does occur albeit unevenly. Rather, it is that at the very moment that such empowerment is attained is also the occasion when a significant loss of 'freedom' from surveillance and control seems to happen. Empowerment is thus apparently bought at a price. The very practices that enable the downtrodden – usually with NGO help—to sometimes 'see off' the forces of *sovereignty* are also an irresistible invitation for processes of *governmentality* to intrude ever more systematically into the lives of marginalized peoples. Empowerment, a bit like NGOs themselves, may be seen to reflect ambivalent, if not contradictory processes.

The preceding analysis may seem unduly pessimistic to those committed to the emancipatory potential of NGOs. Yet, the point of this paper is certainly not to dismiss this actor as an instrument of political rationalities beyond its control. Indeed, NGOs *can* play their part in the promotion of 'an ethic of permanent resistance' to the social control associated with governmentality. Nor is it necessary that such resistance take the form of armed insurrection—a mainstay of certain sorts of political opposition (e.g., Communist, Islamic). True, contemporary neo-liberal governmentality poses new challenges to those who would resist. Specifically, it promotes the selective 'withdrawal of the state' from regulatory matters in favour of enhanced self-regulation by individuals whose lives acquire a 'specific entrepreneurial form'. In effect, neo-liberalism is a rationality that seeks to turn the social arena into just another focus of economic activity as individuals and sometimes communities assume 'responsibility' for their own actions, and failures.

EXERCISE

1. Define biosafety and non-government organisations (NGOs).
2. Explain the need of biosafety in the present day biotechnology dominated scientific field.
3. Name organization for economic cooperation and development (OECD) highlights for effective participation in biosafety for transgenic organisms.
4. Mention the importance of public and non-government organizations', (NGOs) participation in biosafety in biotechnology.
5. Name different types of biodiversity present in the biosphere. Mention their applications in the human society.
6. List out the importance of NGOs in biotechnological field for effective biosafety and biodiversity.
7. Critically comment on the physical containment and biological containment used in the research and testing.
8. Discuss how convention on biological diversity (CBD) on biosafety entered into effect, and its impact on biotechnology.
9. Describe tools and mechanisms of biosafety information that are used for the effective dissemination of informations to the public.
10. Write a note on advantages of public education and public participation in adopting types of biosafety followed for the effective environment and health.
11. Give an account of the merits of the public education and public participation in adopting different types of biosafety.
12. Mention different problems associated with the transgenic organisms, and explain environmental and food safety issues in biotechnology.
13. What do you mean by population explosion? What is its importance in the research and development?
14. Distinguish between government and non-government organisations. Add a note on the roles of non-government organization.
15. Briefly explain the biosafety protocol or UN cartagena biosafety protocol (CBP), and its key issues addressed for achieving maximum safety.
16. What do you mean by low resource farmers in the biotechnology fields? Explain the importance of the effective control of such farmers interest.
17. Differentiate between the different levels of public participation required in the biotechnology research and development for better future.
18. List out the achievements of different national non-government organization (NGOs) in the biotechnological applications of the products.
19. What do you mean by cartagena biosafety protocol (CBP)?
20. Explain the applications of biosafety organization at the international level, and their importance in the biotechnological processes for producing biohazardous organisms.
21. Define "Living Modified Organisms" (LMOs) and their potential danger to the living world.
22. Explain the problems that have to be addressed by public non-government organisations (NGOs).
23. Illustrate how future plans are necessary for better biosafety measurement needed in the biotechnology fields that always has a risk to biodiversity.
24. How non-government organization (NGOs) will work for ensuring the biosafety in the different fields in the biotechnological research?

28

Indian Biodiversity Act

28.1. INTRODUCTION

Biodiversity or biological diversity is the variability among all living organisms (microorganisms, plants, and animals) existing on earth in various ecosystems and ecological complexes. This diversity is the basis of continuous evolution of life forms, and in turn maintains the life-sustaining systems of the biosphere. The conservation of all biological diversity is a common concern of human kind, and it is vital to anticipate, prevent, and tackle the causes of loss or reduction of biological resources. The dependence of human beings on biological diversity is undoubted, as evident in everyday life. The food, fibre, fuel, fodder, shelter, health, and other needs of the growing world population are dependent on various components of biodiversity. It is also recognized that plant genetic resources for food and agriculture are a common concern of all countries, and most countries depend largely on plant genetic resources that have originated elsewhere.

With the extinction of perhaps 50,000 species per year, we are at a critical juncture in history. Ominous storm clouds have gathered to threaten humanity's most basic natural resource of all—the diversity of living and non-living components. Biodiversity encompasses the variety of all life on earth. India has documented over 45,000 species of flora and 81,000 known species of fauna, and contains with its borders two of the world's 10 biogeographic zones. The country is one of the world's 12 mega diverse countries of biodiversity of the world. With only 2.5% of the land area, India already accounts for 7.8% of the global recorded species. India is also rich in traditional and indigenous knowledge, both coded and informal. Contained within the subcontinent are tropical wet evergreen forests, deserts, alpine vegetation and vast coastal systems. Coupled with a corrupt and unaccountable bureaucracy and impoverished local populations, the region is nevertheless the repository of vast traditional knowledge and thus makes for biodiversities delight.

India, like many other developing countries, is home to a rich variety of genetic resources. Among the living organisms that make-up its varied ecosystems, Indian scientists have identified some 126,000 endemic species including 45,000 plants and 81,000 animals. This natural wealth includes many varieties of crops and animals that indigenous and local communities have developed over the centuries. Today, many Indian farmers continue these traditions, developing seed by careful identification and propagation. The varieties that have emerged contain a unique pool of genes well-adapted to local conditions.

India and adjoining countries with diversity of people, culture and livelihoods are endowed with rich biological diversity and genetic resources. The country in the past millennium contributed immensely towards the development of bio-resource and the management of environment for the welfare of the society and posterity. India has leaped high among the developing countries, with the richness of resources and their sustainable management to be placed among the developed nations within foreseeable future. With the global prospects and perspectives, the subject of biological diversity and its management, the country has reached a new height.

A legally binding agreement, Convention on Biological Diversity (CBD), was adopted by the United Nations Conference on Environment and Development, held at Rio de Janeiro in June 1992. The objectives of the CBD are the conservation of biological diversity; the sustainable use of its components; and the fair and equitable sharing of the benefits arising out of the utilization of genetic resources, including appropriate access to genetic resources and appropriate transfer of relevant technologies, taking into account all rights over those resources and technologies, and by appropriate funding. The convention reaffirmed that states have sovereign rights over their biological resources, and that the states are responsible for conserving these resources and using the same in a sustainable manner. The contracting parties to the CBD are, therefore, required to integrate considerations of conservation and sustainable use of biological diversity into relevant sectoral or cross-sectoral plans, programmes, and polices.

India is a party to the convention on biological diversity (1992). Recognizing the sovereign rights of states to use their own biological resources, the convention expects the parties to facilitate access to genetic resources by other parties subject to national legislation and on mutually agreed upon terms (Article 3 and 15 of CBD). Article 8(j) of the Convention on biological diversity recognizes contributions of local and indigenous communities to the conservation and sustainable utilization of biological resources through traditional knowledge, practices, and innovations, and provides for equitable sharing of benefits with such people arising from the utilization of their knowledge, practices, and innovations.

Biodiversity is a multidisciplinary subject involving diverse activities and actions. The stakeholders in biological diversity include the central government, state governments, institutions of local self-governmental organizations, industry, etc. One of the major challenges before India lies in adopting an instrument, which helps realise the objectives of equitable sharing of benefits enshrined in the convention on biological diversity.

After an extensive and intensive consultation process involving the stakeholders, the central government has brought biological diversity act, 2002, with the following salient features:

- To regulate access to biological resources of the country, with the purpose of securing equitable share in benefits arising out of the use of biological resources, and associated knowledge related to biological resources.
- To conserve and sustainably use biological diversity.
- To respect and protect knowledge of local communities related to biodiversity.
- To secure sharing of benefits with local people as conservers of biological resources and holders of knowledge and information relating to the use of biological resources.
- Conservation and development of areas of importance from the standpoint of biological diversity by declaring them as biological diversity heritage sites.

- Protection and rehabilitation of threatened species.
- Involvement of the institutions of state governments in the broad scheme of the implementation of the biological diversity act through constitution of committees.

Patent and its Disadvantages to Biodiversity

Exclusive economic rights can be secured by patenting. Through the recent emergence of a global patenting system, commercial enterprises and particularly transnational corporations are increasingly able to appropriate local genetic resources and traditional knowledge to create commodities attractive to world markets. It can be argued that the deficiencies of intergovernmental regulations dealing with intellectual property rights (IPR) and the conflicts between them have not provided sufficient safeguards against the widespread theft of genetic resources. Problems at the international level have led to delays in national legislation, as policy makers await developments in the international arena.

Biopiracy is becoming a threat to the Indian biodiversity, as patent offices in the USA, the European Union and Japan continue to grant patents to slightly modified agricultural genetic resources and medicinal products grown, developed, and used in large parts of India. The Indian Government's Council for Scientific and Industrial Research (CSIR) is becoming more vigilant. The patent unit analyses and assesses the potential of patent claims, collects authoritative documentation, and files challenges. It has successfully mounted challenges to patents such as the one granted by the United States Patent and Trademark Office (USPTO) on the healing properties of turmeric.

At the international as well as national level, efforts are being made to preserve and protect local genetic resources from commercial exploitation. The CBD, for example, allows contracting parties to develop a legal framework and take administrative measures to preserve, protect and promote indigenous and local knowledge and practices and to restrict 'bio-prospecting'. However, such measures do not safeguard farmers' and breeders' rights. A clear and strict international convention is required and in this context the provisions of the CBD, WIPO, TRIPS, and USA patent law should be re-examined.

National Level Protection of Biodiversity

Biodiversity acts as major resource for natural evolution. The revolutionary new law is in the making. After years of procrastination and some frankly unimaginative moves, the Union Ministry of Environment and Forests has finally come up with a draft legislation on biodiversity, which could go a long way in protecting India's enormous biological resources from being squandered, destroyed, and stolen. The proposed biological diversity act contains the core of a regime to check the runaway theft of the country's genetic wealth, as also the framework for ensuring that both domestic and foreign users of this wealth do so in a manner, which is sustainable and fair.

Agricultural biotechnology encompasses three areas:

- Diagnostic techniques that use an understanding of molecular biology to make conventional plant breeding more efficient.
- Multiplication techniques, such as tissue culture that can rapidly reproduce planting material or help eliminate viruses and other disease-causing organisms.
- Genetic engineering, which involves the transfer of genes from a wide range of sources. Most of the controversy about biotechnology centres on genetic engineering and its products, transgenic varieties, and cloning.

Gene Sources

The gene source is major requirement for the development of biotechnology. The genes may come from any source. The more spectacular and controversial cases involve wide transfers, such as the introduction of a gene from a flounder to a potato—for cold tolerance. Many other examples involve less distance (e.g., a nematode resistance gene from rice transferred to potatoes), and a significant number of cases involve movement of genes within the same species (but with more precision than conventional plant breeding allows). India is one of the world's top twelve megadiversity countries. It possesses a wealth of knowledge associated with biodiversity—be it the orally held knowledge of folk healers or herders, or the traditional knowledge codified in *Ayurvedic, Sidha,* or *Yunani* texts. Its biodiversity resources are far better known scientifically than those of other tropical megadiversity countries such as Brazil or Indonesia. As a result, India has developed a number of excellent biodiversity databases such as the Flora of Karnataka, Traditional Knowledge Digital Library, or the National Register of Green Grassroots Innovations and Traditional Knowledge.

There exist, therefore, rich possibilities of building upon country's biodiversity resources and associated knowledge to promote biodiversity-based enterprises in the modern, as well as traditional sectors; to develop biotechnology industries at the cutting edge of new technologies; as well as to encourage local level value addition to biodiversity resources. Important new markets are also emerging for produce of organic agriculture. Taking advantage of these markets will require development of good databases on agro-ecosystems, including incidence of pests and diseases.

Dominance of Private Companies

The majority of the techniques for gene transfer, and many of the most widely used genes in current transgenic varieties are owned by private firms—most of them are the few large multinationals that dominate the field. All of the transgenic varieties currently grown on a commercial scale, with the exception of some in China, are privately owned. Nevertheless, there is also significant investment in public biotechnology research. Public research organisations in the North and South are developing transgenic varieties using techniques and genes licensed or donated by the private sector as well as sources and methods in the public domain.

The same situation prevails in India. The Indian government has passed three parliamentary acts over the past few years in an attempt to protect the nation's biological diversity and the interests of its researchers, plant breeders, and farmers. This three pronged strategy involves:

1. the Protection of Plant Varieties and Farmers' Rights Act 2001;
2. the Biodiversity Act 2001; and
3. the Geographical Indication of Goods (Registration and Protection) Act 1999.

Varieties and farmers' rights

The Protection of Plant Varieties and Farmers' Rights Act 2001 aims to protect plant varieties developed through public and private sector research and developed and conserved by farmers and traditional communities. It provides legal rights to farmers to save, use, share or sell their farm seed and stimulates plant breeders and researchers to develop new and improved varieties. Generally, this law envisions that farmers will be treated like commercial breeders and receive the same kind of protection. The act prescribes the establishment of the *Plant Varieties Protection Authority* that not only registers the new varieties developed by breeders and farmers, but also ensures fair and equitable benefit sharing and financial compensation.

The act has been criticized for its failure to provide a holistic framework to recognize the variety of stakeholders engaged in agricultural management and seed improvement that have rights to their resources. Secondly, it encourages farmers to get a variety registered, and so fundamentally accepts the registration criteria of the UPOV convention, which is mainly devised for commercial breeders. Thirdly, in the case of benefit sharing, the claimants can neither stop the registration of the variety nor claim IPR on their own variety. They are merely awarded monetary compensation instead. Fourthly, the bill does not discuss the crucial issue of pricing mechanisms for high quality seeds and planting materials, which is a major problem for farmers. It lacks a broader provision on the need for impact assessment of varieties offered for registration. Finally, it has been criticized for containing provisions that allow the registration of essentially derived plant varieties.

Biodiversity act

India's interpretation of the CBD is reflected in the Biodiversity Act 2001. It provides for the establishment of a National Biological Authority (NBA), with extensive powers to protect biological resources. Foreign agents require NBA approval in order to access biological resources or inventions derived from them, and provisions for equitable benefit sharing are clearly stipulated. NBA approval must also be obtained before biological resources can be exported, and proposals have been made to set up biodiversity funds and management committees at national, state, and local levels.

Critics of the bill argue that it does not assert national sovereignty over biodiversity, even though in the context of the CBD the sovereignty provision is an important check to foreign patent seekers. Neither does it provide measures to limit the potential environmental and health risks associated with the introduction of genetically modified organisms (GMOs). However, the business community is critical of the strict licensing requirements of the NBA, arguing that restrictions on foreign collaborative ventures will inhibit the growth of the Indian biotechnology industry. When it was passed on Dec 11, Union Environment and Forests Minister, T.R. Baalu claimed that the legislation would regulate access to genetic resources and associated knowledge by foreign individuals and institutions, and ensure equitable sharing of benefits arising out of the use of resources and knowledge with the country and its people.

He also said that the act provided safeguards to protect the interests of local people, growers, and cultivators of biological diversity, as well as Indian researchers through a new National Biodiversity Authority (NBA), supported by state level boards and management committees that would regulate access to plant and animal genetic resources. "The NBA's approval will be required before obtaining any form of intellectual property rights on an invention based on a biological resource from India or on a traditional knowledge, and it will deal with all cases of access by foreigners," Baalu said.

Indian citizens and companies are allowed free access to biological resources within the country for research purposes, but are barred from transferring findings to foreign entities without the NBA's approval. But all these provisions succeed in burying biodiversity under a mountain of bureaucracy that can only serve to alienate ordinary farmers from their resources, while making international biopiracy easier, says leading activists Suman Sahai of the Gene Campaign and Vandana Shiva of the Research Foundation for Science Technology and the Environment (RFSTE).

"By excluding agriculture from the Act's purview, global corporations can still gain access to valuable biological resources," said Shiva. But an impoverished local farmer, who allows his cow to graze freely on the commons, could find himself penalized for inadvertently destroying a herb considered to be a valuable biological resource, she said.

Geographical indication

The Geographical Indication of Goods (Registration and Protection) *Act* was adopted in 2000 and aims to provide a comprehensive framework to facilitate the registration, conservation, and protection of goods with a unique geographical identity. The act provides for the establishment of a *Geographical Indication Registry* and an *Appellate Board* to take necessary action against infringement.

Biodiversity in India and its Importance

India is gifted with varieties of agroclimatic regions. This confirms that India remains a biomass based civilization. Across the length and breadth of our country, a large proportion of the people subsist as ecosystem people, greatly dependent on local natural living resources to fulfill their manifold requirements. Many cook with fuelwood gathered from forests or scrublands; graze their cattle, sheep, or goat on natural vegetation, collect tree leaves to manure their fields; and employ herbal medicines to heal themselves or their livestock. Many tribal families of Central India live on *Mahua* flowers for several months of the year. Yet others earn a substantial proportion of their small income by collecting forest produce such as canes, beedi leaves, sal seeds, or wild mango fruit for sale in markets. Many landless, and therefore the weakest of rural populace supplement their incomes by weaving baskets or mats in the months, when there is no employment as farm labourers. Yet others depend entirely on fishing. There is thus a substantial dependence on natural living resources. This relationship is especially strong in case of women, who often assume the major responsibility for the collection of water, fuelwood, dung, and fodder.

Cultivation and animal husbandry is even more significant to the livelihoods of a majority of Indians. These plants and animals in turn interact with a number of species of pollinators, pests, parasites, weeds and fodder plants. The farmers, farm labourers, shepherds, and cowherds therefore not only relate to a whole spectrum of varieties or land races of these plants and animals, but to a large number of other plants and animals affecting them as well. This relationship between the human and the natural world is reflected in the diversity of livelihoods present across the country. This diversity in turn has supported the continued protection of biological diversity and the various habitats where it occurs.

Indian tribes or rural population have greater sense of protection for their environment. The significant majority of the Indians are intimately tied to their natural setting in the pursuit of their livelihoods. But Indians not only relate to plants and animals of immediate utility or nuisance value. In almost every one of the study sites people venerate and protect plants like peepal or banyan and animals like peafowl or hanuman langur. In several sites, the protection extends to whole patches of forests or grasslands, or pools along streams as well. However, the number of species thus related to in religious or cultural contexts is relatively small.

A vast majority of the one to five hundred species of plants and animals recorded, as known to people from the various sites, are species of practical significance in the lives of people. So are almost all of the species listed as being of conservation interest to the people. Not only is the diversity of natural biological communities on retreat all over, so is that of crops and livestock. This loss of crop and livestock diversity is in part driven by economic forces, with modern high yielding varieties fetching better returns and attracting farmers to switch over. But such a switchover is also prompted by government subsidies for irrigation, fertilizers, and pesticides. The higher economic returns from monocultures of modern varieties often turn out to be non-sustainable, when soils deteriorate under chemicalized agriculture, pests explode with extensive areas coming under single varieties; or when subsidies are withdrawn. The farmers of Dhikonia from Baran district of Rajasthan are now facing these problems. This is a dry tract, where a traditional variety of wheat called *"Katya"* was being cultivated. Farmers now find the yields from hybrid wheats that replaced *katya* to be lower than what *katya* produced. But now they cannot bring back *katya* cultivation on soils degraded by chemicalised agriculture.

Huge impact has been noticed with altered cultivation and agriculture practice. Replacement of the diversity of traditional crops has other implications. Thus, in and around Bhitar Kanika Wildlife Sanctuary in Orissa, the replacement of traditional tall varieties by dwarf hybrid rice has led to a reduction in availability of thatching material. In turn, this had led to an overharvest of the leaves of a wild ground palm. But while traditional crop and livestock diversity has indeed been eroded in many places, this is not without exception. In Kolar district of Karnataka, farmers continue to grow traditional rice varieties on a small scale, as they consider these more nourishing for their children. On Majuli island on Brahmaputra river in Assam, over 20 traditional rice varieties continue to be cultivated, as the hybrids failed to do better in local soils with problems of silt and low chemical inputs. At the same time, the agricultural and animal husbandry experts have completely ignored on-farm conservation, solely concentrating on the maintenance of germplasm in *ex situ* collections.

The steep change in agriculture and other activities resulted in huge impact. The ongoing erosion of India's biological diversity is also promoted by the narrow focus of the official management, as well as conservation efforts on a handful of plant and animal species—teak and pine, tiger and Siberian cranes. Thus, only a handful of the fifty odd species of medicinal plants and other non-timber forest produce collected from the study site of Mala in Karnataka are subject to forest department regulation in the form of auctions for contractors. The vast majority is collected by locals, sold to agents that ultimately deliver them through unorganized markets to the pharmaceutical or biocosmetics industry, or to the individual consumers. At the same time, the official conservation efforts such as National Parks pay attention to only a few charismatic animals, be they rhinoceros or crocodile, with almost total neglect of the great variety of other living organisms, even of considerable immediate economic significance such as medicinal plants.

Multinational companies dominate the IPR field. The Indian public is by now used to daily news of how foreigners have been quietly taking our crops and medicinal plants, and traditional knowledge related to these, and claiming their ownership after some (often minimal) modifications. Or how they have been using or intending to use the Indian public as guinea pigs, as with the infamous terminator seeds. Last year, the Indian government won a significant victory in getting the patent on turmeric in the USA revoked. It now plans to challenge the patent on Basmati, which has been obtained by an American company. After strong protest from NGOs and farmers, it has also taken a tough stand against Monsanto Corporation—the originator of the deadly terminator gene (which will not allow farmers to regrow their seeds). These, however, are piecemeal approaches to a deeper problem. So far, we have no legal and administrative arrangements, which could prevent the theft of the resources and knowledge in the first place. The proposed Act may just provide the framework for such arrangements.

Discussions on such a law have been going on for years in India and elsewhere, but they received a major boost when the international Convention on Biological Diversity (CBD) came into force in 1993. Ratified by almost every country of the world, the CBD formally assigns ownership over genetic resources to the country within which they were found, and stipulates that if any other country wants to take these resources, they could do so only with "prior informed consent". In other words, if an American company wants to take an Indian plant and experiment with it, or take knowledge related to biodiversity, it needs to inform India of its desire to do so, and needs to have India's permission before the transfer can take place.

The CBD goes beyond this, and also stipulates that where the knowledge and practices of local communities (tribal and others) are involved (as indeed they are with the development of *basmati* and the uses of turmeric, amongst thousands of other such practices), their permission also would need to be taken by outsiders wanting to use this knowledge. This revolutionary concept gives the chance to the rural communities for the first time to benefit from the widespread use of their skills and knowledge, which has before this simply been taken away by outsiders without any corresponding benefits flowing back.

However, the CBD is only applicable if the signatory countries follow up with their own domestic legislation. For the last 5 years, since the CBD came into force, it has been dilly-dallying over formula-ting appropriate legislation. The matters to be dealt with in this law are naturally quite complex, which makes development of such a law especially difficult. However, much time was lost by strange inaction by a previous set of bureaucrats at the Ministry of Environment and Forests (MoEF), and by an amazingly weak draft developed by the Swaminathan Committee. It was only after the current Minister for Environment, Suresh Prabhu, called a national consultation that things once again started moving. Based on the comments received at this consultation and after, a smaller expert group redrafted the proposed Act in October. The result is a much more comprehensive, balanced, and radical document.

Elements of the act

Similar to the CBD, the proposed Indian Biodiversity Act aims to achieve three things:

1. Conservation of biodiversity
2. Sustainable use of biological resources
3. Equitable sharing of benefits arising from such use

To achieve these, the Act contains several important clauses. It:

- prohibits the transfer of Indian genetic material outside the country, without specific approval of the Indian Government through a due process;
- stipulates that anyone wanting to take a patent or other intellectual property right (IPR) over such material, or over related knowledge, will have to seek permission in advance;
- provides for the levying of appropriate fees and royalties on such transfers and IPRs;
- regulates access to such material by Indian national also to ensure that there is some control over over-exploitation (e.g., of medicinal plants), and that there is some sharing of benefits to all concerned parties; however, it provides some relaxation in the case of research;
- provides for measures to conserve and sustainably use biological resources, including habitat and species protection, conservation in gene banks, environmental impact assessments of all projects, which could harm biodiversity, and so on;
- empowers local communities to have a say in the use of resources and knowledge within their jurisdiction, and to enter into negotiations with parties who want to use these resources and knowledge;
- provides for the development of an appropriate legislation or administrative steps, including registration, to protect indigenous and community knowledge;
- empowers governments to declare biodiversity heritage sites as areas of special measures for conservation and sustainable use of biological resources, as also notify threatened species to control their collection and use;
- stipulates that risks associated with biotechnology (including the use of genetically modified organisms) will be regulated or controlled through appropriate means; and
- provides for the designation of repositories of biological resources, at national and other levels.

The Act envisages the creation of funds at local, state, and national levels, which will be used to support conservation and benefit-sharing activities. These funds will be generated from fees, royalties, donations, etc.

The Act proposes to set up bodies at three levels to carry out the above functions. At the national level, there will be a National Biodiversity Authority (NBA), which will screen proposals for the transfer of genetic resources abroad, advise the central government of measures for conservation, sustainable use, and benefit-sharing, suggest the use of the National Biodiversity Fund, and oppose, where necessary, IPRs in India and abroad which violate the Act's provisions. The NBA will consist of a chairperson who has eminence in the field, representatives of relevant government agencies, and non-governmental members including members of local communities and a representative of industry. At the state level, there will be State Biodiversity Boards (SBB), which will oversee use and conservation of biodiversity within state jurisdiction; for instance, they can specify limits of which and how many plants a pharmaceutical company can harvest from its forests. SBBs will also manage the state biodiversity funds. At local levels, there will be biodiversity management committees, which will have a voice in regulating the transfer, use, and conservation of resources and knowledge at community and individual level. These committees or other local bodies will also have a say in the utilisation of the various Funds set up under the act.

Importantly, the Act provides citizens with the power to approach courts if they detect violations. One of the most regressive aspects of the earlier (Swaminathan Committee) draft was that it did not provide such locus standi, but strong citizen's pressure made the MoEF put it squarely in.

The penalties provided in the proposed Act could be stiff: imprisonment for a term up to five years, or fines up to Rs. 10 lakhs.

Difficulties in implementation

There remain many issues which will have to be sorted out in implementing the Act.

- What kind of benefits are appropriate when a commercial product comes out of traditional knowledge?
- A 50:50 share of profits between the company and the community which held the knowledge?
- The transfer of relevant technologies to this community?
- Participation in R&D?
- Exemption from the application of IPRs?
- Revival of tenurial rights to land and resources?
- Non-monetary awards?
- Others?
- By what mechanism would beneficiaries be identified, and benefits be transferred?
- What if more than one community holds the same knowledge, should benefits be shared with all?
- What in the case of innovations made by single individuals, or families, within a community: should they get the benefits, or the whole community?

A recent case of benefit-sharing will illustrate the complexities involved. In the mid-1990s, the Tropical Botanical Garden Research Institute, Kerala, entered into an agreement with the local "Kani" tribe to share the benefits of a new herbal medicine ('Jeevani') developed from knowledge of a plant (*Trichopus zeylanicus*) provided by the tribals. The Kanis are to be given an initial fee and then regular royalties from the sale of the drug. This is perhaps the first formal agreement of this kind in India (there are examples of similar deals in some other countries), and highlights the desire of the director of TBGRI, Dr. P. Pushpangadan to reverse the earlier unfairness in which the company would simply have walked away with all the profits. However, the arrangement contains a number of question marks. For several months its development ground to a halt, because the plant required was on the forest department land, and the department, fearing over-exploitation, stopped its extraction. This shows the need for some tenurial rights to be given to the Kanis for sustainable extraction, coupled with perhaps an overseeing function by the forest department and/or NGOs. Secondly, there was the question of who should get the benefits—the tribal individuals who alerted the TBGRI scientists in the first place, or the entire tribe? This has been partially solved by starting a trust of which all Kanis are potential members, and the whole tribe can, in theory, decide what to do with the money. Third, however, some other tribes in the area may claim to have similar knowledge. Should they also receive benefits if this claim is valid? These and other questions have to be sorted out for the long-term sustainability of the agreement.

There are equally vexed issues in the case of conservation and sustainable use of biological resources. Will the Act help in reigning in a runaway economic roadroller, which is threatening to turn every natural resource into raw material and every villager into cheap labour? Can biodiversity-rich areas be protected against mining, dams, industries, pollution? Like its predecessors (the Wildlife Protection Act, the Environment Protection Act, etc.), will the Biodiversity Act perhaps just be another means for the committed environmentalist and the affected community to fight against such destruction? But even that is a vital new step. Critically, the Act provides for the integration of biodiversity concerns into every economic sector's plans, and also makes mandatory environmental impact assessments of any activities, which could have negative impacts on biodiversity, including biotechnological developments.

Several specific parts of the proposed Act need further amplification in the form of rules or guidelines. The NBA is supposed to consult local bodies when granting permission for access to biological resources within their jurisdiction, but the modalities by which this can be done are to be worked out. The expert group, which drafted the Act, had incorporated a provision to tax industries which are based on biodiversity (e.g., pharmaceutical, seed, and perfumery companies), but this has been deleted from the Ministry's final draft, because of the objections by the Law Department. Hopefully this clause, which will help to generate resources necessary for conservation and benefit-sharing, and also ensure that industry pays some of the true cost of the resources it uses, can come into the rules that will be developed under the Act.

Another aspect of the Act which needs further development is the concept of Biodiversity Heritage Sites, currently not well-defined. Further rules will need to specify what kinds of areas can be declared as such, both to avoid duplication with the protected areas declared under the Wildlife Protection Act, and also to ensure that areas with tremendous human-developed biodiversity (e.g., traditional agricultural areas) receive protection and benefits from it.

By no means should the Act become a means to alienate local communities from their resources and knowledge, as unfortunately the Wildlife Act has become in many areas. But it is equally true that communities in many places are no longer as conservation-oriented as they may have been earlier. If community level bodies (e.g., panchayats, or forest protection committees, or new Biodiversity Management Committees) are really to use the powers given to them under this Act and others (such as panchayat legislation developed in 1996 and 1997), their capacity to manage resources may need revival, and their ability to negotiate with outsiders may need to be built up. Educational and training programmes for all sectors of society will need to be reoriented to become more sensitive to biodiversity issues. This, especially, includes the judiciary at district and other levels, which needs to be convinced about the seriousness of biodiversity related violations if the Act has to have teeth.

Whatever the complexities, it is also true that we will never get around to solving them, unless forced to do so. Once a legislation with a broad framework of conservation, sustainable use, and benefit-sharing is in place, both government and citizens will have to come up with innovative solutions. The next few years will undoubtedly see the refinement of these concepts and practices. First, however, the biggest challenge that Suresh Prabhu and his officials face is to steer the draft Act through Cabinet (he will probably face resistance from fellow ministers who may, mistakenly, see some encroachment into their territory by the Act), and then through Parliament. Prabhu is reportedly keen to introduce the bill into the winter session of Parliament. Whether he is able to do or not, the current draft certainly points India in the right direction, and it is perhaps only a matter of time before we go in for such bold steps to stop the currently unsustainable and unfair use of our biological resources and knowledge.

In this Act, some of the very essential definitions are defined clearly:

- "Benefit claimers" means the conservers of biological resources, their by-products, creators and holders of knowledge and information related to the use of such biological resources, and innovations and practices associated with such use and application.
- "Biological diversity" means the variability among living organisms from all sources and the ecological complexes of which they are part, and includes diversity within species or between species and of ecosystems.
- "Biological resources" means plants, animals and microorganisms or parts thereof, their genetic material and by-products (excluding value added products) with actual or potential use or value, but does not include human genetic material.
- "Bio-survey and bio-utilisation" means survey or collection of species, subspecies, genes, components and extracts of biological resource for any purpose, and includes characterisation, inventorisation, and bioassay.

- "Commercial utilisation" means uses of biological resources for commercial utilisation such as drugs, industrial enzymes, food flavours, fragrance, cosmetics, emulsifiers, oleoresins, colours, extracts and genes used for improving crops and livestock through genetic intervention, but does not include conventional breeding or traditional practices used in any agriculture, horticulture, poultry, diary farming, animal husbandry or bee keeping.
- "Fair and equitable benefit sharing" means sharing of benefits, as determined by the National Biodiversity Authority.
- "Local bodies" means panchayats and municipalities, by whatever name called.
- "Sustainable use" means the use of components of biological diversity in such manner and at such rate that does not lead to the long-term decline of the biological diversity, thereby maintaining its potential to meet the needs and aspirations of present and future generations.

India is party to the Convention on Biological Diversity (CBD) 1992, which recognizes the sovereign rights of states to use their own Biological Resources present in the country. In order to help in realizing the objectives of CBD, India has enacted an umbrella legislation called the Biological Diversity Act 2002 (No. 18 of 2003) aimed at conservation of biological resources and associated knowledge, as well as facilitating access to them in a sustainable manner and through a just process. Prior to 2002, there was no legal protection for biodiversity in India, and the Act was intended to provide some space for biodiversity conservation. The Biological Diversity Act (BDA) was formulated after India became signatory to the CBD. The draft legislation was developed through an intensive consultation process involving all stakeholders such as the central government, state governments, institutions of local self-government, scientific and technical institutions, experts, non-governmental organizations, industry, etc. The Act was passed by the Parliament in December 2002 [Publication of the Biological Diversity Act, 2002 No. 18 of 2003. Ministry of Law and Justice (Legislative Department), Government of India]. The objectives of the Act are 'to provide for conservation of biological diversity, sustainable use of its components, and equitable sharing of the benefits arising out of the use of biological resources, and for matters connected there-with or incidental thereto'.

Mechanisms of Implementation of Biological Diversity Act (BDA)

In India, for the effective implementation of the BDA, the central government would undertake activities to develop national strategies, plans and programmes for conservation and sustainable use of biological resources, with the following proposed institutional mechanisms. It would take measures for identifying and monitoring biodiversity-rich areas and notify threatened species. It would also undertake promotion of incentives for research, training, public awareness and education with respect to biodiversity, and make assessment of environment impact of any activity likely to have adverse impact on biological diversity. It would regulate, manage or control the risks associated with the use and release of living modified organisms resulting from biotechnology, likely to have adverse impact on conservation and sustainable use of biodiversity and human health. It may also declare some resources to be exempted from the provisions of this Act, including resources normally traded as commodities.

National Biodiversity Authority

National biodiversity authority means the National Biodiversity Authority established under section 8 of the Act. *State Biodiversity Board* means the State Biodiversity Board established under section 22 of the Act. In exercise of the powers conferred by sub-section(1)(4) of section 8 of the Biological Diversity Act, 2002 (18 of 2003), the Central Government has established a body called the National Biodiversity Authority, on and from the 1st day of October, 2003.

It is proposed to have National Biodiversity Authority (NBA), State Biodiversity Boards (SBB) and Biodiversity Management Committees (BMC) for effective implementation of the Act.

The NBA will deal with matters relating to requests for access by foreign individuals, institutions or companies, and those relating to transfer of results of research to any foreigner. Imposition of terms and conditions to secure fair and equitable sharing of benefits arising out of utilization of biological resources and approvals for seeking any form of Intellectual Property Rights (IPR) in or outside India for an invention based on research or information pertaining to a biological resource or knowledge associated thereto obtained from India, would also be dealt with by the NBA.

SBB would be constituted for every state in India to deal with matters relating to access by Indians for commercial purposes and restrict any activity that violates the objectives of conservation, sustainable use and equitable sharing of benefits.

Institutions of self-government in their respective areas would constitute a BMC for conservation, sustainable use, documentation of biodiversity and chronicling of knowledge relating to biodiversity. The NBA shall consult BMC and SBB on matters related to use of biological re-sources and associated knowledge within their jurisdiction. It is also proposed to set-up 'Biodiversity Funds' at central, state and local levels. The monetary benefits, fees and royalties received as a result of approvals by NBA will be deposited in the 'National Biodiversity Fund'. The Fund will be used for conservation and development of areas from where resources have been accessed, including management and conservation of heritage sites wherever applicable.

Traditional knowledge associated with bio-logical resources is proposed to be protected. It is also proposed that the state governments will notify national heritage sites, which are important from the standpoint of biodiversity, in consultation with institutions of local self-governments.

Access to Biological Diversity and their Regulation

The sections 3-7 of chapter II of the act deal with the regulation of access to biological diversity. Section 3 of the Act restricts certain persons not to undertake biodiversity-related activities without the approval of the NBA as stated below:

1. No person referred to in sub-section shall without the previous approval of the National Biodiversity Authority obtain any biological resources occurring in India or knowledge associated thereto for research or for commercial utilization or for biosurvey and bioutilization.
2. The persons, who shall be required to take the approval of the National Biodiversity Authority under sub-section, are the following:
   - A person who is not a citizen of India.
   - A citizen of India, who is a non-resident as defined in clause (30) of section 2 of the Income Tax Act, 1961.
   - A body corporate, association or organization.
     — Not incorporated or registered in India.
     — Incorporated or registered in India under any law for the time being in force, which has any non-Indian participation in its share capital or management.

Section 4 provides conditions for transfer of the results of research related to biodiversity. It states that 'no person shall, without the previous approval of the National Biodiversity Authority, transfer the results of any research relating to any biological resources occurring or obtained from India for monetary consideration or otherwise to any person who is not a citizen of India or a body corporate or organization, which is not registered or incorporated in India or which has any non-Indian participation in its share capital or management'.

However, 'transfer' does not include publication of research papers or dissemination of knowledge in any seminar or workshop, if such publication is according to the guidelines issued by the central government. Exemption to provisions of sections 3 and 4 has been provided in section 5. Sub-section (1) of section 5 states that 'the provisions of sections 3 and 4 shall not apply to collaborative research projects involving transfer or exchange of biological resources or information relating thereto between institutions, including government-sponsored institutions of India, and such institutions in other countries, if such collaborative research projects satisfy the conditions specified in sub-section (3):

- Collaborative research projects should conform to the policy guidelines issued by the central government in this behalf.
- They should be approved by the central government. All collaborative research projects other than those referred to in sub-section which are based on agreements concluded before the commencement of this Act and in force shall be void.'

Section 6 deals with the application for IPR. The section states that 'no person shall apply for any intellectual property right by whatever name called in or outside India for any invention based on any research, or information on a biological resource obtained from India, without obtaining the previous approval of the NBA before making such application'. This permission, however, may be obtained after the acceptance of the patent, but before scaling the patent by the patent authority concerned, in case a person applies for a patent.

Main Functions of the Authority

The duties of the NBA are defined under section 18 of the Act. It would regulate activities, issue guidelines for access to, and equitable benefit sharing and it may grant approval for undertaking any activity referred to in sections 3, 4 and 6. The NBA would advise the Central Government on matters relating to the conservation of biodiversity, sustainable use of its components, and equitable sharing of benefits arising out of the utilization of biological resources; advise the state governments in the selection of important areas of biodiversity to be notified as heritage sites and measures for the management of such heritage sites; and perform other functions as may be necessary to carry out the provisions of this act.

The NBA may also take measures necessary to oppose the grant of IPR in any country outside India on behalf of the central government, on any biological resource obtained from India or knowledge associated with biological resource which is derived from India.

The NBA shall determine the benefit sharing subject to any regulations made in this behalf, which shall be given effect in all or any of the following manner:

- Grant of joint ownership of intellectual property rights to the National Biodiversity Authority, or where benefit claimers are identified, to such benefit claimers.
- Transfer of technology.
- Location of production, research and development units in such areas which will facilitate better living standards to the benefit claimers.
- Association of Indian scientists, benefit claimers and the local people with research and development in biological resources and biosurvey and bioutilization.
- Setting up of venture capital fund for aiding the cause of benefit claimers.
- Payment of monetary compensation and other non-monetary benefits to the benefit claimers as the national.

Biodiversity authority may deem fit. It is further provided in the section that where biological resource or knowledge was a result of access from specific individual or group of individuals or organizations, the NBA may direct that the amount

be paid directly to such individuals or group of individuals or organizations in accordance with the terms of any agreement and in such manner as it deems fit. The main functions of the authority are:

- to lay down procedures and guidelines to govern the activities provided under section 3, 4, and 6. (permission to foreigners/NRI's foreign companies):
  — for obtaining any biological resource (section-3);
  — for transferring the results of any research (section-4);
  — certain collaborative research projects exempted (section-5).

- To advice the Government of India. Specific areas mentioned as per the Act are the following:
  — Notifications of threatened species (section-38);
  — Designate institutions as repositories for different categories of biological resources (section-39); and
  — Exempt certain biological resources, normally traded as commodities (section-40).

- To encourage setting up state biodiversity boards.
- To build up database and documentation system.
- To create awareness through mass media.
  — Training of personnel
  — Necessary measures in the areas of intellectual properties rights

The Act mandates the establishment of a National Biodiversity Authority (NBA); State Biodiversity Boards (SBB); and at local levels the Biodiversity Management Committees (BMC). The setting up of the BMCs could have enabled local communities to have some voice in the conservation, sustainable use, and equitable benefit-sharing of biological resources. But as per the rules, the role of the BMCs is merely limited to preparing People's Biodiversity Registers (PBRs) that document local knowledge and bioresources. This immensely undermines the rights of local communities who are the most important stakeholders when it comes to conservation of biological resources. Documentation, without any legal protection, is also an invitation to exploitation. The BMCs would be preparing the PBRs, but where is the power to ensure that they will not be openly accessible for theft or piracy. The Biological Diversity Act visualizes the establishment of a National Biodiversity Authority (NBA), State Biodiversity Boards (SBB) and Biodiversity Management Committees (BMC) at the level of all local bodies, namely, Gram, Taluk and Zilla Panchayats, as well as Municipalities and Corporations. The NBA working with SBBs and BMCs will have the responsibility for and authority to:

- decide upon the admissibility of all patent and other Intellectual Property Rights (IPR) applications based on Indian biodiversity resources, and associated knowledge in consultation with relevant local Biodiversity Management Committees.
- decide upon applications to access biodiversity resources and associated knowledge for commercial use, in consultation with relevant local Biodiversity Management Committees.
- decide upon appropriate benefit sharing arrangements in relation to IPR applications, in consultation with relevant local Biodiversity Management Committees.
- issue guidelines on appropriate collection fees and other benefit sharing arrangements, in relation to applications to access biodiversity resources and associated knowledge for commercial use in consultation with relevant local Biodiversity Management Committees.
- decide on admissibility of joint research proposals involving foreign agencies.
- decide on priorities and appropriate actions for conservation and sustainable use of natural populations of biodiversity resources.
- decide on priorities and appropriate actions for maintenance of health of natural ecosystems.
- decide on priorities and appropriate actions for conservation of domesticated biodiversity.
- decide on priorities and appropriate actions for constitution of heritage sites.
- decide on priorities and appropriate actions for identification of threatened species.
- promote scientific research pertaining to biodiversity and associated knowledge.
- promote public awareness pertaining to biodiversity and associated knowledge.

Evidently, the NBA, SBBs and BMCs would need a well-organized information system on India's biodiversity resources and associated modern as well as traditional, codified as well as oral knowledge to do justice to these ambitious objectives. Such a system will have to deal with a whole range of spatial scales from local to national as well as link properly with global databases. It will also have to address issues such as linking to information on the large holdings of biological specimens of Indian origin located in herbaria and museums abroad.

India with its emerging strengths in Information Technology (IT) as well as biotechnology is in an excellent position to turn this array of significant challenges into welcome opportunities. This calls for networking of country's existing biodiversity databases to take advantage of synergies, and to link all of these to activities leading to value addition. As a part of this process, the existing biodiversity databases will need to be considerably augmented and strengthened and new ones created. We will have to come up with novel ways of bringing on board the substantial knowledge base of country's barefoot ecologists and grass-roots innovators. We also need to devise a country wide decentralized system of monitoring biodiversity. Such a decentralized system could serve to enhance the quality of education by engaging teachers and students in first hand understanding of biodiversity and associated knowledge and in creating, using, and managing electronic databases, including those employing Indian languages.

28.2. NATIONAL BIODIVERSITY AUTHORITY

Official Members

1. Shri Vishwanath Anand, Chair Person, National Biodiversity Authority.
2. S. Chatterjee, Joint Secretary, Ministry of Tribal Affairs.
3. The Additional Director General (Forests), Ministry of Environment and Forests.
4. D.D.Verma, Joint Secretary, (CS), Ministry of Environment and Forests.
5. Shashi Mishra, Additional Secretary, Ministry of Agriculture.
6. Shri U.N. Behara, Joint Secretary, Department of Biotechnology.
7. Dr. S.P. Seth, Joint Secretary, Department of Ocean Technology.
8. Shri Satish Chandra, Joint Secretary, Ministry of Agriculture.
9. Shri Tara Dutt, Joint Secretary, Department of Indian Systems of Medicine and Homeopathy.
10. Hari Gopal, Joint, Department of Science and Technology.
11. Shri Sudhir Kumar, Joint Secretary Ministry of Science and Technology, Non-Official Members.
12. Prof. Raghvender Gadekar, Center for Biological Science.
13. Prof. Anil Gupta, Indian Institute of Management, Ahmedabad.
14. Dr. P. Pushpangadan, Director, National Botanical Research Institute.
15. Prof. L. Kannan, Director Research, Centre for Advanced Study in Marine Biology.
16. Dr A.K. Ghosh, Director, Centre for Environment and Development.

Indian Biodiversity Database

Department of Biotechnology, Government of India at the Centre for Ecological Sciences, Indian Institute of Science, Bangalore along with the Ministry of Environment and Forests, Government of India and eminent scientists, researchers, are planning to make Indian biodiversity database. India is one of the world's top twelve megadiversity countries, rich in biodiversity resources. It also possesses a wealth of knowledge associated with biodiversity, be it the orally held knowledge of folk healers or herders, or the traditional knowledge codified in *Ayurvedic, Sidha* or *Yunani* texts. Its biodiversity resources are far better known scientifically than those of other tropical megadiversity countries such as Brazil or Indonesia. As a result, India has developed a number of excellent biodiversity databases such as the Flora of Karnataka (Ganeshaiah *et. al.*, 2002), Traditional Knowledge Digital Library (NISCAIR, 2002) or the National Register of Green Grassroots Innovations and Traditional Knowledge (NIF, 2002). There exist therefore rich possibilities of building upon country's biodiversity resources and associated knowledge, to promote biodiversity-based enterprises in the modern as well as traditional sectors; to develop biotechnology industries at the cutting edge of new technologies as well as to encourage local level value addition to biodiversity resources. Important new markets are also emerging for produce of organic agriculture. Taking advantage of these markets will require development of good databases on agro-ecosystems, including incidence of pests and diseases.

Firstly, the scope of the databases that need to be thus brought together would have to go beyond the taxoncentric databases, which is the exclusive focus of many efforts including the Global Biodiversity Information Facility (GBIF). Thus, it visualizes linking of data on medicinal uses and chemical composition to taxonomic data on medicinal plants. We also need to link taxonomic data on medicinal plants to data on geographical distribution, abundance, and harvest levels on land under different forms of ownership and access regulations to support development of strategies for conservation and sustainable use of these plant populations. We, therefore, suggest that the networking effort brings under its purview databases that will deal with the whole range of categories of entities listed in **Table 28.1**.

Table 28.1. A possible framework for definition of entities of biodiversity databases.

Class of entities	Sub-class	Examples
Geographic	Physical	Mountain ranges, river basins, wetlands
	Political	Local bodies, states, UTs, GoI
	Ecological	Agro-ecological zones, biomes, landscape elements
	Management regimes	Reserved forests, wildlife sanctuaries, sacred groves
Biological units	Genes	Bt genes in Bt cotton
	Taxonomic groups	Species, orders
	Natural/cultured	Varieties registered under Protection of Plant Varieties and Farmers' Rights Act
	Management status	Species listed under Schedules of WildLife Act, CITES, or in Red Data Books, sacred species
People	Political units	Citizens of India, citizens of particular states or Panchayats
	Membership of organizations	University faculties, Staff of R and D labs
	Occupational groups	Hakims, Vaids, Biotechnologists
	Social groups	Indian communities as recognized by Anthropological Survey, Scheduled tribes
	Holders of material property rights	Holders of rights of collection of forest produce from particular localities
	Holders of intellectual property rights	Holders of patents
	Linguistic units	Speakers of different languages and dialects
Organizations	Relation to government	Government agencies, NGOs, Inter-governmental agencies
	Objective	Commercial, Not-for-profit, Scientific
	Nationality	Indian, foreign
Biological populations	Population levels	Population levels of particular species in specific localities
	Manipulations	Harvests or production under cultivation, including specific techniques employed, of particular produce of particular species in specific localities
	Transport	Transport of the produce of particular species in specific localities to other specific localities
	Marketing	Marketing of the produce of particular species in specific localities in other specific localities

cont....

Table 28.1 contd.

Class of entities	Sub-class	Examples
Biological materials	Processing	Preparation of acetone extracts from particular plant species
	Value addition	Preparation of plant based drugs or cosmetics
	Technology	Stabilization of alkaloids extracted from Neem
	End products	Particular molecules isolated from biological sources
	Uses/disservices	Therapeutic, cosmetic, allergenic
Knowledge	Nature of knowledge	Satellite imageries, folk taxonomies, medicinal properties of particular species
	Source of knowledge	Scientific research, classical tradition, individual hakim
	Medium	Scientific journals, palm-leaf manuscripts, oral traditions
	Generation of knowledge	New collaborative research, new Indian research, grass-roots innovations
	Transmission of knowledge	Websites, scientific publications, scientific meetings, orally from mother to daughter
	Access to knowledge	Public domain, patented, trade secrets
	Rights over knowledge	Individual, community, corporate
	Validation	Raw, Validated through different techniques
Disputes	Disputes over access to material or knowledge resources	Refusal of permission to an application for collaborative research

SUMMARY

India is one of the world's top twelve megadiversity countries, rich in biodiversity resources. We also possess a wealth of knowledge associated with biodiversity, be it the orally held knowledge of folk healers or herders, or the traditional knowledge codified in *Ayurvedic, Sidha* or *Yunani* texts. Its biodiversity resources are far better known scientifically than those of other tropical megadiversity countries such as Brazil or Indonesia. As a result, India has developed a number of excellent biodiversity databases such as the Flora of Karnataka, Traditional Knowledge Digital Library or the National Register of Green Grassroots Innovations and Traditional Knowledge. There exist therefore rich possibilities of building upon country's biodiversity resources and associated knowledge; to promote biodiversity-based enterprises in the modern, as well as traditional sectors; to develop biotechnology industries at the cutting edge of new technologies as well as to encourage local level value addition to biodiversity resources. Important new markets are also emerging for produce of organic agriculture. Taking advantage of these markets will require development of good databases on agro-ecosystems, including incidence of pests and diseases.

As we have seen, India is exploring ways of protecting and developing its natural resources. However, the Government needs to incorporate explicit and robust measures into national law to restrict the patenting of new products that are made with traditional knowledge or that utilize material derived from living organisms originating in India. In this process the role of indigenous and local communities must be recognized by the equitable sharing of the benefits arising from the utilization of their knowledge, innovations and practices. Before getting into the critical issues with the Biological Diversity Rules, it is important to recall how the Biological Diversity Act came into being in the first place. Was it only out of the growing concern to conserve India's rich biodiversity? The process started only after India becoming a signatory to the convention on biological diversity in 1992. Even then it took a good 10 years for the Act to be notified. There is no denying that the efforts of government officials, NGOs and academicians have also contributed to highlighting and pressing for the need for the conservation of biodiversity. However, the gaps in the Act and subsequently (now) in the Rules indicate that the real push was an international treaty obligation.

The Biological Diversity Act 2002, recently enacted by the Indian parliament, is a significant, indeed, a pioneering piece of legislation. It responds to a number of new, emerging concerns, firstly, the result of new developments in technology, in particular, biotechnology and information technology, and, secondly, the ongoing degradation of the environment, inevitably accompanied by an erosion of biological diversity. These developments imply that all organisms, even seemingly insignificant ones like microbes, worms, weeds and mice, are potentially resources of considerable economic value, worthy of efforts at conservation, scientific investigation, and of securing rights over the associated intellectual property. This has prompted the development of two often conflicting international agreements, the Trade Related Intellectual Property Rights provisions (TRIPS) of GATT and the Convention on Biological Diversity (CBD). The latter has two notable stipulations. One is the sovereign right of countries of origin over their genetic and biological diversity resources. The other is the acceptance of the need to share benefits

flowing from commercial utilization of biological diversity resources with holders of traditional knowledge and practices of conservation and sustainable utilization of these resources. There is as yet no proper resolution at the international level of how these will be implemented in view of the fact that the normal intellectual property rights and TRIPS provisions do not stipulate any sharing of benefits for holders of knowledge in public domain, nor the sovereign right of countries of origin over their genetic and biological diversity resources. TRIPS even includes intellectual property rights over micro-organisms and plant varieties. The Biological Diversity Act 2002 is a part of the Indian attempt to make some progress and to operationalize the two important provisions of the convention on biological diversity.

This Act aims to promote conservation, sustainable use and equitable sharing of benefits of India's biodiversity resources, including habitats, cultivars, domesticated stocks and breeds of animals and micro-organisms. With this in view it provides for the establishment of a National Biodiversity Authority, State Biodiversity Boards and Biodiversity Management Committees at the level of Panchayats (village committees) and Municipalities. It was initially designed as an umbrella act, and as a herald of a new age it would have overridden many of the earlier acts such as the Forest Act designed in the colonial era. As passed, however, it only has the status of a complementary act and will have to be operated side by side with a whole range of other acts, including, in particular, those pertaining to forest, wild life, *Panchayati Raj* (village governance) institutions, plant varieties and farmers' rights, and patents. There are a number of potential conflicts in the working of these various acts that need to be resolved carefully to ensure that the Biological Diversity Act 2002 can effectively address the many new and significant challenges resulting from scientific and technological developments and from the growing strength of India's panchayati raj institutions. In particular, one needs to guard against the many entrenched interests ensuring that the National Biodiversity Authority and the various State Biodiversity Boards end up being ineffectual because of the provision in the Act that they must accept all directives of the Central and State Governments. There are other possibili-ties, too, of loopholes that may render the act toothless. For example, any biological resource that is considered a commodity, or biological material that is blended and mixed, may be exempted from the provisions of the act. Finally, there is always a danger that the regulations may merely breed harassment and corruption, rather than effective action. It is to be hoped that public awareness and pressure will overcome these hurdles. This Act is an important step in the attempts to assert the sovereign rights of the people of India over their genetic and biological diversity resources, and to claim a share of benefits flowing from efforts are being made to bring together gram sabhas and other local bodies to pass resolutions against these rules. Letters have been written to the MoEF, the media has been alerted in some parts. But, is this all enough. Specifically, will the ministry's response be sufficient only to revise the Rules, or address core issues with the Act itself. Several NGOs and civil society representatives actually argue that the Biological Diversity Act, 2002 endorses intellectual property rights on the biological resources and knowledge of India by allowing for screening of patent applications on the same. In providing for a regulatory system for access, the law also facilitates biotrade and the increasing privatisation of biological resources and traditional knowledge.

EXERCISE

1. Define biodiversity.
2. Explain the importance of biodiversity of India in the present day biotechnological requirements for the research and developments.
3. Give an account on differents bill passed by the Indian government for the protection and conservation of the biodiversity.
4. Enumerate the different principles of the Indian biodiversity act that are proposed for the protection of Indian biodiversity.
5. Describe briefly about the types of organisms distributed in the different geographical regions of India, and their importance in the near future.
6. Elucidate the applications and problems associated with the Indian biodiversity act that are proposed for the protection of Indian biodiversity.
7. Describe the various types of efforts made by the Indian government for the identification and conservation of Indian biodiversity.
8. Write a note on international and regional agreement/treaties which helped to raise the Indian biodiversity act and are significant for the biotechnology.
9. Mention how geographical indications are useful for the biotechnology business, and their importance in the biotechnology.
10. What is meant by animal biodiversity? Giving a few examples, state how is it important for biotechnology?
11. List out the methods adopted for the sustainable use and conservation of biodiversity that were observed in the India during recent past.
12. Distinguish between the sustainable use and conservation in the biotechnological methods of maintaining resources.
13. Explain how the Indian biodiversity act is stronger for the prevention of loss of bio-diversity and exploitment by foreign researchers?
14. How is the Indian biodiversity act helpful for the biotechnological researcher for the gain of his share in the competition oriented field?
15. Give a note on the ethical issues raised by the intellectual property rights in the field of biotechnology developments.
16. Name the important role of the National Biodiversity Authority that made the news in the recent past in biotechnology research and development in India.
17. Briefly explain access to biological diversity and their regulation in India and other developed countries have an impact on the diversity erosion.
18. Write about the NBA working with SBBs and BMCs that have the responsibility for and authority to biodiversity protection.
19. Critically comment on the Indian biodiversity data-base and its improvements in the Indian scenario in biotechnology research and developments.

Glossary of Terms

A Adenine, a nitrogenous base of nucleic acid.

Ab Antibody produced against specific antigen.

Aback Backwards movement.

Abacterial Absence of bacteria, or sterile condition.

Abatement Lessening of pain.

Abasia Unable to walk because of loss of motor coordination.

Abdomen Area of body between diaphragm and pelvic bones.

Aberrant A growth that invariably deviates either from the normal or usual type; as exceptional growths (aberrations) which take place in some tissue cultures.

Activated Sludge Process Aerobic sewage treatment process using aerobic microorganisms present in sewage sludge to break down organic matters in sewage.

ABO Human antigenic system designating the blood groups A, B, AB and O.

Absolute Plating Efficiency The percentage of cells that give rise to the colonies on the medium.

Accession Seed or plant sample, strain or population held in either a 'gene bank' or a 'breeding programme' for proper usage and conservation.

Acid Phosphatase Enzyme group that catalyzes the hydrolysis of certain organic phosphates. Activator A substance or physical agent that stimulates the transcription of a specific *gene* or *operon.*

Acetyl CoA A condensation product of coenzyme A and acetic acid (attributing acetyl moiety); an intermediate in the transfer of 2-carbon fragments in the process of their gaining entry to tricarboxylic acid (TCA) cycle.

Active Site The region of a protein that gets bound to substrate molecules and facilitates a particular 'chemical conversion'. The fundamental underlying *theory of enzyme activity* assumes formation of an enzyme-substrate complex *via* binding at the active site.

Active Immunity Immunity acquired because of the individual's own reactions to the corresponding *pathogenic micro-organisms* or their respective *antigens,* attributable to the presence of antibody in response to an antigenic stimulus.

Acquired Immunity The ability of an individual to produce specific antibodies in response to antigens to which body has been previously exposed based on the development of a memory response.

Acquired Immune Deficiency Syndrome (AIDS) An infectious disease caused by HIV retrovirus, characterized by the loss of normal functions, followed by various infections.

Adaptive Enzymes An enzyme that is duly formed in response to an outside stimulus during adaptation. This terminology has been duly replaced by 'inducible enzymes'.

Adipocyte Fat cell in animals.

Adult Stem Cell An undifferentiated cell found in a differentiated tissue that can renew itself and (with certain limitations) differentiate to yield all the specialized cell types of the tissue from which it originated.

Aerate To mix or supply with air or gas. The particular process is known as 'aeration'.

Aerobic or Aerobiotic It specifically refers to an *organism* which either lives in or a phenomenon taking place in the presence of molecular oxygen (*i.e.*, Oz-gas).

Adjuvants Substances that enhance immunological response to vaccine and may be incorporated to vaccines to slow down absorption and increase effectiveness: substances that enhance the action of an *antigen* or a *drug.*

Affinity It refers to the precise strength of binding of each separate antibody combining site with its corresponding antigenic determinant.

Agglutination Clumping or aggregation of cells due to the reaction of surface bound antigens with homologous antibodies.

Allele A pair of genes that controls a specific character/trait inherited from each parents.

Allergen A substance, usually a protein, capable of inducing a specific immune hypersensitivity response, often resulting in immunoglobulin E production.

Allogeneic Transplant Transplant using tissue from a donor individual not genetically identical to the recipient.

Allograft is a graft taken from an individual and transferred to a member of the same species; *e.g.*, a human organ to another human.

Allosteric Refers to a binding site in a protein usually an enzyme. The catalytic function of an enzyme can be duly modified by the introduction of small molecules, not only confined to the 'active site', but also at a 'spatially distinct (allosteric) site' having an altogether different specificity. An *allosteric effector* is a molecule bound at such a site that enhances or lowers the *activity of the enzyme.*

Allosteric Enzymes Enzymes with a binding and catalytic site for the substrate and a different' site where a *modulator* (*i.e., allosteric factor)* usually acts.

Agrobacterium tumefaciens The bacterium which is responsible for causing crown-gall disease in plants, besides inducing tumour to form. The Ti plasmid of *A. tumefaciens* is known to initiate and cause the disease; and a part of it is invariably utilized as *vector* in the ensuing genetic modification of higher plant cells.

Alkaline Phosphatase An enzyme that catalyzes hydrolyses of monophosphate ester moiety from specific compounds, such as : 3'- and 5'- terminal ends of nucleic acids (DNA and RNA), and nucleotides.

Ammonification The release of ammonia (NH3) gas from the nitrogenous organic matter by microbial action.

Amphotropic Retrovirus A retrovirus (*q.v.*) that will grow in the cells from which it was isolated and also in cells from a wide range of other species.

Amplify To increase the number of copies of a gene or DNA sequence.

Anaphylactic Shock Physiological shock resulting from an anaphylactic hypersensitivity reaction, invariably followed by death in extremely severe instances *e.g.*, penicillin-anaphylactic shock.

Anaphylactic Hypersensitivity An observed exaggerated immune response either to foreign protein or other substances involving *degranulation of mast-cells* and subsequent release of histamine.

Anchorage Dependent Cell A cell which will exclusively grow and multiply when in close contact with a *suitable solid support i.e.*, several animal cell-lines.

Anthropogenic Originating in or caused by human activity.

Anti-Idiotype Antibody An antibody having the properties of an *antigen.*

Antibody A glycoprotein molecule produced in the body in response to the introduction of an *antigen* or *hapten* that can react with the antigen. It is also termed as immunoglobulin, that is essentially a part of the serum fraction of the blood formed in response to the antigenic stimulus, and which eventually reacts with antigens with utmost specificity.

Antibody-Mediated Immunity Immunity caused by the activation of the B-lymphocyte population thereby giving rise to the production of *several categories of antibodies.*

Antibody-Dependent Cytotoxic Hypersensitivity It is a Type II hypersensitivity in which an *antigen* present on the surface of a cell combines with an antibody, resulting in the death of that cell by stimulating phagocytic attack or initiating the complement pathway.

Antiserum A serum that specifically contains a mixture of antibodies raised against a particular *antigen* is referred to as antiserum to that *antigen.*

Antitoxin An antibody to a *toxin* capable of reacting with that 'poison' and also neutralizing the toxin produced.

Autoimmunity *Immunity* or *hypersensitivity* to certain constituent in one's own body *i.e.,* immunereaction with self-antigens.

Antifoam A compound added to bioreactors so as to reduce the production of foam *viz.,* surface active agents.

Antioxidant A substance which is sometimes ad-ded to the solutions to either *prevent* or *inhibit* oxidative process.

Aseptic Technique Procedures employed to prevent the introduction of *bacteria, fungi, viruses, mycoplasma* or other organisms into cell tissue and organ cultures.

Aneuploid The condition which exists when the nucleus of a cell does not contain an exact multiple of the haploid number of chromosomes.

Aspirate To draw something in or out, up or through using suction or a vacuum (*i.e.,* negative pressure); as aspiration (vacuum) may be employed in the 'disinfection process' to draw *disinfectant* right into the *surface layers of plant tissue.*

Assay The 'substance' to be analyzed *quantitatively* or the process of examining or testing it either by chemical means or by other suitable means.

Attenuation A process of minimizing the virulence of a pathogen.

Attenuated Vaccine A vaccine produced by using an attenuated strain or virus or bacterium.

Autocrine This refers to the ability of cytokines to show the effect on the cell that produced it.

Autoclave A device in which materials are sterilised using steam under high pressure.

Autoimmune Diseases A constellation of different diseases all characterized by the failure of the body to distinguish "self" from "non-self" causing the body to attack its own tissues.

Autologous Transplant Transplant using tissue from the same individual, or a twin.

Auxotroph Mutant that differs from the wild type in requiring a nutritional supplement for' specific growth *viz.,* as in a deficient mutant.

Avidity Total binding between a heterogeneous antiserum and a multi-valent antigen, regardless of how many different specific antibodies each with a precise affinity, contribute to the total binding. This particular measurement is known as the avidity of the antiserum.

Axenic Culture A culture without foreign or undesired life forms. An axenic culture may also include the purposeful co-cultivation of different types of cells, tissues or organisms.

B-Lymphocyte White blood cells (WBC) that are capable of producing specific immunoglobulins. In fact, their surfaces do carry specific *immunoglobulin antigen-binding receptor sites.*

Bacteria Single celled, ubiquitous prokaryotic organism having peptidoglycan cell wall.

Bacteriophage A virus infecting bacterium that replicates with the cells of bacteria.

Base Pairing The particular H-bonding taking place between *purines* and *pyrimidines* in double-stranded nucleic acids (*i.e.,* DNA and RNA).

Batch Culture A cell suspension grown in a liquid medium of a known volume. Organisms in this system invariably exhibit a *'sigmoid'* type growth curve. The *inocula* of successive subcultures 'are of similar size, and the cultures usually contain almost an equivalent quantum of *'cell-mass'* at the end of each passage. Cultures frequently exhibit five distinct phases per passage, namely:

(i) *lag phase;* (ii) *exponential growth phase;* (iii) *linear-growth phase;* (iv) *deceleration phase;* and (v) *stationary phase.*

Bacteriostatic A chemical entity or other substance that does not kill but prevents the possible bacterial growth and multiplication.

Bactericide An agent or a substance that usually kills bacteria (rather rapidly); and this activity is known as bactericidal activity.

Bacterium A single-celled prokaryotic (*q.v.*) organism.

Bacteriophage A virus that infects bacteria; also called phage.

Baculovirus A group of viruses that infect insects and can be used as vectors (*q.v.*) to produce foreign proteins in insect cells.

Bioassay A biological assay carried out on the *'living cells'* or on a *'living organism'.*

Biocentric Regarding all forms of life as worthy of moral respect and consideration, and viewing human beings as one species among others.

Biochip A hi-tech device that essentially combines a small scale 'biosensor' with an integrated circuit.

Bioconversion The actual transformation of matter from one form to another by living organisms or enzymes.

Biodegradation A recognized and accepted pheno-menon whereby a material is broken down into its smaller chemical fragments by the action of living organisms *e.g.,* biodegradable polymers, biodegradable packing components etc.

Biodiversity The number and types of organisms in a region or environment. Includes both species diversity and genetic diversity within the species.

Biolistics A technique that injects DNA into the cells using tiny bullets composed of gold or tungsten particles with DNA attached on the surface.

Biological Safety Cabinet/Biosafety Cabinet Specially constructed cabinets which are desi-gned to protects workers and the environment from dangerous agents, especially bacteria and viruses.

Biomass The cell mass produced during fermentation or the total weight of living matter in a population.

Biophilia Affiliation with and affection for the diversity of life forms, sometimes regarded as an evolved, innate characteristic of human beings.

Biopolymer An altogether different type of large molecules duly formed by organisms *i.e.* nucleic acids, proteins, lipids and polysaccharides.

Biopiracy Illegal commercial development of naturally occurring biological materials such as plant parts or genetic cell lines.

Bioprospecting The legal collection of biological samples that can help medicinal and scientific research with affecting natural sources.

Bioreactor A vessel, preferably made up of SS, employed to perform a biological reaction *i.e.,* a *reactor* used for the culture of aerobic cells, or to the columns or packed beds of immobilized cells or enzymes.

Biosensor A highly specific and sophisticated device that usually makes use of an agent either having a biological origin or a biological principle, for the assay of a chemical compound *viz., tissues, immunosystems, isolated enzymes, organelles,* and *whole cells as catalysts.* The 'catalyst' is immobilized and used in conjunction with a physicochemical device.

Biosynthesis Refers to *'biological synthesis' i.e.,* the building or forming of biological compounds in a living organism.

Biotechnology The meticulous application of *organisms, biological systems,* or *biological phenomena* specifically either to manufacturing units or to service industries. However, this very definition has been logistically extended to include any process or device wherein organisms, tissue cells, organelles, isolated enzymes, design and usage of 'bioreactors', fermentors, downstream processing, and above all the analytical and control equipment associated with biological manufacturing process. In simple sense A set of biological techniques developed through basic research and now applied to research and product development. In particular, the use of recombinant DNA techniques.

Biotype A certain group of organisms having the same genetic characteristic features.

Biotin [Vitamin H; Co-carboxylase; ClOH1603N2S; MW 244.31]: It is a component of vitamin B complex. It serves as a prosthetic moiety for carboxylase-enzymes. Usually present in all living cells bound to proteins or polypeptides, and is of utmost importance in carbohydrate, protein, and fat metabolism. It is essentially component of most plant tissue culture media.

Blastocyst A preimplantation embryo of 30-150 cells.

Bt, *Bacillus thuringiensis* A soil bacterium that produces a toxin that is deadly to some insects. Many strains exist, each with great specificity as to the type of insect it can affect.

Buffer An admixture of various chemical substances (mostly inorganic) in an aqueous medium, or system that prevents/resists change in pH (despite addition of small quantum of acid or base) and withstand shock; as buffered media can appreciably resist pH drift. The extent to which pH change can be resisted logically represents as a measure of the solution's buffering capacity.

Budding A form of asexual reproduction wherein a *daughter cell* develops from a small out-growth or protrusion of the *mother (parent) cell.*

Callus (*plural-calluses* or calli) An actively growing undifferentiated (parenchymous) tissue formed in higher plants in response to wounded surface or certain infections. It usually represents a disorganized tumour-like masses of plant cells which are formed in a culture medium. It proliferates in an irregular tissue mass which varies extensively in several aspects *e.g.,* appearance, texture, rate of growth. Callogenesis—refers to the phenomenon of callus formation *i.e.,* a function of the tissue type (*species* and *explant*) as well as the composition of the medium.

Canadian Nutrient File A compilation of nutrient values for foods available in Canada, produced by Health Canada.

Capsid A protein coat of a virus enclosing the naked nucleic acid (*i.e.,* DNA and RNA).

Capsomere The individual protein subunits which essentially form the *capsid* of a virus.

Carcinogen An agent capable of initiating development of malignant (cancerous) tumours.

Carrageenan Sulphated cell-walled polysaccharide found in certain *red algae.* Contains repeating sulphated disaccharides of galactose and anhydrogalactose.

Carrier A matrix strategically placed within a bioreactor or cell-culture system to serve as a support upon which the active biomass or cells grow and get immobilized ultimately.

Carbon Source A source for non-metallic element carbon (C) *e.g.,* as organic substances, sugars, com-steep liquor, maltose, glucose, glucose-syrup, whey powder, molasses (containing 8-10% residual sucrose) etc., that are usually taken up and metabolized by organisms in culture media or by plant-tissue cultures.

Casein Hydrolysate (Edamin) A milk protein (casein) digestive product very much composed of amino acids (casamino acids) and other substances. The resulting complex (*i.e.,* unidentified product) is sometimes used as an additive (0.02-2.10%) in nutrient solutions employed in various culture media and plant tissue culture as a non-specific source of organic nitrogen.

Cell The smallest structural unit of living organisms that is able to grow and reproduce independently.

Cell Count The number of cells per unit suspension volume or cells weight. Tissue is treated with chromic acid (5-8%) or pectinase (0.25%) for up to 15 minutes normally followed by mechanical dispersion and subsequently the cell numbers are determined with the help of a haemocytometer.

Cell Disrupter A procedure employed to liberate the contents of cells.

Cell Hybridisation The formation of synkaryons viable cell hybrids produced through cell fusion. These hybrids may be identified by their enhanced chromosome number in comparison to the parent cells and the possession of typical characters found in one or of the parent cells.

Cell Line A cell line is permanently established culture which will proliferate indefinitely giving suitable fresh medium and space. The developmental history or descent through cell division is from a *single original cell.* Any deviation in the culture technique may favour one cell line over another.

Cell Lines Cultures of disaggregated tissue that can be maintained and propagated for use in research. The length of time cells will survive in culture varies. Some cell lines are *immortalized*; that is, they can be maintained essentially indefinitely, for one of a variety of reasons. Embryonic stem cells and embryonic germ cells are immortal because they express telomerase, one of the factors necessary for cells to propagate normally.

Cell-Mediated Hypersensitivity Delayed hyper-sensitivity reaction involving T lymphocytes, and taking place almost 24-72 hours after due exposure to the antigen.

Cell-Mediated Immunity A highly specific acquired immunity involving T-cells, primarily responsible for resistance to infectious diseases caused by certain bacteria and viruses that reproduce within the host cells.

Cell Number The absolute number or approxi-mation of the number of cells per unit area of a culture or medium volume.

Cell Selection Selection within a group of gene-tically different cells, and usually involves competition between cells often under certain stress. The selection criteria may involve *three* main variants, namely: (*a*) cell viability; (*b*) biochemical activity; (*c*) another basis for choice. Normally, the selected cells or cell lines are carefully relocated to either fresh medium for continued selection process or duly exposed to an enhanced level of the stress agent. However, the ultimate aim is usually to regenerate plants from those select cells with a possibility that the plants may exhibit the specific traits selected for at the cellular level.

Cell Suspension Cells and small aggregates of cells suspended in a liquid medium *e.g.,* cell suspension cultures. Callus or explants derived from cell suspensions are transferred to liquid medium, and the cultures are subsequently agitated to a mechanical shaker. Importantly, the ensuing single cells as well as small cell clusters are employed for a number of purposes in plant tissue culture *viz.,* single-cell cloning.

Cell Bank A facility for preserving and keeping cells frozen at extremely low temperatures. These cells are used for investigating hereditary diseases, human aging, and cancer. Collection of banked cells are kept by the National Institute of Health (the Human Genetic Mutant Cell Repository and the Aging Cell Repository) and at the Cornell Institute for Medical Research.

Cell Counter (Electronic) An electronic instrument employed to count blood cells, employing either an bioelectrical resistance or an optical grating technique *e.g.,* Dow-cytometry.

Cell Cycle The series of events that take place during the growth and development of a cell *e.g.,* meiosis and mitosis.

Cell Growth Cycle The order of physical and biochemical events which take place during the growth of cells. In tissue-culture studies the cycle changes are divided into specific periods or phases, namely: (*i*) DNA synthesis (or S period); (*ii*) G2 period (or gap); (*c*) M or mitotic period; and (*d*) G1 period.

Cell-Kill In antineoplastic therapy, the number of malignant tumour cells destroyed by a treatment.

Cell Kinetics The study of cells and their subse-quent growth and division. Study of these factors has enormously led to much better understanding of cancer cells and has been found to be useful in developing newer chemotherapeutic methods.

Cell Mass In embryology, the mass of cells that develops into an organ or structure.

Cell Organelles Any of the structures in the cyto-plasm of a cell *viz.,* mitochondria, endoplasmic reticulum, golgi complex, ribosomes, lysosomes and centriole.

Cell Sorting A technique used to separate cells with a surface antigen from those without it.

Characterized DNA DNA which has been sequen-ced and for which there is an understanding of the gene products.

Chelate A chemical compound (liquid) with which metal atoms may be combined in such a manner so as to prevent them (bivalent, trivalent metal ions) from precipitating out of solution, and thus rendering them unavailable to plants, such as: ethylenediamine tetraacetic acid (EDTA), its disodium salt (NaTEDTA), w'rich undergoes complexation to Fe^{2+} (ferrous ion)/Fe^{3+} (ferric ion) present in nutrient solutions employed for plant tissue culture. The phenomenon is usually termed as chelation.

Chemotherapeutant A chemical frequently employed to pretreat diseased source plants prior to *excision* or *incorporation* into the media to support certain specific therapeutic objective, for instance malachite green, virazole (ribavirin) are employed to eliminate completely the virus present in the meristemtip culture. The pheno-menon is known as chemotherapy.

Chromosome A structure in the cell, consisting of DNA and proteins, that carries the organism's genes.

Chimera (Chimaera) A plant or tissue composed of more than one kind of genetic tissue.

Chimeric Gene An artificial gene produced by combining the DNA sequences from several different sources.

Chromatid Single chromosome containing only one DNA complex.

Chromatins The DNA-protein complex that constitutes a chromosome.

Chromosome The deoxyribonucleic acid (DNA) bearing structure that carries the inheritable characteristics of an organism.

Clinical Trial Research to test the safety and efficacy of new treatments or to compare the effects of different treatments in patients or healthy volunteers.

Clonal Selection Theory The theory accounts for exclusive antibody formation during foetal development wherein complete set of lymphocytes are developed. Each lymphocyte consisting.

Clone As a noun: a group of genes, cells or organisms derived from a common ancestor and genetically identical. As a verb: to generate replicas of DNA sequences or whole cells using genetic manipulation techniques.

Conjugative Plasmid A plasmid (*q.v.*) which codes for its own transfer between bacterial cells by the process of conjugation ('mating').

Construct As a noun: genetically manipulated DNA.

Containment Prevention of the spread of genetically manipulated organisms outside the laboratory. Physical containment is accomplished by the use of special procedures and facilities. Biological containment is accomplished by the use of particular strains of the organism that have a reduced ability to survive or reproduce in the open environment.

Containment Level The degree of physical containment provided by a laboratory or facility, which depends on the design of the facility, the equipment installed, and the procedures used.

GMAC physical containment levels are numbered from 1 to 4,4 being the highest level.

Confined Field Trial Field trial carried out with specific restrictions on location, plot size, etc.

Conjugate A reagent that is formed by covalently coupling two molecules together, such as fluorescein coupled to an immunoglobulin molecule.

Constant Regions The relatively invariant parts of immunoglobulin heavy and light chains, and chains of the T-cell receptor.

Contact Hypersensitivity A delayed intlammatory reaction on the skin seen in type IV hyper-sensitivity. Co-stimulation. The signals required for the activation of a lymphocyte, in addition to the antigen-specific signal delivered via their antigen receptors. CD28 is an important co-stimulatory molecule for T cells and CD40 for B cells.

Cosmology A theory or picture of the overall structure, composition, and contents of the universe, whether based on religious, philosophical, or scientific ideas, or some combination of these.

Cross Reaction The sharing of antigenic determinants of two different antigens.

Colony Stimulating Factors (CSFs) A group of cytokines which control the ditlerentiation of haemopoietic stem cells.

Cryopreservation The process of freezing biolo-gical materials in such a way that they can be stored for long periods of time, then thawed for use.

CTLA-4(CD152) A down regulatory signaling molecule of T cells that competes with CD28 for ligation by B7 on antigen-presenting cells.

Culture Medium A specifically prepared nutrient solution (substrate) which may be chemically defined for growing plant tissues and micro-organisms *in vitro.*

Cybrid A plant *or* cell cytoplasmic hybrid (*i.e., heteroplast*) having essentially the nucleus *of* one and cytoplasmic organelles *of* another, or *of* both cells or plants.

Cyclophosphamide A cytotoxic drug frequently used A5 an immuncsuppressive.

Cyclosporin A T-cell suppressive drug that is particularly useful in suppression of graft rejection.

Cytolysis Cell dissolution or disintegration.

Cytokines A generic term far soluble molecules which mediate interactions between cells.

Cytotoxic A chemical or other agent which is toxic to cells.

Cytotoxic T-Cells A class *of* T lymphocytes that are able to kill cells as part *of* the cellmediated immune response.

D Genes Sets of gene segments lying between the V and J genes in the immunoglobulin heavy chain genes, and in the T-cell receptor and J chain genes which arc recombined with V and J genes during ontogeny.

Decay Accelerating Factor (DAF) A cell surface molecule on mammalian cells which limits activation and deposition of complement C3b.

Deacceleration Phase The declining growth rate phase following the linear phase and preceeding the stationary phase in most batch-supension cultures.

Decontaminate To free from contamination or surface sterilization.

Defective Virus A virus that is unable to reproduce in its host without the presence of another ('helper') virus.

Deficiency In genetics, a chromosome aberration which comes into being from the actual loss *of a* gene or series *of* genes by deletion.

Deficiency Disease A disease resulting from a diet that invariably lacks essential nutrients *e.g., rickettsia*—from vitamin D deficiency, *vision impairment*—from vitamin A deficiency; *improper blood-coagulation*- from vitamin K deficiency etc.

Defined Culture Medium A culture medium where-in all the components are quantified and known.

Deletion The lost portion either of a chromosome or of a nucleotide sequence in nucleic acids *e.g.,* DNA, RNA.

Defensins A group of small antibacterial proteins produced by neutrophils.

Degranulation Exocytosis of granules from cells such as mast cells and basophils.

Delayed Hypersensitivity T-Cell A class *of* T-cells.

Deliberate Release Intentional release of a genetically modified organism into the open environment.

Dendritic Cells A set of cells present in tissues, which capture antigens and migrate to the lymph nodes and spleen, where they arc panicularly active in presenting the processed antigen to T-cells. Dendritic cells can be derived from either the lymphoid or mononuclear phagocyte lineages.

De Novo (Latin: *anew, from the very beginning).* Arising sometimes spontaneously from either unknown or very simple precursors.

DNA Deoxyribonucleic acid, the molecule which carries the genetic information for most organisms; consists of four bases and a sugar-phosphate backbone.

Donor The organism or cell from which DNA is derived for insertion into another organism (the host).

Differentiation The resumption of meristematic activity mostly by mature cells *via* a reversal *of* the process *of* cell or tissue differentiation.

Dehumidifier A equipment or device or apparatus that aids in the removal *of* moisture from the air.

Deionized Water Water obtained after due passage *via.* an ion exchange device to get rid *of* either soluble inorganic salt (minerals) or certain organic salts. The phenomenon is termed as deionization.

Derepression The mechanism by which repression is invariably eleviated. Whenever, a repressing metabolite is removed it usually results into an enhanced level *of protein* or *enzyme.*

Diffusional Limitation The phenomenon by which the rate of diffusion of substrate into and product out of an *enzyme, cell, tissue* or above all the aggregate *of immobilized cells* becomes limiting.

Diffusion The movement *of* molecules across a concentration gradient specifically from the area *of higher concentration* to the area *of lower concentration.*

Direct Embryogenesis *Embryoid formation* directly upon the outer surface *of* either somatic or zyogtic embroys, or on seeding plant tissues in culture medium without essentially heating an intervening callus phase.

Direct Organogenesis *Organ formation* directly upon the surface *of* comparatively bigger intact explants mostly without an intervening callus phase.

Dismutages An *enzyme* of great value specifically in organ transplantation, and also in the treatment of 'heat attack' where tissues have been adequately deprived of blood for a short duration. Nevertheless, it prevents reperfusion injury by permitting the oxygen deprived tissues so as to enable their due recovery to the normal state in a more defined, elaborated, and orderly manner.

Dilution Rate Refers to the continuous fermentation whereby a measure of the rate at which the prevailing medium is duly replaced with the fresh medium appropriately. The dilution rate is reciprocal of the *hydraulic retention time.*

Diploid Cell Cell having its chromosome in homologous pairs, and thus having two distinct copies of each autosomal genetic locus.

Double Diffusion Test *Immuno diffusion test* wherein *antigen* and *antibody* predominantly diffuse towards each other from separate wells cut into agar gel.

Downstream Processing Separation and purification of products from a fermentation process produced in a large scale fermentors.

Drosophila A genus of flies whose genetics has been extensively studied.

DNA Ligases Enzymes that specifically bind together the ensuing DNA fragements.

Dual Culture A specialized culture system which invariably includes plant tissue and one organism (*e.g., nematode species)* or micro-organism (*e.g., fungus).* In actual practice, dual cultures are mostly employed to investigate and study the host-parasite interactions or the generation of axenic cultures for a variety of purposes. Normally the microorganisms selected is an obligate parasite.

Dys-Functional Immunity An immune response which causes essentially an allergic reaction or the apparent lack of immunse response resulting in failure to protect the body agirinst a host of infections or toxic agents.

Eastern Originating in or related to the cultures of Asia; Eastern Religions include *Buddhism, Hinduism, Taoism, Shinto, Confucianism,* and others.

Ecocentric Ethically evaluating actions or policies in terms of their effects on ecological communities or ecosystems.

Ectoderm The outermost of the three primary layers of an embryo; produces the nervous system, the epidermis and epidermal derivatives, and the lining of various body cavities such as the mouth.

Ectopic Tissue Tissue that has formed abnormally temporally or spatially.

EG Cells Embryonic germ cells. These cells are found in a specific part of the embryo/fetus called the gonadal ridge, and normally develop into mature gametes.

Embryo Organisms in the early stages of growth and development. In animals, embryos are characterized by the cleavage of the fertilized eggs to many cells, the laying down of the three germ layers, and formative steps in organ development. Although there is some discussion about the characteristics marking the switch from embryo to fetus, in human beings, "embryo" generally refers to the time from implantation to about eight to twelve weeks after conception.

Embryo-Rescue The process in plant breeding whereby tissue from young embryo plants is excised and propagated *in vitro* for subsequent growth as differentiated plants.

Endoderm One of the three primary layers of an embryo; it is the source of the digestive tract and other internal organs.

Enzyme Anyone of a number of specialized proteins produced in the living cells which particularly speed-up, augment, and catalyze the 'rate of specific chemical reactions' even at an extremely low concentrations (*e.g.,* organic catalyst), but is not consumed in the reaction.

Enzyme Induction Regulation in the biosynthesis of specific enzymes (largely accelerated under certain conditions) depending exclusively upon the nature of the nutrients.

Enzyme Inhibition The rate of an *enzyme-catalyzed reaction* may sometimes be controlled in a specific manner by inhibitors as well as inactivators. Substances like acids, alkalis, detergents, urea, proteases etc., serve as enzyme inhibitors and distinctly lower enzyme-catalyzed reaction; however, enzyme inhibition could be either reversible or irreversible. Importantly, the enzyme activation contrarily gives rise to an enhanced rate of an enzyme-catalyzed reaction. Therefore, both inhibition and activation of enzymes by key metabolites predominantly provides the usual means of metabolite control.

Enzyme Activity An expression of the ability of a given enzyme preparation to catalyze a specific reaction effectively.

Enzyme Immobilization It is invariably described as the process whereby an enzyme is converted from a *homogeneous catalyst* into a *heterogeneous catalyst*. It is separated into an altogether distinct and characteristic phase *i.e.,* nonnally water-insoluble and quite often of high molecular weight, that is found to be absolutely separate from the bulk substrate-containing phase. Enzyme immobilization is physically confined and even localized in a particular defined region.

Enzyme-Linked Immunosorbent Assay (ELISA) The excellent technique is invariably employed for detecting and quantifying specific serum antibodies as well as antigens exclusively based upon tagging the antigen-antibody complex with a specific substrate that may be enzymatically converted to readily quantifiable product by a specific enzyme.

Enzyme Immunoassay (EIA) An assay involving the substitution of an enzyme for a radiolabeled product.

Enzymeimmunodiffusion Test A specific test wherein an *antigen* itself is an enzyme and the antigen-antibody complex exhibits distinct enzyme activity. Thus, this test is very useful in detecting invisible or non-stainable antigen-antibody aggregates.

Enhancer Element Refers to the short regions of DNA that possess the effect of enhancing the levels of transcription, but unlike promoters, they invariably affect genes from the chromosomalloci at a relatively longer distance.

Environmental Chamber (Incubator) It is a controlled environment cabinet wherein temperature, light quality, intensity, duration, relative humidity and controlled air flow.

Epigenetic Gene Control It may be regarded as an unique *newer concept*. It essentially describes that the control of gene expression in higher eukaryotic organisms is related to the methylation of *cytosine residues* in gene coding sequence.

Epitope Minimal molecular grouping to which the immune system will respond, for example by producing antibodies. A protein generally contains several epitopes or specific antibody binding sites on the surface of *antigen,* also antigenic determinants.

Erythroprotein Refers to the glucoprotein usually produced in the kidney, and that regulates the production of RBC present in the bone marrow.

ES Cells Embryonic stem cells.

Escherichia coli (E. coli) A bacterium that inhabits the intestinal tract of humans (and other animals).

Escherichia coli K12 A strain of *E. coli* that has been maintained in culture in laboratories for many years. It has lost the ability to colonise the intestinal tract of humans and animals, is well-characterised genetically, and is often used for molecular cloning work.

Escherichia coli B Another well-characterised laboratory strain of *E. coli.*

Ethics 1. Rules defining how persons ought to behave toward one another or toward other beings (sometimes a synonym for morals), 2. The study of, or theories about, such rules.

Etiology The elaborated study of the particular causation of a 'disease'.

Eukaryotes Cellular organisms having a mem-brane-bound nucleus within which genome of the cell is stored as chromosomes composed to DNA *Le.,* algae, fungi, protozoa, plants and animals.

Eukaryotic Belonging to the group of organisms whose cells contain a true nucleus. Eukaryotic organisms include animals, plants and fungi.

Exoenzymes Enzymes that mostly occur either attached to the outer surface or, in the *peri plasmic space* or, released into the medium surrounding a cell.

Exponential Phase Refers to the specific phase in culture wherein the cells usually undergo the maximum rate of cell division.

Expression Manifestation of a characteristic that is specified by a gene; often used to mean the production of a protein by a gene that has been inserted into a host organism.

Fab (Antigen Binding Fragment) Represents either of the two identical fragments generated in the event when an immunoglobin (Ig) is cleaved by papain; and the *antigen-binding-portion* of an antibody, including the hypervariable region.

Faith-Based Motivated by religious beliefs or values, or undertaken by a religious group or institution.

Fc (Crystallisable Fragment) The remainder of the molecule when an immunoglobulin (Ig) undergoes cleavage and the fab fragment gets separated The crystallisable portion of an *Ig molecule* made up of the *constant region,* and the end of an Ig that in turn binds with the complement.

F-Factor Refers to the plasmid that confers the ability to conjugate upon the bacterial cells, and carries the transfer genes.

Fab Batch System An anaerobic bioreactor operated in a batch mode to which substrate either in *solid* or in concentrated *liquid* form is incorporated several times in the course of the run.

Facultative Anaerobe Usually refers to an *organism,* or a *bacterium,* or a *fungus* that may adapt its metabolism to survive and grow in the presence or the absence of oxygen.

Fetus Organisms in later stages of development. In human beings, approximately eight to twelve weeks after conception.

Fibroblast Resident cell of connective tissue that are derived from mesoderm.

Fixative A compound which precisely fixes or sets or stabilizes other compounds or structures securely so as to retain their structural integrity. The phenomenon of fixation makes use of chemical agents to prepare tissues or cells for microscopy rather permanently.

Floccule The coalescence (aggregation) of micro-organisms or colloidal particles floating in or on a liquid. Flocculation may be seen in certain contaminated liquid media showing up as a cloud.

Fouling The deposition of the material both in the 'wrong place' and at the 'wrong time'. In bioreactors and cooling systems the fouling of heat-exchanger surfaces by *wrong biomass, protein, polysaccharides* or other similar materials invariably occurs.

Frame Shift Mutation Deletion or insertion of a number of bases not divisible by three in an open reading frame of a DNA-sequence.

Freeze Drying (Lypholization) The process of drying in a frozen state under vaccum *e.g.,* tissues are freeze-dried to obtain a dry weight or to preserve them for analysis. Lypholization of a specimen is often carried out with liquid N_2, and subsequently causing sublimation of water from the specimen under vacuum. Thus, the proteins are maintained in a reasonably native form, and can easily be dehydrated *viz.,* snake venom, sperm, biological fluids etc.

Fusogen A fusion-inducing agent employed for *protoplast agglutination* in carrying out the somatic hybridization studies *e.g.,* polyethylene glycol (PEG).

Fusion The combination of two distinct cells or macromolecules into a single integrated unit (or entity).

Folic Acid (Vitamin M) [C19H19N,O6; 441.40] A member of the vitamin B group, and is found in green leaves which also exhibit certain extent of coenzyme activities. It is occasionally included as an ingredient for the *plant tissue culture media.*

Fungi Non-photosynthetic eukaryotic organisms, including moulds, that feed on organic matter.

Fusion Joining of the cell membranes of two cells to create a daughter cell that contains the genetic material from both parent cells.

GA-BSL Biosafety Level for genetic modification of animal.

GE Genetically Engineered (see GM).

GI-BSL Biosafety Level for genetic modification of insects.

GM-BSL Biosafety Level for general genetic modification work.

GP-BSL Biosafety Level for genetic modification of plants.

GMAC Genetic modification advisory committee.

GM Genetically Modified; in this context, an organism into whose genome has been deliberately inserted one or more pieces of new DNA (also called Genetically Engineered, or GE).

GMO Genetically modified organism.

Gamete A reproductive (egg or sperm) cell.

Gelation A gelatinous and proteinaceous material produced by boiling bones (animal carcas) with water, which partially hydrolyzes the collagen of animal connective tissues. It is employed skill fully to get or solidify nutrient solutions required for plant-tissue culture.

Gelrite Refers to the brand name of a synthetic refined polysaccharide (*Pseudomonas-de*rived) invariably employed either as a gelling agent or as an agar substitute.

Genotype The genetic make-up of an organism, as distinguished from its physical appearance (the phenotype).

Gene The fundamental physical and functional unit of heredity. A gene is a specific stretch of DNA located in a particular position on a particular chromosome that encodes a specific functional product (*i.e.*, a protein or RNA molecule).

Generation Time The time between successive generations of *cells* or *organisms* within a population.

Gene Therapy The replacement of a defective gene in a person or other animal suffering from a genetic.

Gene Amplification Selective replication of DNA-sequence within a cell, producing a multiple extra copies of that sequence.

Gene Cloning The insertion of a DNA-sequence into a vector that may eventually be propagated in a host organism thereby producing a huge number of copies of the sequence.

Gene Expression The effective utilization of information in a gene *via* transcription and translation ultimately giving rise to the generation of a *'protein'* and therefore, the appearance of the phenotype determined by that gene.

Gene Flow The movement of genes from one population to another.

Gene Gun A device for propelling DNA molecules into living cells.

Gene Isolation The meticulous removal of genetic information, in the form of a DNA-sequence, from a selected organism in order to study its structure or insert it into a vector in the course of 'gene manipulation'.

Gene Magnification Phenomenon in which the number of copies of the ribosomal genes is increased in mutants of drosophila. Such mutants are termed Bobbed, since they also have reduced number of bristles.

Gene Map (Genetic Map) Order of nucleotides and genes that is deduced from genetic recombination experiments. Individual genes can be assigned to specific loci within a genome or a chromosome as a result of gene mapping.

Gene Pool The total of all genes in an interbreeding population at a given time. Gene redundancy. Presence in a cell of many copies of a single gene.

Gene Product The biochemical material, either RNA or protein, resulting from expression of a gene.

Gene Regulatory Protein Protein that functions by altering the transcriptional state of a specific gene. The repressor protein that regulates the activity of the lac operon in *E. coli* is one such protein, and they are widely believed to be an important aspect of eukaryotic gene regulation also.

Gene Sequencing A technique used to determine the sequence of bases in the length of DNA coding for particular polypeptide.

Gene Splicing The enzymatic attachment of one gene or part of a gene to another; also removal of introns and splicing of exons during mRNA synthesis.

Gene Stacking Simultaneous presence of more than one transgene in an organism, usually a GM organism.

Genetic Engineering (Recombinant DNA Technology) It refers to the specific technology involving man-made changes in the genetic constitution of cells (apart from some selective breeding). Genetic engineering invariably uses a vector (*e.g.*, Ti plasmid of *Agrobacterium tumefaciens*) for transferring useful genetic information usually from a donor organism into *a cell* or *organism* which does not possess it.

Genetic Manipulation A technology used to alter the genetic material of living cells or organisms in order to make them capable of producing new substances or performing new functions.

Genetic Selection Refers to selection of genes, clones etc., by man either within populations or between populations or species. The useful purpose is to cause a change in the various specific phenotypic character. Genetic selection normally results in differential success rates of the various genotypes thereby reflecting several variables including selection pressure along with the genetic variability in populations.

Genetic Transformation The precise transfer of extracellular DNA (*genetic information)* among and between species *viz.,* usage of viral or bacterial vectors.

Genetic Variation The actual differences in individuals derived from the same genotype in distinction to differences caused by the environment. Genetic variance designates the very proportion of phenotypic variance caused by differences in the genetic make-up of an individual.

Genome The complete genetic constitution of an organism or individual.

Genotoxic Carcinogenic and toxic to the chromosomes.

Germplasm Refers to the reproductive body tissues distinct from somatic *i.e.,* non-productive, tissues. The basis of heredity of an organism *i.e.,* the *'genetic material'* is usually passed on through the previous generation(s).

Germ Cells Cells comprising actual reproductive components of an organism (specifically, eggs and sperm, and their precursors).

Germline Cells Gametes and the cells from which they are derived. The genetic material of germline cells, unlike that of somatic cells (*q.v.*), can be passed to succeeding generations.

Gibberellic Acid [Gibberellin; GA3; C19H22O6; MW 346.37]. It represents one of the gibberellins, a group of plant-growth hormones essentially promoting *cell division, elongation,* and *germination.* It is isolated from the fungal pathogen *Giberella jujikuroi,* and is extensively used in plant-tissue culture.

Glycosylation Transfer of glucose residue invariably from nucleotide sugar derivative.

Glycolysis. The conversion of monosaccharide to pyruvate *via* the glycolytic pathway [*i.e.,* Embden-Meyerhoff Pathway (EMP-Cycle) in carbohydrate metabolism] in the cytosol.

Glyoxylate Cycle Metabolic pathway present in bacteria and in the glyoxysome 4-C-decarboxylic acid.

GMOs Genetically modified organisms (microorganisms, plants or animals).

Growth Factor A protein that stimulates cell divi-sion when it binds to its specific cell-surface receptor.

HAT Medium A highly selective growth medium for animal tissue cells that essentially comprise of *hypoxanthine, aminopterin* (a folate antagonist), and *thymidine.* It is exclusively used for the selection of hybrid somatic cell lines, such as: production of monoclonal antibodies.

Filter High Efficiency Particulate Air Filter (HEPA) A filter capable of screening out particles larger than 0.3 m. They are frequently employed in Laminar Air Flow Cabinets (Hoods) as well as sterile zones for carrying out transfers/operations in highly aseptic conditions.

HIV Human immunodeficiency virus (a retro-virus).

Hairpin Loop A region of double helix formed by base pairing very much within a single strand of DNA or RNA that has precisely folded back on itself.

Hapten A substance that elicits antibody formation only when combined with other molecules or particles which may specifically react with performed antibodies.

Haploid A single set of homologous chromosomes that have half of the normal diploid number of chromosome.

Haemeagglutination The agglutination or clumping of blood cells *e.g., erythrocytes.*

Haemeagglutination Inhibition Refers to the inhibition of haemeagglutination (*i.e.,* antibody mediated clumpting of RBC), invariably by means of specific immunoglobulins or enzymes, employed to establish/determine whether a patient has been exposed to a specific virus.

Hematopoietic Stem Cell Refers to a particular kind of stem cell that can restore blood.

Helper Virus A virus which, when used to infect cells already infected by a defective virus (*q.v.*), enables the latter to multiply by supplying something the defective virus lacks.

Hepatitis B Viron (Dane Particle) It has 42 nm diameter having an outer wall enclosing inner 27 m core particle comprising of the circular DNA. Aggregates of the envelope proteins are usually found in plasma and are referred to as **hepatitis B surface antigen (HBsAg).**

Heterokaryon A cell containing genetically different nuclear. Found naturally in several fungal species, and can be produced by cell-fusion technique experimentally.

Histones Proteins found in the nuclei of most-eukaryotic cells where they are eventually complexed to DNA in chromatin and chromosomes.

Histocompatibility Antigens Genetically determined antigens located on the lipoprotein membranes of the nucleated cells of most tissues that cause an immune response when grafted on to a genetically different individual and thus determine the capability of tissues in transplantation.

Holism The belief that a complex whole has properties or values that cannot be reduced to the properties or values of its parts, and that an individual thing cannot be adequately valued or understood without reference to the whole(s) of which it is a part.

Host A cell or organism into which foreign DNA is introduced to enable production of proteins or further quantities of the DNA.

Host Range For a virus, the range of species that can be infected by that virus.

Host-Vector System Combination of host (*q.v.*) and the vector (*q.v.*) used for *adult stem cell*. Any stem cell taken from mature tissue, regardless of the age of the donor.

Hybridoma A hybrid cell, used in production of monoclonal antibodies (*q.v.*), which is produced by fusing an antibody-producing cell (a B-lymphocyte) with a tumour cell.

Hypertrophy Refers to an abnormal increase in cell size thereby leading to irregular *growth* or *swelling*. It is invariably regarded as a disease or other stress-induced response.

Hypersensitivity An exaggerated immunological response upon exposure to a particular antigen.

Hypervariable Region A region 'of immunoglobulin (Ig) which essentially accounts for the specificity of antigen-antibody reactions, *e.g.,* specified terminal regions of the fab fragments.

Hybridoma A cell hybrid wherein a tumour cell forms one of the original source cells.

In situ In the original, natural or intact position *viz.,* location of the explants on the mother plant prior to excision.

In vitro (Latin = in glass) Experimentation on organisms or portions thereof in a test tube or any other glassware, *viz.,* growing under artificial conditions as in a tissue culture.

In vivo (Lain = in life) Experimentation on organisms under absolute natural parameters very much within intact living organisms, *viz.,* laboratory animals, voluntary humans etc.

IBC Institutional biosafety committee.

Immobilization A physical or chemical phenomenon used to fix microorganisms, enzymes, and cultures of animal and plant cells, or organelles derived from these sources either upon a solid support or have them trapped in a solid matrix.

Immunodiagnostic An analytical test method exclusively based upon the highly specific interaction between the *antibody* and an *antigen.* The interaction is linked to a method of quantifying the result.

Immortalized Cell Line See cell lines.

Incubator A sophisticated apparatus providing absolute identically controlled environmental conditions (*e.g.,* light, temperature, humidity, photoperiod) most suitable for the growth of microorganisms, yeasts, plant cells or plant cultures (incubation), *viz.,* BOD-Incubator. Indirect Embryogenesis. Refers to the formation upon the callus tissues meticulously derived from somatic or zygotic embryos, seeding plants or other tissues in culture.

Indirect Organogenesis The organ formation upon callus tissue derived from explants. Induction Media. Refers to the media which may afford variation or mutation in the tissues exposed to it.

Informed Consent Autonomous authorization of a medical invention or involvement in research based on substantial understanding.

Intellectual Property (IP) Legal rights associated with inventions, artistic expressions and other products of the imagination (*e.g.,* patent, copyright and trade-mark law).

Insulin A polypeptide hormone usually found in both *vertebrates* and *invertebrates* secreted by B-lymphocytes (cells) of the endocrine pancreas in response to high blood sugar levels, it gives rise to the causation of hypoglycaemia.

Interleukins (ILs) A particular variety of substances (entities) that are produced by leucocytes. Their activity usually get triggered off specifically during inflammatory responses. Interleukins (ILs) represent the larger class of lymphokines.

Intrinsic, inherent Belonging to something independent of its relationship to other things; A thing has intrinsic or inherent value when it is valuable in and for itself.

Ionizing Radiation Refers to such radiations as: γ-radiation and X-ray radiation that specifically gives rise to toxic free radicals which mostly attribute chemical reactions critically disruptive to the *biochemical organization of micro-organisms.*

Isotypes Antibodies differing in heavy chain constant regions intimately associated with different class and subclasses of immunoglobulins (Ig).

Judeo-Christian Originating in or related to the religious and ethical teachings and practices of Judaism and Christianity, particularly those that are common to the two religions.

Jumping Gene A populist terminology for 'transposon' *i.e.,* a genetic unit such as DNA sequence that is transferred from one cell's genetic material to another.

Jiffy Pots A brand name for peat pots that are sometimes employed for *in vitro* transplantation.

Karyotype A complete set of metaphase chromosomes of a cell or organism arranged in decreasing order of chromosomal length.

Karyotyping The meticulous identification and classification of organisms, cells or tissues invariably based on their actual chromosome content.

Killer T-Cells A class of T lymphocytes (cells) that actually participate in killer activity during cell-mediated immune (CMI) response.

Kilobase (Kb) A sequence of 1000 bases or base-pairs of DNA, other nucleic acids or nucleo-tides.

Kinases Enzymes which specifically catalyze the transfer of the a-phosphate of a nucleotide 5'-triphosphate to the 5'-OR terminus of a DNA or RNA molecule.

Kinin The original class name for substances promoting cell-division to which the prefix *cyto* has been added duly (*e.g.,* cytokinins) to distinguish them clearly from kinins found in animal systems.

Knockout Animals Transgenic animals in which an original replaced or "knocked-out" of action by a new gene that doesn't function.

Knockout Mouse A mouse that has been genetically modified by deletion or inactivation of a specific gene.

Krebs Cycle The *citric-acid cycle* or *carboxylic-acid cycle* occurring in 'carbohydrate metabolism'.

LD50 The dose of a toxin or infectious agent which will kill half of a population of organisms.

Lateral Gene Transfer Gene transfer between closely related and very distantly related micro-organisms; an integral part of species evolution in microbial communities.

Laminar Flow Layered monodirectional Air Flow In a laminar airflow (LAF) cabinet air is forced through filters and specially designed ducting to produce an airflow with a minimal lateral disturbance. This has the effect of producing 'layers' of air where the adjacent layers do not mix except on the molecular scale. In the vertical LAF cabinet, this airflow is in the form of a narrow 'curtain' of air passing down the inner front surface of the cabinet: in a horizontal LAP cabinet the air forms a 'block' the size of the front opening blown towards the operator.

Lag Phase Mostly used to describe initial phase of the growth of most batch propagated cell suspension culture wherein the inoculated cells in fresh medium get adapted to the altogether new-environment and get ready to undergo division.

Ligase An enzyme which specifically catalyzes the condensation phenomenon of two molecules coupled to the breakdown of a pyridine triphosphate.

Ligation The process of having direct linkage of the nucleic acid fragments.

Layering *In vitro* layering essentially involves the horizontal placement on agar of cultured shoots (with or without) leaves or nodal segments to promote auxillary bud proliferation.

Lecithin A naturally occurring, choline-containing phospholipid present in animal and plant tissues. Chemically, lecithins are very much akin to fat, but they are found to be rich in phosphorus and nitrogen *e.g.,* yellow yok of an egg; soyalecithin. These are mostly used in food products as an 'emulsifier'.

Limiting Factor An environmental variable whose absolute level at a given time limits the growth or other activity of an organism.

Linear Phase The constant increase in 'cell numbers' following the exponential growth phase' inkers. These are short DNA double strands (*i.e.,* decameric oligonucleotides), which contain sites for the action of one or more restriction enzymes.

Liposome A spontaneously formed layered lipid vesicle in an aqueous medium used as a DNA rector in cell hybridization research activities.

Lyase An enzyme that catalyzes the addition of groups to double bonds or the formation of double bond-lyases as one of the major moieties.

Lyophilize or Freeze Dry To freeze rapidly then dehydrate under high vacuum; and the process termed as 'lyophilization'.

Lysis Cell rupture or destruction *via* enzymatic action.

Lymphocyte White cells of the blood, mediator of specific immunity, derived from the stem :ells of the lymphoid series. Two main categories are: T- and B-Lymphocytes have been duly recognized; the former subdivided into subsets (*viz.,* helpers, suppressors, and cytotoxic T::ells), responsible both for CMI as well as for regulation of the activation of B-cells, whereas the latter solely responsible for antibody production.

Lysosomes Organelles essentially containing hydrolytic enzymes involved in autolytic and digestive processes.

Lysogeny The ability of certain specific *'phages'* to survive in a bacterium as a result of the integration of their respective DNA into the corresponding host chromosome.

Lymphokines A group of biologically active extracellular proteins duly generated by activated T- lymphocytes involved in CMI; the chemical mediators of cellular immunity.

MDA Microdroplet array technique.

MOH Ministry of health.

MOM Ministry of manpower.

Macrolides A group of antibiotics having macrolytic structures *e.g.,* erythromycin, clarithromycin.

Macerase A brand name for *pectinase,* which is used mostly in the isolation of intact cells and protoplasts of higher plants.

Macerate To disintegrate or separate tissues *via* cutting, soaking, enzymatic or other action, resulting in cell dissociation.

Macromolecules Biological terminology referring to proteins, nucleic acids, and carbohydrates. Maternal Inheritance. Inheritance controlled by extra chromosomal (cytoplasmic) hereditary determinants.

Maintenance Energy Energy needed by an organism or a cell culture to survive in a viable condition.

Malignant A primary tumour essentially having the inherited potential to invade locally and also to cause metastasis, *i.e.,* movement of bacteria or body cells (esp. cancer cells) from one part of the body to another.

Marker An 'allele', *i.e.,* one of the two or more different genes containing specific inheritable characteristics that occupy corresponding positions (loci) on paired chromosomes, whose actual inheritance is under observation in a cross.

Marker Gene A gene whose gene product is easily detected (*e.g.,* a gene coding for an antibiotic).

Matrix Ground substance wherein *'things'* are duly embedded. A common usage is for a loose meshwork within which the cells are usually embedded.

Maximum Oxygen Transfer Rate It refers to a measure of the fastest possible rate of uptake of oxygen (O_2) by a growing culture expressed in terms of the volumetric mass transfer coefficient times the oxygen solubility.

Maximum Specific Growth Rate (m_{max}) The theoretical maximum rate of growth of a culture under optimal conditions of nutrient supply.

Macrophages (or Mononuclear Phagocytes) These refer to large, active phagocytic cells found in spleen, liver, lymph nodes, and blood. They represent important factors in non-specific immunity.

Major Histocompatibility Complex (MHC) The genetic region in human beings and mice that controls invariably not only tissue compatibility but also the development and activation of part of the prevailing immune system.

Mast Cells Cells that contain granules of heparin, histamine and serotonin especially in connective tissues involved in hypersensitivity reactions.

Memory Cells Clones of lymphocytes having receptors of high affinity for a particular antigenic molecule; cells that are sensitised duly during the secondary response.

Meristem A localized region of continuing mitotic cell divisions (*meristametic cells)* obtained due to protoplasmic synthesis and tissue initiation. Interestingly, from these undifferentiable tissue new cells arise which eventually differentiates into the specialized tissues.

Mesenchymal Stem Cell A particular kind of stem cell that may give rise to tissues of mesodermal origin, including muscle, bone and related tissues.

Mesoderm One of the three primary layers of an embryo; produces muscle, bone and other related tissues.

Metabolite. A substance that acts as a substrate for or is produced by a metabolic process or enzyme reaction.

Metaphysics The study of, or a theory about, the nature, structure, and composition of reality that goes beyond what can be known by ordinary experience or scientific knowledge, or that analyzes such general and fundamental concepts as time, space, matter, causality, mind, etc.

Microdroplet Array [or Multiple Drop Array (MDA) or Hanging Droplet Technique (HDT)] Introduced by Potrykus *et al.* (1979), this technique is commonly employed to evaluated sufficient numbers of 'media modifications', employing small amounts of medium into which are placed small numbers of cells. Droplets of liquid culture (*i.e.,* medium and suspended cells or protoplasts) are arranged carefully on the lid of a petridish, inverted over the bottom half of the dish containing a solution with a lower osmotic pressure, and the dish is now sealed. The cells or protoplasts form a monolayer at the droplet meniscus and can easily be examined.

Michaelis Constant (Km) A kinetic parameter used to characterize an enzyme defined as the con-centration of a substrate that allows half-maximal rate of reaction.

Microorganism An organism that can be seen only with the aid of a microscope.

Micropropogation Refers to propagation in culture by axillary or adventitious means. It is a general terminology used for vegetative (*i.e., asexual) in vitro* propagation.

Micro Injection The specific insertion of a subs-tance into a cell through a microelectrode. To extrude the substances *via* the very five electrode tips, either using hydrostatic pressure or electric current.

Minimum Inoculation Size The smallest inoculums that may be used successfully for subculture.

Monoclonal Derived exclusively from a single cell; pertaining to a single clone. Monoclonal Antibodies (MABs). Produced from hybrid cells using hybridoma technology. In other words, an antibody produced from hybrid cells employing hybridoma technology.

Monoclonal Antibody An antibody that is derived from a single clone (*q.v.*) of hybridoma (*q.v.*) cells and recognizes only one antigen.

Morals Often a synonym for ethics, but may particularly connote customary beliefs about what constitutes proper behavior.

Mutagen A chemical or physical treatment or agent capable of producing genetic mutation. It may cause insertion, deletion, or alteration of a base or part of the nucleic acid chain (DNA).

Mycelia The interwoven mass of discrete fungal hyphae.

Mutability The propensity of an individual's genes or genotype to undergo heritable mutation. Nick. A point specific in a double-stranded DNA molecule where there is no phosphodiester bond existing between the adjacent nucleotides of one strand.

NEA National environment agency.

Natural Complex A complicated, non-synthetic, and invariably unidentified addendum in plant tissue culture media, *viz.,* orange or tomato juice or coconut milk.

Neoplasm Localized cell-multiplication or tumour, a collection of cells that have undergone genetic transformation. These cells usually differ in structure and function from the original cell type.

Nick Closing Enzyme A form of DNA polymerase which is capable of restoring base sequences. A nick is strategically introduced adjacent to an incorrect base-pair. The base is subsequently removed and the correct base in inserted.

Nif-Genes The complex of genes in nitrogen fixing bacteria, which code for the proteins required for the nitrogen fixation, specifically the enzyme nitrogenase.

Nitrocellulose Paper Paper having a high non-specific absorbing power for biological macro-molecule.

Nonsense Codon (or Nonsense Triplet) The three codons, UAA (ochre), UAG (amber), and UGA (opal), which do not code for an amino acid but serve as signals for the termination of protein synthesis.

Novel Foods Products that have never been used as a food, foods which result from a process that has not previously been used for food; or foods that have been modified by genetic engineering.

N-Terminal The end of a protein or polypeptide chain that contains a free amino group ($-NH_2$); therefore, the abbreviation N.

Nucleases Enzymes that catalyze hydrolysis of the phosphodiester bonds of polynucleotides.

Nuclease SI A very specific nuclease which essentially degrades single-stranded nucleic acids or splits short single-stranded stretches in a DNA sequence, but fails to attack any double stranded structure. It is used invariably in gene manipulation for converting sticky ends of duplex DNA to form blunt ends or to trim off single-stranded ends of nucleic acids after due conversion of single-stranded cDNA to the double-stranded form.

Nucleic Acids Refer to linear polymer of nucleotides, usually linked together by 3', 5'-phospodiester bonds. In deoxyribonucleic acid (DNA), the sugar moiety is deoxyribose, and the corresponding bases of the nucleotides are invariably: *adenine, cytosine, guanine,* and *thymine.* The ribonucleic acid (RNA) has ribose as the sugar, and uracil replaces *thymine* amongst the four bases stated earlier.

Nucleoprotein A conjugated protein closely associated with nucleic acid.

Nucleotide Transferase These enzymes do catalyze attachment of the mononucleotide triphosphate to the 3'-OH function of the initiator polymer, with the release of pyrophosphate.

Oncogene An activated (modified) cellular gene which causes normal cells to become cancerous.

Oocyte A cell that divides to form the female reproductive cell.

Ontology The study of, or a theory about, what sorts of things there are and what characteristics they possess insofar as they exist, apart from their particular or individual.

Open Continuous Culture A cell suspension culture with a continuous influx of fresh medium, maintained at constant volume by the efflux of cells and spent medium.

Operator The specific regulatory section present in a nucleic acid having binding capacity for a repressor protein that categorically blocks the synthesis of mRNAs.

Operon A cluster or group of structural gene whose coordinated expression is duly controlled by a regulator gene.

Organ Culture The growth in an aseptic culture of plant organs *e.g.,* roots or shoots, starting with organ segments and maintaining the characteristic features of the specific organ(s).

Organized Growth The *in vitro* development of organized explants *viz.,* shoot tips or meristem tips or floral buds or organ segments or their *de novo* formation *from* unorganized tissues.

Organogenesis The initiation (*de novo)* and growth of organs (invariably roots and shoots) from cells or tissues; as in organ culture. Organs may be formed upon the very surface of the explants (direct organogenesis) or upon an in-tervening callus phase (indirect organogenesis).

Organoid A anomalous (*i.e.,* unusual) organ-like structural growth formed in culture *e.g.,* as could be seen on leaves, roots, or callus.

Ortet The original *'mother plant'* or *'donor plant'* from which vegetatively propagated plants are derived meticulously.

Osmolarity The overall molar concentration of the solutes affecting the osmotic potential of a solution or nutrient medium.

Oxidative Phosphorylation A metabolic sequence of reactions taking place within a membrane wherein an electron is transferred from a reduced coenzyme by a series of electron carriers thereby strategically establishing an electrochemical gradient across the membrane that virtually drives the formation of ATP from ADP and nitrogen phosphate by chemiosmosis.

Opsonins Refer to intrinsic blood factors that help in establishing a link between particles to be ingested and the macrophage.

Opsonisation The process whereby a cell becomes more susceptible to phagocytosis and lytic digestion when a surface antigen gets combined with an antibody or other serum component.

Outcrossing Mating between different genotypes (organisms with different hereditary constitutions).

Oxygenase An enzyme enabling either a system or an organism to utilize atmospheric oxygen.

PCR See Polymerase chain reaction.

Packaging In the process of virus replication, the assembly of the components of the virus to form the complete virus particle.

Packed Cell Volume (PCV) Refers to a 'quantitative method' of estimating cell growth. It is based on the total cell volume in an aliquot of cell culture. The aliquot is duly centrifuged for 5 min. at 2000 rpm after which the packed cell volume (PCV) is expressed as a percentage (%) of the aliquot volume.

Paratope An important site in the antibody molecule which essentially combines with the antigen.

Parahormone A substance with hormone-like properties which is not a secretory product *e.g.,* ethylene, CO_2.

Parasexual Hybridization Refers to genetic recombination by means other than *via* fertilization of germ cells (parasexual) leading to the formation of individuals or hybrid cells; as could be seen in hybrid cells or plants derived from somatic cell fusion.

Particle Radiation Refers to α-particles (+ vely charged) and β-particles (-vely charged), electrons, protons, and neutrons. These particles are usually employed to produce mutant cells or organisms in plant tissue culture.

Parts Per Million (PPM) This terminology has largely been replaced by equivalent terms, such as: for solids: mg per liter; and for liquids and gases: micro liter per liter.

Patent A limited term monopoly, usually 20 years, granted to inventors of new, useful and non-obvious ideas with industrial application.

Pathogen An organism that causes disease.

Peptide A compound or a *'chemical entity'* existing of two or more amino acids covalently linked by a 'peptide bond' (-CO-NH-) peptide, that are formed with three or more amino acids are usually termed as 'polypeptides'.

Phage See bacteriophage.

Pharming A term derived from pharmaceutical and farming that refers to the growing of genetically engineered organisms for the production of pharmaceutical compounds.

Plasmid A small, self-replicating molecule of DNA which contains a specific origin of replication. Plasmids are often used as cloning vectors (*q.v.*).

Pluripotent Referring to cells able to give rise to virtually any tissue type, but not to a functioning organism.

Polymerase Chain Reaction A technique for generating, *in vitro*, an increased quantity of a target segment of DNA.

Precipitin Test (or Microprecipitin Test) A 'serological assay' wherein visible particulate matters (*i.e.,* precipitates) are duly formed by the interaction of soluble *antigen* and *antibody*. The precipitin test detects and identifies 'antigens'.

Prefilter A *coarse filter* (*i.e.,* Furnace Filter) *e.g.,* those used in a laminar air-flow cabinet to screen large particles before air is forced *via* a much *finer filter* (*i.e.,* HEPA-Filter).

Primordial Germline Cells The source of embryonic germ cells. In normal development, these are the cells that give rise to eggs or sperm.

Pure Culture A culture that contains cells of one type only; the progeny of a single cell *i.e.* genetically identical.

Precautionary Principle A regulatory mechanism for managing environmental and health risk, arising from incomplete scientific knowledge of the impact of a proposed activity or technology.

Protein Digest The enzyme hydrolysis of proteins to give rise to their building block components, amino acids and short-chain peptides (*i.e.,* chains comprising of two to several amino acids normally without enzymatic function).

Prion An infectious agent of unknown etiology that causes spongiform encephalopathies of humans and animals.

Prokaryotic Belonging to the group of microorganisms whose DNA is not enclosed within a nuclear membrane.

Promoter A DNA sequence, located in front of a gene that controls expression of the gene. It is the sequence to which RNA polymerase binds to initiate transcription.

Protein A molecule composed of amino acids.

Protoplast A plant or bacterial cell that has had the outer cell wall removed.

Protoclone Distinct phenotypic regenerates from a plant protoplast. A clone initiated from a protoplast or protoplast-fusion product.

Protocol A sequence of events, activities, techniques or procedures linked to accomplish a definite purpose and objective (goal); as a schedule or protoplast-fusion product.

Protoplast Fusion The coalescence of plasmalemma and cytoplasm of two or more protoplasts in contact with one another. Initial adhesion is indeed a random process but coalesence may be promoted *via* various means (*e.g.,* induced fusion). Nevertheless, when adhesion takes place between adjacent protoplasts during enzymatic wall-degradation or between freshly isolated protoplasts in the absence of a fusion agent, it is known as 'spontaneous fusion'.

Pure Line All cell or individual members that are essentially *homozygous* for one or more characters or genes and will yield more cells or organisms with the character(s) under consideration.

Primary Immune Response The first ever immune response to a particular antigen that has a characteristically long lag period and a low liter of antibody production.

Quiescent Quiet, at rest, not necessarily dormant and having the potential for the resumed activity; may also apply to cells unlikely to divide the non-meristematic cells.

rBST (recombinant) Bovine somatotropin, a genetically modified version of a hormone that controls a cow's rDNA DNA formed by joining, *in vitro*, segments of DNA from different organisms.

RNA Ribonucleic acid, a molecule similar to DNA, whose functions include decoding the instructions for protein synthesis that are carried by the genes; comprises the genetic material of some viruses metabolism and increased milk production. Its use is approved in the USA but not in Canada.

Radiation Rays of heat, light or particles represented in the wave form.

Ramet An individual of a clone. It may also refer to the vegetatively propagated offspring of an ortet.

Radioimmunoassay (RIA) A highly sensitive serological technique invariably employed to assay specific antigens or antibodies, using a specific radioactive label.

Radial Immunodiffusion Radial diffusion is solely based upon the principle that a quantitative relationship prevails between the amount of nitrogen placed in a well where the antibody is already incorporated in the gel, and a ring of precipitation results; generally employed in the assay of serum proteins.

Receptor Cell-surface protein to which molecules such as hormones and growth factors bind to exert their effects on the cell, or to which viruses bind to gain entry to the cell.

Recombinant Organisms, cells, viruses etc. which contain recombinant DNA.

Recombinant Any organism whose genotype has come into being (arisen) as a consequence of recombination; besides, any nucleic acid that has arisen as a result of recombination. Recombinant DNA. The *hybrid DNA* generated specifically by joining pieces of DNA from various sources, invariably designated as rDNA; and the phenomenon is termed as 'recombinant DNA technology' (or Genetic Engineering).

Recombinant DNA (rDNA) DNA molecules created by splicing together two or more different pieces of DNA.

Recombination The occurrence or production of progeny with combinations of genes other than those that occurred in the parents.

Recon or Reconstructed Cell A viable cell hybrid, cybrid or a transformed cell accomplished as a result of 'genetic engineering'.

Reculture The aseptic transfer of a pure culture to another fresh sterilized nutrient medium: also referred to as a 'subculture'.

Regime or Regimen Refers to a systematic treatment.

Redifferentiation Cell or tissue reversal in differentiation from one specific type to another type of cell or tissue.

Rejuvenation Refers to the treatment which leads to *culture invogoration* (*viz.,* subculture) or *revival* (*viz.,* dormancy termination).

Regulatory Genes Genes that serve as a regulatory function; genes that do not code for specific peptides but instead regulate the expression of structural genes.

Replica Plating A technique whereby various types of mutants may be isolated from a population of bacteria grown under nonselective conditions, based upon plating cells from each colony onto multiple plates and thereby noting the exact positions of inoculation.

Replication Reproduction or multiplication of microorganisms, duplication of nucleic acid from a template.

Repression Altered gene expression causing the failure of a particular protein synthesis (anabolism).

Resistance Transfer Factor A plasmid, present in certain types of organism, for instance: *E. coli,* which may. impart specific resistance to antibiotics in animals that are eventually exposed to them.

Restriction Endonucleases Enzymes that specifically cleave DNA at a definite information bearing sequences of nucleotides, and thereby allow an insertion of foreign sequences. Restriction Sites. Specific sites representing categorically the loci at which the 'foreign DNA' gets integrated.

Restriction Fragment Length Polymorphism (RFLP) Fragments of differing lengths of DNA that are produced by cutting with restriction endonucleases enzymes.

Retroviral Vector A retrovirus, which is used to introduce foreign DNA into animal cells, usually by replacing part of the viral genome with the foreign DNA of interest.

Retrovirus A virus that uses the enzyme reverse transcriptase to copy its RNA genome into DNA, which then integrates into the host cell genome.

Rhizogenesis Refers to the formation of roots and their growth as one may come across in root development *de novo* from callus.

Ribosomal RNA (rRNA) RNA of various sizes that make up part of the ribosomes, constituting up to 90% of the total RNA of a cell; single-strand RNA having essentially helical regions duly formed by base-pairing between complementary regions within the strand.

Rights Moral or legal claims to be treated in a certain way, to be entitled to certain goods, or to be allowed to perform certain actions.

Rotator A wheel-like device for slowly (*ca.* I rpm) rotating and gently agitating cultures usually in a vertical plane.

Rotary Shaker A platform shaker having a *'circular motion'* employed particularly for shaking culture flasks at variable speeds.

S-Phase The cell cycle phase during which RNA synthesis takes place.

Sarcoma A solid tumor growing from derivatives of tissue such as embryonal connective tissue, bone, muscle and fat.

Scale-up Expansion of laboratory experiments to full-sized industrial processes.

Schiff's Reagent An admixture of pararosaniline hydrochloride (*i.e.,* an aniline dye) and sodium bisulphite ($NaHSO_3$) invariably employed in staining chromosomes and other nuclear substances present for detection and identification respectively.

Secondary Pests Those species within an ecosystem that are normally kept in check by natural enemies, but which, following certain agronomic practices (*e.g.,* application of pesticides against primary pest), reach densities that cause economic losses.

Selection Culture Makes use of difference(s) in the environmental parameters or more commonly in culture medium composition so that preferred variant cells or cell-lines are predominantly favoured over other variants or the wild-type.

Selection Pressure It is a measure of the effectiveness of natural or experimental selection in altering the genetic composition of a population.

Selection Unit Single cells or small clusters, units of optimum size for isolating and regenerating variants or mutants; the minimum number of cells effective in the screening process.

Selectable Markers Refer to nucleic acid sequences that are phenotypically easy to recognize *i.e.,* antibiotic resistance genes of *E. coli.* plasmids, auxotrophic markers etc.

Semicontinuous Culture The maintenance of cells in a culture vessel in an actively dividing state by draining periodically the medium and replenishing by adding fresh medium.

Serial Float Culture Sunderland's technique of floating anthers (*i.e,* part of a flower's stamen containing pollen) on liquid medium and subculturing them to a new medium at an interval of several days, whereby anther dehiscence, pollen release, and development take place-ultimately enhancing the anther productivity.

Shake Culture An agitated suspension culture, usually an erlenmeyer flask containing the culture is attached to a horizontal or platform shaker or agitated with a magnetic stirrer to provide adequate aeration for cells in the liquid medium.

Secondary Immune Response The response of an individual to the second or subsequent contact with a specific antigen, characterized by a short lag-period, and the production of a high antibody titre.

Sharps Sharp laboratory items such as syringe needles, scalpel and razor blades, and broken glass.

Shot-Gun Cloning The production of a large random collection of cloned fragments of the DNA of an organism, from which genes of interest can later be selected.

Somaclone A plant regenerated from a tissue culture originating from somatic tissue.

Somatic Cell Any cell of a multicellular organism other than germline cells or in general, a cell with the capacity to reproduce itself, and to produce distinct differentiated tissue.

Somatic Cells Refers to cells of the body excluding germ (reproductive) cells.

Somatic Cell Variant (or Embryoid) An organized embryonic structure morphologically similar to a zygotic embryo but initiated from somatic (non-zygotic) cells. Ultimately, these develop into plantlets *in vitro via* developmental processes which are very much akin to zygotic embryos.

Somatic Hyrbid Refers to a cell or plant-product of somatic cell fusion; caused as a result of cell or protoplast fusion and implying genomic integration. The phenomenon is termed as somatic hybridization.

Somatic Organogenesis The production of roots, shoots or other organs upon the somatic tissues of explants (direct organogenesis) or by induction on callus generated by explants (indirect organogenesis).

Spent Medium Medium discarded when a culture is subcultured. The implication is that the medium has been duly depleted of nutrients, dehydrated or accumulated toxic metabolic products.

Spiritual Involving a person's relationship to ultimate value, reality, and meaning; sometimes a synonym for "religious," but often used to contrast with institutional religious affiliations, outward practices, or doctrines.

Spontaneous Fusion Uninduced protoplast fusion that may occur between freshly isolated protoplasts or following adhesion of adjacent cells during enzymatic cell wall degradation.

Spontaneous Variation Refers to the variation in plant populations derived from respective tissue cultures not previously exposed to mutagens but taking place as a consequence of the culture parameters.

Stationary Culture A non-agitated culture *i.e.,* an 'antonym' of shake-culture.

Stages of Culture (I-IV) Refer to : Stage-I: Aseptic explanation or establishment of the explant in culture; State-II: Multiplication of the propagates; State-ID: Rooting of the propagates and preparation for transplant to soil; State-IV: Establishment of State II or III propagates *in vitro* in soil.

Structural Gene A gene whose product is an enzyme, structural protein, tRNA, or rRNA as opposed to regulator gene whose product prominently regulates the transcription of structural gene.

Stock Solution A solution invariably concentrated (10 to 100 times the final medium concentration) of select medium constituents that may be mixed together for attaining perfect compatibility and to avoid precipitation; and usually prepared before hand to save time during preparation of medium. In usual practice, the stock solutions are either frozen (in deep-freezer) or stored in the refrigerator to avoid deterioration in strength and quality. Portions of stock solutions are used conveniently as and when required for preparation of media.

Subculture (Passage) A culture derived from another culture or the aseptic division and transfer of a culture or a portion of that culture (inoculation) to a fresh nutrient medium. Invariably, sub-culturing is Gamed out at a predetermined time intervals, the length of duration is known as the *subculture interval* or *passage time*.

Subline A cell line regenerated from a unique cell line of a hybrid callus colony.

Suspension Culture Cells and group of cells (aggregates) that are adequately dispersed in an aerated, usually agitated, liquid culture medium.

Supraoptimum An amount greater than required as an inhibitory concentration of an exogenous growth factor.

Surface Sterilization The removal of plant surface micro-flora prior to aseptic excision of explants. Surface sterilization is accomplished by immersing of tissue in one of several sterilants *e.g.,* calcium hypochlorite, sodium hypochlorite, hydrogen peroxide, mercuric chloride, silver nitrate, bromine water for an empirically determined period of time.

Substantial Equivalence An assessment approach which is an early criterion in the regulatory decision tree. It states conditions under which it can be assumed that a new crop or novel food poses no more risks than a non-modified couterpart that is already considered safe.

Synchronized Cells Synchronized mitosis in a group of cells in culture by natural or artificial means.

Synchronous Culture A microbial or plant-cell culture treated in such a manner so as to have most cells or individuals in the same stage of development or mitosis. It may be accomplished in several ways including: *via temperature variation* and *nutrient limitation.*

Ti Plasmid A portion of the genome of the bacterium *Agrobacterium tumefaciens i.e.,* the agent directly involved in crown gall disease. This specific plasmid happens to be an useful (experimental) vector for the transfer of genetic information into the plant cells.

Tissue Culture *In vitro* growth of tissue cells in nutrient medium.

Tissue Explant An excised section of plant tissue usually employed to initiate a culture.

T-Cells T lymphocytes that are differentiated in the thymus and are virtually important in cell mediated immunity (CMI); besides, in the regulation of antibody-mediated immunity (AMI).

Totipotency The potential or inherent capacity of a plant cell or tissue to develop into an entire plant when stimulated appropriately. It critically implies that all the information necessary for growth and reproduction of the organism is very much contained in the cell.

Totipotent Refers to cells able to give rise to virtually any tissue type and in some cases, as shown experimentally in mice, to a functioning organism.

Traditions Teachings and practices (written or unwritten) that are handed down from generation to generation and especially in the case of religious traditions, formative of communities which regard them as authoritative.

Transcription The synthesis of RNA (ribonucleic acid) molecules (mRNA, rRNA, tRNA) concerned in translating the structure of DNA into the structure of protein molecules or others.

Transduction Introduction of foreign genetic material into a cell by the help of virus or, in the particular instance of bacteria by means of phages.

Transformation Uptake of free nucleic acids or plasmids into a cell.

Transgene A gene from one organism inserted into the genome of another.

Transgenic (organism) An organism whose cells, including the germline cells, contain foreign DNA; transgenic animals are produced by the insertion of the foreign DNA into the newly fertilised egg or embryo or an organisms that consists of (foreign) genetic materials from a different species or genus.

Translation Refers to the synthesis of a protein upon mRNA.

Trophoblast The outer layer of cells of the mammalian blastocyst that gives rise to the placenta.

Toxic Poisonous in nature; as are certain chemical substances (toxicants) or any specific substance present in excess which tends to be detrimental to the normal plant function or growth.

Toxin A poisonous substance, produced mainly by microorganisms, but also by some fungi, plants and animals.

T Suppressor Cells A class of T-cells that usually suppress the activates of B-cells in the antibody mediated immunity.

Triplet Code Refers to the 'genetic code' *i.e.,* three sequential nucleotides in mRNA are required to code for a specific amino acid.

Tumour Inducing Principle (TIP) The plasmid carried by *Agrobacterium tumefaciens* the crown gall organism. Through incorporation into the host genome the host tissue is duly transformed into the tumour tissue.

Tumour Suppressor Gene A type of gene in which inactivating mutations contribute (or antioncogene) to tumour development.

Turbidostat An open continuous culture system wherein the inflow of fresh medium is controlled by the turbidity of the culture, a function of the amount of cell growth. Thus, balancing the fresh medium inflow which being a regulated outflow of cells and spent medium thereby restoring the original turbidity level.

Ultrasonic Cleaner A device to include high frequency vibration of materials, removing adhering substances from surfaces by mechanical action. This specific device is useful for cleaning glassware and for also disinfecting plant materials.

Unorganized Growth *In vitro* formation of tissues with a few differentiated cell types and lacking recognizable structure.

Unwinding Enzymes Refer to such enzymes that specifically catalyze separation of complementary strands of DNA.

Vaccine Any antigenic preparation administered to stimulate the recipients immune defence mechanisms with respect to a given pathogen or toxic agent.

Vector Plasmid or virus after ligation with a foreign DNA is employed to introduce new information into the cell.

Vegetative Cells Refers to such cells that are intimately engaged in nutrition and growth; they do not act as specialized reproductive or dormant forms.

Viroid A disease-causing agent of plants, which is smaller than a virus and consists of a naked RNA molecule.

Virulence Ability of an organism to cause disease.

Virus A submicroscopic infectious particle, containing genetic material (DNA or RNA) and protein, which can replicate only within the cell of an organism (plant, animal or bacteria).

Virus Elimination Chemotherapy, thermotherapy and meristem or meristem tip culture, used alone or in combination have been used successfully for the elimination of systemic viruses from plants.

Virus-Free (or Virus Tested) A plant that appears healthy and repeatedly gives negative tests for the presence of one or more identifiable viruses. Such a plant may subsequently employed as a stack or donor plant (explant source) solely for propagation purposes, which may be duly certified as virus tested (or certified virus tested).

Worldview A comprehensive understanding of reality and our place within it, which includes and integrates factual beliefs, moral values, and ultimate goals.

Wild Types The *genotype* or *phenotype* of an organism predominating in the control (standard or wild) population, in its natural environment.

Water of Hydration The quantum of water chemically bound to a substance. The amount could be variable and should be taken into account when solutions of salts are prepared; as in the instance of medium preparation.

Xenograft A tissue graft between animals of different species.

Xenotransplantation Grafting tissue or organs from one animal species to another; *e.g.,* grafting pig organs to a human being.

Xylem The vascular tissue of plants that conduct water and dissolved minerals from roots to the aerial parts of plants.

Yeast Artificial Chromosome A stretch of DNA that contains all the elements required to propagate a chromosome in yeast and that is used to clone foreign DNA fragments in yeast cells.

Yeast Extract An unidentified complex Vitamin B complex used as an addendum to certain plant tissue-culture media.

Yield Refers usually to the 'productivity assessment'.

Zoonosis A disease of animals that can be transmitted to humans.

Zoospore A motile, flagessaed spore.

Zygote The cell produced by the union of the male and female gametes or the fertilized egg or ovum.

Appendix

APPENDIX—A

Commonly Used Abbreviations

Ab—Antibody
AIDS—Acquired immune deficiency syndrome
ACP—Acyl carrier protein
ADH—Alcohol dehydrogenase
AMP, ADP, ATP—Adenosine 5' -mono di-, and tri-phosphate
APCs—Antigen-presenting cells
cAMP—Cyclic AMP
cDNA—Complementary DNA
COA, COA-SH—Coenzyme A
ConA—Concanavalin A
CR—Complement receptors
DEAE-cellulose—Diethylaminoethyl cellulose
DNA—Deoxyribonucleic acid
DNase—Deoxyribonuclease
DNP—2,4-Dinitrophenol
DTH—Delayed type hypersensitivity
EBV—Epstein-Barr virus
EDTA—Ethylenediaminetetraacetic acid
EPR—Electron paramagnetic resonance
ESR—Electron spin resonance
FAD—Flavin adenine dinucleotide
FADH$_2$—Flavin adenine dinucleotide reduced
Met—N-formyl methionine
FMN—Flavin mononucleotide

FMNH$_2$—Flavin mononucleotide reduced
GDH—Glutamate dehydrogenase
GLC—Gas-liquid chromatography
Glu—Glutamine
cGMP—Cyclic GMP
GS—Glutamine synthetase
HbO$_2$—Oxyhemoglobin
HEPA—High efficiency particulate air filter
His—Histidine
HnRNA—Heterogeneous nuclear RNA
Ig—Immunoglobulin
Kb—Kilobase
LAF—Laminar airflow
LDH—Lactate dehydrogenase
Leu—Leucine
MAC—Membrane attack complex
MHC—Major histocompatibility complex
NAD$^+$—Nicotinamide adenine dinucleotide
NADH—Nicotinamide adenine dinucleotide (reduced)
NADP$^+$—Nicotinamide adenine dinucleotide phosphate
NADPH—Nicotinamide adenine dinucleotide phosphate (reduced)
NMR—Nuclear magnetic resonance
OD—Opticail density
Pi—Inorganic phosphate
PEP—Phosphoenol pyruvate
RNA—Ribonucleic acid
mRNA—Messenger RNA
rRNA—Ribosomal RNA
sRNA—(Soluble RNA)
tRNA—Transfer RNA
RNase—Ribonuclease
TCA—Tricarboxylic acid
TLC—Thin layer chromatography
UV—Ultraviolet
UMP—Uridine monophosphate
UDP—Uridine diphosphate
UTP—Uridine triphosphate
Val—Valine
Met—Methionine
Phe—Phenylalanine
PPi—Inorganic pyrophosphate
Ser—Serine
TPP—Thiamine pyrophosphate
Trp—Tryptophan

APPENDIX—B

Commonly used Terms in Stem Cell Biology

AGM—The region where the aorta, gonads, and fetal kidney mesh.

ALS—Amyotrophic lateral sclerosis. Also known as Lou Gehrig's disease.

BME—Beta-mercaptoethanol.

BMP-1 to BMP-9—Bone morphogenetic proteins that are signaling molecules.

BRCA1—Breast cancer gene 1.

BRCA2—Breast cancer gene 2.

C/EBC—CCAAT/Enhancer binding protein.

CD4—Helper T-cells that are instrumental in initiating an immune response by supplying help in the form of special cytokines to both CD8 cytotoxic T-cells and B-cells.

CD8—Cytotoxic (killer) T-cells that are capable of killing infected cells once activated by cytokines secreted by antigen-specific CD4 helper T-cells.

CMV—Cytomegalovirus.

EBs—Embryoid bodies.

EG—Embryonic germ cell.

ES—Embryonic stem cell.

FACS—Fluorescence-activated cell sorting.

Fas receptor (CD95)—Fatty acid synthase.

FGF-1 to FGF-10—Fibroblast growth factor 1 to 10. A growth factor molecule.

GATA4—Transcription factor. Important in embryonic stem differentiation into yolk sac endoderm.

GATA6—Important for embryonic stem cell differentiation into heart smooth muscle.

GCSF—Granulocytecolony stimulating factor.

Gdf-5—Growth/differentiation factor-5. A growth factor molecule.

GDNF—Glial cell-derived neurotrophic factor. A growth factor molecule.

GFP—Green fluorescent protein.

Gp130—Glycoprotein. Signal transducing receptor of cytokines.

Gsc—Goosecoid. A signaling molecule.

hCNS-SC—Human central nervous system stem cell.

Hesx1—Pituitary transcription factor.

Hex—Hexosaminidase. Enzyme for processing lipid (fat).

HGF—Hepatic growth factor molecule. Also a scatter factor.

HLAs—Human leukocyte antigens.

Hoxa-d—Homeobox-containing a to d. A transcription factor.

HPC—Hematopoietic progenitor cell.

HSC—Hematopoietic stem cell.

ICM—Inner cell mass.

IVF—*In vitro* fertilization.

LIF—Leukemia inhibitory factor. A growth factor molecule.

Lim1—A transcription factor molecule.

Mac-1 (CD11b)—Antigen found in blood cells. Indicative of murine and progenitor cells.

MPC—Mesenchymal progenitor cell.

MR4—Metabolic regulator. Important for electron transport and ATP synthesis.

MSC—Mesenchymal stem cell.

Myf-5—Myogenic regulatory factor molecule.

NK—Natural killer lymphocytes.

NSC—Neural stem cell.

Oct4—Octamer binding gene. Important for germ cell generation.

Otx2—A transcription factor molecule.

Pax-1 to Pax-9—Paired box 1-9. A transcription factor molecule.

PDGF—Platelet-derived growth factor.

PDX-1—A transcription factor molecule.

PECAM 1—Platelet. Endothelial cell adhesion molecule.

SDF-1/CXCR4—Stromal-derived factor and its receptor.

SHH—Sonic hedgehog.

SMA—Alpha-smooth muscle actin.

SP—Side population stem cell.

Stat 3—Signal transducers and activators of transcription 3.

T3—Triiodothyronine. A thyroid hormone important for hematopietic cells.

TGF-β1 to TGF-β5—Transforming growth factors.

TPO/mpl—Thrombopoietin and receptor.

VEGF—Vascular endothelial growth factor.

Wnt1—A signaling molecule.

XIST—X-inactive specific transcript. Uncertain function.

APPENDIX—C

International Organizations

Advisory Committee on Animal Feeds (ACAF)
American Association for the Advancement of Science (AAAS)
Animal and Plant Health Inspection Service (APHIS)
Biosafety Working Group (BSWG)
Biotechnology, Biologics, and Environmental Protection unit (BBEP)
Centers for Disease Control and Prevention (CDC)
Convention on Biological Diversity (CBD)
Center for Biologics Evaluation and Research (CBER)
Center for Bioethics and Human Dignity (CBHD)
Center for the Study of Environmental Change (CSEC)
Consultative Group on International Agricultural Research (CGIAR)
Convention on Biological Diversity (CBD)
Commission on Genetic Resources for Food and Agriculture (CGRFA)
Community Biodiversity and Development Conservation Program (CBDC)
Copyright Enforcement Advisory Council (CEAC)
Council for International Organizations of Medical Sciences (CIOMS)
Department of Energy (DOE)
ELSI Research Planning and Evaluation Group (ERPEG)
Ethics Advisory Board (EAB)
European Union (EU)
Federal Plant Pest Act (FPPA)
Federal Plant Quarantine Act (FPQA)
Food and Agriculture Organization (FAO)
Federal Bureau of Investigation (FBI)
General Agreement on Tariffs and Trade (GATT)
Genetic Resources Action International (GRAIN)
Global Environment Facility (GEF)
Global Plan of Action (GPA)
Human Genome Organization (HUGO)
Indian Council for Agricultural Research (ICAR)
Indian Performing Rights Society Limited (IPRS)
Industrial Toxicology Research Institute (ITRC)
Indigenous Peoples Council on Biocolonialism (IPCB)
International Agricultural Research Center (IARC)
International Association of Plant Breeders for the Protection of Plant Varieties (ASSINSEL)
International Bioethics Committee (IBC-UNESCO)
International Chamber of Commerce (ICC)
International Development Research Center (IDRC)
International Federation of Organic Agricultural Movements (IFOAM)

International Plant Protection Convention (IPPC)
International Potato Center (CIP)
International Rice Research Institute (IRRI)
International Searching Authority (ISA)
International Service for the Acquisition of Agri-biotech Applications (ISAAA)
International Service for National Agricultural Research (ISNAR)
International Standards for Phytosanitary Measure (ISPM)
International Treaty on Plant Genetic Resources for Food and Agriculture (ITPGRFA)
International Union for the Protection of New Varieties of Plants (UPOV)
International Union for Conservation of Nature and Natural Resources (IUCN)
National Agricultural Research System (NARS)
National Bioethics Advisory Commission (NBAC)
National Human Genome Research Institute (NHGRI)
National Institutes of Health (NIH)
National Institute of Allergy and Infectious Disease (NIAID)
National Science Foundation (NSF)
Occupational Safety and Health Administration (OSHA)
Organization for Economic Cooperation and Development (OECD)
Patent Cooperation Treaty (PCT)
People for the Ethical Treatment of Animals (PETA)
Plant genetic resources for food and agriculture (PGRFA)
Pharmaceutical Research and Manufacturers Association (PhRMA)
Phonographic Performance Limited (PPL)
Plant Genetic Resources for Food and Agriculture (PGRFA)
Recombinant DNA Advisory Committee (RAC)
Review Committee on Genetic Manipulation (RCGM)
Royal Society for the Protection of Birds (RSPB)
Rural Advancement Foundation International (RAFI)
Society for Copyright Regulations of Indian Producers of Films and Television (SCRIPT)
Search for Extraterrestrial Intelligence (SETI)
Trade Related Aspects of Intellectual Property (TRIPS)
The United Nations Development Programme (UNDP)
U.S. Patent and Trademark Office (USPTO)
United States Conference of Catholic Bishops (USCCB)
United States Department of Agriculture (USDA)
United Nations Conference on Environment and Development (UNCED)
United Nations Conference on Trade and Development (UNCTAD)
United Nations Educational, Scientific and Cultural Organization (UNESCO)
United Nations Environment Program (UNEP)
United Nations Human Rights Commission (UNHRC)
United Nations Industrial Development Organization (UNIDO)
World Health Organization (WHO)
World Intellectual Property Organization (WIPO)
WTO (World Trade Organization)

APPENDIX—D

Factors which could Influence the Results of Animal Experiments

Genetic Quality
- Strain/stock
- Breeding system
- Quality breeder/supplier

Biological Status
- Sex
- Age
- Body weight

Health Status
- Quality breeder/supplier
- Constant level of quality
- Hygiene barrier in maintenance

Nutrition
- Quality supplier
- Constant composition (lot-nr.)
- Quality drinking water

Maintenance
- Cage-type (dimensions)
- Bedding
- Number of animals per cage
- Animal room-ventilation
- Temperature
- Relative humidity
- Lighting
- Noise
- Other animals

Transportation
- Means of transportation
- Transport cage
- Food supply

Animals Care
- Qualification of animal care taker

Experimental Techniques
- Qualification of animal technician
- Standardisation of techniques
- Time of intervention

APPENDIX—E

Alternate Names for Growth Factors

Activins (A, AB, B)
- FSH releasing protein (FRP)

Colony-stimulating factor 1 (CSF-1)
- Macrophage colony-stimulating factor (M-CSF)

Epidermal growth factor (EGF)
- Urogastrone

Erythropoietin (epo; EP)
- Hemopoietine
- Erythrocyte stimulating factor (ESF)

Fibroblast growth factors (FGF)
- Acidic fibroblast growth factor (aFGF)
- Basic fibroblast growth factor (bFGF) Brain-derived growth factor (BNDF; BDGF) Heparin binding growth factor (HBFG) Endothelial cell growth factor (ECGF) Retina-derived growth factor (RDGF) Eye-derived growth factor (EDGF)
- Kidney angiogenic factor (KAF)
- Adrenal growth factor (AGF)
- Corpus luteum angiogenic factor (CLAF) Ovarian growth factor (OGF)
- Placental angiogenic factor (P AG)
- Hepatocyte growth factor (HGF)
- Myogenic growth factor (MGF)
- Cartilage-derived growth factor (CDGF) Bone growth factor (BGF)
- Seminiferous growth factor (SGF)
- Prostatropin (PGF)
- Tumor-derived growth factor (TDGF)
- Hepatoma-derived growth factor (HDGF)
- Melanoma-derived growth factor (MDGF)
- Mammary tumor-derived growth factor (MTGF)

Gastrin-releasing peptide (GRP)
- Mammalian bombesin

Granulocyte colony-stimulating factor (G-CSF)
- Macrophage|granulocyte inducer type 1, granulocyte (MG-1G)
- Pluripotent colony-stimulating factor (pluripoetin)
- Granulocyte-macrophage colony-stimulating factor-p (GM-CSF *P)*

Granulocyte-macrophage colony-stimulating factor (GM-CSF)
- Macrophage-granulocyte inducer
- Colony-stimulating factor 2 (CSF-2)

Inhibin

Insulin-like growth factor I (IGF-I)
- Somatomedin A
- Somatomedin C
- Basic somatomedin

Insulin-like growth factor II (IGF-II)
- Multiplication stimulating activity (MSA)

Interferon-α (IFN-α)
- Leukocyte (Le) interferon
- Type I interferon

Interferon-β (IFN-β)
- Fibroblast (F) interferon
- Type I interferon

Interferon-γ (IFN-γ)
- Immune interferon
- T interferon
- Type II interferon

Interleukin-1 (IL-1)
- Interleukin-I α
- Interleukin-1 β
- Lymphocyte activating factor (LAF)

Interleukin-2 (IL-2)
- T-cell growth factor (TCG F)

Interleukin-3 (IL-3)
- Mast cell growth factor
- P-cell stimulating factor Multi-colony-stimulating factor
- Burst promoting activity
- WEHI-3 hematopoietic growth factor
- Thy-1 inducing factor
- Histamine cell-producing stimulating factor
- 20 α-dehydrogenase-inducing factor

Interleukin-4 (IL-4)
- B-cell stimulatory factor 1 (BSF-1)
- T-cell growth factor (TCGF)
- B-cell growth factor I (BCGF-I)
- Mast-cell growth factor (MCGF)

Interleukin-5 (IL-5)
- T-cell replacing factor (TRF-I)
- B-cell growth factor II (BCGF II)
- B-cell differentiation factor *f1* (BCDFf1)
- Eosinophil differentiation factor (EDF)
- IgA enhancing factor (IgA-EF)
- B-cell maturation factor (BMF)
- B-cell growth and differentiation factor (BGDF)

Interleukin-6 (IL-6)
- B-cell stimulatory factor 2 (BSF-2)
- Interferon-β 2
- 26-kDa protein
- Hepatocyte stimulating factor
- Hybridoma/plasmacytoma growth factor
- Interleukin-HPI
- Macrophage granulocyte inducer type 2

Lymphotoxin
- Tumor necrosis factor-β (TNF-β)

Mullerian inhibiting substance (MIS)
- Anti-MUllerian hormone (AMH)
- Mullerian inhibiting factor (MIF)

Nerve growth factor (NGF)

Platelet-derived endothelial cell growth factor (PD-ECGF)

Platelet-derived growth factor (PDGF-AA, PDGF-BB, PDGF-AB)

Transforming growth factor-α: (TGF-α:)
- Sarcoma growth factor (SGF)

Transforming growth factor-β (TGF-β)
- Cartilage inducing factor-A (CIF-A)
- Cartilage inducing factor-B (CIF-B)
- BSC-I growth inhibitor (BSC-I GI)
- Differentiation inhibitor (DI)
- Polyergin

Tumor necrosis factor (TNF) Cachectin

Tumor necrosis factor-α: (TNF-α:)

APPENDIX—F

Chromosomal Locations of Human Growth Factors/Growth Factor Receptors

Chromosome 1
- Nerve growth factor (p22.1)
- Transforming growth factor-p2 (q41)

Chromosome 2
- Inhibin-a (distal portion of long arm)
- Inhibin-PB (near centromere on short arm)
- Interleukin-1p (2q13-2q21)
- Transforming growth factor-a (pl1-p13)

Chromosome 4
- Basic fibroblast growth factor
- Epidermal growth factor (q2S-q27)

Chromosome 5
- Acidic fibroblast growth factor
- Colony-stimulating factor 1 (q33.1)
- Colony-stimulating factor 1 receptor (q33.2-33.3) Granulocyte-macrophage colony-stimulating factor (q2331)
- Interleukin-3 (q23-q31)
- Interleukin-4 (q23-31)
- Interleukin-S (q23.3-32)
- Platelet-derived growth factor receptor (p-sub-unit)

Chromosome 6
- Interferon-y receptor

Chromosome 7
- Epidermal growth factor receptor (p14-p12)
- Erythropoietin (q11-q22)
- Inhibin-PA
- Interleukin-6 (p21)
- Platelet-derived growth factor A chain

Chromosome 9
- Interferon-a
- Interferon-p

Chromosome 11
- Insulin-like growth factor II

Chromosome 12
- Insulin-like growth factor I
- Interferon-y

Chromosome 14
- Transforming growth factor-p3 (q24)

Chromosome 17
- Granulocyte colony-stimulating factor (q21-q22)
- Nerve growth factor receptor (q12-q22)

Chromosome 18
- Gastrin-releasing peptide (q21)

Chromosome 19
- Mullerian inhibiting substance
- Transforming growth factor-p1 (q13)

Chromosome 21
- Interferon-alP receptor

Chromosome 22
- Platelet-derived growth factor B chain

APPENDIX—G

Chromosomal Locations of Mouse Growth Factors/Growth Factor Receptors

Chromosome 1
- Transforming growth factor-α

Chromosome 2
- Interleukin-1 α
- Interleukin-1 β

Chromosome 3
- Epidermal growth factor
- Nerve growth factor

Chromosome 4
- Interferon-α
- Interferon-β

Chromosome 5
- Erythropoietin
- Interleukin-6

Chromosome 7
- Transforming growth factor-p1

Chromosome 10
- Interferon-γ
- Interferon-γ receptor

Chromosome 11
- Epidermal growth factor receptor
- Granulocyte colony-stimulating factor
- Granulocyte-macrophage collony-stimulating factor Interleukin-3
- Interleukin-S
- Nerve growth factor receptor

Chromosome 12
- Transforming growth factor-β

Chromosome 16
- Interferon-α receptor

Chromosome 18
- Colony stimulating growth factor 1 receptor

APPENDIX—H

CD (Cluster of Differentiation) Markers

CD marker	Mol. Wt. kDa	Cell source
CD1a-1e	49-55	T-cells
CD3	50	T-cells
CD4	55	T-cells
CD8	36	T-cells
CD10	100	B-cells
CD11a	180	T-cells/B-cells
CD25	55	Activated T-cells/B-cells
CD35	260	T-cells/B-cells
CD114	110	Granulocytes
CD115	150	Macrophages

APPENDIX—I

Decimal Prefixes and Multiples for Area/Length, Mass Weight, and Volume

Prefix	Symbol	Multiple	Magnitude	Common name
yotta	Y	10^{24}	1 000 000 000 000 000 000 000 000	heptillion
zetta	Z	10^{21}	1 000 000 000 000 000 000 000	hexillion
exa	E	10^{18}	1 000 000 000 000 000 000	quintillion
peta	P	10^{15}	1 000 000 000 000 000	quadrillion
tera	T	10^{12}	1 000 000 000 000	trillion
giga	G	10^{9}	1 000 000 000	billion
mega	M	10^{6}	1 000 000	million
kilo	K	10^{3}	1000	thousand
hecto	h	10^{2}	100	hundred
deca	da	10^{1}	10	ten
unit*	—	—	1	one
deci	d	10^{-1}	0.1	tenth
centi	c	10^{-2}	0.01	hundredth
milli	m	10^{-3}	0.001	thousandth
micro	μ(mu)	10^{-6}	0.000 001	millionth
nano	n	10^{-9}	0.000 000 001	billionth
pico	p	10^{-12}	0.000 000 000 001	trillionth
femto	f	10^{-15}	0.000 000 000 000 001	quadrillionth
atto	a	10^{-18}	0.000 000 000 000 000 001	quintillionth
zepto	z	10^{-21}	0.000 000 000 000 000 000 001	hexillionth
yocto	y	10^{-24}	0.000 000 000 000 000 000 000 001	heptillionth

*unit = meter (m) for area/length, gram (g) for mass/weight, and liter (L) for volume (capacity).

APPENDIX—J

Structures and One- and Three-Letter Abbreviations of Twenty Common Amino Acids

It is very helpful to memorize these abbreviations and to become familiar with the physical properties of the amino acids. The percentages refer to the relative abundance of each amino acid in proteins.

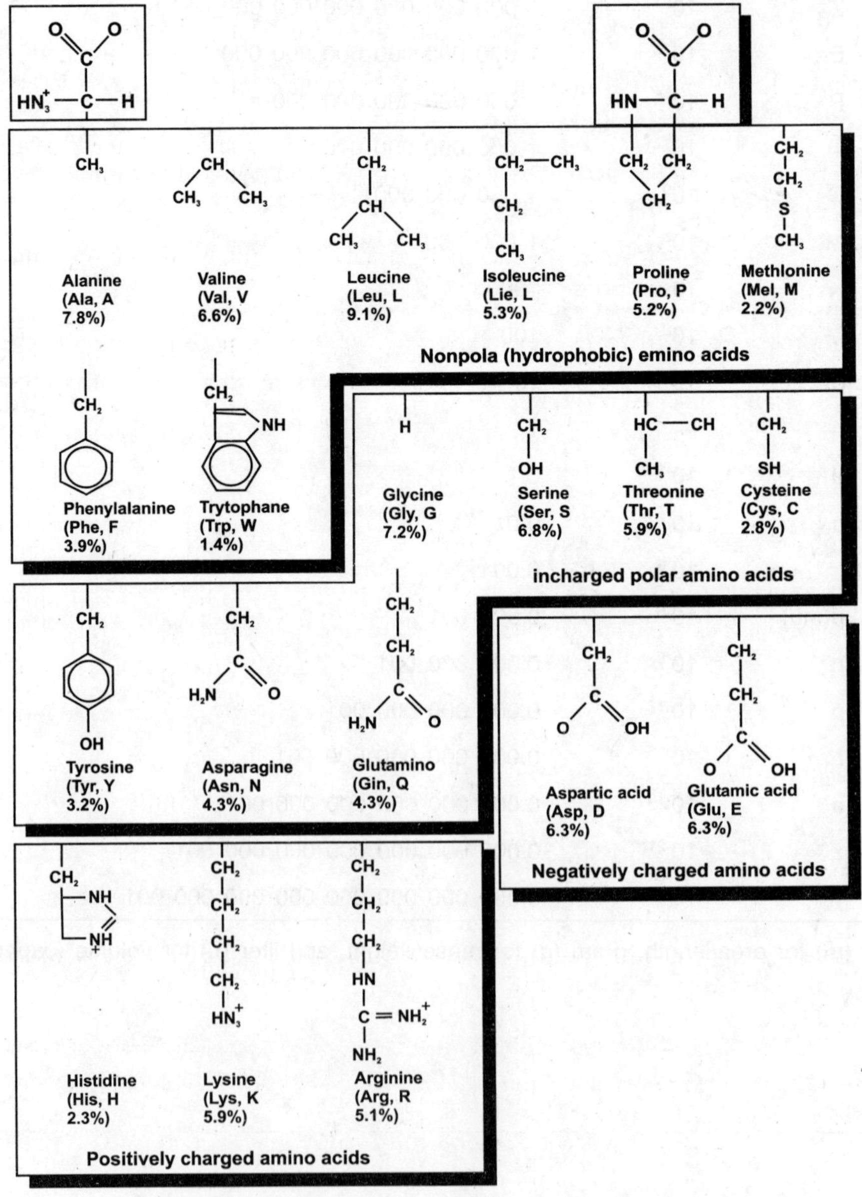

APPENDIX—K

Eight Major Goals of Human Genome Project (1998-2003) from Scientific Point of View

1.	Human DNA sequence	• Finish the complete human genome sequence by the end of 2003. • Achieve coverage of at least 90% of the genome in a working draft based on mapped clones by the end of 2001. • Make the sequence totally and freely accessible.
2.	Sequencing technology	• Continue to increase the throughput and reduce the cost of current sequencing technology. • Support research on novel technologies that can lead to significant improvements in sequencing technology. • Develop effective methods for the development and introduction of new sequencing technologies.
3.	Human genome sequence variation	• Develop technologies for rapid, large-scale identification and/or scoring of single-nucleotide polymorphisms and other DNA sequence variants. • Identify common variants in the coding regions of the majority of identified genes during this five-year-period. • Create a SNP map of at least 100,000 markers. • Create public resources of DNA samples and cell lines.
4.	Functional genomics technology	• Generate sets of full-length cDNA clones and sequences that represent human genes and model organisms. • Support research on methods for studying functions of nonprotein-coding sequences. • Develop technology for comprehensive analysis of gene expression. • Improve methods for genomewide mutagenesis. • Develop technology for large-scale protein analyses.
5.	Comparative genomics	• Complete the sequence of the roundworm *C. elegans* genome and the fruitfly *Drosophila* genome. • Develop an intergrated physical and genetic map for the mouse, generate additional mouse cDNA resources, and complete the sequence of the mouse genome by 2008.

6.	Ethical, legal, and social issues	• Examine issues surrounding completion of the human DNA sequence and the study of genetic variation. • Examine issues raised by the integration of genetic technologies and information into health care and public health activities. • Examine issues raised by the integration of knowledge about genomics and gene-environment interactions in non-clinical settings. • Explore how new genetic knowledge may interact with a variety of philosophical, theological, and ethical perspectives. • Explore how racial, ethnic, and socio-economic factors affect the use, understanding, and interpretation of genetic information, the use of genetic services, and the development of policy.
7.	Bioinformatics and computational biology	• Improve content and utility of databases. • Develop better tools for data generation, capture, and annotation. • Develop and improve tools and databases for comprehensive functional studies. • Develop and improve tools for representing and an alyzing sequence similarity and variation. • Create mechanisms to support effective approaches for producing robust, exportable software that can be widely shared.
8.	Training and manpower	• Nurture the training of scientists skilled in genomics research. • Encourage the establishment of academic career paths for genomic scientists. • Increase the number of scholars who are knowledgeable in both genomic and genetic sciences and in ethics, law, or the social sciences.

APPENDIX—L

Various Genomes

Comparison of the sizes of various genomes. The x-axis on each graph represents a 10-fold change in scale. For bacterial genomes, the genome size ranges from a mere 580,000 bp (M. genitalium, with 470 protein-coding genes is the smallest sequenced genome) to cyanobacteria with genome sizes of 13 Mb. This is a 22-fold range. For eukaryotic genomes, there range is from the 8 Mb of some fungi to 686 Gb for some amoebac. This range is over 75,000-fold and has been called the C value paradox. The C value is the total amount of DNA in the genome, and the paradox is the relation between complexity of a eukaryote and its amount of genomic DNA.

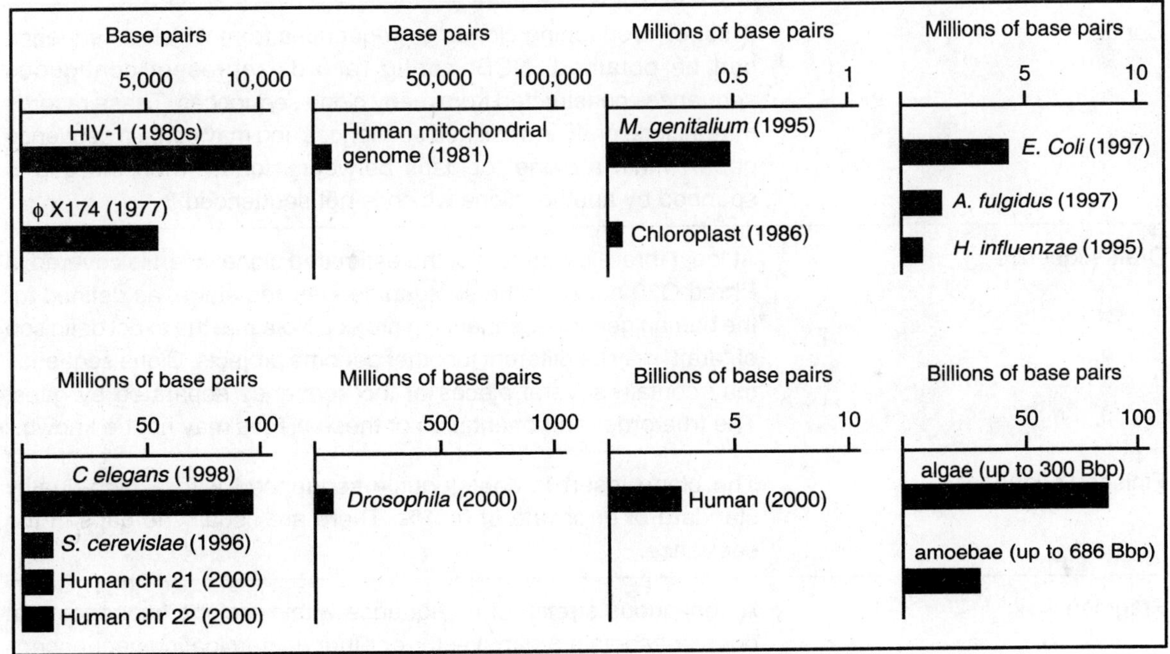

APPENDIX—M

Terminology Used in Genome-Sequencing Projects

Term	Definition
BAC end sequence	The ends of a bacterial artificial chromosome (BAC) have been sequenced and submitted to GenBank; the internal BAC sequence may not be available. When both end sequences from the same BAC are available, this information can be used to order contigs into scaffolds.
Contig	A set of overlapping clones or sequences from which a sequence can be obtained. NCBI contig records represent contiguous sequences constructed from many clone sequences. These records may include draft and finished sequences and may contain sequence gaps (within a clone) or gaps between clones when the gap is spanned by another clone which is not sequenced.
Draft sequence	At least three-to-four fold of the estimated clone insert is covered in Phred Q20 bases in the shotgun sequencing stage, as defined for the human genome-sequencing project. Note that the exact definition of "draft" may be different for other genome projects. Clone sequence may contain several pieces of the sequence separated by gaps. The true order and orientation of these pieces may not be known.
Finished sequence	The clone insert is contiguously sequenced with a high-quality standard of error rate of 0.01%. There are usually no gaps in the sequence.
Fragment	A contiguous stretch of a sequence within a clone sequence that does not contain a gap, vector or other contaminating sequence.
Meld	When two or more fragments overlap in the entire alignable region, these sequences are merged together to make a single longer sequence.
Order and orientation	Sequence overlap information is used to order and orient (ONO) fragments within a large clone sequence.
Scaffold	Ordered set of contigs placed on the chromosome.

APPENDIX—N

Twenty Most Sequenced Organisms in GenBank (Release 135.0, April 2003)

Entries	Bases	Species	Common name
6,574,171	9,743,398,611	*Homo sapiens*	Human
4,839,873	5,815,119,777	*Mus musculus*	Mouse
702,180	5,565,107,808	*Rattus norvegicus*	Rat
481,072	725,258,089	*Danio rerio*	Zebrafish
350,741	695,203,285	*Drosophila melanogaster*	Fruit fly
221,048	620,437,677	*Oryza sativa (japonica cultivar-group)*	Rice
730,719	402,816,176	*Zea mays*	Corn
502,590	394,829,161	*Arabidopsis thaliana*	Thale cress
567,863	386,355,126	*Brassica oleracea*	Broccoli
13,379	326,808,471	*Macaca mulatta*	Rhesus monkey
427,323	317,823,422	*Gallus gallus*	Chicken
499,207	294,135,749	*Ciona intestinalis*	Ascidian
390,231	223,252,029	*Bos taurus*	Cow
199,302	222,094,921	*Caenorhabditis elegans*	Worm
418,485	207,706,076	*Triticum aestivum*	Wheat
166,243	191,992,299	*Pan troglodytes*	Chimpanzee
189,208	170,811,939	*Tetraodon nigroviridis*	Pufferfish
273,621	166,171,916	*Xenopus laevis*	Frog
282,062	156,196,664	*Hordeum vulgare subsp. vulgare*	Barley
186,121	154,322,129	*Medicago truncatula*	Legume

APPENDIX—O

Human mtDNA

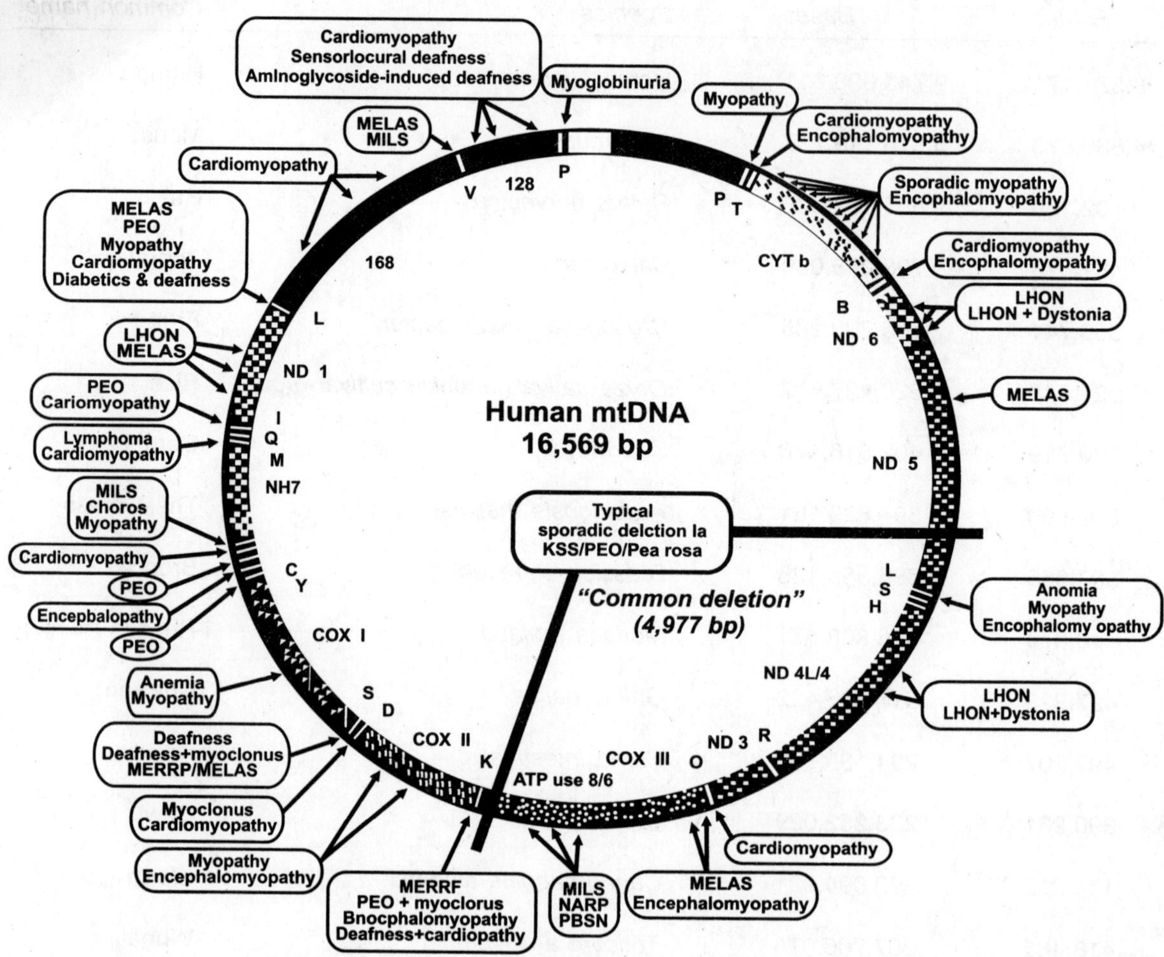

Morbidity map of the human mitochondrial genome. Abbreviations are for the genes encoding seven subunits of complex I (ND), three subunits of cytochrome coxidase (COX), cytochrome b (Cyt b), and the two subunits of ATP synthase (ATPase 6 and 8). 12S and 16S refer to ribosomal RNAs; 22 transfer RNAs are identified by the one-letter codes for the corresponding amino acids. FBSN, familial bilateral striatal necrosis; KSS, Kearns-Sayre syndrome;LHON, Leber hereditary optic neuropathy; MELAS, mitochondrial encephalomyopathy, lactic acidosis,and strokelike episodes; MERRF, myoclonic epilepsy with ragged-red fibers; MILS, maternally inherited Leigh syndrome; NARP, neuropathy, ataxia, retinitis pigmentosa; PEO, progressive external ophthalmoplegia. From DiMauro and Schon (2001). Used with permission.

APPENDIX—P

General Web Resources for Study of Human Diseases

Site	Description	URL
Diseases, disorders and related topics	Karolinska Institute (Stockholm)	http://www.mic.ki.se/Diseases/
The frequency of inherited disorders database (FIDD)	From the Institute of Medical Genetics University of Wales College of Medicine	http://archive.uwcm.ac.uk/uwcm/mg/fidd/
Gene cards	A database of human genes, their products, and their involvement in diseases	http://bioinfo.weizmann.ac.il/cards/
Genes and disease (NCBI)	Organized by chromosome, provides descriptions of 60 diseases	http://www.ncbi.nlm.nih.gov/disease/
Gene clinics	A clinical information resource from the University of Washington, Seattle	http://www.geneclinics.org/
Genetic alliance	International coalition of individuals, profesionals and genetic support organizations	http://www.geneticalliance.org/;search form http://www.geneticalliance.org/diseaseinfo/ search.html
Inherited disease genes identified by positional cloning	From the National Human Genome Research Institute (NHGRI) at the NIH	http://genome.nhgri.nih.gov/clone/
The national information center for children and youth with disabilities	An information and referral center in the United States	http://www.nichcy.org/
National organization for rare disorders (NORD)	Federation of voluntary health organizations dedicated to helping people with rare "orphan" diseases and assisting the organizations that serve them	http://www.rarediseases.org/
Online mendelian inheritance in man (OMIM)	Over 12,000 entries	http://www.ncbi.nlm.nih.gov/entrez/ query.fcgi?db-OMIM

APPENDIX—Q

Noncoding Genes in Human Genome

RNA gene	Number of noncoding genes	Number of related genes	Function
tRNA	497	324	Protein synthesis
SSU (185) RNA	0	40	Protein synthesis
5.85 rRNA	1	11	Protein synthesis
LSU (285) rRNA	0	181	Protein synthesis
5S RNA	4	520	Protein synthesis
U1	16	134	Spliceosome component
U2	6	94	Spliceosome component
U4	4	87	Spliceosome component
U4atac	1	20	Minor (U11/U12) spliceosome component
U5	1	31	Spliceosome component
U6	44	1135	Spliceosome component
U6atac	4	32	Minor (U11/U12) spliceosome component
U7	1	3	Histone mRNA 3' processing
U11	0	6	Minor (U11/U12) spliceosome component
U12	1	0	Minor (U11/U12) spliceosome component
SRP (7SL) RNA	3	773	Component of signal recognition particle
RNAse P	1	2	tRNA 5' end processing
RNAse MRP	1	6	rRNA procesing
Telomerase RNA	1	4	Template for addition of telomeres
hY1	1	353	Component of Ro RNP, function unknown
hY3	25	414	Component of Ro RNP, function unknown
hY4	3	115	Component of Ro RNP, function unknown
hY5 (4.55 RNA)	1	9	Component of Ro RNP, function unknown
Vault RNAs	3	1	Component of 13 Mda vault RNP
7SK	1	330	Unknown
H19	1	2	Unknown
Xist	1	0	Initiation of X chromosome inactivation
Known C/D snoRNAs	69	558	Pre-rRNA processing or site-specific ribose methylation of rRNA
Known H/ACA snoRNAs	15	87	Pre-rRNA processing or site-specific pseu-douridylation of rRNA

APPENDIX—R

Organisms Represented in UniGene

Organism	Species	Abbreviation	Common name
Chordata	*Bos taurus*	Bt	Cow
	Ciona intestinalis	Cin	Sea squirt
	Danio rerio	Dr	Zebrafish
	Gallus gallus	Gga	Chicken
	Homo sapiens	Hs	Human
	Mus musculus	Mm	Mouse
	Oryzias latipes	Ola	Japanese medaka
	Rattus norvegicus	Rn	Rat
	Silurana tropicalis	Str	Western clawed frog
	Sus scrofa	Ssc	Pig
	Xenopus laevis	Xl	African clawed frog
Arthropoda	*Anopheles gambiae*	Aga	Malaria mosquito
	Drosophila melanogaster	Dm	Fruit fly
Nematoda	*Caenorhabditis elegans*	Cel	Worm
Embryophyta	*Arbidopsis thaliana*	At	Thale cress
	Glycine max	Gma	Soybean
	Hordeum vulgare	Hv	Barley
	Lycopersicon esculentum	Les	Tomato
	Medicago truncatula	Mtr	Barrel medic
	Oryza sativa	Os	Rice
	Solanum tuberosum	Stu	Potato
	Sorghum bicolor	Sbi	Sorghum
	Triticum aestivum	Ta	Wheat
	Zea mays	Zm	Corn
Chlorophyta	*Chlamydomonas reinhardtii*	Cre	Green alga
Mycetozoa	*Dictyostelium discoideum*	Ddi	Slime mold

APPENDIX—S

Algorithms for Finding Genes in Eukaryotic DNA

A summary of sites is given at http://linkage.rockefeller.edu/wli/gene/

Program	Description	URL
AAT	Analysis and automation tool; web-based server	http://genome.cs.mtu.edu/aat.html
FGeneH	Predicts exons using linear discriminant functions	http://genomic.sanger.ac.uk/gf/gf.html
FgeneSH	Ab initio gene finder	http://www.softberry.com/berry.phtml
Gene Finder	For human, mouse, *Arabidopsis* and fission yeast	http://argon.cshl.org/genefinder/
GeneParser2	Identification cf protein-coding regions in genomic DNA	http://beagle.colorado.edu/%7Eeesnyder/GeneParser.html
Genie	Based on HMMs	http://www.cse.ucsc.edu/~dkulp/cgi-bin/genie
GenLang	Syntactic pattern recognition system; uses computational linguistics to find genes	http://www.cbil.upenn.edu/genlang/genlang.home.html
Genscan	Based on HMMs; rule based rather than homology based	http://genes.mit.edu/GENSCAN.html
GenTerpret	From RabbitHutch Biotechnology Corporation	http://www.rabbithutch.com/
GlimmerM	From TIGR	http://www.tigr.org/software/glimmerm/
GlimmerM web server	For *Arabidopsis thaliana*, *Oryza sativa* (rice), *Plasmodium falciparum* (malaria), other7	http://www.tigr.org/tdb/glimmerm/glmr.form.html
GRAIL	One of the most widely used algorithms	http://compbio.ornl.gov/tools/index.html
MORGAN	A decision tree system for finding genes in vertebrate DNA	http://www.tigr.org/~salzberg/morgan.html
PROCRUSTES	Gene recognition via spliced alignment	http://www-hto.usc.edu/software/procrustes/index.html
VEIL	HMM for finding genes in vertebrate DNA	http://www.cs.jhu.edu/labs/compbio/veil.html
Xpound	A probabilistic model for detecting coding regions	http://bioweb.pasteur.fr/seqanal/interfaces/xpound-simple.html

Index